**SECOND EDITION**

# Essentials of College Mathematics

**CHERYL CLEAVES**

*Southwest Tennessee Community College*

**MARGIE HOBBS**

*The University of Mississippi*

PEARSON

Prentice
Hall

Upper Saddle River, New Jersey
Columbus, Ohio

*To my brother, Jim Smith*

*To my brother, Byron Johnson*

**Library of Congress Cataloging-in-Publication Data**

Cleaves, Cheryl.
  Essentials of college mathematics / Cheryl Cleaves, Margie Hobbs.--2nd ed.
    p. cm.
  Rev. ed. of: Essentials of technical mathematics. c2002.
  Includes bibliographical references and index.
  ISBN 0-13-171480-5
    1. Mathematics--Textbooks. I. Hobbs, Margie J. II. Cleaves, Cheryl S.
  Essentials of technical mathematics. III. Title.

  QA39.3.C54 2006
  510--dc22                                      2005048888

**Senior Acquisitions Editor:** Gary Bauer
**Editorial Assistant:** Jacqueline Knapke
**Production Editor:** Louise N. Sette
**Production Supervision:** Ann Imhof, Carlisle Publishers Services
**Design Coordinator:** Diane Ernsberger
**Photo Research:** Ann Brunner
**Cover Designer:** Bryan Huber
**Production Manager:** Pat Tonneman
**Marketing Coordinator:** Leigh Ann Sims

This book was set in Times Roman by Carlisle Communications, Ltd. It was printed and bound by Banta Book Group. The cover was printed by Coral Graphic Services, Inc.

Pearson Education Ltd.                         Pearson Education Australia Pty. Limited
Pearson Education Singapore Pte. Ltd.     Pearson Education North Asia Ltd.
Pearson Education Canada, Ltd.            Pearson Educación de Mexico, S.A. de C.V.
Pearson Education—Japan                  Pearson Education Malaysia Pte. Ltd.

8  16
ISBN 0-13-171480-5

# *Preface*

In *Essentials of College Mathematics,* Second Edition, we have preserved all the features that made the first edition one of the most appropriate texts on the market for a comprehensive study of mathematics in career programs. We continue to use real-life situations as a context for applied problems.

## *Changes in the Second Edition*

The second edition incorporates many valuable suggestions made by users of the first edition. As a result, we have placed greater emphasis on problem solving, included new material, and rearranged some topics. Content changes include the following:

- Topics have been rearranged to distinguish more clearly between linear and nonlinear equations. Graphing linear equations, equations of lines, and systems of linear equations now appear before quadratic equations are introduced.
- Many related formulas have been brought together to form more logical units of study. For example, Heron's formula for finding the area of a triangle has been moved from one of the trigonometry chapters to the chapter that introduces the area of a triangle. Both radian and degree measures of angles are introduced in the geometry chapter, and the formulas for finding arc length and the area of a sector are presented for both types of angle measures. Students can better correlate the two notations for angle measure.
- Evaluating and rearranging formulas has been moved to the chapter on linear equations. This will help students more immediately understand the usefulness of algebra.
- Geometry (perimeter, area, and volume) has been collected in the geometry chapter (Chapter 12). This provides a more organized body of study for students and allows easier reference. It also is a more efficient presentation of the material.
- U.S. customary measures are introduced in Chapter 2, "Review of Fractions and Percents." This change allows students to experience authentic applications of fractions more quickly.
- "Interpreting and Analyzing Data" has been renamed "Statistics" and is now Chapter 15. Students are more familiar with this terminology. Because these topics can be incorporated into a technical program at many different points in a program of study, this will provide the flexibility of using this material at any point after the chapter on percents.
- Period and phase shift have been added to Chapter 14, "Trigonometry with Any Angle." These concepts enable students to experience applications in electronics and other technical fields.

To help students connect math with their career path and develop collaborative skills we have added these new features to the book:

- **Focus on Careers:** This feature opens each chapter with an interesting overview of selected careers and their job outlook, so students will become familiar with the type of career information that is provided on the Internet from the U.S. Bureau of Labor Statistics (http://www.bls.gov/oco/).

- **Teamwork Exercises:** These activities provide students the opportunity to develop and refine team interaction skills. They use skills that have been identified by employers as important skills for employees.
- **Career Coding:** Examples, section self-study exercises, and chapter review exercises have been coded using 14 different career categories. This will strengthen the students' ability to make connections between mathematical concepts and career applications. An index of applications by career code follows the table of contents.

| | |
|---|---|
| HLTH/N | Allied Health/Nursing/EMS |
| BUS | Business/Accounting/Personal Finance/Real Estate |
| HELPP | Helping Professions/Education/Criminal Justice/Fire Fighting |
| HOSP | Hospitality/Culinary/Food Technology |
| CAD/ARC | CAD/Drafting/Architecture/Surveying |
| COMP | Computer Technologies |
| AG/H | Agriculture/Horticulture/Landscaping |
| CON | Construction Trades |
| INDTR | Industrial Trades/Welding/Machine Tool/Industrial Maintenance |
| INDTEC | Industrial Technology/Manufacturing/Machine Technology/Engineering Technology |
| AVIA | Aviation Technology |
| ELEC | Electronics Technology |
| TELE | Telecommunication |
| AUTO | Auto/Diesel Technology |

- **Cumulative Practice Tests:** To help students assess their understanding of the material as the course progresses, new cumulative tests have been added. Four tests are included after selected chapters to enhance students' assimilation of mathematical concepts.

Our goal is to present a systematic framework for successful learning in mathematics that will strengthen students' *mathematical sense* and give students a greater appreciation for the power of mathematics in everyday life and in the workplace. The new material in this edition has been added to broaden the usefulness of the text. Many of the explanations have been enhanced with carefully constructed visualizations. Exercises have been updated and new ones added.

## Commitment to Improving Mathematics Education

The authors continue to be active in the development, revision, and implementation of the standards (*Beyond Crossroads*) of the American Mathematical Association of Two-Year Colleges (AMATYC). We enthusiastically promote the standards and guidelines encouraged by AMATYC, NCTM, MAA, and the SCANS document.

## Calculator Usage

Calculator tips appropriate for both scientific and graphing calculators are periodically included. These generic tips guide students to use critical thinking to determine how their calculator operates without referring to a user's manual.

We continue to emphasize the calculator as a tool that *facilitates* learning and understanding. Assessment strategies are included throughout the text and supplementary materials to enable students to test their understanding of a concept independently of their calculator.

## Additional Resources

A variety of instructional tools are available with adoption of this text including a printed Instructor's Resource Manual, printed Test Item File, TestGen computerized test generation software, a Student Solutions Manual, and a Companion Website at **www.prenhall.com/cleaves.**

The Instructor's Resource Manual, Test Item File, and TestGen computerized test generator are also downloadable from our Instructor Resource Center. Go to **www.prenhall.com,** click the **Instructor Resource Center** link, and then click **Register Today** for an instructor access code. Within 48 hours after registering you will receive a confirming e-mail including an instructor access code. Once you have received your code, go to the site and log on for full instructions on downloading the materials you wish to use.

## Acknowledgments

A project such as this does not come together without help from many people. Our first avenue for input is through students and faculty who use the text. Their comments and suggestions have been invaluable.

We wish to express thanks to all the people who helped make this edition a reality. In particular, we thank Gary Bauer, Senior Acquisitions Editor, whose belief in our work and support of our ideas have been a major factor in the success of this book. We thank Louise Sette, Prentice Hall production editor. We also thank Ann Imhof and Kelly Mulligan of Carlisle Publishers Services.

The teaching of mathematics over time produces a wealth of knowledge about instructional strategies and specific content. We are grateful for the many valuable suggestions received in these areas. We wish to thank the following individuals:

Behnaz Rouhani, Athens Area Technical Institute (GA)
Brent Hamilton, North Iowa Area Community College
David C. Shellabarger, Lane Community College (OR)
Edwin G. Landauer, Clackamas Community College (OR)
Joseph Sukta, Moraine Valley Community College (IL)
Jimmie A. Van Alphen, Ozarks Technical Community College (MO)
Karen Newson, Triangle Tech (PA)
Milton Clark, Florence Darlington Technical College (SC)
Nicole Muth, Lakeshore Technical College (WI)
Scott Randby, University of Akron (OH)
Terry B. Gaalswyk, Western Iowa Technical Community College

Cheryl Cleaves
Margie Hobbs

## To the Student

The mathematics you learn from this book will serve you well and will help you advance your career goals. We have given much thought to the best way to teach mathematics and have done extensive research on how students learn. We have provided a wide variety of features and resources so that you can customize your study to your needs and circumstances. The following features are key to helping you learn the mathematics in this text.

**Table of Contents.**   The table of contents is your "roadmap" to this text. Study it carefully to determine how the topics are arranged. This will aid you in relating topics to each other.

**Glossary/Index.**   An extensive glossary/index is an important part of every mathematics book. Use the index to cross-reference topics and to locate other topics that relate to the topic you are studying.

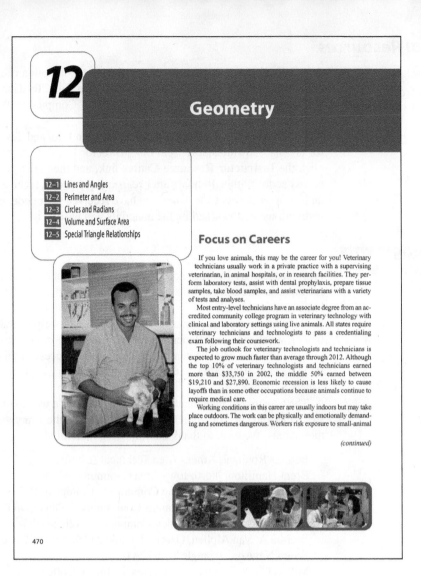

**12**

## Geometry

12–1  Lines and Angles
12–2  Perimeter and Area
12–3  Circles and Radians
12–4  Volume and Surface Area
12–5  Special Triangle Relationships

### Focus on Careers

If you love animals, this may be the career for you! Veterinary technicians usually work in a private practice with a supervising veterinarian, in animal hospitals, or in research facilities. They perform laboratory tests, assist with dental prophylaxis, prepare tissue samples, take blood samples, and assist veterinarians with a variety of tests and analyses.

Most entry-level technicians have an associate degree from an accredited community college program in veterinary technology with clinical and laboratory settings using live animals. All states require veterinary technicians and technologists to pass a credentialing exam following their coursework.

The job outlook for veterinary technologists and technicians is expected to grow much faster than average through 2012. Although the top 10% of veterinary technologists and technicians earned more than $33,750 in 2002, the middle 50% earned between $19,210 and $27,890. Economic recession is less likely to cause layoffs than in some other occupations because animals continue to require medical care.

Working conditions in this career are usually indoors but may take place outdoors. The work can be physically and emotionally demanding and sometimes dangerous. Workers risk exposure to small-animal

*(continued)*

470

**Focus on Careers.**   Each chapter opens with interesting information about a selected career. The article includes a description of the type of work done in this career, the expected salary, and the job prospects. A good resource for investigating other careers is the website for the U.S. Department of Labor, Bureau of Labor Statistics (http://www.bls.gov/oco/).

**Learning Outcomes.**   A learning outcome is what you should be able to do when you master a concept. These outcomes can guide you through your study plan. Each section begins with a statement of learning outcomes that shows you what you should look for and learn in that section. If you read and think about these outcomes before you begin the section, you will know what to look for as you work through the section. Section Self-Study Exercises are organized by learning outcomes, and the Chapter Review of Key Concepts lists the learning outcomes for your review.

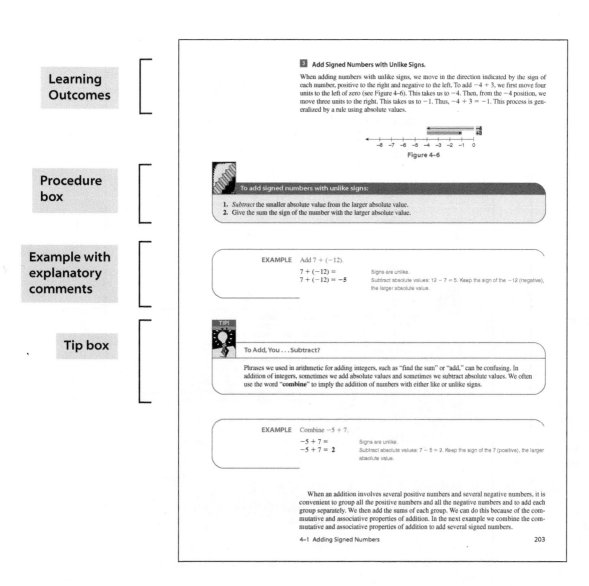

**3** Add Signed Numbers with Unlike Signs.

When adding numbers with unlike signs, we move in the direction indicated by the sign of each number, positive to the right and negative to the left. To add $-4 + 3$, we first move four units to the left of zero (see Figure 4–6). This takes us to $-4$. Then, from the $-4$ position, we move three units to the right. This takes us to $-1$. Thus, $-4 + 3 = -1$. This process is generalized by a rule using absolute values.

Figure 4–6

**To add signed numbers with unlike signs:**

1. *Subtract* the smaller absolute value from the larger absolute value.
2. Give the sum the sign of the number with the larger absolute value.

**EXAMPLE**   Add $7 + (-12)$.

$7 + (-12) =$          Signs are unlike.
$7 + (-12) = -5$      Subtract absolute values: $12 - 7 = 5$. Keep the sign of the $-12$ (negative), the larger absolute value.

**TIP!**

**To Add, You . . . Subtract?**

Phrases we used in arithmetic for adding integers, such as "find the sum" or "add," can be confusing. In addition of integers, sometimes we add absolute values and sometimes we subtract absolute values. We often use the word "**combine**" to imply the addition of numbers with either like or unlike signs.

**EXAMPLE**   Combine $-5 + 7$.

$-5 + 7 =$          Signs are unlike.
$-5 + 7 = 2$      Subtract absolute values: $7 - 5 = 2$. Keep the sign of the 7 (positive), the larger absolute value.

When an addition involves several positive numbers and several negative numbers, it is convenient to group all the positive numbers and all the negative numbers and to add each group separately. We then add the sums of each group. We can do this because of the commutative and associative properties of addition. In the next example we combine the commutative and associative properties of addition to add several signed numbers.

4–1 Adding Signed Numbers                                                                203

**Procedures Boxes.**   Each learning outcome has one or more procedures boxes. These boxes provide rules or procedures presented as numbered steps. A procedures box may also present a mathematical property, formula, or fact.

**Tip Boxes.**   These boxes give helpful hints for doing mathematics, and they draw your attention to important observations and connections that you may have missed in an example.

**Use of Color in the Text.**   As you read the text and work through the examples, notice the items shaded with color or gray. These will help you follow the logic of working through the example. Color also highlights important items and boxed features such as the Tips, Learning Outcomes, and rules, procedures, and formulas.

**Six-Step Problem Solving Example**

**Tip box**

**Career coding in examples**

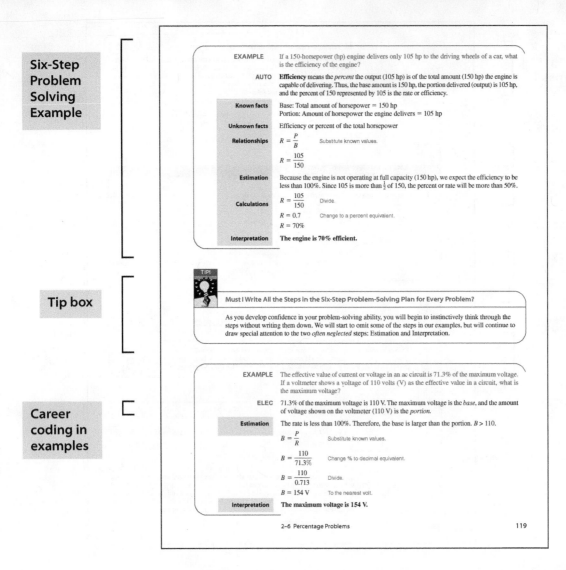

EXAMPLE  If a 150-horsepower (hp) engine delivers only 105 hp to the driving wheels of a car, what is the efficiency of the engine?

AUTO  **Efficiency** means the *percent* the output (105 hp) is of the total amount (150 hp) the engine is capable of delivering. Thus, the base amount is 150 hp, the portion delivered (output) is 105 hp, and the percent of 150 represented by 105 is the rate or efficiency.

**Known facts**  Base: Total amount of horsepower = 150 hp
Portion: Amount of horsepower the engine delivers = 105 hp

**Unknown facts**  Efficiency or percent of the total horsepower

**Relationships**  $R = \dfrac{P}{B}$    Substitute known values.

$R = \dfrac{105}{150}$

**Estimation**  Because the engine is not operating at full capacity (150 hp), we expect the efficiency to be less than 100%. Since 105 is more than $\frac{1}{2}$ of 150, the percent or rate will be more than 50%.

**Calculations**  $R = \dfrac{105}{150}$    Divide.

$R = 0.7$    Change to a percent equivalent.

$R = 70\%$

**Interpretation**  **The engine is 70% efficient.**

TIP!

**Must I Write All the Steps in the Six-Step Problem-Solving Plan for Every Problem?**

As you develop confidence in your problem-solving ability, you will begin to instinctively think through the steps without writing them down. We will start to omit some of the steps in our examples, but will continue to draw special attention to the two *often neglected* steps: Estimation and Interpretation.

EXAMPLE  The effective value of current or voltage in an ac circuit is 71.3% of the maximum voltage. If a voltmeter shows a voltage of 110 volts (V) as the effective value in a circuit, what is the maximum voltage?

ELEC  71.3% of the maximum voltage is 110 V. The maximum voltage is the *base*, and the amount of voltage shown on the voltmeter (110 V) is the *portion*.

**Estimation**  The rate is less than 100%. Therefore, the base is larger than the portion. $B > 110$.

$B = \dfrac{P}{R}$    Substitute known values.

$B = \dfrac{110}{71.3\%}$    Change % to decimal equivalent.

$B = \dfrac{110}{0.713}$    Divide.

$B = 154 \text{ V}$    To the nearest volt.

**Interpretation**  **The maximum voltage is 154 V.**

2–6 Percentage Problems    119

**Six-Step Approach to Problem Solving.**   Successful problem solvers use a systematic, logical approach. We use a six-step approach to problem solving. This approach gives you a system for solving a variety of math problems. You will learn how to organize the information given and how to develop a logical plan for solving the problem. You are asked to analyze and compare and to estimate as you solve problems. Estimation helps you decide whether your answer is reasonable. You will learn to interpret the results of your calculations within the problem's context, a skill you will use on your job.

**Career-Coded Examples and Exercises.**   Applied problems focus on a wide variety of careers available as a course of study at your community college, technology center, or university. These careers are grouped into 14 categories, and the examples and exercises are coded to these categories as appropriate. An index of applications is provided for your convenience.

**Using Your Calculator.**   Calculators are useful in all levels of mathematics. Some tips introduce easy-to-follow calculator strategies. The tips show you how to analyze the procedure and set up a problem for a calculator solution; a sample series of keystrokes is often included. In addition, the tips help you determine how your type of calculator operates for various mathematical processes.

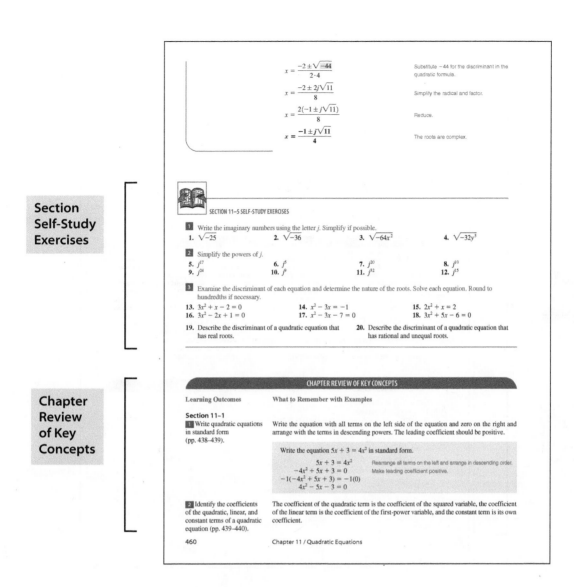

$$x = \frac{-2 \pm \sqrt{-44}}{2 \cdot 4}$$

Substitute −44 for the discriminant in the quadratic formula.

$$x = \frac{-2 \pm 2j\sqrt{11}}{8}$$

Simplify the radical and factor.

$$x = \frac{2(-1 \pm j\sqrt{11})}{8}$$

Reduce.

$$x = \frac{-1 \pm j\sqrt{11}}{4}$$

The roots are complex.

**SECTION 11–5 SELF-STUDY EXERCISES**

**1** Write the imaginary numbers using the letter $j$. Simplify if possible.

**1.** $\sqrt{-25}$ **2.** $\sqrt{-36}$ **3.** $\sqrt{-64x^2}$ **4.** $\sqrt{-32y^5}$

**2** Simplify the powers of $j$.

**5.** $j^{17}$ **6.** $j^5$ **7.** $j^{20}$ **8.** $j^{10}$
**9.** $j^{24}$ **10.** $j^9$ **11.** $j^{32}$ **12.** $j^{15}$

**3** Examine the discriminant of each equation and determine the nature of the roots. Solve each equation. Round to hundredths if necessary.

**13.** $3x^2 + x - 2 = 0$ **14.** $x^2 - 3x = -1$ **15.** $2x^2 + x = 2$
**16.** $3x^2 - 2x + 1 = 0$ **17.** $x^2 - 3x - 7 = 0$ **18.** $3x^2 + 5x - 6 = 0$

**19.** Describe the discriminant of a quadratic equation that has real roots.

**20.** Describe the discriminant of a quadratic equation that has rational and unequal roots.

**CHAPTER REVIEW OF KEY CONCEPTS**

| Learning Outcomes | What to Remember with Examples |
|---|---|
| **Section 11–1**<br>**1** Write quadratic equations in standard form (pp. 438–439). | Write the equation with all terms on the left side of the equation and zero on the right and arrange with the terms in descending powers. The leading coefficient should be positive. |

Write the equation $5x + 3 = 4x^2$ in standard form.

$$5x + 3 = 4x^2$$ Rearrange all terms on the left and arrange in descending order.
$$-4x^2 + 5x + 3 = 0$$ Make leading coefficient positive.
$$-1(-4x^2 + 5x + 3) = -1(0)$$
$$4x^2 - 5x - 3 = 0$$

| | |
|---|---|
| **2** Identify the coefficients of the quadratic, linear, and constant terms of a quadratic equation (pp. 439–440). | The coefficient of the quadratic term is the coefficient of the squared variable, the coefficient of the linear term is the coefficient of the first-power variable, and the constant term is its own coefficient. |

460    Chapter 11 / Quadratic Equations

**Section Self-Study Exercises.** These practice sets are keyed to the learning outcomes and appear at the end of each section. Use these exercises to check your understanding of the section. The answers to every exercise are at the end of the text, so you can get immediate feedback on whether you understand the concepts.

**Chapter Review of Key Concepts.** Each chapter includes a summary in the form of a two-column chart. The first column lists the learning outcomes of the chapter. The second column gives the procedures and examples for each outcome. Page references are included to facilitate your preview or review of the chapter.

Career
Coding in
Chapter
Review
Exercises

Team
Problem-
Solving
Exercises

Practice
Test

**152. INDTR** According to specifications, a machined part may vary from its specified measure by ±0.4% and still be usable. If the specified measure of the part is 75 in. long, what is the range of measures acceptable for the part?

**153. INDTEC** A wet casting weighing 145 kg has a 2% weight loss in the drying process. How much does the dried casting weigh?

**154. BUS** A book that did sell for $18.50 now sells for 20% more. How much does the book now sell for?

**155. COMP** A laptop computer originally priced at $2,400 now sells for $300 more. What is the percent of the increase?

**156. INDTEC** A 20-in. bar of iron measures 20.025 in. when it is heated. What is the percent of increase?

**157. HLTH/N** A dieter went from 168 lb to 160 lb in 1 week. To the nearest tenth of a percent, what was the percent of weight loss?

**158. BUS** Workers took a 10% pay cut to help their company stay open during economic hard times. What is the reduced annual salary of a worker who originally earned $35,000?

**159. COMP** The Bureau of Labor Statistics predicts that the computer and data processing services industry will have the fastest wage and salary employment growth from 2000 to 2010. The industry had 2,095,000 jobs in 2000 and expects to have 3,900,000 jobs in 2010. What is the percent increase?

**160. HLTH/N** According to the Bureau of Labor Statistics the residential care industry is predicted to be the second highest growth industry from 2000 to 2010. The number of jobs is expected to increase from 806,000 to 1,318,000. What is the expected percent of increase?

### TEAM PROBLEM-SOLVING EXERCISES

1. Your team is preparing a report that is to be printed on both sides of the paper. It is customary to put odd-numbered pages on the front and even-numbered pages on the back. A new chapter is started on the front of a sheet of paper, even if this creates a preceding blank page. Assign page numbers to the document based on these guidelines.

   Chap. 1, 15 pages    Chap. 2, 17 pages
   Chap. 3, 24 pages    Chap. 4, 15 pages

   **(a)** What page number will start Chapter 2?
   **(b)** How many pages are in the document?
   **(c)** How many blank pages are in the document?

2. Your team wants to analyze the percent of income that is spent on housing and transportation. Assume you have a family annual income of $35,500.
   **(a)** In a year $6,900 is spent for a home mortgage, $950 for property taxes, $380 for homeowner's insurance, $2,400 for utilities, and $200 for maintenance and repair. What percent, to the nearest tenth percent of the family's annual income, is spent for housing?
   **(b)** In a year $3,420 is spent on automobile financing, $1,152 on gasoline, $625 on insurance, and $150 on maintenance and repair. What percent, to the nearest tenth percent of the family's annual income, is spent for transportation?

### PRACTICE TEST

1. Write 0.68 as a fraction
2. Reduce $\frac{18}{24}$ to lowest terms.
3. Find the measure of the line segment $AB$ in Fig. 2–32.
4. Write $\frac{7}{20}$ as a decimal.
5. Write $\frac{51}{8}$ as a decimal.
6. Write $5\frac{3}{8}$ as an improper fraction.
7. Which fraction is smaller, $\frac{4}{5}$ or $\frac{7}{10}$?

Perform the indicated operations.

8. $\dfrac{7}{12} + \dfrac{5}{6}$

9. $3\dfrac{4}{15} + 4\dfrac{3}{10}$

10. $\dfrac{5}{6} \times \dfrac{3}{10}$

11. $2\dfrac{2}{9} \times 1\dfrac{3}{4}$

12. $\dfrac{7}{8} - \dfrac{1}{4}$

13. $6\dfrac{1}{4} - 2\dfrac{3}{4}$

14. $\dfrac{5}{12} \div \dfrac{5}{6}$

15. $7\dfrac{1}{2} \div \dfrac{5}{9}$

16. $5\dfrac{2}{3} \div 1\dfrac{1}{9}$

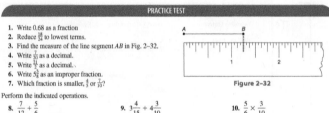

**Figure 2–32**

Practice Test      139

**Chapter Review Exercises.** An extensive set of exercises appears at the end of each chapter so you can review all the learning outcomes presented in the chapter. These exercises, organized by section, may be assigned as homework, or you may want to work them on your own for additional practice. Answers to the odd-numbered exercises are given at the end of the text, and worked-out solutions appear in a separate Student Solutions Manual available for purchase. Your instructor has the solutions to the even-numbered exercises in the Instructor's Resource Manual.

**Team Problem-Solving Exercises.** Employers value an employee's ability to interact with others in a team environment. These exercises will allow you to develop and refine your team-interaction skills.

**Practice Test.** The practice test at the end of each chapter lets you check your understanding of the chapter learning outcomes. You should be able to work each problem without referring any examples in your text or your notes. Take this test before you take the class test to check and verify your understanding of the chapter material. Answers to the odd-numbered exercises appear at the end of the text, and their solutions appear in a separate Student Solutions Manual. Your instructor has the solutions to the even-numbered exercises in the Instructor's Resource Manual.

Describe the roots of the equations without solving.
**19.** $x^2 - 8x + 12 = 0$          **20.** $3x^2 - 2x + 3 = 0$

Graph the quadratic equations by examining properties. Show the axis of symmetry, vertex, and solutions.
**21.** $y = x^2 + 2x + 1$          **22.** $y = x^2 + 4x + 4$

### CAREER APPLICATION: CIVIL ENGINEERING— FREEWAY SUPPORTS

Safe, dependable, durable freeway-support design is extremely important because highways carry an ever-increasing number of heavy vehicles (such as tandem tractor-trailer trucks) and heavier weight loads. Civil engineers try to design supports that hold up during temperature extremes, earthquakes, and high wind conditions (for example, tornadoes and hurricanes). When designing a rectangular box support, researchers found that support is optimal when the depth is at least one-half of the width and the volume is at least 10 ft$^3$ per ft of support height.

Civil engineers lobby a state government to allot construction money for a freeway passing through an earthquake fault zone at a higher, safer volume rate of 12.5 ft$^3$ per ft of support height. At intersections, rectangular box supports 16 ft high are required.

**Exercises**

Use the information for the higher, safer volume rate to answer the questions. Round answers to the nearest tenth.

1. Draw a diagram of a rectangular box support with a 16-ft height. Label the sides $W$ for width and $L$ for length.

2. What is the required volume of this safer 16-ft-high support?

3. Use $L$ to write an expression for the width if it is to be half the length.

4. Use the volume formula $V = LWH$ to write an equation to find the width and depth of the support. Solve the equation, and state your answers rounded to the nearest tenth of a foot. What is the volume of this support?

5. If a support has a width three-fourths the size of the length, find its dimensions and volume.

6. This freeway will be in an earthquake fault zone, so state civil engineers decide to make the depth equal to the width of the support to withstand maximum vibration stress. Find the dimensions and volume of each support.

7. Why do the dimensions you found in Exercise 4 have a volume of exactly 200 ft$^3$, while the dimensions you found in Exercises 5 and 6 have volumes over and under the prescribed 200 ft$^3$?

8. A recent computer modeling simulation found that supports with the smallest cross-sectional perimeter per volume are best able to endure vibration stress, which means that supports in earthquake zones should be in the shape of a cylinder (a cylinder has a circular cross-sectional perimeter). Use $V = \pi r^2 h$ to find the radius of the support with a height of 16 ft and a volume of 200 ft$^3$. Use the $\pi$ key on your calculator.

9. Find the cross-sectional perimeters of the supports designed in Exercises 4, 5, 6, and 8. Does the circular cross section have the smallest perimeter? The perimeter or circumference of a circle is $C = 2\pi r$.

**Answers**

1. Support diagram:
2. At 12.5 ft$^3$ per ft of height, 200 ft$^3$ is allotted.
3. $W = (\frac{1}{2})L$ or $W = \frac{L}{2}$ or $W = 0.5L$.
4. The length should be 5.0 ft and the width 2.5 ft, with a volume of 200.0 ft$^3$.
5. From the equation $16(0.75L)(L) = 200$, the length should be 4.1 ft, the width should be 3.1 ft, and the height should be 16 ft. These rounded dimensions have a volume of 203.4 ft$^3$.

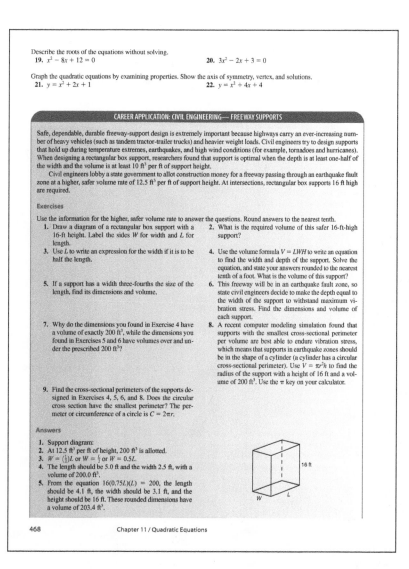

**Career Applications.** The career applications included at the end of each chapter resulted from interviews and research in the workplace. They demonstrate how widespread math applications are in the workplace and the world around you. They provide opportunities to solve real-world problems, and they demonstrate the ways that you regularly use the math concepts you are learning.

**Cumulative Practice Tests.** Practice tests for a group of chapters are included after chapters 4, 8, 11, and 14. These tests will help you prepare for mid-course or end-of-course exams. Periodically reviewing previously learned material will help you retain the concepts for a longer period of time.

**Student Solutions Manual.** This manual can be purchased at your college bookstore or from online bookstores. It gives you extra *learning insurance* to help you master learning outcomes in the text. The manual contains worked-out solutions to the odd-numbered exercises in the Chapter Review Exercises and the Practice Test for each chapter of the text. Answers to these exercises appear in the back of your text, but using the manual to study the worked-out solutions reinforces your problem-solving skills and your understanding of the concepts.

**Companion Website.** This free website, available at **www.prenhall.com/cleaves**, provides even more practice with the math concepts presented in the form of short quizzes for each section of the text. These quizzes are immediately graded, and you have the opportunity to send the results to your instructor via email.

We wish you much success in your study of mathematics. Many of the improvements for this book were suggested by students such as yourself. If you have suggestions for improving the presentation, please give them to your instructor or email the authors at ccleaves@bellsouth.net or margiehobbs@bellsouth.net.

Cheryl Cleaves
Margie Hobbs

# Contents

# List of Career Applications

## ELECTRONICS TECHNOLOGY (ELEC)

## HELPING PROFESSIONS/EDUCATION/ CRIMINAL JUSTICE/FIRE FIGHTING (HELPP)

## ALLIED HEALTH/NURSING/EMS (HLTH/N)

## HOSPITALITY/CULINARY/FOOD TECHNOLOGY (HOSP)

## INDUSTRIAL TECHNOLOGY/ MANUFACTURING/MACHINE TECHNOLOGY/ ENGINEERING TECHNOLOGY (INDTEC)

## INDUSTRIAL TRADES/WELDING/MACHINE TOOL/INDUSTRIAL MAINTENANCE (INDTR)

## TELECOMMUNICATIONS (TELE)

# 1

# Review of Basic Concepts

## Focus on Careers

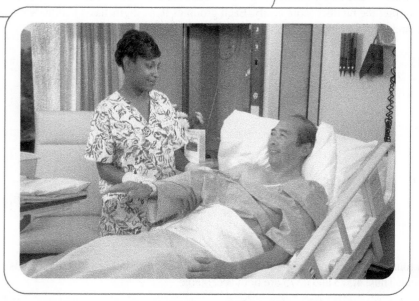

The *Occupational Outlook Handbook* is an excellent source of career information. It provides information for making decisions about careers. The *Handbook* is revised every two years and describes what workers do on the job and their working conditions. It also gives the training and education needed, earnings, and expected job prospects for many careers and occupations. The *Handbook* can be accessed on the Internet at http://www.bls.gov/oco/home.htm.

Nursing is one of the largest health-care occupations with 2.3 million jobs. The associate degree (ADN programs) is one path for becoming a nurse and takes two to three years to complete. Job opportunities for this occupation are expected to be very good in the next 10 years. In 2002 the median (middle) annual earnings for registered nurses were $48,090.

Although most nurses work in hospitals, some work in offices, nursing care facilities, and patients' homes. Nurses should be caring and responsible individuals. They must be detail oriented and be able to correctly assess patients' conditions.

Source: *Occupational Outlook Handbook,* 2004–2005 Edition, U.S. Department of Labor, Bureau of Labor Statistics.

When we study a subject for the first time or review in some detail a subject we studied a while ago, we begin with the basics. Often, as we examine the basics of a subject, we discover—or rediscover—many pieces of useful information. In this sense, mathematics is no different from any other subject. We begin with a study of whole numbers and decimals and the basic operations that we perform with them.

## 1–1 | Basic Operations with Whole Numbers and Decimals

*Learning Outcomes*

**1** Compare whole numbers.

**2** Write fractions with power-of-10 denominators as decimal numbers.

**3** Compare decimal numbers.

**4** Round a whole number or a decimal number to a specified place value.

**5** Add and subtract whole numbers and decimals.

**6** Multiply and divide whole numbers and decimals.

Our system of numbers, the **decimal-number system,** uses 10 figures called **digits:** 0, 1, 2, 3, 4, 5, 6, 7, 8, 9. A **whole number** is made up of one or more digits. When a number contains two or more digits, each digit must be in the correct place for the number to have the value we intend it to have.

The place values are arranged in **periods,** or groups of three (Fig. 1–1). The first group of three is called **units,** the second group of three is called **thousands,** the third group is called **millions,** and the fourth group is called **billions.** Commas are used to mark off these periods. The commas make larger numbers easier to read because we can locate specific place values and interpret the number more easily. Each group of three digits has a hundreds place, a tens place, and a ones place.

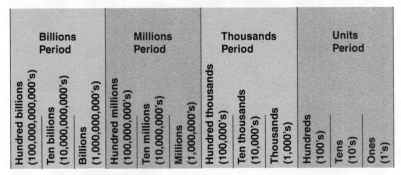

**Figure 1–1** Whole-number place values and periods.

In four-digit numbers, the comma separating the units period from the thousands period is optional. Thus, 4,575 and 4575 are both acceptable.

### **1** Compare Whole Numbers.

Whole numbers can be arranged on a **number line** to show a visual representation of the relationship of numbers by size. The most common arrangement is to begin with zero and place numbers on the line from left to right as they get larger.

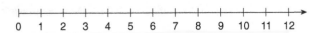

All numbers have a place on the number line and the numbers continue indefinitely without end. A term that is often used to describe this concept is **infinity** and the symbol is ∞.

Whole numbers can be compared by size by determining which of the two numbers is larger or smaller. If two numbers are positioned on a number line, the smaller number is positioned to the left of the larger number. The order relationship can be written in a mathematical statement called an **inequality.** An inequality shows that two numbers are not equal;

that is, one is larger than the other. Symbols for showing inequalities are the **less than** symbol < and the **greater than** symbol >.

5 < 7     Five is less than seven.
7 > 5     Seven is greater than five.

### To compare whole numbers:

1. Mentally position the numbers on a number line.
2. Select the number that is farther to the left to be the smaller number.
3. Write an inequality using the *less than* symbol.

smaller number < larger number

or

Write an inequality using the *greater than* symbol.

larger number > smaller number

**TIP!**

### Which Way Does the Inequality Symbol Point?

In using an inequality symbol to relate two numbers, the point of the *less than* symbol is directed toward the *left* like the arrowhead on the *left* of the number line. The point of the *greater than* symbol is directed toward the *right* like the arrowhead on the *right* of the number line.

3 < 5     6 < 14     4 > 2     9 > 7

**EXAMPLE**  Write an inequality comparing the numbers 12 and 19:

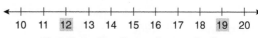

Mentally position the numbers on the line.

12 is the smaller number.        12 is to the *left* of 19.
**12 < 19 or 19 > 12**           Use appropriate inequality symbol.

Numbers are used to show *how many* and to show *order.* **Cardinal numbers** show *how many* and **ordinal numbers** show *order* or position (such as first, second, third, fourth, etc.). For example, in the statement "three students are doing a presentation," three is a cardinal number (showing how many). In the statement "Margaret is the third tallest student in the class," third is an ordinal number (showing order).

### 2  Write Fractions with Power-of-10 Denominators as Decimal Numbers.

Numbers that are parts of a whole number are called **fractions.** In fraction notation, we write one number over another number.

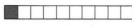

1 of 10 parts

$\dfrac{1}{10}$  numerator
denominator

The bottom number, the **denominator,** represents the number of parts that a whole unit contains. The top number, the **numerator,** represents the number of parts being considered.

A special type of fraction is called a **decimal fraction.** Other fractions will be examined in Chapter 2.

A decimal fraction is a fraction whose denominator is 10 or some power of 10, such as 100 or 1,000. Often the terms decimal fraction, **decimal number,** and **decimal** are used interchangeably. In fraction notation, 3 out of 10 parts is written as $\frac{3}{10}$. In decimal notation, the denominator 10 is not written but is implied by position on the place-value chart (Fig. 1–2). A decimal point (.) separates whole-number amounts on the left and fractional parts on the right. The fraction $\frac{3}{10}$ can be written in decimal notation as 0.3.

3 out of 10

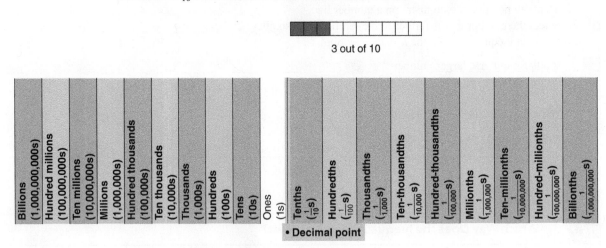

**Figure 1–2** Place-value chart for whole numbers and decimals.

To extend the place-value chart to include parts of whole amounts, we place a decimal point (.) after the ones place. The place on the right of the ones place is called the *tenths place.* A **decimal point** is placed between the ones place and the tenths place to distinguish between whole amounts and fractional amounts.

TIP!

**Informal Use of the Word *Point***

Informally, the decimal point is sometimes read as "point." Thus, 3.6 is read "three and six tenths" or "three point six." The decimal 0.0162 can be read as "one hundred sixty-two ten-thousandths," or "point zero one six two," or "zero point zero one six two." This informal process is often used in communication to ensure that numbers are not miscommunicated.

TIP!

**Unwritten Decimal Points**

When we write whole numbers, using numerals, we usually omit the decimal point; the decimal point is understood to be at the end of the whole number. Therefore, any whole number, such as 32, can be written without a decimal (32) or with a decimal (32.).

Fractions like $\frac{1}{10}$ and $\frac{75}{100}$ have denominators that are powers of 10. Any fraction whose denominator is 10, 100, 1,000, 10,000, and so on, can be written as a decimal number without making any calculations.

Chapter 1 / Review of Basic Concepts

**To write a fraction whose denominator is 10, 100, 1,000, 10,000, and so on, as a decimal:**

1. Use the denominator to find the number of decimal places.

$$10 \rightarrow 1 \text{ place}$$
$$100 \rightarrow 2 \text{ places}$$
$$1,000 \rightarrow 3 \text{ places}$$
$$10,000 \rightarrow 4 \text{ places}$$

Write $\frac{17}{1,000}$ as a decimal.

0.___    Three decimal places are needed.

2. Place the numerator so that the last digit is in the farthest place on the right.

0._1 7

3. Fill in any blank spaces with zeros.

0.017

---

**EXAMPLE**   Write $\frac{3}{10}$, $\frac{25}{100}$, $\frac{425}{100}$, and $\frac{3}{1,000}$ as decimal numbers.

$\frac{3}{10}$ is written **0.3.**          One decimal place

$\frac{25}{100}$ is written **0.25.**        Two decimal places

$\frac{425}{100}$ is written **4.25.**        Two decimal places

$\frac{3}{1,000}$ is written **0.003.**      Three decimal places

---

**TIP!**

**Do Ending Zeros Change the Value of a Decimal Number?**

When we attach zeros on the *right* end of a decimal number, we do not change the value of the number.

$$0.5 = 0.50 = 0.500 \qquad \frac{5}{10} = \frac{50}{100} = \frac{500}{1,000}$$

See equivalent fractions on page 67.

---

**3**   **Compare Decimal Numbers.**

As with whole numbers, we often need to compare decimals by size. To make valid comparisons, we must compare like amounts. Whole numbers compare with whole numbers, tenths compare with tenths, thousandths compare with thousandths, and so on.

**To compare decimal numbers:**

1. Compare whole-number parts.
2. If the whole-number parts are equal, compare digits place by place, starting at the tenths place and moving to the right.
3. Stop when two digits in the same place are different.
4. The digit that is larger determines the larger decimal number.

**EXAMPLE**  Compare the numbers 32.47 and 32.48 to see which is larger.

32.47    Look at the whole-number parts. They are the same.
32.48    Look at the tenths place for each number. Both numbers have a 4 in the tenths place.
         Look at the hundredths place. They are different and 8 is larger than 7.

**32.48 is the larger number.**

**EXAMPLE**  Write two inequalities for the numbers 0.4 and 0.07.
Since the whole-number parts are the same (0), we compare the digits in the tenths place. 0.4 is larger because 4 is larger than 0.

**$0.4 > 0.07$    or    $0.07 < 0.4$**

Another procedure for comparing decimals is to affix an appropriate number of zeros to make each number have the same number of decimal places.

$$0.4 \; = \; 0.40 \qquad \frac{4}{10} \; = \; \frac{40}{100}$$

Now, compare 0.40 and 0.07. The larger number is 0.40. Then, $0.40 > 0.07$.

**TIP!**

**Common Denominators in Decimals**

The denominator of an equivalent decimal fraction is determined by the number of decimal places in a number. Decimal fractions have a common denominator if they have the same number of digits to the right of the decimal point. See common denominators of fractions on page 74.

**4  Round a Whole Number or a Decimal Number to a Specified Place Value.**

**Rounding** a number means finding the closest *approximate number* to a given number. For example, if 37 is rounded to the nearest ten, is 37 closer to 30 or 40? Locate 37 on the number line.

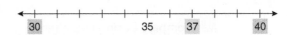

37 is closer to 40 than 30. Thus, 40 is a better approximation to the nearest ten for 37. Another way to say this is that 37 rounded to the nearest ten is 40.

When rounding a number to a certain place value, we must make sure that we are as accurate as our employer wants us to be. Generally, the size of the number and its use dictate the decimal place to which it should be rounded.

**TIP!**

**Procedures Do Not Substitute for Understanding Concepts**

Once you understand that rounding means to find the closest approximate number to a given number, we can use procedures or rules to facilitate rounding. Procedures are meaningless and are easily confused or forgotten if you do not understand the concept.

Chapter 1 / Review of Basic Concepts

## To round a whole or decimal number to a given place value:

1. Locate the digit that occupies the rounding place. Then examine the digit to the immediate right.
2. If the digit to the right of the rounding place is 0, 1, 2, 3, or 4, do not change the digit in the rounding place. If the digit to the right of the rounding place is 5, 6, 7, 8, or 9, add 1 to the digit in the rounding place.
3. Replace all digits to the *right* of the digit in the rounding place with zeros if they are to the left of the decimal point. Drop digits that are to the right of the digit in the rounding place *and also* to the right of the decimal point.

---

**EXAMPLE**

**AG/H**

Oregon has a land area of 96,187 square miles. What would be a reasonable approximate number for this land area?

96,187 rounds to the following approximate numbers:

  96,190 to the nearest ten
  96,200 to the nearest hundred
  96,000 to the nearest thousand
100,000 to the nearest ten thousand

Deciding to which place to round a value is a judgment depending on what use you will make of the rounded or approximate value.

**Both 96,000 and 100,000 are reasonable approximations.**

---

**EXAMPLE**

Round 46.897 to the hundredths place.

46.8⑨7    9 is in the hundredths place.
46.8⑨7    The next digit to the right is 7, so add 1 to 9. (9 + 1 = 10 and 89 + 1 = 90.)
             Drop the 7 in the thousandths place because it is to the right of the rounding place and also to the right of the decimal.

**46.90**

---

**TIP!**

### Nine Plus One Still Equals Ten

When the digit in the rounding place is 9 and must be rounded up, it becomes 10. The 0 replaces the 9 and the 1 is carried to the next place to the left.

---

**EXAMPLE**

Round $293.48 to the nearest dollar.

$29③.48    When we round to the nearest dollar, we are rounding to the *ones* place.

**$293**

---

**EXAMPLE**

Round $71.8986 to the nearest cent.

$71.8⑨86    One cent is 1 hundredth of a dollar, so to round to the nearest cent is to round to the *hundredths* place.

**$71.90**

---

### Exact Amount Versus Approximate Amount

When an amount has been rounded, it is no longer an exact amount. The rounded amount is now an **approximate amount.**

**5** **Add and Subtract Whole Numbers and Decimals.**

In an addition problem, the numbers being added are called **addends** and the answer is called the **sum** or **total.**

Two useful properties to know about addition are that it is **commutative** and **associative.** By commutative, we mean that it does not matter in what *order* we add numbers. We can add 7 and 6 in any order and still get the same answer:

$$7 + 6 = 13 \qquad 6 + 7 = 13$$

By *associative,* we mean that we can *group* numbers together any way we want when we add and still get the same answer. To add $7 + 4 + 6$, we can group the $4 + 6$ to get 10; then we add the 10 to the 7:

$$7 + (4 + 6) = 7 + 10 = 17$$

Or we can group the $7 + 4$ to get 11; then we add the 6 to the 11:

$$(7 + 4) + 6 = 11 + 6 = 17$$

Addition is a **binary operation;** that is, the rules of addition apply to adding *two* numbers at a time. The associative property of addition shows how addition is extended to more than two numbers.

### Using Symbols to Write Rules and Definitions

Many rules and definitions can be written symbolically. Symbolic representation allows a quick recall of the rule or definition.

**Commutative Property of Addition:** Two numbers may be added in any order and the sum remains the same.

$$a + b = b + a \quad \text{where } a \text{ and } b \text{ are numbers.}$$

**Associative Property of Addition:** Three numbers may be added using different groupings and the sum remains the same.

$$a + (b + c) = (a + b) + c \quad \text{where } a, b, \text{ and } c \text{ are numbers.}$$

The associative property of addition also allows other possible groupings and extends to more than three numbers.

$$7 + 4 + 6 = 13 + 4 = 17$$
$$3 + 5 + 7 + 9 = 8 + 16 = 24$$

Adding zero to any number results in the same number. This property is called the **zero property of addition** and zero is called the **additive identity.**

---

**Zero property of addition:**

Adding zero to any number results in the same number.

$$n + 0 = n \quad \text{or} \quad 0 + n = n$$
$$5 + 0 = 5 \quad \text{or} \quad 0 + 5 = 5$$

---

When adding numbers of two or more digits, the same place values must be aligned under one another so that all the ones are in the far right column, all the tens in the next column, and so on.

**To add numbers of two or more digits:**

1. Arrange the numbers in columns so that the ones place values are in the same column.
2. Add the ones column, then the tens column, then the hundreds column, and so on, until all the columns have been added. **Carry** whenever the sum of a column is more than one digit. This is also called **regrouping.**

---

**EXAMPLE**  Shipping fees are often charged by the total weight of the shipment. Find the total weight of this order: nails, 250 pounds (lb); tacks, 75 lb; brackets, 12 lb; and screws, 8 lb. Arrange in columns and add.

```
   11
  250      The sum of the digits in the right column is 15. Record the 5 in the ones column and carry
   75      the 1 to the tens column.
   12      Add the tens column. 1 + 5 + 7 + 1 = 14. Carry the 1.
 +  8
  345
```

**The total weight is 345 lb.**

---

When adding, decimal numbers are aligned so that all decimal points fall in the same vertical line. Aligning decimal points has the same effect as using *like* denominators when adding (or subtracting) fractions. Like denominators are discussed in Chapter 2.

**To add decimals:**

1. Arrange the numbers so that the decimal points are in one vertical line.
2. Add each column.
3. Align the decimal for the sum in the same vertical line.

**EXAMPLE**   Add 42.3 + 17 + 0.36.

42.3
17       Note that the decimal in 17 is understood to be at the right end.
0.36

42.30    If we prefer, we may write each number so that all have the same number of decimal places
17.00    by attaching zeros on the right.
0.36

**59.66**

Estimating is an increasingly important skill to develop and it relates in part to your *number sense.* You aren't born with number sense. It must be developed. We develop and strengthen our number sense by estimating and doing mental calculations. In making calculations it is important to **estimate** and **check** your work.

In estimating or rounding we often refer to a nonzero digit. A **nonzero digit** is a 1, 2, 3, 4, 5, 6, 7, 8, or 9. That is, it is a digit that is not zero.

### To estimate the answer to an addition problem:

1. Round each addend to a specific place value or to a number with one nonzero digit.
2. Add the rounded addends.

### To check an addition problem:

1. Add the numbers a second time and compare with the first sum.
2. Use a different order or grouping if convenient.

**EXAMPLE**   Elston Home Renovators spent the following amounts on a job: $16,466.15, $23,963.10, and $5,855.20. Estimate the total amount by rounding to thousands. Then find the exact amount and check your answer.

| Thousands Place | Estimate | Exact | Check |
|---|---|---|---|
| $16,466.15 | $16,000 | $16,466.15 | $16,466.15 |
| 23,963.10 | 24,000 | 23,963.10 | 23,963.10 |
| 5,855.20 | 6,000 | 5,855.20 | 5,855.20 |
| | **$46,000** | **$46,284.45** | **$46,284.45** |

The estimate and exact answer are close. The exact answer is reasonable.

**TIP!**

## Decimal Point and Zeros on the Calculator

The decimal key $\boxed{\cdot}$ is most often located near the number keys on a calculator. This key is pressed when the decimal point appears in the number being entered.

- Does the zero to the left of the decimal have to be entered before the decimal in a number like 0.5?
  Check it out: Add 3 + 0.2 on the calculator.

  Options:    3 $\boxed{+}$ 0 $\boxed{\cdot}$ 2 $\boxed{=}$ $\Rightarrow$    3.2 entering the zero

                    3 $\boxed{+}$ $\boxed{\cdot}$ 2 $\boxed{=}$ $\Rightarrow$    3.2 not entering the zero

- Do zeros that follow decimals have to be entered?
  Check it out: Add $3.00 + $1.50 on the calculator.

  Options:    3 $\boxed{\cdot}$ 0 0 $\boxed{+}$ 1 $\boxed{\cdot}$ 50 $\boxed{=}$ $\Rightarrow$    4.5 entering ending zeros

                    3 $\boxed{+}$ 1 $\boxed{\cdot}$ 5 $\boxed{=}$ $\Rightarrow$    4.5 not entering ending zeros

Ending zeros to the right of the decimal are usually dropped in a calculator display unless the calculator is set to display a specific number of decimal places.

Subtraction is the **inverse operation** of addition. In addition, we add numbers to get their total (such as 5 + 4 = 9), but to solve the subtraction problem 9 − 5 = ? we ask, "What number must be added to 5 to give us 9?" The answer is 4 because 4 added to 5 gives a total of 9. When we subtract two numbers, the answer is called the **difference** or **remainder**. The initial quantity is the **minuend**. The amount being subtracted from the initial quantity is the **subtrahend**.

Subtraction is *not* commutative. 8 − 3 = 5, but 3 − 8 does not equal 5; that is, 3 − 8 ≠ 5. The symbol ≠ is read "is not equal to."

Subtraction is not associative.

**EXAMPLE**    Show that 9 − (5 − 1) does not equal (9 − 5) − 1.

$$9 - (5 - 1) = 9 - 4 = \mathbf{5}, \text{ but } (9 - 5) - 1 = 4 - 1 = \mathbf{3}$$

When subtracting two or more numbers, if no grouping symbols are included, perform subtractions from left to right.

**EXAMPLE**    Subtract 8 − 3 − 1.

$$8 - 3 - 1 = 5 - 1 = \mathbf{4}$$    Subtract from left to right.

**Subtracting zero from a number results in the same number:**

$$n - 0 = n, \qquad 7 - 0 = 7$$

## Subtraction and Zeros

Subtracting a number from zero is not the same as subtracting zero from a number; that is, $7 - 0 = 7$, but $0 - 7$ does not equal 7.

### To subtract numbers of two or more digits:

1. Arrange the numbers in columns, with the minuend at the top and the subtrahend at the bottom.
2. Make sure the ones digits are in a vertical line on the right.
3. Subtract the ones column first, then the tens column, the hundreds column, and so on.
4. To subtract a larger digit from a smaller digit in a column, **regroup** by subtracting 1 from the digit in the next column to the left. This is the equivalent to *one* group of 10; thus, add 10 to the digit in the given column, Then, continue subtracting. The concept of *regrouping* is also referred to as **borrowing.**

EXAMPLE  Subtract $9,327 - 3,514$.
Arrange in columns.

$$
\begin{array}{r}
^{8\ 13}\\
9,3\,27\\
-\ 3,5\,14\\
\hline
5,8\,13
\end{array}
$$

Arrange in columns.

In the hundreds place, 5 is more than 3. Regroup by subtracting 1 group of 10 from 9.

$9 - 1 = 8$, $10 + 3 = 13$.

## Words and Phrases That Imply Subtraction

These phrases indicate subtraction in applied problems:

how many are left     how many more
how much less         how much larger
how much smaller

Also, some applied problems require more than one operation. Many problems that require more than one operation involve parts and a total. If we know the total and all the parts but one, we can add all the known parts and subtract the result from the total to find the missing part.

EXAMPLE  The Froehlichs left Memphis and drove 356 mi on the first day of their vacation. They drove 426 mi on the second day. If they are traveling to Albuquerque, which is 1,050 mi from Memphis, how many more miles do they have to drive?
The phrase *how many more* indicates subtraction.

$$1{,}050 \text{ mi} = \text{total miles}$$
$$356 + 426 + \text{mi left to drive} = 1{,}050 \text{ mi}$$

To find the miles left to drive, add $356 + 426$ and subtract the result from 1,050.

$$356 + 426 = \boxed{782} \qquad 1{,}050 - \boxed{782} = \textbf{268 miles}$$

## To subtract decimals:

1. Arrange the numbers so that the decimal points align vertically.
2. Subtract each column beginning at the right.
3. Interpret blank places as zeros.
4. Place the decimal in the difference in the same vertical line.

**EXAMPLE**    Subtract 7.18 from 15.

Take care to align the decimals properly.

$$\begin{array}{r} 15. \\ -\ 7.18 \\ \hline \end{array}$$   Because 15 is a whole number, its decimal point is placed after the 5.

$$\begin{array}{r} 15.00 \\ -\ 7.18 \\ \hline \textbf{7.82} \end{array}$$   To subtract, we put zeros in the tenths and hundredths places of 15 and then borrow.

When a worker machines an object using a blueprint as a guide, a certain amount of variation from the blueprint specification is allowed for the machining process. This variation is called the **tolerance.** Thus, if a blueprint calls for a part to be 9.47 in. with a tolerance of ± 0.05 (read **plus or minus** five hundredths), this means that the actual part can be 0.05 in. *more* or 0.05 in. *less* than the specification. To find the *largest* acceptable measure of the object, we add $9.47 + 0.05 = 9.52$ in. To find the *smallest* acceptable size of the object, we subtract $9.47 - 0.05 = 9.42$ in. *The dimensions 9.52 in. and 9.42 in. are called the* **limit dimensions** *of the object.* That is, 9.52 in. is the largest acceptable measure, and 9.42 in. is the smallest acceptable measure.

**EXAMPLE**

**CON**

Find the limit dimensions of an object with a blueprint specification of 8.097 in. and a tolerance of ± 0.005 in. (This is often written 8.097 ± 0.005.)

8.097 in. = the blueprint specification for the dimension of an object.
± 0.005 in. = tolerance of object's dimension.
Smallest dimension for object = blueprint specification − tolerance.
Largest dimension for object = blueprint specification + tolerance.

**Estimation**

The tolerance of the part is very small—just five thousandths—so the limit dimensions for the part should be only a few thousandths of an inch smaller or larger than the blueprint specification.

$8.097 - 0.005 = 8.092$
$8.097 + 0.005 = 8.102$

**The smallest acceptable dimension for the object is 8.092 in. and the largest acceptable dimension is 8.102 in.**

Estimating a subtraction problem is similar to estimating an addition problem. The numbers in the problem are rounded before the subtraction is performed.

## To estimate the difference:

1. Round each number to the desired place value or to a number with one nonzero digit.
2. Subtract the rounded numbers.

To check subtraction, we can use the inverse relationship between addition and subtraction. If $9 - 5 = 4$, then $4 + 5$ should equal 9.

## To check a subtraction problem:

1. Add the subtrahend and difference.
2. Compare the result of Step 1 with the minuend. If the two numbers are equal, the subtraction is correct.

---

**EXAMPLE**  Estimate by rounding to hundreds, then find the exact difference, and check.

$$427.45 - 125$$

|  | Estimate | Exact | Check |
|---|---|---|---|
| 427.45 | 400 | 427.45 | 125.00 |
| − 125.00 | − 100 | − 125.00 | + 302.45 |
|  | **300** | **302.45** | **427.45** |

---

**EXAMPLE**  Find the difference between $53,943.76 and $34,256.45 using a calculator.
One option:

53943 $\boxed{\cdot}$ 76 $\boxed{-}$ 34256 $\boxed{\cdot}$ 45 $\boxed{=}$    $\boxed{\text{ENTER}}$ or $\boxed{\text{EXE}}$ may replace $\boxed{=}$.

Calculator display: **19687.31**

---

**TIP!**

### What Happens If the Numbers Are Entered in the Wrong Order?

Rework the previous example by entering the smaller number first.

$$34256.45 \boxed{-} 53943.76 \boxed{=}$$

The display shows $-19687.31$ as the difference. Is this correct? No. Why not? The larger number should be entered first.

Since the digit portion of the display is the same as the correct answer, is it acceptable to just ignore the negative sign? No. 19687.31 and $-19687.31$ are not equal. Negative numbers are discussed in Chapter 4.

Two cuts are made from a 72-in. pipe (see Fig. 1–3). The two lengths cut from the pipe are 28 in. and 15 in. How much of the pipe is left after these cuts are made?

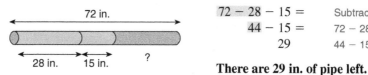

72 in.

28 in.　15 in.　?

$72 - 28 - 15 =$    Subtract from left to right.
$44 - 15 =$    $72 - 28 = 44$
$29$    $44 - 15 = 29$

**There are 29 in. of pipe left.**

**Figure 1–3** Lengths cut from pipe.

When more than one step is required to solve a problem, there is usually more than one way to solve the problem. In the previous example we could have found the total of the cuts first and then subtracted.

28 in. $+$ 15 in. $=$ 43 in.    Sum of two cuts.
72 in. $-$ 43 in. $=$ 29 in.    Remaining length of pipe.

### 6   Multiply and Divide Whole Numbers and Decimals.

**Multiplication** is repeated addition. If we have three \$10 bills, we have \$10 $+$ \$10 $+$ \$10, or \$30. Using multiplication, we see that this is the same as 3 times \$10, or \$30.

When we multiply two numbers, the first number is called the **multiplicand,** and the number we multiply by is called the **multiplier.** Either number is referred to as a **factor.** The answer or result of multiplication is called the **product.**

$$2 \qquad \times \qquad 3 \qquad = \qquad 6$$
multiplicand        multiplier        product
or factor          or factor

TIP!

**Various Notations for Multiplication**

Besides the familiar $\times$ or "times" sign, parentheses ( ), a raised dot ($\cdot$), and an asterisk ($*$) are also used to show multiplication. Parentheses are most often used as notation for multiplication.

$$2(3) = 6, \qquad (2)(3) = 6, \qquad 2 \cdot 3 = 6, \qquad 2 * 3 = 6$$

Multiplication is commutative and associative, just like addition. The **commutative property of multiplication** permits two numbers to be multiplied in any order. In symbols, $a(b) = b(a)$.

$$4(5) = 20, \qquad 5(4) = 20$$

When more than two numbers are multiplied, the numbers must be grouped, and the **associative property of multiplication** permits the numbers to be grouped in any way. In symbols, $a(b \cdot c) = (a \cdot b)c$.

$$2(3 \cdot 5) \quad \text{or} \quad (2 \cdot 3)5$$
$$2(15) \qquad\qquad (6)5$$
$$30 \qquad\qquad\quad 30$$

**EXAMPLE**  Multiply 3(2)(9)

$$3(2)(9)$$  Group any two factors.
$$(3 \cdot 2)(9)$$  Multiply grouped factors.
$$6\,(9)$$  Multiply the factors: 6 and 9.

**54**

---

**The product of a number and zero is zero:**

$$n \times 0 = 0, \qquad 0 \times n = 0, \qquad 4 \times 0 = 0, \qquad 0 \times 4 = 0$$

This is called the **zero property of multiplication.**

---

**EXAMPLE**  Multiply 2(5)(0)(7)

$$(2)(5)(0)(7)$$  Group factors.
$$(2 * 5)(0 * 7)$$  Multiply each group of factors.
$$(10)(0)$$  Multiply products.

**0**

---

**To multiply factors of two or more digits:**

1. Arrange the factors one under the other.
2. Multiply each digit in the multiplicand by each digit in the multiplier. The product of the multiplicand and each digit in the multiplier gives a **partial product.**
   (a) To start, multiply the ones digit in the multiplier by the multiplicand from right to left.
   (b) Align each partial product with its first digit directly under its multiplier digit.
3. Add the partial products.

---

**EXAMPLE**  Multiply (204)(103).

```
      204
  ×   103
      612        Multiply: 3 × 204 = 612. Align 612 under 3 in the multiplier.
    0 00         Multiply: 0 × 204 = 000. Align 000 under 0 in the multiplier.
   20 4          Multiply: 1 × 204 = 204. Align 204 under 1 in the multiplier.
  21,012         Add the partial products as they are aligned.
```

Partial products 000 and 204 could be combined on a single line.

```
      204
  ×  ˙103
      612
   20 40         0 × 204 = 0. Align under 0 in multiplier.
  21,012         1 × 204 = 204. Align under 1 in multiplier on the same line as the previous partial product.
```

Since decimals are fractions, multiplication of decimals causes us to rethink or expand our basic number sense of multiplication. The product of any number and a decimal number less than 1 is less than the original number.

## To multiply decimal numbers:

1. Align the numbers as if they were whole numbers and multiply.
2. Count the total number of digits to the right of the decimal in each factor.
3. Place the decimal in the product so that the number of decimal places is the sum of the number of decimal places in the factors.

---

**EXAMPLE**   Multiply $1.36 \times 0.2$.

$$\begin{array}{r} 1.36 \\ \times \ \ 0.2 \\ \hline \mathbf{0.272} \end{array}$$

Note that in multiplication the decimals do *not* have to be in a straight line.

The product has three places to the right of the decimal. Place a zero in the ones place so that the decimal point will not be overlooked.

---

**EXAMPLE**   Multiply $0.309 \times 0.17$.

$$\begin{array}{r} 0.309 \\ \times \ \ 0.17 \\ \hline 2163 \\ 309 \ \ \ \\ \hline \mathbf{0.05253} \end{array}$$

No decimals are placed in the partial products.

We did not have enough digits in the product for five decimal places, so we inserted a zero on the *left*.

---

**EXAMPLE**

**AG/H**

7.82 m
(outside
diameter)

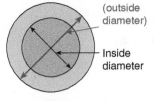

Inside
diameter

**Figure 1–4**

The outside diameter of a flower bed is 7.82 meters (m) (see Fig. 1–4). If the brick walk surrounding the bed is 1.56 m thick, find the inside diameter of the flower bed.

The diameter of a circle is the distance across the center of the circle.
7.82 m = outside diameter of the flower bed
1.56 m = width of brick walk surrounding the flower bed
Outside diameter − two widths (one on each end of the inside diameter) of the walk = inside diameter

**Estimation**

The outside diameter is nearly 8 m, and the total of the two widths to be subtracted is about 3 m, so the inside diameter should be about 5 m.

$$7.82 - (2 \times 1.56) = 7.82 - 3.12 = 4.70$$

**The inside diameter of the flower bed is 4.70 m.**

---

The **distributive property of multiplication** means that multiplying a sum or difference by a factor is equivalent to multiplying each term of the sum or difference by the factor.

1. Add or subtract the numbers within the grouping.
2. Multiply the result of Step 1 by the factor outside the grouping.

or

1. Multiply each number inside the grouping by the factor outside the grouping.
2. Add or subtract the products from Step 1.

Symbolically, $a(b + c) = ab + ac$    or    $a(b - c) = ab - ac$.

TIP!

### Additional Notations for Multiplication

- Parentheses show multiplication when the distributive property is used: $a(b + c)$ means $a \times (b + c)$.
- The letters represent numbers.
- Letters written together with no operation sign between them imply multiplication: $ab$ means $a \times b$; $ac$ means $a \times c$.

**EXAMPLE**    Multiply $3(2 + 4)$.

Multiplying first gives:          Adding first gives:
$$3(2 + 4) =$$                      $$3(2 + 4) =$$
$$3(2) + 3(4) =$$                   $$3(6) = \mathbf{18}$$
$$6 + 12 = \mathbf{18}$$

**EXAMPLE**    Multiply $2(6 - 5)$.

Multiplying first gives:          Subtracting first gives:
$$2(6 - 5) =$$                      $$2(6 - 5) =$$
$$2(6) - 2(5) =$$                   $$2(1) = \mathbf{2}$$
$$12 - 10 = \mathbf{2}$$

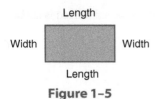

**Figure 1–5**

Perimeter of rectangle.

The distributive property is found in many formulas. One example is the formula for the perimeter of a rectangle. A **rectangle** is a four-sided geometric shape whose opposite sides are equal in length, and each corner makes a square corner (see Fig. 1–5). The **perimeter** of a rectangle is the distance around the figure. We use the perimeter when we fence a rectangular yard, install baseboard in a rectangular room, frame a picture, or outline a flower bed with landscaping timbers.

The **formula** for finding the perimeter of a rectangle is

$$P = 2(l + w) \qquad \text{or} \qquad P = 2l + 2w$$

EXAMPLE

AG/H

Find the number of feet of fencing needed to enclose a rectangular pasture that is 1,784.6 feet (ft) long and 847.3 ft wide.

$P = 2(l + w)$      or      $P = 2l + 2w$

$P = 2(1,784.6 + 847.3)$          $P = 2(1,784.6) + 2(847.3)$

$P = 2(2,631.9)$                $P = 3,569.2 + 1,694.6$

$P = 5,263.8$                  $P = 5,263.8$

**The amount of fencing needed is 5,263.8 ft.**

## To estimate the answer for a multiplication problem:

1. Round both factors to a chosen or specified place value or to one nonzero digit.
2. Then multiply the rounded numbers.

## To check a multiplication problem:

1. Multiply the numbers a second time and check the product.
2. Interchange the factors if convenient.

EXAMPLE

AG/H

Find the approximate and exact costs of 48 flower bulbs if each bulb costs $2.15. Estimate the cost by rounding each factor to a number with one nonzero digit. Then find the exact cost and check your work.

48 = total number of flower bulbs
$2.15 = cost of each bulb

Total cost of flower bulbs = number of bulbs × cost of each bulb

**Estimation**

If 50 bulbs were purchased at $2 each, the total cost would be $100. So the exact cost should be close to $100.

48 × $2.15 = $103.20     Number of bulbs × Cost of each bulb

**The total cost of 48 flower bulbs is $103.20, which is approximately $100, as noted in the estimation.**

      The **area** of a geometric shape is the number of square units needed to cover the shape. We use area when we find the amount of carpet needed to cover a rectangular floor, the amount of paint needed to cover a wall, the amount of asphalt needed to pave a parking lot, the amount of fertilizer needed to treat a yard, or the amount of material needed to produce a rectangular

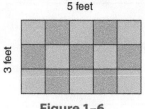

5 feet

3 feet

**Figure 1–6**
Area of a rectangle.

sign. Area is measured in **square units.** One square foot (1 ft²) indicates a square that measures 1 foot on each side.

5 ft long by 3 ft wide = 15 ft²

To find the area of a rectangle, we multiply the length by the width (see Fig. 1–6). We can state this in symbols using a formula.

$A = lw$    Area of a rectangle = length times width

---

EXAMPLE

AG/H

Maintenance Consultants needs to apply fertilizer to a customer's lawn. The lawn is 223.4 ft long and 132.8 ft wide. The fertilizer costs $0.004 per square foot to apply. What is the cost of applying the fertilizer?

Lawn is in the shape of a rectangle.
Length of the lawn = 223.4 ft.
Width of the lawn = 132.8 ft.
Fertilizer costs $0.004 per square foot to apply.
Area of a rectangle is the product of the length times the width ($A = lw$).
Total cost = total number of square feet × cost per square foot.

Estimation

A lawn 200 ft by 100 ft contains 20,000 ft², and at a cost of $0.004 per square foot, the total cost is approximately $80. Note: the cost will be significantly higher because we rounded both length and width *down*.

$A = lw$
$A = 223.4(132.8)$
$A = 29,667.52$ ft²

cost = 29,667.52($0.004)
cost = $118.67      Rounded.

**The total cost of applying fertilizer to the lawn is $118.67, which is significantly higher than the estimated cost of $80, as we anticipated.**

---

When either or both factors of a multiplication problem end in zeros, a shortcut process such as the one in the following example can be used.

---

EXAMPLE

Multiply $2,600 \times 70$.

1.     2600
     × 70          Separate the ending zeros from the other digits.

2.     2600
     × 70          Multiply the other digits as if the zeros were not there (26 × 7 = 182).
     182

3.     2600          Attach the zeros to the basic product. Note that the number of zeros affixed to
     × 70           the basic product is the same as the sum of the number of zeros at the end
     **182,000**     of each factor.

---

This process is sometimes necessary whenever a multiplication problem is too long to fit into a calculator. Many basic calculators have only an eight-digit display window. The problem 26,000,000 × 3,000 would not fit into many basic calculators. This shortcut allows us to work the problem with or without a calculator.

**EXAMPLE** Multiply $26,000,000 \times 3,000$.

$$
\begin{array}{r}
26,000,000 \\
\times \quad 3,000 \\
\hline
78,000,000,000
\end{array}
$$

Separate ending zeros and multiply $26 \times 3$.
Attach 9 zeros.

**TIP!**

### The Mind Is Often Quicker Than the Fingers

Don't use your calculator as a crutch. It is a tool! When multiplying 2,500 times 30, you can multiply 25 times 3 mentally. $25 \times 3 = 75$. Then, attach three zeros to that product.

$$2,500 \times 30 = 75,000$$

This skill does not come automatically. You have to practice it! Like playing a musical instrument or mastering a sport, you don't develop skill by watching. You have to practice!

**Division** is the **inverse operation** of multiplication. Since $4 \times 7 = 28$, then 28 divided by 7 is 4 and 28 divided by 4 is 7. The number being divided is called the **dividend.** The number divided by is the **divisor.** The result is the **quotient.**

Division is *not* commutative. $12 \div 6 = 2$, but $6 \div 12$ does not equal 2; that is, $6 \div 12 \neq 2$.

Division is *not* associative. $(12 \div 6) \div 2 = 2 \div 2 = 1$. But $12 \div (6 \div 2) = 12 \div 3 = 4$. That is, $(12 \div 6) \div 2 \neq 12 \div (6 \div 2)$.

**TIP!**

### To Write Division Symbolically:

**1.** To use the *divided by* symbol ($\div$), write the dividend first.

$$28 \div 7 = 4 \longleftarrow \text{quotient}$$
dividend $\longrightarrow$ $\quad$ $\longleftarrow$ divisor

**2.** To use the *long-division symbol* ($\overline{)}$), write the dividend under the bar.

$$\begin{array}{r} 4 \longleftarrow \text{quotient} \\ 7\overline{)28} \end{array}$$
divisor $\longrightarrow$ $\quad$ $\longleftarrow$ dividend

**3.** To use the *division bar* or *slash symbol,* write the dividend on top or first.

$\longleftarrow$ dividend $\longrightarrow$
$$\frac{28}{7} = 4 \longleftarrow \text{quotient} \quad \text{or} \quad 28/7 = 4 \longleftarrow \text{quotient}$$
$\longleftarrow$ divisor $\longrightarrow$

## Which Number Goes First When Dividing?

When using the "divided by" symbol, the division bar, or the slash as division, the dividend is written first or on top. The divisor is written second or on the bottom.

Technology tools such as the calculator and computer normally use the "divided by" or slash symbol.

The divisor is written first *only* when using the long division symbol.

Long division, like long multiplication, involves using one-digit multiplication facts repeatedly to find the quotient.

When the quotient is not a whole number, the quotient may have a *whole-number part* and a **remainder.** When a dividend has more digits than a divisor, parts of the dividend are called **partial dividends,** and the quotient of a partial dividend and the divisor is called a **partial quotient.**

## To divide whole numbers:

1. Beginning with its leftmost digit, identify the first group of digits of the dividend that is larger than or equal to the divisor. This group of digits is the first *partial dividend.*
2. For each partial dividend in turn, beginning with the first:
   (a) Divide the partial dividend by the divisor. Write this partial quotient above the rightmost digit of the partial dividend.
   (b) Multiply the partial quotient by the divisor. Write the product below the partial dividend, aligning places.
   (c) Subtract the product from the partial dividend. Write the difference below the product, aligning places. The difference must be less than the divisor.
   (d) Next to the ones place of the difference, write the next digit of the dividend. This is the new partial dividend.
3. When all the digits of the dividend have been used, write the final difference in Step 2c as the remainder (unless the remainder is 0). The whole-number part of the quotient is the number written above the dividend.

**EXAMPLE**   To divide 881 by 35:

$$35\overline{)88\ 1}$$   The first partial dividend is 88.

$$\begin{array}{r} 2 \\ 35\overline{)88\ 1} \\ 70 \\ \hline 18 \end{array}$$   The partial quotient for $88 \div 35$ is 2. Multiply $2 \times 35 = 70$. Then subtract $88 - 70 = 18$. The difference 18 is less than the divisor 35.

$$\begin{array}{r} 2 \\ 35\overline{)881} \\ 70 \\ \hline 181 \end{array}$$   1 from the dividend is written next to 18 to form the next partial dividend.

$$\begin{array}{r} 25 \\ 35\overline{)881} \\ 70 \\ \hline 181 \\ 175 \\ \hline 6 \end{array}$$   The partial quotient for $181 \div 35$ is 5. The product of $5 \times 35$ is 175. The difference of $181 - 175$ is 6. The remainder is 6.

881 divided by 35 = **25 R6.**

### Importance of Placing the First Digit Carefully

The correct placement of the first digit in the quotient is critical. If the first digit is out of place, all digits that follow will be out of place, giving the quotient too few or too many digits.

**EXAMPLE**   To divide $5\overline{)2{,}535}$.

$$
\begin{array}{r}
5 \\
5\overline{)2{,}535} \\
2\,5 \\
\hline
03
\end{array}
$$

5 divides into 25 five times. Write 5 over the last digit of the 25. Subtract and bring down the 3.

$$
\begin{array}{r}
\mathbf{507} \\
5\overline{)2{,}535} \\
2\,5 \\
\hline
035 \\
35 \\
\hline
0
\end{array}
$$

5 divides into 3 zero times. Write 0 over the 3 of the dividend. Bring down the next digit, which is 5. Divide 5 into 35: $35 \div 5 = 7$. Write 7 over the 5 of the dividend. Multiply $7 \times 5 = 35$. Subtract.

Special properties are associated with division involving zeros.
The quotient of zero divided by a nonzero number is zero:

$$0 \div n = 0, \qquad n\overline{)0}, \qquad \frac{0}{n} = 0$$

$$0 \div 5 = 0, \qquad 5\overline{)0}, \qquad \frac{0}{5} = 0$$

The quotient of a number divided by zero is **undefined** or **indeterminate:**

$$n \div 0 \text{ is } \textit{undefined}, \qquad 0\overline{)n}\,^{\text{undefined}}, \qquad \frac{n}{0} \text{ is undefined}$$

$$12 \div 0 \text{ is } \textit{undefined}$$

$$0 \div 0 \text{ is } \textit{indeterminate}$$

Dividing any nonzero number by itself yields 1:

$$n \div n = 1, \qquad \text{if } n \text{ is not equal to zero;} \qquad 12 \div 12 = 1$$

Dividing any number by 1 yields the same number:

$$n \div 1 = n, \qquad 5 \div 1 = 5$$

Since division is *not* associative, if there are no grouping symbols, we divide from left to right.

**EXAMPLE**   Divide $16 \div 4 \div 2$.

$$16 \div 4 \div 2 = 4 \div 2 = 2$$   Divide from left to right.

## To divide decimal numbers written in long-division form:

1. Move the decimal in the divisor so that it is on the right side of all digits. (By moving the decimal, you are multiplying by 10, 100, 1,000, and so on.)
2. Move the decimal in the dividend to the right as many places as the decimal was moved in the divisor. Attach zeros if necessary. (This is multiplying the dividend by the same number as was used in Step 1.)
3. Write the decimal point in the answer directly above the new position of the decimal in the dividend. (Do this *before* dividing.)
4. Divide as you would in whole numbers.

---

**EXAMPLE**   Divide $4.8 \div 6$.

$$6\overline{)\overset{\cdot}{4.8}}$$   Insert a decimal point above the decimal point in the dividend.

When the divisor is a whole number, the decimal is understood to be to the right of 6 and is not moved. The decimal is placed in the quotient directly above the decimal in the dividend.

$$6\overline{)\overset{\mathbf{0.8}}{4.8}}$$   Divide.

---

**EXAMPLE**   Divide $3.12 \div 1.2$.

$$1.2\overline{)3.12}$$   Move the decimal in the divisor and dividend.

$$12\overline{)\overset{\cdot}{31.2}}$$   Write the decimal in the quotient, then divide.

$$
\begin{array}{r}
2.6 \\
12\overline{)31.2} \\
\underline{24} \\
7\,2 \\
\underline{7\,2}
\end{array}
$$

---

## To round a quotient to a place value:

1. Divide to one place past the desired rounding place.
2. Attach zeros to the dividend after the decimal if necessary to carry out the division.
3. Round the quotient to the place specified.

---

**EXAMPLE**   Divide and round the quotient to the nearest tenth.

$$3.2\overline{)15.27}$$

$$
\begin{array}{r}
4.77 \\
3.2\overline{\smash{)}15.2\,70} \\
\underline{12\,8} \\
2\,4\,7 \\
\underline{2\,2\,4} \\
2\,30 \\
\underline{2\,24} \\
6
\end{array}
$$

Since we are rounding to the nearest tenth, divide to the hundredths place.

**4.77 rounds to 4.8.**

Estimating division is similar to estimating other operations. The numbers are rounded *before* the calculation is made.

**To estimate division:**

1. Round the divisor and dividend to one nonzero digit.
2. Find the first digit of the quotient.
3. Attach a zero in the quotient for each remaining digit in the dividend.

Because division and multiplication are inverse operations, division is checked by multiplication.

**To check division:**

1. Multiply the divisor by the quotient.
2. Add any remainder to the product in Step 1.
3. The result of Step 2 should equal the dividend.

**EXAMPLE**    Estimate, find the exact answer, and check $913 \div 22$.

Estimate:
$$
\begin{array}{r}
40 \\
20\overline{\smash{)}900}
\end{array}
$$
20 divides into 90 four whole times. Attach a zero after 4.

Exact:
$$
\begin{array}{r}
41\ \text{R}11 \\
22\overline{\smash{)}913} \\
\underline{88} \\
33 \\
\underline{22} \\
11
\end{array}
\qquad \text{or} \qquad
\begin{array}{r}
41.5 \\
22\overline{\smash{)}913.0} \\
\underline{88} \\
33 \\
\underline{22} \\
11\,0 \\
\underline{11\,0}
\end{array}
$$

Check:
$$
\begin{array}{r}
41 \\
\times\ 22 \\
\hline
82 \\
\underline{82} \\
902
\end{array}
\qquad
\begin{array}{r}
902 \\
+\ 11 \\
\hline
913
\end{array}
\qquad \text{or} \qquad
\begin{array}{r}
41.5 \\
2\,2 \\
\hline
83\,0 \\
\underline{830} \\
913.0
\end{array}
$$

**The answer checks.**

We use averages to make comparisons, such as when we compare the average mileage different cars get per gallon of gasoline. There are several types of averages.

In most courses students take, their numerical grade is determined by a process called **numerical averaging.** This average is also called the **arithmetic average,** or **mean.** If the grades are 92, 87, 76, 88, 95, and 96, we can find the average grade by adding the grades and dividing by the number of grades. Because we have six grades, we divide the sum by 6.

$$\frac{92 + 87 + 76 + 88 + 95 + 96}{6} = \frac{534}{6} = 89$$

## To find the average of a group of numbers or like measures:

1. Add the numbers or like measures.
2. Divide the sum by the number of addends.

**EXAMPLE**  A car involved in an energy efficiency study had the following miles per gallon (mpg) listings for five tanks of gasoline: 21.7, 22.4, 26.9, 23.7, and 22.6 mpg. Find the average miles per gallon for the five tanks of gasoline.

Estimate:  The low value is 21.7 and the high value is 26.9. The average will be between the two values.

Exact:  $$\frac{21.7 + 22.4 + 26.9 + 23.7 + 22.6}{5} = \frac{117.3}{5} = 23.46$$

$$= 23.5 \frac{mi}{gal} \qquad \text{Rounded.}$$

**The average is 23.5 $\frac{mi}{gal}$.**

**TIP!**

## Rounding Answers When Averaging

If the numbers being averaged are expressed to the same place value, the answer is usually rounded to that place value.

SECTION 1–1 SELF-STUDY EXERCISES

**1**  Write two inequalities to compare each pair of numbers.

**1.** 6 and 8
**2.** 42 and 32
**3.** 196 and 148
**4.** 2,802 and 2,517
**5.** 7,809 and 8,902
**6.** 44,000 and 42,999

7. **BUS** A house that sold for $183,500 four years ago has just sold for $198,500. Write two inequalities to compare the housing prices.

8. **AG/H** Touliatas Nursery sold 786 flats of annual bedding plants and 583 flats of perennial bedding plants. Write two inequalities to compare the number of plants of each type.

9. **HOSP** The Orlando Renaissance Resort sold 758 rooms for a horticulture convention and 893 rooms for a motorcycle trade show. Write two inequalities to compare the number of rooms sold for each of the events.

10. **COMP** The first day a spam filter was installed at Phoenix College, 5,982 spam e-mails were blocked. On the second day 2,807 spam emails were blocked. Write two inequalities to compare the number of blocked spam e-mails.

**2** Write the fractions as decimal numbers.

11. $\dfrac{5}{10}$    12. $\dfrac{23}{100}$    13. $\dfrac{7}{100}$    14. $\dfrac{683}{100}$    15. $\dfrac{79}{1000}$    16. $\dfrac{468}{1000}$

**3** Compare the number pairs and identify the larger number.

17. 3.72, 3.68    18. 7.08, 7.06    19. 0.23, 0.3

Arrange the numbers in order from smallest to largest.

20. 1.9, 1.87, 1.92
21. 72.1, 72.07, 73
22. **INDTEC** Two micrometer readings are recorded as 0.837 in. and 0.81 in. Which is larger?
23. **INDTEC** A micrometer reading for a part is 3.85 in. The specifications call for a dimension of 3.8 in. Which is larger, the micrometer reading or the specification?
24. **INDTR** A washer has an inside diameter of 0.33 in. Will it fit a bolt that has a diameter of 0.325 in.?
25. **INDTEC** Aluminum sheeting can be purchased in thicknesses of 0.04 in. or 0.035 in. Which sheeting is thicker?
26. **CON** If No. 14 copper wire has a diameter of 0.064 in., and No. 10 wire has a diameter of 0.09 in., which has the larger diameter?
27. **HLTH/N** A nurse recorded the weights of two patients as 64.8 kilograms and 72.3 kilograms. Which weight is greater?

Write two inequalities for each pair of numbers.

28. 4.2 and 3.8
29. 1.68 and 1.6
30. **INDTEC** A 100-watt bulb that burns continuously for two minutes uses 0.003 kilowatt-hours of electricity, and a 800-watt toaster uses 0.026 kilowatt-hours for a piece of toast. Which uses the greater number of kilowatt-hours?
31. To change centimeters to inches, multiply by 0.394, and to change kilometers to miles, multiply by 0.621. Which factor is larger?

**4** Round to the place value indicated.

32. Nearest hundred: 468
33. Nearest ten thousand: 429,207
34. Nearest billion: 82,629,426,021
35. Nearest ten million: 297,384,726
36. **CON** A micrometer measure is listed as 0.7835 in. Round this measure to the nearest thousandth.
37. **AUTO** To the nearest tenth, what is the current of a 2.836-amp (A) motor?
38. **HOSP** If round steak costs $2.78 per pound, what is the cost per pound to the nearest dollar?
39. **CON** The average response times (in seconds) for drivers braking when they first see a road hazard were measured as follows: driver A, 0.0275; driver B, 0.0264; driver C, 0.0234; driver D, 0.0284; and driver E, 0.0379. Round each response time to hundredths to identify the drivers whose response times were most similar.

**5** Write in columns and add.

40. 4,582 + 86,724 + 482 + 5,826
41. 6,017 + 893 + 15 + 82
42. 17 + 5,804 + 23,907 + 405
43. 4.2 + 3.6 + 7.9
44. 12.8 + 13.52 + 7.86
45. 83.37 + 42 + 1.6 + 3
46. **CON** A 0.103-in.-thick pipe has an inside diameter of 2.871 in. Find the outside diameter of the pipe. (Hint: The thickness of the pipe is on both sides of the inside diameter.)
47. **ELEC** The total current in amps in a parallel circuit is found by adding the individual currents. If a circuit has individual currents of 3.98 A, 2.805 A, and 8.718 A, find the total current.

**48. BUS** A part-time hourly worker earned $25.97 on Monday, $7.48 on Tuesday, $5.88 on Wednesday, $65.45 on Thursday, and $76.47 on Friday. Find the total week's wages.

**49. CON** A four-sided residential lot that measures 100.8 ft, 87.3 ft, 104.7 ft, and 98.6 ft is to be fenced. How many feet of fencing are required?

**50. HLTH/N** A patient's normal body temperature registered at 98.2° and his temperature rose 2.7°. What was his increased body temperature?

**51. BUS** Your investment portfolio totals $25,915.53 at the beginning of the year and it increases by $2,418.48 over the one-year period. What is the value of your portfolio at the end of the year?

Estimate the sum by rounding to hundreds. Then find the exact sum and check.

| **52.** | **53.** | **54.** | **55.** | **56.** |
|---|---|---|---|---|
| 34.07 | 52,843 | 0.935 | 24,003 | 24.381 |
| 15.962 | 17,497 | 12.4 | 5,874 | 1.1 |
| 5.81 | 13,052 | 152.07 | 319,467 | 17.92 |
| + 0.523 | + 821 | + 18 | + 52,855 | + 38 |

**57. CON** Palmer Associates provided the following prices for items needed to build a sidewalk: concrete, $2,583.45; wire, $43.25; frame material, $18.90; labor, $798. Estimate the cost by rounding each amount to the nearest ten. Find the exact total.

**58. BUS** Antonio's expenses for one semester are: food, $1,500; lodging, $1,285; books, $288; supplies, $130; transportation, $162. Estimate his expenses by rounding each amount to the nearest hundred. Calculate the exact amount.

**59. CON** A hardware store filled an order for nails: 25 lb, $2\frac{1}{2}$-in. common; 16 lb, 4-in. common; 12 lb, 2-in. siding; 24 lb, $2\frac{1}{2}$-in. floor brads; 48 lb, 2-in. roofing; and 34 lb, $2\frac{1}{2}$-in. finish. Find the total weight.

**60. AG/H** If four containers have a capacity of 12 gal, 27 gal, 55 gal, and 21 gal, can 100 gal of fuel be stored in these containers? (Find the total capacity of the containers first.)

**61. BUS** A printer has three printing jobs that require the following numbers of sheets of paper: 185, 83, and 211. Will one ream of paper (500 sheets) be enough to finish the three jobs?

**62. CON** How many feet of fencing are needed to enclose the area shown in Fig. 1–7?

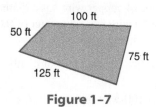

**Figure 1–7**

Subtract.

**63.** $3,672 - 2,652$

**64.** $946 - 831$

**65.** $53,867 - 831$

**66. CON** If a mason orders 75 bags of cement for a job and uses only 53, how many bags are left?

**67. BUS** An inventory sheet shows that 468 outlet boxes were in stock on March 1. Sales during March were 127. How many outlet boxes were left at the end of the month?

Subtract.

**68.** Subtract 24.38 from 316.2.

**69.** Subtract 13.5 from 21.

**70.** Subtract 67.2 from 378.

**71.** Find the difference between 42 and 37.6.

**72. HTH/N** Nurse Jennings recorded a temperature of 103.6°F for his patient. If the patient's normal body temperature was 98.6°F, how many degrees of fever did the patient have?

**73. INDTEC** One box of rivets weighs 52.6 lb and another box weighs 37.5 lb. How much more does the first box of rivets weigh?

**74. INDTEC** According to a blueprint, the length of an object is 12.09 in. If the tolerance is ±0.01 in., what are the limit dimensions of the object?

**75. CON** Two lengths of copper tubing measure 63.6 cm and 3.77 cm. What is the difference in their lengths?

**76. CAD/ARC** Find the limit dimensions of an object whose blueprint dimension is 4.195 in. ± 0.006 in.

**77. CON** A bricklayer laid 1,283 bricks on one day. A second bricklayer laid 1,097 bricks. How many more bricks did the first bricklayer lay?

**78. INDTEC** A stockroom has 285.8 in. of bar stock and 173.5 in. of round stock. How many inches of bar stock remain after the object in Fig. 1–8 is made from this stock?

**79. CON** In Exercise 78, how many inches of round stock remain after the object in Fig. 1–8 is made?

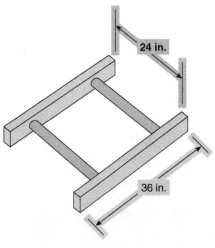

**Figure 1–8**

**80. CON** Find the missing dimension in Fig. 1–9.

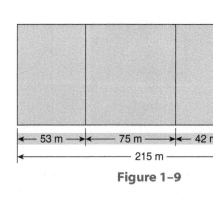

**Figure 1–9**

---

**6** Multiply the following.

**81.** (3)(7)(9)(2)

**82.** 90,000
× 7,000

**83.** 503
× 204

**84. BUS** Helene Wright counted 8 unopened boxes of washers. Each box contained 512 washers. What total number of washers will be shown on the inventory sheet?

**85. AUTO** Bill Wepner repaired 6 automobiles a week over a period of 14 weeks. How many automobiles did Bill repair during this time period?

**86. HELPP** Each officer in the Public Safety office wrote, on average, 25 tickets a week. If there are 7 officers, how many tickets were written over a 4-week period?

**87. BUS** Jossie Moore is planning to sell candy bars in her Smart Shop. She receives 12 boxes, and each box contains 24 candy bars. If Jossie sells the bars for $1 each, how much money will she get for all the bars?

**88.** 37.7
× 1.5

**89.** 9.27
× 0.35

**90.** 0.215
× 0.27

**91.** 0.271
× 0.32

**92.** 73.806(2.305)

**93.** 1.9067 · 0.2013

**94.** 8.2037 × 0.602

**95.** 42(0.73)

**96. CON** A plastic pipe has an inside diameter of 4.75 in. Find the outside diameter if the pipe wall is 0.25 in. thick.

**97. CON** Electrical switches cost $5.70 wholesale. If the retail price is $7.99 each, how much would an electrician save over retail by buying 12 switches at the wholesale price?

**98. CON** How much No. 24 electrical wire is needed to make eight pieces each 18.9 in. long?

**99. BUS** A retailer purchases 15 cases of potato chips at $8.67 per case. If the chips are sold at $12.95 per case, how much profit does the retailer make?

**100. CON** A contractor purchases 500 cubic yards (yd³) of concrete at $21.00 per cubic yard, 24 yd³ of sand at $5.00 per cubic yard, and 36 yd³ of fill dirt at $3.00 per cubic yard. What is the total cost of the materials?

Find the perimeter of the rectangular crop fields in Figs. 1–10 and 1–11.

**101. AG/H**

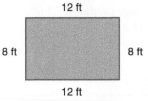

**Figure 1–10**

**102. AG/H**

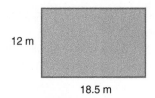

**Figure 1–11**

Find the number of square feet in each rectangular garden illustrated in Figs. 1–12 and 1–13. Check your work.

**103. CON**

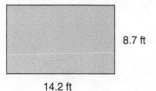

8.7 ft

14.2 ft

**Figure 1–12**

**104. CON**

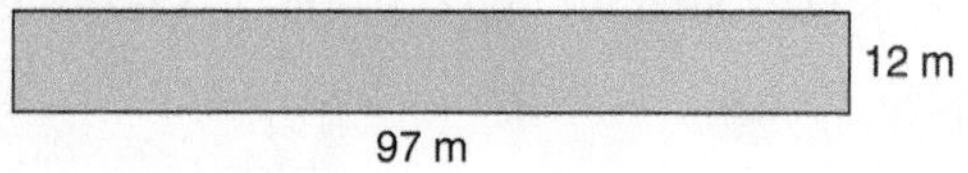

12 m

97 m

**Figure 1–13**

Perform the long-division problems.

**105.** $5\overline{)215}$

**106.** $37\overline{)1,739}$

**107. CON**  In a building where 46 outlets are installed, 1,472 ft of cable are used. What is the number of feet of cable used per outlet?

**108. INDTR**  Twelve water tanks are constructed in a welding shop at a total contract price of $14,940. What is the price per tank?

**109. INDTEC**  Five equally spaced holes are drilled in a piece of $\frac{1}{4}$-in. flat metal stock. The centers of the first and last holes are 2 in. from the end (see Fig. 1–14). What is the distance between the centers of any two adjacent holes? (*Caution:* How many equal center-to-center distances are there?)

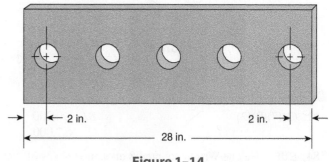

← 2 in.        2 in. →

28 in.

**Figure 1–14**

Divide.

**110.** $6.3\overline{)68.67}$    **111.** $0.23\overline{)0.0437}$    **112.** $8 \div 0.32$

**113. CON**  A 7.3-ft-long pipe weighs 43.8 lb. What is the weight of 1 ft of pipe?

**114. CON**  A room requires 770.5 ft² of wallpaper, including waste. How many whole single rolls are needed for the job if a roll covers 33.5 ft² of surface?

**115. CON**  The feed per revolution of a drill is 0.012 in. A hole 7.2 in. deep will require how many revolutions of the drill?

**116. INDTR**  A piece of channel iron 5.6 ft long is cut into eight pieces. Assuming that there is no waste, what is the length of each piece?

Divide and round the quotient to the place indicated.

**117.** Nearest hundredth: $25\overline{)3.897}$

**118.** Nearest whole number: $4.1\overline{)34.86}$

**119. INDTR**  If 12 lathes cost $6,895, find the cost of each lathe to the nearest dollar.

**120. ELEC**  If 12 electrolytic capacitors cost $23.75, find the cost of one capacitor to the nearest cent.

**121. CON**  Three bricklayers laid 3,210 bricks on a job in one day. What was the average number of bricks laid by each bricklayer?

**122. CON**  A developer divides a tract of land into 14 equally valued parcels. If the tract is valued at $147,000, what is the value of each parcel?

**123. INDTR**  A shipment of 150 machine parts costs $15,737.50. If this includes a $25 shipping charge, find the cost of each part.

**124. INDTR**  If a shop manager earns $23,400 annually, what is the monthly salary?

Find the average, then round to the same place value used in the problems.

**125.** Test scores: 86, 73, 95, 85

**126. AG/H**  Weight of cotton bales: 515 lb, 468 lb, 435 lb, 396 lb

**127. BUS**  Monthly income: $873.46, $598.21, $293.85, $546.83, $695.83, $429.86, $955.34, $846.95, $1,025.73, $1,152.89, $957.64, $807.25

**128. AG/H**  Average rainfall: 1.25 in., 0.54 in., 0.78 in., 2.35 in., 4.15 in., 1.09 in.

**129. CON**  Amperes of current: 3.0 A, 2.5 A, 3.5 A, 4.0 A, 4.5 A

*Learning Outcomes*

**1** Simplify expressions that contain exponents.

**2** Square numbers and find the square roots of numbers.

**3** Use powers of 10 to multiply and divide.

### **1** Simplify Expressions That Contain Exponents.

The product of repeated factors can be written in shorter form using natural-number exponents. **Natural numbers** are also called **counting numbers** and include all whole numbers except zero. Exponents that are not natural numbers will have a different interpretation.

In the example, $4 \times 4 \times 4 = 4^3$, the 4 is called the **base** and is the repeated factor. The 3 is the **exponent** and indicates the number of times the factor is repeated. The expression $4^3$ is read "four **cubed**" or "four to the third **power**" or "four raised to the third *power*." In expressions where 2 is an exponent, such as $4^2$, the expression is usually read as "four **squared**"; however, it may also be read as "four to the second *power*."

The expression $4^3$ is written in **exponential notation**. The number 64 is written in **standard notation** and is called the **power**.

### To change from exponential notation to standard notation:

1. Use the base as a factor as many times as indicated by the exponent.
2. Perform the multiplication.

---

**EXAMPLE** Identify the base and exponent of the expressions, and write in standard notation.

(a) $5^3$ (b) $1.5^2$

(a) $5^3$ 5 is the base; 3 is the exponent.
$5^3 = \mathbf{5 \times 5 \times 5} = \mathbf{125}$     Standard notation

(b) $1.5^2$ 1.5 is the base; 2 is the exponent.
$1.5^2 = \mathbf{1.5 \times 1.5} = \mathbf{2.25}$     Standard notation

---

### Any number with an exponent of 1 is the number itself:

$$a^1 = a, \quad \text{for any base } a \quad 8^1 = 8 \quad 2.3^1 = 2.3$$

---

### **2** Square Numbers and Find the Square Roots of Numbers.

The result of using a number as a factor 2 times is a **square** number or a **perfect square.** In the expression $7^2 = 49$, the 49 is a perfect square. This terminology evolved from a common formula for finding the area of a square. $A = s^2$, where $A$ represents the area and $s$ represents the length of one side. When we square 3, we write $3^2 = 3 \times 3 = 9$. We say 9 is the square of 3.

**EXAMPLE** Find the square.

(a) 2    (b) 7    (c) 3.2

(a) 2;    $2^2 = 2 \times 2 = \mathbf{4}$
(b) 7;    $7^2 = 7 \times 7 = \mathbf{49}$
(c) 3.2;    $3.2^2 = 3.2 \times 3.2 = \mathbf{10.24}$

The inverse operation of squaring is taking the square root of a number. The **principal square root** of a perfect square is the number that was used as a factor twice to equal that perfect square. The principal square root of 9 is 3 because $3^2$ or $3 \times 3 = 9$.

The **radical sign** $\sqrt{\phantom{x}}$ indicates that the square root is to be taken of the number under the bar. This bar serves as a grouping symbol just like parentheses. The number under the bar is called the **radicand.** The entire expression is called a **radical expression.**

radical sign ──┐ ┌── bar
$$\sqrt{25} = 5 \leftarrow \text{principal square root}$$
radicand ──┘

**To find the square root of a perfect square by estimation:**

1. Select a trial estimate of the square root.
2. Square the estimate.
3. If the square of the estimate is less than the original number, adjust the estimate to a larger number. If the square of the estimate is more than the original number, adjust the estimate to a smaller number.
4. Square the adjusted estimate from Step 3.
5. Continue the adjusting process until the square of the trial estimate is the original number.

**EXAMPLE** Find $\sqrt{256}$.

Select 15 as the estimated square root: $15^2$ or $15 \times 15 = 225$. The number 225 is less than 256, so the square root of 256 must be larger than 15. We adjust the estimate to 17: $17^2$ or $17 \times 17 = 289$. The number 289 is more than 256, so the square root of 256 must be smaller than 17. Now adjust the estimate to 16.

**Because $16^2 = 256$, 16 is the square root of 256.**

The squares of the numbers 1 through 10 are 1, 4, 9, 16, 25, 36, 49, 64, 81, and 100. These are the only whole numbers from 1 through 100 that are perfect squares.

**3  Use Powers of 10 to Multiply and Divide.**

A **nonzero digit** is any digit except zero. That is, 1, 2, 3, 4, 5, 6, 7, 8, and 9 are nonzero digits. **Powers of 10** are numbers whose only nonzero digit is 1. Thus, 10, 100, 1,000, and so on are powers of 10 because each value can be written in exponential form with a base of 10.

Compare the number of zeros in standard notation with the exponent in exponential notation.

| | | |
|---|---|---|
| One million | $1,000,000 = 10^6$ | 6 zeros |
| One hundred thousand | $100,000 = 10^5$ | 5 zeros |
| Ten thousand | $10,000 = 10^4$ | 4 zeros |
| One thousand | $1,000 = 10^3$ | 3 zeros |
| One hundred | $100 = 10^2$ | 2 zeros |
| Ten | $10 = 10^1$ | 1 zero |
| One | $1 = 10^0$ | 0 zeros |

The exponents in powers of 10 indicate the number of zeros used in standard notation. Zero exponents will be discussed later in Chapter 4.

**EXAMPLE**  Express as powers of 10.

(a) 10,000,000        (b) 100,000,000        (c) 100,000,000,000

**(a)**  10,000,000         $= 10^7$
**(b)**  100,000,000        $= 10^8$
**(c)**  100,000,000,000  $= 10^{11}$

**EXAMPLE**  Express in standard notation.

(a) $10^5$      (b) $10^{13}$

**(a)**  $10^5$  $= 100,000$
**(b)**  $10^{13} = 10,000,000,000,000$

In some applications, powers of 10 are used to simplify multiplication and division problems. Compare the examples to find the pattern for multiplying a number by a power of 10.

| | | |
|---|---|---|
| $5 \times 100 = 500$ | $5 \times 10^2 = 500$ | Decimal point moved two places to the right and two zeros attached. |
| $27 \times 1,000 = 27,000$ | $27 \times 10^3 = 27,000$ | Decimal point moved three places to the right and three zeros attached. |

**To multiply a number by a power of 10:**

1. Move the decimal point in the number to the *right* as many places as the number of zeros in 10, 100, 1,000, and so on.
2. Attach zeros on the right if necessary.

**EXAMPLE**  Multiply $237 \times 100$.

$237 \times 100 = 237.00 = \textbf{23,700}$     Attach two zeros to the *right* of the 7 in 237, and insert the appropriate comma.

**EXAMPLE**  Multiply $36.2 \times 1,000$.

$$36.2 \times 1,000 = 36.200 = \mathbf{36,200}$$  Move the decimal point three places to the *right*. Two zeros need to be attached.

Examine these divisions.

$$32 \div 10 = 3.2$$  Decimal point moved one place to the *left*.
$$78.9 \div 100 = 0.789$$  Decimal point moved two places to the *left*.
$$52,900 \div 1,000 = 52.9$$  Decimal point moved three places to the *left*.

### To divide a number by a power of 10:

1. Move the decimal to the *left* as many places as the divisor has zeros.
2. Attach zeros to the left if necessary.
3. You may drop zeros to the right of the decimal point if they follow the last nonzero digit of the quotient.

Compare this rule with the rule for multiplying decimal numbers by 10, 100, 1,000, and so on. For multiplication, the decimal shifts to the right; for division, the decimal shifts to the left.

**EXAMPLE**  If 100 lb of floor cleaner costs \$63, what is the cost of 1 lb?
Divide \$63 by 100.

$$63 \div 100 = 0.63 = \$0.63$$  Decimal is after 3 in 63. Move decimal two places to the left.

**The cost of 1 lb is \$0.63.**

Some scientific and graphing calculators have specific power keys, such as keys for squares and cubes. They also have a "general power" key that can be used for all powers. Even though the "general power" key can be used to square and cube numbers, the "square" and "cube" keys require fewer keystrokes.

**TIP!**

### Labels on Calculator Keys Are Not Universal

To show calculator steps, we often identify a common label for a function in a box. The exact label will vary with the specific calculator model. Some calculators also provide many functions on menus instead of on specific keys. Even though we use a key notation in this text to show a calculator function, this function may actually appear on a menu. Check your calculator manual for exact location and labeling of functions.

**Common Labels for Power Keys:**   $\boxed{x^2}$ $\boxed{x^3}$ $\boxed{\wedge}$ $\boxed{x^y}$

**Common Labels for Square Root Keys:**   $\boxed{\sqrt{\phantom{x}}}$ $\boxed{\sqrt{x}}$

EXAMPLE    Use the calculator to find $35^2$.

35 $\boxed{x^2}$ $\boxed{=}$     On some calculators pressing the equal key may not be required.

**1,225**

EXAMPLE    Use the calculator to find $2.3^5$.

2 $\boxed{\cdot}$ 3 $\boxed{\wedge}$ 5 $\boxed{=}$     $\boxed{\wedge}$ and $\boxed{x^y}$ are the most common labels for the general power key.

**64.36343**

EXAMPLE    **(a)** Evaluate $\sqrt{529}$.          **(b)** Evaluate $\sqrt{10.89}$.

$\boxed{\sqrt{}}$ 529 $\boxed{=}$                $\boxed{\sqrt{}}$ 10 $\boxed{\cdot}$ 89 $\boxed{=}$

**23**                          **3.3**

SECTION 1–2 SELF-STUDY EXERCISES

**1** Give the base and exponent of the expressions.

**1.** $4^3$                **2.** $9^4$                **3.** $2.7^9$                **4.** $15^2$

Simplify the exponential expressions.

**5.** $10^3$               **6.** $2^4$                **7.** $3.4^2$               **8.** $15^1$
**9.** $8^1$                **10.** $9^2$

Express with an exponent of 1.

**11.** 8                  **12.** 14.5               **13.** 12                 **14.** 23

**15.** Explain how an exponent of 1 relates to the base.     **16.** Explain what an exponent of 3 means.

**2** Write in standard notation.

**17.** $8^2$               **18.** $18^2$             **19.** $1.4^2$             **20.** $13^2$
**21.** $100^2$             **22.** $121^2$            **23.** $114^2$             **24.** $3.8^2$

Square the numbers.

**25.** 8                  **26.** 12                 **27.** 18                 **28.** 101

Perform the operations.

**29.** $\sqrt{25}$         **30.** $\sqrt{49}$         **31.** $\sqrt{81}$         **32.** $\sqrt{36}$
**33.** $\sqrt{196}$        **34.** $\sqrt{225}$        **35.** $\sqrt{121}$        **36.** $\sqrt{144}$

**37.** What does it mean to square a number?     **38.** What does it mean to take the square root of a number?

**3** Multiply the whole numbers using powers of 10.

**39.** $10 \times 10^2$    **40.** $12 \times 10^5$    **41.** $2 * 10^4$          **42.** $102 * 100$

Divide the whole numbers using powers of 10.

**43.** $250 \div 10$       **44.** $210 \div 10$       **45.** $\dfrac{300}{10^2}$     **46.** $2,500 \div 10$

**47.** What does the exponent of a power of 10 mean when you are multiplying?

**48.** What does the exponent of a power of 10 mean when you are dividing?

Use a calculator to evaluate.

**49.** $15^2$

**50.** $7^3$

**51.** $5^7$

**52.** $12^4$

**53.** $\sqrt{324}$

**54.** $\sqrt{784}$

**55.** $\sqrt{1,089}$

**56.** $\sqrt{196}$

---

## 1–3 | Order of Operations and Problem Solving

*Learning Outcomes*

**1** Apply the order of operations to a series of operations.

**2** Evaluate a formula.

**3** Solve applied problems using problem-solving strategies.

**1** **Apply the Order of Operations to a Series of Operations.**

Whenever several mathematical operations are performed, the proper **order of operations** must be followed.

**To apply the order of operations:**

1. **Parentheses (grouping symbols):** Perform operations within parentheses (or other grouping symbols), beginning with the innermost set of parentheses; or, apply the distributive property.
2. **Exponents and roots:** Evaluate exponential operations and find square roots in order from left to right.
3. **Multiply and divide** in order from left to right.
4. **Add and subtract** in order from left to right.

To summarize, use the following key words:

Parentheses (grouping), Exponents (and roots), Multiplication and Division, Addition and Subtraction

Other grouping symbols are **brackets** [ ], **braces** { }, and a **bar.** The bar can combine with other symbols like the radical sign and be used as a grouping symbol, $\sqrt{4 + 5} = \sqrt{9} = 3$.

**A Memory Aid for the Order of Operations**

To remember the order of operations, use the sentence, "<u>P</u>lease <u>E</u>xcuse <u>M</u>y <u>D</u>ear <u>A</u>unt <u>S</u>ally."

- Parentheses (grouping)
- Exponents (roots)
- Multiply/Divide
- Add/Subtract

**EXAMPLE** Evaluate $3(2 + 3)$.

Parentheses may show both a grouping and multiplication. When a number is multiplied by a sum or a difference, the distributive property can also be applied. Using the distributive property:

$$3(2 + 3) = 3(2) + 3(3) \quad \text{Distribute.}$$
$$= 6 + 9 \quad \text{Add.}$$
$$= \mathbf{15}$$

Using the order of operations:

$$3(2 + 3) = 3(5)$$      Do operation in parentheses first.

$$= 15$$      Multiply.

---

**EXAMPLE**    Simplify $4^2 - 5(2) \div (4 + 6)$ by performing the operations in the correct order.

| | | |
|---|---|---|
| $4^2 - 5(2) \div (4 + 6)$ | Do operations within parentheses first: $4 + 6 = 10$. | P |
| $4^2 - 5(2) \div 10$ | Evaluate exponential notation: $4^2 = 16$. | E |
| $16 - 5(2) \div 10$ | Multiply: $5(2) = 10$. | M D |
| $16 - 10 \div 10$ | Divide: $10 \div 10 = 1$. | M D |
| $16 - 1$ | Subtract last. | A S |
| $16 - 1 = \mathbf{15}$ | | |

---

**TIP!**

## Parentheses Indicate Multiplication, the Distributive Property, or a Grouping

Parentheses can indicate multiplication or an operation that should be done first. If the parentheses contain an operation, they indicate a grouping. Otherwise, they indicate multiplication. The expression $5(2)$ indicates multiplication, while $(4 + 6)$ indicates a grouping.

Many calculators have parentheses keys $\boxed{(}\,\boxed{)}$. These keys are used for all types of grouping symbols.

---

**EXAMPLE**    Evaluate $5 \times \sqrt{16} - 5 + [15 - (3 \times 2)]$.

| | | |
|---|---|---|
| $5 \times \sqrt{16} - 5 + [15 - (3 \times 2)]$ | Work innermost grouping: $3 \times 2$. | P |
| $5 \times \sqrt{16} - 5 + [15 - 6]$ | Work remaining grouping: $15 - 6$. | P |
| $5 \times \sqrt{16} - 5 + 9$ | Find square root: $\sqrt{16} = 4$. | E |
| $5 \times 4 - 5 + 9$ | Multiply: $5 \times 4$. | M D |
| $20 - 5 + 9$ | Add and subtract from left to right. | A S |
| $15 + 9 = \mathbf{24}$ | | |

---

**EXAMPLE**    Evaluate $3.2^2 + \sqrt{21 - 5} \times 2$.

| | |
|---|---|
| $3.2^2 + \sqrt{21 - 5} \times 2$ | Do operations within grouping first; the bar of the radical symbol is a grouping symbol: $21 - 5 = 16$. |
| $3.2^2 + \sqrt{16} \times 2$ | Evaluate exponent and square root from left to right: $3.2^2 = 10.24$; $\sqrt{16} = 4$. |
| $10.24 + 4 \times 2$ | Then multiply: $4 \times 2 = 8$. |
| $10.24 + 8$ | Add last. |
| $10.24 + 8 = \mathbf{18.24}$ | |

A division bar serves as a grouping symbol in the same way parentheses do.

**EXAMPLE** Makesha Lee took six tests and scored 87, 92, 76, 85, 95, and 89. Find Makesha's average score.

$$\frac{87 + 92 + 76 + 85 + 95 + 89}{6} =$$ The division bar serves to group the addends. Perform operations that are grouped first.

$$\frac{524}{6} =$$ Perform division.

$$87.3$$ Found.

**The average score is 87.**

## 2 Evaluate a Formula.

**Formulas** are procedures that have been used so frequently to solve certain types of problems that they have become the accepted means of solving these problems. Most formulas are expressed with one or more letter terms rather than words, and these procedures are written as symbolic equations, such as $P = 2(l + w)$, the formula for the perimeter of a rectangle. Electronic spreadsheets and calculator and computer programs are developed through formulas.

The most common use of formulas is for finding missing values. If we know values for all but one letter of a formula, we can find the missing value. To **evaluate** a formula is to substitute known values for letters of the formula and perform the indicated operations to find the missing value.

### To evaluate a formula:

1. Write the formula.
2. Rewrite the formula substituting known values for letters of the formula.
3. Perform the indicated operations, applying the order of operations.
4. Interpret the solution within the context of the formula.

Some of the first formulas that we use are the formulas for basic geometric shapes called polygons. A **polygon** is a plane or flat, closed figure described by straight-line segments and angles. Polygons have different numbers of sides and different properties. Some common polygons are the parallelogram, rectangle, and square (Fig. 1–15).

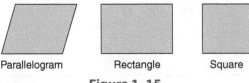

Parallelogram        Rectangle        Square

**Figure 1–15**

The **base** of any polygon is the horizontal side or a side that would be horizontal if the polygon's orientation is modified.

The **adjacent side** of any polygon is the side that has an end point in common with the base.

A **parallelogram** is a four-sided polygon with opposite sides that are parallel.

A **rectangle** is a parallelogram with angles that are all right angles.

A **square** is a parallelogram with all sides of equal length and with all right angles. A **square** can also be described as a rectangle with all sides of equal length.

The **perimeter** is the total length of the sides of a plane figure. As we saw earlier, some common applications for perimeter are finding the amount of trim molding for a room, determining the amount of fencing for a yard, finding the amount of edging for a flower bed, and so on.

A general procedure for finding the perimeter of any shape is to add the lengths of the sides. However, shortcuts based on the properties of the shape are given as formulas.

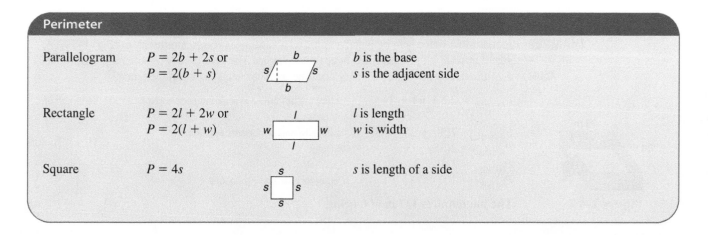

| Perimeter | | | |
|---|---|---|---|
| Parallelogram | $P = 2b + 2s$ or $P = 2(b + s)$ | | $b$ is the base $s$ is the adjacent side |
| Rectangle | $P = 2l + 2w$ or $P = 2(l + w)$ | | $l$ is length $w$ is width |
| Square | $P = 4s$ | | $s$ is length of a side |

The perimeter of a parallelogram, like the perimeter of a rectangle, is the sum of its four sides. The sides of the parallelogram are called **base** and **adjacent side** (instead of length and width). Notice the locations of the base and adjacent side in Fig. 1–16.

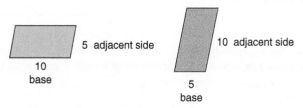

5  adjacent side

10  adjacent side

10
base

5
base

**Figure 1–16**

**EXAMPLE** Find the perimeter of a parallelogram with a base of 16 in. and an adjacent side of 8 in. (Fig. 1–17).

Visualize the parallelogram.

8 in.

16 in.

**Figure 1–17**

$P_{\text{parallelogram}} = 2(b + s)$     Select the perimeter of parallelogram formula.
$P_{\text{parallelogram}} = 2(16 + 8)$     Substitute: $b = 16$, $s = 8$.
$P_{\text{parallelogram}} = 2(24)$     Evaluate.

$\boldsymbol{P_{\text{parallelogram}} = 48 \text{ in.}}$

**EXAMPLE**

**BUS**

A shop that makes custom picture frames has an order for a frame whose outside measurements are 42 in. by 30 in. (Fig. 1–18). How many inches of picture frame molding are needed for the job?

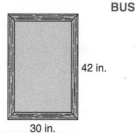

42 in.

30 in.

**Figure 1–18**

$P_{\text{rectangle}} = 2(l + w)$     The figure is a rectangle. Select the appropriate formula.
$P_{\text{rectangle}} = 2(42 + 30)$     Substitute and evaluate the formula.
$P_{\text{rectangle}} = 2(72)$
$P_{\text{rectangle}} = 144$

When analyzing the dimensions, inches are added to inches, so the result is written in inches. Then the inches are multiplied by a number and the final result is inches.

**Thus, 144 in. are needed for the project.**

In some applications, we need to decrease the total perimeter to account for doorways and other openings in the perimeter.

**EXAMPLE**

**AG/H**

A chain-link fence is to be installed around a yard measuring 25 m by 30 m (Fig. 1–19). A gate 3 m wide will be installed over the driveway. The gate comes preassembled from the manufacturer. How much fencing does the installer need to put up the fence?

25 m

30 m

$P_{\text{rectangle}} = 2(l + w) - 3$    Select the perimeter of a rectangle formula. Subtract the length of the gate.

$P_{\text{rectangle}} = 2(30 + 25) - 3$    Substitute and perform operations.

$P_{\text{rectangle}} = 2(55) - 3$    Multiply.

$P_{\text{rectangle}} = 110 - 3$    Subtract.

$P_{\text{rectangle}} = 107$    Perimeter is a linear measure.

**Figure 1–19**    **The job requires 107 m of fencing.**

**TIP!**

**Using Subscripts**

Subscripted words, letters, or numbers are a handy way to provide additional information. Because we will examine the formulas for the perimeter of several different shapes, we sometimes use subscripts to distinguish among them. For instance, the perimeter of a square, rectangle, or parallelogram can be indicated as $P_{\text{square}}$, $P_{\text{rectangle}}$, or $P_{\text{parallelogram}}$, respectively.

**EXAMPLE**

**CON**

How much aluminum edge molding is needed to surround a stainless steel kitchen sink that measures 40 cm on each side (Fig. 1–20)?

40 cm

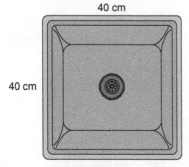

40 cm

$P_{\text{square}} = 4s$    The figure is a square. Select the appropriate formula.

$P_{\text{square}} = 4(40 \text{ cm})$    Substitute and evaluate the formula.

$P_{\text{square}} = 160 \text{ cm}$

**Figure 1–20**    **The sink requires 160 cm of molding.**

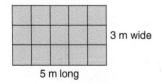

3 m wide

5 m long

**Figure 1–21**

The area of a polygon is the amount of surface of a plane figure. Area is expressed in square units. For example, if a rectangle is 3 m wide and 5 m long, there are 15 m$^2$ in the area (Fig. 1–21).

$$\text{area} = 15 \text{ m}^2$$

## Writing Square Units

The exponent 2 following a unit of measure indicates *square measure* or area. Thus, 15 m² = 15 square meters, 23 ft² = 23 square feet, 120 cm² = 120 square centimeters, and so on. This is a shortcut way to express a measure that has *already* been "squared."

When do you need the area of a polygon? Some common applications for area are finding the amount of carpeting needed to cover a floor, the amount of paint needed to paint a surface, the amount of fertilizer needed to treat a lawn, and so on.

## Area

| | | | |
|---|---|---|---|
| Rectangle | $A = lw$ |  | $l$ is length<br>$w$ is width |
| Square | $A = s^2$ | 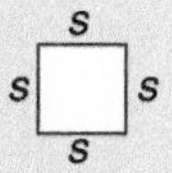 | $s$ is length of a side |
| Parallelogram | $A = bh$ | 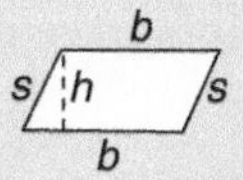 | $b$ is base<br>$h$ is height |

**EXAMPLE**

**CON**

A carpet installer is carpeting a room measuring 16 ft by 20 ft. Projecting out from one wall is a fireplace whose hearth measures 3 ft by 6 ft (Fig. 1–22). How many square feet of carpet does the installer require for the job? How much is wasted?

First, compute the area of the room without considering the area of the hearth. This is the amount of carpet needed. Second, compute the area of the fireplace hearth. This is the amount wasted.

Let $A_{room}$ = the area of the room and $A_{hearth}$ = the area of the hearth.

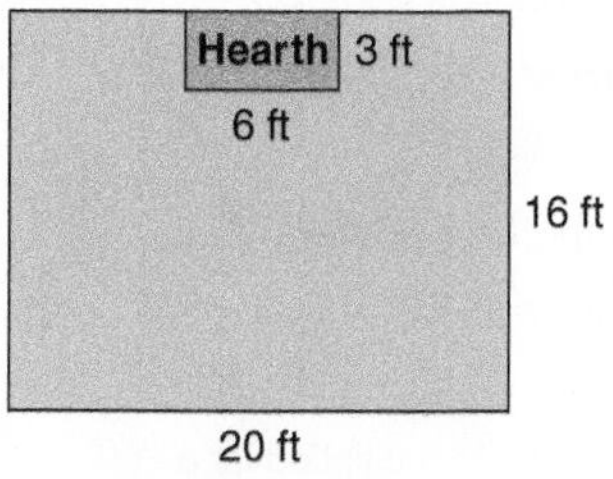

| | |
|---|---|
| $A_{room} = lw$ | Select the appropriate formula and substitute. |
| $A_{room} = 20 \times 16$ | Multiply. |
| $A_{room} = 320$ ft² | Area is a square measure. |
| $A_{hearth} = lw$ | Select the appropriate formula and substitute. |
| $A_{hearth} = 3 \times 6$ | Multiply. |
| $A_{hearth} = 18$ ft² | Area is a square measure. |

**Figure 1–22**

**This job requires 320 ft² of carpet. Of this, 18 ft² is waste from the hearth.**

The preceding example brings up some interesting questions. Is carpet purchased in square feet? If carpet can only be purchased in 12-ft or 15-ft widths, will there be additional waste? Is it more practical to purchase extra carpet that will be wasted or to spend extra labor costs to install the carpet with several seams? You may want to investigate common practices in the industry.

Asphalt roofing shingles are sold in bundles. The number of bundles needed to cover a square (100 ft²) depends on the overlap when the shingles are installed. If the overlap allows 4 in. of each shingle to be exposed, then four bundles are needed per square. With a 5-in. exposure, 3.2 bundles are needed per square. Figure the number of bundles for a 4-in. exposure and for a 5-in. exposure for a roof measuring 30 ft × 20 ft.

$$A_{\text{roof}} = 30 \times 20$$
$$A_{\text{roof}} = 600 \text{ ft}^2$$

$600 \text{ ft}^2 \div 100 \text{ ft}^2 = 6 \text{ squares}$  Find the number of squares (100 ft²) by dividing by 100 ft².

$6 \times 4 = 24 \text{ bundles}$  Find the number of bundles for a 4-in. exposure.

$6 \times 3.2 = 19.2 \text{ } or \text{ } 20 \text{ bundles}$  Find the number of bundles for a 5-in. exposure.

**Therefore, 24 bundles of shingles are required for a 4-in. exposure, and 20 bundles (from 19.2 bundles) are required for a 5-in. exposure.**

Lynn Fly, a vinyl floor installer, discovers that one of the squares in the flooring pattern is damaged. He decides to cut out the damaged square and replace it. The damaged square measures 20 cm on each side. What is the area of the square to be replaced?

$A_{\text{square}} = s^2$  Select the appropriate formula.

$A_{\text{square}} = (20 \text{ cm})^2$  Substitute measurement of one side and square it.

$A_{\text{square}} = 400 \text{ cm}^2$

**The vinyl square measures 400 cm².**

TIP!

## Dimension Analysis

When we evaluate formulas, the measuring units are sometimes omitted from the written steps. However, it is very important to use the correct unit in your calculations so that the unit in the solution is correct. Compare examples involving perimeter and area.

**Example (molding for sink, p. 40)**

$P_{\text{square}} = 4s$

$P_{\text{square}} = 4(40)$  The measuring unit is centimeters; the measure is multiplied by a number.

$P_{\text{square}} = \textbf{160 cm}$  The measuring unit in the solution is centimeters.

Perimeter is always a linear measure, which means the measuring unit should be to the first power.

**Example (vinyl flooring in the preceding example)**

$A_{\text{square}} = s^2$

$A_{\text{square}} = (20)^2$  The measuring unit is centimeters and a measure is squared or a measure is multiplied by a measure.

$A_{\text{square}} = 400 \text{ cm}^2$  The measuring unit in the solution is square centimeters.

Area is always a square measure, which means that measures of area are to the second power or have an exponent of 2.

Since a rectangle is a parallelogram, the areas of both types of polygons are related.

The **height** of a parallelogram is the perpendicular distance between two parallel sides. Height is also called **altitude.**

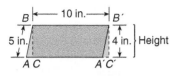

**Figure 1–23**

If triangle $ABC$ ($\triangle ABC$) in Fig. 1–23 were transposed to the right side of the parallelogram ($\triangle A'B'C'$), we would have a rectangle.

The base of the original parallelogram is the same as the length of the newly formed rectangle. Similarly, the height of the parallelogram is the same as the width of the rectangle. Thus, the area of the rectangle and parallelogram is $10 \times 4$ or 40 in$^2$.

---

**EXAMPLE**

Find the area of a parallelogram with a base of 16 in., an adjacent side of 8 in., and a height of 7 in. (Fig. 1–24).

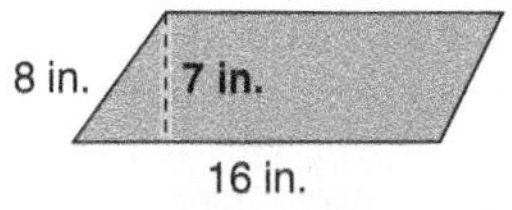

Visualize the parallelogram.

8 in. / 7 in.

16 in.

**Figure 1–24**

$A_{\text{parallelogram}} = bh$      Substitute.
$A_{\text{parallelogram}} = 16 \times 7$      Multiply.
$A_{\text{parallelogram}} = \textbf{112 in}^2$

---

### 3   Solve Applied Problems Using Problem-Solving Strategies.

Problem solving is an important skill in the workplace and in everyday life. In developing good problem-solving skills, it is helpful to use a systematic problem-solving plan. A plan gives you a framework for approaching problems.

You will encounter many different plans for solving problems. They help you organize the problem details so that you can find an appropriate procedure for solving the problem. Our plan, like others, is very structured. As you develop confidence and skill in solving problems, you will probably use less structured and more intuitive approaches.

**Six-Step Problem-Solving Plan**
1. **Unknown Facts.** What facts are missing from the problem? What are you trying to find?
2. **Known Facts.** What relevant facts are known or given? What facts must you bring to the problem from your own background?
3. **Relationships.** How are the known facts and the unknown facts related? What formulas or definitions can you use to establish a model?
4. **Estimation.** What are some of the characteristics of a reasonable solution? For instance, should the answer be more than a certain amount or less than a certain amount?
5. **Calculations.** Perform the operations identified in the relationships.
6. **Interpretation.** What do the results of the calculations represent within the context of the problem? Is the answer reasonable? Have all unknown facts been found? Do the unknown facts check in the context of the problem?

---

**EXAMPLE**

**BUS**

The 7th Inning buys baseball cards from eight different vendors. In November, the company purchased 8,832 boxes of cards. If an equal number of boxes was purchased from each vendor, how many boxes of cards were supplied by each vendor?

**Unknown fact**

Number of boxes of cards supplied by each vendor

**Known facts**

8,832 = total number of boxes purchased
8 = total number of vendors

An equal number of boxes was purchased from each vendor.

**Relationships**

total boxes purchased ÷ number of vendors = boxes purchased from each vendor

**Estimation**

If 8,000 boxes were purchased in equal amounts from eight vendors, then 1,000 were purchased from each vendor. Since more than 8,000 boxes were purchased, then more than 1,000 were purchased from each vendor.

**Calculations**

$$\begin{array}{r} 1,104 \\ 8\overline{)8,832} \end{array}$$

**Interpretation**

**Each vendor supplied 1,104 boxes of cards.**

In identifying the relationships of a problem, it is often helpful if you look for key words or phrases that give you clues about the mathematical operations involved in the relationships (Table 1–1). Key words give clues as to whether one quantity is added to, subtracted from, or multiplied or divided by another quantity. For example, if a problem tells you that Carol's salary in 2006 exceeds her 2005 salary by $2,500, you know that you should add $2,500 to her 2005 salary to find her 2006 salary.

**Table 1–1    Key Words and What They Generally Imply in Word Problems**

| Addition | Subtraction | Multiplication | Division | Equality |
|---|---|---|---|---|
| The sum of | Less than | Times | Divide(s) | Equals |
| Plus/total | Decreased by | Multiplied by | Divided by | Is/was/are |
| Increased by | Subtracted from | Of | Divided into | Is equal to |
| More/more than | Difference between | The product of | Half of (divided by 2) | The result is |
| Added to | Diminished by | Twice (2 times) | Third of (divided by 3) | What is left |
| Exceeds | Take away | Double (2 times) | Per | What remains |
| Expands | Reduced by | Triple (3 times) | How big is each part? | The same as |
| Greater than | Less/minus | Half of ($\frac{1}{2}$ times) | How many parts can be made from | Gives/giving |
| Gain/profit | Loss | Third of ($\frac{1}{3}$ times) | | Makes |
| Longer | Lower | | | Leaves |
| Older | Shrinks | | | |
| Heavier | Smaller than | | | |
| Wider | Younger | | | |
| Taller | Slower | | | |
| Larger than | Shorter | | | |

More than one relationship may be needed to find the unknown facts. When this is the case, you usually need to read the problem several times to find all the relationships and plan your solution strategy.

**EXAMPLE**

**BUS**

Carlee Anne McAnally needs to ship 78 crystal vases. With standard packing to prevent damage, 5 vases fit in each available box. How many boxes are required to pack the vases?

**Unknown fact**    Number of boxes required to ship the vases

**Known facts**    total vases to be shipped = 78
number of vases per box = 5

**Relationships**    total boxes needed = total number of vases ÷ number per box
total boxes needed = 78 ÷ 5

**Estimation**    70 ÷ 5 = 14     Round down the dividend.
80 ÷ 5 = 16     Round up the dividend.

Since 78 is between 70 and 80, the number of boxes needed is between 14 and 16.

**Calculation**    78 ÷ 5 = 15 R3

**Interpretation**    **16 boxes are needed;** 15 boxes will contain 5 vases each, and 1 box will contain 3 vases. The box with 3 vases will need extra packing.

**TIP!**

### Using Guessing and Checking to Solve Problems

An effective strategy for solving problems involves guessing. Make a guess that you think may be reasonable, and check to see if the answer is correct. If your guess is not correct, decide if it is too high or too low. Make another guess based on what you learned from your first guess. Continue until you find the correct answer.

Let's try guessing in the previous example. We found that we could pack 70 vases in 14 boxes and 80 vases in 16 boxes. Since we need to pack 78 vases, how many vases can we pack with 15 boxes? $15 \times 5 = 75$. Still not enough. Therefore, we will need 16 boxes, but the last box will not be full.

You can probably think of other ways to solve this problem. Any plan that leads to a correct solution is acceptable. Some plans will be more efficient than others, but you develop your problem-solving skills by pursuing a variety of strategies.

### SECTION 1–3 SELF-STUDY EXERCISES

 Use the order of operations to evaluate each problem.

**1.** $5^2 + 4 - 3$  
**2.** $4^2 + 6 - 4$  
**3.** $4(3) - 9 \div 3$  
**4.** $5 \cdot 2.9 - 4 \div 2$  
**5.** $25 \div 5 \cdot 4.8$  
**6.** $64 \div 4(2)$  
**7.** $48 \div 8 \times 3$  
**8.** $15 - 2 \times 3$  
**9.** $4^2 - (4)(3) + 6$  
**10.** $17 - 4 \times 2$  
**11.** $6 \times \sqrt{36} - 2 \times 3$  
**12.** $3 \cdot \sqrt{81} - (3)(4)$  

**13.** $4^2 \cdot 3^2 + (4 + 2)(2)$  
**14.** $2^2 \cdot 5^2 + (2 + 1)(3)$  
**15.** $54 - 3^3 - \dfrac{8}{2}$  
**16.** $156 - 2^3 - \dfrac{9}{3}$  
**17.** $3 - 2 + 3 \times 3 - \sqrt{9}$  
**18.** $2 - 1 + (4)(4) - \sqrt{4}$  

**19.** $2^4 \times (7 - 2) \times 2$  
**20.** $3^4 \times (9 - 3) \times 3$  
**21.** $124 - 8 \cdot 7 + 12$  

Use a calculator to perform the operations.

**22.** $4^3 + 14 - 8$  
**23.** $2^3 + 12 - 7$  
**24.** $2 \times \sqrt{16} + (8 - \sqrt{25})$  
**25.** $4 \times \sqrt{49} + (9 - \sqrt{64})$  
**26.** $3(2^2 + 1) - 30 \div 3$  
**27.** $4(3^2 + 2) - 60 \div 12$  
**28.** $25 - 5^2 \div (7 - 2)$  
**29.** $36 - 6^2 \div (8 - 2)$  
**30.** $2(3.1^2 + 2) - \sqrt{7.29}$  

**2**

**31.** Find the perimeter and area of the parallelogram in Fig. 1–25.

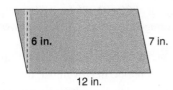

**Figure 1–25**

**32.** Find the perimeter and area of the parallelogram in Fig. 1–26.

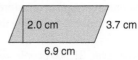

**Figure 1–26**

Solve the problems involving perimeter.

**33. CON** An illuminated sign in the main entrance of a hospital is a parallelogram with a base of 48 in. and an adjacent side of 30 in. How many feet of aluminum molding are needed to frame the sign?

**34. AUTO** A customized van has a window cut in each side in the shape of a parallelogram with a base of 20 in. and an adjacent side of 11 in. How many inches of trim are needed to surround the two windows?

**35. CON** A contemporary building has a window in the shape of a parallelogram with a base of 50 in. and an adjacent side of 30 in. How many inches of trim are needed to surround the window?

**36. CON** A table for a reading lab has a top in the shape of a parallelogram with a base of 36 in. and an adjacent side of 18 in. How many inches of edge trim are needed to surround the tabletop?

**37.** Find the perimeter and area of the rectangle in Fig. 1–27.

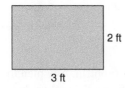

3 ft

2 ft

**Figure 1–27**

**38.** Find the perimeter and area of the rectangle in Fig. 1–28.

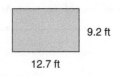

9.2 ft

12.7 ft

**Figure 1–28**

**39. CON** A rectangular parking lot is 340 ft by 125 ft. Find the perimeter of the parking lot.

**40. CON** A room is 15 ft by 12 ft. How many feet of chair rail are needed for the room? Disregard openings.

**41. CON** How many feet of quarter-round molding are needed to finish around the baseboard after sheet vinyl flooring is installed if the room is 16 ft by 18 ft and there are three 3-ft-wide doorways?

**42. CON** The swimming pool in Fig. 1–29 measures 32 ft by 18 ft. How much fencing is needed, including material for a gate, if the fence is to be built 7 ft from each side of the pool?

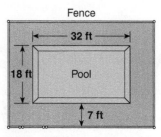

Fence

32 ft

18 ft    Pool

7 ft

**Figure 1–29**

**43.** Find the perimeter and area of Fig. 1–30.

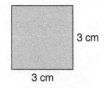

3 cm

3 cm

**Figure 1–30**

**44.** Find the perimeter and area of Fig. 1–31.

8.9 cm

8.9 cm

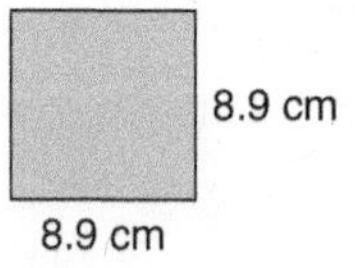

**Figure 1–31**

**45. CON** The square parking lot of a doctor's office is to have curbs built on all four sides. If the lot is 150 ft on each side, how many feet of curb are needed? Allow 10 ft for a driveway into the parking lot.

**46. CON** A border of 4-in. × 4-in. wall tiles surrounds the floor of a shower stall that is 48 in. × 48 in. How many tiles are needed for this border? Disregard spaces for grout (connecting material between the tiles).

**3**

**47. BUS** If you have 348 packages of Halloween candy to rebox for shipment to a discount store and you can pack 12 packages in each box, how many boxes will you need?

**48. BUS** If American Communications Network (ACN) has an annual payroll of $5,602,186 for its 214 employees, what is the average salary of an ACN employee?

**49. BUS** In a recent year, 21,960 people visited Bio Fach, Germany's biggest ecologically sound consumer goods trade fair. This figure was up from the 18,090 the previous year and 16,300 two years ago. What was the increase in visitors to Bio Fach over the two-year period?

**50. BUS** The On-The-Square Card and Gift Shop buys cards from six vendors. In November, the company purchased 6,480 boxes of cards. If the shop purchased an equal number of boxes from each vendor, how many boxes of cards did each vendor supply?

| Learning Outcome | Key Concepts and Examples |
|---|---|

**Section 1–1**

**1** Compare whole numbers (pp. 2–3).

Mentally position the numbers on a number line. The leftmost number is the smaller number. Write an inequality using either the *less than* < or *greater than* > symbol.

> Write two inequalities comparing 5 and 9.
> 5 < 9    or    9 > 5

**2** Write fractions with power-of-10 denominators as decimal numbers (pp. 3–5).

Write the numerator as the decimal number with the same number of decimal digits as there are zeros in the denominator.

> Express $\dfrac{53}{1,000}$ as a decimal: 0.053.

**3** Compare decimal numbers (pp. 5–6).

Compare decimal numbers by comparing the digits in the same place beginning from the left of each number.

> Which decimal is larger, 0.23 or 0.225? Both numbers have the same digit, 2, in the tenths place. 0.23 is larger because it has a 3 in the hundredths place, while 0.225 has a 2 in that place.

**4** Round a whole number or a decimal number to a specified place value (pp. 6–8).

**Whole number: 1.** Locate the rounding place.   **2.** Examine the digit to the right.   **3.** Round down if the digit is less than 5.   **4.** Round up if the digit is 5 or more.

**Decimal number:** Round as in whole numbers; however, digits on the right side of the digit in the rounding place are dropped rather than replaced with zeros.

> Round 3,624 to the nearest tens place: 2 is in the tens place, 4 is the digit to the right; 4 is less than 5, so round down. The rounded value is 3,620.
> Round 5.847 to the nearest tenth: 8 is in the tenths place; 4 is less than 5, so leave 8 as is and drop the 4 and 7. The rounded value is 5.8.

**5** Add and subtract whole numbers and decimals (pp. 8–15).

Addition is a binary operation that is commutative and associative.

Zero property of addition (additive identity): $0 + n = n + 0 = n$

**Add whole numbers:**
Arrange addends in columns of like places. Add each column beginning with the ones place. Carry when necessary.

> | | |
> |---|---|
> | $5 + 7 = 7 + 5 = 12$ | Commutative property of addition |
> | $(3 + 2) + 6 = 3 + (2 + 6) = 11$ | Associative property of addition |
> | $0 + 3 = 3 + 0 = 3$ | Zero property of addition |
>
> | | |
> |---|---|
> | 4,824 | Addend |
> | + 745 | Addend |
> | 5,569 | Sum |

**Add decimal numbers:**
Add decimal numbers by arranging the addends so that the decimal points are aligned.

> Add: 43.35 + 3.7 + 0.462
>
> | | |
> |---|---|
> | 43.35 | Addend |
> | 3.7 | Addend |
> | + 0.462 | Addend |
> | 47.512 | Sum |

**Estimate and check addition:**
Estimate by rounding the addends before finding the sum. Check addition by adding a second time.

> Estimate the sum by rounding to the nearest hundred: $483 + 723$;
> $500 + 700 = 1,200$. Exact sum: $1,206$.

Subtraction is a binary operation that is *not* commutative or associative. Addition and subtraction are inverse operations.

Zero property of subtraction: $n - 0 = n$

**Subtract whole numbers:**
Arrange numbers in columns of like places. Subtract each column, beginning with the ones place. Borrow when necessary.

> If $5 + 4 = 9$, then $9 - 5 = 4$ or $9 - 4 = 5$.     Inverse operations
> $7 - 0 = 7$     Zero property of subtraction
> $\begin{array}{r} 4,227 \\ -\ 745 \\ \hline 3,482 \end{array}$

**Subtract decimal numbers:**
Subtract decimal numbers by arranging the minuend and subtrahend so that the decimal points are aligned.

> Subtract: $53.824 - 4.0423$
>
> $\begin{array}{rl} 53.824 & \text{Minuend} \\ -\ 4.0423 & \text{Subtrahend} \\ \hline 49.7817 & \text{Difference} \end{array}$

**Estimate and check subtraction:**
Estimate by rounding the minuend and subtrahend before finding the difference. Check subtraction by adding the difference and the subtrahend. The result should equal the minuend.

> Estimate the difference by rounding to the nearest hundred: $783 - 423$
> $800 - 400 = 400$. Exact difference: $783 - 423 = 360$. Check: $360 + 423 = 783$.

**6** Multiply and divide whole numbers and decimals (pp. 15–26).

Multiplication is a binary operation that is commutative and associative.

Zero property of multiplication: $n \times 0 = 0 \times n = 0$

**Multiply whole-number factors:**
Arrange numbers in columns of like place values. Multiply the multiplicand by each digit in the multiplier. Add the partial products.

> $5(7) = 7(5) = 35$;     Commutative property of multiplication
> $(3 \times 2) \cdot 6 = 3 \cdot (2 \times 6) = 36$:     Associative property of multiplication
> $5 \times 0 = 0 \times 5 = 0$     Zero property of multiplication
> $\begin{array}{rl} 259 & \text{Multiplicand} \\ \times\ \ 23 & \text{Multiplier} \\ \hline 777 & \text{Partial product} \\ 5\ 18\ \ & \text{Partial product} \\ \hline 5,957 & \text{Product} \end{array}$

**Multiply decimal factors:**
Multiply the decimal numbers as in whole numbers, and count the number of decimal digits in both numbers. Place the decimal in the product to the left of the same number of digits, counting from the right.

Multiply: $3.25 \times 0.53$

| | |
|---|---|
| $3.25$ | Two decimal places |
| $\times 0.53$ | Two decimal places |
| $9\ 75$ | |
| $1\ 62\ 5$ | |
| $1.72\ 25$ | Four decimal places |

**Apply the distributive property:**
$a(b + c) = ab + ac$
$a(b - c) = ab - ac$

$3(5 + 6) = 3(5) + 3(6)$
$3(11) \quad = 15 + 18$
$\qquad 33 = 33$

**Estimate and check multiplication:**
Round the factors, then multiply. Check multiplication by multiplying a second time.

Estimate the product by rounding to one nonzero digit: $483 \times 72$;
$500 \times 70 = 35,000$. Exact product: $34,776$.

**Use symbols to indicate division:**
The division $a$ divided by $b$ can be written as

$$a \div b \qquad b\overline{)a} \qquad \frac{a}{b} \qquad \text{or} \qquad a/b$$

The divisor, $b$, cannot be zero.

Write 12 divided by 4 in four ways.

$$12 \div 4 \qquad 4\overline{)12} \qquad \frac{12}{4} \qquad 12/4$$

**Divide whole numbers:**
Align the numbers properly in long division.

$$
\begin{array}{r}
20\ \text{R}15 \\
23\overline{)475} \\
\underline{46}\phantom{5} \\
15 \\
\underline{0} \\
15
\end{array}
$$

## Divide decimal numbers:

When dividing by a decimal number, move the decimal in the divisor to the right end; move the decimal in the dividend the same number of places to the right. Place the decimal in the quotient. Divide one place past the specified place value. Attach zeros in the dividend if necessary. Round to the specified place.

| Divide: | Divide and round to tenths: |
|---|---|
| $\begin{array}{r} 3.8 \\ 2.1\overline{\smash{)}7.9\,8} \\ \underline{6\,3} \\ 1\,6\,8 \\ \underline{1\,6\,8} \end{array}$ | $\begin{array}{r} 1.70 \approx 1.7 \\ 15\overline{\smash{)}25.60} \\ \underline{15} \\ 10\,6 \\ \underline{10\,5} \\ 10 \end{array}$ |

## Estimate and check division:

To estimate division, round the dividend and divisor before finding the quotient. Find the first digit of the quotient, and add a zero for each remaining digit in the dividend.

To check division, multiply the quotient by the divisor and add the remainder. The result should equal the dividend.

Estimate $2{,}934 \div 42$.

$$\begin{array}{r} 70 \\ 40\overline{\smash{)}3{,}000} \end{array}$$

Exact quotient = 69 R36
Check:
$69 \times 42 = 2{,}898$
$2{,}898 + 36 = 2{,}934$

## Find the numerical average:

Find the sum of the values and divide the sum by the number of values.

Find the average of 74, 65, and 85.

| | |
|---|---|
| $74 + 65 + 85 = 224$ | Find the sum of the values. |
| $224 \div 3 = 74.6$ or 75 (rounded) | Divide by the number of values. |

## Section 1–2

**1** Simplify expressions that contain exponents (p. 31).

To simplify an exponential expression, use the base as a factor the number of times indicated by the exponent.

$a^1 = a$, for any base $a$

| | |
|---|---|
| $5^3 = 5 \times 5 \times 5 = 125$ | Five is used as a factor 3 times. |
| $7^1 = 7$ | Any number raised to the first power is the number. |

**2** Square numbers and find the square roots of numbers (pp. 31–32).

Squaring and finding square roots are inverse operations.

$$7^2 = 49$$
$$\sqrt{49} = 7$$

**3** Use powers of 10 to multiply and divide (pp. 32–35).

To multiply a decimal number by a power of 10, move the decimal to the *right* as many digits as the power of 10 has zeros. Attach zeros if necessary.

To divide a decimal by a power of 10, move the decimal to the *left*. Note that this process is just the opposite of that for multiplication.

> Multiply:
> $27 \times 10^3 = 27,000$
> $18 \times 100 = 1,800$
> $23.52 \times 1,000 = 23,520$
>
> Divide:
> $3.52 \div 10 = 0.352$
> $400 \div 10^2 = 4$
> $23,000 \div 10^3 = 23$

## Section 1–3

**1** Apply the order of operations to a series of operations (pp. 36–38).

**Order of operations:**

1. Parentheses or groupings, innermost first.
2. Exponential operations and roots from left to right.
3. Multiplication and division from left to right.
4. Addition and subtraction from left to right.

> Simplify.
> | | |
> |---|---|
> | $3^2 + 5(6 - 4) \div 2$ | Work inside parentheses. |
> | $3^2 + 5(2) \div 2$ | Raise to power. |
> | $9 + 5(2) \div 2$ | Multiply. |
> | $9 + 10 \div 2$ | Divide. |
> | $9 + 5$ | Add. |
> | $14$ | |

**2** Evaluate a formula (pp. 38–43).

1. Write the formula.
2. Rewrite the formula substituting known values for letters of the formula.
3. Perform the indicated operations, applying the order of operations.
4. Interpret the solution within the context of the formula.

> The formula for the distance ($d$) an object falls is $d = 0.5gt^2$. Find the distance if gravity ($g$) is 32 ft per second squared and seconds ($t$) is 3. Feet per second squared is $\frac{\text{ft}}{\text{s}^2}$.
>
> | | |
> |---|---|
> | $d = 0.5gt^2$ | Dimension analysis: $\frac{\text{ft}}{\text{s}^2}(\text{s}^2) = \text{ft}$ |
> | $d = 0.5(32)(3)^2$ | Raise to power. |
> | $d = 0.5(32)(9)$ | Multiply. |
> | $d = 144$ ft | |

**3** Solve applied problems using problem-solving strategies (pp. 43–45).

Six-Step Problem-Solving Plan

> A shipment of textbooks to a college bookstore is sent in two boxes. One box weighs 9 lb, and the second box weighs 14 pounds more than the first box. What is the total weight of the boxes?

| | |
|---|---|
| **Unknown facts** | Weight of second box<br>Total weight of the two boxes together |
| **Known facts** | First box weighs 9 lb.<br>Second box weighs 14 lb more than first box. |
| **Relationships** | weight of first box $= 9$<br>weight of second box $= 9 + 14$<br>total weight $= 9 + 9 + 14$ |
| **Estimation** | Since the second box weighs more than twice as much as the first, the two boxes together weigh approximately $3 \times 9$ or 27 lb. |
| **Calculation** | $9 + 9 + 14 = 32$ |
| **Interpretation** | The two boxes weigh 32 lb together, and 9 lb and 23 lb separately. |

## CHAPTER REVIEW EXERCISES

### Section 1–1

Write as decimal numbers.

1. (a) $\dfrac{3}{10}$    (b) $\dfrac{15}{100}$    (c) $\dfrac{4}{100}$

2. (a) $\dfrac{75}{1,000}$    (b) $\dfrac{21}{10}$    (c) $\dfrac{652}{100,000}$

3. Round to the indicated place.
   (a) Nearest hundred: 468
   (b) Nearest ten thousand: 49,238
   (c) Nearest tenth: 41.378
   (d) Nearest hundredth: 6.8957
   (e) Nearest ten-thousandth: 23.46097

4. Round to the indicated place.
   (a) Nearest ten: 98
   (b) Nearest ten: 94
   (c) Nearest thousand: 25,786
   (d) Nearest hundredth: 0.0736
   (e) Nearest whole number: 7.93
   (f) Nearest whole number: 1.876

5. Which of these decimal numbers is larger: 4.783 or 4.79?

6. Which of these decimal numbers is smaller: 0.83 or 0.825?

7. Write these decimal numbers in order of size from smallest to largest: 0.021, 0.0216, 0.02.

8. Two measurements of an object are recorded. If the measures are 4.831 in. and 4.820 in., which is larger?

9. The decimal equivalent of $\frac{7}{8}$ is 0.875. The decimal equivalent of $\frac{6}{7}$ is approximately 0.857. Which fraction is larger?

10. Two parts are machined from the same stock. They measure 1.023 in. and 1.03 in. after machining. Which part has been machined more; that is, which part is now smaller?

11. Add.
   (a) $8 + 5 + 3 + 6 + 2 + 4$
   (b) $7 + 4 + 3 + 2 + 5 + 4$

12. (a) $6.2 + 32.7 + 46.82 + 0.29 + 4.237$
   (b) $86.3 + 9.2 + 70.02 + 3 + 2.7$

13. An air conditioner uses 10.4 kW (kilowatts), a stove uses 15.3 kW, a washer uses 2.9 kW, and a dryer uses 6.3 kW. What is the total number of kilowatts used?

14. A do-it-yourself project requires \$57.32 for concrete, \$74.26 for fence posts, and \$174.85 for fence boards. Estimate the cost by rounding to numbers with one nonzero digit, then find the exact cost.

15. Subtract.
   (a) $21.34 - 16.73$
   (b) $15.934 - 12.807$
   (c) $284.73 - 79.831$
   (d) $13,342 - 1,202$

16. Estimate by rounding to hundreds, then find the exact answer.
   (a) $\$12,346.87 - \$4,468.63$
   (b) $3,495 - 3,090$
   (c) $6,767 - 478$
   (d) $293.86 - 148$

17. A blueprint calls for the length of a part to be 8.296 in. with a tolerance of ±0.005 in. What are the limit dimensions of the part?

18. For a moving sale, a family sold a sofa for \$75 and a table for \$25. If a newspaper ad for the sale cost \$12.75, how much did the family clear on the two items sold?

Use Fig. 1–32 for Exercises 19–22.

**19.** Find the length of *A* if *D* = 4.237 in., *B* = 1.861 in., and *C* = 1.946 in.

**20.** What is the dimension of *A* if *E* = 4.86 in., and *B* = 1.972 in.?

**21.** Give the limit dimensions of *D* if *D* = 8.935 in. with a ± 0.005-in. tolerance.

**22.** What dimension should be listed for *C* if *D* measures 3.7 in. and *E* measures 1.6 in.?

**Figure 1–32**

Multiply.

**23.** (2)(6)(7)      **24.** $6 \times 3 \times 2 \times 4$      **25.** (305)(45)      **26.** $236 \cdot 244$      **27.** $56{,}002 \cdot 7{,}040$

**28.** A college bookstore sold 327 American history textbooks for $39 each. How much did the bookstore receive for the books?

**29.** A mail-order supplier sells computer keyboards for $67 each. How much would a business pay for 21 keyboards?

**30.** An automotive tire dealer ran a special on heavy-duty, deluxe whitewall truck tires. If the dealer sold 105 tires for $112 each, how much did the dealer take in on the sale?

**31.** If a wholesaler ordered 144 computers for $305 each, how much did the dealer pay for the order?

**32.** A luxury car dealer pays a sound system installer $33.25 per hour. How much is the sound system installer paid for 37 hours of work? Estimate by rounding to a number with one nonzero digit, then find the exact answer.

**33.** A worker is offered a job that pays $365 per week. If the worker takes the job for 36 weeks, how much will the worker earn? Estimate by rounding to tens, then find the exact answer.

**34.** Find the area of a field 234.6 ft by 123.2 ft. Estimate the area by rounding to one nonzero digit, then find the exact answer. Check your answer. Express the area in square feet ($A = lw$).

**35.** A parcel of land measures 1,940.7 ft by 620.4 ft. Estimate the area by rounding to hundreds, then find the exact area. Express the area in square feet ($A = lw$).

**36.** A piecework employee averages 178.6 pieces per day. If the employee earns $0.28 per item, how much is earned in 5 days?

**37.** If a steel tape expands 0.00014 in. for each inch when heated, how much will a tape 864 in. long expand?

Divide.

**38.** $29.25 \div 0.36$      **39.** $325 \div 25$      **40.** $364.8 \div 6$      **41.** $30{,}126 \div 15$      **42.** $10{,}160 \div 20$

**43.** A group of 27 volunteers is seeking contributions to send a first-grade class to the circus. There are 632 envelopes for the collection to be divided equally among the 27 volunteers. How many will each receive? How many will be left over?

**44.** A school marching band is in a formation of 7 rows, each with the same number of students. If the band has 56 members, how many are in each row?

**45.** Find the average measure for 42.34 ft, 38.97 ft, 51.95 ft, and 61.88 ft. Round to the nearest hundredth.

**46.** Five light fixtures cost $74.98, $23.72, $51.27, $125.36, and $85.93. Find the average cost of the fixtures to the nearest cent.

**Section 1–2**

**47.** Give the base and exponent of each expression, then simplify.
(a) $7^3$      (b) $2.3^4$      (c) $8^4$

**48.** Give the base and exponent of each expression, then simplify.
(a) $5^6$      (b) $1.2^2$      (c) $10^6$

**49.** Evaluate:
(a) $1^2$      (b) $125^2$      (c) $5.6^2$      (d) $21^2$

**50.** Find the square root.
(a) $\sqrt{2500}$      (b) $\sqrt{1.44}$
(c) $\sqrt{289}$      (d) $\sqrt{81}$

**51.** Express as powers of 10.
(a) 10      (b) 1,000
(c) 10,000      (d) 100,000

**52.** Multiply by using powers of 10.
(a) $3 \times 100$      (b) $75 \times 10{,}000$   (c) $2.2 \times 1{,}000$
(d) $5 \times 100$      (e) $40.6 \times 10$

**53.** Divide by using powers of 10.
(a) $700 \div 100$      (b) $40.56 \div 1{,}000$
(c) $60.5 \div 100$      (d) $23{,}079 \div 10{,}000$

Evaluate.

**54.** $2 + 3 \cdot 3 \div 3$

**55.** $4^2 \cdot (12 - 7) - 8 + 3$

**56.** $18 \div 6 - 3$

**57.** $5 + 21 \div 3 \cdot 7$

**58.** $82 + 4 \div 2 \times 5$

**59.** $21 + 7 \cdot 2 - 5 \cdot 4$

**60.** $15 - 6 \cdot 2 + 3$

**61.** $18 - 5 \cdot 2 + 7$

**62.** $24 \div 4 - 18 \div 6$

**63.** $5 - 2 \cdot 2 + 12$

**64.** $26 + 8 \div 2 - 3 \cdot 3$

**65.** $3.1 \cdot 4 \cdot \sqrt{16} - 6^2$

**66.** $\sqrt{12.25} \times (4 - 2) + 8$

**67.** $5.2^3 - \sqrt{81} \times (2 + 1)$

**68.** $2^4 \div 2 - \sqrt{10 - 1}$

**69.** If you have 584 packages of hard candy to rebox for shipment to a discount store and you can pack 12 packages in each box, how many boxes will you need?

**70.** If European Internet Service (EIS) has an annual payroll of $7,460,174,000 for its 194,582 employees, what is the average salary of an EIS employee?

**71.** Cottage House Antique Mall has 88 booths that rent for $1.40 per square foot. If 48 of the booths have 100 ft² and 40 booths have 110 ft², what is the total rental from the booths?

**72.** Find the average of 42, 68, 72, and 96.

**73.** Judy Ackerman purchases 5 pairs of shoes for $43, $68, $72, $59, and $21. What is the average cost of each pair of shoes?

**74.** Makisha Brown records the high temperatures for the week of August 13 to be 78°, 72°, 86°, 88°, 90°, 85°, and 82°. What is the average daily temperature?

Find the perimeter of Figs. 1–33 through 1–38.

**75.**

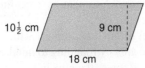

**Figure 1-33**

**76.**

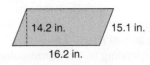

**Figure 1-34**

**77.**

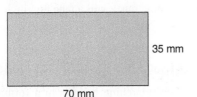

**Figure 1-35**

**78.**

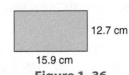

**Figure 1-36**

**79.**

**Figure 1-37**

**80.**

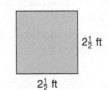

**Figure 1-38**

**81.** The Tennessee Highway Department has signs in the form of a parallelogram. One set of parallel sides each measures 15 ft and one set of parallel sides each measures 18 ft. Find the perimeter of the sign.

**82.** Antique tiles were often made in the form of a square that is 6 in. on each side. What is the perimeter of a tile?

**83.** A rectangular tablecloth measures 84 in. by 60 in. What length of lace is required to trim the edges of the cloth?

**84.** Find the number of feet of roll fencing needed to fence a square storage area measuring $15\frac{1}{2}$ ft on a side. A pre-assembled gate 4 ft wide will be installed.

Find the area of Figs. 1–39 through 1–44.

**85.**

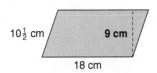

**Figure 1-39**

**86.**

**Figure 1-40**

**87.**

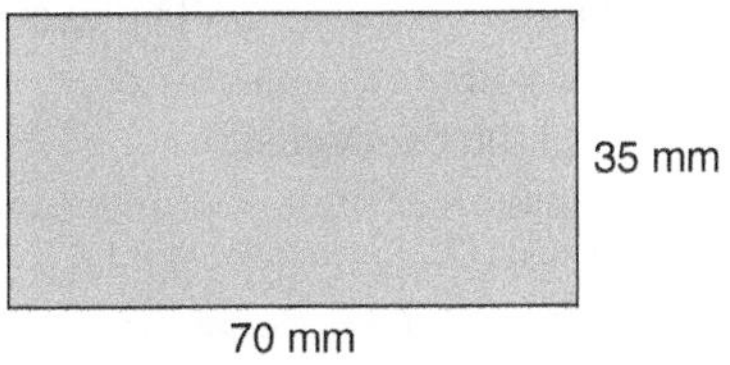

35 mm

70 mm

**Figure 1–41**

**88.**

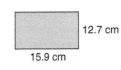

12.7 cm

15.9 cm

**Figure 1–42**

**89.**

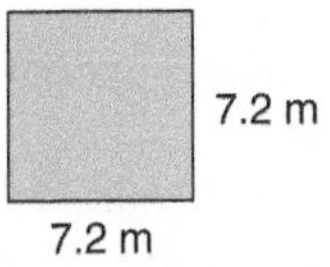

7.2 m

7.2 m

**Figure 1–43**

**90.**

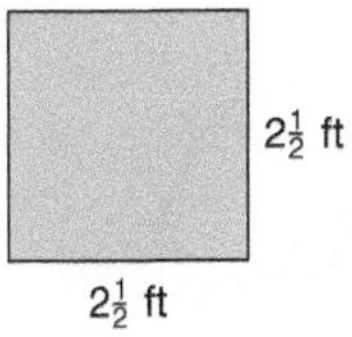

$2\frac{1}{2}$ ft

$2\frac{1}{2}$ ft

**Figure 1–44**

**91.** A parking lot for a new hospital is in the shape of a parallelogram measures 275 ft by 150 ft and has a height of 120 ft. How many square feet need to be paved?

**93.** An office 18 ft by $16\frac{1}{2}$ ft is to be carpeted. How many square yards of carpeting are needed?

**92.** A hall wall with no windows or doors measures 25 ft long by 8 ft high. Find the number of square feet to be covered if paneling is installed on the two walls.

**94.** Vincent Ores, a contractor, is to brick the storefront of a landscape service that has a doorway measuring 7 ft by 6 ft. How many bricks are needed if the storefront is 20 ft by 12 ft and the bricks cover at the rate of 6 per square foot using $\frac{1}{2}$-in. mortar joints?

**95.** A roof measuring 16 ft by 20 ft is to be covered with asphalt roofing cement. How much would the project cost if the asphalt roofing cement spreads at the rate of 150 square feet per gallon and costs $4.75 per gallon? The cement is purchased by the gallon only.

---

## TEAM PROBLEM-SOLVING EXERCISES

**1.** Your team is to design a rectangular playground that has 2,160 yd$^2$ of space.
   **(a)** Examine different options for the length and width of the playground if one piece of equipment requires at least 15 yd of length. Consider only whole-number options.
   **(b)** Give three practical options for the dimensions of the playground and explain why you selected each of these options.
   **(c)** Of all possible rectangular designs, which design requires the least amount of fencing? Consider only whole-number options.

**2.** An important practice in minimizing calculation errors when using a calculator is to anticipate some characteristics of your answer.
   **(a)** Find the squares of five different decimal numbers when each has one decimal place.
   **(b)** Find the squares of five different decimal numbers when each has two decimal places.
   **(c)** Find the squares of five different decimal numbers when each has three decimal places.
   **(d)** Examine the patterns established with the squares in parts a, b, and c. Write a statement describing each of these patterns.

---

## PRACTICE TEST

**1.** Which number is smaller, 5.09 or 5.1?

**3.** Round 48.3284 to the nearest tenth.

**2.** Round 4.018 to the nearest hundredth.

**4.** Round $4.834 to the nearest cent.

Perform the indicated operations.

**5.** $37 + 158 + 764 + 48$

**7.** $\$13,207(702)$

**9.** $3^2 + 5^3$

**11.** $3 \cdot 6^2 - 4 \div 2$

**6.** $\$61,532 - \$47,245$

**8.** $\$25,600 \div 12$

**10.** $46 \times 10^3$

**12.** $5^3 - (3 + 2) \times \sqrt{9}$

13. If a man has a bill for $165 and his paycheck is $475, estimate how much of his paycheck is left after paying the bill by rounding to tens. Find the exact amount left.

14. A corporation buys 45 DVD players for $335 each. Use numbers with one nonzero digit to estimate the total cost. Find the exact cost.

15. A softball coach paid $126 for 9 pizzas for a party after a successful season. Estimate the cost of each pizza using numbers with one nonzero digit. Find the exact cost per pizza.

16. A mathematics professor promises to give 2 extra points for each set of exercises a student works. If one student works 17 sets of exercises, how many points should be given?

17. Find the product: $42.73 \times 1,000$.

18. Divide: $25\overline{)27.75}$.

19. Round answer to the nearest tenth: $7.2\overline{)83.41}$.

20. Divide: $52.38 \div 10,000$.

21. Find the average of these test scores: 82, 95, 76, 84, 72, and 91. Round to the nearest whole number.

22. Estimate the sum by rounding to the nearest whole number: $3.85 + 7.46$.

23. Estimate the difference by rounding each number to the nearest tenth: $0.87 - 0.328$.

24. The blueprint specification for a machined part calls for its thickness to be 1.485 in. with a tolerance of ±0.010 in. Find the limit dimensions of the part.

25. Heating oil costs $1.75 per gallon. What is the cost of 10,000 gal?

26. A construction job requires 16 pieces of steel, each 7.96 ft long. What length of steel is needed?

Find the perimeter and area of Figs. 1–45 through 1–47.

27.

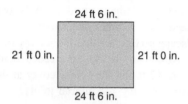

24 ft 6 in.
21 ft 0 in. 21 ft 0 in.
24 ft 6 in.

**Figure 1–45**

28.

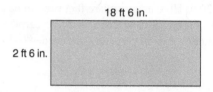

18 ft 6 in.
2 ft 6 in.

**Figure 1–46**

29.

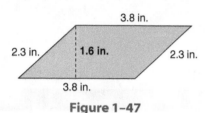

3.8 in.
2.3 in. **1.6 in.** 2.3 in.
3.8 in.

**Figure 1–47**

## CAREER APPLICATION: DC CIRCUITS

In electronics one of the basic types of circuits is the **direct current (dc) circuit.** One main component of a dc circuit is the **electromotive force (emf)** or **source** of energy. The most common source of energy in a dc circuit is a battery. The force or energy, $\mathcal{E}$, is also called the **voltage** and is measured in units of **volts, V.**

A second common component of a dc circuit is a **resistance.** The resistance will most often be a **resistor.** Resistance is measured in units of **ohms, $\Omega$.**

In a simple circuit diagram, symbols are used to indicate the energy source and the resistors. Figure 1–48 illustrates these symbols.

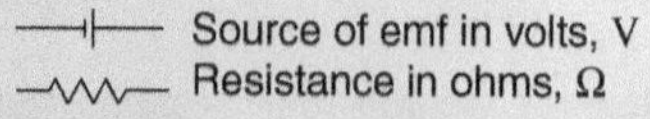

⊣⊢ Source of emf in volts, V
⌇⌇⌇ Resistance in ohms, $\Omega$

**Figure 1–48** Circuit symbols.

An important relationship among the **voltage,** $V$, the current, $I$, and the resistance, $R$, is called **Ohm's law.** The **current,** $I$, is measured in units called **amperes, or amps, A.**

**Ohm's law:** $V = IR$    where $V$ = voltage in volts, $I$ = current in amps, and
           $R$ = resistance in ohms

This version of Ohm's law is used to find the voltage when the current and resistance are known.

**EXAMPLE**  Find the voltage in the circuit in Figure 1–49.

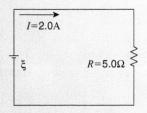

**Figure 1–49**

$V = IR$        Substitute 2.0 A for $I$ and 5.0 Ω for $R$.
$V = 2.0\,(5.0)$    Multiply.
$V = 10.0$      Interpret units: voltage is measured in volts.
$V = 10.0\ \text{V}$

**The voltage is 10.0 V.**

Some variations of Ohm's law that can be used to find the current or resistance are:

$I = \dfrac{V}{R}$    Used to find the current when the voltage and resistance are known

$R = \dfrac{V}{I}$    Used to find the resistance when the voltage and current are known

A circuit can have more than one resistor. If the resistors in a circuit are arranged one after another, they are said to be **in series.** The **total resistance** in a circuit in series is the sum of all the resistors. Figure 1–50 illustrates a circuit with three resistors labeled $R_1$, $R_2$, and $R_3$. The arrow indicates the direction of the flow of current.

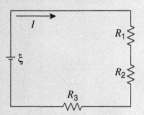

**Figure 1–50**  Resistors connected in series.

**EXAMPLE**  Find the total resistance in the circuit in Figure 1–50 if $R_1 = 4.0\ \Omega$, $R_2 = 3.0\ \Omega$, and $R_3 = 2.0\ \Omega$.

total resistance $= R_1 + R_2 + R_3$    Substitute values.
$= 4.0 + 3.0 + 2.0$    Add.
$= 9.0$    Interpret units.
$= 9.0\ \Omega$

**The total resistance is 9.0 Ω.**

## Exercises

Use Figure 1–51 and find the missing components in Exercises 1–6.

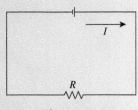

**Figure 1–51**

1. Find the voltage when $I = 5.0$ A and $R = 3.0\ \Omega$.
2. Find the voltage when $I = 4.2$ A and $R = 3.7\ \Omega$.
3. Find the current when $V = 10.0$ V and $R = 2.0\ \Omega$.
4. Find the current when $V = 15.4$ V and $R = 2.2\ \Omega$.
5. Find the resistance when $I = 10.0$ A and $V = 2.0$ V.
6. Find the resistance when $I = 9.0$ A and $V = 20.7$ V.

Use Figure 1–52 in Exercises 7–10.

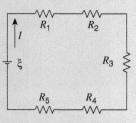

**Figure 1–52**

7. Find the total resistance if $R_1 = 2.3\ \Omega$, $R_2 = 3.8\ \Omega$, $R_3 = 4.5\ \Omega$, $R_4 = 2.0\ \Omega$, and $R_5 = 3.1\ \Omega$.
8. Find the total resistance if $R_1 = 3.7\ \Omega$, $R_2 = 5.9\ \Omega$, $R_3 = 2.5\ \Omega$, $R_4 = 4.2\ \Omega$, and $R_5 = 1.0\ \Omega$.
9. Find $R_1$ if the total resistance is $19.4\ \Omega$, $R_2 = 4.3\ \Omega$, $R_3 = 1.5\ \Omega$, $R_4 = 5.2\ \Omega$, and $R_5 = 3.8\ \Omega$.
10. Find $R_1$ if the total resistance is $23.2\ \Omega$, $R_2 = 5.3\ \Omega$, $R_3 = 1.7\ \Omega$, $R_4 = 2.2\ \Omega$, and $R_5 = 8.4\ \Omega$.

**Answers**

1. 15.0 V    2. 15.54 V    3. 5.0 A    4. 7.0 A    5. 0.2 $\Omega$
6. 2.3 $\Omega$    7. 15.7 $\Omega$    8. 17.3 $\Omega$    9. 4.6 $\Omega$    10. 5.6 $\Omega$

# 2

# Review of Fractions and Percents

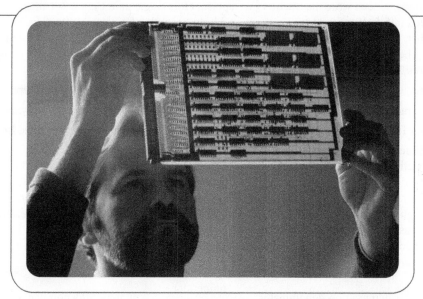

## Focus on Careers

Forty-two percent of all engineering technicians are electrical and electronics engineering technicians. An associate degree or extensive job training provides the best opportunities for engineering technicians.

Engineering technicians may assist engineers and scientists to solve technical problems in manufacturing, construction, and maintenance. Some technicians inspect products and processes to ensure quality control. Others work in product design, development, or production.

Nearly a half-million jobs are available in this career, and median earnings range from $36,850 to $51,650, depending on the type of technician. The median annual earnings of electrical and electronics engineering technicians were $42,950 in 2002. Federal government jobs in this career paid the highest annual earnings and jobs in the telecommunication industry paid the second highest.

Certification for engineering technicians is available at various levels from the National Institute for Certification in Engineering Technologies (NICET). In addition to passing a written examination, job-related experience and a supervisory evaluation and recommendation are required for certification.

Source: *Occupational Outlook Handbook,* 2004–2005 Edition, U.S. Department of Labor, Bureau of Labor Statistics.

The language of mathematics is important in the study of mathematics. Mathematical symbols help us describe situations in shortcut fashion, and terminology helps us understand mathematical concepts. In this chapter we study how fractions relate to whole numbers and decimals, and how we add, subtract, multiply, and divide fractions.

## 2-1 | Equivalent Fractions and Decimals

*Learning Outcomes*

1 Write equivalent fractions with different denominators.

2 Write improper fractions as whole numbers or mixed numbers.

3 Write whole numbers or mixed numbers as improper fractions.

4 Write decimals as fractions and fractions as decimals.

5 Compare fractions, mixed numbers, and decimals.

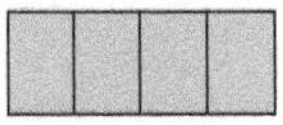

In Chapter 1, we examined a special type of fraction called a *decimal fraction.* A **fraction** is a number that can be expressed as the quotient of two whole numbers.

$1 = \frac{4}{4}$

**Figure 2–1**

### 1 Write Equivalent Fractions with Different Denominators.

Symbolically, a **common fraction** is written as $\frac{a}{b}$ or $a/b$, where $a$ and $b$ are whole numbers and $b$ cannot equal zero ($b \neq 0$). The symbol $\neq$ is read "is not equal to." If one unit is divided into four parts, we can write the fraction $\frac{4}{4}$ to represent this single unit. Figure 2–1 illustrates a unit divided into four equal parts. The fraction $\frac{4}{4}$ is an example of a *common fraction.* The bottom number, the **denominator,** indicates *how many equal parts* one whole unit has been divided into. The top number, the **numerator,** tells *how many of these parts* are being considered. Figures 2–2 and 2–3 illustrate the fractions $\frac{1}{4}$ and $\frac{3}{7}$.

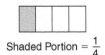

Shaded Portion = $\frac{1}{4}$

**Figure 2–2**

The unit is the *standard* amount when writing fractions. Thus, $\frac{4}{4}$ and $\frac{3}{3}$ each represent one unit. Fractions that represent less than one unit (less than 1), for example $\frac{1}{4}$ or $\frac{3}{7}$, are called **proper fractions.** Fractions that represent one or more units, for example $\frac{7}{5}$, are called **improper fractions.** Figures 2–4 and 2–5 illustrate the improper fractions $\frac{7}{5}$ and $\frac{10}{5}$.

Shaded Portion = $\frac{3}{7}$

**Figure 2–3**

Fractions and division are related. In the fraction $\frac{10}{5}$ (two units), $10 \div 5 = 2$. This relationship to division is important to our understanding of fractions.

The numerator and denominator of a fraction are most often separated by a horizontal bar, although sometimes a slash is used. This horizontal bar or slash is the **fraction line,** and it also serves as a division symbol.

When using fraction terminology or notation to describe division, *the numerator is divided by the denominator in all cases.*

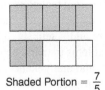

Shaded Portion = $\frac{7}{5}$

**Figure 2–4**

$$\text{numerator} \div \text{denominator} \qquad \frac{\text{numerator}}{\text{denominator}} \qquad \text{numerator/denominator}$$

An improper fraction can be written as a whole number when the denominator divides evenly into the numerator ($\frac{8}{4} = 2$). When the fraction is more than one unit and the denominator cannot divide evenly into the numerator, the improper fraction can be written as a combination of a whole number and a fractional part, such as $\frac{9}{4} = 2\frac{1}{4}$.

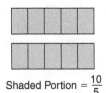

Shaded Portion = $\frac{10}{5}$

**Figure 2–5**

A **mixed number** consists of both a whole number and a fraction. The whole number and fraction are added together. *Example:* $3\frac{2}{5}$ means 3 whole units and $\frac{2}{5}$ of another unit or $3\frac{2}{5} = 3 + \frac{2}{5}$.

Fractions indicate division. Multiplication and division are inverse operations. Let's look at some relationships involving multiplication and division.

If we count by 3s, such as 3, 6, 9, 12, 15, 18, we obtain natural numbers that are *multiples* of 3 because each is the product of 3 and a natural number; that is,

$$3 = 3 \times 1, \qquad 6 = 3 \times 2, \qquad 9 = 3 \times 3$$
$$12 = 3 \times 4, \qquad 15 = 3 \times 5, \qquad 18 = 3 \times 6$$

A **multiple** of a natural number is the product of that number and a natural number.

Show that 2, 4, 6, 8, and 10 are multiples of 2 by writing each as the product of 2 and a natural number.

$$2 = 2 \times 1, \quad 4 = 2 \times 2, \quad 6 = 2 \times 3, \quad 8 = 2 \times 4, \quad 10 = 2 \times 5$$

Natural numbers that are multiples of 2 are **even numbers.** Natural numbers that are not multiples of 2 are **odd numbers.**

EXAMPLE  Find five multiples of 16.

$$1 \times 16 = 16, \quad 2 \times 16 = 32, \quad 3 \times 16 = 48, \quad 4 \times 16 = 64, \quad 5 \times 16 = 80$$

**16, 32, 48, 64, and 80 are multiples of 16.**

We say that a number is **divisible** by another number if the quotient has no remainder or if the dividend is a multiple of the divisor.

We want to be able to determine divisibility by **inspection.** This means that we can examine the number being divided (dividend) and decide if it is divisible by a divisor without actually having to perform the division.

EXAMPLE  Is 35 divisible by 7?

35 is divisible by 7 if $35 \div 7$ has no remainder or if 35 is a multiple of 7.

$$35 \div 7 = 5 \quad \text{or} \quad 35 = 5 \times 7$$

**Yes, 35 is divisible by 7.**

Rules or tests can help us decide by inspection if certain numbers are divisible by other numbers.

## Tests for divisibility:

A number is divisible by

1. 2 if the last digit is an even number (0, 2, 4, 6, or 8).
2. 3 if the sum of its digits is divisible by 3.
3. 4 if the last two digits form a number that is divisible by 4.
4. 5 if the last digit is 0 or 5.
5. 6 if the number is divisible by *both* 2 *and* 3.
6. 7 if the division has no remainder.
7. 8 if the last three digits form a number divisible by 8.
8. 9 if the sum of its digits is divisible by 9.
9. 10 if the last digit is 0.

EXAMPLE  Use the tests for divisibility to identify which number in each pair is divisible by the given divisor.

| Numbers | Divisor | Answer |
|---|---|---|
| 874 or 873 | 2 | **874;** the last digit is an even digit (4). |
| 275 or 270 | 2 | **270;** the last digit is an even digit (0). |

| | | |
|---|---|---|
| 427 or 423 | 3 | **423;** the sum of the digits is divisible by 3: $4 + 2 + 3 = 9$. |
| 5,912 or 5,913 | 4 | **5,912;** the last two digits form a number divisible by 4: $12 \div 4 = 3$. |
| 80 or 82 | 5 | **80;** the last digit is 0. |
| 56 or 65 | 5 | **65;** the last digit is 5. |
| 804 or 802 | 6 | **804;** the last digit is even and the sum of the digits is divisible by 3: $8 + 0 + 4 = 12$. |
| 58 or 56 | 7 | **56;** it divides by 7 with no remainder. |
| 3,160 or 3,162 | 8 | **3,160;** the last three digits form a number divisible by 8: $160 \div 8 = 20$. |
| 477 or 475 | 9 | **477;** the sum of the digits is divisible by 9: $4 + 7 + 7 = 18$. |
| 182 or 180 | 10 | **180;** the last digit is 0. |

The **natural numbers** are the counting numbers—1, 2, 3, 4, 5, 6, 7, 8, 9, 10, 11, 12, and so forth. The natural numbers can also be described as the whole numbers excluding zero.

Any natural number can be expressed as the product of two natural numbers. These two natural numbers are called a **factor pair** of the number.

Every natural number greater than 1 has at least one factor pair, the number itself and 1.

1 and 3 form a factor pair for 3: $1 \times 3 = 3$.

1 and 5 form a factor pair for 5: $1 \times 5 = 5$.

Many natural numbers have more than one factor pair. For example, list all number pairs of 12. Start with the pair $1 \times 12$. Every number has a factor pair of the number and 1. Next, examine each number from 2 to 11.

| | | | |
|---|---|---|---|
| 2 | 12 is divisible by 2 | $12 \div 2 = 6$ | $2 \times 6$ is a factor pair of 12. |
| 3 | 12 is divisible by 3 | $12 \div 3 = 4$ | $3 \times 4$ is a factor pair of 12. |
| 4 | 12 is divisible by 4 | $12 \div 4 = 3$ | $4 \times 3$ is a factor pair of 12. |
| 5 | 12 is not divisible by 5 | | |
| 6 | 12 is divisible by 6 | $12 \div 6 = 2$ | $6 \times 2$ is a factor pair of 12. |
| 7 | 12 is not divisible by 7 | | |
| 8 | 12 is not divisible by 8 | | |
| 9 | 12 is not divisible by 9 | | |
| 10 | 12 is not divisible by 10 | | |
| 11 | 12 is not divisible by 11 | | |

Are the factor pairs $2 \times 6$ and $6 \times 2$ different pairs? No. Both pairs have the same numbers, and since multiplication is commutative, they count as one pair. Similarly, $3 \times 4$ and $4 \times 3$ are the same factor pair. Is it necessary to examine every number less than 12? No. Once we get the first repeat, $4 \times 3$, we can assume we have found all the factor pairs of the number.

The factor pairs of 12 are $1 \times 12$, $2 \times 6$, and $3 \times 4$.

**To find all factor pairs of a natural number:**

1. Write the factor pair of 1 and the given natural number.
2. Check to see if the number is divisible by 2. If so, write the factor pair of 2 and the quotient of the given number and 2.
3. Check the next number for divisibility; if the given number is divisible by the next number, write the factor pair.
4. Continue Step 3 until you reach a number that already has been found in a previous factor pair.

**EXAMPLE** List all the factor pairs of 18.

$1 \times 18$

$2 \times 9$     18 is divisible by 2: $18 \div 2 = 9$.

$3 \times 6$     18 is divisible by 3: $18 \div 3 = 6$.

4 or 5     18 is not divisible by 4 or 5.

6     18 is divisible by 6; 6 was found in the factor pair $3 \times 6$, so we stop.

**The factor pairs of 18 are 1 and 18, 2 and 9, and 3 and 6.**

Once we have listed all factor pairs of a number, we can list all factors of a number. From the factor pairs, we list every different factor that appears in any factor pair. The factors of 12 are 1, 2, 3, 4, 6, and 12.

**EXAMPLE** List all the factor pairs of 48; then write each distinct factor in order from smallest to largest.

$1 \times 48$

$2 \times 24$     48 is divisible by 2: $48 \div 2 = 24$.

$3 \times 16$     48 is divisible by 3: $48 \div 3 = 16$.

$4 \times 12$     48 is divisible by 4: $48 \div 4 = 12$.

5     48 is not divisible by 5.

$6 \times 8$     48 is divisible by 6: $48 \div 6 = 8$.

7     48 is not divisible by 7.

8     48 is divisible by 8; 8 was found in the factor pair $6 \times 8$, so we stop.

**Factors: 1, 2, 3, 4, 6, 8, 12, 16, 24, 48**

When all the factor pairs of natural numbers are listed, some numbers have only one pair of factors, the number itself and 1. These numbers form a special set of numbers called prime numbers. A **prime number** is a whole number greater than 1 that has only one factor pair, the number itself and 1. Note that 1 is not a prime number because it has only one factor.

**EXAMPLE** Identify the prime numbers by examining the factor pairs of the numbers.

**(a)** 8    **(b)** 1    **(c)** 3    **(d)** 9    **(e)** 7

**(a)** **8 is *not* a prime number** because its factor pairs are $1 \times 8$ and $2 \times 4$.

**(b)** **1 is *not* a prime number** because a prime must be greater than 1 or because it does not have a factor pair.

**(c)** **3 is a prime number** because it has only one factor pair, $1 \times 3$.

**(d)** **9 is *not* a prime number** because its factor pairs are $1 \times 9$ and $3 \times 3$.

**(e)** **7 is a prime number** because it has only one factor pair, $1 \times 7$.

A **composite number** is a whole number greater than 1 that is not a prime number. In the preceding example, 8 and 9 are composite numbers. A composite number has at least one factor pair other than itself and 1.

**EXAMPLE** Identify the composite numbers by examining the factor pairs of the numbers.

(a) 4 (b) 10 (c) 13 (d) 12 (e) 5

**(a)** **4 is a composite number** because its factor pairs are $1 \times 4$ and $2 \times 2$.
**(b)** **10 is a composite number** because its factor pairs are $1 \times 10$ and $2 \times 5$.
**(c)** **13 is *not* a composite number** because its only factor pair is $1 \times 13$. It is a prime number.
**(d)** **12 is a composite number** because its factor pairs are $1 \times 12$, $2 \times 6$, and $3 \times 4$.
**(e)** **5 is *not* a composite number** because its only factor pair is $1 \times 5$. It is a prime number.

TIP!

### Prime Numbers Less Than 50

We can find all the prime numbers that are 50 or less using an ancient technique developed by the mathematician Eratosthenes. This technique is called **sieve of Eratosthenes.**

Step 1. List the numbers from 1 through 50.

Step 2. Eliminate numbers that are not prime using the systematic process:
**(a)** 1 is not prime. Eliminate 1.
**(b)** 2 is prime. Eliminate all multiples of 2.
**(c)** 3 is prime. Eliminate all multiples of 3.
**(d)** 4 has already been eliminated.
**(e)** 5 is prime. Eliminate all multiples of 5.
**(f)** 6 has already been eliminated.
**(g)** 7 is prime. Eliminate all multiples of 7.

Step 3. Circle remaining numbers as prime numbers.

| | | | | | | | | | |
|---|---|---|---|---|---|---|---|---|---|
| ~~1~~ | 2 | 3 | ~~4~~ | 5 | ~~6~~ | 7 | ~~8~~ | ~~9~~ | ~~10~~ |
| 11 | ~~12~~ | 13 | ~~14~~ | ~~15~~ | ~~16~~ | 17 | ~~18~~ | 19 | ~~20~~ |
| ~~21~~ | ~~22~~ | 23 | ~~24~~ | ~~25~~ | ~~26~~ | ~~27~~ | ~~28~~ | 29 | ~~30~~ |
| 31 | ~~32~~ | ~~33~~ | ~~34~~ | ~~35~~ | ~~36~~ | 37 | ~~38~~ | ~~39~~ | ~~40~~ |
| 41 | ~~42~~ | 43 | ~~44~~ | ~~45~~ | ~~46~~ | 47 | ~~48~~ | ~~49~~ | ~~50~~ |

All numbers not already eliminated are prime. Why? The numbers 8, 9, and 10 have already been eliminated as multiples of 2, 3, and 5, respectively. Multiples of 11 that are less than 50 have already been eliminated: $11 \times 2 = 22$, $11 \times 3 = 33$, $11 \times 4 = 44$. The product $11 \times 5 = 55$ is greater than 50. Similarly, all other composite numbers have already been eliminated.

A composite number can be expressed as a product of prime numbers. **Prime factorization** refers to writing a composite number as the product of *only* prime numbers. In this case, the factors are **prime factors.**

### To find the prime factors of a composite number:

1. Test each prime number to see if the composite number is divisible by the prime.
2. Make a factor pair using the first prime number that passes the test in Step 1.
3. Carry forward the prime factors and test the remaining factors by repeating Steps 1 and 2.

**EXAMPLE**   Find the prime factorization of 30.

$$30 = 2 \times 15$$

first prime

30 is divisible by 2. Factor 30 into a factor pair using its smallest prime factor, 2.

$$30 = 2 \times 3 \times 5$$

last primes

Carry the prime factor 2 forward. Factor the composite number 15 using its smallest prime factor, 3. Because 5 is also prime, the factoring is complete.

**The prime factorization of 30 is $2 \times 3 \times 5$.**

---

**EXAMPLE**   Find the prime factorization of 16.

$$16 = 2 \times 8$$

first prime

Factor 16 into two factors using its smallest prime factor.

$$16 = 2 \times 2 \times 4$$

second prime

Factor 8 into two factors using its smallest prime factor.

$$16 = 2 \times 2 \times 2 \times 2$$

last two primes

Factor 4 into two factors using its smallest prime factor.

**The prime factorization of 16 is $2 \times 2 \times 2 \times 2$. We can write this expression in exponential notation as $2^4$.**

---

The **least common multiple (LCM)** of two or more natural numbers is the smallest number that is a multiple of each number. It is divisible by each number.

To find the least common multiple of 3 and 5, examine the multiples of each number.

multiples of 3:   3, 6, 9, 12, 15, 18, 21, 24, 27, 30, 33, 36, 39, . . .

multiples of 5:   5, 10, 15, 20, 25, 30, 35, 40, . . .

The common multiples in these lists that are less than 40 are 15 and 30; 15 is the *least common multiple* of 3 and 5.

Prime factorization can also be used to find the least common multiple of two or more numbers.

**To find the least common multiple of two or more natural numbers using the prime factorization of the numbers:**

1.  List the prime factorization of each number using exponential notation.
2.  List the prime factorization of the least common multiple by including the prime factors appearing in *each* number. If a prime factor appears in more than one number, use the factor with the *largest* exponent.
3.  Write the resulting expression in standard notation.

---

**EXAMPLE**   Find the least common multiple of 12 and 40 by prime factorization.

$$12 = 2 \times 2 \times 3 \quad\quad = 2^2 \times 3$$   Prime factorization of 12.

$$40 = 2 \times 2 \times 2 \times 5 = 2^3 \times 5$$   Prime factorization of 40.

$$\text{LCM} = 2^3 \times 3 \times 5$$   Prime factorization of LCM.

$$\textbf{LCM} = \textbf{120}$$   LCM in standard notation.

The **greatest common factor (GCF)** of two or more numbers is the largest factor common to each number. Each number is divisible by the GCF.

Let's take the numbers 30 and 42. The prime factors are

$$30 = \boxed{2 \times 3} \times 5 \qquad 42 = \boxed{2 \times 3} \times 7$$

The *common* prime factors of both 30 and 42 are 2 and 3, which represent the composite factor 6. The *greatest* common factor is the product of the common prime factors, $2 \times 3 = 6$.

**To find the greatest common factor (GCF) of two or more natural numbers:**

1. List the prime factorization of each number using exponential notation when appropriate.
2. List the prime factorization of the greatest common factor by including each prime factor appearing in *every* number. If a prime factor appears more than one time in any number (that is, the exponent is greater than 1), use the factor with the *smallest* exponent. If there are no common prime factors, the GCF is 1.
3. Write the resulting expression in standard notation.

---

**EXAMPLE**  Find the greatest common factor of 15, 30, and 45.

$15 = 3 \times 5 \quad\ = \boxed{3} \times \boxed{5}$    Prime factorization of 15.
$30 = 2 \times 3 \times 5 = 2 \times \boxed{3} \times \boxed{5}$    Prime factorization of 30.
$45 = 3 \times 3 \times 5 = \boxed{3^2} \times \boxed{5}$    Prime factorization of 45.
$\text{GCF} = \boxed{3} \times \boxed{5}$    Common prime factors.
$\mathbf{GCF = 15}$    GCF in standard notation.

---

**EXAMPLE**  Find the greatest common factor of 10, 12, and 13.

$10 = 2 \times 5 \quad\ = 2 \times 5$    Prime factorization of 10.
$12 = 2 \times 2 \times 3 = 2^2 \times 3$    Prime factorization of 12.
$13 = 13 \qquad\ = 13$    Prime factorization of 13.
$\mathbf{GCF = 1}$    No common prime factors.

---

**TIP!**

**LCM Versus GCF**

The least common multiple (LCM) and greatest common factor (GCF) are easily confused. Because multiples of a number are the products of the number and any natural number, multiples will be as large as or larger than the original number. The LCM is the *smallest* of the "same or larger" common multiples of the original numbers.

Because factors of a number are the same as or smaller than the given number, the GCF is the *largest* of the "same or smaller" common factors of the original numbers.

---

There are many different ways to express the same value in fractional form. For example, the whole number 1 can be written as $\frac{1}{1}, \frac{4}{4}, \frac{7}{7}, \frac{15}{15}$, and so on.

Fractions are equivalent if they represent the same value. Compare the illustrations in Figure 2–6. In Figure 2–6, line *a* is one whole unit divided into only 1 part ($\frac{1}{1}$). Line *b* is one

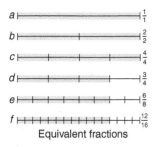

Equivalent fractions

**Figure 2–6**

unit divided into 2 parts ($\frac{2}{2}$). Lines $c$ and $d$ are divided into 4 parts ($\frac{4}{4}$), line $e$ is divided into 8 parts ($\frac{8}{8}$), and line $f$ is divided into 16 parts ($\frac{16}{16}$).

Again, look at lines $d$, $e$, and $f$. Line $d$ is divided into 4 parts and 3 of them are shaded. Line $e$ is divided into 8 parts and 6 of them are shaded. Line $f$ is divided into 16 parts and 12 of them are shaded.

Look at the shaded portions of lines $d$, $e$, and $f$. They are the same length, even though they are divided into a different number of parts. Thus, $\frac{3}{4}$, $\frac{6}{8}$, and $\frac{12}{16}$ are **equivalent fractions** since they represent the same shaded part of one whole unit. That is,

$$\frac{3}{4} = \frac{6}{8} = \frac{12}{16}$$

*Equivalent fractions* are in the same "family of fractions." The first member of the "family" is the fraction in **lowest terms;** that is, no natural number divides evenly into *both* the numerator and the denominator except the number 1. Other "family members" are found by multiplying both the numerator and the denominator by the same natural number.

---

**EXAMPLE** Find five fractions that are equivalent to $\frac{1}{2}$.

$\dfrac{1 \times 2}{2 \times 2}$    or    $\dfrac{1}{2} \times \dfrac{2}{2} = \dfrac{\mathbf{2}}{\mathbf{4}}$      $\dfrac{1 \times 5}{2 \times 5}$    or    $\dfrac{1}{2} \times \dfrac{5}{5} = \dfrac{\mathbf{5}}{\mathbf{10}}$

$\dfrac{1 \times 3}{2 \times 3}$    or    $\dfrac{1}{2} \times \dfrac{3}{3} = \dfrac{\mathbf{3}}{\mathbf{6}}$      $\dfrac{1 \times 6}{2 \times 6}$    or    $\dfrac{1}{2} \times \dfrac{6}{6} = \dfrac{\mathbf{6}}{\mathbf{12}}$

$\dfrac{1 \times 4}{2 \times 4}$    or    $\dfrac{1}{2} \times \dfrac{4}{4} = \dfrac{\mathbf{4}}{\mathbf{8}}$

Other equivalent fractions can be found.

---

**TIP!**

### Multiplication, Division, and 1

In the preceding example $\frac{1}{2}$ is multiplied by a fraction whose value is 1, and 1 times any number does not change the value of that number. Written symbolically,

$$\frac{n}{n} = 1 \text{ when } n \neq 0 \qquad \text{and} \qquad 1 \times n = n$$

Fractions in the same family can be generated by multiplying the fraction by 1 in the form of $\frac{2}{2}$, $\frac{3}{3}$, $\frac{4}{4}$, and so on.

---

The concept presented in the preceding example and tip is referred to as the **fundamental principle of fractions.** If the numerator and denominator of a fraction are multiplied by the same nonzero number, the value of the fraction remains unchanged.

### To change a fraction to an equivalent fraction with a specified larger denominator:

1. Divide the specified larger denominator by the original denominator.
2. Multiply the original numerator and denominator by the quotient found in Step 1. That is, multiply by 1 in the form of $\frac{n}{n}$ when $n \neq 0$.

**EXAMPLE**  Change $\frac{5}{8}$ to an equivalent fraction whose denominator is 32.

$$\frac{5}{8} = \frac{?}{32} \qquad 32 \div 8 = 4 \text{ and } 4 \times 5 = 20$$

$$\frac{5}{8} \times \frac{4}{4} = \frac{\mathbf{20}}{\mathbf{32}} \qquad \text{Apply the fundamental principle of fractions.}$$

Because each fraction has an unlimited number of equivalent fractions, we usually work with fractions in *lowest terms*. When we find an equivalent fraction with smaller numbers and there are no common factors in the numerator and denominator, we have **reduced to lowest terms.**

**To change a fraction to an equivalent fraction with a smaller denominator or to reduce a fraction to lowest terms:**

1. Find a common factor greater than 1 for the numerator and denominator.
2. Divide both the numerator and the denominator by this common factor.
3. Continue until the fraction is in lowest terms or has the desired smaller denominator.

*Note:* To reduce to lowest terms in the fewest steps, find the greatest common factor (GCF) in Step 1.

**EXAMPLE**  Reduce $\frac{8}{10}$ to lowest terms.

Prime factors of 8: $2 \times 2 \times 2$ or $2^3$

Prime factors of 10: $2 \times 5$

The greatest common factor (GCF) is $2$.

$$\frac{8 \div 2}{10 \div 2} \qquad \text{or} \qquad \frac{8}{10} \div \frac{2}{2} = \frac{\mathbf{4}}{\mathbf{5}}$$

**TIP!**

**Reducing and the Properties of 1**

To reduce the fraction $\frac{8}{10}$, we divide by the whole number 1 in the form of $\frac{2}{2}$. That is, $\frac{8}{10} \div \frac{2}{2} = \frac{4}{5}$. A nonzero number divided by itself is 1, and to divide a number by 1 does not change the value of the number. Symbolically,

$$\frac{n}{n} = 1 \text{ when } n \neq 0 \qquad \text{and} \qquad n \div 1 = n \qquad \text{or} \qquad \frac{n}{1} = n$$

**EXAMPLE**  Reduce $\frac{18}{24}$ to lowest terms.

Prime factors of 18: $2 \times 3 \times 3$    or    $2 \times 3^2$

Prime factors of 24: $2 \times 2 \times 2 \times 3$    or    $2^3 \times 3$

The GCF is $2 \times 3$ or 6.

$$\frac{18 \div 6}{24 \div 6} \qquad \text{or} \qquad \frac{18}{24} \div \frac{6}{6} = \frac{\mathbf{3}}{\mathbf{4}}$$

Chapter 2 / Review of Fractions and Percents

## Do You Have to Use the GCF to Reduce to Lowest Terms?

A fraction can be reduced to lowest terms in the fewest steps by using the *greatest common factor*; however, it still can be reduced to lowest terms using any common factor; this just takes a few more steps.

$$\frac{18 \div 2}{24 \div 2} = \frac{9}{12} \qquad \text{Reduce with common factor 2.}$$

$$\frac{9 \div 3}{12 \div 3} = \frac{3}{4} \qquad \text{Reduce with common factor 3.}$$

The *U.S. customary rule* is divided into inches and parts of an inch and is used to measure length. Each inch is subdivided into fractional parts, usually 8, 16, 32, or 64.

The rule in Figure 2–7 shows each inch divided into 16 equal parts, so each part is $\frac{1}{16}$ in.; that is, the first mark from the left edge represents $\frac{1}{16}$ in. The left end of the rule represents zero (0).

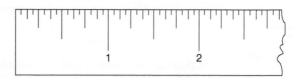

**Figure 2–7**  U.S. customary rule.

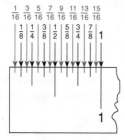

**Figure 2–8**  1 in.

Figure 2–8 labels division marks for the parts of 1 in. The second mark from the left edge represents $\frac{2}{16}$ or $\frac{1}{8}$ in. This mark is slightly longer than the first mark.

The fourth mark from the left is labeled $\frac{1}{4}$; that is, $\frac{4}{16} = \frac{1}{4}$. In each case, fractions are always reduced to lowest terms. Notice that the $\frac{1}{4}$ mark is slightly longer than the $\frac{1}{8}$ mark.

The division marks are different lengths to make the rule easier to read. The shortest marks represent fractions that, in lowest terms, are sixteenths $\left(\frac{1}{16}, \frac{3}{16}, \frac{5}{16}, \frac{7}{16}, \frac{9}{16}, \frac{11}{16}, \frac{13}{16}, \frac{15}{16}\right)$. The fractions that reduce to eighths are slightly longer than the sixteenths marks $\left(\frac{1}{8}, \frac{3}{8}, \frac{5}{8}, \frac{7}{8}\right)$. The marks representing fractions that reduce to fourths are slightly longer than the eighths $\left(\frac{1}{4}, \frac{3}{4}\right)$. The marks for one-half $\left(\frac{1}{2}\right)$ are longer than the fourths, and the inch marks are the longest.

**EXAMPLE**  Measure line segment *AB* (Figure 2–9).

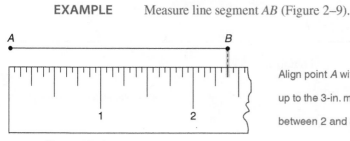

Align point *A* with zero. Line segment *AB* goes past the 2-in. mark but not up to the 3-in. mark. Therefore, the measure of *AB* will be a mixed number between 2 and 3. Point *B* is $\frac{3}{8}$ in. past 2.

**Figure 2–9**  Line segment *AB*.

**AB is $2\frac{3}{8}$ in.**

### Judge to the Closest Mark

A line segment may not always align exactly with a division mark. Use eye judgment to decide which mark is closer to the end of the line segment.

**EXAMPLE**   Measure line segment *CD* (Figure 2–10).

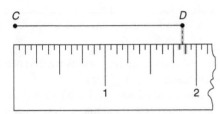

Point *D* aligns between $1\frac{13}{16}$ and $1\frac{7}{8}$. Measurements are always approximations; using our best eye judgment, point *D* seems to be halfway between $1\frac{13}{16}$ and $1\frac{7}{8}$.

**Figure 2–10**

**We say *CD* is $1\frac{13}{16}$ in. or $1\frac{7}{8}$ in. *to the nearest sixteenth of an inch.***

In practice, measurements are considered acceptable if they are within a desired *tolerance*. In the preceding example, the smallest division is $\frac{1}{16}$, so an appropriate tolerance would be plus or minus one-half of one-sixteenth, or $\pm\frac{1}{32}$ ($\frac{1}{2}$ of $\frac{1}{16} = \frac{1}{32}$). That is, the acceptable measure can be $\frac{1}{32}$ more than or $\frac{1}{32}$ less than the ideal measure.

If the ideal measure is halfway between $1\frac{13}{16}$ and $1\frac{7}{8}$ (or equal to $1\frac{27}{32}$) and the tolerance is $\frac{1}{32}$ in., the interval of acceptable values is from $1\frac{27}{32} - \frac{1}{32}$ to $1\frac{27}{32} + \frac{1}{32}$. The acceptable interval is from $1\frac{26}{32}$ to $1\frac{28}{32}$, or $1\frac{13}{16}$ to $1\frac{7}{8}$.

**2**   **Write Improper Fractions as Whole Numbers or Mixed Numbers.**

Earlier we noted that an improper fraction is a fraction whose value is equal to or greater than one unit, such as $\frac{6}{3}$, $\frac{15}{7}$, or $\frac{5}{5}$. These fractions can be changed to equivalent whole numbers or mixed numbers.

### To write an improper fraction as a whole or mixed number:

1. Perform the division indicated (numerator ÷ denominator).
2. Express any remainder of the division as a fraction or decimal equivalent.

**EXAMPLE**   Write $\frac{15}{3}$ as a whole or mixed number.

$$\frac{15}{3} \text{ means } 15 \div 3 \quad \text{or} \quad 3\overline{)15} \quad \begin{array}{r} 5 \\ \underline{15} \\ 0 \end{array} \quad \text{Then, } \frac{15}{3} = 5.$$    Divide the numerator by the denominator.

If there is no remainder from the division, the improper fraction converts to a whole number.

## Writing a Whole or Mixed Number Is Different from Reducing

Do not confuse converting an improper fraction to a whole or mixed number with reducing a fraction to lowest terms. An improper fraction is in lowest terms if its numerator and denominator have no common factor. Therefore, the improper fraction $\frac{10}{7}$ is written in lowest terms. The improper fraction $\frac{10}{4}$ is not in lowest terms. It will reduce to $\frac{5}{2}$, which is in lowest terms.

When writing an improper fraction as a whole or mixed number, we must make sure that the fraction is in lowest terms. We can reduce to lowest terms either before dividing or after dividing.

**EXAMPLE**   Write $\frac{28}{8}$ as a mixed number.

$$\frac{28}{8} = \frac{7}{2} \qquad 2\overline{)7} = 3\frac{1}{2}$$
$$\underline{6}$$
$$1$$

Fraction is reduced before dividing.

or

$$\frac{28}{8} \qquad 8\overline{)28} = 3\frac{4}{8} = 3\frac{1}{2}$$
$$\underline{24}$$
$$4$$

Fraction is reduced after dividing.

**3**   Write Whole Numbers or Mixed Numbers as Improper Fractions.

Some computation processes require mixed numbers to be expressed as equivalent improper fractions.

### To write a mixed number as an improper fraction:

1. Multiply the denominator of the fractional part by the whole number.
2. Add the numerator of the fractional part to the product; this sum becomes the numerator of the improper fraction.
3. The denominator of the improper fraction is the same as the denominator of the fractional part of the mixed number.

**EXAMPLE**   Change $6\frac{2}{3}$ to an improper fraction.

$$6\frac{2}{3} = \frac{(3 \times 6) + 2}{3} = \frac{20}{3}$$

Multiply the denominator times the whole number and add the numerator.

Whole numbers can also be written as improper fractions by writing the whole number as a fraction with a denominator of 1.

**EXAMPLE**   Change 8 to fifths.

$$8 = \frac{8}{1}$$   Write 8 as a fraction with a denominator of 1.

$$\frac{8 \times 5}{1 \times 5} = \frac{40}{5}$$   Multiply by 1 in the form $\frac{5}{5}$.

### 4   Write Decimals as Fractions and Fractions as Decimals.

When a decimal is written as a fraction, the number of digits that follow the decimal point determines the denominator of the fraction.

**To write a decimal number as a fraction or mixed number in lowest terms:**

1. Write the digits without the decimal point as the numerator.
2. Write the denominator as a power of 10 with as many zeros as there are places after the decimal point.
3. Reduce and, if the fraction is improper, convert to a mixed number. See Outcome 2 of this section.

**EXAMPLE**   Write as a fraction 0.4 and 0.075.

$$0.4 = \frac{4}{10} = \frac{2}{5}$$   Tenths indicates a denominator of 10. Reduce to lowest terms.

$$0.075 = \frac{75}{1,000} = \frac{3}{40}$$   Thousandths indicates a denominator of 1,000. Reduce to lowest terms.

Writing decimal numbers as fractions was introduced in greater detail in Chapter 1, Section 1, Outcome 2.

**EXAMPLE**   Write $0.33\frac{1}{3}$ as a fraction.

$$0.33\frac{1}{3} = \frac{33\frac{1}{3}}{100} = 33\frac{1}{3} \div 100$$   Count places for digits only; that is, do not count the fraction $\frac{1}{3}$ as a place.

$$33\frac{1}{3} \div 100 = \frac{100}{3} \div \frac{100}{1} = \frac{\overset{1}{\cancel{100}}}{3} \times \frac{1}{\underset{1}{\cancel{100}}} = \frac{1}{3}$$   Change division to multiplication. Reduce and multiply. Multiplication and division of fractions is introduced in Section 2–3.

**What Place Does $\frac{1}{3}$ Hold If the Decimal Is $0.33\frac{1}{3}$?**

When changing decimals to fractions, we divide by the place value of the last digit. Is $\frac{1}{3}$ in $0.33\frac{1}{3}$ in the hundredths or thousandths place? Hundredths. The fraction attaches to the last digit.

$0.12\frac{1}{2}$ is read "twelve and one-half hundredths"      $0.008\frac{1}{3}$ is read "eight and one-third thousandths"

With increased calculator and computer use, we find it convenient to work with decimals.

The bar separating the numerator and denominator of a fraction indicates division: $\frac{2}{5}$ also means $2 \div 5$ or $5\overline{)2}$.

If we show the decimal after the 2 and attach a zero in the tenths place, we can divide.

$$\begin{array}{r} 0.4 \\ 5\overline{)2.0} \end{array}$$

Therefore, $\frac{2}{5} = 0.4$.

## To convert a fraction to a decimal number:

1. Place a decimal point after the numerator.
2. Divide the numerator by the denominator using long division.
3. Attach zeros after the decimal point in the dividend as needed for division.

**EXAMPLE**   Change $\frac{7}{8}$ to a decimal.

$7 \div 8$      or

$$\begin{array}{r} 0.875 \\ 8\overline{)7.000} \\ \underline{6\,4}\phantom{00} \\ 60\phantom{0} \\ \underline{56}\phantom{0} \\ 40 \\ \underline{40} \end{array}$$

Attach zeros and divide until the division terminates; that is, it has no remainder.

When we change some fractions to decimals, the quotient does not terminate.

**EXAMPLE**   Write $\frac{1}{3}$ and $\frac{4}{11}$ as decimals to the nearest thousandth.

$1 \div 3$      or

$$\begin{array}{r} 0.3333 \\ 3\overline{)1.0000} = 0.333 \\ \underline{9}\phantom{.0000} \\ 10\phantom{000} \\ \underline{9}\phantom{000} \\ 10\phantom{00} \\ \underline{9}\phantom{00} \\ 10\phantom{0} \\ \underline{9}\phantom{0} \\ 1 \end{array}$$    Rounded.

$4 \div 11$      or

$$\begin{array}{r} 0.3636 \\ 11\overline{)4.0000} = 0.364 \\ \underline{3\,3}\phantom{0000} \\ 70\phantom{000} \\ \underline{66}\phantom{000} \\ 40\phantom{00} \\ \underline{33}\phantom{00} \\ 70\phantom{0} \\ \underline{66}\phantom{0} \\ 4 \end{array}$$    Rounded.

When fractions are changed into decimals that do not terminate, the decimals are called **repeating decimals.** Repeating decimals can be written with a line over the digit(s) that repeat or three dots at the end to indicate that they repeat. The decimal equivalent can be rounded to any desirable place.

$$\frac{1}{3} = 0.\overline{3} \quad \text{or} \quad 0.333333\ldots \qquad \frac{4}{11} = 0.\overline{36} \quad \text{or} \quad 0.3636\ldots$$

Another option for expressing the decimal equivalent is to carry the division to a specified number of places and to write the remainder as a fraction. In the preceding example the mixed decimal equivalents to the hundredths place would be $0.33\frac{1}{3}$ and $0.36\frac{4}{11}$. The rounded decimal equivalents are **approximate equivalents,** and the repeating decimals and mixed decimals are **exact equivalents.**

## To write a mixed number as a decimal number:

1. The whole-number part remains the same.
2. Write only the fraction part as a decimal by dividing the numerator by the denominator.

---

**EXAMPLE**   Change $3\frac{2}{5}$ to a mixed decimal.

$$2 \div 5 \quad \text{or} \quad 5\overline{)2.0}^{\;0.4} \qquad \text{Write the fraction part as an equivalent decimal.}$$

**Then, $3\frac{2}{5} = 3.4$.**

---

### 5   Compare Fractions, Mixed Numbers, and Decimals.

**Like fractions** are fractions that have the same denominator. When fractions are not like fractions, they can be changed to equivalent fractions with like denominators. We call these like denominators **common denominators.**

The **least common denominator (LCD)** for two or more fractions is the *least common multiple* (LCM) of the denominators.

The least common denominator can be often found by inspection. By inspection, we mean examine each denominator and intuitively (or mentally) determine the LCD. For fractions with larger denominators, you may need to use the procedure for finding the LCM discussed in Outcome 1 of this section.

---

**EXAMPLE**   Find the least common denominator for the fractions $\frac{5}{12}, \frac{4}{15}, \frac{3}{8}$.

| 12 | 15 | 8 | Find the LCM using the prime factorization of the denominators. |
|---|---|---|---|
| $2 \times 6$ | $3 \times 5$ | $2 \times 4$ | |
| $2 \times 2 \times 3$ | | $2 \times 2 \times 2$ | |
| $2^2 \times 3$ | | $2^3$ | |

$$\begin{aligned} \text{LCM or LCD} &= 2^3 \times 3 \times 5 \\ &= 8 \times 3 \times 5 \\ &= \mathbf{120} \end{aligned}$$

Chapter 2 / Review of Fractions and Percents

## Alternative Procedure for Finding the LCM or LCD

We can also find the LCM or LCD of several fractions by dividing duplicated factors and then multiplying.

1. Arrange the denominators horizontally.
2. Divide by any prime factor that divides evenly into *at least two* denominators.
3. The LCM or LCD is the product of the primes and remaining factors.

Look at the denominators from the preceding example.

| | | | | |
|---|---|---|---|---|
| 2 | 8 | 12 | 15 | Divide by the prime factor 2. |
| 2 | 4 | 6 | 15 | Divide by the prime factor 2. |
| 3 | 2 | 3 | 15 | Divide by the prime factor 3. |
| | 2 | 1 | 5 | |

Primes: 2, 2, 3

Remaining factors: 2, 1, 5

LCM $= 2 \cdot 2 \cdot 3 \cdot 2 \cdot 1 \cdot 5$

   $= 120$

If we are to compare fractions, the fractions must have the same denominators. To compare $\frac{3}{7}$ and $\frac{5}{7}$, which have the same denominators, we compare the numerators, and $\frac{3}{7}$ is smaller than $\frac{5}{7}$.

### To compare fractions:

1. Find the least common denominator (LCD) that is also the least common multiple (LCM).
2. Change each fraction to an equivalent fraction with the least common denominator (LCD) as its denominator.
3. Compare the numerators.

**EXAMPLE CON**  Is it possible to have a pipe with an outside diameter of $\frac{5}{8}$ in. and an inside diameter of $\frac{21}{32}$ in. (see Fig. 2–11)?

To answer this question, we need to compare the two fractions $\frac{5}{8}$ and $\frac{21}{32}$.

$$\frac{5}{8} = \frac{5 \times 4}{8 \times 4} = \frac{20}{32}$$   Change $\frac{5}{8}$ to 32nds.

Is $\frac{20}{32}$, which is equivalent to $\frac{5}{8}$, larger than $\frac{21}{32}$? No, so $\frac{5}{8}$ **in. cannot be the outside diameter of a pipe with an inside diameter of $\frac{21}{32}$ in.**

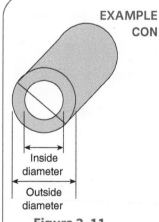

Inside
diameter

Outside
diameter

**Figure 2–11**

**EXAMPLE CON**
Two drill bits have diameters of $\frac{3}{8}$ in. and $\frac{5}{16}$ in., respectively. Which drill bit makes the larger hole?

$$\frac{3}{8} = \frac{6}{16} \qquad \frac{5}{16} = \frac{5}{16}$$

The least common denominator is 16. Change $\frac{3}{8}$ to 16ths.

Compare the numerators: $\frac{6}{16}$ is larger than $\frac{5}{16}$ because 6 is larger than 5, so $\frac{3}{8}$, which is equivalent to $\frac{6}{16}$, is larger than $\frac{5}{16}$.

**The drill bit with a $\frac{3}{8}$-in. diameter will drill the larger hole.**

**TIP!**

### Comparing Fractions Using Decimal Equivalents

We can compare fractions by changing them to equivalent fractions with *any* common denominator, but using the least common denominator gives us smaller numbers with which to work. We can also compare fractions by changing them to decimal equivalents. Review comparing decimals in Chapter 1, Section 1, Outcome 3.

To compare a fraction and a decimal we change one number so that both numbers are either in fraction form or decimal form. Unless otherwise specified, you may use either form.

**EXAMPLE HELPP**
The specifications of a document camera state the length is 15.8 in. A carrying case is $15\frac{7}{8}$ in. long on the inside. Will the camera fit inside the case?

Change $15\frac{7}{8}$ to a decimal.

Change the fractional part of the mixed number to an equivalent decimal.
$7 \div 8 = 0.875$

$$15\frac{7}{8} = 15.875$$

Compare decimals.

$$15.875 > 15.8$$

**The case measurement, 15.875 in., is greater than 15.8 in., so the camera will fit into the case.**

### SECTION 2–1 SELF-STUDY EXERCISES

**1** Show that each number is a multiple of the first number by writing each as the product of the first number and a natural number.

**1.** 5, 10, 15, 20, 25, 30      **2.** 9, 18, 27, 36, 45, 54      **3.** 10, 20, 30, 40, 50, 60

Write five multiples of the given number.

**4.** 12      **5.** 13      **6.** 3

Use the tests for divisibility to determine which number is divisible by the given number. Explain.

**7.** 2,434 by 6      **8.** 230 by 5      **9.** 2,434 by 4

List all the factor pairs for each number.

**10.** 4      **11.** 15      **12.** 24      **13.** 30      **14.** 36

List all factors of each number.

**15.** 46        **16.** 52        **17.** 81        **18.** 85        **19.** 98

Identify as prime numbers or composite numbers by examining the factor pairs of each number.

**20.** 11     **21.** 39     **22.** 16     **23.** 51     **24.** 52     **25.** 53

Find the prime factorization.

**26.** 18     **27.** 20     **28.** 27     **29.** 29     **30.** 40

Find the least common multiple.

**31.** 2 and 3        **32.** 10 and 12        **33.** 12 and 24
**34.** 20, 25, and 35        **35.** 3, 9, and 27        **36.** 2, 8, and 16

Find the greatest common factor.

**37.** 9 and 18        **38.** 40 and 55
**39.** 2, 8, and 16        **40.** 18, 30, and 36
**41.** 20, 25, and 35        **42.** 6, 15, and 18
**43.** Find five fractions that are equivalent to $\frac{4}{5}$.        **44.** Find five fractions that are equivalent to $\frac{7}{10}$.

Find equivalent fractions with the indicated denominators.

**45.** $\dfrac{4}{7} = \dfrac{?}{21}$        **46.** $\dfrac{9}{11} = \dfrac{?}{44}$        **47.** $\dfrac{1}{3} = \dfrac{?}{15}$

Reduce the fractions to lowest terms.

**48.** $\dfrac{6}{10}$      **49.** $\dfrac{12}{16}$      **50.** $\dfrac{10}{32}$      **51.** $\dfrac{16}{32}$      **52.** $\dfrac{28}{32}$

Measure line segments 53–58 in Fig. 2–12 to the nearest sixteenth of an inch (tolerance $= \pm\frac{1}{32}$ in).

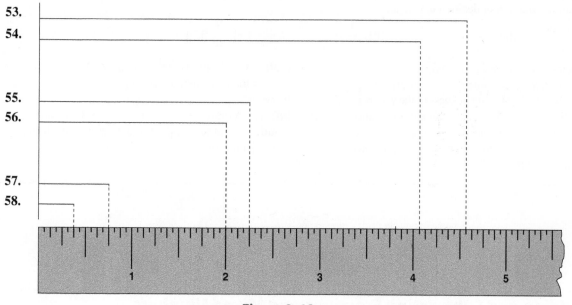

**Figure 2–12**

**2**   Write the improper fractions as whole or mixed numbers.

**59.** $\dfrac{12}{5}$      **60.** $\dfrac{10}{7}$      **61.** $\dfrac{12}{12}$      **62.** $\dfrac{32}{7}$      **63.** $\dfrac{24}{6}$

**3**   Write the mixed numbers as improper fractions.

**64.** $1\dfrac{7}{8}$      **65.** $6\dfrac{5}{12}$      **66.** $9\dfrac{5}{8}$      **67.** $3\dfrac{7}{8}$      **68.** $7\dfrac{5}{12}$

2-1 Equivalent Fractions and Decimals

**4** Write the decimals as their fraction or mixed-number equivalents, and reduce answers to lowest terms.

**69.** 0.1          **70.** 0.25          **71.** 0.025          **72.** 3.9

**73.** 0.875        **74.** 0.375         **75.** 0.625          **76.** 0.43

**77. ELEC** An electronic instrument weighs 0.83 lb. Write this as a fraction of a pound.

**78. CON** Some sheet metal is 0.3125 in. thick. What is the thickness expressed as a fraction?

Write as decimal numbers.

**79.** $\dfrac{3}{10}$          **80.** $\dfrac{21}{100}$          **81.** $3\dfrac{7}{8}$          **82.** $1\dfrac{7}{16}$          **83.** $4\dfrac{9}{16}$

Write these fractions and mixed numbers as decimals to the nearest hundredth.

**84.** $\dfrac{4}{9}$          **85.** $1\dfrac{3}{7}$          **86.** $3\dfrac{5}{11}$

**87. BUS** The property tax rate is $45 per $1,000 of the assessed value. Express the tax rate as a decimal.

**88. CON** A plan allows a gap of $\frac{1}{8}$ in. between vinyl flooring and the wall for expansion. What is the gap measure in decimal notation?

**5** Find the least common denominator.

**89.** $\dfrac{5}{8}, \dfrac{4}{9}$          **90.** $\dfrac{3}{10}, \dfrac{4}{15}$          **91.** $\dfrac{9}{10}, \dfrac{4}{25}$          **92.** $\dfrac{7}{12}, \dfrac{9}{16}, \dfrac{5}{8}$          **93.** $\dfrac{2}{3}, \dfrac{5}{12}, \dfrac{7}{8}$

Show which fraction is larger.

**94.** $\dfrac{2}{3}, \dfrac{3}{5}$          **95.** $\dfrac{5}{12}, \dfrac{7}{16}$          **96.** $\dfrac{15}{32}, \dfrac{29}{64}$          **97.** $\dfrac{7}{12}, \dfrac{9}{16}$          **98.** $\dfrac{3}{8}, \dfrac{4}{5}$

Show which common fraction or decimal is smaller.

**99.** $\dfrac{4}{5}$, 0.82          **100.** $\dfrac{5}{12}$, 0.42          **101.** $\dfrac{1}{2}$, 0.65          **102.** $\dfrac{3}{4}$, 0.34

**103. CON** A hollow-wall fastener has a grip range up to $\frac{7}{16}$ in. Is it long enough to fasten a thin sheet metal strip to a plywood wall if the combined thickness of the wall is $\frac{3}{8}$ in.?

**104. CON** A range top is $29\frac{1}{8}$ in. long by $19\frac{1}{2}$ in. wide. Is it smaller than an existing opening $29\frac{3}{16}$ in. long by $19\frac{9}{16}$ in. wide?

**105. INDTEC** Brenda Jinkins needs to cut a piece of sheet metal slightly longer than the $10\frac{21}{32}$ in. required in the plans and to trim it down to size. Brenda cuts the piece $10\frac{3}{4}$ in. Is it too short?

**106. INDTR** A plastic anchor for a No. 6 × 1-in. screw requires that at least a $\frac{3}{16}$-in.-diameter hole be drilled. Will a $\frac{1}{4}$-in. drill bit be large enough?

---

## 2-2 | Adding and Subtracting Fractions and Mixed Numbers

*Learning Outcomes*    **1** Add fractions and mixed numbers.

**2** Subtract fractions and mixed numbers.

**1** **Add Fractions and Mixed Numbers.**

Adding fractions requires that all fractions being added have the same denominator. Trying to add unlike fractions is like trying to add unlike objects or measures. Before adding unlike fractions, we change the fractions to equivalent fractions with a common denominator.

1. If the denominators are not the same, find the least common denominator.
2. Change each fraction not already expressed in terms of the common denominator to an equivalent fraction with the common denominator.
3. Add the numerators only.
4. The common denominator is the denominator of the sum.
5. Reduce the sum to lowest terms and change improper fractions to whole or mixed numbers.

**EXAMPLE**   Find the sum of $\frac{3}{8} + \frac{1}{8}$.

Because the denominators are the same, start with Step 3 of the addition procedure.

$$\frac{3}{8} + \frac{1}{8} = \frac{4}{8} = \mathbf{\frac{1}{2}} \qquad \text{Add numerators and reduce.}$$

**EXAMPLE**   Add $\frac{5}{32} + \frac{3}{16} + \frac{7}{8}$.

The least common denominator may be found by inspection. Both 8 and 16 divide evenly into 32, so we use 32 as the common denominator.

$$\frac{5}{32} = \frac{5}{32}, \qquad \frac{3}{16} = \frac{6}{32}, \qquad \frac{7}{8} = \frac{28}{32} \qquad \text{Change each fraction to an equivalent fraction whose denominator is 32.}$$

$$\frac{5}{32} + \frac{6}{32} + \frac{28}{32} = \frac{39}{32} \qquad \text{Add the numerators.}$$

$$\frac{39}{32} = \mathbf{1\frac{7}{32}} \qquad \text{Change to a mixed number.}$$

**EXAMPLE
CON**   A plumber uses a $\frac{9}{16}$-in.-diameter copper tube wrapped with $\frac{5}{8}$-in. insulation (Fig. 2–13). What size hole must he bore in the stud (wall support) to install the insulated pipe?

To find the total diameter of the pipe and insulation, add $\frac{9}{16} + \frac{5}{8} + \frac{5}{8}$. The thickness of the insulation is added twice because it counts in the total diameter of the pipe and insulation two times.

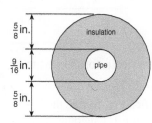

**Figure 2–13**

$$\frac{9}{16} = \frac{9}{16} \qquad \text{The LCD is 16. Change each fraction to 16ths.}$$

$$\frac{5}{8} = \frac{10}{16}$$

$$+\frac{5}{8} = \frac{10}{16} \qquad \text{Add numerators. Keep the common denominator.}$$

$$\frac{29}{16} = 1\frac{13}{16} \qquad \text{Change to a mixed number.}$$

**The total diameter is $1\frac{13}{16}$ in., and the diameter of the hole must be at least this large.**

## To add mixed numbers:

1. Add the whole-number parts.
2. Add the fractional parts and reduce to lowest terms.
3. Change improper fractions to whole or mixed numbers.
4. Add whole-number parts.

---

**EXAMPLE**  Add $5\frac{2}{3} + 7\frac{3}{8} + 4\frac{1}{2}$.

$$5\frac{2}{3} = 5\frac{16}{24}$$  The LCD is 24. Change fractions to equivalent fractions.

$$7\frac{3}{8} = 7\frac{9}{24}$$  Add whole numbers.

$$+\ 4\frac{1}{2} = 4\frac{12}{24}$$  Add fractional parts.

$$\overline{\qquad\qquad 16\frac{37}{24}}$$  $\frac{37}{24} = 1\frac{13}{24}$  Change improper fraction to a mixed number.

$$16 + 1\frac{13}{24} = \mathbf{17\frac{13}{24}}$$  Add whole-number parts.

---

**TIP!**

### Writing Whole Numbers in Mixed-Number Form

When you add mixed numbers and whole numbers, think of the whole number as a mixed number with zero as the numerator in the fraction.

$$16 \quad = 16\frac{0}{24}$$
$$+\ 1\frac{13}{24} = 1\frac{13}{24}$$
$$\overline{\qquad\qquad 17\frac{13}{24}}$$

---

**EXAMPLE**
**CAD/ARC**  Find the largest acceptable measurement of a part if the blueprint calls for the part to be 2 in. long and the tolerance is $\pm\frac{1}{8}$ in.

*Tolerance* is the amount a part can vary from the blueprint specification. To find the largest acceptable measure, we add. Review tolerance in Chapter 1, Section 1, Outcome 5.

$$2 + \frac{1}{8} = 2\frac{1}{8}$$

**The largest acceptable measure is $2\frac{1}{8}$ in.**

---

Chapter 2 / Review of Fractions and Percents

**2** **Subtract Fractions and Mixed Numbers.**

The steps for subtracting fractions and mixed numbers are very similar to the steps for adding fractions and mixed numbers.

**To subtract fractions:**

1. If the denominators are not the same, find the least common denominator.
2. Change each fraction not expressed in terms of the common denominator to an equivalent fraction having the common denominator.
3. Subtract the numerators.
4. The common denominator will be the denominator of the difference.
5. Reduce the difference to lowest terms.

---

**EXAMPLE** Subtract $\dfrac{3}{8} - \dfrac{7}{32}$.

$$\dfrac{3}{8} = \dfrac{12}{32}$$

Change $\frac{3}{8}$ to an equivalent fraction with a denominator of 32.

$$-\dfrac{7}{32} = \dfrac{7}{32}$$

Subtract numerators and keep the common denominator.

$$\dfrac{5}{32}$$

---

**To subtract mixed numbers:**

1. If the fractional parts of the mixed numbers do not have the same denominator, change them to equivalent fractions with a common denominator.
2. When the fraction in the minuend is larger than the fraction in the subtrahend, go to Step 6.
3. When the fraction in the subtrahend is larger than the fraction in the minuend, regroup (borrow) by taking one whole number from the whole-number part of the minuend. This makes the whole number 1 less.
4. Change the whole number that was borrowed to an improper fraction with the common denominator. For example, $1 = \dfrac{3}{3}, 1 = \dfrac{8}{8}, 1 = \dfrac{n}{n}$, where $n$ is the common denominator.
5. Add the borrowed fraction $\left(\dfrac{n}{n}\right)$ to the fraction already in the minuend.
6. Subtract the fractional parts and the whole-number parts.
7. Reduce the difference to lowest terms.

---

**EXAMPLE** Subtract $15\dfrac{7}{8} - 4\dfrac{1}{2}$.

$$15\dfrac{7}{8} = 15\dfrac{7}{8}$$

LCD is 8. Change $\frac{1}{2}$ to $\frac{4}{8}$.

$$-4\dfrac{1}{2} = 4\dfrac{4}{8}$$

Subtract fractions.
Subtract whole numbers.

$$11\dfrac{3}{8}$$

---

2–2 Adding and Subtracting Fractions and Mixed Numbers

**EXAMPLE** Subtract $15\frac{3}{4}$ from $18\frac{1}{2}$.

$$18\frac{1}{2} = 18\frac{2}{4} = 17\frac{4}{4} + \frac{2}{4} = 17\frac{6}{4}$$

$$- 15\frac{3}{4} = 15\frac{3}{4} = 15\frac{3}{4} \qquad = 15\frac{3}{4}$$

$$2\frac{3}{4}$$

LCD = 4. Change $\frac{1}{2}$ to $\frac{2}{4}$.

Regroup. $18 - 1 = 17$, $1 = \frac{4}{4}$, $\frac{4}{4} + \frac{2}{4} = \frac{6}{4}$

Subtract fractions.
Subtract whole numbers.

---

**EXAMPLE**
**HOSP**

$127\frac{1}{2}$ lb of sugar is used from an inventory of $433\frac{3}{8}$ lb. How many pounds of sugar remain in inventory?

$$433\frac{3}{8} = 433\frac{3}{8} = 432\frac{11}{8}$$

$$- 127\frac{1}{2} = 127\frac{4}{8} = 127\frac{4}{8}$$

$$305\frac{7}{8}$$

$433 - 1 = 432$, $1 = \frac{8}{8}$, $\frac{8}{8} + \frac{3}{8} = \frac{11}{8}$

Subtract fractions. Subtract whole numbers.

**$305\frac{7}{8}$ lb of sugar remain in inventory.**

---

**EXAMPLE** Subtract 27 from $45\frac{1}{3}$.

$$45\frac{1}{3} = 45\frac{1}{3}$$

$$- 27 \quad = 27\frac{0}{3}$$

$$18\frac{1}{3}$$

Write 27 as a mixed number. Subtract.

---

**EXAMPLE**
**TELE**

How many feet of coaxial cable are left on a 100-ft roll if $27\frac{1}{4}$ ft are used from the roll?

$$100 = 100\frac{0}{4} = 99\frac{4}{4}$$

$$- 27\frac{1}{4} = 27\frac{1}{4} = 27\frac{1}{4}$$

$$72\frac{3}{4}$$

Write 100 as a mixed number. Regroup.

Subtract.

**$72\frac{3}{4}$ ft of cable are left on the roll.**

**EXAMPLE**
**INDTR**

Three lengths measuring $5\frac{1}{4}$ in., $7\frac{3}{8}$ in., and $6\frac{1}{2}$ in. are cut from a 64-in. bar of angle iron. If $\frac{3}{16}$ in. is wasted on each cut, how many inches of angle iron remain?

Visualize the problem by making a sketch (see Fig. 2–14). Then find the total amount of angle iron used. This includes the three lengths and the waste for three cuts.

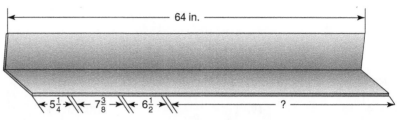

**Figure 2–14**

Total removed and wasted = 3 lengths + 3 cuts

| Three Lengths | Three Cuts | |
|---|---|---|

$$5\frac{1}{4} + 7\frac{3}{8} + 6\frac{1}{2} + \frac{3}{16} + \frac{3}{16} + \frac{3}{16} =$$

Change fractions to equivalent fractions with LCD of 16 and add 5 + 7 + 6 = 18
4 + 6 + 8 + 3 + 3 + 3 = 27.

$$5\frac{4}{16} + 7\frac{6}{16} + 6\frac{8}{16} + \frac{3}{16} + \frac{3}{16} + \frac{3}{16} = 18\frac{27}{16}$$

Write answer as standard mixed number.

$$18\frac{27}{16} = 18 + \frac{27}{16} = 18 + 1\frac{11}{16} = 19\frac{11}{16}$$

Total amount removed and wasted.

amount of angle iron remaining = $\dfrac{\text{beginning}}{\text{length}} - \dfrac{\text{total iron removed}}{\text{and wasted}}$

$$64 - 19\frac{11}{16} =$$   Regroup and subtract.

$$63\frac{16}{16} - 19\frac{11}{16} = 44\frac{5}{16}$$   Total remaining.

**$44\frac{5}{16}$ in. of angle iron remain.**

SECTION 2–2 SELF-STUDY EXERCISES

**1** Add; reduce answers to lowest terms and write any improper fractions as whole or mixed numbers.

**1.** $\dfrac{5}{16} + \dfrac{1}{16}$

**2.** $\dfrac{1}{2} + \dfrac{1}{8} + \dfrac{3}{4}$

**3.** $\dfrac{1}{8} + \dfrac{1}{2}$

**4.** $\dfrac{3}{8} + \dfrac{5}{32} + \dfrac{1}{4}$

**5.** $\dfrac{5}{16} + \dfrac{1}{4}$

**6.** $\dfrac{15}{16} + \dfrac{1}{2}$

**7.** $\dfrac{3}{32} + \dfrac{5}{64}$

**8.** $\dfrac{7}{8} + \dfrac{3}{5}$

**9.** $\dfrac{3}{4} + \dfrac{8}{9}$

**10.** $\dfrac{7}{8} + \dfrac{5}{24}$

**11.** $\dfrac{3}{5} + \dfrac{4}{5}$

**12.** $\dfrac{5}{7} + \dfrac{4}{21}$

**13. CON** What is the thickness of a countertop made of $\frac{7}{8}$-in. plywood and $\frac{1}{16}$-in. Formica?

**14. INDTR** Three pieces of steel are joined together. What is the total thickness if the pieces are $\frac{1}{2}$ in., $\frac{7}{16}$ in., and $\frac{29}{32}$ in.?

15. **BUS** Three books are placed side by side. They are $\frac{5}{16}$ in., $\frac{7}{8}$ in., and $\frac{3}{4}$ in. wide. What is the total width of the books if they are polywrapped in one package?

16. **AUTO** Find the outside diameter of a hose (Fig. 2–15) whose wall is $\frac{1}{2}$ in. thick if its inside diameter is $\frac{7}{8}$ in.

17. **INDTEC** What length bolt is needed to fasten two pieces of metal each $\frac{7}{16}$ in. thick if a $\frac{1}{8}$-in. lock washer and a $\frac{1}{4}$-in. nut are used?

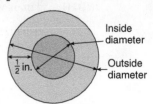

Inside diameter

$\frac{1}{2}$ in.

Outside diameter

**Figure 2–15**

Add; reduce answers to lowest terms and write any improper fractions as whole or mixed numbers.

18. $2\frac{3}{5} + 4\frac{1}{5}$

19. $1\frac{5}{8} + 2\frac{1}{2}$

20. $3\frac{3}{4} + 7\frac{3}{16} + 5\frac{7}{8}$

21. $\frac{1}{6} + \frac{7}{9} + \frac{2}{3}$

22. $2\frac{1}{4} + 2\frac{9}{16}$

23. $1\frac{5}{16} + 4\frac{7}{32}$

24. $3\frac{1}{4} + 1\frac{7}{16}$

25. $4\frac{1}{2} + 9$

26. **CON** The studs (interior supports) of an outside wall are $5\frac{3}{4}$ in. thick. The inside wallboard is $\frac{7}{8}$ in. thick, and the outside covering is $2\frac{3}{16}$ in. thick. What is the total thickness of the wall?

27. **CAD/ARC** A blueprint calls for a piece of bar stock $3\frac{7}{8}$ in. long. If a tolerance of $\pm\frac{1}{16}$ in. is allowed, what is the longest acceptable measurement for the bar stock?

28. **HOSP** If $4\frac{3}{8}$ gal of water are used to dilute $7\frac{1}{4}$ gal of juice, how many gallons are in the mixture?

29. **CON** How much bar stock is needed to make bars of the following lengths: $10\frac{1}{4}$ in., $8\frac{7}{16}$ in., $5\frac{15}{32}$ in.? Disregard waste.

30. **HLTH/N** Three pieces of bandage material each measuring $7\frac{5}{8}$ in. are needed to complete a job. How much bandage material is needed?

31. **HOSP** If $5\frac{1}{8}$ cups of water are mixed with $\frac{3}{4}$ cup of Kool Aid, how many cups are in the mixture?

Subtract; reduce when necessary.

32. $\frac{9}{16} - \frac{3}{8}$

33. $\frac{7}{16} - \frac{3}{8}$

34. $\frac{5}{8} - \frac{1}{2}$

35. $\frac{5}{32} - \frac{1}{64}$

Subtract; reduce when necessary.

36. $23\frac{3}{16} - 5\frac{7}{16}$

37. $9\frac{1}{4} - 4\frac{5}{16}$

38. $9\frac{1}{32} - 3\frac{3}{8}$

39. $14\frac{1}{7} - 12\frac{3}{7}$

40. **AG/H** A flower bed includes $7\frac{7}{8}$ in. of base fill. If the bed is to be 18 in. thick, how thick must the topsoil be?

41. **INDTR** A casting is machined so that $22\frac{1}{5}$ lb of metal remain. If the casting weighed $25\frac{3}{10}$ lb, how many pounds were removed by machine?

42. **AUTO** A bolt $2\frac{5}{8}$ in. long fastens two pieces of metal 1 in. and $1\frac{7}{32}$ in. thick. If a $\frac{3}{32}$-in.-thick lock washer and a $\frac{1}{8}$-in.-thick washer are used, what thickness is the nut if it is even with the end of the bolt? The measure of a bolt length does not include the bolt head.

43. **CAD/ARC** Find the missing length in Fig. 2–16.

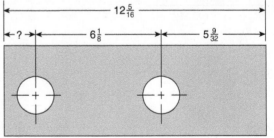

$12\frac{5}{16}$

? $6\frac{1}{8}$ $5\frac{9}{32}$

**Figure 2–16**

*Learning Outcomes*

1 Multiply fractions and mixed numbers.

2 Raise a fraction to a power.

3 Divide fractions and mixed numbers.

When multiplying a fraction by a fraction, we are finding *a part of a part*.

## 1  Multiply Fractions and Mixed Numbers.

In the multiplication, $\frac{1}{2} \times \frac{1}{2}$ is $\frac{1}{2}$ of $\frac{1}{2}$ (Fig. 2–17). The word "of" is the clue that we must multiply to find the part we are looking for.

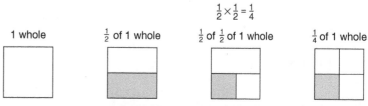

$$\frac{1}{2} \times \frac{1}{2} = \frac{1}{4}$$

1 whole     $\frac{1}{2}$ of 1 whole     $\frac{1}{2}$ of $\frac{1}{2}$ of 1 whole     $\frac{1}{4}$ of 1 whole

**Figure 2–17**

Adding or subtracting fractions and mixed numbers requires a common denominator. In multiplying fractions, we do *not* change fractions to equivalent fractions with a common denominator. Look at two more examples of taking a part of a part (Figures 2–18 and 2–19).

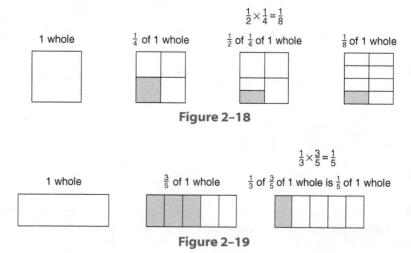

$$\frac{1}{2} \times \frac{1}{4} = \frac{1}{8}$$

1 whole     $\frac{1}{4}$ of 1 whole     $\frac{1}{2}$ of $\frac{1}{4}$ of 1 whole     $\frac{1}{8}$ of 1 whole

**Figure 2–18**

$$\frac{1}{3} \times \frac{3}{5} = \frac{1}{5}$$

1 whole     $\frac{3}{5}$ of 1 whole     $\frac{1}{3}$ of $\frac{3}{5}$ of 1 whole is $\frac{1}{5}$ of 1 whole

**Figure 2–19**

### To multiply fractions:

1. Multiply the numerators of the fractions to get the numerator of the product.
2. Multiply the denominators to get the denominator of the product.
3. Reduce the product to lowest terms.

---

**EXAMPLE**    Find $\frac{1}{2}$ of $\frac{1}{4}$.

$$\frac{1}{2} \times \frac{1}{4} = \frac{\mathbf{1}}{\mathbf{8}}$$
    Multiply numerators.
    Multiply denominators.

---

**EXAMPLE**    Find $\frac{1}{3}$ of $\frac{3}{5}$.

$$\frac{1}{3} \times \frac{3}{5} = \frac{3}{15} = \frac{\mathbf{1}}{\mathbf{5}}$$
    Multiply numerators.
    Multiply denominators. Reduce.

---

### Reduce, or Cancel, Before Multiplying

In the preceding example, reducing is possible. When multiplying fractions, we can reduce common factors before multiplying them.

$$\frac{1}{3} \times \frac{3}{5} = \frac{1 \times \overset{1}{\cancel{3}}}{\underset{1}{\cancel{3}} \times 5} = \frac{1}{5} \qquad \frac{1 \times 3}{3 \times 5} = \frac{1 \times 3}{5 \times 3} = \frac{1}{5} \times \frac{3}{3} = \frac{1}{5} \times 1 = \frac{1}{5}$$

In this example a numerator and a denominator both have a common factor of 3, so the common factor can be reduced before multiplying. Reducing applies the principles $\frac{n}{n} = 1$ and $1 \times n = n$. This process is also referred to as *canceling*.

---

**EXAMPLE**   Multiply $\dfrac{2}{3} \times \dfrac{5}{9} \times \dfrac{1}{6}$.

$$\frac{\overset{1}{\cancel{2}}}{3} \times \frac{5}{9} \times \frac{1}{\underset{3}{\cancel{6}}} = \frac{5}{81} \qquad$$ 2 is a common factor of both a numerator and a denominator. Reduce before multiplying.

---

### Reduce from Any Numerator to Any Denominator

Common factors that are reduced can be diagonal to each other, one above the other, or separated by another fraction, but one factor *must* be in the numerator and the other in the denominator.

$$\frac{2}{\underset{1}{\cancel{3}}} \times \frac{\overset{1}{\cancel{3}}}{5} = \frac{2}{5} \qquad \frac{\overset{3}{\cancel{6}}}{\underset{4}{\cancel{8}}} \times \frac{3}{5} = \frac{9}{20} \qquad \frac{\overset{1}{\cancel{2}}}{7} \times \frac{1}{3} \times \frac{5}{\underset{4}{\cancel{8}}} = \frac{5}{84}$$

---

**EXAMPLE**   Multiply $\dfrac{2}{5} \times \dfrac{10}{21} \times \dfrac{6}{12}$.

$$\frac{\overset{1}{\cancel{2}}}{\underset{1}{\cancel{5}}} \times \frac{\overset{2}{\cancel{10}}}{21} \times \frac{\overset{1}{\cancel{6}}}{\underset{1}{\underset{2}{\cancel{12}}}} = \frac{2}{21} \qquad$$ 5 and 10 are diagonal to each other.
6 is above 12.
2 and 2 are separated by another fraction.

Other patterns of reducing could also have been used.

---

When multiplying mixed numbers or combinations of whole numbers, fractions, and mixed numbers, we first change each whole number or mixed number to an improper fraction. Then we proceed as in multiplying fractions.

## To multiply mixed numbers, fractions, and whole numbers:

1. Change each mixed number or whole number to an improper fraction.
2. Reduce as much as possible.
3. Multiply numerators.
4. Multiply denominators.
5. Convert the answer to a whole or mixed number if possible.

**EXAMPLE**   Multiply $2\frac{1}{2} \times 5\frac{1}{3}$.

$2\frac{1}{2} \times 5\frac{1}{3} =$     Change each mixed number to an improper fraction.

$\frac{5}{2} \times \frac{16}{3} =$     Reduce.

$\frac{5}{\underset{1}{2}} \times \frac{\overset{8}{16}}{3} =$

$\frac{5}{1} \times \frac{8}{3} =$     Multiply numerators. Multiply denominators.

$\frac{40}{3} =$     Change the product to a mixed number.

$\mathbf{13\frac{1}{3}}$

**EXAMPLE AG/H**   Bedding plants are to be planted $3\frac{5}{8}$ in. apart. What length planter is needed for 9 plants (Fig. 2–20)?

$3\frac{5}{8} \times 10 = \frac{29}{\underset{4}{8}} \times \frac{\overset{5}{10}}{1} = \frac{145}{4} = 36\frac{1}{4}$     10 equal spaces of $3\frac{5}{8}$ in. are required.

$3\frac{5}{8}$                    $3\frac{5}{8}$

**Figure 2–20**

**A $36\frac{1}{4}$ in. planter will be needed.**

EXAMPLE
CON

Bricks that are $2\frac{1}{4}$ in. thick form a brick wall with $\frac{3}{8}$-in. mortar joints (Fig. 2–21). What is the height of the wall above the foundation after nine courses?

Use the Six-Step Problem-Solving Plan.

**Unknown facts**

Height of the wall after nine courses of brick have been laid

**Known facts**

$2\frac{1}{4}$ in.    thickness of each brick

$\frac{3}{8}$ in.    thickness of each mortar joint

9    number of courses (or rows) of brick and mortar joints

**Relationships**

Height of wall = thickness of each mortar joint × number of mortar joints + thickness of each brick × number of rows of brick

**Estimation**

Each brick is a little more than 2 in. thick, so the wall should be at least $2 \times 9$ or 18 in. high. Since the mortar joint is not quite $\frac{1}{2}$ in. and the fractional portion of the brick's thickness is less than $\frac{1}{2}$ in., the combined thickness of the brick and mortar joint must be less than 3 in. So the total height of the wall should be less than $3 \times 9$ or 27 in. Thus, we estimate the wall height to be between 18 and 27 in.

**Calculation**

$$\left(9 \times 2\frac{1}{4}\right) + \left(9 \times \frac{3}{8}\right)$$    Write as improper fractions.

$$\left(\frac{9}{1} \times \frac{9}{4}\right) + \left(\frac{9}{1} \times \frac{3}{8}\right)$$    Multiply.

$$\frac{81}{4} + \frac{27}{8}$$    Change to mixed numbers.

Foundation

**Figure 2–21**

$$20\frac{1}{4} + 3\frac{3}{8}$$    Write fractions with a common denominator.

$$20\frac{2}{8} + 3\frac{3}{8} = 23\frac{5}{8}$$    Add.

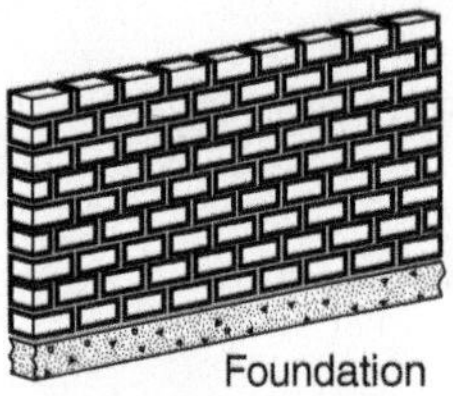

**Interpretation**

The wall will be $23\frac{5}{8}$ in. high.

---

**2**   Raise a Fraction to a Power.

### To raise a fraction or quotient to a power:

1. Raise the numerator to the power.
2. Raise the denominator to the power. The denominator cannot be zero.

Symbolically,

$$\left(\frac{a}{b}\right)^n = \frac{a^n}{b^n} \qquad b \neq 0$$

**EXAMPLE** Raise the fractions to the indicated power.

(a) $\left(\dfrac{2}{3}\right)^2$   (b) $\left(\dfrac{1}{2}\right)^3$

(a) $\left(\dfrac{2}{3}\right)^2 = \dfrac{2^2}{3^2} = \dfrac{4}{9}$   Raise the numerator to the power.
Raise the denominator to the power.

(b) $\left(\dfrac{1}{2}\right)^3 = \dfrac{1^3}{2^3} = \dfrac{1}{8}$   $(1)(1)(1) = 1;\ (2)(2)(2) = 8$

**3** **Divide Fractions and Mixed Numbers.**

If we compare $12 \div 3$ and $\frac{1}{3} \times 12$, we find that both answers are 4. That is, $12 \div 3 = \frac{1}{3} \times 12$ or $12 \times \frac{1}{3}$. Not only is there a relationship between multiplication and division, there is also a relationship between numbers like 3 and $\frac{1}{3}$. Pairs of numbers like $\frac{1}{3}$ and 3 are called *reciprocals*.

Two numbers are **reciprocals** if their product is 1. Thus, $\frac{1}{3}$ and 3 are reciprocals because $\frac{1}{3} \times 3 = 1$, and $\frac{2}{3}$ and $\frac{3}{2}$ are reciprocals because $\frac{2}{3} \times \frac{3}{2} = 1$. The **multiplicative inverse** of a number is its reciprocal. A number times its multiplicative inverse is 1, the **multiplicative identity.**

**To find the reciprocal of a number:**

1. Write the number in fractional form.
2. Interchange the numerator and denominator so that the numerator is the denominator and the denominator is the numerator.

**TIP!**

**Reciprocals and Inverting**

Interchanging the numerator and denominator of a fraction is commonly called **inverting** the fraction.

**EXAMPLE** Find the reciprocal of $\frac{4}{7}$, $\frac{1}{5}$, 3, $2\frac{1}{2}$, 0.8, 1, and 0.

The reciprocal of $\frac{4}{7}$ is $\frac{7}{4}$ or $\mathbf{1\frac{3}{4}}$.   Interchange the numerator and denominator.
Write in improper-fraction or mixed-number form.

The reciprocal of $\frac{1}{5}$ is $\frac{5}{1}$ or **5.**   Write in improper-fraction or whole-number form.

The reciprocal of 3 is $\frac{1}{3}$.   Write 3 as an improper fraction. $3 = \frac{3}{1}$.

The reciprocal of $2\frac{1}{2}$ is $\frac{2}{5}$.   Write $2\frac{1}{2}$ as an improper fraction. $2\frac{1}{2} = \frac{5}{2}$.

The reciprocal of 0.8 is $\frac{5}{4}$ or **1.25.**   Write 0.8 as a common fraction. $0.8 = \dfrac{8}{10} = \dfrac{4}{5}$.

The reciprocal of 1 is **1.**   Write 1 as an improper fraction. $1 = \frac{1}{1}$.

**0 has no reciprocal.**   $0 = \frac{0}{1}$ and $\frac{1}{0}$ is undefined.

Let's review the terminology of division.

$$15 \div 3 = 5$$
dividend · divisor · quotient

To identify the divisor, remember that the symbol ÷ is always read "divided by."

## To divide fractions:

1. Change the division to an equivalent multiplication by replacing the divisor with its reciprocal and replacing the division notation with multiplication notation.
2. Perform the resulting multiplication.

---

**EXAMPLE** Find $\frac{5}{8} \div \frac{2}{3}$.

$$\frac{5}{8} \div \frac{2}{3} = \frac{5}{8} \times \frac{3}{2} = \frac{15}{16}$$    Change division to an equivalent multiplication.

---

**Put Rules into Your Own Words**

Some common phrases for stating the division of fractions rule are:

- Invert the divisor and multiply.
- Invert the second number and multiply.
- Invert the number after the division sign and multiply.

A rule in your own words is often easier for you to remember. Be sure the words guide you to an appropriate process.

---

**EXAMPLE CON** An auger bit advances $\frac{1}{16}$ in. for each turn (see Figure 2–22). How many turns are needed to drill a hole $\frac{5}{8}$ in. deep? ($\frac{5}{8}$ in. can be divided into how many $\frac{1}{16}$-in. parts?)

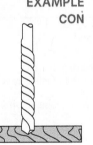

$$\frac{5}{8} \div \frac{1}{16} = \frac{5}{\underset{1}{8}} \times \frac{\overset{2}{16}}{1} = 10$$    Multiply by the reciprocal of the divisor.

**Ten turns are needed.**

**Figure 2–22**

---

To divide mixed numbers and whole numbers, we first write the mixed numbers or whole numbers as improper fractions and then follow the rule for dividing fractions.

Chapter 2 / Review of Fractions and Percents

**To divide mixed numbers, fractions, and whole numbers:**

1. Change each mixed number or whole number to an improper fraction.
2. Convert to an equivalent multiplication problem using the reciprocal of the divisor.
3. Multiply according to the rule for multiplying fractions.

---

**EXAMPLE**  Find $2\frac{1}{2} \div 3\frac{1}{3}$.

$2\frac{1}{2} \div 3\frac{1}{3} =$   Change each mixed or whole number to an improper fraction.

$\frac{5}{2} \div \frac{10}{3} =$   Change division to equivalent multiplication.

$\frac{\overset{1}{\cancel{5}}}{2} \times \frac{3}{\underset{2}{\cancel{10}}} =$   Reduce and multiply.

$\mathbf{\frac{3}{4}}$

---

**EXAMPLE**  Find $5\frac{3}{8} \div 3$.

$5\frac{3}{8} \div 3 =$   Change mixed number and whole number to improper fractions.

$\frac{43}{8} \div \frac{3}{1} =$   Change division to equivalent multiplication.

$\frac{43}{8} \times \frac{1}{3} =$   Multiply.

$\frac{43}{24} = \mathbf{1\frac{19}{24}}$

---

**EXAMPLE**
**BUS**

A developer subdivides $5\frac{1}{4}$ acres into lots; each lot is $\frac{7}{10}$ of an acre. How many lots are made?

$$5\frac{1}{4} \div \frac{7}{10} = \frac{21}{4} \div \frac{7}{10} = \frac{\overset{3}{\cancel{21}}}{\underset{2}{\cancel{4}}} \times \frac{\overset{5}{\cancel{10}}}{\underset{1}{\cancel{7}}} = \frac{15}{2} = 7\frac{1}{2}$$

**Seven lots are made and each is $\frac{7}{10}$ of an acre. The $\frac{1}{2}$ lot is left over or combined with one of the other lots.**

---

2-3 Multiplying and Dividing Fractions and Mixed Numbers

**EXAMPLE**
**HLTH/N**

How many Velcro fasteners, each requiring $1\frac{3}{4}$ ft of Velcro, can be made from a roll containing 100 ft of Velcro? See Fig. 2–23.

$$100 \div 1\frac{3}{4} = \frac{100}{1} \div \frac{7}{4} = \frac{100}{1} \times \frac{4}{7} = \frac{400}{7} = 57\frac{1}{7}$$

$1\frac{3}{4}$ ft.

**Figure 2–23**

**Fifty-seven fasteners can be cut to the desired length. The extra $\frac{1}{7}$ of the desired length is considered waste.**

**EXAMPLE**
**INDTEC**

A piece of trophy column stock that is $21\frac{1}{2}$ in. long is cut into four equal trophy columns (Fig. 2–24). If $\frac{1}{16}$ in. is wasted on each cut, find the length of each piece:

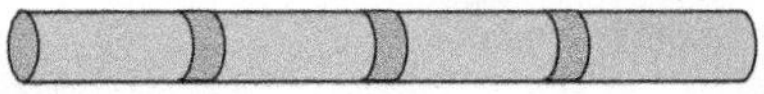

**Figure 2–24**

Use the Six-Step Problem-Solving Strategy.

| | |
|---|---|
| **Unknown facts** | Length of cuts to be made. |
| **Known facts** | 4      Note the number of pieces needed. |
| | 3      Note the number of cuts to be made (4 − 1 = 3). |
| | $\frac{1}{16}$ in.      Note the waste for each cut. |
| | $21\frac{1}{2}$ in.      Note the length of trophy column stock. |
| **Relationships** | Length of each piece = (total length − 3 cuts × amount wasted for each cut) ÷ 4 pieces of stock needed |
| **Estimation** | If the stock is 20 in. long and cut into 4 pieces and if waste is disregarded, each piece will be 5 in. |
| **Calculation** | Three cuts are to be made and each cut wastes $\frac{1}{16}$ in. Find the amount wasted. |

$$\frac{1}{16} \times 3 = \boxed{\frac{3}{16}} \qquad \text{Total waste.}$$

$$21\frac{1}{2} - \boxed{\frac{3}{16}} \qquad \begin{array}{l}\text{Subtract to find the amount of stock that will be left}\\\text{to divide equally into four trophy columns.}\end{array}$$

$$21\frac{1}{2} = 21\frac{8}{16} \qquad \text{Align vertically and use common denominators.}$$

$$\begin{array}{r} -\dfrac{3}{16} = \dfrac{3}{16} \\ \hline \end{array}$$

$$\boxed{21\frac{5}{16}} \text{ in.} \qquad \text{Amount of stock left to be divided.}$$

Find the length of each trophy column.

$$\boxed{21\frac{5}{16}} \div 4 = \frac{341}{16} \div \frac{4}{1} = \frac{341}{16} \times \frac{1}{4} = \frac{341}{64} = 5\frac{21}{64}$$

| | |
|---|---|
| **Interpretation** | **Each trophy column is $5\frac{21}{64}$ in. long.** |

**1** Multiply and reduce answers to lowest terms. Write improper fractions as whole or mixed numbers.

1. $\dfrac{3}{4} \times \dfrac{1}{8}$

2. $\dfrac{1}{2} \times \dfrac{7}{16}$

3. $\dfrac{5}{8} \times \dfrac{7}{10}$

4. $\dfrac{2}{3} \times \dfrac{7}{8}$

5. $\dfrac{1}{2} \times \dfrac{3}{4} \times \dfrac{8}{9}$

6. $\dfrac{3}{8} \times \dfrac{5}{6} \times \dfrac{1}{2}$

7. $7 \times 3\dfrac{1}{8}$

8. $\dfrac{3}{5} \times 125$

9. $2\dfrac{3}{4} \times 1\dfrac{1}{2}$

10. $9\dfrac{1}{2} \times 3\dfrac{4}{5}$

11. $\dfrac{1}{5} \times 7\dfrac{5}{8}$

12. $\dfrac{2}{3} \times 3\dfrac{1}{4}$

13. **AUTO** A fuel tank that holds 75 liters (L) of fuel is $\dfrac{1}{4}$ full. How many liters of fuel are in the tank?

14. **CON** If steps are 12 risers high and each riser is $7\dfrac{1}{2}$ in. high, what is the total rise of the steps?

15. **INDTEC** An alloy, which is a substance composed of two or more metals, is $\dfrac{11}{16}$ copper, $\dfrac{7}{32}$ tin, and $\dfrac{3}{32}$ zinc. How many kilograms of each metal are needed to make 384 kg of alloy?

**2** Raise the fractions to the indicated power.

16. $\left(\dfrac{1}{4}\right)^2$

17. $\left(\dfrac{3}{5}\right)^3$

18. $\left(\dfrac{1}{3}\right)^3$

19. $\left(\dfrac{7}{9}\right)^2$

**3** Give the reciprocal.

20. $\dfrac{5}{8}$

21. $2\dfrac{1}{5}$

22. 8

23. 0.9

24. 1.8

Divide and reduce answers to lowest terms. Convert improper fractions to whole or mixed numbers.

25. $\dfrac{1}{2} \div \dfrac{7}{12}$

26. $\dfrac{4}{5} \div \dfrac{8}{9}$

27. $\dfrac{11}{32} \div \dfrac{3}{8}$

28. $\dfrac{3}{4} \div \dfrac{3}{8}$

29. $\dfrac{7}{8} \div \dfrac{3}{16}$

30. $10 \div \dfrac{3}{4}$

31. $3\dfrac{1}{8} \div \dfrac{1}{4}$

32. $2\dfrac{1}{2} \div 4$

33. $1\dfrac{1}{7} \div \dfrac{2}{7}$

34. $3\dfrac{3}{4} \div 1\dfrac{1}{2}$

35. **CON** A truck will hold 21 yd$^3$ (cubic yards) of gravel. If an earth mover has a shovel capacity of $1\dfrac{3}{4}$ yd$^3$, how many shovelfuls are needed to fill the truck?

36. **CON** Three shelves of equal length are cut from a 72-in. board. If $\dfrac{1}{8}$ in. is wasted on each cut, what is the maximum length of each shelf? (Two cuts are made to divide the entire board into three equal lengths.)

37. **CAD/ARC** If $\dfrac{1}{8}$ in. represents 1 ft on a drawing, find the dimensions of a room that measures $2\dfrac{1}{2}$ in. by $1\dfrac{7}{8}$ in. on the drawing. (How many $\dfrac{1}{8}$s are there in $2\dfrac{1}{2}$; how many $\dfrac{1}{8}$s are there in $1\dfrac{7}{8}$?)

38. **INDTEC** A segment of I-beam is $10\dfrac{1}{2}$ ft long. Into how many whole $2\dfrac{1}{4}$-ft pieces can it be divided? Disregard waste.

39. **CON** How many $17\dfrac{5}{8}$-in. strips of quarter-round molding can be cut from a piece $132\dfrac{3}{4}$ in. long? Disregard waste.

40. **INDTEC** How many $9\dfrac{1}{4}$-in. drinking straws can be cut from a $216\dfrac{1}{2}$-in. length of stock? How much stock is left over?

---

## 2–4 | The U.S. Customary System of Measurement

*Learning Outcomes*

**1** Convert one unit of measure to another using unity ratios.

**2** Convert one unit of measure to another using conversion factors.

**3** Add and subtract U.S. customary measures.

**4** Multiply and divide U.S. customary measures.

**5** Change one U.S. customary rate measure to another.

The **U.S. customary system of measurement** evolved from the English or British system of measurement, and many of the original units in this system are now obsolete. In our discussion of the U.S. customary system of measurement, we will include only selected measuring units.

**1** **Convert One Unit of Measure to Another Using Unity Ratios.**

### Length

Four basic units in the U.S. customary system are commonly used to measure length. They are the inch, the foot, the yard, and the mile. Table 2–1 gives the relationships among these measurements of length.

### Weight or Mass

The terms *weight* and *mass* are commonly used interchangebly. In technical, engineering, and scientific applications, the *mass* of an object is the quantity of material that makes up the object. The *weight* is a measure of the earth's gravitational pull on the object. Three commonly used measuring units for weight or mass in the U.S. customary system are the ounce, pound, and ton. Table 2–2 gives the relationships among these measures of weight or mass.

### Capacity or Volume

The U.S. customary system includes units for both liquid and dry capacity measures; however, the dry capacity measures are seldom used. It is more common to express dry measures in terms of weight than in terms of capacity.

Common U.S. customary units of measure for capacity or volume are the ounce, cup, pint, quart, and gallon. Table 2–3 gives the relationships among the liquid measures for capacity or volume. In the U.S. customary system the term *ounce* represents both weight and liquid capacity. The measures are different and have no common relationship. The context of the problem will suggest whether the unit for weight or capacity is meant.

Using the relationship between two units of measure, we can form a ratio in two different ways that has a value of 1. We call this type of ratio a **unity ratio.**

A **ratio** is a fraction. A unity ratio, then, is a fraction with one unit of measure in the numerator and a different, but equivalent, unit of measure in the denominator. Some examples of unity ratios are

$$\frac{12 \text{ in.}}{1 \text{ ft}}, \quad \frac{1 \text{ ft}}{12 \text{ in.}}, \quad \frac{3 \text{ ft}}{1 \text{ yd}}, \quad \frac{1 \text{ mi}}{5,280 \text{ ft}}$$

---

**Table 2–1    U.S. Customary Units of Length or Distance**

| | |
|---|---|
| 12 inches (in.)[a] = 1 foot (ft)[b] | 36 inches (in.) = 1 yard (yd) |
| 3 feet (ft) = 1 yard (yd) | 5,280 feet (ft) = 1 mile (mi) |

[a]The symbol ″ means inches (8″ = 8 in.) or seconds (60″ = 60 seconds).
[b]The symbol ′ means feet (3′ = 3 ft) or minutes (60′ = 60 minutes).

---

**Table 2–2    U.S. Customary Units of Weight or Mass**

16 ounces (oz) = 1 pound (lb)
2,000 pounds (lb) = 1 ton (T)

---

**Table 2–3    U.S. Customary Units of Liquid Capacity or Volume**

| | |
|---|---|
| 3 teaspoons (t) = 1 tablespoon (T) | 2 tablespoons (T) = 1 ounce (oz) |
| 8 ounces (oz) = 1 cup (c) | 4 cups (c) = 1 quart (qt) |
| 2 cups (c) = 1 pint (pt) | 4 quarts (qt) = 1 gallon (gal) |
| 2 pints (pt) = 1 quart (qt) | |

In each unity ratio, the value of the numerator equals the value of the denominator. A ratio with the numerator and denominator equal has a value of 1. When we convert from one unit of measure to another, we use a unity ratio that contains the original unit and the new unit.

**EXAMPLE**   Write two unity ratios that relate the pair of measures.

(a) ounces and pounds                (b) cups and pints

(a) The relationship between ounces and pounds is 1 lb contains 16 oz. The unity ratios involving ounces and pounds are

$$\frac{1 \text{ lb}}{16 \text{ oz}} \quad \text{and} \quad \frac{16 \text{ oz}}{1 \text{ lb}}$$

(b) The relationship between cups and pints is 1 pint contains 2 cups. The unity ratios involving cups and pints are

$$\frac{1 \text{ pt}}{2 \text{ c}} \quad \text{and} \quad \frac{2 \text{ c}}{1 \text{ pt}}$$

Unity ratios help us convert from one unit of measure to another.

### To change from one U.S. customary unit of measure to another using unity ratios:

1. Set up the original amount as a fraction with the original unit of measure in the numerator.
2. Multiply this fraction by a unity ratio with the original unit in the denominator and the new unit in the numerator.
3. Reduce like units of measure and all numbers wherever possible.

**EXAMPLE**   Find the number of inches in 5 ft.

Multiply 5 ft by a unity ratio that contains both inches and feet.
  Because 5 ft is a whole number, we write it with 1 as the denominator.

$$\frac{5 \text{ ft}}{1}\left(\underline{\quad\quad}\right)$$      Place the original unit with the 5 in the *numerator* of the first fraction.

$$\frac{5 \text{ ft}}{1}\left(\frac{\quad}{\text{ft}}\right)$$      We are changing *from* feet, so we place ft in the *denominator* of the unity ratio, which is shown in parentheses. This allows us to reduce the units later.

$$\frac{5 \text{ ft}}{1}\left(\frac{\text{in.}}{\text{ft}}\right)$$      To change *to* inches, place inches in the *numerator* of the unity ratio.

$$\frac{5 \cancel{\text{ ft}}}{1}\left(\frac{12 \text{ in.}}{1 \cancel{\text{ ft}}}\right) = 60 \text{ in.}$$      Place in the unity ratio the numerical values that make these two units of measure equivalent (1 ft = 12 in.). Complete the calculation, reducing wherever possible.

5 ft = **60 in.**

**EXAMPLE**   How many pints are in 4.5 quarts?

$$\frac{4.5 \cancel{\text{ qt}}}{1}\left(\frac{2 \text{ pt}}{1 \cancel{\text{ qt}}}\right) = 4.5\,(2 \text{ pt}) = 9 \text{ pt}$$      From quart (denominator) to pint (numerator).

4.5 qt = **9 pt**

When working with units of measure, it is very important to include the measuring unit in our analysis. Measurements are also referred to as **dimensions,** and the systematic examination of the appropriate measuring units of a solution is referred to as **dimension analysis.**

Sometimes it is necessary to convert a U.S. customary unit to a unit that is *not* the next larger or smaller unit of measure.

### Changing to Any Larger or Smaller Unit

To change from a U.S. customary unit to one other than the next larger or smaller unit, proceed as before, but multiply the original amount by as many unity ratios as needed to attain the new U.S. customary unit.

For instance, to change from yards to inches: yards → feet → inches.
To change from gallons to ounces: gallons → quarts → pints → cups → ounces.

**EXAMPLE** How many inches are in $2\frac{1}{3}$ yd?

$$2\frac{1}{3}\text{ yd} = \frac{7}{3}\text{ yd}$$

Write $2\frac{1}{3}$ as an improper fraction.

$$\frac{7\text{ yd}}{3}\left(\frac{\text{ft}}{\text{yd}}\right)\left(\frac{\text{in.}}{\text{ft}}\right)$$

Multiply the improper fraction by two unity ratios. To change from yards to inches, first change from yards to feet, and then from feet to inches. Place the original unit in the *numerator* of the improper fraction, $\frac{7}{3}$.

$$\frac{7\cancel{\text{yd}}}{\cancel{3}_1}\left(\frac{\cancel{3}^1\cancel{\text{ft}}}{1\cancel{\text{yd}}}\right)\left(\frac{12\text{ in.}}{1\cancel{\text{ft}}}\right) = 84\text{ in.}$$

Insert the appropriate numerical values for each unity ratio (3 ft = 1 yd; 12 in. = 1 ft ), and multiply, reducing wherever possible.

$$2\frac{1}{3}\text{ yd} = \textbf{84 in.}$$

**Alternative method**

If we use the relationship for inches and yards, 36 in. = 1 yd, we need only one unity ratio for the calculation. That is, we convert $2\frac{1}{3}$ yd ($\frac{7}{3}$ yd) to inches as follows:

$$\frac{7\text{ yd}}{3}\left(\frac{36\text{ in.}}{1\text{ yd}}\right)$$

Set up the unity ratio using 36 in. = 1 yd.

$$\frac{7\cancel{\text{yd}}}{\cancel{3}_1}\left(\frac{\overset{12}{\cancel{36}}\text{ in.}}{1\cancel{\text{yd}}}\right) = 84\text{ in.}$$

Reduce units and numbers; then multiply.

$$2\frac{1}{3}\text{ yd} = \textbf{84 in.}$$

### Focus on One Thing at a Time

Sometimes the steps in a multistepped problem can be overwhelming. It is often helpful to focus on one aspect of the problem at a time. In the example changing yards to inches, focus first on just the units of measure or dimensions.

$$\frac{\cancel{\text{yd}}}{1}\left(\frac{\cancel{\text{ft}}}{\cancel{\text{yd}}}\right)\left(\frac{\text{in.}}{\cancel{\text{ft}}}\right)$$

Reduce as appropriate. Yards reduce to 1. Feet reduce to 1. The only measuring unit left is inches. Therefore, the result will be in inches.

Next, focus on the numbers.

$$\frac{7\cancel{\text{yd}}}{3}\left(\frac{3\cancel{\text{ft}}}{1\cancel{\text{yd}}}\right)\left(\frac{12\text{ in.}}{1\cancel{\text{ft}}}\right) \rightarrow \frac{7}{\cancel{3}}\left(\frac{\cancel{3}}{1}\right)\left(\frac{12}{1}\right) = 84$$

Putting the number and unit together, you have 84 in.

**TIP!**

**Estimation and Dimension Analysis**

When estimating unit conversions, first see if the new unit is larger or smaller than the original unit.

**Larger to Smaller**

Each larger unit can be divided into smaller units. Larger-to-smaller conversions mean *more* smaller units. *More* implies multiplication.

To convert a U.S. customary unit to a desired *smaller* unit: *Multiply* the number of larger units by the number of smaller units that equals 1 larger unit.

| | |
|---|---|
| 2 yd = ____ ft | The smaller unit is feet: 3 ft = 1 yd. |
| 2 yd × 3 ft = 6 ft | Multiply number of yards by 3 ft. |
| 2 yd = 6 ft | Dimension analysis: $\frac{2\text{ yd}}{1} \times \frac{3\text{ ft}}{1\text{ yd}} = 6$ ft |

The key word clues are:

$$\text{larger to smaller unit} \rightarrow \text{obtain more units} \rightarrow \text{multiply}$$

**Smaller to Larger**

Several small units combine to make one large unit. Thus, smaller-to-larger conversions mean *fewer* large units. *Fewer* implies division.

To convert a U.S. customary unit to a *larger* unit: *Divide* the original unit by the number of smaller units that equals 1 desired larger unit.

| | |
|---|---|
| 12 ft = ____ yd | The larger unit is yards: 1 yd = 3 ft. |
| 12 ft ÷ 3 ft = 4 | Divide by 3 ft to get yards. |
| 12 ft = 4 yd | Dimension analysis: $\frac{12\text{ ft}}{1} \times \frac{1\text{ yd}}{3\text{ ft}} = 4$ yd |

Using key word clues:

$$\text{smaller to larger unit} \rightarrow \text{fewer units} \rightarrow \text{divide}$$

Estimation can catch errors in setting up the problem, but it will not likely catch calculation errors.

**2** **Convert One Unit of Measure to Another Using Conversion Factors.**

Unity ratios can be used to develop conversion factors. With conversion factors, you *always* multiply to change from one measuring unit to another.

**To develop a conversion factor for converting from one measure to another:**

1. Write a unity ratio that changes the given unit to the new unit.
2. Change the fraction (or ratio) to its decimal equivalent by dividing the numerator by the denominator.

**EXAMPLE** Develop two conversion factors relating pounds and ounces.

| **Pounds to Ounces** | **Ounces to Pounds** |
|---|---|
| $\frac{\text{pounds}}{1}\left(\frac{\text{ounces}}{\text{pounds}}\right)$ | $\frac{\text{ounces}}{1}\left(\frac{\text{pounds}}{\text{ounces}}\right)$ |
| $\frac{\text{pounds}}{1}\left(\frac{16\text{ ounces}}{1\text{ pound}}\right)$ | $\frac{\text{ounces}}{1}\left(\frac{1\text{ pound}}{16\text{ ounces}}\right)$ |

$$\frac{16}{1} = 16 \qquad \frac{1}{16} \text{ or } 1 \div 16 = 0.0625$$

**pounds $\times$ 16 = ounces     ounces $\times$ 0.0625 = pounds**

## To change from one U.S. customary unit of measure to another using conversion factors:

1. Select the appropriate conversion factor.
2. Multiply the original measure by the conversion factor.

**EXAMPLE**  Use a conversion factor from Table 2–4 to convert 56 ounces to pounds.

ounces $\times$ 0.0625 = pounds    Conversion factor for ounces to pounds.
56 $\times$ 0.0625 = 3.5 pounds    Multiply.

**56 ounces is 3.5 pounds.**

### Table 2–4    U.S. Customary Conversion Factors

| | | | |
|---|---|---|---|
| | TO CHANGE | | |
| | **From** | **To** | **Multiply By** |
| *Length or Distance* | | | |
| 12 inches (in.) = 1 foot (ft) | feet | inches | 12 |
| | inches | feet | 0.0833333 |
| 3 feet (ft) = 1 yard (yd) | yards | feet | 3 |
| | feet | yards | 0.3333333 |
| 36 inches (in.) = 1 yard (yd) | yards | inches | 36 |
| | inches | yards | 0.0277778 |
| 5,280 feet (ft) = 1 mile (mi) | miles | feet | 5,280 |
| | feet | miles | 0.0001894 |
| *Weight or Mass* | | | |
| 16 ounces (oz) = 1 pound (lb) | pounds | ounces | 16 |
| | ounces | pounds | 0.0625 |
| 2,000 pounds (lb) = 1 ton (T) | pounds | tons | 2,000 |
| | tons | pounds | 0.0005 |
| *Liquid Capacity or Volume* | | | |
| 8 ounces (oz) = 1 cup (c) | cups | ounces | 8 |
| | ounces | cups | 0.125 |
| 2 cups (c) = 1 pint (pt) | cups | pints | 2 |
| | pints | cups | 0.5 |
| 2 pints (pt) = 1 quart (qt) | quarts | pints | 2 |
| | pints | quarts | 0.5 |
| 4 quarts (qt) = 1 gallon (gal) | gallons | quarts | 4 |
| | quarts | gallons | 0.25 |

Additional conversion factors are found on the inside covers of the text.

Measures that use two or more units are called **mixed measures.** A mixed measure is in **standard notation** if the number associated with each unit of measure is smaller than the number required to convert to the next larger unit. The number in the largest unit of measure given may or may not be converted as desired.

## To express mixed measures in standard notation:

1. Start with the smallest unit of measure and determine if there are enough units to make one or more of the next larger unit.
2. Regroup to make as many of the larger units as possible.
3. Combine like units.
4. Repeat the process with each given measuring unit.

**EXAMPLE**  Express (a) 8 lb 20 oz and (b) 1 gal 5 qt in standard notation.

(a)  8 lb 20 oz                                                            20 oz = 1 lb 4 oz

    8 lb 20 oz = 8 lb + 1 lb 4 oz = **9 lb 4 oz**      Standard notation.

(b)  1 gal 5 qt                                                            5 qt = 1 gal 1 qt

    1 gal 5 qt = 1 gal + 1 gal 1 qt = **2 gal 1 qt**      Standard notation.

**EXAMPLE**  Express 2 yd 4 ft 16 in. in standard notation.

2 yd 4 ft 16 in. = 2 yd 4 ft + 1 ft 4 in.                Regroup inches. 16 in. = 1 ft 4 in.

            = 2 yd 5 ft 4 in. = 2 yd + 1 yd 2 ft 4 in.      Regroup feet. 5 ft = 1 yd 2 ft

            = **3 yd 2 ft 4 in.**                            Standard notation.

**TIP!**

## Standard Conventions

We can say that 3 ft 15 in. is 4 ft 3 in. in standard notation. Why not change 4 ft 3 in. to 1 yd 1 ft 3 in.? There may be situations when 1 yd 1 ft 3 in. is the desirable form; however, in general, we keep the same units of measure in standard notation as used in the original measure.

### 3  Add and Subtract U.S. Customary Measures.

We can add U.S. customary measures *only* when their units are the same. Measures with the same units are **like measures.** Measures with different units are **unlike measures.**

## To add unlike U.S. customary measures:

1. Convert each measure to a measure with a common U.S. customary unit.
2. Add.

**EXAMPLE**   Add 3 ft + 2 in.

Because 2 in. is a fraction of a foot, we can avoid working with fractions by converting the larger unit (feet) to the smaller unit (inches).

$$3 \text{ ft} = \frac{3 \text{ ft}}{1} \left( \frac{12 \text{ in.}}{1 \text{ ft}} \right) = 36 \text{ in.} \qquad \text{Convert ft to in., and add like measures.}$$

36 in. + 2 in. = **38 in.**

## To add mixed U.S. customary measures:

1. Align the measures vertically so the common units are written in the same vertical column.
2. Add.
3. Express the sum in standard notation.

**EXAMPLE**   Add and write the answer in standard form: 6 lb 7 oz and 3 lb 13 oz.

$$\begin{array}{r} 6 \text{ lb } \; 7 \text{ oz} \\ + \, 3 \text{ lb } 13 \text{ oz} \\ \hline 9 \text{ lb } 20 \text{ oz} \end{array} \qquad \text{Write in standard notation, 20 oz = 1 lb 4 oz.}$$

**Thus, 9 lb 20 oz = 10 lb 4 oz.**

## To subtract unlike U.S. customary units:

1. Convert to a common U.S. customary unit.
2. Subtract.

**EXAMPLE**   Subtract 15 in. from 2 ft.

Changing 15 in. to feet gives us a mixed number, so it is more convenient to convert 2 ft to inches.

2 ft = 24 in.          Convert ft to in. and subtract.
24 in. − 15 in. = **9 in.**

## To subtract mixed U.S. customary measures:

1. Align the measures vertically so that the common units are written in the same vertical column.
2. Subtract.

Chapter 2 / Review of Fractions and Percents

> **EXAMPLE** Subtract 5 ft 3 in. from 7 ft 4 in.
>
> 7 ft 4 in.     Align like measures in a vertical line, and then subtract.
> − 5 ft 3 in.
> **2 ft 1 in.**

When we subtract mixed measures, we use our knowledge of regrouping to subtract a larger unit from a smaller one.

**To subtract a larger U.S. customary unit from a smaller unit in mixed measures:**

1. Align the common measures in vertical columns.
2. Regroup by subtracting one unit from the next larger unit of measure in the minuend, convert to the equivalent smaller unit, and add it to the smaller unit.
3. Subtract the measures.

> **EXAMPLE** Subtract 3 lb 12 oz from 7 lb 8 oz.
>
> 7 lb  8 oz
> −3 lb 12 oz
>
> We always begin subtraction with the smallest unit, which should be on the *right*. 12 oz cannot be subtracted from 8 oz.
>
> 7 lb 8 oz  = 6 lb 16 oz + 8 oz =     6 lb 24 oz      Rewrite 7 lb as 6 lb 16 oz. Add 16 oz to
> 3 lb 12 oz =                      − 3 lb 12 oz      8 oz to get 24 oz.
>                                    **3 lb 12 oz**

### 4  Multiply and Divide U.S. Customary Measures.

Suppose a tank of weed killer contains 21 gal 3 qt. What is the total amount of weed killer in eight tanks? To find the total amount of weed killer, we multiply 21 gal 3 qt by 8.

**To multiply a U.S. customary measure by a number:**

1. Multiply the numbers associated with each unit of measure by the given number.
2. Write the resulting measure in standard notation.

> **EXAMPLE** A tank holds 21 gal 3 qt of weed killer. How much do eight containers hold?
>
> 21 gal 3 qt     Multiply each unit of measure by 8.
> ×    8
> ――――――――
> 168 gal 24 qt     Write in standard notation, 24 qt = 6 gal.
>
> 168 gal 24 qt = 168 gal + 6 gal = 174 gal in standard notation
>
> **The eight containers hold 174 gal.**

Length measures multiplied by like length measures produce square units. Square measures indicate areas (Fig. 2–25).

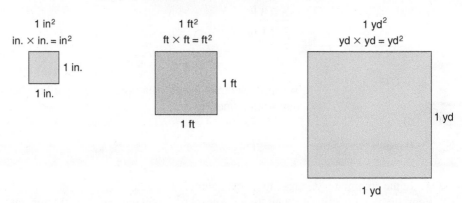

1 in²
in. × in. = in²

1 in.

1 in.

1 ft²
ft × ft = ft²

1 ft

1 ft

1 yd²
yd × yd = yd²

1 yd

1 yd

**Figure 2–25**

**Area** is a measure of a surface instead of a length. Area indicates the number of squares required to cover a surface.

$$\text{length} \times \text{length} = \text{length}^2 \text{ or area}$$

**To multiply a length measure by a like length measure:**

1. Multiply the numbers associated with each like unit of measure.
2. The unit of measure of the product is a square measure or an area.

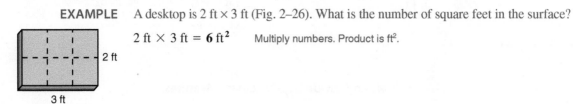

EXAMPLE    A desktop is 2 ft × 3 ft (Fig. 2–26). What is the number of square feet in the surface?

2 ft × 3 ft = **6 ft²**    Multiply numbers. Product is ft².

2 ft

3 ft

**Figure 2–26**

We are frequently required to divide measures by a number.

**To divide a U.S. customary measure by a number that divides evenly into each measure:**

1. Divide the numbers associated with each unit of measure by the given number.
2. Write the resulting measure in standard notation.

**EXAMPLE** How much milk is needed for a half-recipe if the original recipe calls for 2 gal 2 qt?

$2\overline{)2\text{ gal }2\text{ qt}}$     Divide each measure by 2.

$$\frac{1\text{ gal }1\text{ qt}}{2\overline{)2\text{ gal }2\text{ qt}}}$$

**The half-recipe requires 1 gal 1 qt of milk.**

If a given number does not divide evenly into each measure, there will be a remainder.

**To divide U.S. customary measures by a number that does not divide evenly into each measure:**

1. Set up the problem and proceed as in long division.
2. When a remainder occurs after subtraction, convert the remainder to the same unit used in the next smaller measure, and add it to the quantity in the next smaller measure.
3. Divide the given number into this next smaller unit.
4. If a remainder occurs when the smallest unit is divided, express the remainder as a fractional part of the smallest unit.

**EXAMPLE** Divide 5 gal 3 qt 1 pt by 3.

$$\begin{array}{c}
\phantom{3)}1\text{ gal}\qquad 3\text{ qt}\qquad 1\tfrac{2}{3}\text{pt} \\
\hline
3\overline{)5\text{ gal}\qquad 3\text{ qt}\qquad 1\text{ pt}} \\
\end{array}$$

| | |
|---|---|
| 3 gal | 5 gal ÷ 3 = 1 gal, remainder 2 gal |
| 2 gal = 8 qt | 2 gal = 8 qt |
| 11 qt | 3 qt + 8 qt = 11 qt |
| 9 qt | 11 qt ÷ 3 = 3 qt, remainder 2 qt |
| 2 qt = 4 pt | 2 qt = 4 pt |
| 5 pt | 1 pt + 4 pt = 5 pt |
| 3 pt | 5 pt ÷ 3 = 1 pt, remainder 2 pt |
| 2 pt | Write the final remainder, 2, as a fraction $\frac{2}{3}$ pints, and add to 1 pint to get $1\frac{2}{3}$ pints. |

**Thus, 1 gal 3 qt $1\frac{2}{3}$ pt is the solution.**

If tubing is manufactured in lengths of 8 ft 4 in. and a part is 10 in. long, how many parts can be cut from the length of tubing if we do not account for waste? To solve such a problem, we express both measures in the same unit, just as in adding and subtracting measures. We generally convert to the *smallest* unit used in the example.

**To divide a U.S. customary measure by a U.S. customary measure:**

1. Convert both measures to the same unit if they are different.
2. Write the division as a fraction, including the common unit in the numerator and the denominator.
3. Reduce the units and divide the numbers.

**EXAMPLE**    If tubing is manufactured in lengths of 8 ft 4 in. and a part is 10 in. long, how many parts can be cut from the length of tubing if we do not account for waste? Divide 8 ft 4 in. by 10 in.

$$8 \text{ ft} = \frac{8 \text{ ft}}{1}\left(\frac{12 \text{ in.}}{1 \text{ ft}}\right) = 96 \text{ in.} \qquad \text{Convert 8 ft to inches. Then add to 4 in.}$$

8 ft 4 in. = 96 in. + 4 in. = 100 in.    Write the mixed measure as a measure with one
                                         measuring unit.

100 in. ÷ 10 in.                         Divide the total length by the length of each part.

If we write this division in fraction form, we can see more easily that the common units reduce. In other words, the answer will be a number (not a measure) telling *how many equal parts* can be cut from the tubing.

$$\frac{100 \text{ in.}}{10 \text{ in.}} = 10$$

**Therefore, 10 parts of equal length can be cut from the tubing.**

**5**    **Change One U.S. Customary Rate Measure to Another.**

A **rate measure** is a ratio of two different kinds of measures. A rate measure is often referred to as a **rate.** Some examples of rates are 55 miles per hour and 20 cents per mile. In each of these rates, the word **per** means *divided by.*

The rate 55 miles per hour means 55 miles ÷ 1 hour or $\frac{55 \text{ mi}}{1 \text{ h}}$. The rate 20 cents per mile means 20 cents ÷ 1 mile or $\frac{20 \text{ cents}}{1 \text{ mi}}$.

Many rate measures involve measures of time. The units we use to measure time are universally accepted. The basic units of time are the year, month, week, day, hour, minute, and second. Table 2–5 gives the relationships among the units for time. These units are often used when working with rates.

---

**Table 2–5    Units of Time**

| | |
|---|---|
| 1 year (yr) = 12 months (mo) | 1 minute (min) = 60 seconds (s)[c] |
| 1 year (yr) = 365 days (da) | 1 millisecond (ms) = $\frac{1}{1,000}$ s |
| 1 week (wk) = 7 days (da) | |
| 1 day (da) = 24 hours (h)[a] | 1 nanosecond (ns) = $\frac{1}{1,000,000,000}$ s |
| 1 hour (h) = 60 minutes (min)[b] | |

[a] The abbreviation for hour can also be hr.
[b] The symbol ' means feet (3' = 3 ft) or minutes (60' = 60 minutes).
[c] The symbol " means inches (8" = 8 in.) or seconds (60" = 60 seconds). The abbreviation for second can also be sec.

---

**To convert one U.S. customary rate measure to another:**

1. Compare the units of both numerators and both denominators to determine which units will change.
2. Multiply each unit that changes by a unity ratio containing the new unit so that the unit to be changed will reduce.

**EXAMPLE**   Change $8\dfrac{\text{pt}}{\text{min}}$ to $\dfrac{\text{qt}}{\text{min}}$.

**Estimation**

Pints to quarts is *smaller* to *larger,* so there will be fewer quarts.

Examine the rates:

numerators—pints change to quarts
denominators—no change

$$\frac{\text{pt}}{\text{min}}\left(\frac{\text{qt}}{\text{pt}}\right) = \frac{\text{qt}}{\text{min}}$$

Develop a unity ratio with pints in the denominator.

$$\frac{\overset{4}{8\,\text{pt}}}{\text{min}}\left(\frac{1\,\text{qt}}{\underset{1}{2\,\text{pt}}}\right) = \frac{4\,\text{qt}}{\text{min}}$$

Insert numbers in the unity ratio and reduce. Multiply.

**Interpretation**    **Thus, $8\dfrac{\text{pt}}{\text{min}}$ equals $4\dfrac{\text{qt}}{\text{min}}$.**

A separate unity ratio is used for each unit in the rate that changes. For example, when both the numerator and the denominator of a rate measure change, we multiply by at least two unity ratios to make the conversion.

**EXAMPLE**   Change 60 miles per hour to feet per second.

$$60 \text{ miles per hour} = 60\frac{\text{mi}}{\text{h}}$$

Write rate as a fraction.

$$\text{feet per second} = \frac{\text{ft}}{\text{s}}$$

Numerators—miles change to feet
Denominators—hours change to seconds

Examine the changes in the measures.

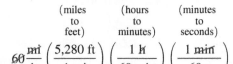

$$60\frac{\text{mi}}{\text{h}}\left(\frac{5{,}280\text{ ft}}{1\text{ mi}}\right)\left(\frac{1\text{ h}}{60\text{ min}}\right)\left(\frac{1\text{ min}}{60\text{ s}}\right)$$

Develop appropriate unity ratios and reduce.

$$\frac{5{,}280\text{ ft}}{60\text{ s}} = 88\frac{\text{ft}}{\text{s}}$$

Divide.

**At 60 mi/h you are traveling 88 ft/s.**

### SECTION 2–4 SELF-STUDY EXERCISES

**1**   Write two unity ratios that relate the *given* pair of measures.

**1.** pints and quarts        **2.** feet and miles        **3.** inches and feet        **4.** feet and yards

Use unity ratios to convert the units.

**5.** 4 ft = _____ in.

**6.** 7 yd = _____ ft

**7.** $2\frac{1}{2}$ mi = _____ yd

**8.** Find the number of yards in 28 ft.

9. Find the number of pounds in $36\frac{4}{5}$ oz.

10. How many ounces are in 45.8 lb?

11. How many quarts are in 5 gal?

12. How many pints are in $6\frac{1}{2}$ qt?

2 Develop two conversion factors for the pairs of units.

13. feet and yards

14. quarts and gallons

Use conversion factors to convert the units of measure.

15. 460 oz is equivalent to how many pints?

16. How many pints are in 46 qt?

17. How many pounds are in 580 oz?

18. If a cabinet is $12\frac{3}{4}$ ft long, how many inches is this?

19. How many feet are in 1.5 mi?

20. How many quarts are in 7 gal?

Express the measures in standard notation.

21. 2 ft 20 in.
22. 1 mi 6,375 ft
23. 2 lb $19\frac{1}{2}$ oz
24. 1 gal 5 qt
25. 2 ft 10 in.
26. 5 lb 25 oz
27. 3 gal 5 qt 48 oz
28. 6 qt 20 oz

3 Add and write the answer in standard notation.

29. 5 oz + 2 lb

30. 4 ft + 7 in.

31. 8 lb 2 oz + 7 lb 9 oz

32. 5 ft 45 in. + 7 ft 30 in.

33. 5 qt 1 pt + 2 qt $1\frac{1}{2}$ pt

34. 8 gal 3 qt + 5 gal 2 qt

35. 7 yd 2 ft + 1 yd 2 ft

36. 4 yd 2 ft 7 in. + 2 yd 1 ft 10 in.

Solve the problems. When necessary, express the answers in standard notation.

37. **CON** A plumber has a 2-ft length of copper tubing and a 7-in. length of copper tubing. What is the total?

38. **COMP** How long are two computer cables together if one is 6 ft and the other is 60 in.?

39. **HOSP** A mixture for hamburgers contains 2 lb 8 oz of ground round steak and 3 lb 7 oz of regular ground beef. How much does the hamburger mixture weigh?

40. **HLTH/N** According to hospital maternity records, one infant twin weighed 6 lb 1 oz and the other weighed 5 lb 15 oz at birth. What was their total weight?

Subtract and write the answer in standard form.

41. 2 ft − 18 in.

42. 2 qt − 3 pt

43. 3 yd − 7 ft

44. 6 lb − 18 oz

45. 3 lb 12 oz − 2 lb 6 oz

46. 12 lb 7 oz − 5 lb 12 oz

47. 2 ft 30 in. − 1 ft 40 in.

48. 5 gal 3 qt 1 pt − 1 gal 3 qt $1\frac{1}{2}$ pt

49. **AUTO** A mechanic has a length of hose 5 ft long. What is its length after 10 in. is cut off?

50. **CON** A cabinetmaker cut 9 in. from a 3-ft shelf in a medical lab. How long was the shelf after it was cut?

51. **COMP** If a computer sorts a list of names in 1 min 30 s and a faster computer does the same job in 45 s, how much time is saved by using the faster computer?

52. **AUTO** A car with a 2.0-L engine accelerates a certain distance in 1 min 27 s. A car with a 5.0-L engine accelerates the same distance in 48 s. How much faster does the car with the 5.0-L engine accelerate?

53. **COMP** A package containing a laser printer weighs 74 lb 3 oz. The container and packing material weigh 4 lb 12 oz. How much does the laser printer weigh?

54. **AG/H** To weigh a baby pig, Stacey held the pig while standing on a scale. If the scale showed 132 lb 6 oz and Stacey weighs 115 lb 8 oz, how much does the pig weigh?

4 Multiply and write the answers for mixed measures in standard notation.

55. 12 mi × 5

56. 18 gal × 6

57. 7 lb 3 oz × 8

58. 7 ft 3 in. × 8

59. **HOSP** Tuna is packed in 1-lb 8-oz cans. If a case contains 24 cans, how much does a case weigh?

60. **AUTO** A car used 1 qt 1 pt of oil each month for 5 months. Find the total amount of oil used.

Multiply.

61. 5 in. × 7 in.

62. 12 ft × 9 ft

63. 15 yd × 12 yd

64. 4 mi × 27 mi

65. **CON** A room is to be covered with square linoleum tiles that are 1 ft by 1 ft. If the room is 18 ft by 21 ft, how many tiles (square feet) are needed?

66. **AG/H** A horticulturist stores a stock solution of fertilizer in two tanks, each with a capacity of 23 gal 9 oz. How much liquid fertilizer is needed to fill both tanks?

**67. ELEC** Latonya has three containers, each containing 1 qt 3 pt of photographic solution. How much total photographic solution is in all three containers?

**68. AG/H** A package of grass seed weighs 1 lb 4 oz. How much would five packages weigh?

Divide.

**69.** 3 days 6 h ÷ 2

**70.** 20 yd 2 ft 6 in. ÷ 2

**71.** 4 yd 1 ft 9 in. ÷ 3

**72. HELPP** Sixty feet of crime-scene tape are required to complete eight jobs. If each job requires an equal amount of tape, find the amount of tape required for one job. (Express your answer in feet and inches.)

**73. AG/H** A vat holding 10 gal 2 qt of defoliant is emptied equally into three tanks. How many gallons and quarts are in each tank?

**74. CON** How many pieces of $\frac{1}{2}$-in. OD (outside diameter) plastic pipe 8 in. long can be cut from a piece 72 in. long?

**75. ELEC** A roll of No. 14 electrical cable 150 ft long is divided into 30 equal sections. How long is each section?

**76. AG/H** A greenhouse attendant has a container with 6 gal 2 qt 10 oz of potassium nitrate solution that will be stored in two smaller containers. How much solution will be stored in each smaller container?

**77. HOSP** For a catering order, Mr. Sonnier prepared 96 lb 12 oz of boiled crawfish. He brought the crawfish to the site in four containers containing equal amounts. How much did the crawfish in each container weigh?

Divide.

**78.** 36 ft ÷ 12 ft

**79.** 6 lb 12 oz ÷ 6 oz

**80.** 2 ft 6 in. ÷ 10 in.

**81. INDTR** How many 6-in. pieces can be cut from 48 in. of pipe?

**82. HOSP** How many 2-lb boxes can be filled from 18 lb of candy?

**83. HOSP** How many 15-oz cans are in a case if the case weighs 22 lb 8 oz?

**84. BUS** How many $8 tickets can be purchased for $72?

**5** Change to the indicated rate measure.

**85.** $\dfrac{45 \text{ lb}}{\text{h}} = \underline{\hspace{1cm}} \dfrac{\text{lb}}{\text{min}}$

**86.** $\dfrac{3 \text{ mi}}{\text{h}} = \underline{\hspace{1cm}} \dfrac{\text{ft}}{\text{h}}$

**87.** $\dfrac{144 \text{ lb}}{\text{min}} = \underline{\hspace{1cm}} \dfrac{\text{oz}}{\text{min}}$

**88.** $\dfrac{30 \text{ gal}}{\text{min}} = \underline{\hspace{1cm}} \dfrac{\text{qt}}{\text{s}}$

**89. INDTEC** A pump that can pump 45 $\frac{\text{gal}}{\text{h}}$ can pump how many quarts per minute?

**90. INDTEC** A pump can dispose of sludge at the rate of 3,200 $\frac{\text{lb}}{\text{h}}$. How many pounds can be disposed of per minute? Round to the nearest tenth.

---

## 2–5 | *Number and Percent Equivalents*

*Learning Outcomes*

**1** Change any number to its percent equivalent.

**2** Change any percent to its numerical equivalent.

The word **percent** means "per hundred" or "for every hundred." Thus, 35 percent means 35 per hundred, or 35 out of every hundred, or $\frac{35}{100}$. Also, 100 percent means 100 out of 100 parts, or $\frac{100}{100}$, or 1 whole quantity. The symbol % is used to represent "percent."

**1 Change Any Number to Its Percent Equivalent.**

We often need to use percents, but the numbers we are given are not expressed as percents. To express a number as a percent, we multiply by 1 in the form of 100%. For example, the fraction $\frac{1}{2}$ can be written as the decimal 0.5. Both $\frac{1}{2}$ and 0.5 change to the same equivalent percent.

$$\frac{1}{2} \cdot 100\% = \frac{1}{\underset{1}{2}} \cdot \frac{\overset{50}{\cancel{100\%}}}{1} = 50\% \qquad 0.5 \cdot 100\% = 50\%$$

**To change any number to its percent equivalent:**

Multiply by 1 in the form of 100%.

We can use this rule to change any type of number—fraction, decimal, whole number, or mixed number—to a percent equivalent.

**TIP!**

## Multiplying by 1 Can Take Many Forms

The multiplicative identity states that $n \times 1 = n$. In fractions, we use the multiplicative identity to change a fraction to an equivalent fraction. We can use the property: $\frac{n}{n} = 1$.

$$\frac{1}{2} \cdot \frac{3}{3} = \frac{3}{6} \qquad \frac{3}{3} = 1.$$

With percents, we use the property of 1 in the form of 100%.

$$\frac{1}{2} \times 100\% = 50\% \qquad 100\% = 1.$$

---

**EXAMPLE** Change the fractions to percent equivalents: $\frac{1}{4}, \frac{1}{3}, \frac{3}{8},$ and $\frac{1}{200}$.

$$\frac{1}{4} \cdot 100\% = \frac{1}{\overset{}{\underset{1}{4}}} \cdot \frac{\overset{25}{\cancel{100}}\%}{1} = \mathbf{25\%}$$

$$\frac{1}{3} \cdot 100\% = \frac{1}{3} \cdot \frac{100\%}{1} = \frac{100\%}{3} = \mathbf{33\frac{1}{3}\%}$$

$$\frac{3}{8} \cdot 100\% = \frac{3}{\underset{2}{8}} \cdot \frac{\overset{25}{\cancel{100}}\%}{1} = \frac{75\%}{2} = \mathbf{37\frac{1}{2}\%}$$

$$\frac{1}{200} \cdot 100\% = \frac{1}{\underset{2}{200}} \cdot \frac{\overset{1}{\cancel{100}}\%}{1} = \mathbf{\frac{1}{2}\%} \qquad \text{$\frac{1}{2}$\% means $\frac{1}{2}$ of every hundredth, or $\frac{1}{2}$ of 1\%.}$$

---

**EXAMPLE** Change the decimals 0.3 and 0.006 to percent equivalents.

$0.3 \cdot 100\% = 030.\% = \mathbf{30\%}$ 　　Use the shortcut procedure to multiply by 100: Move the decimal point two places to the right.

$0.006 \cdot 100\% = 000.6\% = \mathbf{0.6\%}$ 　　0.6% means 0.6 of every hundredth, or 0.6 of 1%.

---

**EXAMPLE** Change the whole numbers 1 and 7 to their percent equivalents.

$1 \cdot 100\% = \mathbf{100\%}$ 　　100 out of 100 or all of 1 quantity.

$7 \cdot 100\% = \mathbf{700\%}$ 　　7 whole quantities, or 7 times a quantity.

**EXAMPLE**  Change the mixed numbers and decimals to their percent equivalents: $1\frac{1}{4}$, $3\frac{2}{3}$, and $5.3$.

$$1\frac{1}{4} \cdot 100\% = \frac{5}{\cancel{4}} \cdot \frac{\overset{25}{\cancel{100\%}}}{1} = \mathbf{125\%}$$    Write $1\frac{1}{4}$ as an improper fraction and multiply.

$$3\frac{2}{3} \cdot 100\% = \frac{11}{3} \cdot \frac{100\%}{1} = \frac{1,100\%}{3} = \mathbf{366\frac{2}{3}\%}$$    Write $3\frac{2}{3}$ as an improper fraction and multiply.

$$5.3 \cdot 100\% = 530.\% = \mathbf{530\%}$$    Multiply by 100 by moving the decimal two places to the right.

### 2  Change Any Percent to Its Numerical Equivalent.

Percents are a convenient way to express the ratio of any quantity to 100. They are excellent time-savers when we make comparisons or state problems on the job. In many calculations we first convert the percents to fraction-, decimal-, whole-, or mixed-number equivalents.

**To change a percent to a numerical equivalent:**

Divide by 1 in the form of 100%.

**TIP!**

**Dividing by 1 Can Take Many Forms**

With whole numbers, $\frac{n}{n} = 1$. To reduce fractions, we use this property again as

$$\frac{6}{10} \div \frac{2}{2} = \frac{6 \div 2}{10 \div 2} = \frac{3}{5}$$

With percents we apply this property using 1 in the form of 100%.

The numerical equivalent of a percent can be expressed in fraction or decimal form as shown in the following example.

**EXAMPLE**  Change the percents to their fraction and decimal equivalents: 75%, 38%, and 5%.

|  | **Fraction equivalent** | **Decimal equivalent** |  |
|---|---|---|---|
| 75% | 75% ÷ 100% | 75% ÷ 100% | For fractions, change division to an equivalent multiplication. Reduce. |
|  | $\dfrac{\overset{3}{\cancel{75\%}}}{1} \cdot \dfrac{1}{\underset{4}{\cancel{100\%}}} = \dfrac{3}{4}$ | $.75 = \mathbf{0.75}$ | For decimals, use the shortcut procedure for dividing by 100%: Move the decimal point two places to the left. |

2–5 Number and Percent Equivalents

$$38\% \qquad 38\% \div 100\% \qquad 38\% \div 100\%$$

$$\frac{\overset{19}{\cancel{38}}\%}{1} \cdot \frac{1}{\underset{50}{\cancel{100}}\%} = \frac{\mathbf{19}}{\mathbf{50}} \qquad .\overset{\frown}{38} = \mathbf{0.38} \qquad \frac{\%}{\%} \text{ reduces to 1.}$$

$$5\% \qquad 5\% \div 100\% \qquad 5\% \div 100\%$$

$$\frac{\overset{1}{\cancel{5}}\%}{1} \cdot \frac{1}{\underset{20}{\cancel{100}}\%} = \frac{\mathbf{1}}{\mathbf{20}} \qquad .\overset{\frown}{05} = \mathbf{0.05}$$

## TIP!

### Division Expressed as Multiplication

As with fractions, we see that it is again convenient to change division to an equivalent multiplication. Is *dividing* by 100% the same as *multiplying* by $\frac{1}{100\%}$? Yes.

$$\text{a percent} \div 100\% = \text{a percent} \div \frac{100\%}{1} = \text{a percent} \cdot \frac{1}{100\%}$$

Some percents change more conveniently to a fraction equivalent, and others change more conveniently to a decimal equivalent. In solving problems, we normally change the percent to the most convenient equivalent for the problem. In the following examples, both fraction and decimal equivalents are given, and you can judge for yourself when fraction equivalents are more convenient than decimal equivalents, and vice versa.

**EXAMPLE**  Change the percents to their fraction and decimal equivalents: $33\frac{1}{3}\%$, $37\frac{1}{2}\%$.

|  | **Fractional equivalent** | **Decimal equivalent** |
|---|---|---|

$$33\frac{1}{3}\% \div 100\% \qquad \frac{\overset{1}{\cancel{100}}\%}{3} \cdot \frac{1}{\underset{1}{\cancel{100}}\%} = \frac{\mathbf{1}}{\mathbf{3}} \qquad 33\frac{1}{3}\% = \mathbf{0.33}\frac{1}{3} \text{ or } \mathbf{0.33} \text{ (rounded)}$$

A decimal point separates whole quantities from fraction parts. Therefore, there is an *unwritten* decimal between 33 and $\frac{1}{3}$. Because $\frac{1}{3}$ does not change to a terminating decimal equivalent, using the decimal equivalent of $33\frac{1}{3}\%$ will create more extensive calculations. Using a rounded decimal equivalent changes the result from an exact to an approximate amount.

|  | **Fractional equivalent** | **Decimal equivalent** |
|---|---|---|

$$37\frac{1}{2}\% \div 100\% \qquad \frac{\overset{3}{\cancel{75}}\%}{2} \cdot \frac{1}{\underset{4}{\cancel{100}}\%} = \frac{\mathbf{3}}{\mathbf{8}} \qquad 37\frac{1}{2}\% = \underset{\frown}{37.5}\% = \mathbf{0.375}$$

Since no rounding was necessary, the decimal equivalent is an exact amount.

**EXAMPLE**   Change 5.25% to its fractional and decimal equivalents.

| **Fractional equivalent** | **Decimal equivalent** |

First, write the percent in fraction form.

$$5.25\% = 5\frac{25}{100}\% = 5\frac{1}{4}\% \qquad\qquad 5.25\% = 5.25\% \div 100\% = \mathbf{0.0525}$$

$$5\frac{1}{4}\% = 5\frac{1}{4}\% \div 100\%$$

$$= \frac{21}{4}\% \cdot \frac{1}{100\%}$$

$$= \mathbf{\frac{21}{400}}$$

**TIP!**

**What Happens to the % (Percent) Sign?**

From multiplying fractions we can reduce or cancel common factors from a numerator to a denominator. Percent signs and other types of labels can also cancel.

$$\frac{\%}{1} \cdot \frac{1}{\%} = 1$$

**EXAMPLE**   Change the percents to their decimal equivalents: $\frac{1}{2}\%$ and 0.25%.

$$\frac{1}{2}\% = 0.5\% = 0.5\% \div 100\% = \mathbf{0.005}$$

$$0.25\% = 0.25\% \div 100\% = \mathbf{0.0025}$$

When a quantity is 100% or more, the fractional and decimal equivalents will be equal to or more than the whole number 1. The numerical equivalents will be whole numbers, mixed numbers, mixed decimals, or rounded decimals. When solving problems, use the more convenient equivalent.

**EXAMPLE**   Change the percents to their whole-number equivalents or to their rounded decimal equivalents: 700%, 375%, $233\frac{1}{3}\%$, and $462\frac{1}{2}\%$. Round decimals to the nearest thousandth when appropriate.

$$700\% \div 100\% = \mathbf{7}$$

$$375\% \div 100\% = \mathbf{3.75}$$

$$233\frac{1}{3}\% \div 100\% = \frac{\overset{7}{\cancel{700\%}}}{3} \cdot \frac{1}{\underset{1}{\cancel{100\%}}} = \frac{7}{3} = \mathbf{2.333}\ (\textbf{rounded})$$

$$462\frac{1}{2}\% = 462.5\%$$

$$462.5\% \div 100\% = \mathbf{4.625}$$

## TIP!

### Do Part of the Calculation on the Calculator and the Other Part Mentally

When you need to find the percent equivalent of a fraction, why not change the fraction to a decimal equivalent by dividing? Then, change the decimal to a percent mentally by multiplying by 100%.

$$\frac{4}{7} = 4 \div 7 = 0.571428571$$

Then, $\frac{4}{7} \approx 57.14\%$   Mentally move the decimal two places to the right and round to the nearest hundredth of a percent.

### SECTION 2–5 SELF-STUDY EXERCISES

1. If $\frac{2}{5}$ of the electricians in a city are self-employed, what percent are self-employed?

2. If $\frac{7}{10}$ of the bricklayers in a city are male, what percent are male?

Change the numbers to their percent equivalents. Round to the nearest tenth of a percent if necessary.

3. $\frac{5}{8}$

4. $\frac{1}{350}$

5. 0.2

6. 0.14

7. 5

8. 8

9. $1\frac{1}{3}$

10. $3\frac{1}{2}$

11. 3.05

12. 7.2

**2** Change to both fractional and decimal equivalents. Round to the nearest ten thousandth if necessary.

13. 36%

14. 75%

15. $6\frac{1}{4}\%$

16. 62.5%

17. $66\frac{2}{3}\%$

18. 0.6%

19. $\frac{1}{5}\%$

20. 0.05%

21. $8\frac{1}{3}\%$

22. 18.75%

Change to equivalent whole numbers.

23. 800%

24. 400%

Change to both mixed-number and mixed-decimal equivalents.

25. 250%

26. 425%

27. 176%

28. 380%

29. $137\frac{1}{2}\%$

30. 387.5%

Change to mixed-number equivalents.

31. $166\frac{2}{3}\%$

32. $316\frac{2}{3}\%$

Change to mixed-decimal equivalents.

33. 115.3%

34. $212\frac{1}{2}\%$

Chapter 2 / Review of Fractions and Percents

**Learning Outcomes**

**1** Identify the portion, base, and rate in percent problems.

**2** Solve percent problems using the percentage formula.

**1 Identify the Portion, Base, and Rate in Percent Problems.**

All problems involving percents have three basic elements: the rate, the base, and the portion. Knowing what each element is and how all three are related helps us solve problems with percents.

The **rate R** is the percent; the **base B** is the original or total amount; the **portion P** is part of the base. In the statement 50% of 80 is 40, the rate is 50%, the base is 80, and the portion is 40.

**TIP!**

**Identify the Rate, Base, and Portion**

The following descriptions may help you recognize the rate, base, or portion more quickly:

*Rate* is usually written as a percent, but it may be a decimal or fraction.

*Base* is the total amount, original amount, or entire amount. It is the amount that the *portion* is a part of. In a sentence the base is often closely associated with the preposition *of*.

*Portion* can refer to the part, partial amount, amount of increase or decrease, or amount of change. It is a portion of the *base*. In a sentence the portion is often closely associated with a form of the verb *is*. The portion can also be referred to as the **percentage.** We choose to use the term *portion* to minimize the confusion between the terms *percent* and *percentage.*

A common memory jogger for finding the percent: $\text{Rate} = \dfrac{\text{is}}{\text{of}}$.

**EXAMPLE**  Identify the given and missing elements for

(a) 20% of 75 is what number?
(b) What percent of 50 is 30?
(c) Eight is 10% of what number?

$$
\begin{array}{ccc}
R & B & P
\end{array}
$$
(a) 20% of 75 is what number?
  percent   total       portion

Use the identifying key words for rate (*percent* or %), base (*total*, *original*, associated with the word *of* ), and portion (*part*, associated with the word *is*).

$$
\begin{array}{ccc}
R & B & P
\end{array}
$$
(b) What percent of 50 is 30?
      percent   total  portion

$$
\begin{array}{ccc}
P & R & B
\end{array}
$$
(c) Eight is 10% of what number?
  portion  percent    total

**2 Solve Percent Problems Using the Percentage Formula.**

The percentage formula, Portion = Rate × Base, can be written as $P = R \times B$. The letters or words represent numbers. When the numbers are put in place of the letters, the formula guides you through the calculations.

The three percentage formulas are

Portion = Rate × Base          $P = R \cdot B$          For finding the portion.

Base $= \dfrac{\text{Portion}}{\text{Rate}}$          $B = \dfrac{P}{R}$          For finding the base.

Rate $= \dfrac{\text{Portion}}{\text{Base}}$          $R = \dfrac{P}{B}$          For finding the rate.

Circles can help us visualize these formulas. The shaded part of the circle in Fig. 2–27 represents the missing amount. The unshaded parts represent the known amounts. If the unshaded parts are *side by side*, *multiply* their corresponding numbers to find the missing number.

If the unshaded parts are *one on top of the other*, *divide* the corresponding numbers to find the missing number.

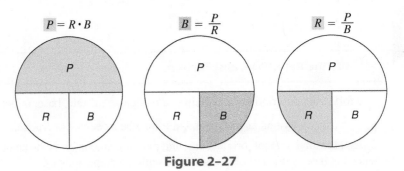

**Figure 2–27**

## To use the percentage formula to solve percentage problems:

1. Identify and classify the two known values and the one missing value.
2. Choose the appropriate percentage formula for finding the missing value.
3. Substitute the known values into the formula. For the rate, use the decimal or fractional equivalent of the percent.
4. Perform the calculation indicated by the formula.
5. Interpret the result. If finding the rate, convert decimal or fractional equivalents of the rate to a percent.

**EXAMPLE**  Solve the problems: (a) 20% of 400 is what number? (b) 20% of what number is 80? (c) 80 is what percent of 400?

**(a)** 20% = rate                    Identify known values and missing value.
400 = base
Portion is missing.                    Choose the appropriate formula.
$P = R \cdot B$                         Substitute values using the decimal equivalent of 20%.
$P = 0.2 \cdot 400$                     Perform calculation.
$P = \mathbf{80}$                       Interpret result.
**20% of 400 is 80.**

**(b)** 20% = rate                    Identify known values and missing value.
80 = portion
Base is missing.                       Choose the appropriate formula.

$B = \dfrac{P}{R}$                      Substitute values.

$B = \dfrac{80}{0.2}$                   Perform calculation.

$B = 400$                     Interpret result.

**20% of 400 is 80.**

**(c)** $80 = $ portion          Identify known values and missing value.

$400 = $ base

Rate is missing.              Choose the appropriate formula.

$R = \dfrac{P}{B}$            Substitute values.

$R = \dfrac{80}{400}$         Perform calculation.

$R = 0.2$ or **20%**          Interpret result. $0.2 = 20\%$.

**80 is 20% of 400.**

---

**EXAMPLE**  A mortgage company requires that the monthly cost of a new home be no more than 28% of the buyer's gross monthly income. Find the monthly income a prospective buyer must have or exceed to qualify for a home that has a monthly cost (including taxes and insurance) of $1,200.

**BUS**   $28\% = $ rate        Identify the known values and the missing value.

$\$1,200 = $ portion

Base is missing.              Choose the appropriate formula.

$B = \dfrac{P}{R}$            Substitute values.

$B = \dfrac{1,200}{0.28}$     Perform calculation.

$B = \$4,285.71$             Interpret result.

**Prospective buyer must have a monthly income of at least $4,286, rounded to the nearest dollar.**

The next example shows how to use the fractional equivalent of a mixed-number percent.

---

**EXAMPLE**   $33\frac{1}{3}\%$ of 282 is what number?

The rate is $33\frac{1}{3}\%$. The key word *of* tells us that 282 is the base. The portion is missing.

$P = RB$          Substitute $\frac{1}{3}$ for $33\frac{1}{3}\%$.   $33\frac{1}{3}\% = \dfrac{33\frac{1}{3}\%}{100\%} = \dfrac{1}{3}$.

$P = \dfrac{1}{3}(282)$      Multiply.

$P = \dfrac{282}{3}$         Divide.

$P = 94$

**$33\frac{1}{3}\%$ of 282 is 94.**

## How Many Digits Do You Use?

Even with a calculator, you may prefer to use the fractional equivalent because the decimal equivalent of $33\frac{1}{3}\%$ is a repeating decimal and requires rounding. Examine the effect of rounding to various places.

$$0.3 \cdot 282 = 84.6$$
$$0.33 \cdot 282 = 93.06$$
$$0.333 \cdot 282 = 93.906$$
$$0.3333 \cdot 282 = 93.9906$$
$$0.333333333 \cdot 282 = 93.999999906$$

Because the desired degree of accuracy varies depending on the problem, it is advisable to find the exact answer on the calculator using the fractional equivalent of the percent. A close approximate answer can be found by using the full calculator value for nonterminating decimal equivalents.

It is advisable to anticipate the approximate size of an answer before making any calculations. This approximation helps you discover mistakes, especially when using a calculator.

TIP!

## Estimating a Percent of a Number

Estimate a percent by comparing it to a percent that you can calculate mentally.

|  | **To find** |  |
|---|---|---|
| 1% of a number | *Multiply by 0.01* | Move decimal two places to the left. |
| 2% of a number | $2 \cdot 1\%$ of a number |  |
| 3% of a number | $3 \cdot 1\%$ of a number |  |
| 10% of a number | *Multiply by 0.1* | Move decimal one place to the left. |
| 5% of a number | $\frac{1}{2}$ of 10% of a number |  |
| 20% of a number | $2 \cdot 10\%$ of a number |  |
| 30% of a number | $3 \cdot 10\%$ of a number |  |
| 50% of a number | $\frac{1}{2}$ *of a number* | Divide by 2. |
| 25% of a number | $\frac{1}{4}$ *of a number* | Divide by 4. |
| 75% of a number | $3 \cdot 25\%$ of a number |  |
| $33\frac{1}{3}\%$ of a number | $\frac{1}{3}$ *of a number* | Divide by 3. |
| $66\frac{2}{3}\%$ of a number | $2 \cdot 33\frac{1}{3}\%$ of a number |  |

**EXAMPLE** Find 10% of $51.00.

10% of $51.00 = **$5.10**.    Mentally, move the decimal point in the number one place to the left.

**EXAMPLE**    Estimate a 15% tip on a restaurant bill of $48.18.

**BUS**    10% of $48.00 = $4.80        Find 10% of the rounded total bill.

$$5\% \text{ of } \$48.00 = \frac{1}{2} \text{ of } \$4.80 = \$2.40$$      $48.18 rounded to $48.00.
                                               Find 5% of the rounded total bill.

15% = 10% + 5%

$4.80 + $2.40 =               Add the amounts for 10% and 5%.

**$7.20**                      15% tip.

---

**EXAMPLE**    A taxi fare is $24.00. Find the amount of a 20% tip.

**BUS**    10% of $24.00 = $2.40        Find 10% of the fare.
2 · $2.40 = $4.80             Double the 10% amount.

**A 20% tip is $4.80.**

---

1% of a number can be found mentally. When using a percent that is less than 1%, the portion is less than 1% of the number. Using 1% of the number as an estimate is very useful in checking the decimal placement.

---

**EXAMPLE**    $\frac{1}{4}$% of 875 is what number?

$\frac{1}{4}$% is less than 1% and 1% of 875 is 8.75. Therefore, $\frac{1}{4}$% of 875 is less than 8.75. Since $\frac{1}{4}$% and 0.25% are equivalent, either can be used to solve this problem. We are given the rate and the base, and need to find the portion.

$P = RB$            Substitute known values for the rate and base.
$P = 0.25\% (875)$     Change rate to the decimal equivalent.
$P = 0.0025 (875)$     Multiply.
$P = 2.1875$

$\frac{1}{4}$**% of 875 is 2.1875.**

---

100% of a number is 1 times the number, or the number itself. When using percents that are larger than 100%, the portion will be more than the original number.

---

**EXAMPLE**    325% of 86 is what number?

100% of 86 is 86. 325% is more than 3 times 100%. 3 · 86 = 258. So, 325% of 86 is more than 258. The rate is 325%, the base is 86, and the portion is missing.

$P = RB$            Substitute known values.
$P = 325\% (86)$      325% = 3.25
$P = 3.25 (86)$       Multiply.
$P = 279.5$

**325% of 86 is 279.5.**

In the following example the rate is more than 100% so we expect the base to be less than the portion.

---

**EXAMPLE**  398.18 is 215% of what number?

We are looking for the base, as indicated by the key word *of*. We are given the portion and the rate. Because the rate is more than 100%, we expect the base to be smaller than the portion.

$B = \dfrac{P}{R}$    Substitute values.

$B = \dfrac{398.18}{215\%}$    215% = 2.15

$B = \dfrac{398.18}{2.15}$    Divide.

$B = 185.2$

**398.18 is 215% of 185.2.**

---

When solving applied problems, our most difficult task is identifying the two given parts and determining which part is missing.

---

**EXAMPLE**  If a type of solder contains 55% tin, how many pounds of tin are needed to make 10 lb of solder?

**INDTR**  First, let's be sure we understand the word *solder*. **Solder** is a mixture of metals. In this example, the *total amount* or the *base* is the 10 lb of solder, and it is made of tin and other metals.

**Known facts**  55% of 10 lb of solder is tin.
Rate: Percent of tin = 55%
Base: Amount of solder = 10 lb

**Unknown fact**  Portion or number of pounds of tin = $P$

**Relationships**  $P = RB$    Substitute known values.
$P = 55\% (10)$

**Estimation**  To estimate, 55% is more than $\frac{1}{2}$ and $\frac{1}{2}$ of 10 = 5. So the amount of tin should be more than 5 lb.

**Calculations**  $P = 55\% (10)$    55% = 0.55
$P = 0.55 (10)$    Multiply.
$P = 5.5$

**Interpretation**  **Thus, 5.5 or $5\frac{1}{2}$ lb of tin are needed to make 10 lb of solder.**
To check the reasonableness of the answer, $5\frac{1}{2}$ lb is a little more than 5.

---

Chapter 2 / Review of Fractions and Percents

| | |
|---|---|
| **EXAMPLE** | If a 150-horsepower (hp) engine delivers only 105 hp to the driving wheels of a car, what is the efficiency of the engine? |
| **AUTO** | **Efficiency** means the *percent* the output (105 hp) is of the total amount (150 hp) the engine is capable of delivering. Thus, the base amount is 150 hp, the portion delivered (output) is 105 hp, and the percent of 150 represented by 105 is the rate or efficiency. |
| **Known facts** | Base: Total amount of horsepower = 150 hp <br> Portion: Amount of horsepower the engine delivers = 105 hp |
| **Unknown facts** | Efficiency or percent of the total horsepower |
| **Relationships** | $R = \dfrac{P}{B}$      Substitute known values. <br><br> $R = \dfrac{105}{150}$ |
| **Estimation** | Because the engine is not operating at full capacity (150 hp), we expect the efficiency to be less than 100%. Since 105 is more than $\frac{1}{2}$ of 150, the percent or rate will be more than 50%. |
| **Calculations** | $R = \dfrac{105}{150}$      Divide. <br><br> $R = 0.7$      Change to a percent equivalent. <br><br> $R = 70\%$ |
| **Interpretation** | **The engine is 70% efficient.** |

**TIP!**

### Must I Write All the Steps in the Six-Step Problem-Solving Plan for Every Problem?

As you develop confidence in your problem-solving ability, you will begin to instinctively think through the steps without writing them down. We will start to omit some of the steps in our examples, but will continue to draw special attention to the two *often neglected* steps: Estimation and Interpretation.

| | |
|---|---|
| **EXAMPLE** | The effective value of current or voltage in an ac circuit is 71.3% of the maximum voltage. If a voltmeter shows a voltage of 110 volts (V) as the effective value in a circuit, what is the maximum voltage? |
| **ELEC** | 71.3% of the maximum voltage is 110 V. The maximum voltage is the *base*, and the amount of voltage shown on the voltmeter (110 V) is the *portion*. |
| **Estimation** | The rate is less than 100%. Therefore, the base is larger than the portion. $B > 110$. <br><br> $B = \dfrac{P}{R}$      Substitute known values. <br><br> $B = \dfrac{110}{71.3\%}$      Change % to decimal equivalent. <br><br> $B = \dfrac{110}{0.713}$      Divide. <br><br> $B = 154 \text{ V}$      To the nearest volt. |
| **Interpretation** | **The maximum voltage is 154 V.** |

**1** Identify the given and missing elements as $R$ (rate), $B$ (base), and $P$ (portion).

1. What percent of 10 is 2?
2. 2 is 20% of what number?
3. Three books is what percent of four books?
4. What percent of 25 students is 5 female students?
5. 6 of 15 motorists is what percent?
6. How many nurses is 20% of the 15 nurses on duty?

**2** Solve using the percentage formula.

7. 30% of 18 is what number?
8. 42% of 600 is what number?
9. 1.8% of 100 is what number?
10. What is $33\frac{1}{3}\%$ of 78?
11. What percent of 18 is 9?
12. What percent of 15.6 is 52?
13. What percent of 88 is 77?
14. 150 is what percent of 100?
15. 125% of what number is 80?
16. 0.01% of what number is 7?
17. **INDTEC** Cast iron contains 4.25% carbon. How much carbon is contained in a 25-lb bar of cast iron?
18. **AG/H** 3,645 rolls of landscape fabric are manufactured during one day. After being inspected, 121 of these rolls are rejected as imperfect. What percent of the rolls is rejected? (Round to the nearest whole percent.)
19. **AUTO** An engine operating at 82% efficiency transmits 164 hp. What is the engine's maximum capacity (base) in horsepower?
20. **AG/H** If wrought iron contains 0.07% carbon, how much carbon is in a 30-lb bar of wrought iron?
21. **ELEC** The voltage of a generator is 120 V. If 6 V are lost in a supply line, what is the rate of voltage loss?
22. **AG/H** A landscape contractor figures it costs $\frac{1}{2}\%$ of the total cost of a job to make a bid. What would be the cost of making a bid on a $115,000 job?
23. **INDTEC** A certain ore yields an average of 67% iron. How much ore is needed to obtain 804 lb of iron?
24. **CON** A contractor makes a profit of $12,350 on a $115,750 job. What is the percent of profit? (Round to the nearest whole percent.)
25. **INDTEC** 385 defective alcohol swabs were produced during a day. If 4% of the alcohol swabs produced were defective, how many alcohol swabs were produced in all?
26. **INDTR** In a welding shop, 104,000 welds are made. If 97% of them are acceptable, how many are acceptable?
27. Estimate a 15% tip on a restaurant bill of $31.15.
28. Estimate a 15% tip for a taxi fare of $43.00.
29. **BUS** Find the monthly income a prospective buyer must have to qualify for a home that has a monthly cost (including taxes and insurance) of $1,100 if the mortgage company requires the monthly cost to be no more than 28% of the buyer's gross monthly income.
30. **BUS** Chloe Duke earns $4,872 monthly and wants to purchase a home, but the mortgage firm requires the monthly payment to be no more than 28% of her gross monthly income. What is the most the monthly mortgage payment can be, including taxes and insurance?
31. **BUS** Find the sales tax and the total bill on an order of office supplies costing $75.83 if the tax rate is 6%. Round to the nearest cent.
32. **AG/H** Materials to landscape a property total $785.84. What is the total bill if the sales tax rate is $5\frac{1}{2}\%$? Round to the nearest cent.
33. **BUS** If the rate of social security tax is 6.2%, find the tax on gross earnings of $375.80. Round to the nearest cent.
34. **BUS** An employee's gross earnings for a pay period are $895.65. The net pay for this salary is $675.23. What percent of the gross pay are the total deductions? Round to the nearest whole percent.
35. **BUS** An employee has a net salary of $576.89 and a gross salary of $745.60. What percent of the gross salary is the total of the deductions? Round to the nearest whole percent.
36. **BUS** Madison Duke earns $2,892 in one pay period. Medicare tax is 1.45% of the earnings. How much is deducted for Medicare tax?
37. **BUS** A real estate salesperson earns 4% commission. What is the commission on a property that sold for $295,800?
38. **INDTEC** A manufacturer gives a 2% discount to customers paying cash. If a parts store pays cash for an order totaling $875.84, what amount is saved? Calculate to the nearest cent.
39. **HLTH/N** Find the cash price for an order of hospital supplies totaling $3,985.57 if a 3% discount is offered for cash orders. Calculate to the nearest cent.
40. **BUS** What commission is earned by a salesperson who sells $18,890 in merchandise if a 5% commission is paid on all sales?

41. **TELE** A telecommunications representative is paid a salary of $140 per week and 7% commission on all sales over $3,200 per week. The sales for a recent week were $7,412. What is the representative's salary for that week?

42. **BUS** A computer store manager is paid a salary of $2,153 monthly plus a bonus of 1% of the net earnings of the business. Find the total salary for a month when the net earnings of the business are $105,275.

43. **HLTH/N** The adult human skeleton consists of 206 bones. The face has 14 bones. What percent of human bones are found in the face? Round to the nearest tenth percent.

44. **HLTH/N** Each lower limb of the adult human skeleton has 31 bones. What percent of the 206 bones are found in both lower limbs? Round to the nearest tenth percent.

45. **HLTH/N** Approximately 60% of an adult man's body is water. A male that weighs 185 lb has approximately how many pounds of water?

46. **HLTH/N** The average American consumes about $3\frac{1}{2}$ lb of sodium each year. If salt is 40% sodium, how many pounds of salt are consumed, on average, each year by an individual?

47. **CON** Home heat loss through poor-fitting doors and windows can account for 15% of a homeowner's energy cost. A January energy bill is $219.42. Find the cost of unnecessary heat loss.

48. **INDTEC** Alumina makes up 45% of high grade aluminum ore. If 48,000 lb of high-grade aluminum ore are mined, how many pounds of alumina can be obtained?

49. **INDTEC** How many pounds of pure sulfur are contained in 3,400 lb of coal that has 5% pure sulfur content?

50. **INDTEC** ACME Products estimates a product costs $42.80 to manufacture. How much did it cost to make the product if a 12% cost overrun occurred?

---

## 2-7 | Increase and Decrease

*Learning Outcomes*

**1** Find the amount of increase or decrease.

**2** Find the new amount in increase or decrease problems.

**3** Find the rate or base in increase or decrease problems.

### 1 Find the Amount of Increase or Decrease.

Percents are often used in problems dealing with increases or decreases.

TIP!

**Relating Increases and Decreases to Percents**

Increases and decreases are applications of the percentage formula.

|  | Rate | Base | Portion |
|---|---|---|---|
| **Increase** | Rate of increase | Original amount | Amount of increase |
| **Decrease** | Rate of discount | Original amount | Amount of decrease |

New amount for increase = Original amount + Amount of increase
New amount for decrease = Original amount − Amount of decrease

---

**EXAMPLE** Medical assistants are to receive a 9% increase in wages per hour. If they were making $19.25 an hour, what is the *amount of increase per hour* (to the nearest cent)? Also, what is the *new wage per hour*?

The original wage per hour is the base, and we want to find the amount of increase (portion).

10% of $19.25 is $1.92. The increase will be less than $1.92. The new wage per hour will be approximately $21.

| | |
|---|---|
| $P = RB$ | *P* represents amount of increase, 9% is the rate of increase, and *B* is the original wage per hour. |
| $P = 9\% \ (\$19.25)$ | $9\% = 0.09$ |
| $P = 0.09 \ (19.25)$ | Multiply. |
| $P = 1.7325$ | $1.73 to the nearest cent is the amount of increase. |

**Interpretation**

**The medical assistants will receive a $1.73 per-hour increase in wages.**

$19.25 + \$1.73 = \$20.98$     New amount = original amount + amount of increase.

**Their new hourly wage will be $20.98.**

---

**EXAMPLE**    Molten iron shrinks 1.2% while cooling. What is the cooled length of a piece of iron if it is cast in a 24-cm pattern?

**INDTEC**    First, we find the amount of shrinkage (amount of decrease, portion). The original amount, 24 cm, is the base.

**Estimation**    The cooled piece will be less than 24 cm.

| | |
|---|---|
| $P = RB$ | *P* is the amount of decrease, 1.2% is the rate of decrease, and the base is the original amount, or 24 cm. |
| $P = 1.2\% \ (24)$ | $1.2\% = 0.012$ |
| $P = 0.012 \ (24)$ | Multiply. |
| $P = 0.288$ | Amount of shrinkage (decrease). |

**Interpretation**    The amount of shrinkage is 0.288 cm, so the length of the cooled piece (new amount) is

**$24 - 0.288 = 23.712$ cm**     New amount = original amount − amount of decrease.

---

2   **Find the New Amount in Increase or Decrease Problems.**

When knowing the amount of increase is not necessary, the new amount can be found directly.

TIP!

**New Amounts and New Rates**

| | Rate | Base | Portion |
|---|---|---|---|
| New amount for increase | New rate *100% + rate of increase* | Original amount | New amount *original + increase* |
| New amount for decrease | New rate *100% − rate of decrease* | Original amount | New amount *original − decrease* |

---

**EXAMPLE**    A 3% error is acceptable for a machine part to be usable. If the part is intended to be 57 cm long, what is the range of measures that is acceptable for this part?

**INDTR**    The machine part can be ±3% from the ideal length of 57 cm. The range of acceptable measures is found by calculating the smallest acceptable measure and the largest acceptable measure. Recall that the symbol ± is read *plus or minus*. It means that the measure can be more (+) or less (−) than the designated amount. In this case, the part can be 3% longer than or shorter than 57 cm and still be usable. The smallest acceptable value is 97% of the ideal length (100% − 3%).

$P = RB$         *P* is the smallest acceptable amount because 97% is the *smallest acceptable percent*. The base is 57 cm.

$P = 97\% (57)$     97% = 0.97
$P = 0.97 (57)$     Multiply.
$P = 55.29$

**Interpretation**   **The smallest acceptable measure is 55.29 cm.**

The largest acceptable value is 103% of the ideal length (100% + 3%).

$P = RB$         *P* is the largest acceptable amount because 103% is the *largest acceptable percent*. The base is 57 cm.

$P = 103\% (57)$     103% = 1.03
$P = 1.03 (57)$     Multiply.
$P = 58.71$

**Interpretation**   **The largest acceptable value is 58.71 cm. The range of acceptable measures is from 55.29 cm to 58.71 cm.**

It is helpful in developing our number sense with percents to think of percents in pairs. 100% is one whole quantity. A percent that is less than 100% represents part of a quantity. For example, 30% represents part of one quantity. Then, 70% is the rest of the quantity if 30% is removed. This percent, 70%, is the complement of 30%. The ***complement of a percent*** is the difference between 100% and a given percent.

### To find the complement of a percent:

Subtract the percent from 100%.

EXAMPLE   Find the complement of (a) 25%, (b) 80%, and (c) 36%.

(a)  $100\% − 25\% = \mathbf{75\%}$     Subtract from 100%.
(b)  $100\% − 80\% = \mathbf{20\%}$     Subtract from 100%.
(c)  $100\% − 36\% = \mathbf{64\%}$     Subtract from 100%.

### To estimate the amount you pay for a sale item:

1. Round the original price and the percent to numbers you can work with mentally.
2. Find the complement of the rounded percent.
3. Relate the complement to 10% by dividing it by 10%.
4. Find 10% of the rounded original price.
5. Multiply the results from Steps 3 and 4.

**EXAMPLE**  Estimate the amount you pay on a $49.99 item advertised at 70% off.

**BUS**  $49.99 rounds to $50.

| | |
|---|---|
| $100\% - 70\% = 30\%$ | Complement of 70% (percent you pay). |
| $30\% \div 10\% = 3$ | Relate to 10%. |
| $10\%$ of $\$50 = \$5$ | Move decimal one place to the left. |
| $3 \cdot \$5 = \mathbf{\$15}$ | Three times 10% of $50. |

---

**EXAMPLE**  Estimate the amount you pay on a $28 item advertised at 18% off.

**BUS**  18% rounds to 20%. $28 rounds to $30.

| | |
|---|---|
| $100\% - 20\% = 80\%$ | Complement of 20% (percent you pay). |
| $80\% \div 10\% = 8$ | Relate to 10%. |
| $10\%$ of $\$30 = \$3$ | Move decimal one place to the left. |
| $8 \cdot \$3 = \mathbf{\$24}$ | Eight times 10% of $30. |

---

**3**  **Find the Rate or Base in Increase or Decrease Problems.**

Many kinds of increase or decrease problems involve finding either the rate or the base.

The rate is the **percent of change** or the **percent of increase or decrease**. The base is still the *original amount*.

---

**EXAMPLE**  A worn brake lining is measured to be $\frac{3}{32}$ in. thick. If the original thickness was $\frac{1}{4}$ in., what is the percent of wear?

**AUTO**  First, the amount of wear is $\frac{1}{4} - \frac{3}{32}$.

$$\frac{8}{32} - \frac{3}{32} = \frac{5}{32} \qquad \text{Find the common denominator. Subtract.}$$

The amount of wear (decrease) is the portion. The base is the original amount.

$$R = \frac{P}{B} \qquad \begin{array}{l} R \text{ is the percent of wear.} \\ P \text{ is the amount of wear, or } \frac{5}{32} \text{ in.} \quad B \text{ is } \frac{1}{4} \text{ in.} \end{array}$$

$$R = \frac{\frac{5}{32}}{\frac{1}{4}} \qquad \text{Divide.}$$

$$R = \frac{5}{32} \div \frac{1}{4} \qquad \text{Change to an equivalent multiplication.}$$

$$R = \frac{5}{\overset{}{\underset{8}{32}}} \cdot \frac{\overset{1}{4}}{1} \qquad \text{Multiply.}$$

$$R = \frac{5}{8} \qquad \text{Change to an equivalent percent.}$$

$$R = \frac{5}{8} \times 100\%$$

$$R = \frac{5}{\underset{2}{8}} \cdot \frac{\overset{25}{\cancel{100}}\%}{1}$$ Reduce and multiply.

$$R = 62\frac{1}{2}\%$$

**Interpretation**    The percent of wear is $62\frac{1}{2}\%$.

### SECTION 2–7 SELF-STUDY EXERCISES

**1** Solve.

1. Find the amount of increase if 432 is increased by 25%.
2. If 78 is increased by 40%, what is the new amount?
3. Find the amount of decrease if 68 is decreased by 15%.
4. If 135 is decreased by 75%, what is the new amount?
5. **HOSP**  Jobs in the food preparation and serving occupation totaled 2,206,000 in 2000. According to the Bureau of Labor Statistics, jobs in this occupation are scheduled to grow by 30.5% by 2010. How many new jobs will there be in 2010? Find the total number of jobs expected in this occupation in 2010.
6. **INDTR**  The Bureau of Labor Statistics notes that the occupation of customer service representative will be a fast-growing occupation from 2000 to 2010. In 2000 the number of jobs was 1,946,000 and an increase of 32.4% is expected. Find the number of new jobs expected and the total number of jobs available in 2010.

**2** Solve.

7. Find the complement of 40%.
8. Find the complement of 18%.
9. Find the complement of 86.3%.
10. Find the complement of $33\frac{1}{3}\%$.
11. **CON**  If 17% extra flooring is needed to allow for waste when boards are laid diagonally, how much flooring should be ordered to cover 2,045 board feet of floor? Answer to the nearest whole board foot.
12. **CON**  A construction company requires 25,400 bricks for a job. If it allows 2% more bricks for breakage, how many bricks must it order?
13. **BUS**  Ciara Walker was earning $49,860 and received a 7% raise. Find her new annual earnings directly.
14. **BUS**  LaTreas Walker received a 3% salary increase on her weekly earnings of $1,982. What are her new weekly earnings?
15. **BUS**  David Dawson earned $4,290 but paid 6.2% of his earnings in social security taxes and 1.45% in Medicare taxes. What were his net earnings after paying social security and Medicare taxes?
16. **BUS**  Megan Anders purchased a swimsuit that was priced at $84.00 with a 30% discount. What was the sale price?

**3** Solve.

17. **HOSP**  The cost of a pound of ground beef increased from $2.36 to $2.53. What is the percent of increase to the nearest whole-number percent?
18. **AG/H**  A landscape contractor estimated that materials, shrubs, saplings, and labor for a job would cost $5,385. An estimate 1 year later for the same job was $7,808, due to inflation. Find the percent of increase due to inflation to the nearest whole number.
19. **AG/H**  A chicken farmer bought 2,575 baby chicks. Of this number, 2,060 lived to maturity. What percent loss was experienced by the chicken farmer?
20. **ELEC**  An electrician recorded costs of $1,297 for a job. If he received $1,232 for the job, what was the percent of money lost on the job? Round to the nearest whole number.
21. **AUTO**  An engine that has a 4% loss of power has an output of 336 hp. What is the input (base) horsepower of the engine?
22. **AG/H**  A contractor figures that 10 yd$^3$ of sand are needed for a job. If a 5% allowance for waste must be included, how much sand must be ordered?

23. **HLTH/N** A floor in a doctor's office that would normally require 2,580 board feet is to be laid diagonally. If a 17% waste allowance is necessary for flooring laid diagonally, how much flooring must be ordered? (Round to the nearest whole number.)

24. **INDTR** A shop manager records a 14% loss on rivets for waste. If the shop needs 25 lb of rivets, how many pounds must be ordered to compensate for loss due to waste?

25. **INDTEC** Steel bars shrink 10% when cooled from furnace temperature to room temperature. If a cooled steel bar is 36 in. long, how long was it when it was formed?

26. **CON** The cost of No. 1 pine studs increased from $3.85 each to $4.62 each. Find the percent of increase.

---

## CHAPTER REVIEW OF KEY CONCEPTS

**Learning Outcomes**
**Section 2–1**

**1** Write equivalent fractions with different denominators. (pp. 60–70)

**What to Remember with Examples**

The *numerator* (top number) of a fraction is the number of parts of the whole amount we are considering. The *denominator* (bottom number) of a fraction is the number of parts a whole amount has been divided into. *Proper fractions* are less than 1. *Improper fractions* are equal to or larger than 1. A *mixed number* consists of a whole number and a common fraction written together and indicates addition of the whole number and fraction. A *decimal fraction* is a fraction whose denominator is 10 or a power of 10 and it can be written in decimal notation with the place value representing the denominator of the fraction.

To change a fraction to an equivalent fraction with a larger denominator: **1.** Divide the larger denominator by the original denominator. **2.** Multiply the original denominator by the quotient found in Step 1.

Change $\frac{5}{9}$ to an equivalent fraction that has a denominator of 36.

$$\frac{5}{9} \times \frac{4}{4} = \frac{20}{36} \qquad 36 \div 9 = 4$$

To reduce a fraction to lowest terms: Divide both the numerator and the denominator by the greatest common factor (GCF).

Reduce $\frac{12}{16}$.

$$\frac{12}{16} = \frac{12}{16} \div \frac{4}{4} = \frac{3}{4} \qquad \text{GCF is 4.}$$

To read the U.S. customary rule: Align the rule along the object (Fig. 2–28). Count the number of whole and fractional inches ($\frac{1}{16}$'s, $\frac{1}{8}$'s, $\frac{1}{4}$'s, and so on) to determine the approximate length. Use eye judgment to estimate closeness of the object to a mark on the rule.

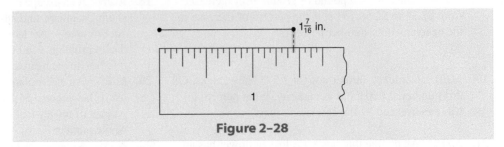

$1\frac{7}{16}$ in.

**Figure 2–28**

**2** Write improper fractions as whole numbers or mixed numbers (pp. 70–71).

To convert an improper fraction to a whole or mixed number: **1.** Divide the numerator by the denominator. **2.** Write any remainder as a fraction with the original denominator as its denominator.

Chapter 2 / Review of Fractions and Percents

Convert $\frac{18}{6}$, $\frac{15}{4}$ to whole or mixed numbers.

$$\frac{18}{6} = 3 \qquad \frac{15}{4} = 3\frac{3}{4}$$

**3** Write whole numbers or mixed numbers as improper fractions (pp. 71–72).

To convert a whole number to an improper fraction: **1.** Write the whole number as the numerator. **2.** Write 1 as the denominator.

To convert a mixed number to an improper fraction: **1.** Multiply the whole number by the denominator. **2.** Add the numerator to the result of Step 1. **3.** Place the sum from Step 2 over the original denominator.

Change 7 and $4\frac{7}{8}$ to improper fractions.

$$7 = \frac{7}{1} \qquad 4\frac{7}{8} = \frac{(8 \times 4) + 7}{8} = \frac{39}{8}$$

**4** Write decimals as fractions and fractions as decimals (pp. 72–74).

To change a decimal to a fraction (or mixed number) in lowest terms: **1.** Write the number without the decimal as the numerator. **2.** Give the denominator the same number of zeros as the decimal has decimal places. **3.** Reduce or convert to a mixed number.

Write 0.23 as a fraction:

$$\frac{23}{100}$$ Two decimal digits so denominator is 100.

To change a fraction to a decimal: Divide the numerator by the denominator by placing a decimal after the last digit of the numerator and attaching zeros as needed.

Change $\frac{5}{8}$ to a decimal:

$$
\begin{array}{r}
0.625 \\
8\overline{)5.000} \\
\underline{4\ 8}\phantom{00} \\
20\phantom{0} \\
\underline{16}\phantom{0} \\
40 \\
\underline{40}
\end{array}
$$
Divide by denominator.

To find the least common denominator, use the process for finding the least common multiple. **1.** List the prime factorization of each number using exponential notation. **2.** List the factors of the LCM by including each prime factor appearing in any of the numbers the greatest number of times that it appears in any factor. **3.** Write the result in standard notation.

Find the least common denominator for $\frac{7}{18}$ and $\frac{5}{24}$.

$$18 = 2 \cdot 3 \cdot 3 = 2 \cdot 3^2$$
$$24 = 2 \cdot 2 \cdot 2 \cdot 3 = 2^3 \cdot 3$$
$$\text{LCM} = \text{LCD} = 2^3 \cdot 3^2 = 2 \cdot 2 \cdot 2 \cdot 3 \cdot 3 = 72$$

The smallest number that can be divided evenly by both 18 and 24 is 72.

**5** Compare fractions, mixed numbers, and decimals (pp. 74–76).

To compare fractions: **1.** Write the fractions as equivalent fractions with common denominators. **2.** Compare the numerators. The larger numerator indicates the larger fraction.

To compare mixed numbers: **1.** Compare the whole-number parts, if different. **2.** If the whole-number parts are equal, write the fractions with common denominators. **3.** Compare the numerators.

Which fraction is smaller, $\frac{2}{5}$ or $\frac{5}{12}$?

$$\frac{2}{5} = \frac{24}{60}$$

$$\frac{5}{12} = \frac{25}{60}$$   Since $\frac{24}{60}$ is smaller than $\frac{25}{60}$, $\frac{2}{5}$ is smaller than $\frac{5}{12}$.

## Section 2–2

**1** Add fractions and mixed numbers (pp. 78–80).

To add fractions: **1.** Find the common denominator. **2.** Change each fraction to an equivalent fraction with the common denominator. **3.** Add the numerators and place the sum over the common denominator. **4.** Reduce the sum if possible. **5.** Convert improper fractions to whole or mixed numbers if desired.

Add:

$$\frac{1}{7} + \frac{3}{7} + \frac{2}{7} = \frac{6}{7} \qquad \frac{5}{8} + \frac{3}{4} = \frac{5}{8} + \frac{6}{8} = \frac{11}{8} = 1\frac{3}{8}$$

To add mixed numbers: **1.** Change each fraction to an equivalent fraction with the LCD. **2.** Place the sum of the numerators over the LCD. **3.** Add the whole-number parts. **4.** Write the improper fraction from Step 2 as a whole or mixed number; add the result to the whole number from Step 3. **5.** Simplify if necessary.

Add $4\frac{3}{4} + 5\frac{2}{8} + 1\frac{1}{2}$.

$$4\frac{3}{4} = 4\frac{6}{8}$$   Change each fraction to an equivalent fraction with the LCD.

$$5\frac{2}{8} = 5\frac{2}{8}$$

$$1\frac{1}{2} = 1\frac{4}{8}$$   Add whole numbers. Add fractions.

$$10\frac{12}{8}$$   Simplify. $\frac{12}{8} = 1\frac{4}{8}$

$$10 + 1\frac{4}{8} = 11\frac{4}{8} = 11\frac{1}{2}$$

**2** Subtract fractions and mixed numbers (pp. 81–83).

To subtract fractions: **1.** Change each fraction to an equivalent fraction that has the LCD as its denominator. **2.** Subtract the numerators. **3.** Place the difference over the LCD. **4.** Reduce if possible.

Subtract $\frac{5}{8} - \frac{7}{16}$.

$$\frac{5}{8} = \frac{10}{16}$$   Convert each fraction to an equivalent fraction with a common denominator.

$$-\frac{7}{16} = \frac{7}{16}$$   Subtract the numerators.

$$\frac{3}{16}$$

To subtract mixed numbers: **1.** Write fractions as equivalent fractions with common denominators. **2.** Borrow from the whole number and add to the fraction if the fraction in the minuend is smaller than the fraction in the subtrahend. **3.** Subtract the fractions. **4.** Subtract the whole numbers. **5.** Simplify if necessary.

Subtract $5\frac{3}{8} - 3\frac{9}{16}$.

$$5\frac{3}{8} = 5\frac{6}{16} = 4\frac{22}{16}$$   Regroup.

$$-3\frac{9}{16} = 3\frac{9}{16} = 3\frac{9}{16}$$   Subtract whole numbers. Subtract fractions.

$$1\frac{13}{16}$$

## Section 2–3

**1** Multiply fractions and mixed numbers (pp. 85–88).

To multiply fractions: **1.** Reduce any numerator and denominator that have a common factor. **2.** Multiply the numerators to get the numerator of the product. **3.** Multiply the denominators to get the denominator of the product. **4.** Be sure the product is reduced.

Multiply $\frac{4}{5} \times \frac{7}{10} \times \frac{15}{35}$.

$$\frac{\overset{2}{\cancel{4}}}{\underset{1}{\cancel{5}}} \times \frac{\overset{1}{\cancel{7}}}{\underset{5}{\cancel{10}}} \times \frac{\overset{3}{\cancel{15}}}{\underset{5}{\cancel{35}}} = \frac{6}{25}$$   Reduce and multiply.

To multiply fractions, whole numbers, and mixed numbers: **1.** Write whole numbers as fractions with denominators of 1. **2.** Write mixed numbers as improper fractions. **3.** Reduce as much as possible. **4.** Multiply the numerators to get the numerator of the product. **5.** Multiply the denominators to get the denominator of the product. **6.** Write the product as a whole number, mixed number, or fraction in lowest terms.

Multiply $4 \times 3\frac{1}{5} \times \frac{2}{7}$.

$$4 \times 3\frac{1}{5} \times \frac{2}{7} = \frac{4}{1} \times \frac{16}{5} \times \frac{2}{7} = \frac{128}{35} = 3\frac{23}{35}$$

**2** Raise a fraction to a power (pp. 88–89).

To raise a fraction or quotient to a power: **1.** Raise the numerator to the power. **2.** Raise the denominator to the power.

Raise $\left(\frac{3}{5}\right)^3$ to the indicated power.

$$\left(\frac{3}{5}\right)^3 = \frac{3^3}{5^3} = \frac{3 \cdot 3 \cdot 3}{5 \cdot 5 \cdot 5} = \frac{27}{125}$$

**3** Divide fractions and mixed numbers (pp. 89–92).

To write the reciprocal of a number: **1.** Express the number as a fraction. **2.** Interchange the numerator and denominator.

Find the reciprocal of $\frac{3}{5}$, 6, $2\frac{3}{4}$, and 0.2.

The reciprocal of $\frac{3}{5}$ is $\frac{5}{3}$; the reciprocal of 6 is $\frac{1}{6}$; the reciprocal of $2\frac{3}{4}$ or $\frac{11}{4}$ is $\frac{4}{11}$.

The reciprocal of 0.2 or $\frac{2}{10}$ is $\frac{10}{2}$ or 5.

To divide fractions: **1.** Replace the divisor with its reciprocal. **2.** Change the division to multiplication. **3.** Multiply.

Divide $\frac{4}{5} \div \frac{8}{9}$.

$$\frac{4}{5} \div \frac{8}{9} = \frac{\overset{1}{\cancel{4}}}{5} \times \frac{9}{\underset{2}{\cancel{8}}} = \frac{9}{10}$$   Change to an equivalent multiplication and multiply.

To divide mixed numbers: **1.** Write each mixed number as an improper fraction. **2.** Change the division to an equivalent multiplication by using the reciprocal of the divisor. **3.** Simplify if possible. **4.** Multiply.

Divide $4\frac{2}{3} \div 1\frac{1}{6}$.

$$4\frac{2}{3} \div 1\frac{1}{6} = \frac{14}{3} \div \frac{7}{6} = \frac{\overset{2}{\cancel{14}}}{\underset{1}{\cancel{3}}} \times \frac{\overset{2}{\cancel{6}}}{\underset{1}{\cancel{7}}} = \frac{4}{1} = 4$$

## Section 2–4

**1** Convert one unit of measure to another using unity ratios (pp. 94–97).

To write two equivalent measures as a unity ratio, write one measure in the numerator and its equivalent measure in the denominator. The ratio has a value of 1.

Write two unity ratios for the equivalent measures: 1 yd = 36 in.

$$\frac{1 \text{ yd}}{36 \text{ in.}} \quad \text{or} \quad \frac{36 \text{ in.}}{1 \text{ yd}}$$

To convert from one U.S. customary unit of measure to another using unity ratios: **1.** Write the original measure in the numerator of a fraction with 1 in the denominator. **2.** Multiply by a unity ratio with the original unit of measure in the denominator and the new unit in the numerator.

To estimate: Multiplication can be used to change to smaller units. There will be more smaller units. Division can be used to change to larger units. There will be fewer larger units.

Change 5 ft to inches.

$$\frac{5 \text{ ft}}{1} \times \frac{12 \text{ in.}}{1 \text{ ft}} = 60 \text{ in.}$$

Change 285 ft to yards.

$$\frac{285 \text{ ft}}{1} \times \frac{1 \text{ yd}}{3 \text{ ft}} = 95 \text{ yd}$$

Change $3\frac{1}{2}$ qt to cups.

$$3\frac{1}{2} \text{ qt} = \frac{7}{2} \text{ qt}$$

$$\frac{7 \text{ qt}}{2} \times \frac{2 \text{ pt}}{1 \text{ qt}} \times \frac{2 \text{ c}}{1 \text{ pt}} = 14 \text{ c}$$

More than one unity ratio may be needed.

**2** Convert one unit of measure to another using conversion factors (pp. 97–99).

When we use conversion factors to convert units, we always multiply. Two equivalent measures have two conversion factors.

To develop a conversion factor: **1.** Write a unity ratio that changes the given unit to the new unit. **2.** Change the fraction (or ratio) to its decimal equivalent by dividing the numerator by the denominator.

Develop two conversion factors relating pints and quarts. 1 qt = 2 pt

| **Pints to Quarts** | | **Quarts to Pints** |
|---|---|---|
| $\frac{\text{pt}}{1}\left(\frac{\text{qt}}{\text{pt}}\right)$ | Write appropriate labels. | $\frac{\text{qt}}{1}\left(\frac{\text{pt}}{\text{qt}}\right)$ |
| $\frac{\text{pt}}{1}\left(\frac{1 \text{ qt}}{2 \text{ pt}}\right)$ | Insert numbers to form a unity ratio. | $\frac{\text{qt}}{1}\left(\frac{2 \text{ pt}}{1 \text{ qt}}\right)$ |
| $\frac{1}{2} = 0.5$ | Convert fraction to whole number, decimal, or mixed decimal equivalent. | $\frac{2}{1} = 2$ |
| $\text{pt} \times 0.5 = \text{qt}$ | | $\text{qt} \times 2 = \text{pt}$ |

To convert from one U.S. customary unit of measure to another using conversion factors: **1.** Select the appropriate conversion factor. **2.** Multiply the original measure by the conversion factor.

Change 9 pt to quarts.

$$9 \text{ pt} \times 0.5 = 4.5 \text{ qt} \qquad \text{Conversion factor is 0.5.}$$

Each measure is converted to the next larger unit of measure when possible.

$$2 \text{ ft } 16 \text{ in.} = 2 \text{ ft} + 1 \text{ ft } 4 \text{ in.} = 3 \text{ ft } 4 \text{ in.} \qquad 16 \text{ in.} = 1 \text{ ft } 4 \text{ in.}$$

**3** Add and subtract U.S. customary measures (pp. 99–101).

To add or subtract U.S. customary measures: **1.** Convert each measure to a measure with a common U.S. customary unit. **2.** Add or subtract.

Add 5 lb and 7 oz.     Convert 5 lb to oz.

$$\frac{5 \ \cancel{lb}}{1} \times \frac{16 \text{ oz}}{1 \ \cancel{lb}} = 80 \text{ oz} \qquad \text{Multiply by unity ratio } \frac{16 \text{ oz}}{1 \text{ lb}}.$$

$$80 \text{ oz} + 7 \text{ oz} = 87 \text{ oz} \qquad \text{Combine like measures.}$$

To add or subtract mixed U.S. customary measures: **1.** Align like measures in columns. **2.** Add or subtract, carrying or regrouping as necessary.

Subtract 2 ft 8 in. from 7 ft.

$$
\begin{array}{rl}
7 \text{ ft} = & 6 \text{ ft } 12 \text{ in.} \\
- & 2 \text{ ft } \ 8 \text{ in.} \\
\hline
& 4 \text{ ft } \ 4 \text{ in.}
\end{array}
$$

Rewrite 7 ft as 6 ft 12 in.
Subtract.

**4** Multiply and divide U.S. customary measures (pp. 101–104).

To multiply a measure by a number: **1.** Multiply the numbers associated with each unit of measure by the given number. **2.** Express the answer in standard notation.

Multiply 2 gal 3 qt by 5.

$$
\begin{array}{r}
2 \text{ gal } \ 3 \text{ qt} \\
\times \qquad \quad 5 \\
\hline
10 \text{ gal } 15 \text{ qt} \\
13 \text{ gal } \ 3 \text{ qt}
\end{array}
$$

15 qt = 3 gal 3 qt

To multiply a length measure by a like length measure: **1.** Multiply the numbers associated with each like unit. **2.** The product will be a square unit of measure.

Multiply 5 yd by 12 yd.

$$5 \text{ yd} \times 12 \text{ yd} = 60 \text{ yd}^2$$

To divide a measure by a number: **1.** Divide the largest measure by the number. **2.** Convert any remainder to the next smaller unit. **3.** Repeat Steps 1 and 2 until no other units are left. **4.** Express any remainder as a fraction of the smallest unit. **5.** The quotient will be a measure.

Divide 7 lb 5 oz by 5.

$$
\begin{array}{r}
1 \text{ lb} \quad 7\frac{2}{5} \text{ oz} \\
\hline
5\overline{)7 \text{ lb} \quad \ 5 \text{ oz}} \\
\underline{5 \text{ lb}} \qquad \quad \\
2 \text{ lb} = \underline{32 \text{ oz}} \\
37 \text{ oz} \\
\underline{35 \text{ oz}} \\
2
\end{array}
$$

To divide a measure by a measure: **1.** Convert the measures to the same unit. **2.** Divide. The quotient is a number that indicates how many.

How many 5-oz glasses of juice can be poured from 1 gallon of juice?

$$\frac{1 \text{ gal}}{1} \times \frac{4 \text{ qt}}{1 \text{ gal}} \times \frac{2 \text{ pt}}{1 \text{ qt}} \times \frac{2 \text{ c}}{1 \text{ pt}} \times \frac{8 \text{ oz}}{1 \text{ c}} = 128 \text{ oz}$$     Change 1 gal to ounces.

$$\frac{128 \text{ oz}}{5 \text{ oz}} = 25\frac{3}{5} \text{ glasses}$$

**5** Change one U.S. customary rate measure to another (pp. 104–105).

**1.** Compare the units of both numerators and both denominators to determine which units will change. **2.** Multiply by the unity ratio (or ratios) containing the new unit so that the unit to be changed will reduce.

Change $10\dfrac{\text{mi}}{\text{hr}}$ to $\dfrac{\text{ft}}{\text{hr}}$.

$$\frac{10 \text{ mi}}{1 \text{ hr}} \text{ to } \frac{\text{ft}}{\text{hr}}$$     Miles change to feet. Hours stay the same.

$$\frac{10 \text{ mi}}{1 \text{ hr}} \left( \frac{5{,}280 \text{ ft}}{1 \text{ mi}} \right) = 52{,}800 \frac{\text{ft}}{\text{hr}}$$

## Section 2–5
**1** Change any number to its percent equivalent (pp. 107–109).

Multiply a number by 1 in the form of 100% to change a number to a percent. For a shortcut, move the decimal two places to the right.

Change $\frac{1}{2}$, 1.2, and 7 to percents.

$$\frac{1}{2} \cdot 100\% = \frac{100\%}{2} = 50\% \qquad 1.2 \cdot 100\% = 120\% \qquad 7 \cdot 100\% = 700\%$$

**2** Change any percent to its numerical equivalent (pp. 109–112).

Divide by 1 in the form of 100% to change a percent to a number. Reduce if possible. For a shortcut, move the decimal two places to the left.

Change 7% to a fraction.

$$7\% \div 100\% = \frac{7}{100}$$

Change $1\frac{1}{4}\%$ to a fraction.

$$1\frac{1}{4}\% \div 100\% = \frac{5\%}{4} \div 100\% = \frac{5\%}{4} \cdot \frac{1}{100\%} = \frac{5}{400} = \frac{1}{80}$$

Convert 3.5% to a decimal.

$$3.5\% \div 100\% = 0.035 = 0.035$$

Convert 245% to a mixed number.

$$245\% \div 100\% = \frac{245\%}{100\%} = 2\frac{45}{100} = 2\frac{9}{20}$$

Convert 124.5% to a decimal.

$$124.5\% \div 100\% = 1.245 = 1.245$$

Convert $100\frac{1}{4}\%$ to a decimal.

$$100\frac{1}{4}\% = 100.25\%$$

$$100.25\% \div 100\% = 1.0025$$

## Section 2-6

**1** Identify the portion, base, and rate in percent problems (p. 113).

Use the key words for rate (*percent* or *%*), base (*total, original,* associated with the word *of*), and portion (*part,* associated with the word *is*).

      *R*      *B*  *P*
**(a)** What percent of 10 is 5?

   *R*           *B*    *P*
**(b)** 25% of what number is 3?

  *P*  *R*           *B*
**(c)** 3 is 20% of what number?

**2** Solve percent problems using the percentage formula (pp. 113–119).

Use key words to identify the known elements and the missing element. Then select the appropriate percentage formula.

$$P = R \cdot B \qquad R = \frac{P}{B} \qquad B = \frac{P}{R}$$

What amount is 5% of $200? *Of* identifies $200 as the base.
The percent or rate is 5%. The missing element is the portion.

| | |
|---|---|
| $P = RB$ | Select the appropriate percentage formula and substitute given amounts. |
| $P = 5\% \ (\$200)$ | Change 5% to a fraction or decimal equivalent by dividing by 100%. |
| $P = 0.05 \ (200)$ | Multiply. |
| $P = \$10$ | Portion. |

What percent of 6 is 2?
*Is* suggests 2 is the portion. *Of* identifies 6 as the base. The rate is missing.

| | |
|---|---|
| $R = \dfrac{P}{B}$ | Select the appropriate percentage formula and substitute given amounts. |
| $R = \dfrac{2}{6}$ | Reduce fraction or change to decimal equivalent by dividing. |
| $R = \dfrac{1}{3}$ | Change to a percent by multiplying by 100%. |
| $R = 33\frac{1}{3}\%$ | Rate $\left(\dfrac{1}{3} \times \dfrac{100\%}{1} = \dfrac{100\%}{3} = 33\frac{1}{3}\%\right)$. |

12 is 24% of what number?
The rate is 24%. 12 is the part or portion and is suggested by *is*. *Of* identifies "what number" as the missing base.

| | |
|---|---|
| $B = \dfrac{P}{R}$ | Select the appropriate percentage formula and substitute given amounts. |
| $B = \dfrac{12}{24\%}$ | Change 24% to a decimal equivalent by dividing by 100%. |
| $B = \dfrac{12}{0.24}$ | Divide. |
| $B = 50$ | Base. |

## Section 2-7

**1** Find the amount of increase or decrease (pp. 121–122).

The original amount is the base. The new amount is the original amount plus the increase or the original amount minus the decrease. Subtract the new amount and the original amount to find the increase or decrease.

Julio made $15.25 an hour but took a 20% pay cut. What was the new hourly pay?

| | Estimation: |
|---|---|
| $P = 20\% \ (\$15.25)$ | 10% of $15 is $1.50. |
| $P = 0.2 \ (15.25)$ | 20% of $15 is $3.00. |
| $P = 3.05$ | Hourly pay cut (decrease) |
| $\$15.25 - \$3.05 = \$12.20$ | Original amount − decrease = new amount |

**2** Find the new amount in increase or decrease problems (pp. 122–124).

Add the percent of increase to 100% or subtract the percent of decrease from 100%. Use this new percent in the percentage formula.

A project requires 5 lb of galvanized nails. If 15% of the nails will be wasted, how many pounds must be purchased?

$P = 115\% (5)$    $B = 5 \text{ lb}, R = 100\% + 15\% = 115\%$
            $P$ is missing.
$P = 1.15 (5)$    Estimation:
            10% of 5 = 0.5.
            More than 0.5 or $\frac{1}{2}$ lb of nails must be added for waste.
            Order > 5.5 lb.
            Interpretation:
$P = 5.75 \text{ lb}$    5.75 lb must be ordered.

**3** Find the rate or base in increase or decrease problems (pp. 124–125).

Subtract the original amount and the new amount to find the amount of increase or decrease. Then use the percentage formula, with $R$ or $B$ as the missing element.

A 5-in. power edger blade now measures $4\frac{3}{4}$ in. What is the percent of wear?

$R = \dfrac{\frac{1}{4}}{5}$    $4\frac{4}{4} - 4\frac{3}{4} = \frac{1}{4}$. The portion is $\frac{1}{4}$. The base or original amount is 5 in. $R$ is missing.

$R = \frac{1}{4} \div 5$

            Estimation:
            Wear is much less than 1 in.; 1 in. $5\frac{1}{5}$ original length.

$R = \dfrac{1}{20}$    $\cdot \frac{1}{5} = 20\%$

$R = \dfrac{1}{20} \times 100\%$    Wear < 20%. (< is read "is less than.")

            Interpretation:
$R = 5\%$    The percent of wear is 5%.

A PC has 20% of its hard drive storage capacity filled. If the PC now has 5.12 G of storage capacity available, what was the original storage capacity?

Current storage capacity available is $100\% - 20\% = 80\%$ of the original storage capacity, or the base. 5.12 G is the portion.

$B = \dfrac{5.12}{80\%}$    Estimation:
            $20\% = \frac{1}{5}$

$B = \dfrac{5.12}{0.8}$    $\frac{1}{5}$ of = 5 1

            $\frac{1}{5}$ of original storage capacity > 1

            Original storage capacity > 5.12 + 1 or
            Original storage capacity > 6.12 G
            Interpretation:
$B = 6.4 \text{ G}$    Original storage capacity was 6.4 G

**Section 2–1**

Write five multiples of each number.

**1.** 10  **2.** 9

Is the number divisible by the given number? Explain.

**3.** 153 by 3  **4.** 8,234 by 4

List all factor pairs for each number; then write the factors in order from smallest to largest.

**5.** 48  **6.** 50

Identify each number as *prime* or *composite*. Explain.

**7.** 18  **8.** 20

Find the prime factorization of each number. Write in factored form and then in exponential notation.

**9.** 98  **10.** 120

Find the least common multiple.

**11.** 18 and 40  **12.** 12 and 18

Find the greatest common factor.

**13.** 12, 18, and 30  **14.** 4, 9, and 16

Find the equivalent fractions using the indicated denominators.

**15.** $\dfrac{5}{8} = \dfrac{?}{24}$  **16.** $\dfrac{3}{7} = \dfrac{?}{35}$  **17.** $\dfrac{5}{12} = \dfrac{?}{60}$  **18.** $\dfrac{4}{5} = \dfrac{?}{40}$

Reduce to lowest terms.

**19.** $\dfrac{6}{12}$  **20.** $\dfrac{26}{64}$  **21.** $\dfrac{6}{8}$  **22.** $\dfrac{75}{100}$

Measure line segments 23–24 in Fig. 2–29 (tolerance $= \pm\frac{1}{32}$ in.).

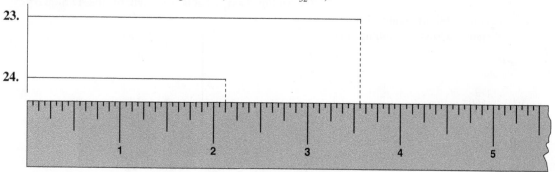

**Figure 2–29**

Write the improper fractions as whole or mixed numbers.

**25.** $\dfrac{18}{5}$  **26.** $\dfrac{27}{6}$

Write the whole or mixed numbers as improper fractions.

**27.** $7\dfrac{1}{8}$  **28.** $5\dfrac{7}{12}$  **29.** $4\dfrac{3}{5}$

Change the whole number to an equivalent fraction using the indicated denominator.

**30.** $5 = \dfrac{?}{3}$  **31.** $2 = \dfrac{?}{10}$

Write as a fraction.

**32.** 0.8  **33.** 0.035  **34.** 0.16  **35.** 0.85  **36.** 2.7

Write as a decimal. Round to the nearest hundredth if necessary.

**37.** $\dfrac{5}{8}$

**38.** $4\dfrac{3}{5}$

**39.** $\dfrac{7}{11}$

Find the least common denominator.

**40.** $\dfrac{7}{8}, \dfrac{2}{3}$

**41.** $\dfrac{5}{12}, \dfrac{3}{10}, \dfrac{13}{15}$

**42.** $\dfrac{1}{12}, \dfrac{3}{8}, \dfrac{15}{16}$

**43. INDTR** Is a $\frac{3}{8}$-in.-thick piece of plasterboard thicker than a $\frac{1}{2}$-in.-thick piece?

**44. AUTO** A $\frac{9}{16}$-in. tube must pass through an opening in a wall. Is a $\frac{3}{4}$-in.-diameter hole large enough?

**45. INDTR** Is a $\frac{19}{32}$-in. wrench larger or smaller than a $\frac{7}{8}$-in. bolt head?

**46. HOSP** A cook top that has a width of $22\frac{1}{4}$ in. needs to fit inside an opening of $22\frac{5}{16}$ in. wide. Will the opening need to be made larger?

Which common or decimal fraction is larger?

**47.** $0.26, \dfrac{3}{4}$

**48.** $0.127, \dfrac{1}{8}$

### Section 2–2

Add; reduce sums to lowest terms and convert improper fractions to mixed numbers or whole numbers.

**49.** $\dfrac{3}{16} + \dfrac{9}{64}$

**50.** $3\dfrac{7}{8} + 7 + 5\dfrac{1}{2}$

**51.** $3\dfrac{7}{8} + 5\dfrac{3}{16} + 1\dfrac{7}{32}$

**52.** $2\dfrac{1}{4} + 3\dfrac{7}{8}$

**53. HELPP** A forest fire advanced $7\frac{5}{8}$ mi one day and $10\frac{7}{16}$ mi the next. How far did the fire advance?

**54. CON** Find the total thickness of a wall if the outside covering is $3\frac{7}{8}$ in. thick, the studs (interior supports) are $3\frac{7}{8}$ in., and the inside covering is $\frac{5}{16}$-in. paneling.

**55. INDTEC** If $7\frac{5}{16}$ in. of a piece of square bar stock is turned (machined) so that it is cylindrical and $5\frac{9}{32}$ in. remains square, what is the total length of the original bar stock?

**56. AVIA** Two carts of airline food weigh $27\frac{1}{2}$ lb and $20\frac{3}{4}$ lb. What is the total weight of the two carts?

**58.** In Fig. 2–31, what is the length of side $A$? Side $B$?

**57.** Figure 2–30 shows $\frac{1}{2}$-in. copper tubing wrapped in insulation. What is the distance across the tubing and insulation?

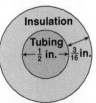

**Figure 2–30**

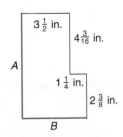

**Figure 2–31**

Subtract; reduce to lowest terms when necessary.

**59.** $\dfrac{5}{9} - \dfrac{2}{9}$

**60.** $\dfrac{11}{32} - \dfrac{5}{64}$

**61.** $122\dfrac{1}{2} - 87\dfrac{3}{4}$

**62. INDTR** Pins of $2\frac{3}{8}$ in. and $3\frac{7}{16}$ in. are cut from a drill rod 12 in. long. If $\frac{1}{16}$ in. of waste is allowed for each cut, how many inches of drill rod are left?

**63. CON** A bolt 2 in. long fastens a piece of $\frac{7}{8}$-in.-thick wood to a piece of metal. If a $\frac{3}{16}$-in.-thick lock washer, a $\frac{1}{16}$-in. washer, and a $\frac{7}{16}$-in.-thick nut are used, what is the thickness of the metal if the nut is even with the end of the bolt after tightening?

**64. TELE** Four lengths measuring $6\frac{1}{4}$ in., $9\frac{3}{16}$ in., $7\frac{1}{8}$ in., and $5\frac{9}{32}$ in. are cut from 48 in. of cable. How much cable remains? Disregard waste.

**65. INDTR** A piece of tapered stock has a diameter of $2\frac{5}{16}$ in. at one end and a diameter of $\frac{55}{64}$ in. at the other end. What is the difference in the diameters?

Multiply and reduce answers to lowest terms. Convert improper fractions to whole or mixed numbers.

**66.** $\dfrac{2}{3} \times \dfrac{5}{8} \times \dfrac{3}{16}$  **67.** $1\dfrac{1}{2} \times \dfrac{4}{5}$  **68.** $3\dfrac{1}{3} \times 4\dfrac{1}{2}$  **69.** $1\dfrac{3}{4} \times 1\dfrac{1}{7}$

**70. CON** In a concrete mixture, $\dfrac{4}{7}$ of the total volume is sand. How much sand is needed for 135 $yd^3$ of concrete?

**71. HOSP** A chef is making a dessert that is $\dfrac{3}{4}$ the original recipe. How much flour should be used if the original recipe calls for $3\dfrac{2}{3}$ cups of flour?

**72. INDTEC** If an alloy is $\dfrac{3}{5}$ copper and $\dfrac{2}{5}$ zinc, how many pounds of each metal are in a casting weighing $112\dfrac{1}{2}$ lb?

**73. CON** A water pipe has an outside diameter of $18\dfrac{3}{4}$ cm. What is the width of eight pipes that have the same diameter?

Raise the fractions to the indicated power.

**74.** $\left(\dfrac{1}{5}\right)^2$  **75.** $\left(\dfrac{3}{4}\right)^2$  **76.** $\left(\dfrac{1}{7}\right)^3$  **77.** $\left(\dfrac{1}{2}\right)^3$

Give the reciprocal.

**78.** $\dfrac{7}{8}$  **79.** 4  **80.** 1.8

Divide and reduce answers to lowest terms. Convert improper fractions to whole or mixed numbers.

**81.** $\dfrac{7}{8} \div \dfrac{3}{4}$  **82.** $8 \div \dfrac{2}{3}$  **83.** $5\dfrac{1}{10} \div 2\dfrac{11}{20}$

**84. CAD/ARC** On a house plan, $\dfrac{1}{4}$ in. represents 1 ft. Find the dimensions of a porch that measures $4\dfrac{1}{8}$ in. by $6\dfrac{1}{2}$ in. on the plan. (How many $\dfrac{1}{4}$s are there in $4\dfrac{1}{8}$; how many $\dfrac{1}{4}$s are there in $6\dfrac{1}{2}$?)

**85. CON** A pipe that is 12 in. long is cut into four equal parts. If $\dfrac{3}{16}$ in. is wasted per cut, what is the maximum length of each pipe? (It takes three cuts to divide the entire length into four equal parts.)

**86. AG/H** If $7\dfrac{1}{2}$ gal of wastewater are distributed equally among five containers, what is the number of gallons per container?

**87. BUS** Fabric that is $22\dfrac{1}{2}$ yd long is cut into lengths of $\dfrac{5}{8}$ yd. How many equal lengths can be made?

## Section 2–4

Write two unity ratios that relate the given pair of measures.

**88.** Hours and days  **89.** Pounds and tons  **90.** Yards and miles

Using unity ratios or conversion factors, convert the given measures to the new units.

**91.** How many ounces are in 5 lb?

**92. HELPP** A fire suit weighing $57\dfrac{3}{5}$ lb weighs how many ounces?

**93.** Find the number of pounds in 680 oz.

**94. HOSP** A can of fruit weighs 22.4 oz. How many pounds is this?

**95. AG/H** How many feet of wire are needed to fence a property line $1\dfrac{1}{4}$ mi long?

Express the measures in standard notation.

**96.** 1 mi 5,375 ft  **97.** 12 lb $17\dfrac{1}{2}$ oz  **98.** 2 gal 7 qt

Add or subtract. Write answers in standard notation.

**99.** $\begin{array}{r} 5\text{ gal 3 qt} \\ +\ 2\text{ gal 3 qt} \\ \hline \end{array}$

**100.** $\begin{array}{r} 7\text{ ft 9 in.} \\ -\ 4\text{ ft 6 in.} \\ \hline \end{array}$

**101.** $\begin{array}{r} 4\text{ lb}\ \ 9\text{ oz} \\ -\ 3\text{ lb 11 oz} \\ \hline \end{array}$

**102. AVIA** Two packages to be sent air express weigh 5 lb 4 oz each. What is the shipping weight of the two packages?

**103. AUTO** A water hose purchased for an RV was 2 ft long. What was its length after 7 in. were cut off?

Multiply and write answers for mixed measures in standard notation.

**104.** $\begin{array}{r} 8\text{ lb 3 oz} \\ \times\ \ 9 \\ \hline \end{array}$

**105.** $\begin{array}{r} 9\text{ in.} \\ \times 7\text{ in.} \\ \hline \end{array}$

**106.** $\begin{array}{r} 10\text{ gal 3 qt} \\ \times\ \ \ \ 7 \\ \hline \end{array}$

Divide.

**107.** 20 yd 2 ft 6 in. ÷ 2

**108.** 5 gal 3 qt 2 pt ÷ 6

**109.** **HELPP** If 18 lb of flame retardant are divided equally into four boxes, express the weight of the contents of each box in pounds and ounces.

**110.** **CON** If 32 equal lengths of pipe are needed for a job and each length is to be 2 ft 8 in., how many feet of pipe are needed for the job?

**111.** 14 ft ÷ 4 ft

**112.** 2 mi 120 ft ÷ 15 ft

**113.** $5\dfrac{\text{mi}}{\text{min}} = \underline{\hspace{1cm}}\dfrac{\text{mi}}{\text{h}}$

**114.** $2{,}520\dfrac{\text{gal}}{\text{h}} = \underline{\hspace{1cm}}\dfrac{\text{qt}}{\text{h}}$

**115.** **HOSP** How many quarts of milk are needed for a recipe that calls for 3 pt of milk?

**116.** **HOSP** How many $\frac{1}{2}$-oz servings of jelly can be made from a $1\frac{1}{2}$-lb container of jelly?

## Section 2–5

Change to percent equivalents.

**117.** $\dfrac{72}{100}$

**118.** $\dfrac{9}{100}$

**119.** 0.7

**120.** 0.35

**121.** $3\dfrac{1}{5}$

Change to fraction equivalents.

**122.** 40%

**123.** $12\dfrac{1}{2}\%$

**124.** $\dfrac{3}{5}\%$

**125.** 275%

**126.** $183\dfrac{1}{3}\%$

Change to decimal equivalents.

**127.** 227.2%

**128.** 73.8%

**129.** 83%

**130.** $37\dfrac{1}{2}\%$

## Section 2–6

Identify the given and missing elements as $R$ (rate), $B$ (base), and $P$ (portion).

**131.** 5% of 180 is what number?

**132.** $15 is what percent of $120?

**133.** 45% of how many dollars is $36?

**134.** How many landscape contractors is 15% of 40 landscape contractors?

Solve using the percentage formula.

**135.** 5% of 480 is what number?

**136.** $62\frac{1}{2}\%$ of 120 is what number?

**137.** 39 is what percent of 65?

**138.** 80% of what number is 116?

**139.** 38.25 is what percent of 250?

**140.** 3 is 0.375% of what number?

**141.** **INDTEC** Of the 2,374 pieces produced by a particular machine, 27 were defective. What percent (to the nearest hundredth of a percent) were defective?

**142.** **ELEC** The voltage loss in a line is 2.5 V. If this is 2% of the generator voltage, what is the generator voltage?

**143.** **CON** An order of lumber totals $348.25. If a 5% sales tax is added to the bill, what is the total bill? Round to the nearest cent.

**144.** **CON** A builder purchases a concrete mixer for $785. The builder does not know the sales tax rate, but the total bill is $828.18. Find the sales tax rate. Round to the nearest tenth of a percent.

**145.** **BUS** A businessperson is charged a $4.96 monthly finance charge on a bill of $283.15. What is the monthly interest rate on the account? Round to the nearest hundredth of a percent.

**146.** **BUS** A salesperson earns a commission of 15% of the total monthly sales. If the salesperson earns $2,145, how much were the total sales?

**147.** **BUS** A salesclerk in a store is paid a salary plus 3% commission. If the salesclerk earns $10.65 commission for a weekend, how much were the sales?

## Section 2–7

**148.** Find the complement of 23%.

**149.** Find the complement of 12.9%.

**150.** **CON** A brickmason received a 12% increase in wages, amounting to $35.40. Find the amount of wages received before the increase. Find the amount of wages received after the increase.

**151.** **BUS** A lathe costing $600 was sold for $516. What was the percent of decrease in the price of the lathe?

152. **INDTR** According to specifications, a machined part may vary from its specified measure by ±0.4% and still be usable. If the specified measure of the part is 75 in. long, what is the range of measures acceptable for the part?

153. **INDTEC** A wet casting weighing 145 kg has a 2% weight loss in the drying process. How much does the dried casting weigh?

154. **BUS** A book that did sell for $18.50 now sells for 20% more. How much does the book now sell for?

155. **COMP** A laptop computer originally priced at $2,400 now sells for $300 more. What is the percent of the increase?

156. **INDTEC** A 20-in. bar of iron measures 20.025 in. when it is heated. What is the percent of increase?

157. **HLTH/N** A dieter went from 168 lb to 160 lb in 1 week. To the nearest tenth of a percent, what was the percent of weight loss?

158. **BUS** Workers took a 10% pay cut to help their company stay open during economic hard times. What is the reduced annual salary of a worker who originally earned $35,000?

159. **COMP** The Bureau of Labor Statistics predicts that the computer and data processing services industry will have the fastest wage and salary employment growth from 2000 to 2010. The industry had 2,095,000 jobs in 2000 and expects to have 3,900,000 jobs in 2010. What is the percent increase?

160. **HLTH/N** According to the Bureau of Labor Statistics the residential care industry is predicted to be the second highest growth industry from 2000 to 2010. The number of jobs is expected to increase from 806,000 to 1,318,000. What is the expected percent of increase?

## TEAM PROBLEM-SOLVING EXERCISES

1. Your team is preparing a report that is to be printed on both sides of the paper. It is customary to put odd-numbered pages on the front and even-numbered pages on the back. A new chapter is started on the front of a sheet of paper, even if this creates a preceding blank page. Assign page numbers to the document based on these guidelines.

   Chap. 1, 15 pages    Chap. 2, 17 pages
   Chap. 3, 24 pages    Chap. 4, 15 pages

   (a) What page number will start Chapter 2?
   (b) How many pages are in the document?
   (c) How many blank pages are in the document?

2. Your team wants to analyze the percent of income that is spent on housing and transportation. Assume you have a family annual income of $35,500.

   (a) In a year $6,900 is spent for a home mortgage, $950 for property taxes, $380 for homeowner's insurance, $2,400 for utilities, and $200 for maintenance and repair. What percent, to the nearest tenth percent of the family's annual income, is spent for housing?

   (b) In a year $3,420 is spent on automobile financing, $1,152 on gasoline, $625 on insurance, and $150 on maintenance and repair. What percent, to the nearest tenth percent of the family's annual income, is spent for transportation?

## PRACTICE TEST

1. Write 0.68 as a fraction
2. Reduce $\frac{18}{24}$ to lowest terms.
3. Find the measure of the line segment $AB$ in Fig. 2–32.
4. Write $\frac{7}{20}$ as a decimal.
5. Write $\frac{21}{2}$ as a decimal.
6. Write $5\frac{3}{8}$ as an improper fraction.
7. Which fraction is smaller, $\frac{4}{5}$ or $\frac{7}{10}$?

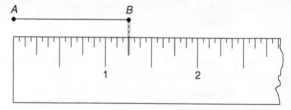

**Figure 2–32**

Perform the indicated operations.

8. $\dfrac{7}{12} + \dfrac{5}{6}$

9. $3\dfrac{4}{15} + 4\dfrac{3}{10}$

10. $\dfrac{5}{6} \times \dfrac{3}{10}$

11. $2\dfrac{2}{9} \times 1\dfrac{3}{4}$

12. $\dfrac{7}{8} - \dfrac{1}{4}$

13. $6\dfrac{1}{4} - 2\dfrac{3}{4}$

14. $\dfrac{5}{12} \div \dfrac{5}{6}$

15. $7\dfrac{1}{2} \div \dfrac{5}{9}$

16. $5\dfrac{2}{3} \div 1\dfrac{1}{9}$

Solve.
17. 40% of 10 X-ray technicians is how many?
19. 25% of how much is 18?

18. What percent of 275 is 33?
20. What percent of 480 is 3?

Solve the problems.
21. Two of the seven security employees at the local community college received safety awards from the governor. Represent the part of the total number of employees who received an award as a fraction.
23. A homemaker has $5\frac{1}{2}$ cups of sugar on hand to make a batch of cookies requiring $1\frac{2}{3}$ cups of sugar. How much sugar is left?
25. A costume maker figures one costume requires $2\frac{2}{3}$ yd of red satin. How many yards of red satin are needed to make three costumes?
27. Find the complement of 88%.
29. $15,000 is invested at 14% per year for 3 months. How much interest is earned on the investment?
31. An electronic parts salesperson earned $175 in commission. If the commission is 7% of sales, how much did the salesperson sell?

33. In 2003, an area vocational school had an enrollment of 325 men and 123 women. In 2004, there were 149 women. What was the percent of increase of women students? Round to the nearest hundredth.
35. During one period, a bakery rejected 372 items as unfit for sale. In the following period, the bakery rejected only 323 items, a decrease in unfit bakery items. What was the percent of decrease? Round to the nearest hundredth.
37. A casting weighed 36.6 kg. After milling, it weighs 34.7 kg. Find the percent of weight loss to the nearest whole percent.
39. The total bill for machinist supplies was $873.92 before a discount of 12%. How much was the discount? Round to the nearest cent.

22. A candy-store owner mixes $1\frac{1}{2}$ lb of caramels, $\frac{3}{4}$ lb of chocolates, and $\frac{1}{2}$ lb of candy corn. What is the total weight of the mixed candy?
24. If $6\frac{1}{4}$ ft of wire is needed to make one electrical extension cord, how many extension cords can be made from $68\frac{3}{4}$ ft of wire?
26. Will a $\frac{5}{8}$-in.-wide drill bit make a hole wide enough to allow a $\frac{1}{2}$-in. (outside diameter) copper tube to pass through?
28. Find the complement of 13%.
30. A casting measuring 48 cm when poured shrinks to 47.4 cm when cooled. What is the percent of decrease?
32. Electronic parts increased 15% in cost during a certain period, amounting to an increase of $65.15 on one order. How much would the order have cost before the increase? Round to the nearest cent.
34. The payroll for a hobby shop for 1 week is $1,500. If federal income and social security taxes average 28%, how much is withheld from the $1,500?

36. Materials to landscape a new home cost $643.75. What is the amount of tax if the rate is 6%? Round to the nearest cent.

38. After soil was excavated for a project, it swelled 15%. If 275 yd³ were excavated, how many cubic yards of soil were there after excavation?
40. A business paid a $5.58 finance charge on a monthly balance of $318.76. What was the monthly rate of interest? Round to the nearest hundredth.

## CAREER APPLICATION: WHAT PERCENT TAX IS REALLY WITHHELD?

True or false? As your income increases and you move to a higher tax bracket, your total taxable income is taxed at the higher rate. Before you answer the question, examine the withholding tax for the various monthly incomes for Exercises 1–19. To determine the amount of income tax to be withheld, use the given recent IRS Table for Percentage Method of Withholding for a single taxpayer paid monthly (IRS Publication 15, Circular E). The directions for calculating the income tax to be withheld are given in the table.

**SINGLE person** (including head of household)—
If the amount of wages
(after subtracting withholding allowances) is:
Not over $221 . . .

The amount of income tax to withhold is:
$0

| Over— | But not over— | | of excess over— |
|---|---|---|---|
| $221 | $808 .. | 10% | $221 |
| $808 | $2,567 .. | $58.70 plus 15% | $808 |
| $2,567 | $5,708 .. | $322.55 plus 25% | $2,567 |
| $5,708 | $12,392 .. | $1,107.80 plus 28% | $5,708 |
| $12,392 | $26,767 .. | $2,979.32 plus 33% | $12,392 |
| $26,767 | . . . . . . | $7,723.07 plus 35% | $26,767 |

For each amount in Exercises 1–19:

**(a)** Enter the tax bracket percent (0%, 10%, 15%, 25%, 28%, 33%, or 35%) for the given income.

**(b)** Calculate the income tax to be withheld for each amount of monthly income that is subject to withholding.

**(c)** Calculate the percent to the nearest whole percent that the withholding tax is of the given income.

| | Monthly Income Subject to Withholding Tax | (a) Tax Bracket Percent | (b) Income Tax to Be Withheld | (c) Actual Percent Withholding Tax Is of Income |
|---|---|---|---|---|
| 1. | $221 | | | |
| 2. | $500 | | | |
| 3. | $1,000 | | | |
| 4. | $1,500 | | | |
| 5. | $2,000 | | | |
| 6. | $2,567 | | | |
| 7. | $4,000 | | | |
| 8. | $5,000 | | | |
| 9. | $5,708 | | | |
| 10. | $7,000 | | | |
| 11. | $10,000 | | | |
| 12. | $12,393 | | | |
| 13. | $15,000 | | | |
| 14. | $20,000 | | | |
| 15. | $25,000 | | | |
| 16. | $26,767 | | | |
| 17. | $30,000 | | | |
| 18. | $50,000 | | | |
| 19. | $100,000 | | | |

**20.** Now, true or false: As your income increases and you move to a higher tax bracket, your total taxable income is taxed at the higher rate.
Explain why you think your answer is correct.

**Answers:**

**1.** a. 0% b. $0 c. 0%
**4.** a. 15% b. $162.50 c. 11%
**7.** a. 25% b. $680.80 c. 17%
**10.** a. 28% b. $1,469.56 c. 21%
**13.** a. 33% b. $3,839.96 c. 26%
**16.** a. 33% b. $7,723.07 c. 29%
**19.** a. 35% b. $33,354.62 c. 33%

**2.** a. 10% b. $27.90 c. 6%
**5.** a. 15% b. $237.50 c. 12%
**8.** a. 25% b. $930.80 c. 19%
**11.** a. 28% b. $2,309.56 c. 23%
**14.** a. 33% b. $5,489.96 c. 27%
**17.** a. 35% b. $8,854.62 c. 30%
**20.** False. Explanations will vary.

**3.** a. 15% b. $87.50 c. 9%
**6.** a. 15% b. $322.50 c. 13%
**9.** a. 25% b. $1,107.80 c. 19%
**12.** a. 28% b. $2,979.32 c. 24%
**15.** a. 33% b. $7,139.96 c. 29%
**18.** a. 35% b. $15,854.62 c. 32%

# 3

# Measurement

## Focus on Careers

Job prospects for brickmasons, block-masons, and stonemasons are expected to be excellent over the next few years. Local contractors, trade associations, or local union-management communities sponsor apprenticeships that usually require three years of on-the-job training. Additionally, a minimum of 144 hours of classroom instruction in each of the three years is required. Subjects such as blueprint reading, mathematics, layout work and sketching are required.

More than one out of four masons are self-employed, and many specialize in contracting to work on small jobs such as patios, walkways, and fireplaces. Some masons become supervisors for masonry contractors and others become owners of businesses employing many workers.

Those choosing this career usually work outdoors and are exposed to all kinds of weather. They stand, kneel, and bend for long periods of time and often have to lift heavy materials. The work may be hazardous; injuries include falls from scaffolds and injuries from tools. Proper safety equipment and practices do minimize these risks.

*(continued)*

In 2002, there were 165,000 brickmasons, blockmasons, and stonemasons. The median hourly earnings for brickmasons and blockmasons were $20.11. The middle 50% earned between $15.36 and $25.32 per hour. Stonemasons had median hourly earnings of $16.36. Apprentices and helpers usually start at a lower wage rate. Pay increases as apprentices gain experience and learn new skills.

Source: *Occupational Outlook Handbook,* 2004–2005 Edition, U.S. Department of Labor, Bureau of Labor Statistics.

---

## 3–1 | Introduction to the Metric System

**Learning Outcomes**

1. Identify uses of metric measures of length, mass, weight, and capacity.
2. Convert from one metric unit of measure to another.
3. Make calculations with metric measures.

The **metric system** is an international system of measurement that uses standard units and powers-of-10 prefixes to indicate other units of measure.

In the metric system, or the **International System of Units (SI),** there is a standard unit for each type of measurement. A **standard unit** is the unit that assumes the position of the ones place on the place-value chart and is the word to which the prefixes attach. In most cases the standard unit is also the base unit for the type of measure. The **base unit** is the unit that is used most often in practice. The **meter** is used for length or distance, the **gram** is used for mass or weight, and the **liter** is used for capacity or volume. A prefix is affixed to the standard unit to indicate a measure greater than the standard unit or less than the standard unit.

### 1 Identify Uses of Metric Measures of Length, Mass, Weight, and Capacity.

The most common prefixes for units *smaller* than the standard unit are

$$\textbf{deci-}\ \frac{1}{10}\ \text{of} \qquad \textbf{centi-}\ \frac{1}{100}\ \text{of} \qquad \textbf{milli-}\ \frac{1}{1,000}\ \text{of}$$

The most common prefixes for units *larger* than the standard unit are

$$\textbf{deka-}\ 10\ \text{times} \qquad \textbf{hecto-}\ 100\ \text{times} \qquad \textbf{kilo-}\ 1,000\ \text{times}$$

| Thousands (1,000) | Hundreds (100) | Tens (10) | Units or ones (1) | Tenths ($\frac{1}{10}$) | Hundredths ($\frac{1}{100}$) | Thousandths ($\frac{1}{1,000}$) |
|---|---|---|---|---|---|---|
| Kilo- | Hecto- | Deka- | STANDARD UNIT | Deci- | Centi- | Milli- |

• Decimal point

**Figure 3–1**

We can compare our decimal place-value chart with the prefixes (Figure 3–1). The standard unit (whether meter, gram, or liter) corresponds to the *ones* place. All the places to the left are powers of the standard unit. That is, the value of *deka-* (some sources use *deca-*) is 10 times the standard unit; the value of *hecto-* is 100 times the standard unit; the value of *kilo-* is 1,000 times the standard unit; and so on. All the places to the right of the standard unit are subdivisions of the standard unit. That is, the value of *deci-* is $\frac{1}{10}$ of the standard unit; the value of *centi-* is $\frac{1}{100}$ of the standard unit; the value of *milli-* is $\frac{1}{1,000}$ of the standard unit; and so on.

EXAMPLE    Give the value of the metric units using the standard unit (gram, liter, or meter).

(a) Kilogram (kg) = 1,000 times 1 gram or **1,000 g**
(b) Deciliter (dL) = $\frac{1}{10}$ of 1 liter or **0.1 L**
(c) Hectometer (hm) = 100 times 1 meter or **100 m**
(d) Dekaliter (dkL) = 10 times 1 liter or **10 L**
(e) Milliliter (mL) = $\frac{1}{1,000}$ of 1 liter or **0.001 L**
(f) Centigram (cg) = $\frac{1}{100}$ of 1 gram or **0.01 g**

There are other metric prefixes for very large and very small amounts. For measurements smaller than one-thousandth of a unit or larger than one thousand times a unit, prefixes that align with periods on a place-value chart are commonly used.

Powers of 10 for decimal places are written with negative exponents. The relationship between the exponent and the value of the number is discussed more completely in Chapter 4.

## Metric Prefixes

| Prefix | Relationship to Standard Unit* |
| --- | --- |
| atto-(a) | quintillionth part ($\times\ 0.000000000000000001$ or $10^{-18}$) |
| femto-(f) | quadrillionth part ($\times\ 0.000000000000001$ or $10^{-15}$) |
| pico-(p) | trillionth part ($\times\ 0.000000000001$ or $10^{-12}$) |
| nano-(n) | billionth of ($\times\ 0.000000001$ or $10^{-9}$) |
| micro-(μ) | millionth of ($\times\ 0.000001$ or $10^{-6}$) |
| milli-(m) | thousandth of ($\times\ 0.001$ or $10^{-3}$) |
| centi-(c) | hundredth of ($\times\ 0.01$ or $10^{-2}$) |
| deci-(d) | tenth of ($\times\ 0.1$ or $10^{-1}$) |
| deka-/deca-(dk) | ten times ($\times\ 10$ or $10^{1}$) |
| hecto-(h) | hundred times ($\times\ 100$ or $10^{2}$) |
| kilo-(k) | thousand times ($\times\ 1,000$ or $10^{3}$) |
| mega-(M) | million times ($\times\ 1,000,000$ or $10^{6}$) |
| giga-(G) | billion times ($\times\ 1,000,000,000$ or $10^{9}$) |
| tera-(T) | trillion times ($\times\ 1,000,000,000,000$ or $10^{12}$) |
| peta-(P) | quadrillion times ($\times\ 1,000,000,000,000,000$ or $10^{15}$) |
| exa-(E) | quintillion times ($\times\ 1,000,000,000,000,000,000$ or $10^{18}$) |

\* See Chapter 4, Section 5 for powers of 10.

All metric units of length, weight, and volume are expressed either as a standard unit or as a standard unit with a prefix. Let's examine some common metric units to develop an intuitive sense of their size.

### Length

*Meter:*    The **meter** is the standard unit for measuring length. Both *meter* and *metre* are acceptable spellings for this unit of measure. A meter is about 39.37 in. It is 3.37 in. longer than a yard (Figure 3–2). We use the meter to measure lengths and distances like room dimensions, land dimensions, lengths of poles, heights of mountains, and heights of buildings. The abbreviation for meter is m.

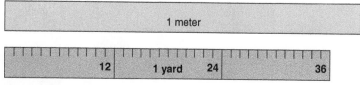

1 meter

1 meter – slightly longer than a yard
(36 inches = 1 yard) (39.37 inches = 1 meter)

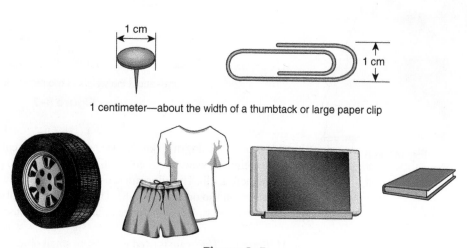

Figure 3–2

*Kilometer:* A **kilometer** is 1,000 m and is used for longer distances. The abbreviation for kilometer is km. The prefix *kilo* means 1,000. We measure the distance from one city to another, one country to another, or one landmark to another in kilometers (Figure 3–3). Driving at a speed of 55 mi (or 90 km) per hour, we would travel 1 km in about 40 sec. An average walking speed is 1 km (approximately 5 city blocks) in about 10 min. (Figure 3–4)

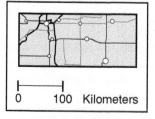

0    100   Kilometers

Figure 3–3

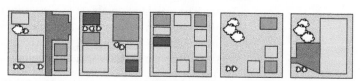

Figure 3–4

*Centimeter:* To measure objects less than 1 m long, we commonly use the **centimeter** (cm). The prefix *centi* means "$\frac{1}{100}$ of," and a centimeter is one-hundredth of a meter. A centimeter is about the width of a thumbtack head, somewhat less than $\frac{1}{2}$ in. (Figure 3–5). We use centimeters to measure medium-sized objects such as tires, clothing, textbooks, and television pictures.

1 cm

1 cm

1 centimeter—about the width of a thumbtack or large paper clip

Figure 3–5

*Millimeter:* Many objects are too small to be measured in centimeters, so we use a **millimeter** (mm), which is "$\frac{1}{1,000}$ of" a meter. It is about the thickness of a plastic credit card or a dime (Figure 3–6). Certain film sizes, bolt and nut sizes, the length of insects, and similar items are measured in millimeters.

Other units and their abbreviations are decimeter, dm; dekameter, dkm; and hectometer, hm.

3–1 Introduction to the Metric System

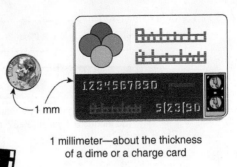

1 mm

1 millimeter—about the thickness
of a dime or a charge card

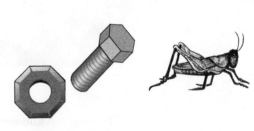

**Figure 3–6**

## Weight or Mass

Mass and weight are often used interchangeably, but in technical or scientific terms they are different. The weight of an object is a measure of the earth's gravitational pull on the object. As an illustration, an object may have a specific weight and mass on Earth. As an object moves away from Earth, as out in space, the mass remains constant (the same), whereas the weight decreases. When the object is in a "weightless" state, it floats freely in space.

*Gram:* A metric unit for measuring mass in the metric system is the **gram**. A gram is the mass of 1 cubic centimeter ($cm^3$) of water at its maximum density. The metric unit for measuring weight is the Newton (N); however, in common usage the gram is used in comparing metric and U.S customary units of weight. A cubic centimeter is a cube whose edges are each 1 centimeter long. It is a little smaller than a sugar cube. The abbreviation g is used for gram. We use grams to measure small or light objects such as paper clips, cubes of sugar, coins, and bars of soap (Figure 3–7).

1 gram—about the weight of two paper clips

**Figure 3–7**

**Figure 3–8**

*Kilogram:* A **kilogram** (kg) is 1,000 grams. Since a cube 10 cm on each edge will be 1,000 $cm^3$, the weight of water required to fill this cube is 1,000 grams or 1 kilogram. A kilogram is approximately 2.2 lb. The kilogram is used to measure the weight of people, books, meat, grain, automobiles, and so on (Figure 3–8). The kilogram is probably the most commonly used metric measure of mass. It is sometimes referred to as the base metric unit for mass.

1 milligram

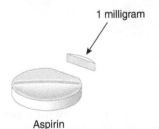

Aspirin

**Figure 3–9**

*Milligram:* The gram is used to measure small objects, and the **milligram** ($\frac{1}{1,000}$ of a gram) is used to measure *very* small objects. Milligrams are too small for ordinary uses; however, pharmacists and manufacturers use milligrams (mg) to measure small amounts of drugs, vitamins, and medications (Figure 3–9).

Other units and their abbreviations are decigram, dg; centigram, cg; dekagram, dkg; and hectogram, hg.

Chapter 3 / Measurement

1 liter—about the volume of a quart of milk or a soft drink in a plastic bottle

**Figure 3–10**

## Capacity or Volume

*Liter:* A **liter** (L) is the volume of a cube 10 cm on each edge. It is the standard metric unit of capacity. Like the meter, it may be spelled *liter* or *litre*, but we use the spelling *liter*. A cube 10 cm on each edge filled with water weighs approximately 1 kg, so 1 L of water weighs about 1 kg. One liter is just a little larger than a liquid quart. Soft drinks are often sold in 2-L bottles, gasoline is sold by the liter at some service stations, and numerous other products are sold in liter containers (Figure 3–10).

*Milliliter:* A liter is 1,000 cm$^3$, so $\frac{1}{1,000}$ of a liter, or a **milliliter,** has the same volume as a cubic centimeter. Most liquid medicine is labeled and sold in milliliters (mL) or cubic centimeters (cc or cm$^3$). Medicines, perfumes, and other very small quantities are measured in milliliters (Figure 3–11).

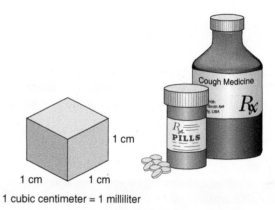

1 cubic centimeter = 1 milliliter

**Figure 3–11**

Other units and their abbreviations are deciliter, dL; centiliter, cL; dekaliter, dkL; hectoliter, hL; and kiloliter, kL.

Other standard units in the metric system exist, but those discussed here are the ones most frequently used.

---

**EXAMPLE** Choose the most reasonable metric measure.

1. Distance from Jackson, Mississippi, to New Orleans, Louisiana
   (a) 322 m      (b) 322 km      (c) 322 cm      (d) 322 mm
2. Weight of an adult woman
   (a) 56 g      (b) 56 mg      (c) 56 kg      (d) 56 dkg
3. Bottle of eye drops
   (a) 30 dL      (b) 30 dkL      (c) 30 L      (d) 30 mL
4. Weight of an aspirin
   (a) 352 mg      (b) 352 dg      (c) 352 g      (d) 352 kg

1. **(b)** 322 km (about 200 mi)
2. **(c)** 56 kg (about 120 lb)
3. **(d)** 30 mL (about 1 oz)
4. **(a)** 352 mg (one regular-strength aspirin)

---

**2** **Convert from One Metric Unit of Measure to Another.**

To understand how we change one metric unit to another, let's arrange the units into a place-value chart like the one we used for decimals (Figure 3–12). The units are arranged from left to right and from largest to smallest.

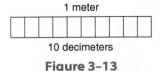

| | | | standard units 1,000 | standard units 100 | standard units 10 | meter, gram, liter | $\frac{1}{10}$ of a standard unit | $\frac{1}{100}$ of a standard unit | $\frac{1}{1,000}$ of a standard unit |
|---|---|---|---|---|---|---|---|---|---|
| | | | Kilo- | Hecto- | Deka- | STANDARD UNIT | Deci- | Centi- | Milli- |

• Decimal point

**Figure 3–12**

1 meter

**Figure 3–13**

10 decimeters

As we move from any place in the chart one place to the *right*, the metric unit changes to the next smaller unit. The larger unit is broken down into 10 smaller units, so we are *multiplying* the larger unit by 10 when we move one place to the right (Figure 3–13).

**To change from one metric unit to *any smaller* metric unit:**

1. Mentally position the measure on the metric-value chart so that the decimal immediately follows the original measuring unit.
2. Move the decimal to the right so that it *follows* the new measuring unit. (Attach zeros if necessary.)

---

**EXAMPLE**  43 dkm = _____ cm?

Place 43 dkm on the metric-value chart so that the last digit is in the dekameters place; that is, the understood decimal that follows the 3 will be *after* the dekameters place (see Figure 3–14). To change to centimeters, move the decimal point so that it follows the centimeters place. Fill in the empty places with zeros (see Figure 3–15).

Note the shortcut to multiplication by powers of 10: Move the decimal point one place to the right for each time 10 is used as a factor.

**43 dkm = 43,000 cm.**

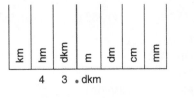

**Figure 3–14**

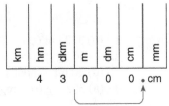

**Figure 3–15**

---

**EXAMPLE**  2.5 dg = _____ mg? (Note the decimal location.)

Place the number 2.5 on the chart so that the decimal point follows the decigrams place (see Figure 3–16). (Note the decimal after the decigrams place.) To change to milligrams, shift

Chapter 3 / Measurement

the decimal two places to the right so that it follows the milligrams place (see Figure 3–17). (Note the decimal after the milligrams place.)

**2.5 dg = 250 mg.**

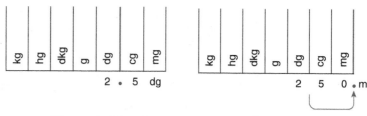

Figure 3–16                    Figure 3–17

As we move one place to the *left* on the metric-value chart, the metric unit changes to the next larger unit. The smaller units are combined into one larger unit 10 times larger than each smaller unit, so we are *dividing* the smaller unit by 10 when we move one place to the left.

**To change from one metric unit to *any larger* metric unit:**

1. Mentally position the measure on the metric-value chart so that the decimal immediately follows the original unit.
2. Move the decimal to the left so that it *follows* the new unit. (Attach zeros if necessary.)

**EXAMPLE**    3,495 L = _____ kL?

Place the number on the chart so the digit 5 is in the liters place; that is, the decimal point follows the liters place (see Figure 3–18). (Note the understood decimal point after the liters place.) To change to kiloliters, move the decimal *three* places to the left so the decimal follows the kiloliters place (see Figure 3–19).

**Thus, 3,495 L = 3.495 kL.**

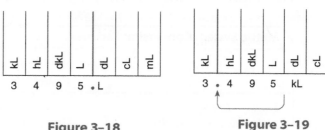

Figure 3–18                    Figure 3–19

**EXAMPLE**    2.78 cm = _____ dkm?

Place the number on the chart so that the decimal follows the centimeters place (see Figure 3–20). (Note the decimal position after the centimeters place.) Move the decimal three places to the left so that it follows the dekameters place (see Figure 3–21). (Note the decimal position after the dekameters place.)

**Therefore, 2.78 cm = 0.00278 dkm.**

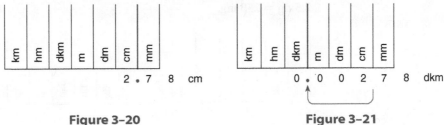

**Figure 3–20**                    **Figure 3–21**

TIP!

### How Far and Which Way?

To determine the movement of the decimal point when changing from one metric unit to another, answer these questions:

1. *How far* is it from the original unit to the new unit (how many places)?
2. *Which way* is the movement on the chart (left or right)?

Change 28.392 cm to m.

*How far* is it from cm to m (Figure 3–22)?   **Two places**

*Which way*?   **Left**

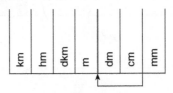

**Figure 3–22**

Move the decimal in the original measure *two places* to the *left*.

28.392 cm = 0.28392 m

Conversion factors can also be used to convert from one metric measure to another.

| Metric System Conversion Factors | | | |
|---|---|---|---|
| | **From** | **To** | **Multiply By** |
| *Length or Distance* | | | |
| 1 kilometer (km) = 1,000 meters (m) | kilometers | meters | 1,000 |
| | meters | kilometers | 0.001 |
| 1 hectometer (hm) = 100 meters | hectometers | meters | 100 |
| | meters | hectometers | 0.01 |
| 1 dekameter (dkm) = 10 meters | dekameters | meters | 10 |
| | meters | dekameters | 0.1 |
| 1 decimeter (dm) = 0.1 meter | decimeters | meters | 0.1 |
| | meters | decimeters | 10 |
| 1 centimeter (cm) = 0.01 meter | centimeters | meters | 0.01 |
| | meters | centimeters | 100 |
| 1 millimeter (mm) = 0.001 meter | millimeters | meters | 0.001 |
| | meters | millimeters | 1,000 |

|  | From | To | Multiply By |
|---|---|---|---|
| **Weight** | | | |
| 1 kilogram (kg) = 1,000 grams (g) | kilograms | grams | 1,000 |
| | grams | kilograms | 0.001 |
| 1 hectogram (hg) = 100 grams | hectograms | grams | 100 |
| | grams | hectograms | 0.01 |
| 1 dekagram (dkg) = 10 grams | dekagrams | grams | 10 |
| | grams | dekagrams | 0.1 |
| 1 decigram (dg) = 0.1 gram | decigrams | grams | 0.1 |
| | grams | decigrams | 10 |
| 1 centigram (cg) = 0.01 gram | centigrams | grams | 0.01 |
| | grams | centigrams | 100 |
| 1 milligram (mg) = 0.001 gram | milligrams | grams | 0.001 |
| | grams | milligrams | 1,000 |
| **Capacity** | | | |
| 1 kiloliter (kL) = 1,000 liters (L) | kiloliters | liters | 1,000 |
| | liters | kiloliters | 0.001 |
| 1 hectoliter (hL) = 100 liters | hectoliters | liters | 100 |
| | liters | hectoliters | 0.01 |
| 1 dekaliter (dkL) = 10 liters | dekaliters | liters | 10 |
| | liters | dekaliters | 0.1 |
| 1 deciliter (dL) = 0.1 liter | deciliters | liters | 0.1 |
| | liters | deciliters | 10 |
| 1 centiliter (cL) = 0.01 liter | centiliters | liters | 0.01 |
| | liters | centiliters | 100 |
| 1 milliliter (mL) = 0.001 liter | milliliters | liters | 0.001 |
| | liters | milliliters | 1,000 |

### 3 Make Calculations with Metric Measures.

We can add or subtract only *like* or *common* measures. For instance, 5 cm and 3 cm are *like* measures. In contrast, 17 kg and 4 hg are *unlike* measures.

**To add or subtract *like* or *common* metric measures:**

1. Add or subtract the numerical values.  $7\,cm + 4\,cm = 11\,cm$
2. Give the answer in the common unit of measure.  $15\,dkg - 3\,dkg = 12\,dkg$

**To add or subtract *unlike* metric measures:**

1. Change the measures to a common unit of measure.
2. Add or subtract the numerical values.
3. Give the answer in the common unit of measure.

**EXAMPLE**  Add 9 mL + 2 cL.

9 mL + 2 cL        Change cL to mL; that is, 2 cL = 20 mL.
9 mL + 20 mL = **29 mL**

**Alternative solution:**

9 mL + 2 cL        Change mL to cL. 9 mL = 0.9 cL.
0.9 cL + 2 cL =

   0.9 cL        Note alignment of decimals.
   2  cL        An understood decimal follows the addend 2.
   **2.9 cL**        2.9 cL = 29 mL

**29 mL or 2.9 cL**

---

**EXAMPLE**  Subtract: 14 km − 34 hm.

  14.0 km        Change hm to km; that is, 34 hm = 3.4 km.
−  3.4 km        Caution: Notice alignment of decimals.
  10.6 km

**The difference is 10.6 km, or 106 hm.**

---

**TIP!**

### Incompatible Measures

Can 5 g be added to 2 cm? Can a common unit be found for centimeters and grams? No, grams measure weight and centimeters measure length (Figure 3–23), so there is no common unit. Thus, we cannot add 5 g + 2 cm.

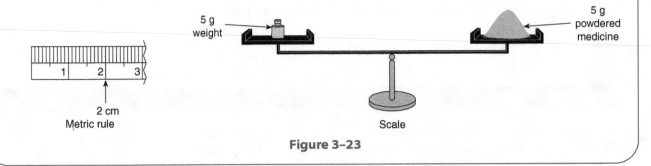

**Figure 3–23**

Multiplication is a shortcut for repeated addition. If we want to find the total length of three pieces of landscape timber that are each 2.7 m long, we can multiply 2.7 m by 3.

### To multiply a metric measure by a number:

**1.** Multiply the numerical values.        2.7 m × 3 = 8.1 m
**2.** Give the answer in the same unit as the original measure.

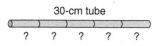

30-cm tube

**Figure 3–24**

Suppose we have a tube that is 30 cm long and we need it cut into five equal parts (Figure 3–24). How long is each part? To solve this problem, we divide 30 cm by 5.

$$\frac{30 \text{ cm}}{5} = 6 \text{ cm}$$

**To divide a metric measure by a number:**

1. Divide the numerical values.
2. Give the quotient in the same unit as the original measure.

$30 \text{ cm} \div 5 = 6 \text{ cm}$

We can divide all measures by a number because we are dividing a quantity into parts. In some situations, we need to divide a measure by a measure. Suppose we have to administer a 250-mg dose of ascorbic acid in 100-mg tablets. How many tablets are needed per dose?

**To divide a metric measure by a like measure:**

1. Divide the numerical values.
2. Give the quotient as a number that tells how many parts there are.
3. If unlike measures are used, first change them to like measures.

$\dfrac{250 \text{ mg}}{100 \text{ mg}} = 2.5 \text{ tablets}$

The answer, 2.5 is labeled tablets rather than mg because we are looking for *how many* tablets are needed.

---

**EXAMPLE**   Solve.

(a) Four micrometers weighing 752 g each will fit into a shipping carton. If the shipping carton and filler weigh 217 g, what is the total weight of the shipment?
(b) A 5-m-long board is to be cut into four equal parts. How long is each part?
(c) How many 25-g packages of seed can be made from 2 kg of seed?

(a) Total weight = carton and filler + 4 micrometers
     = 217 g + (4 × 752 g)        Multiply.
     = 217 g + 3,008 g        Add.
     = 3,225 g, or 3.225 kg

**The total weight is 3,225 g, or 3.225 kg.**

(b) $\dfrac{5 \text{ m}}{4} = 1.25 \text{ m}$        Divide length in meters by number of parts. Result is expressed in meters.

**Each part is 1.25 m long.**

(c) $\dfrac{2 \text{ kg}}{25 \text{ g}} = \dfrac{2,000 \text{ g}}{25 \text{ g}}$        Convert kg to g; then divide. Reduce units.

     $= 80$        Result expresses the number of packages.

**80 packages of seed can be made.**

---

3–1  Introduction to the Metric System

SECTION 3–1 SELF-STUDY EXERCISES

**1**  Give the value of the metric units in standard units.

1. (a) kilometer (km)
   (c) decigram (dg)
   (e) hectogram (hg)

   (b) dekaliter (dkL)
   (d) millimeter (mm)
   (f) centiliter (cL)

Choose the most reasonable metric measure.

2. Height of a 16-year-old patient
   (a) 1.5 km          (b) 1.5 m
   (c) 1.5 cm          (d) 1.5 mm
4. Distance from Los Angeles to San Francisco
   (a) 800 m           (b) 800 mm
   (c) 800 km          (d) 800 cm
6. Overnight accumulation of snowfall
   (a) 8 cm            (b) 8 m
   (c) 8 km            (d) 8 dkm
8. Weight of the average male adult
   (a) 70 mg           (b) 70 kg
   (c) 70 g            (d) 70 dg
10. Weight of a teaspoonful of sugar
   (a) 2 g             (b) 2 kg
   (c) 2 mg            (d) 2 dkg

3. Diameter of a dime
   (a) 1.5 m           (b) 1.5 cm
   (c) 1.5 mm          (d) 1.5 km
5. Length of a pencil
   (a) 20 km           (b) 20 cm
   (c) 20 m            (d) 20 mm
7. Width of a home DVD case
   (a) 12 m            (b) 12 km
   (c) 12 cm           (d) 12 mm
9. Weight of a can of tuna
   (a) 184 mg          (b) 184 kg
   (c) 184 g           (d) 184 dg
11. Weight of a vitamin C tablet
   (a) 250 g           (b) 250 mg
   (c) 250 kg          (d) 250 dkg

**2**  Change to the measure indicated. When using the metric-value chart, place the decimal immediately *after* the measuring unit.

12. 7 g = _____ dg
15. 0.25 km = _____ hm
18. 14.2 dg = _____ cg
21. 0.25 km = _____ m
24. 8.33 L = _____ mL
27. Change 2.36 hL to liters.
30. Change 13 dkm to centimeters.
33. 101 mg = _____ cg
36. 192.5 g = _____ dkg
39. 2,743 mm = _____ m

13. 58 km = _____ hm
16. 21 dkL = _____ L
19. 4 L = _____ cL
22. 8 hg = _____ g
25. 2 km = _____ mm
28. Change 0.467 dkm to centimeters.
31. 28 m = _____ dkm
34. 60 hm = _____ km
37. 17 cm = _____ dm

14. 8 hL = _____ dkL
17. 8.5 cm = _____ mm
20. 8 m = _____ mm
23. 10.25 hm = _____ cm
26. 0.7 g = _____ cg
29. Change 3.8 kg to decigrams.
32. 238 hL = _____ kL
35. 29 dkL = _____ hL
38. How many deciliters are in 2.25 cL?

**3**  Add or subtract as indicated.

40. 7 hL + 5 hL
43. 5 cL + 9 mL
46. 1 g − 45 cg

41. 15 cg − 9 cg
44. 4 m + 2 L
47. 3 cg − 5 mL

42. 2 dm + 4 cm
45. 14 kL − 39 hL
48. 7 km + 2 m

49. **HLTH/N**  A patient absorbs 175 mL of fluid through an IV. If the IV bag has 825 mL left, how much fluid was in the bag to begin with?

50. **HOSP**  653 dkL of orange juice concentrate are removed from a vat containing 8 kL of the concentrate. How much concentrate remains in the vat?

Multiply.

51. 43 m × 12

52. 3.4 m × 12

53. 50.32 dm × 3

54. **CAD/ARC**  A plot of ground is divided into seven plots, each with road frontage of 138.5 m. What is the total road frontage of the plot of ground?

55. **AVIA**  Earth consists of a series of relatively thin plates which are in constant motion. These plates move at different velocities. The Australian plate moves about 60 mm per year. How many millimeters will the plate move in 78 years?

Divide.

**56.** $48 \text{ g} \div 8$  **57.** $39 \text{ m} \div 3$  **58.** $\dfrac{54 \text{ cL}}{6}$

**59. INDTEC** A block of silver weighing 978 g is cut into six equal pieces. How much does each piece weigh?

**60. INDTEC** Two pieces of steel each 12 m long are cut into a total of 60 equal pieces. How long is each piece?

**61.** $2.5 \text{ cg} \div 0.5 \text{ cg}$

**62.** $3 \text{ m} \div 10 \text{ cm}$

**63. HLTH/N** How many 250-mL prescriptions can be made from a container of 4 L of decongestant?

**64. AG/H** How many 500-g containers are needed to hold 40 kg of grass seed?

Solve.

**65. INDTEC** A length of satin fabric 30 dm long is cut into 4 equal pieces. How long is each piece?

**66. HLTH/N** An IV bag holding 250 mL of an antibiotic is calibrated (marked off) into five equal sections. How many milliliters are represented by each section?

**67. INDTEC** How many containers of jelly can be made from 8,500 L of jelly if each container holds 4 dL of jelly?

**68. HELPP** How many 2-kg vials of fire retardant can be obtained from 38 kg of fire retardant?

---

## 3–2 | Time, Temperature, and Other Measures

**Learning Outcomes**

1 Convert from one unit of time to another.
2 Make calculations with measures of time.
3 Convert a Fahrenheit temperature to a Celsius temperature.
4 Convert a Celsius temperature to a Fahrenheit temperature.
5 Examine other useful measures.

As we examined rate measures in Chapter 2, Section 4, Outcome 5, we introduced the units of time. We convert from one unit of time to another just as we convert other compatible units.

### 1 Convert from One Unit of Time to Another.

The relationships among various units of time are given in Chapter 2, Section 4, Outcome 5. We can use either unity ratios or conversion factors to convert from one unit of time to another.

---

**EXAMPLE** Convert 3 h to minutes.

**Estimation**

Hours to minutes → larger to smaller → more minutes

Use a unity ratio to make the conversion:

$3 \text{ h} \left( \dfrac{\text{min}}{\text{h}} \right)$  Use an appropriate unity ratio.

$3 \text{ h} \left( \dfrac{60 \text{ min}}{1 \text{ h}} \right) = 3 \times 60 \text{ min} = 180 \text{ min}$

**Interpretation**

**There are 180 min in 3 h.**

Or use a conversion factor to make the conversion:

$1 \text{ h} = 60 \text{ min}$  Conversion factor = 60.

$3 \text{ h} \times 60 \text{ min per hour} = 180 \text{ min}$  Multiply hours by 60 min.

---

### 2 Make Calculations with Measures of Time.

Suppose we have a 1-h meeting and five items to be covered on our agenda. If we are to devote the same amount of time to each item, how much time should we give to each item?

$$1 \text{ h} \div 5 = \frac{1}{5} \text{ h}$$

How long is $\frac{1}{5}$ h? Is this the usual way to express an amount of time? No. When less than 1 h is involved, we often change to minutes.

$$\frac{1}{\cancel{5}} \cancel{\text{h}} \left( \frac{\overset{12}{\cancel{60} \text{ min}}}{1 \cancel{\text{h}}} \right) = 12 \text{ min}$$

---

**EXAMPLE**    A technician can assemble one precut picture frame in 7 min. How long will it take to assemble 25 frames?

**Estimation**    If the frame could be assembled in 6 min, then the technician could assemble 10 per hour (6 min. × 10 = 60 min = 1 h). Thus, it would take 2 h to assemble 20 frames and $2\frac{1}{2}$ h to assemble 25 frames. Since it actually takes 7 min per frame, it will take more than $2\frac{1}{2}$ h.

$$\frac{7 \text{ min}}{\text{frame}} \times 25 \text{ frames} = 175 \text{ min}$$

Alternative method for converting to hours and minutes:

$$175 \text{ min} \times \frac{1 \text{ h}}{60 \text{ min}} = 2.91\overline{6} \text{ h} \qquad \text{Convert min to h.}$$

If we want to know how many hours and minutes, 2 h = 120 min:

$$175 \text{ min} - 120 \text{ min} = 55 \text{ min}$$

**Interpretation**    **It will take 2 h 55 min to assemble 25 frames.**

$$0.91\overline{6} \text{ h} \times \frac{60 \text{ min}}{1 \text{ h}} = 54.99\overline{9}, \text{ or } 55, \text{ min} \qquad \begin{array}{l}\text{To calculate minutes from the decimal part of an hour,} \\ \text{use the decimal portion of } 2.91\overline{6} \text{ h.}\end{array}$$

---

### 3 Convert a Fahrenheit Temperature to a Celsius Temperature.

The **Kelvin** scale is one scale used to measure temperature in the metric system of measurement. Units on the scale are abbreviated with a capital $K$ (without the symbol ° because these units are called *kelvins*) and are measured from absolute zero, the temperature at which *all* heat is said to be removed from matter. Another metric temperature scale is the **Celsius** scale (abbreviated °C), which has as its zero the freezing point of water. The Kelvin and Celsius scales are related such that absolute zero on the Kelvin scale is the same as $-273$°C on the Celsius scale. Each unit of change on the Kelvin scale is equal to 1 degree of change on the Celsius scale; that is, the size of a kelvin and a Celsius degree is the same on both scales.

The U.S. customary system temperature scale that starts at absolute zero is called the **Rankine** scale. It is related to the more familiar **Fahrenheit** scale, which places the freezing point of water at 32°. One degree of change on the Rankine scale equals 1 degree of change on the Fahrenheit scale. Absolute zero (the zero for the Rankine scale) corresponds to 460 degrees *below* zero ($-460$°) on the Fahrenheit scale.

The Celsius and Fahrenheit scales are the most common temperature scales used for reporting air and body temperatures. The formulas for converting temperatures using these two scales are more complicated than the previous ones because 1 degree of change on the Celsius scale does *not* equal 1 degree of change on the Fahrenheit scale (Figure 3–25).

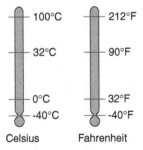

**Figure 3–25**

Chapter 3 / Measurement

## To convert Fahrenheit degrees to Celsius degrees:

Use the formula: $°C = \dfrac{5}{9}(°F - 32)$,   where $°C$ = degrees Celsius and $°F$ = degrees Fahrenheit.

---

**EXAMPLE**   Change 212°F (the boiling point of water) to degrees Celsius.

$°C = \dfrac{5}{9}(\boxed{°F} - 32)$          Substitute 212 for °F in the formula.

$°C = \dfrac{5}{9}(\boxed{212} - 32)$          Work within grouping; subtract $212 - 32$.

$°C = \dfrac{5}{9}(180)$          Multiply. $\dfrac{5}{\underset{1}{\cancel{9}}} \times \dfrac{\overset{20}{\cancel{180}}}{1} = 100$.

$°C = 100$

**212°F = 100°C.**

---

**EXAMPLE**   According to Dr. Shotwell, an antifungal powder containing tolnaftate may be stored at a room temperature of 77°F. Change 77°F to degrees Celsius.

**Estimation**   The Celsius temperature is a smaller value than the Fahrenheit temperature for values near 77°F. From Figure 3–25 we see that 90°F is approximately 32°C. Then 77°F will be less than 32°C.

$°C = \dfrac{5}{9}(\boxed{°F} - 32)$          Substitute 77 for °F.

$°C = \dfrac{5}{9}(\boxed{77} - 32)$          Work grouping. $77 - 32 = 45$.

$°C = \dfrac{5}{9}(45)$          Multiply. $\dfrac{5}{\underset{1}{\cancel{9}}} \times \dfrac{\overset{5}{\cancel{45}}}{1}$

$°C = 25$

**Interpretation**   **77°F = 25°C.**          Based on the estimation the solution is reasonable.

---

**4**   **Convert a Celsius Temperature to a Fahrenheit Temperature.**

## To convert Celsius degrees to Fahrenheit degrees:

Use the formula: $°F = \dfrac{9}{5}°C + 32$, where $°F$ = degrees Fahrenheit and $°C$ = degrees Celsius.

**EXAMPLE**    Change 100°C to degrees Fahrenheit.

$$°F = \frac{9}{5} \; °C + 32 \qquad \text{Substitute 100 for °C.}$$

$$°F = \frac{9}{5}(100) + 32 \qquad \text{Multiply first. } \frac{9}{\overset{}{\underset{1}{5}}}\left(\frac{\overset{20}{\cancel{100}}}{1}\right) = 180.$$

$$°F = 180 + 32 \qquad \text{Add.}$$

$$°F = 212$$

**100°C = 212°F.**

---

**EXAMPLE**    The label on a dropper bottle of ofloxacin ophthalmic solution warns that the medicine must not be stored at a temperature above 25°C. Change 25°C to degrees Fahrenheit.

**Estimation**    In Figure 3–25 we see that the Fahrenheit temperature for values near 25°C is less than 90°F.

$$°F = \frac{9}{5} \; °C + 32 \qquad \text{Substitute 25 for °C.}$$

$$°F = \frac{9}{5}(25) + 32 \qquad \text{Multiply. } \frac{9}{\overset{}{\underset{1}{5}}}\left(\frac{\overset{5}{\cancel{25}}}{1}\right).$$

$$°F = 45 + 32 \qquad \text{Add.}$$

$$°F = 77$$

**Interpretation**    **25°C = 77°F.**      Based on the estimate the solution is reasonable.

---

**5**   **Examine Other Useful Measures.**

All SI metric measures use the same system of prefixes for multiples and submultiples. As with the kilogram, the base unit for a type of measure may not always be the standard unit that is used as the stem word for the SI prefixes. Table 3–1 summarizes the SI base units and some other commonly used metric units.

**Table 3–1   SI Metric Units**

| Base Unit and Abbreviation | Standard Unit When Different from the Base Unit | Type of Measure |
|---|---|---|
| meter (m) | | length |
| kilogram (kg) | gram (g) | mass |
| liter (L) | | volume or capacity |
| second (s) | | time |
| kelvin (K) | | temperature |
| ampere (A) | | electric current |
| candela (cd) | | light intensity |
| mole (mol) | | molecular substance |
| newton (N) | | force |
| joule (J) | | energy |
| watt (W) | | power |
| square meter (m²) | | area |
| meter per second (m/s) | | speed |
| cubic meter (m³) | | volume |
| volt (V) | | voltage |

| Base Unit and Abbreviation | Standard Unit When Different from the Base Unit | Type of Measure |
| --- | --- | --- |
| ohm (Ω) | | resistance |
| hertz (Hz) | | frequency |
| farad (F) | | capacitance |
| henry (H) | | inductance |
| coulomb (C) | | charge |

Metric prefixes for very large and small units are given in Secion 1, page 144.

**EXAMPLE** 12 nanoseconds (ns) is what part of a second?

1 ns = 1 billionth of a second, or 0.000000001 s
12 ns = 12 × 0.000000001 = **0.000000012 s**

Using unity ratios:

$$12 \text{ ns} \times \frac{0.000000001 \text{ s}}{1 \text{ ns}} = \textbf{0.000000012 s}$$

**EXAMPLE** Give the meaning of the following units.

**(a)** 115 μA        **(b)** 3.2 MW

**(a)** μA means microamperes, microamps, or millionths of an amp.

115 μA = **115 microamps, or 0.000115 A**

**(b)** MW means megawatts, or 1 million watts.

3.2 MW = **3.2 megawatts or 3,200,000 W**

SECTION 3–2 SELF-STUDY EXERCISES

**1** Use ratios or conversion factors to convert the measures of time. Round to hundredths.

**1.** How many days are in 36 h?

**2.** Find the number of minutes in 580 s.

**3.** How many minutes are in 2.5 h?

**4.** A physical test took 5.3 min to perform. How many seconds is this?

**5.** If you studied for 3.5 h, how many minutes did you study?

**6.** A process takes 182 s to perform. How many minutes is this?

**7.** Convert 7.2 h to minutes.

**8.** Convert 88 s to minutes.    $\frac{88}{60} = 1.47 \text{ min.}$

**2** Make the conversions.

**9. INDTEC** A conveyor belt can move 72 lb of rice per minute. How many pounds can be moved on the conveyor belt in 1 hour?

**10. AVIA** Aircraft fuel moves through a pipeline at 40 gal per minute. How many gallons per hour can move through the pipeline?

**11. BUS** A dog kennel uses 30 lb of dog food per day. How many pounds are used in a year?

**12. BUS** A toll station can accommodate on average 28 vehicles per minute. How many vehicles can be accommodated in an hour?

**13. AUTO** A car traveling at the rate of 60 $\frac{\text{mi}}{\text{h}}$ is traveling how many miles per second?

**14. CON** A pump that can pump 125 $\frac{\text{gal}}{\text{h}}$ can pump how many gallons per minute?

**15. INDTR**  A pump can dispose of sludge at the rate of 3,600 $\frac{lb}{h}$. How many pounds can be disposed of per minute?

**16. HELPP**  If water flows through a pipe at the rate of 96 $\frac{gal}{min}$, how many gallons will flow per second?

**3**  Change the Fahrenheit temperatures to Celsius.

| | | | | |
|---|---|---|---|---|
| **17.** 95°F | **18.** 32°F | **19.** 113°F | **20.** 41°F | **21.** 59°F |
| **22.** 50°F | **23.** 149°F | **24.** 122°F | **25.** 176°F | **26.** 248°F |

**4**  Change the Celsius temperatures to Fahrenheit.

| | | | | |
|---|---|---|---|---|
| **27.** 70°C | **28.** 15°C | **29.** 45°C | **30.** 50°C | **31.** 20°C |
| **32.** 215°C | **33.** 310°C | **34.** 410°C | **35.** 185°C | **36.** 0°C |

**5**

**37. TELE**  One hertz is a frequency of one cycle per second. Frequencies of radio and television waves are often measured in kilohertz or megahertz. How many hertz are in 500 MHz?

**38. ELEC**  The henry is a large unit. Inductances in circuits are often measured in millihenrys (mH) or microhenrys (μH). How many henrys are in 420 mH?

**39. ELEC**  The watt (W) is the unit used for measuring electrical power. An electrical device that used 1,400 kW has used how many watts?

**40.**  45 ms is what part of a second?

---

| **3-3** | **Metric–U.S. Customary Comparisons** |
|---|---|

*Learning Outcome*  **1**  Convert between U.S. customary measures and metric measures.

Converting a measure from the U.S. customary system to the metric system, and vice versa, is often necessary because both systems are used in the United States. Most industries use one system exclusively, making conversions from one system to another uncommon. However, an industry that uses the U.S. customary system in the United States often needs to convert to the metric system to market its products in other countries.

**1**  **Convert Between U.S. Customary Measures and Metric Measures.**

To convert measures from the U.S. customary system to the metric system, and vice versa, we need only one conversion relationship for each type of measure (length, weight, and capacity). The following equivalency relationships have been rounded to the nearest ten thousandth of a unit:

| | |
|---|---|
| For length: | 1 m = 1.0936 yd |
| For weight: | 1 kg = 2.2046 lb |
| For capacity: | 1 L = 1.0567 liquid qt (liquid measure) |

Additional conversion relationships, such as centimeters to inches and kilometers to miles, can be derived from these relationships using unity ratios.

The most common procedure for converting between the U.S. customary and metric systems is to use conversion factors. Conversion factors always involve multiplication, whereas unity ratios may involve multiplication or division. Many conversion factors are given in the table below. Calculators sometimes have conversion factors programmed into the calculator for greater accuracy.

**U.S. Customary and Metric Conversion Factors**

| | From | To | Multiply By |
|---|---|---|---|
| *Length* | | | |
| 1 meter = 39.37 inches | meters | inches | 39.37 |
| | inches | meters | 0.0254 |

| | From | To | **Multiply By** |
|---|---|---|---|
| 1 meter = 3.2808 feet | meters | feet | 3.2808 |
| | feet | meters | 0.3048 |
| 1 meter = 1.0936 yards | meters | yards | 1.0936 |
| | yards | meters | 0.9144 |
| 1 centimeter = 0.3937 inch | centimeters | inches | 0.3937 |
| | inches | centimeters | 2.54 |
| 1 millimeter = 0.03937 inch | millimeters | inches | 0.03937 |
| | inches | millimeters | 25.4 |
| 1 kilometer = 0.6214 miles | kilometers | miles | 0.6214 |
| | miles | kilometers | 1.6093 |
| *Weight* | | | |
| 1 gram = 0.0353 ounce | grams | ounces | 0.0353 |
| | ounces | grams | 28.3286 |
| 1 kilogram = 2.2046 pounds | kilograms | pounds | 2.2046 |
| | pounds | kilograms | 0.4536 |
| *Liquid Capacity* | | | |
| 1 liter = 1.0567 quarts | liters | quarts | 1.0567 |
| | quarts | liters | 0.9463 |

*1 gram = .0353 oz*

**EXAMPLE**  Change 50 ft to meters.

feet to meters:  Conversion factor is 0.3048.

$50 \times 0.3048 = \mathbf{15.24\ m}$  Multiply.

Body mass index (BMI) is the standard unit for measuring a person's degree of obesity or emaciation. BMI is body weight in kilograms (kg) divided by height in meters squared, or $BMI = w/h^2$.

According to federal guidelines in the U.S., a BMI greater than 25 means that you are overweight. Surveys show that 59% of men and 49% of women have BMIs greater than 25. Extreme obesity is defined as a BMI greater than 40.

### To calculate your BMI:

1. Multiply your weight in pounds by 0.4536 to convert to kilograms.
2. Convert your height to inches.
3. Multiply the inches by 0.0254 to get meters.
4. Square the number found in Step 3.
5. Divide your weight in kilograms by the result from Step 4, and round to the nearest whole number. The result is your BMI.

**EXAMPLE**  Alexa May is 5 feet 6 inches and weighs 138 pounds. Find her body mass index.

**HLTH/N**  
$138 \times 0.4536 = 62.5968$   Change pounds to kilograms.  
$5\text{ ft }6\text{ in.} = 5 \times 12 + 6$   Change feet to inches.  
$\qquad = 60 + 6$  
$\qquad = 66\text{ in.}$  
$66 \times 0.0254 = \boxed{1.6764}$   Change inches to meters.

3–3 Metric–U.S. Customary Comparisons

$$BMI = \frac{w}{h^2} \qquad w = 62.5968, \; h = 1.6764.$$

$$BMI = \frac{62.5968}{(1.6764)^2}$$

$$BMI = \frac{62.5968}{2.81031696}$$

$$\textbf{BMI} = \textbf{22.27392885}$$

$$\textbf{BMI} = \textbf{22} \qquad \text{Rounded to nearest whole number.}$$

## SECTION 3–3 SELF-STUDY EXERCISES

**1**   Change to the units indicated.

**1.** 9 m to inches
**2.** 120 m to yards
**3.** 42 km to miles
**4.** 6 L to liquid quarts
**5.** 10 qt to liters
**6.** 27 kg to pounds
**7.** 50 lb to kilograms
**8.** 7 in. to centimeters
**9.** 18 ft to meters
**10.** 500 km to miles
**11.** 28 in. to millimeters
**12.** 36 in. to centimeters
**13.** 48 in. to meters
**14.** 5,280 ft to meters
**15.** 100 yd to meters
**16.** 200 mi to kilometers
**17.** 50 L to quarts
**18.** 100 qt to liters

**19. ELEC**   A spool of wire contains 100 ft of wire. How many meters of wire are on the spool?

**20. INDTR**   A 60-lb sheet of metal weighs how many kilograms?

**21. CAD/ARC**   Two cities 150 mi apart are how many kilometers apart?

**22. AG/H**   A 30-m-wide field is how many yards wide?

**23. HOSP**   A tourist in Europe traveled 200 km, 60 km, and 120 km by car. How many total miles was this?

**24. HLTH/N**   A patient in therapy jogged 5 km, 4 km, and 3 km. How many miles did the patient jog?

**25.** A container holds 12 quarts. How many liters will the container hold?

**26. ELEC**   A spool of electrical wire contains 100 m of wire. How many feet of wire are on the spool?

**27. HLTH/N**   Gary Druckemiller is 6 feet 7 inches and weighs 192 pounds. What is his body mass index?

**28. HLTH/N**   Jo Ella Stearns weighs 121 pounds and is 5 feet 8 inches. What is her body mass index?

## 3–4   Accuracy, Precision, and Error

*Learning Outcomes*

**1**   Determine the significant digits of a number.
**2**   Find the precision and greatest possible error of a measurement.
**3**   Determine the relative error and the percent error of a measurement.
**4**   Determine an appropriate approximation of measurement calculations.

### 1   Determine the Significant Digits of a Number.

Approximate numbers that represent measured values may have varying degrees of accuracy. The **accuracy** of a measurement refers to how close the measured value is to the true or accepted value. **Precision** is the degree to which the measuring process gives consistent results. (See Figure 3–26.)

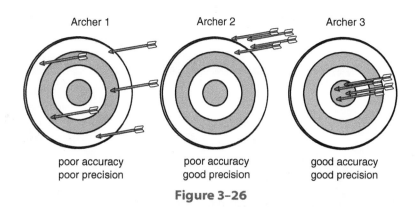

Archer 1     Archer 2     Archer 3

poor accuracy
poor precision

poor accuracy
good precision

good accuracy
good precision

**Figure 3–26**

One indicator of the degree of precision of a measurement is the number of **significant digits** in a measure. The numbers 2,500; 250; 25; 2.5; 0.25 and 0.025 all have two significant digits. When a number contains no zeros, all the digits are significant. However, zeros are special digits. Sometimes zeros are significant digits, and sometimes they are just placeholders.

## To determine the significant digits of a number:

**For whole numbers:**

1. Start with the leftmost nonzero digit.
2. Count each digit (including zeros) through the rightmost nonzero digit.

**For decimals and mixed decimals:**

1. Start with the leftmost nonzero digit.
2. Count each digit (including zeros) through the last digit.

**Determine the number of significant digits in the following:**

200 has 1 significant digit.
28 has 2 significant digits.
1,320 has 3 significant digits.
2,005 has 4 significant digits.

0.07 has 1 significant digit.
2.0 has 2 significant digits.
1.20 has 3 significant digits.
0.01250 has 4 significant digits.

---

**2**   **Find the Precision and Greatest Possible Error of a Measurement.**

The concept of **precision** is associated with approximate numbers. Exact numbers have infinite precision. The average of the measurements of 2.3 cm, 4.25 cm, and 5.125 cm can be no more precise than the least precise of the measurements (tenths).

For the calculations $\dfrac{2.3 + 4.25 + 5.125}{3}$, we would round the results to the nearest tenth.

The number 3 is an exact number (exactly three measures) and has an infinite precision, so tenths is still the least precise.

$$\frac{2.3 + 4.25 + 5.125}{3} = \frac{11.675}{3} = 3.891666667 = 3.9 \text{ (rounded to tenths)}$$

The **greatest possible error** of a measurement is half of the precision of the measurement.

Find the greatest possible error of a measurement of $1\frac{3}{4}$ ft.

1. Determine the precision of the measurement of the smallest subdivision.

Precision $= \frac{1}{4}$ ft

2. The greatest possible error is one-half of the precision.

$\frac{1}{2} \times \frac{1}{4} = \frac{1}{8}$ ft

---

**EXAMPLE**    Find the greatest possible error of the measurements.

(a) $2\frac{5}{8}$ in.          (b) 3.5 cm.

(a) $2\frac{5}{8}$ in.                    Precision $= \frac{1}{8}$ in.

$\frac{1}{2} \times \frac{1}{8} = \frac{1}{16}$ **in.**          Greatest possible error is one-half the precision.

(b) 3.5 cm                    Precision $= 0.1$ cm.

$0.5(0.1) = \textbf{0.05 cm.}$          Greatest possible error is one-half the precision.

---

**3**  **Determine the Relative Error and the Percent Error of a Measurement.**

Because factories and other industrial concerns are extremely interested in issues related to quality control, most regularly measure parts to ensure that they fall within preestablished tolerance limits for quality. Every measurement taken has some amount of error. There are three ways of reporting the error: as an absolute error, a relative error, and a percent of error. Reporting the error as an absolute error does not give much information, so relative error and percent error are frequently calculated.

- The absolute value of the difference between the observed measurement and the true value is the **absolute error.**
- The quotient of the absolute error and the true value is the **relative error.**
- The relative error converted to a percent is the **percent error.**

---

**EXAMPLE**    The blueprint for a part calls for it to be 32.112 mm. A measurement of an actual part is recorded as 32.155 mm. Find the absolute error, relative error, and percent error.

**INDTR**                    absolute error $= \mid$ observed value $-$ true value $\mid$

absolute error $= \mid 32.155 - 32.112 \mid$

**absolute error $= 0.043$ mm**

$$\text{relative error} = \frac{\text{absolute error}}{\text{true value}}$$

$$\text{relative error} = \frac{0.043}{32.112}$$

**relative error = 0.001339**

$$\text{percent error} = \text{relative error} \times 100\%$$

$$\text{percent error} = 0.001339 \times 100\%$$

**percent error = 0.134%**     Rounded.

## 4 Determine an Appropriate Approximation of Measurement Calculations.

It is often necessary to round the results of calculations. In some instances specific industry standards are appropriate. In other cases, specific directions will be given. However, in many instances, you will determine the appropriate rounding place. When adding or subtracting measurements, we examine the precision of the measurements being added or subtracted.

### To round the sum or difference of measurements of different precisions:

1. If necessary, change all measurements to a common unit of measure.
2. Add or subtract.
3. Round the result to have the same precision (place value) as the *least precise* measurement.

---

**EXAMPLE**     Add the measurements: 12.5 m, 38 cm, 2.9 m, 43.25 cm.

| | |
|---|---|
| 12.5 m | Change all measurements to meters. |
| 38 cm = $38 \times 0.01$ m = 0.38 m | |
| 2.9 m | |
| 43.25 cm = $43.25 \times 0.01$ m = 0.4325 m | Add the like measurements. |

$$
\begin{array}{r}
12.5 \\
0.38 \\
2.9 \\
+ \quad 0.4325 \\
\hline
16.2125
\end{array}
$$

The least precise measurement is tenths.

**16.2125 m rounds to 16.2 m.**

---

When multiplying or dividing measurements, the product or quotient can be no more accurate than the *least accurate* measurement.

### To round the product or quotient of measurements of different accuracy:

1. Multiply or divide the measurements.
2. Round the product or quotient to have the same number of significant digits as the *least accurate* measurement (fewest significant digits).

EXAMPLE    Multiply the measurements: (150 m)(105 m)(50 m).

$$(150 \text{ m})(105 \text{ m})(50 \text{ m}) = 787{,}500 \text{ m}^3 \qquad (m)(m)(m) = m^3$$

50 m is the least accurate measurement, with one significant digit.

**787,500 m³ rounds to 800,000 m³.**

## SECTION 3–4 SELF-STUDY EXERCISES

**1**  Indicate the number of significant digits in each number.

**1.** 583,000        **2.** 702,500        **3.** 0.0057        **4.** 82.07        **5.** 7.200
**6.** 860            **7.** 5,080          **8.** 2.0091        **9.** 572,080        **10.** 1,010

**2**  Find the greatest possible error of each measurement.

**11.** $7\frac{3}{16}$ in.        **12.** 7.2 mm        **13.** $5\frac{1}{4}$ in.        **14.** 19 oz

**15.** 8 L        **16.** $5\frac{1}{8}$ in.        **17.** $12\frac{1}{32}$ in.        **18.** $1\frac{5}{8}$ oz

**19.** 6.2 mi        **20.** 7.9 cm        **21.** 8.17 dg        **22.** 5.3 cg
**23.** 6.7 cL        **24.** 1.4 km        **25.** 12.7 km        **26.** 125.3 km

**3**

**27.** A measurement is observed to be 52.02 cm, against a true measure of 52 cm. Find the absolute error, relative error, and percent error.

**28. INDTR**  A blueprint for a machine part specifies a measure of 52.2 mm. An actual part has a measurement of 52.4 mm. Find the absolute error, relative error, and percent error.

**29. CAD/ARC**  The blueprint for a part specifies a measurement of 48.7 cm. An actual part has a measurement of 50.2 cm. Find the absolute error, relative error, and percent error.

**30. AUTO**  The blueprint for a racing steering wheel specifies a measurement of 14.25 in. An actual wheel has a measurement of 14.54 in. Find the absolute error, relative error, and percent error.

**31. INDTEC**  A blueprint for the gasoline tank of an industrial lawn mower specifies that it holds 50.05 L of gasoline. Upon close measurement, an actual tank is found to hold 50.25 L of gasoline. Find the absolute error, relative error, and percent error.

**32. AUTO**  An actual gear is measured to be 15.4 in. in diameter. The blueprint specifies that the diameter of the part be 15 in. Find the absolute error, relative error, and percent error.

**4**  Add. Give the sum using the appropriate precision.

**33.** 12.84 m, 13.5 m, 182.3 m, 6.732 m
**35.** 14.3 cm, 0.7 m, 4.9 m, 23.45 cm

**34.** 0.93 g, 1.8 g, 42.7 g, 18.932 g
**36.** 12.78 g, 0.23 kg, 4.752 kg, 0.3466 kg

Give the product or quotient using the appropriate number of significant digits.

**37.** (202 m)(45 m)(100 m)
**39.** 1,500 g ÷ 35 g

**38.** (125 m)(403 m)(340 m)
**40.** 1,968 km ÷ 30 km

---

*Learning Outcomes*

**1** Read a metric rule.

**2** Read a slide caliper.

**3** Read a micrometer.

**4** Read circular, uniform, and nonuniform scales.

Many measuring instruments are available for making measurements of all types. On some measuring instruments we read the measurement directly from a scale, whereas on other measuring instruments that have electronic sensors we read the measurement on a digital display. Whether the measurement is made manually or electronically, all measurements are *approximate* values.

### **1** Read a Metric Rule.

Many standard rulers have both a U.S. customary scale and a metric scale. The metric rule is usually calibrated in centimeters or millimeters. The **metric rule** illustrated in Figure 3–27 shows centimeters (cm) as the major divisions, represented by the longest lines. Each centimeter is divided into 10 mm, which are the shortest lines. A line slightly longer than the millimeter line divides each centimeter into two equal parts of 5 mm each.

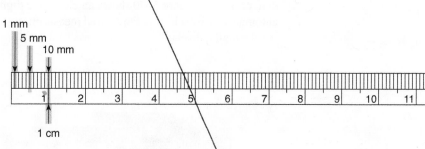

**Figure 3–27** The Metric Rule.

The metric rule is read like the U.S. customary rule with the exception that only two measures are indicated, centimeters and millimeters. Other measures are calculated in relation to millimeters or centimeters.

Since 1 mm is $\frac{1}{10}$ cm, we can write metric measures in decimals. For example, a measure of 3 cm and 4 mm is written as either 3.4 cm or 34 mm.

**EXAMPLE** Find the length of line segment *AB* (Figure 3–28) to the nearest millimeter.

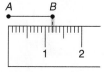

The line segment *AB* extends two marks past 1 cm.

**Figure 3–28**

**Line segment *AB* is 12 mm or 1.2 cm to the nearest millimeter.**

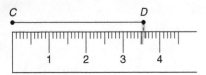

**EXAMPLE**    Find the length of the line segment *CD* (Figure 3–29).

C                                D

```
|‖‖‖‖‖‖‖‖‖‖‖‖‖‖‖‖‖‖‖‖‖‖‖‖‖‖‖‖‖‖‖‖‖‖‖|
      1       2       3       4
```

**Figure 3–29**

The end of line segment *CD* falls approximately halfway between the 35-mm mark and the 36-mm mark, by eye judgment, measuring about 35.5 mm. Both 35 and 36 mm would be acceptable measures of the line segment *CD*.

**35 mm, 35.5 mm, and 36 mm are acceptable approximations for line segment *CD*.**

### 2  Read a Slide Caliper.

The U.S. customary and metric rules are used to make measurements of items that do not require a high degree of precision. The **slide caliper** is just one of several measuring instruments used to make precise measurements.

One of the oldest precision measuring tools that measures by "feel" is a caliper. To find the measure of an object using a slide caliper, we may make the reading from a digital display, a dial, or a scale. Figure 3–30 shows an electronic digital caliper. Measurements are calculated automatically by selecting the desired measuring system (U.S. customary or metric) and reading the digital display.

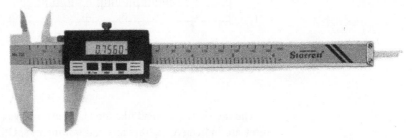

**Figure 3–30**    Digital caliper.

A dial caliper is specifically purchased for making measurements in the U.S. customary or metric system (Figure 3–31). The procedures for reading dials are given in Learning Outcome 8.

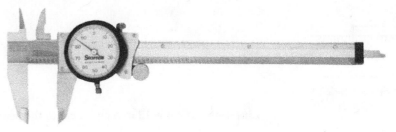

**Figure 3–31**    Dial caliper.

A slide caliper that requires the user to read the measure from a scale is available with scales of different graduations. Scales graduated in 32nds of an inch, 64ths of an inch, or metric units, are the most common (Figure 3–32).

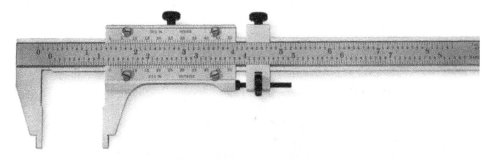

**Figure 3–32** Slide caliper.

For a higher degree of precision the vernier caliper is used. A **vernier caliper** has an additional graduated scale that allows reading to the nearest thousandth of an inch in the U.S. customary system or to the nearest hundredth of a millimeter in the metric system. To get a good reading with any precision tool involves skill and experience. Figure 3–33 shows a vernier caliper with both U.S. customary and metric scales.

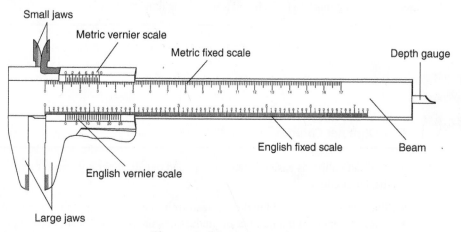

**Figure 3–33** Vernier caliper.

The vernier caliper can be used to take inside, outside, or depth measurements. Outside measurements are made by closing the large jaws of the tool around the outside of the object. To measure the inside of an object, place the small jaws inside the object. The depth of an object is measured by inserting the depth gauge into the object.

One type of vernier caliper has two metric scales and two U.S. customary scales. Figure 3–34 shows the two metric scales on the *upper part* of the beam. The fixed scale is graduated in centimeters and subdivided into millimeters. Measurements taken with the metric scales are recorded in millimeters (mm). Each numbered graduation on the fixed scale represents 10 mm, and the unnumbered graduations represent 1 mm. The other metric scale, called the **vernier scale,** has 11 long marks (0, 1, 2, . . . , 9, 10) graduated in tenths of millimeters (0.10 mm) that are subdivided into halves of tenths $\left(\frac{1}{2} \times \frac{1}{10} = \frac{1}{20}\right)$, or five-hundredths (0.05). The precision of the metric vernier scale is 0.05 mm.

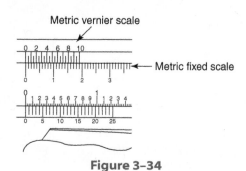

**Figure 3–34**

## To read a vernier caliper metric scale:

1. On the fixed metric scale (top part of the fixed beam) determine the numbered graduation that is the tens place of the measurement to the left of the zero graduation on the vernier metric scale. This number is the tens place of the measurement.

2. Determine the number of unmarked graduations between the numbered graduation on the fixed scale and the zero (0) mark on the vernier scale. This is the ones place of the measurement.

3. If the zero graduation on the vernier metric scale is aligned directly above a line on the fixed scale, this indicates there are 0.00 (zero hundredths), so read the measurement from the fixed scale. Write the measurement as the whole number read from the fixed scale followed by .00 (zero hundredths).

3a. If the zero graduation on the vernier scale is not directly aligned with a graduation on the fixed scale, record the number of whole millimeters in the measurement; then find the graduation on the vernier scale that most nearly aligns with *any* graduation on the fixed scale. If a long mark represents the best alignment, record the number of tenths of millimeters and a zero in the hundredths place.

3b. If a short mark on the vernier scale represents the best alignment, record the number of tenths of millimeters represented by the long mark to the left and a five in the hundredths place.

**TIP!**

### One Digit per Column

The vernier caliper is easier to read if you read one digit per column for each of the four places—tens, ones, tenths, hundredths.

- Tens place (or tens and hundreds): read from the numbered graduation on the fixed beam.
- Ones place: count the number of graduations on the fixed beam between the numbered graduation and zero on the vernier scale.
- Tenths place: read from the numbered graduation on the vernier scale that most closely aligns with *any* graduation on the fixed scale.
- Hundredths place: read as zero if the two beams align on a long graduation of the vernier scale, and read as 5 if the two beams align on a short graduation of the vernier scale.

**EXAMPLE**  Read the measurement in millimeters (to the nearest hundredth) on the vernier caliper in Figure 3–35.

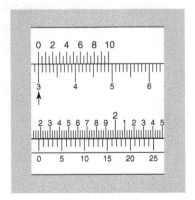

**Figure 3–35**

The zero graduation on the vernier scale is the graduation that most nearly aligns with a graduation on the fixed beam, and it aligns with the long mark numbered 3.
**The measurement is 30.00 mm.**

EXAMPLE   Read the measurement in millimeters (to the nearest hundredth) on the vernier caliper in Figure 3–36.

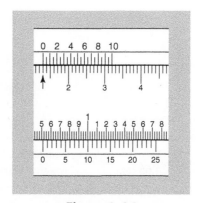

**Figure 3–36**

The zero graduation on the vernier scale is between 1 and 2 on the fixed beam and is the graduation that most nearly aligns with a graduation on the fixed beam. It aligns with the short mark 3 units to the right of the graduation marked 1.  **The measurement is 13.00 mm.**

EXAMPLE   Read the measurement in millimeters (to the nearest hundredth) on the vernier caliper in Figure 3–37.

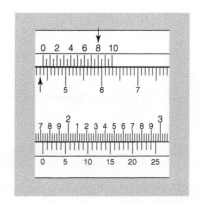

**Figure 3–37**

| | |
|---|---|
| 4__.__ __ mm | The zero graduation on the vernier scale is between the numbers 4 and 5 on the fixed beam. The number 4 is not visible. |
| 43.__ __ mm | The first graduation to the left of the zero is 3 marks to the right of 4. |
| 43.8__ mm | The graduation on the vernier scale that best aligns with a mark on the fixed beam is 8. |
| 43.80 mm | Since 8 is a long mark, the hundredths place is zero. |

**The measurement is 43.80 mm.**

**EXAMPLE**   Read the measurement in millimeters (to the nearest 0.05 mm) on the vernier caliper in Figure 3–38.

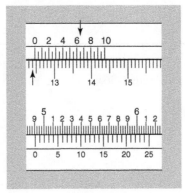

**Figure 3–38**

| 124.__ __ mm | The zero on the vernier scale is between 12 and 13 on the fixed beam and is 4 graduations to the right of 12. |
| 124.65 mm | The short mark to the right of the 6 on the vernier scale aligns best with a mark on the fixed scale. |

**The measurement is 124.65 mm.**

The U.S. customary scales on the vernier caliper are located at the bottom of the scale (Figure 3–39). The fixed scale is on the *lower part* of the fixed beam and is divided into whole inches and tenths (0.100 in., 0.200 in., . . . ) of an inch. These graduations are numbered. Each tenth is subdivided into 4 parts, so each part represents $\frac{1}{4}$ of a tenth-inch ($\frac{1}{4} \times \frac{1}{10} = \frac{1}{40}$ or 0.025 in.). The vernier scale (movable scale at the bottom) is divided into 25 parts, and each part represents one thousandth of an inch. The precision of the U.S. customary scale is 0.001 in.

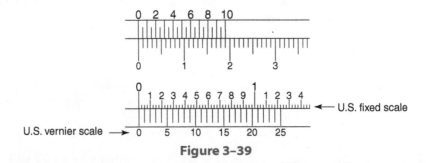

**Figure 3–39**

### To read a vernier caliper U.S. customary scale:

1. Determine the number of whole inches by reading the number on the longest mark to the immediate left of the zero graduation on the vernier scale.
2. Read the number of tenths of inches by reading the number on the short mark that is immediately to the left of the zero mark on the vernier scale. Read each unnumbered graduation as 0.025 in.
3. Read each mark on the vernier scale as one-thousandth of an inch (0.001 in.) and add its value to the thousandths read from the fixed beam.
4a. If the zero graduation on the vernier scale directly aligns with a graduation on the fixed scale, read the measurement from the fixed scale and attach zeros in the hundredths and thousandths places to show the precision level.
4b. If the zero graduation on the vernier scale does not align with a graduation on the fixed scale, find the graduation on the vernier scale that most nearly aligns with *any* graduation on the fixed scale. Add that value in thousandths of an inch to the measurement.

**EXAMPLE**   Read the measurement in inches shown on the vernier caliper in Figure 3–40.

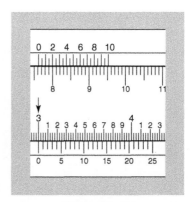

**Figure 3–40**

Zero aligns best with the mark numbered 3, **so the measurement is 3.000 in.**

**EXAMPLE**   Read the measurement in inches shown on the vernier caliper in Figure 3–41.

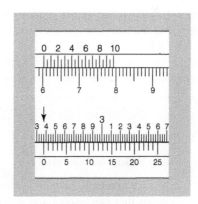

**Figure 3–41**

| | | |
|---|---|---|
| 2.    in. | The first longest mark to the left of zero is numbered 2 (not shown). |
| 0.3   in. | The shorter numbered mark to the left is 3. |
| 0.075 in. | The number of 0.025-in. graduations is 3, so add 3 × 0.025 in. = 0.075 in. |
| 0.000 in. | Zero on the vernier scale aligns directly with a mark on the fixed beam, so read 0.000 in. from the vernier scale. |
| ————— | |
| 2.375 in. | *(Note the vertical alignment of decimal points.)* |

**The measurement is 2.375 in.**

**EXAMPLE**   Read the measurement in inches shown on the vernier caliper in Figure 3–42.

**Figure 3–42**

| 1.2 in. | Read 1.2 in. from the numbered graduations on the fixed beam. |
|---|---|
| 0.050 in. | There are 2 unnumbered graduations to the left of zero, so add 2 × 0.025 in. = 0.050 in. |
| 0.013 in. | The 13 mark on the vernier scale best aligns with a mark on the fixed scale, so add 0.013 in. |
| 1.263 in. | *(Note the vertical alignment of decimal points.)* |

**The measurement is 1.263 in.**

TIP!

### Proper Use of a Vernier Caliper

- Hold the caliper so that it is perpendicular (at right angles) to the object being measured.
- When measuring the diameter of a round object, be sure to position the caliper so the greatest distance across the piece is measured.
- When measuring round parts or to ensure uniform thickness, take two or more measurements; then calculate and use their average.

### 3  Read a Micrometer.

A **micrometer** is a type of caliper used in technical fields that require more precise measurements than can be obtained with a vernier caliper. Some industries use computerized micrometers that provide measurements as digital readouts.

A micrometer is designed to make measurements in *either* metric or U.S. customary units of measure. The metric micrometer is graduated to provide readings to the nearest hundredth of a millimeter (0.01 mm). The U.S. customary "mike" provides readings to the nearest thousandth of an inch (0.001 in.).

Figure 3–43 illustrates the parts of an outside micrometer.

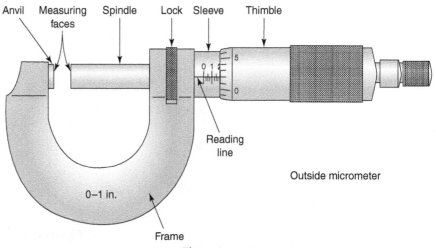

Figure 3-43

The frame of the micrometer is the curved rigid portion. Micrometers come in various sizes, and the size of the micrometer is generally identified on the frame. The anvil is located on the upper end of the frame. The spindle is the movable part that is adjusted to the size of the object being measured. The object being measured fits snugly between the anvil and the spindle.

The sleeve is the graduated portion of the micrometer that is attached to the frame. It has 40 vertical graduation marks, with every fourth mark labeled with a number from 1 to 10. Each labeled graduation represents one-tenth of an inch (0.1 in.). Each vertical mark represents 0.025 in. Thus, the three marks between each labeled vertical mark represent 0.025, 0.050, and 0.075 in., respectively.

The thimble of the micrometer is graduated into 25 parts, each representing one-thousandth of an inch (0.001 in.). These marks are labeled every five marks for ease in reading.

The micrometer is placed so the object to be measured is between the anvil and the spindle. The thimble is turned slowly until it fits snugly on the object. *The threads of the spindle located inside the thimble are very delicate and can be damaged if the thimble is forced.* Some micrometers use a ratchet design that prevents the thimble from being turned with damaging force. Some micrometers have a lock that keeps the micrometer from slipping while the measurement is read.

### To read an outside micrometer:

1. Place the object to be measured snugly between the anvil and the spindle.
2. Determine the inch portion of the measure by reading the smallest number given in the range of the micrometer. This range is found on the frame.
3. Read the tenth portion of the measurement by reading the last number that is showing on the sleeve of the micrometer.
4. Determine *part* of the thousandths measure by reading the number of marks that are showing to the right of the tenths number. Each mark represents 0.025.
5. Read *additional* thousandths from the thimble of the micrometer. The reading is made where the horizontal line on the sleeve meets the thimble. Give the reading to the closest mark.
6. Add the readings in 2 through 5. (Steps 2 and 3 represent whole numbers and tenths, respectively. Steps 4 and 5 represent thousandths and will fall in the same columns when the numbers are added.)

EXAMPLE    Determine the reading on the micrometer in Figure 3–44.

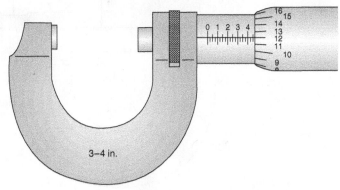

**Figure 3–44**

| | |
|---|---|
| 3. | The inch measurement is given to be 3 in. |
| 0.4 | The tenths measurement is 0.4 in. |
| 0.050 | The sleeve reading is two marks (0.050) past the 4. |
| <u>0.012</u> | The thimble reading is 12 (0.012). |
| 3.462 in. | *(Note the vertical alignment of decimal points.)* |

**The measurement is 3.462 in.**

A **vernier micrometer** has a vernier scale that allows readings to the nearest ten-thousandth of an inch (0.0001 in.). The graduations of the vernier scale are located on the top of the sleeve or barrel of the micrometer, and they run lengthwise on the barrel. There are 10 spaces indicated on the vernier scale, and each space represents one-tenth of a space on the thimble, or 0.0001 in. (One-tenth of one thousandth is one ten-thousandth.) Locate the vernier scale on Figure 3–45.

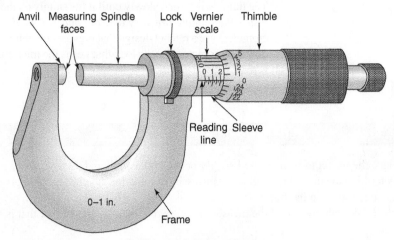

Outside vernier micrometer

**Figure 3–45**

**To read an outside vernier micrometer:**

1. Follow the first four steps for reading an outside micrometer.
2. When determining the thimble reading, use the smaller number that the measure falls between.
3. Locate the number on the vernier scale that aligns with *any* mark on the thimble. This indicates the number of ten-thousandths.
4. Add the readings in each step.

**EXAMPLE**    Determine the reading on the 2- to 3-in. vernier micrometer in Figure 3–46.

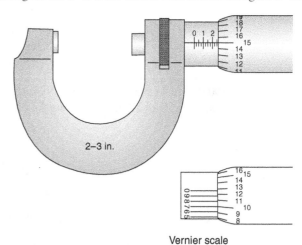

Vernier scale

**Figure 3–46**

| 2. | The inch measurement is given to be 2 in. |
|---|---|
| 0.2 | The tenth reading is 0.2 in. |
| 0.050 | The sleeve reading is two marks past the 2; thus, it is 0.050 in. |
| 0.015 | The thimble reading is 15; thus, it is 0.015. |
| <u>0.0007</u> | On the vernier scale the number 7 aligns with the 10 on the thimble; thus, it is 0.0007. |
| 2.2657 in. | *(Note the vertical alignment of decimal points.)* |

**The measurement is 2.2657 in.**

Several measuring instruments have scales similar to those on the outside micrometer and the outside vernier micrometer. Two are shown in Figure 3–47. The basic principles for reading these instruments are similar to those for reading the micrometer. Figure 3–47 illustrates a depth micrometer and an inside micrometer.

Depth micrometer

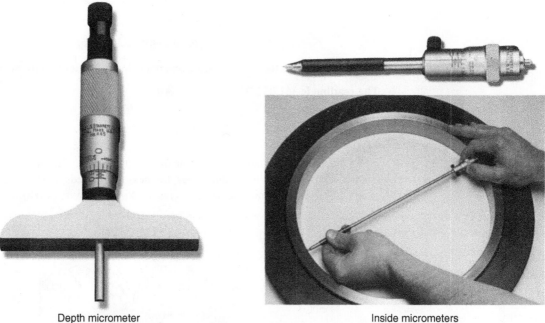

Depth micrometer                    Inside micrometers

**Figure 3–47**    Other types of micrometers.

3-5  Reading Measuring Instruments

#### 4 Read Circular, Uniform, and Nonuniform Scales.

**Circular scales** are prevalent in our environment and workplace. Whether reading a water or gas meter dial or a dial used to test the depth of cut on a flat metal bar, the first thing we need to determine is the scale being used and the basic unit for that scale. For example, a water meter may be composed of six circular dials such as the one in Figure 3–48. Notice that each dial has ten graduations on its scale and represents a power of ten. To read the dial, start with the 100,000 scale and continue clockwise to the 1-ft³ scale. If a dial indicator is between digits, read the smaller digit. The reading for the water meter in Figure 3–48 is 381,835 ft³.

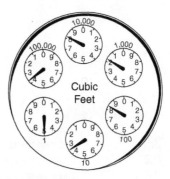

**Figure 3–48**   Water meter.

EXAMPLE   Read the water meter shown in Figure 3–49.

**Figure 3–49**   Water meter.

Begin reading with the 100,000 scale and read clockwise through the scale marked 1.

**The reading is 341,052 ft³.**

An electric meter is another example of a circular scale. The electric meter in Figure 3–50 has four dials; the first reads thousands, the next is hundreds, the third is tens, and the fourth is units (or ones). The measurement is given in kilowatt hours (kWh).

**Figure 3–50**   Electric meter.

**EXAMPLE**   Read the electric meter shown in Figure 3–51.

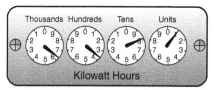

Read and record the thousands dial. 6
Read and record the hundreds dial.  3
Read and record the tens dial.       8
Read and record the ones dial.       1

**Figure 3–51**   Electric meter.

**The meter reads 6,381 kWh.**

**Figure 3–52**

Some dial gauges have a large circular scale and a smaller circular scale on the inside face. Such a metric gauge is shown in Figure 3–52. Each graduation of the large dial represents 0.01 mm. Thus, when the needle registers at or below 35 on the right of zero, the object being measured is 35 × 0.01 mm. The small needle records the number of complete revolutions made by the large needle when making the measurement. A complete revolution corresponds to 1.00 mm. The small needle is usually read first, followed by the large needle.

**EXAMPLE**   Read the metric dial in Figure 3–52.

The small needle indicates +5:        $5 \times 1.00$ mm = + 5.00 mm
The large needle indicates +21:       $21 \times 0.01$ mm = + 0.21 mm
Total reading:                                        **+5.21 mm**

**Figure 3–53**

A U.S. customary dial is read much like the metric dial. Examine the U.S. customary dial in Figure 3–53. Each complete revolution of the large needle corresponds to 0.100 in. and the small needle on the left indicates the number of complete revolutions made by the large needle. Each graduation on the large scale represents 0.001 in.

**EXAMPLE**   Read the U.S. customary dial in Figure 3–53.

The small needle reads 5:             $-5 \times 0.100$ in. = $-0.500$ in.
The large needle reads −6:            $-6 \times 0.001$ in. = $-0.006$ in.
Total reading:                                      $-0.506$ in.

**The reading is −0.506 in.**

The volt-ohm meter (VOM) is an example of a uniform scale. This instrument is used in electrical circuits to measure voltage in volts (V) and resistance in ohms ($\Omega$). The voltage scales are uniform, but the resistance scale is nonuniform. On a **uniform scale** all the graduations are equally spaced, and each subdivision represents the same number of units.

3-5  Reading Measuring Instruments

179

Read the voltage scale shown in Figure 3–54.

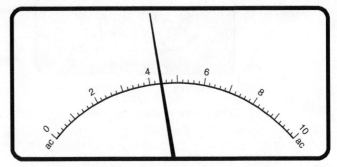

**Figure 3–54** Volt-ohm meter (VOM) scales.

Each long unnumbered mark represents an odd whole number of volts. Each short un-numbered mark represents 0.2 V. The needle is on the second graduation to the right of 4 ($2 \times 0.2 = 0.4$).

**The reading is 4.4 V.**

An ohmmeter is an example of a **nonuniform scale**; that is, the spacing between numbered graduations is not the same. In Figure 3–55, each unit between 0 and 5 has 5 subdivisions. Therefore, each subdivision represents $\frac{1}{5} \times 1\ \Omega$, or $0.2\ \Omega$. Between 5 and 10, each large division is subdivided into 2 parts, so each subdivision represents $\frac{1}{2} \times 1\ \Omega$, or $0.5\ \Omega$. Between 10 and 20, each division represents $1\ \Omega$. The values of the subdivisions for the other divisions of the scale are determined in a similar manner. The ranges and value of each subdivision are shown in the table.

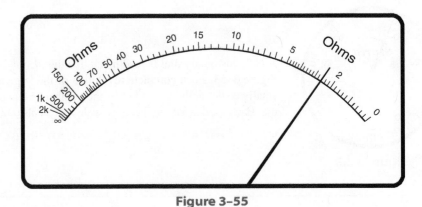

**Figure 3–55**

| Range | Value of each subdivision |
| --- | --- |
| 0–5 Ω | 0.2 Ω |
| 5–10 Ω | 0.5 Ω |
| 10–20 Ω | 1 Ω |
| 20–30 Ω | 2 Ω |
| 30–100 Ω | 5 Ω |
| 100–150 Ω | 10 Ω |
| 200–500 Ω | 100 Ω |

**EXAMPLE** Read the scale shown in Figure 3–55.

The needle is on the third subdivision to the left of 2. Each subdivision in that range represents 0.2 Ω.

**The scale reads 2.6 Ω.**

**Thickness gauges** come in sets of various thicknesses of metal, with the measurements indicated on the gauge. These sets of gauges are available in different combinations of sizes and numbers and to varying degrees of precision. The gauges can be produced to allow measurements to the nearest two-millionth of an inch (0.000002 in.). Figure 3–56 illustrates two different sets of thickness gauges. These gauges are used in different combinations to determine the appropriate measurement. The sum of the measures of the gauges used represents the measurement of the object.

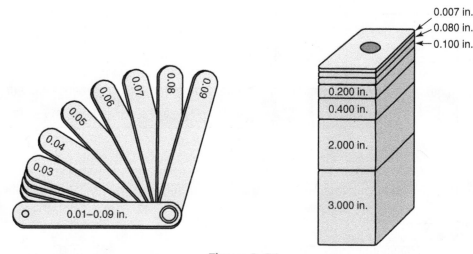

**Figure 3–56**

Suppose that a set of thickness gauges comes in series of ten-thousandths (0.0001 to 0.0009), thousandths (0.001 to 0.009), hundredths (0.01 to 0.09), and tenths (0.1 to 0.9). A measurement of 0.2493 in. can be made by selecting the 0.2, 0.04, 0.009, and 0.0003 gauges.

To measure an object that is less than 1 in., select the largest tenth-inch gauge that is less than the object. Then combine the tenth-inch gauge with the largest gauge from the hundredth-inch series that is less than the object. Continue with the largest appropriate thousandth-inch and ten-thousandth-inch gauges. To determine the measure of the object, add the measures of each of the gauges used.

**EXAMPLE** The following gauges were used to measure an object to the nearest hundred-thousandth of an inch: 0.2, 0.04, 0.006, 0.0001, and 0.00008. What is the measure of the object?

To find the measure, add the measures of the gauges used.

$$
\begin{array}{r}
0.2 \\
0.04 \\
0.006 \\
0.0001 \\
\underline{0.00008} \\
\mathbf{0.24618 \text{ in.}}
\end{array}
$$

**1** Measure line segments 1–10 in Figure 3–57 to the nearest millimeter. Express answers in millimeters or centimeters.

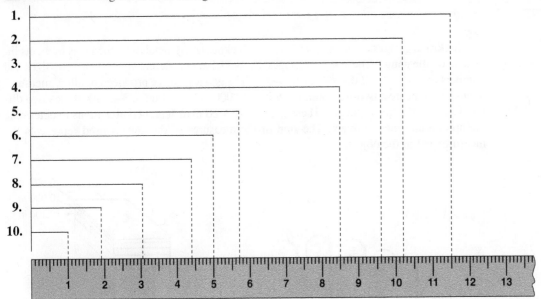

1.
2.
3.
4.
5.
6.
7.
8.
9.
10.

**Figure 3–57**

**2** Read the measurement shown on the vernier calipers in Figures 3–58 to 3–61 in millimeters and in inches.

11.

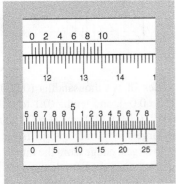

**Figure 3–58**

12.

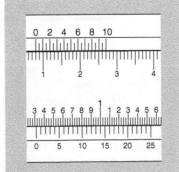

**Figure 3–59**

13.

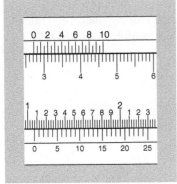

**Figure 3–60**

14.

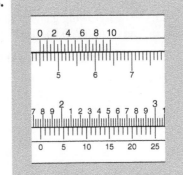

**Figure 3–61**

**3** Determine the measurements indicated on the micrometer readings in Figures 3–62 to 3–66.

15.

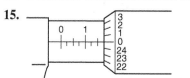

**Figure 3–62**  Frame size 0–1 in.

16.

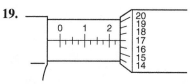

**Figure 3–63**  Frame size 4–5 in.

17.

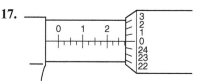

**Figure 3–64**  Frame size 2–3 in.

18.

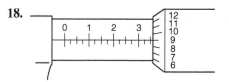

**Figure 3–65**  Frame size 4–5 in.

19.

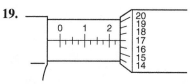

**Figure 3–66**  Frame size 1–2 in.

Determine the measurements indicated on the vernier micrometers in Figs. 3–67 to 3–71.

20.

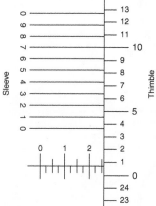

**Figure 3–67**  Frame size 0–1 in.

21.

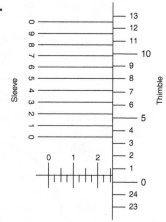

**Figure 3–68**  Frame size 2–3 in.

22.

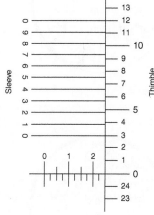

**Figure 3–69**  Frame size 1–2 in.

23.

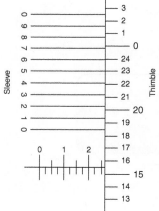

**Figure 3–70**  Frame size 4–5 in.

24.

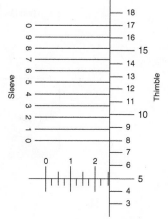

**Figure 3–71**  Frame size 2–3 in.

**4** Write the readings for the scales in Figures 3–72 to 3–77.

Read each water meter:

**25.**

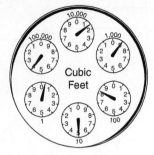

**Figure 3–72** December 19.

**26.**

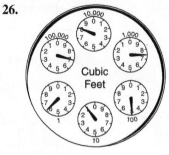

**Figure 3–73** August 12.

**27.**

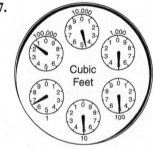

**Figure 3–74** June 13.

**28.**

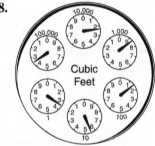

**Figure 3–75** October 18.

**29.**

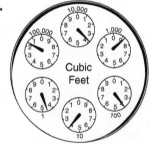

**Figure 3–76** November 5.

**30.**

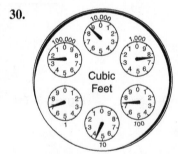

**Figure 3–77** January 3.

Read each electric meter in Figures 3–78 and 3–79.

**31.**

**Figure 3–78**

**32.**

**Figure 3–79**

Read each metric dial in Figures 3–80 to 3–83.

**33.**

Figure 3–80

**34.**

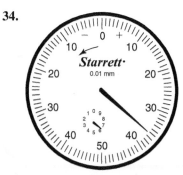

Figure 3–81

**35.**

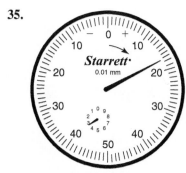

Figure 3–82

**36.**

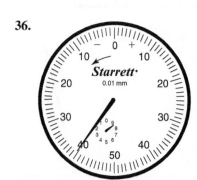

Figure 3–83

Read each voltmeter or ohmmeter in Figures 3–84 to 3–87.

**37.**

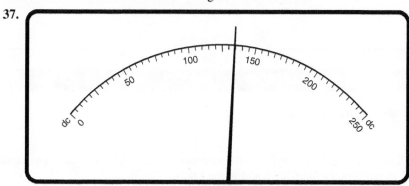

Figure 3–84

**38.**

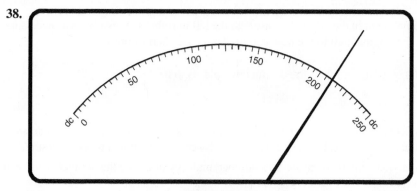

Figure 3–85

**39.**

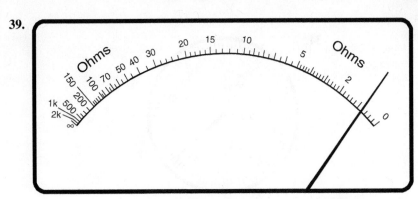

**Figure 3–86**

**40.**

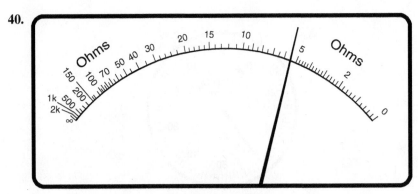

**Figure 3–87**

**41.** Find the measure of an object that is flush (even in height) with the following thickness gauges:

       0.4 in.     0.003 in.     0.000002 in.

**42.** What thickness gauges would be needed to make a measure of 1.03456 in.?

---

## CHAPTER REVIEW OF KEY CONCEPTS

| Learning Outcomes | What to Remember with Examples |
|---|---|

### Section 3–1

**1** Identify uses of metric measures of length, mass, weight, and capacity (pp. 143–147).

Associate small, medium, and large objects with appropriate measuring units of comparable size. Powers-of-10 prefixes are used, such as *kilo* for 1,000.

> Perfume, mL; soda, L; travel, km; racetrack, m; vitamin, mg; potatoes, kg; eye drops, mL

**2** Convert from one metric unit of measure to another (pp. 147–151).

Move the decimal point in the original measure to the left or right as many places as necessary to move from the original unit to the new unit on the metric chart of prefixes.

> Change 5.04 cL to liters. Move the decimal two places to the left.
>
> 5.04 cL = 0.0504 L

**3** Make calculations with metric measures (pp. 151–153).

To add or subtract metric measures: **1.** Change measures to measures with like units if necessary. **2.** Add or subtract the numerical values. **3.** Give the answer in the common unit of measure.

To multiply or divide a metric measure by a number, multiply or divide the numbers and keep the same unit.

To divide a metric measure by a measure: **1.** Change measures to measures with like units if necessary. **2.** Divide the numbers, canceling the units of measure. The answer will be a number.

$$7 \text{ km} + 34 \text{ m} = \qquad\qquad\qquad 7 \text{ mg} \times 3 = 21 \text{ mg}$$
$$7{,}000 \text{ m} + 34 \text{ m} = 7{,}034 \text{ m} \qquad 5 \text{ m} \div 25 \text{ cm}$$

or $\qquad\qquad\qquad\qquad\qquad\qquad\qquad$ or

$$7 \text{ km} + 0.034 \text{ km} = 7.034 \text{ km} \qquad \frac{500 \text{ cm}}{25 \text{ cm}} = 20$$

## Section 3–2

**1** Convert from one unit of time to another (p. 155).

Use unity ratios or conversion factors to convert one unit of time to another.

How many days are in 3,420 hours?

$$1 \text{ day} = 24 \text{ h}$$

$$3{,}420 \text{ h} \times \frac{1 \text{ day}}{24 \text{ h}} = 142.5 \text{ days}$$

**2** Make calculations with measures of time (p. 156).

Measures of time that are the same unit of measure can be added or subtracted by adding or subtracting the quantities and keeping the same unit of measure. A time measure can be multiplied or divided by a number by multiplying or dividing the measure by the number and keeping the same measuring unit.

A printer prints 1 page in 30 s. How long does it take to print 45 pages?

$1 \text{ min} = 60 \text{ s}; 1 \text{ h} = 60 \text{ min}$

$30 \text{ s} \times 45 = 1{,}350 \text{ s}$ $\qquad\qquad$ Seconds to complete job.

$1{,}350 \text{ s} \times \dfrac{1 \text{ min}}{60 \text{ s}} = 22.5 \text{ min}$ $\qquad$ Minutes to complete job.

$22.5 \text{ min} \times \dfrac{1 \text{ h}}{60 \text{ min}} = 0.375 \text{ h}$ $\qquad$ Portion of hour to complete job.

The printer can print 45 pages in 22.5 min or 0.375 h.

**3** Convert a Fahrenheit temperature to a Celsius temperature (pp. 156–157).

To change degrees Fahrenheit to degrees Celsius use the formula $°C = \dfrac{5}{9}(°F - 32)$

Change 50°F to °C.

$°C = \dfrac{5}{9}(°F - 32)$ $\qquad\qquad$ Substitute 50 for °F.

$°C = \dfrac{5}{9}(50 - 32)$ $\qquad\qquad$ Perform operation in parentheses.

$°C = \dfrac{5}{9}(18)$ $\qquad\qquad\qquad$ Multiply.

$°C = 10$

**4** Convert a Celsius temperature to a Fahrenheit temperature (pp. 157–158).

To change degrees Celsius to degrees Fahrenheit use the formula $°F = \dfrac{9}{5}°C + 32$

Change 28°C to °F.

$°F = \dfrac{9}{5}°C + 32$ $\qquad\qquad$ Substitute 28 for °C.

$°F = \dfrac{9}{5}(28) + 32$ $\qquad\qquad$ Multiply.

$$°F = \frac{252}{5} + 32 \qquad \text{Divide.}$$

$$°F = 50.4 + 32 \qquad \text{Add.}$$

$$°F = 82.4$$

**5** Examine other useful measures (pp. 158–159).

All SI metric measures use the same system of prefixes for multiples and submultiples. Table 3–1 on page 158 summarizes the SI base units and some other commonly used metric units.

What is the meaning of 3 μA?

$$1\ \mu A = 1\ \text{millionth of an ampere, or}\ 0.000001\ A,\ \text{or}\ 10^{-6}\ A$$
$$3\ \mu A = 3 \times 0.000001 = \textbf{0.000003 A, or } \mathbf{3 \times 10^{-6}\ A}$$

Using unity ratios:

$$3\ \mu A \times \frac{0.000001\ A}{1\ \mu A} = 0.000003\ A$$

## Section 3–3

**1** Convert between U.S. customary measures and metric measures (pp. 160–162).

**1.** Set up the original amount as a fraction with the original unit in the numerator and 1 in the denominator. **2.** Multiply by a unity ratio with the original unit in the denominator and the new unit in the numerator.

How many feet are in 14 m?

$$\frac{14\ \cancel{m}}{1} \times \frac{3.28\ \text{ft}}{1\ \cancel{m}} = 45.92\ \text{ft} \qquad \text{Use the conversion factor 1 m = 3.28 ft.}$$

## Section 3–4

**1** Determine the significant digits of a number (pp. 162–163).

For whole numbers: **1.** Start with the leftmost nonzero digit. **2.** Count each digit through the rightmost nonzero digit.

258 has three significant digits.
1,050 has three significant digits.

For decimal numbers: **1.** Start with the leftmost nonzero digit. **2.** Count each digit through the last digit.

0.023 has two significant digits.
1.50 has three significant digits.

**2** Find the precision and the greatest possible error of a measurement (pp. 163–164).

**1.** Determine the precision of the measurement of the smallest subdivision. **2.** Find half of the precision.

The precision of $2\frac{5}{8}$ in. is $\frac{1}{8}$ in.

$$\frac{1}{2} \times \frac{1}{8} = \frac{1}{16}\ \text{in.}$$

The greatest possible error is $\frac{1}{16}$ in.

**3** Determine the relative error and the percent error of a measurement (pp. 164–165).

To find the absolute error, find the absolute value of the difference between the observed measurement and the true value.

To find the relative error, find the quotient of the absolute error and the true value.

To find the percent error, change the relative error to a percent by multiplying by 1 in the form of 100%.

The actual measure of a widget is 15.963 cm and the blueprint specifies 15.94 cm for the widget.

$$\text{absolute error} = |15.963 - 15.94|$$

$$= 0.023 \text{ cm}$$

$$\text{relative error} = \frac{0.023}{15.94} = 0.001443$$

$$\text{percent error} = 0.001443 \times 100\% = 0.1443\%$$

**4** Determine an appropriate approximation of measurement calculations (pp. 165–166).

To round the sum or difference of measurements of different precisions: **1.** If necessary, change all measurements to a common unit of measure. **2.** Add or subtract. **3.** Round the result to have the same precision (place value) as the *least precise* measurement.

Add the measurements: 34.5 m, 15 cm, 4.6 mm.

Change all measurements to meters.

$$34.5 \text{ m} = 34.5 \text{ m}$$
$$15 \text{ cm} = 15 \times 0.01 \text{ m} = 0.15 \text{ m}$$
$$4.6 \text{ mm} = 4.6 \times 0.01 \text{ m} = 0.046 \text{ m}$$

Add the like measurements.

$$\begin{array}{r} 34.5 \\ 0.15 \\ \underline{0.046} \\ 34.696 \end{array}$$

The least precise measurement is tenths.

**34.696 m rounds to 34.7 m.**

To round the product or quotient of measurements of different accuracy: **1.** Multiply or divide the measurements. **2.** Round the product or quotient to have the same number of significant digits as the *least accurate* measurement (fewest significant digits).

Multiply the measurements: $(12.3 \text{ cm})(5 \text{ cm})(0.25 \text{ cm})$.

$$(12.3 \text{ cm})(5 \text{ cm})(0.25 \text{ cm}) = 15.375 \text{ cm}^3 \qquad \text{(cm)(cm)(cm)} = \text{cm}^3$$

5 cm is the least accurate measurement with one significant digit.

**15.375 cm³ rounds to 20 cm³.**

## Section 3–5

**1** Read a metric rule (pp. 167–168).

See Figure 3–88. Align the rule along the object. Count the number of millimeters and/or centimeters (10 mm = 1 cm) to determine the approximate length. Use eye judgment to estimate closeness of the object to a mark on the rule.

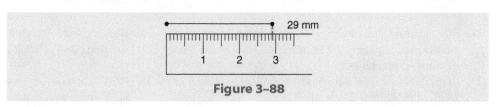

**Figure 3–88**

**2** Read a slide caliper (pp. 168–174).

A slide caliper is a measuring instrument for making precise measurements. Refer to page 170 for the procedure and illustrations for reading a vernier caliper metric scale. Refer to page 172 for the procedure and illustrations for reading the vernier caliper U.S. customary scale.

**3** Read a micrometer (pp. 174–177).

The micrometer is a type of caliper that allows more precise measurements than the vernier caliper. Refer to page 175 for the procedure and an illustration for reading an outside micrometer. Refer to page 176 for the procedure and an illustration for reading an outside vernier micrometer.

**4** Read circular, uniform, and nonuniform scales (pp. 178–181).

Circular scales are dials and an application may require taking a reading from one or more dials. Refer to pages 178–179 for illustrations of reading circular scales.

Uniform scales have graduations that are equally spaced and each subdivision represents the same number of units. Refer to page 180 for illustrations of uniform scales.

Nonuniform scales have graduations that are not equally spaced and subdivisions may represent the different number of units. Refer to page 180 for illustrations of nonuniform scales.

---

## CHAPTER REVIEW EXERCISES

**Section 3–1**

Give the prefix that relates each number to the standard unit in the metric system.

**1.** 1,000 times

**2.** $\frac{1}{10}$ of

**3.** $\frac{1}{1,000}$ of

**4.** 10 times

Give the value of the prefixes based on a standard measuring unit.

**5.** dekameter (dkm)
**8.** centigram (cg)

**6.** hectogram (hg)
**9.** kiloliter (kL)

**7.** milligram (mg)
**10.** deciliter (dL)

Choose the most reasonable answer.

**11.** Height of the Washington Monument
   **(a)** 200 m
   **(b)** 200 cm
   **(c)** 200 mm
   **(d)** 200 km

**12.** Height of Mt. Rushmore
   **(a)** 1.6 km
   **(b)** 1.6 m
   **(c)** 1.6 cm
   **(d)** 1.6 mm

**13.** Weight of a man's shoe
   **(a)** 0.25 g
   **(b)** 0.25 mg
   **(c)** 0.25 kg

Change to the unit indicated.

**14.** 0.4 dkm = _____ hm
**16.** 4 m = _____ dm
**18.** 5 cm = _____ mm
**20.** 42.7 cm = _____ dkm
**22.** 2.7 hg = _____ dg

**15.** 67.1 m = _____ dkm
**17.** 2.3 m = _____ mm
**19.** 0.123 hm = _____ mm
**21.** 41,327 dkm = _____ km
**23.** 3,000,974 cg = _____ kg

Perform the operations indicated.

**24.** 25 mm − 14 mm

**25.** 12 g + 5 m

**26.** 43 dkg × 7

**27.** 6.83 cg × 9

**28.** $\frac{18 \text{ cm}}{9}$

**29.** 34 hL ÷ 4

**30.** 2.4 m ÷ 5 cm

**31.** **BUS** Fabric must be purchased to make seven garments, each requiring 2.7 m of fabric. How much fabric must be purchased?

**32.** **BUS** Candy weighing 526 g is mixed with candy weighing 342 g. What is the weight of the mixture?

**33.** **HOSP** A recipe calls for 5 mL of vanilla flavoring and 24 cL of milk. How much liquid is this?

**34.** **BUS** 20 boxes, each weighing 42 kg, are to be moved. How much weight must be moved?

**35.** **CON** A metal rod 42 m long is cut into seven equal pieces. How long is each piece?

**36.** **HOSP** How many containers of jelly can be made from 8,548 L of jelly if each container holds 4 dL of jelly?

**37.** **BUS** A bolt of fabric contains 6.8 dkm. If a shirt requires 1.7 m, how many shirts can be made from the bolt?

## Section 3-2

Use unity ratios or conversion factors to convert the measures of time.

**38.** How many minutes are in 2.4 h?

**39.** How many days are in 72 h?

**40.** **HLTH/N** A doctor ordered a surgery patient not to drive for 3 weeks. How many days is this?

**41.** Convert 158 min to hours. Round to hundredths.

**42.** **HLTH/N** A bone marrow patient remained hospitalized for 72 days. How many weeks is this?

**43.** **HLTH/N** A heart surgery patient was held in intensive care for 96 hr. How many days is this?

**44.** There are 96 days remaining in a year. How many weeks remain?

**45.** **BUS** If you have worked for your employer for 39 months, how many years have you worked?

Make the temperature conversions.

**46.** 86°F = _____ °C

**47.** 95°C = _____ °F

**48.** The freezing point of benzene is 5°C. What is this temperature on the Fahrenheit scale?

**49.** **HLTH/N** The label on a container of the acid-reducer famotidine states that storage above 40°C should be avoided. What is 40°C on the Fahrenheit scale?

**50.** **AUTO** Summer road surface temperatures of 122°F give what reading on the Celsius scale?

**51.** **HOSP** Candy that should reach a cooking temperature of 365°F should reach what temperature using the Celsius scale?

**52.** **AVIA** A jet plane that travels at 250 m/s travels how many kilometers per second?

**53.** **HLTH/N** A hospital patient has a temperature reading of 37.9°C. What is the temperature in degrees Fahrenheit?

## Section 3-3

Make the conversions.

**54.** 7 m = _____ inches

**55.** 215 m = _____ yards

**56.** 69 km = _____ miles

**57.** 15 L = _____ liquid quarts

**58.** 12 qt = _____ liters

**59.** 32 kg = _____ pounds

**60.** 10 lb = _____ kilograms

**61.** 9 in. = _____ centimeters

**62.** 21 ft = _____ meters

**63.** **CON** How many meters long is 200 ft of pipe?

**64.** **CON** Concrete weighing 90 lb weighs how many kilograms?

**65.** **HLTH/N** Wanda Williams is 5 ft 9 in. and weighs 142 lb. What is her body mass index?

**66.** **CON** A room 10 m wide is how many feet wide?

## Section 3-4

Determine the number of significant digits of each number.

**67.** 304,243

**68.** 2,401,000

**69.** 4.010

**70.** 0.023

Find the greatest possible error of each measurement.

**71.** $2\frac{1}{2}$ in.

**72.** $7\frac{3}{8}$ in.

**73.** 15.3 cm

**74.** 7.5 oz

**75.** **BUS** A gas pump registers 4.95 gal when the measuring standard of 5.00 gal is used. What is the percent error?

**76.** **HLTH/N** A pharmacy balance reads 2.03 g when a standard 2.00-g weight is used in calibration. What is the percent error?

**77.** How many significant digits are there in 0.5010?

**78.** How many significant digits are there in 203.07?

**79.** What is the greatest possible error of the measurement 4.7 cg?

**80.** What is the greatest possible error of the measurement $2\frac{5}{32}$ in.?

**81.** Add and write the sum using appropriate precision:

$$4.2 \text{ m} + 508 \text{ cm} + 31.72 \text{ m} + 5.46 \text{ m}$$

**82.** Multiply and write the product using the appropriate number of significant digits.

$$(132 \text{ m})(80 \text{ m})(27.5 \text{ m})$$

**Section 3–5**

Measure line segments 83–86 in Figure 3–89 to the nearest millimeter.

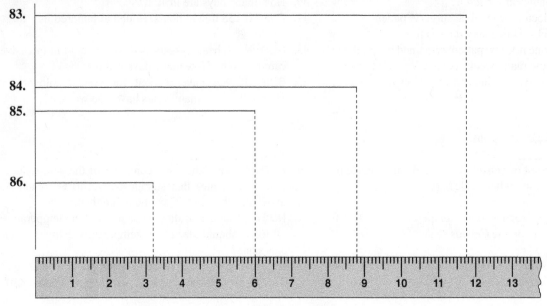

**Figure 3–89**

Read the measurements shown on metric and U.S. customary scales on the vernier calipers in Figures 3–90 and 3–91.

**87.**

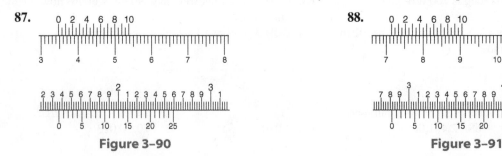

**88.**

**Figure 3–90**

**Figure 3–91**

Determine the measurement indicated on the micrometer readings in Figures 3–92 and 3–93.

**89.**

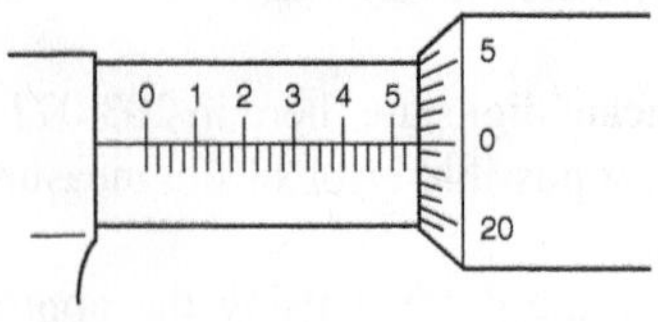

**Figure 3–92** Frame size 2–3 in.

**90.**

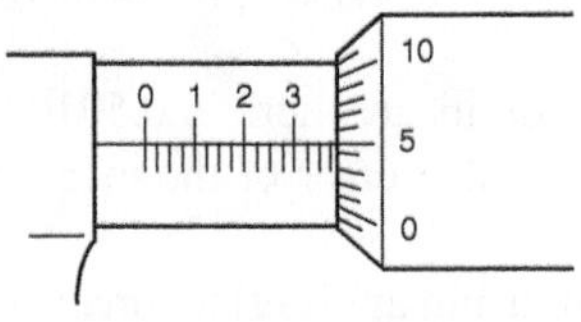

**Figure 3–93** Frame size 4–5 in.

Determine the measurements indicated on the vernier micrometer readings in Figures 3–94 and 3–95.

**91.**

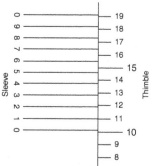

**Figure 3–94**  Frame size 4–5 in.

**92.**

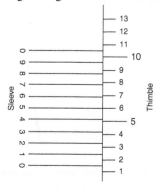

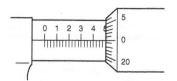

**Figure 3–95**  Frame size 0–1 in.

**93. INDTEC** A blueprint reading indicates the thickness of a casting to be 12.40 cm. The actual part is measured and found to be 12.92 cm. Find the absolute error, relative error, and percent error.

**94. INDTR** The true measure of a machine tool is 18.0 in., but its measurement is recorded as 18.09 in. Find the absolute error, relative error, and percent error.

**95.** Read the water meter in Figure 3–96.

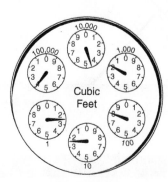

**Figure 3–96**

**96.** Read the electric meter in Figure 3–97.

**Figure 3–97**

**97.** Read the metric dial in Figure 3–98 (the arrow near the zero indicates the initial deflection of the needle).

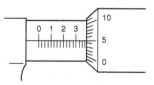

**Figure 3–98**

**98.** Read the voltmeter in Figure 3–99.

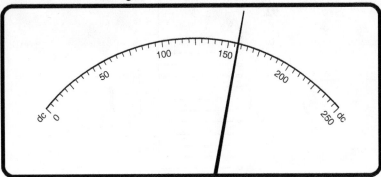

**Figure 3–99**

**99.** Read the ohmmeter in Figure 3–100 (in ohms, Ω).

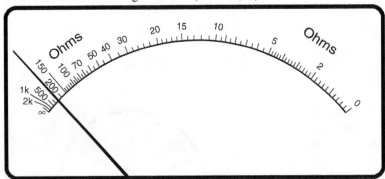

**Figure 3–100**

## TEAM PROBLEM-SOLVING EXERCISES

A **lumen** is the most common measurement of light output, or luminous flux. The lumen rating of a light source is a measure of the total light output of the light source. Light sources are labeled with an output rating in lumens. For example, an R30, 65-W indoor flood lamp may have a rating of 750 lumens. A **watt** is the unit used to measure electrical power. It defines the rate of energy consumption by an electrical device when it is in operation. The energy cost of operating an electrical device is calculated as its wattage time in hours of use.

The measure of electrical energy from which electricity billing is determined is the kilowatt-hour (kWh). For example, a 100-W bulb operated for 1,000 h would consume 100 kWh (100 W × 1,000 h) = 100,000 Wh, or 100 kWh.

**1.** A conference room has five R30, 65-W indoor flood lamps. How many lumens are produced if all 5 lamps are lighted?

**2.** How many kilowatt-hours are consumed if the 5 lamps are lighted for 36 h?

**3.** At the electrical billing rate of $0.12/kWh, how much will it cost to operate the five lamps over 1,000 h?

## PRACTICE TEST

Change to the metric unit indicated.

**1.** 298 m = _____ km

**2.** 8 dm = _____ mm

**3.** 5.2 dL of liquid are poured from a container holding 10 L. How many liters of liquid remain in the container?

**4.** A bar of soap weighs 175 g. How much do 15 bars of soap weigh (in kilograms)?

**5.** 75 mi = _____ km

**6.** 25 kg = _____ lb

**7.** 4 L = _____ pt

**8.** How many liters of weed killer are contained in a 55-gal drum?

**9.** Change 48°F to degrees Celsius.

**10.** Change 70°C to degrees Fahrenheit.

**11.** Find the number of minutes in 42 h.

**12.** How many hours are in 5,700 min?

**13.** A GIS Specialist worked 47 h a week for 5 consecutive weeks. How many hours did she work?

**14.** A machine was in use for 20 h a day for the 30 days in June. How many hours was the machine producing?

**15.** The number 840 has how many significant digits?

**16.** A tool measures 3.5 cm. What is the greatest possible error?

**17.** A hip replacement part measures 24.75 cm and the blueprint specifies the part is 24.72 cm. What is the relative error and percent error?

**18.** Find the sum of 32.5 cg, 4.76 cg, 503.2 cg, and 16.3 cg and use an appropriate approximation to report the sum.

Measure line segments 19–20 in Figure 3–101 to the nearest millimeter.

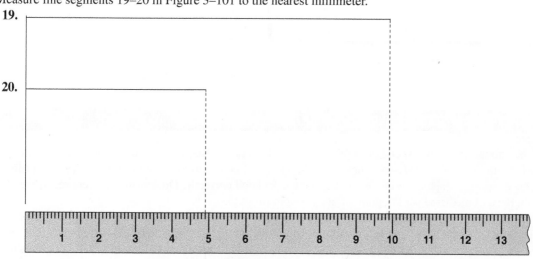

**19.**

**20.**

**Figure 3–101**

**21.** Determine the measurement indicated on the micrometer in Figure 3–102.

**Figure 3–102** Frame size 2–3 in.

**22.** Read the electric meter in Figure 3–103.

Kilowatt Hours

**Figure 3–103**

**23.** Read the U.S. customary dial in Figure 3–104.

**Figure 3–104**

**24.** Read the voltmeter in Figure 3–105.

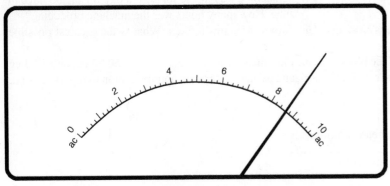

**Figure 3–105**

## CAREER APPLICATION: NURSING—DRUG DOSAGE AND IV CONCENTRATION

When nurses administer patient drugs, they spend most of their time calculating correct dosage and concentration, both of which depend on the patient's body weight. Even minor errors in dosage or concentration can have drastic (or even life-threatening) consequences, so it is imperative that the dosages be calculated correctly. Drug manufacturers recommend dosages based on milligrams of medicine per kilogram of body weight (mg/kg).

For example, the recommended dosage of *Lasix*, a common drug used in congestive heart failure, is 1 mg of *Lasix* per kg of body weight (1 mg/kg). A patient's weight is typically measured in pounds, so nurses must convert pounds to kilograms by dividing the weight in pounds by 2.2 (1 kg = 2.2 lb). Conversely, to change a patient's weight from kilograms to pounds, nurses multiply the weight in kilograms by 2.2.

Although nurses administer metric doses in hospitals, they usually convert to the U.S. customary system for patient at-home-care instructions. For example, *Children's Tylenol*, a common drug used to treat fever and pain, contains 80 mg of medicine per 5 cubic centimeters (cc) of volume. Knowing that 5 cc equals 1 teaspoon (tsp), nurses calculate how many tsp should be given at home to a patient whose body weight requires a stronger or weaker dose of *Tylenol*.

Some medicines are given through an intravenous (IV) drip. The rate of the IV drip determines the concentration of medicine given. For example, an infusion of *Dobutamine*, a cardiac drug used to improve a patient's heart muscle contractility, at 1.0 cc/hr equals a concentration of 10 micrograms of medicine per kilogram of body weight per minute (10 mcg/kg/min). Nurses then calculate how fast the drip needs to be if the doctor increases the dosage to 14 mcg/kg/min. Doctors often change the IV concentration order, which means the amount of medicine dissolved in the IV liquid is increased (for a higher concentration) or decreased (for a lower concentration).

### Exercises

Use the preceding information for the exercises. Round all answers to the nearest tenth.

**1.** A premature baby boy weighing 4.5 lb needs a dose of *Lasix*. Convert his weight to kilograms and calculate how many milligrams of the drug he should receive.

**2.** Find the dosage of *Lasix* needed for his 3.4-lb twin sister.

**3.** Calculate the dosage of *Lasix* needed for a 67-year-old man weighing 192 lb.

**4.** The recommended dosage of *Children's Tylenol* is 10 mg/kg. How many mg of *Tylenol* should be given to an 18-month-old child who weighs 24 lb and has an ear infection?

**5.** Use your answer from Exercise 4 to tell the parent this child's dosage in teaspoons.

**6.** If a *Dobutamine* IV drip is flowing at 1.0 cc/hr, which equals 10 mcg/kg/min, how fast would it need to flow if the doctor ordered it increased to 20 mcg/kg/min?

**7.** For a *Dobutamine* IV drip flowing at 1.0 cc/hr (10 mcg/kg/min), how fast would it need to flow if the doctor ordered it decreased to 8 mcg/kg/min?

**8.** If a doctor ordered the concentration of a *Dobutamine* IV drip doubled, how would a nurse change the amount of medicine added to a bag of IV liquid?

**Answers**

1. For a weight of about 2 kg, 2 mg of *Lasix* should be given.
2. For a weight of about 1.5 kg, 1.5 mg of *Lasix* should be given.
3. For about 87.3 kg of weight, 87.3 mg of *Lasix* should be given.
4. For about 10.9 kg of weight, 109.0 mg of *Tylenol* should be given.
5. *Children's Tylenol* comes 80 mg/l tsp, so about 1.4 or $1\frac{1}{2}$ tsp should be given.
6. The concentration has been doubled, so the IV should drip at twice the rate, or 2.0 cc/hr.
7. The IV should drip at 0.8 cc/hr.
8. The amount of medicine added to the bag of IV liquid should be doubled.

# 4

# Signed Numbers and Powers of 10

## Focus on Careers

Geographic Information Systems (GIS) specialists work with computer programs and software to create and maintain data or maps that can be combined with geographically referenced data. Different types of data, such as socioeconomic, demographic, political, land use, land cover, environmental, and transportation networks, are related through the use of GIS software.

GIS specialists must have knowledge of geography, written and oral communication, mathematics, critical thinking, information gathering, computer science, and systems evaluation. These specialists usually work in offices that are clean, well lighted, and well ventilated. Most of the GIS specialist's time is spent in front of a computer screen; however, some GIS employees collect data in the field.

*(continued)*

Jobs in this field are expected to grow by about 65% over the next 10 years. Hourly wages for jobs in California range from $20.20 to $39.26, with the average being $29.97. GIS specialists usually work 40 hours a week, but longer hours and weekend hours are not uncommon.

Many 2-year colleges offer GIS certificate programs as well as degree programs. GIS courses are offered through distance learning over the Internet by many colleges. Graduates with a GIS major and significant study in business, sociology, political science, or economics often have an advantage over job seekers who have only a GIS major. GIS specialists are expected to keep up with new technology and practices in this relatively new and fast-growing field through continuing education.

Source: California Employment Development Department.

When a temperature is colder than zero degrees, how can we express this temperature numerically? When the selling price of an item is less than the cost of an item, how can we express the profit made on the sale numerically? When a withdrawal on a bank account exceeds the amount of money in the account, how can we express the account balance numerically? These situations demonstrate a need to express numbers that are less than zero.

# 4-1 | Adding Signed Numbers

*Learning Outcomes*

**1** Compare signed numbers.

**2** Add signed numbers with like signs.

**3** Add signed numbers with unlike signs.

## 1 Compare Signed Numbers.

Many physical phenomena have values that are less than zero. Another type of number is needed to express these values. When the opposite of each natural number is included, the set of whole numbers can be expanded to form the set of **integers.** Figure 4–1 shows how the set of whole numbers is extended to include all integers. A number line for integers continues indefinitely in *both* directions. Numbers get smaller as we proceed to the left and larger as we proceed to the right.

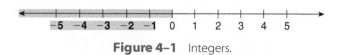

**Figure 4–1** Integers.

The **opposite** of a number is the same number of units from zero, but in the opposite direction. Figure 4–2 shows a number line with numbers and their opposites.

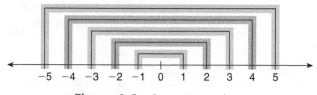

**Figure 4-2** Opposite numbers.

The relationship among the three sets of numbers—natural numbers, whole numbers, and integers—is shown in Figure 4–3.

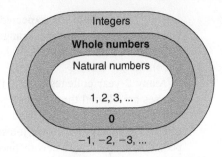

**Figure 4–3**   Relating natural numbers, whole numbers, and integers.

### Are Positive Signs Necessary?

As you can see from the number line in Figure 4–1, it is not necessary to include the positive sign for positive values. You may sometimes want to include the sign to draw attention to it or to emphasize its direction. Negative signs, however, must always be included.

Integers are just one type of signed numbers. Fractions, mixed numbers, and decimals also have values less than zero. **Signed numbers** are all types of numbers that are either positive or negative or 0. We begin our study of signed numbers with integers and expand to other types of signed numbers later in the chapter.

If a number is positioned to the left of another number on a number line, it represents a smaller value. Larger values are positioned to the right of a number on the number line. We show how two numbers compare in size by using two symbols: $<$ and $>$. The symbol $<$ is used when the first number is smaller than or "is less than" the second number. For example, "$-7$ is less than 5" is written $-7 < 5$. When the first number is larger than or "is greater than" the second number, the symbol $>$ is used. For example, "5 is greater than $-7$" is written $5 > -7$.

**EXAMPLE**   Use the symbols $<$ and $>$ to show the relationship between the two numbers.

**(a)** 7 ___ 9        **(b)** 0 ___ $-1$        **(c)** $-4$ ___ $-2$        **(d)** $-2$ ___ 0

**(a)** $7 < 9$          7 is smaller—to the left of 9 on the number line.
**(b)** $0 > -1$         0 is larger—to the right of $-1$ on the number line.
**(c)** $-4 < -2$       $-4$ is smaller—to the left of $-2$ on the number line.
**(d)** $-2 < 0$         $-2$ is smaller—to the left of 0 on the number line.

The distance between each pair of consecutive integers is the same for all integers located on the number line. This concept of *distance between numbers* on the number line is related to the concept of *absolute value*. The **absolute value** of a number is often described as the number of units or distance the number is from zero. The symbol for absolute value is $|\ |$; $|3|$ is read "the absolute value of 3."

Distance is a physical property, and it cannot have a negative value. The absolute value of a number is always a nonnegative value.

### What Does Nonnegative Mean?

Are positive numbers also nonnegative? Yes, every positive number is nonnegative.
Are any nonnegative numbers not positive?
Yes, zero is nonnegative *and* nonpositive. Therefore, when we say *nonnegative*, we mean positive or zero.
Similarly, *nonpositive* is negative or zero.

Chapter 4 / Signed Numbers and Powers of 10

## Write Facts Symbolically

Positive numbers are represented symbolically as $a > 0$, where $a$ is any number greater than zero. Negative numbers are represented by $a < 0$, where $a$ is any number less than zero. The symbols $\geq$ which means "is greater than or equal to" and $\leq$ which means "is less than or equal to" are used to represent nonnegative and nonpositive numbers respectively. Practice reading symbolic statements in words.

| | | | |
|---|---|---|---|
| $a > 0$ | positive numbers | $a \geq 0$ | nonnegative numbers (positives and zero) |
| $a < 0$ | negative numbers | $a \leq 0$ | nonpositive numbers (negatives and zero) |

Then, the *definition of the absolute value* of a number can be written symbolically and read in words.

$|a| = a$, for $a > 0$     The absolute value of a positive number is equal to itself.

$|a| = -a$, for $a < 0$     The absolute value of a negative number is equal to its opposite.

$|0| = 0$     The absolute value of zero is zero.

**EXAMPLE**    Give the absolute value of the following quantities: (a) $|9|$ (b) $|-4|$

(a) $|9| = 9$     9 is positive. Its absolute value is the number itself.

(b) $|-4| = 4$     $-4$ is negative. Its absolute value is the opposite of $-4$.

## Two Components of a Signed Number

Integers and other signed numbers tell us two things. For the integer $-8$, the negative sign tells us the number is to the left of zero and is referred to as the *directional sign* of the number. The 8 tells us how many units the number is from zero; it is the *absolute value* of the number.

Every number except zero has an opposite. The opposite of a number is also called the **additive inverse** of the number. A number and its additive inverse (opposite) have the same absolute value but different directional signs.

**EXAMPLE**    Find the opposite of each number and show your answer on the number line.
(a) $-8$    (b) 6    (c) 0

(a) The opposite of $-8$ is 8.

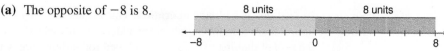

(b) The opposite of 6 is $-6$.

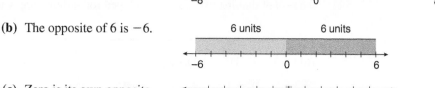

(c) Zero is its own opposite.

### 2 Add Signed Numbers with Like Signs.

Since our earliest experiences with addition we have added positive numbers. In the addition problem $3 + 2 = 5$, all three numbers are positive. Numbers are understood to be positive when no sign is given. To emphasize that numbers are positive we can write the problem as $+3 + +2 = +5$. We can also write the problem using parentheses to separate the two plus signs, $+3 + (+2) = +5$. The plus symbol serves two purposes in mathematical expressions. It is used as the directional sign for a positive number, and it is used to show addition. In the expression $+3 + (+2) = +5$, the plus sign in front of the parentheses shows the operation of addition. The plus sign inside the parentheses identifies the number as positive.

Adding integers builds on our knowledge of addition of positive numbers.

To add 3 and 2 on the number line we begin at zero and move three units to the right (positive direction). This move takes us to $+3$. From the $+3$ position, we move two units to the right. This second move takes us to the $+5$ position (Figure 4–4). In other words, $3 + 2 = 5$.

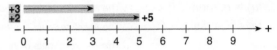

**Figure 4–4** Adding using a number line.

To add two negative integers on the number line, $(-3) + (-2)$, we begin at zero and move three units to the left (negative direction). This takes us to $-3$ on the number line. Then, from $-3$ we move two units to the left. This second move takes us to $-5$ (Figure 4–5). Thus, $-3 + (-2) = -5$.

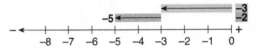

**Figure 4–5** Adding two negative numbers using a number line.

When adding the two positive numbers or the two negative numbers, the sum could have been obtained using the following rule.

## To add numbers with like signs:

1. Add the absolute values of the numbers.
2. Give the sum the common or like sign.

**EXAMPLE**  Add $-12$ and $-5$.

$$-12 + (-5) = -17 \qquad \text{Add absolute values. Keep common negative sign.}$$

Addition is a **binary operation.** This means that only two numbers are used in the operation. If more than two numbers are added, two are added and then the next number is added to the sum of the first two. Rules developed for addition apply to two numbers at a time.

**EXAMPLE**  Add $-3 + (-4) + (-1)$.

$$-3 + (-4) + (-1) = \qquad \text{Add first two numbers.}$$

$$-7 + (-1) = -8 \qquad \text{Add sum to remaining number.}$$

**Add Signed Numbers with Unlike Signs.**

When adding numbers with unlike signs, we move in the direction indicated by the sign of each number, positive to the right and negative to the left. To add $-4 + 3$, we first move four units to the left of zero (see Figure 4–6). This takes us to $-4$. Then, from the $-4$ position, we move three units to the right. This takes us to $-1$. Thus, $-4 + 3 = -1$. This process is generalized by a rule using absolute values.

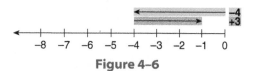

**Figure 4–6**

## To add signed numbers with unlike signs:

1. *Subtract* the smaller absolute value from the larger absolute value.
2. Give the sum the sign of the number with the larger absolute value.

---

**EXAMPLE**  Add $7 + (-12)$.

$7 + (-12) =$          Signs are unlike.

$7 + (-12) = \mathbf{-5}$          Subtract absolute values: $12 - 7 = 5$. Keep the sign of the $-12$ (negative), the larger absolute value.

---

**TIP!**

## To Add, You . . . Subtract?

Phrases we used in arithmetic for adding integers, such as "find the sum" or "add," can be confusing. In addition of integers, sometimes we add absolute values and sometimes we subtract absolute values. We often use the word "**combine**" to imply the addition of numbers with either like or unlike signs.

---

**EXAMPLE**  Combine $-5 + 7$.

$-5 + 7 =$          Signs are unlike.

$-5 + 7 = \mathbf{2}$          Subtract absolute values: $7 - 5 = 2$. Keep the sign of the 7 (positive), the larger absolute value.

---

When an addition involves several positive numbers and several negative numbers, it is convenient to group all the positive numbers and all the negative numbers and to add each group separately. We then add the sums of each group. We can do this because of the commutative and associative properties of addition. In the next example we combine the commutative and associative properties of addition to add several signed numbers.

4–1 Adding Signed Numbers

203

**EXAMPLE**    Add $8 + (-9) + 13 + (-15)$.

$8 + (-9) + 13 + (-15) =$    Signs are unlike. Rearrange and group numbers with like signs.

$[8 + 13] + [-9 + (-15)] =$    The brackets, [ ], show the grouping and separate the operational plus sign from the directional minus sign of $-9$. Add groups of numbers with like signs.

$21 + (-24) = -3$    Add resulting sum using the rule for adding numbers with unlike signs.

The properties of addition that involve the number zero apply to integers as well as to whole numbers, fractions, and decimals. When zero is added to a number, the number is not changed. We say that zero is the **additive identity.** The sum of a number and its opposite is zero. We call the opposite of a number the **additive inverse** of that number. We can also write these definitions using symbols.

Zero is the *additive identity* because $a + 0 = a$ for all values of $a$.
The opposite of a number is the *additive inverse* of the number because $a + (-a) = 0$ for all values of $a$.

**EXAMPLE**    Add: **(a)** $-2 + 0$    **(b)** $5 + (-5)$    **(c)** $0 + (-3) + 4$    **(d)** $5 + (-2) + (-5)$

**(a)** $-2 + 0 = -2$    Additive identity.
**(b)** $5 + (-5) = 0$    Additive inverse.
**(c)** $0 + (-3) + 4 =$    Additive identity.
$-3 + 4 = 1$    Add numbers with unlike signs.
**(d)** $5 + (-2) + (-5) =$    Commutative and associative properties.
$5 + (-5) + (-2) =$    Additive inverse.
$0 + (-2) = -2$    Additive identity.

### SECTION 4–1 SELF-STUDY EXERCISES

**1** Use the symbol $>$ or $<$ to show the relationship between each pair of numbers.

**1.** $5$ ___ $8$ 
**2.** $-3$ ___ $-2$ 
**3.** $-7$ ___ $-9$ 
**4.** $0$ ___ $5$
**5.** $-8$ ___ $0$ 
**6.** $12$ ___ $7$ 
**7.** $-3$ ___ $2$ 
**8.** $5$ ___ $-7$

Write the absolute value.

**9.** $|7|$ 
**10.** $|-17|$ 
**11.** $|8|$ 
**12.** $|-7|$

Write the opposite of each number.

**13.** $42$ 
**14.** $-17$ 
**15.** $-78$ 
**16.** $57$

**2** Add.

**17.** $8 + 15$ 
**18.** $-9 + (-7)$ 
**19.** $-5 + (-8)$
**20.** $(-12) + (-17)$ 
**21.** $-24 + (-31)$ 
**22.** $-78 + (-46)$
**23.** $7 + 10 + 12$ 
**24.** $-5 + (-8) + (-7)$ 
**25.** $12 + 87$
**26.** $-21 + (-38)$ 
**27.** $-32 + (-16)$ 
**28.** $(-58) + (-103)$
**29.** $-5 + (-12) + (-36)$ 
**30.** $52 + (+92) + (+88)$ 
**31.** $-107 + (-502) + (-396)$

**3** Add.

**32.** $5 + (-3)$

**33.** $-3 + 5$

**34.** $8 + (-4)$

**35.** $-8 + 4$

**36.** $9 + (-3)$

**37.** $3 + (-9)$

**38.** $-4 + (+2)$

**39.** $-3 + 7$

**40.** $4 + (-6)$

**41.** $-4 + 6$

**42.** $-18 + 8$

**43.** $32 + (-72)$

**44.** $21 + (-14)$

**45.** $17 + (-4) + 3 + (-1)$

**46.** $-3 + 2 + (-7)$

**47.** $47 + (-82) + 2$

**48.** $14 + (-6) + 1$

**49.** $-7 + (-3) + (-1)$

**50.** $4 + 2 + (-3) + 10$

**51.** $2 + (-1) + 8$

**52.** $-2 + 1 + (-8) + 12$

**53.** $15 + (-15)$

**54.** $-7 + 7$

**55.** $92 + (-92)$

**56.** $(-396) + (396)$

**57.** $-3 + 0$

**58.** $0 + (-7)$

**59.** $6 + (-3) + 5 + (-8)$

**60.** $12 + (-5) + 8 + (-18)$

**61.** $4 + (-7) + 6 + (-15)$

**62.** $18 + (-7) + (-9) + (-16)$

**63.** $8 + (-4) + (-13) + 22$

**64.** $18 + (-3) + 5 + (-32)$

**65.** $12 + (-15) + 73 + (-46)$

**66.** $(-83) + (-21) + 42 + (-36)$

**67.** $(-5) + (-18) + 32 + (-7)$

**68.** $15 + (-32) + 46 + (-19)$

**69. BUS** You open a bank account by depositing $242. You then write checks for $21, $32, and $123. What is your new account balance?

**70. BUS** A stock priced at $42 has the following changes in one week: $-3, +8, -6, -7, +2$. What is the value of the stock at the end of the week?

**71. AVIA** Liquid hydrogen, fuel in the space shuttle, must be kept below $-253°C$. A tank of liquid hydrogen is stored at $-320°C$, but the temperature is raised by $25°C$. What is the new temperature of the hydrogen?

**72.** The temperature at the South Pole is recorded as $-38°F$ at midnight. The temperature rises $15°$ by noon. Determine the temperature reading at noon.

---

## 4-2 | Subtracting Signed Numbers

*Learning Outcomes*

**1** Subtract signed numbers.

**2** Combine addition and subtraction.

**1** Subtract Signed Numbers.

**Addition and subtraction are related.**

Subtracting a number is the *same as* adding the opposite of the number.

| | |
|---|---|
| $6 - 9$ | Show direction signs of the numbers. |
| $+6 - (+9)$ | Subtract +9. |
| $+6 + (-9)$ | Add the opposite of +9 or add −9. |

Let's write this relationship as a rule.

**To subtract signed numbers:**

1. Change the subtraction sign to addition.
2. Change the sign of the second number (subtrahend) to its opposite.
3. Apply the appropriate rule for adding signed numbers.

**EXAMPLE** Subtract: **(a)** $2 - 6$ **(b)** $-9 - 5$ **(c)** $12 - (-4)$ **(d)** $-7 - (-9)$

| **(a)** | $2 - 6 =$ | Subtract positive 6 from positive 2. |
|---|---|---|
| | $2 - (+6) =$ | Write sign of subtrahend. |
| | $2 + (-6) =$ | Add the opposite of +6, which is −6, to 2. |
| | $2 + (-6) = \mathbf{-4}$ | Apply the rule for adding numbers with unlike signs. |

**(b)** 
$$-9 - 5 =$$        Subtract positive 5 from negative 9.

$$-9 - (+5) =$$        Write sign of subtrahend.

$$-9 + (-5) =$$        Add the opposite of +5, which is −5, to −9.

$$-9 + (-5) = \mathbf{-14}$$        Apply the rule for adding numbers with like signs.

**(c)** 
$$12 - (-4) =$$        Subtract negative 4 from positive 12.

$$12 + (+4) =$$        Add the opposite of −4, which is +4, to 12.

$$12 + 4 = \mathbf{16}$$        Apply the rule for adding numbers with like signs.

**(d)** 
$$-7 - (-9) =$$        Subtract negative 9 from negative 7.

$$-7 + (+9) =$$        Add the opposite of −9, which is +9, to −7.

$$-7 + 9 = \mathbf{2}$$        Apply the rule for adding numbers with unlike signs.

**TIP!**

### Writing Subtractions as Equivalent Additions

When writing a subtraction as an equivalent addition, you make *two* changes.

1. Change the operation from subtraction to addition.
2. Change the subtrahend to its opposite.

| | | | |
|---|---|---|---|
| $2 - (+6)$ | $-9 - (+5)$ | $12 - (-4)$ | $-7 - (-9)$ |
| $2 + (-6)$ | $-9 + (-5)$ | $12 + (+4)$ | $-7 + (+9)$ |
| ↑ ↑ | ↑ ↑ | ↑ ↑ | ↑ ↑ |
| 1.  2. | 1.  2. | 1.  2. | 1.  2. |

Subtractions involving zero and opposites are interpreted first as equivalent addition problems. Since zero has no opposite, +0 and −0 are still 0.

**EXAMPLE**   Evaluate: **(a)** $8 - 0$   **(b)** $0 - 15$   **(c)** $-32 - (-32)$   **(d)** $-15 - 15$

**(a)** 
$$8 - 0 =$$        Subtract zero from positive 8.

$$8 + (0) =$$        Add zero to positive 8.

$$8 + (0) = \mathbf{8}$$        Zero added to any number is that number (additive identity).

**(b)** 
$$0 - 15 =$$        Subtract positive 15 from zero.

$$0 - (+15) =$$        Write sign of subtrahend.

$$0 + (-15) =$$        Add the opposite of +15, which is −15, to zero.

$$0 + (-15) = \mathbf{-15}$$        Any number added to zero (additive identity) is that number.

**(c)** 
$$-32 - (-32) =$$        Subtract negative 32 from negative 32.

$$-32 + (+32) =$$        Add the opposite of −32, which is +32, to −32.

$$-32 + 32 = \mathbf{0}$$        A number added to its opposite (additive inverse) is zero.

**(d)** 
$$-15 - (+15) =$$        Subtract positive 15 from negative 15.

$$-15 + (-15) =$$        Add the opposite of +15, which is −15, to −15.

$$-15 + (-15) = \mathbf{-30}$$        Apply the rule for adding numbers with like signs.

### 2   Combine Addition and Subtraction.

When we express the sum and difference of signed numbers with all the appropriate operational and directional signs, we have an expression with both an operational sign and a directional sign between every two numbers. We have four different possibilities when two signs are written together. The following tip allows us to simplify these four possibilities.

## Omitting Signs

Writing mathematical expressions that include every operational and directional sign is cumbersome. In general practice, we omit as many signs as possible. When two signs are written between two numbers, we can write a simplified expression with only one sign.

|  | Double Signs |  | Single Sign |
|---|---|---|---|
| **Like Signs** | Plus, Plus: $+3 + (+ \ 5)$ | Add $+3$ and $+5$. | $3 + 5$ |
|  | Minus, Minus: $+3 - (- \ 5)$ | Change to addition. $+3 + (+5)$ | $3 + 5$ |
| **Unlike Signs** | Plus, Minus: $+3 + (- \ 5)$ | Add $+3$ and $-5$. | $3 - 5$ |
|  | Minus, Plus: $+3 - (+ \ 5)$ | Change to addition. $+3 + (-5)$ | $3 - 5$ |

To generalize, *two like signs between signed numbers,* whether both plus or both minus, translate to adding a positive number. Use just one plus sign:

$$+ + \rightarrow + \qquad\qquad - - \rightarrow + \qquad\qquad \textbf{like signs} \rightarrow +$$

*Two unlike signs between signed numbers,* either a plus/minus or a minus/plus, translate to adding a negative number. Use just one minus sign:

$$+ - \rightarrow - \qquad\qquad - + \rightarrow - \qquad\qquad \textbf{unlike signs} \rightarrow -$$

## To add and subtract more than two numbers:

1. Rewrite the problem so that all integers are separated by only one sign.
2. Add the series of signed numbers.

---

**EXAMPLE**  Evaluate: **(a)** $3 - (-5) - 6$   **(b)** $-8 + 10 - (-7)$

**(a)** $3 - (-5) - 6 =$   Rewrite with only one sign between signed numbers.

$\quad\; 3 + 5 \; - 6 =$   Add numbers with like signs.

$\qquad\qquad 8 \; - 6 = 2$   Apply the rule for adding numbers with unlike signs.

**(b)** $-8 + 10 - (-7) =$   Rewrite with only one sign between signed numbers.

$\quad\; -8 + 10 + 7 =$   Add numbers with like signs.

$\qquad\quad -8 + 17 = 9$   Apply the rule for adding numbers with unlike signs.

---

The key to solving applied problems with signed numbers is to identify which numbers are positive and which are negative.

> Positive key words: profits, gains, money in the bank, temperatures above zero, receipts, income, winnings, and so on.

> Negative key words: losses, deficits, checks that cleared the bank, temperatures below zero, drops, declines, payments, and so on.

**EXAMPLE** A landscaping business makes a profit of $345 one week, has a loss of $34 the next week, and makes a profit of $235 the third week. What is the total profit?

**BUS** The total profit is the sum of the weekly profits and losses.

$345 − $34 + $235          Interpret profits as positive and losses as negative.
profit    loss    profit

345 + 235 − 34             Rearrange and add positives.

580 − 34 = 546             Apply rule for adding numbers with unlike signs.

**The total profit for the**          Interpret answer.
**three weeks is $546.**

---

**EXAMPLE** In a recent year, 98°F was the highest temperature in Boston, and −2°F was the lowest temperature. What was the temperature range for the city that year? (The range is the difference between the highest and lowest values.)

98 − (−2) =               Subtract −2 from 98.
    98 + 2 = 100          Rewrite with only one sign between integers, and apply the
                          appropriate rule for adding integers.

**The temperature range for**          Interpret answer.
**Boston that year was 100°F.**

---

**SECTION 4–2 SELF-STUDY EXERCISES**

**1** Subtract.

1. $-3 - 9$
2. $8 - 2$
3. $9 - 15$
4. $(-3) - (-7)$
5. $-11 - 14$
6. $(-6) - (-3)$
7. $5 - (-3)$
8. $8 - 11$
9. $-8 - 1$
10. $11 - (-2)$
11. $(-8) - (-7)$
12. $(-15) - (-7)$
13. $(20) - (-42)$
14. $(-38) - (-27)$
15. $(-42) - (+16)$
16. $(-21) - (+36)$
17. $(-18) - (+15) - (-18)$
18. $(-12) - (+21) - (+72)$
19. $14 - (-21) - 17$
20. $(-14) - (-21) - 24$
21. $-142 - (+46) - (-21)$
22. $217 - (-38) - (+172)$
23. $802 + (196) - (-415)$
24. $72 - (-23) - (+198)$
25. $15 - 0$
26. $0 - 8$
27. $-12 - 0$
28. $0 - (-8)$
29. $0 - (-7)$
30. $10 - 0$
31. $28 - (-28)$
32. $-46 - 46$
33. $7 - (-7)$
34. $-18 - 18$
35. $5 - 5$
36. $21 - 21$

**2** Evaluate.

37. $5 + 8 - 9$
38. $6 - 4 + 5$
39. $9 + 2 - 5$
40. $-1 + 1 - 4$
41. $5 + 3 - 7$
42. $7 + 3 - (-4)$
43. $-8 - 2 - (-7)$
44. $-3 + 4 - 7 - 3$
45. $2 - 4 - 5 - 6 + 8$
46. $8 - 3 + 2 - 1 + 7$
47. $-5 - 3 + 8 - 2 + 4$
48. $6 - (-3) + 5 - 6 - 9$
49. $-8 + 2 - 7 + 14$
50. $3 - 5 + 8 - 11 - 15$
51. $5 - 2 + 9 - 6 - 12 + 32$
52. $18 + 12 - 16 - 32 - 81$
53. $-28 + 32 - 17 - 32$
54. $78 - 64 + 32 - 78$
55. $52 - 96 - 102 + 86$
56. $42 - 17 + 86 - 191$
57. $77 - 96 - 102 + 86$
58. $149 - 23 + 6 - 82$
59. $54 - 87 + 33 - 21$
60. $228 - 21 + 33 - 54$

**61.** The temperatures for Bowling Green, Kentucky, ranged from 102°F to −5°F. What was the temperature range for the city?

**62.** New Boston, Texas, registered −8°F as its lowest temperature one year and 99°F as its highest temperature for the same year. What was the temperature range for New Boston?

**63.** **BUS** Computing Solutions records a profit of $28,296 one quarter (three months), a loss of $1,896 for the second quarter, a profit of $52,597 for the third quarter, and a profit of $36,057 for the fourth quarter. What is their net profit for the year?

**64.** Explain the difference between subtracting zero from a number and subtracting a number from zero.

---

## 4-3 | Multiplying and Dividing Signed Numbers

*Learning Outcomes*

**1** Multiply signed numbers.

**2** Evaluate powers of signed numbers.

**3** Divide signed numbers.

Multiplying absolute values of signed numbers is exactly the same as multiplying whole numbers. However, the rules for multiplying signed numbers must also include the assignment of the proper sign to the product.

### 1 Multiply Signed Numbers.

Just as with whole numbers, multiplication of signed numbers is a binary operation that is commutative and associative; that is, two factors can be multiplied at a time in any order. More than two factors can be grouped in any manner. These examples show how we assign signs to the product of two signed numbers:

| **Like Signs** | **Unlike Signs** |
|---|---|
| $4(6) = 24,$  $(-3)(-7) = 21$ | $-10(2) = -20,$  $8(-2) = -16$ |

**To multiply two signed numbers:**

1. Multiply the absolute values of the numbers.
2. If the factors have like signs, the sign of the product is positive.
3. If the factors have unlike signs, the sign of the product is negative.

---

**EXAMPLE**  Multiply: **(a)** $-12(-2)$  **(b)** $10 \cdot 3$  **(c)** $25(-3)$  **(d)** $-5 * 7$

**(a)** $-12(-2) = \mathbf{24}$   Like signs give a positive product.
**(b)** $10 \cdot 3 = \mathbf{30}$   Like signs give a positive product.
**(c)** $25(-3) = \mathbf{-75}$   Unlike signs give a negative product.
**(d)** $-5 * 7 = \mathbf{-35}$   Unlike signs give a negative product.

---

When the number 1 is multiplied by a number, the result is the number. The number 1 is the **multiplicative identity** because $a \cdot 1 = 1 \cdot a = a$ for all values of $a$.

Multiplication, like addition, is a binary operation, so when we multiply three or more factors, we multiply two at a time.

EXAMPLE  Multiply: **(a)** $4(-2)(6)$  **(b)** $-3(4)(-5)$  **(c)** $-2(-8)(-3)$  **(d)** $-2(-3)(-4)(-1)$

**(a)** $\boxed{4(-2)}\ (6) =$          Multiply the first two factors and apply the rule for factors with unlike signs.

$\boxed{-8}\ (6) = \mathbf{-48}$          Multiply and apply the rule for factors with unlike signs.

**(b)** $\boxed{-3(4)}\ (-5) =$          Multiply the first two factors and apply the rule for factors with unlike signs.

$\boxed{-12}\ (-5) = \mathbf{60}$          Multiply and apply the rule for factors with like signs.

**(c)** $\boxed{-2(-8)}\ (-3) =$          Multiply the first two factors and apply the rule for factors with like signs.

$\boxed{16}\ (-3) = \mathbf{-48}$          Multiply and apply the rule for factors with unlike signs.

**(d)** $\boxed{-2(-3)}\ \boxed{(-4)(-1)} =$          Multiply the first two and the last two factors.

$\boxed{6}\ \boxed{(4)} = \mathbf{24}$          Multiply and apply the rule for factors with like signs.

If we examine the multiplications in the preceding example more closely, we see that the number of negative factors affects the sign of the answer.

|  | Number of Negative Factors | Sign of Product |
|---|:---:|:---:|
| $4(-2)(6)$ | 1 | $-$ |
| $-3(4)(-5)$ | 2 | $+$ |
| $-2(-8)(-3)$ | 3 | $-$ |
| $-2(-3)(-4)(-1)$ | 4 | $+$ |

**To determine the sign of the product when multiplying three or more factors:**

1. The sign of the product is *positive* if the number of negative factors is *even*.
2. The sign of the product is *negative* if the number of negative factors is *odd*.

EXAMPLE  Multiply: **(a)** $-2(6)(-1)(-3)$  **(b)** $(2)(-5)(1)(-3)$

**(a)** $-2(6)(-1)(-3) = \mathbf{-36}$          Multiply absolute values. The odd number of negative factors makes the product negative.

**(b)** $(2)(-5)(1)(-3) = \mathbf{30}$          Multiply absolute values. The even number of negative factors makes the product positive.

The *zero property of multiplication* extends to integers. The product of zero and any number is zero: $x \cdot 0 = 0$. If we have two or more factors and one factor is zero, we can immediately write the product as zero without having to work through the steps.

EXAMPLE  Multiply: **(a)** $3(-21)(2)(0)$  **(b)** $-9(-2)(8)(-1)$

**(a)** $3(-21)(2)(0) = \mathbf{0}$          Zero is a factor.

**(b)** $-9(-2)(8)(-1) = \mathbf{-144}$          Zero is not a factor.

Applied problems often require multiplication of integers.

**EXAMPLE**  In a 3-week period, a technology stock declined approximately 2 points each week. How many points did the stock decline in the 3 weeks?

**BUS**  Because there are equal declines each week, we multiply the amount of weekly decline times the number of weeks: $-2(3) = -6$. Thus, **the stock declined (negative) a total of 6 points over the three-week period**.

### 2 Evaluate Powers of Signed Numbers.

Raising a number to a natural-number power is an extension of multiplication, so determining the sign of the result is similar to multiplying several integers. Observe the pattern for determining the sign of the result:

**Positive Base**

$(+4)^2 = (+4)(+4) = +16$
$(+4)^3 = (+4)(+4)(+4) = +64$

**Negative Base**

$(-4)^2 = (-4)(-4) = +16$
$(-4)^3 = (-4)(-4)(-4) = -64$

**To raise signed numbers to a natural-number power, use the following patterns:**

1. A positive number raised to any natural-number power is positive.
2. Zero raised to any natural-number power is zero.
3. A negative number raised to an even natural-number power is positive.
4. A negative number raised to an odd natural-number power is negative.

**EXAMPLE**  Evaluate the powers: **(a)** $4^3$ **(b)** $0^8$ **(c)** $(-2)^4$ **(d)** $(-3)^5$

| | |
|---|---|
| **(a)** $4^3 = 4(4)(4) = \mathbf{64}$ | Base is positive. |
| **(b)** $0^8 = \mathbf{0}$ | Base is zero. |
| **(c)** $(-2)^4 = (-2)(-2)(-2)(-2) = \mathbf{16}$ | Base is negative, exponent is even. |
| **(d)** $(-3)^5 = (-3)(-3)(-3)(-3)(-3) = \mathbf{-243}$ | Base is negative, exponent is odd. |

**EXAMPLE**  Security systems sometimes use a four-digit code to activate the system. How many different codes can be made with four digits? Identify codes that may be impractical.

**ELEC**  The number system has 10 digits: 0, 1, 2, 3, 4, 5, 6, 7, 8, and 9, so we can fill each one of the four slots of the code in 10 different ways. We allow the same digit to be used repeatedly. (This is called *sampling with replacement*.) To find the total number of different codes, we multiply $10 \cdot 10 \cdot 10 \cdot 10$. This product can be written as $10^4$, or 10,000. **There are 10,000 ways to make a four-digit security code. Some codes, such as 0000, may be impractical**.

**TIP!**

### Negative Base Versus an Opposite

$(-2)^4$ is not the same expression as $-2^4$. *To raise a negative number to a power requires parentheses.* $(-2)^4$ is a negative base raised to a power. $-2^4$ is the opposite of $2^4$.

**Negative Base**

$(-2)^4 = (-2)(-2)(-2)(-2) = 16$

**Opposite**

$-2^4 = -(2)(2)(2)(2) = -16$

Sometimes the values of two expressions are equal, but the interpretation is different. The expressions $(-3)^5$ and $-3^5$ give the same result because the exponent is an odd number.

$$(-3)^5 = (-3)(-3)(-3)(-3)(-3) = -243 \qquad -3^5 = -(3)(3)(3)(3)(3) = -243$$

## 3 Divide Signed Numbers.

The rules for determining the sign when dividing signed numbers are similar to the rules for multiplying signed numbers.

### To divide two signed numbers:

1. Divide the absolute values of the numbers.
2. If the numbers have like signs, the sign of the quotient is positive.
3. If the numbers have unlike signs, the sign of the quotient is negative.

**EXAMPLE**    Divide: **(a)** $\dfrac{-8}{-2}$  **(b)** $\dfrac{6}{3}$  **(c)** $\dfrac{10}{-2}$  **(d)** $\dfrac{-9}{1}$

**(a)** $\dfrac{-8}{-2} = \mathbf{4}$    Like signs give a positive quotient.

**(b)** $\dfrac{6}{3} = \mathbf{2}$    Like signs give a positive quotient.

**(c)** $\dfrac{10}{-2} = \mathbf{-5}$    Unlike signs give a negative quotient.

**(d)** $\dfrac{-9}{1} = \mathbf{-9}$    Unlike signs give a negative quotient.

Division with zero works the same for integers as it does for whole numbers.

### To evaluate division with zero:

Zero divided by any nonzero number is zero.

$$\frac{0}{n} = 0; \qquad n \neq 0 \qquad \frac{0}{-5} = 0$$

Division by zero is either undefined or indeterminate.

$$\frac{n}{0} \quad \text{and} \quad \frac{-5}{0} \text{ are undefined} \qquad \frac{0}{0} \text{ is indeterminate}$$

**EXAMPLE** Evaluate: **(a)** $\dfrac{-3}{0}$ **(b)** $\dfrac{0}{0}$ **(c)** $\dfrac{0}{-5}$

**(a)** $\dfrac{-3}{0}$ **is undefined**

**(b)** $\dfrac{0}{0}$ **is indeterminate**

**(c)** $\dfrac{0}{-5} = \mathbf{0}$

### SECTION 4–3 SELF-STUDY EXERCISES

**1** Multiply.

1. $5 \cdot 8$
2. $-4(-3)$
3. $7 * 5$
4. $(-3)(-7)$
5. $-8(-3)$
6. $(-2)(-3)$
7. $5(-3)$
8. $(-2)(5)$
9. $-4 * 8$
10. $-3 \cdot 4$
11. $-7 * 8$
12. $6(-4)$

13. **BUS** Madison Duke had four checks returned for insufficient funds, and her bank charged her a $28 service charge for each check. Use multiplication of signed numbers to show how these transactions affected her checking account balance.

14. **BUS** Chloe Duke made seven withdrawals of $40 each from her checking account. Use signed numbers to show how these transactions affected her checking account balance.

15. $8(-3)(-2)(7)$
16. $5(-2)(3)(2)$
17. $6(1)(-3)(-2)$
18. $4(0)(-12)(3)$
19. $15(-2)(-3)$
20. $5(2)(-3)(0)$
21. $-3(2)(-7)(-1)$
22. $9(-1)(3)(-2)$
23. $(-7)(-5)(-6)$
24. $(-3)(-9)(-12)(-7)$
25. $7(-3)(-10)(12)(-8)$
26. $(7)(8)(-5)(-3)$
27. $(-3)(-8)(-2)(5)$
28. $-8(0)$
29. $5(0)$
30. $0(-12)$
31. $(-15)(0)$
32. $18(0)$
33. $0 \cdot 3$
34. $0(-15)$
35. $0(-17)$
36. $-28 \cdot 0$
37. $46 \cdot 0$
38. $5(1)(-3)(2) + 5(0)$
39. $-8(1)(3)(-7) + 7(0)$
40. $5(0) + 2(-3)(-7)$
41. $4(-2) + 3(0)(-5)$
42. $(-3)(-6)(-8)(0)$
43. $2(-3) - 5(2) + 7(0)(-4)$

44. Review the definitions for additive inverse and additive identity and write a similar definition for multiplicative inverse.

45. Illustrate the commutative property of multiplication using one positive and one negative integer.

**2** Evaluate.

46. $(-3)^2$
47. $(-2)^3$
48. $(-5)^2$
49. $0^{10}$
50. $-2^3$
51. $(-8)^3$
52. $(5)^4$
53. $3^4$
54. $7^4$
55. $(-42)^2$
56. $(5)^1$
57. $(-28)^1$
58. $12^1$
59. $(-8)^1$
60. $-10^1$
61. $-8^1$
62. $(-3)^3$
63. $-11^2$
64. $(-11)^2$
65. $-5^3$

**3** Divide.

66. $-15 \div 5$
67. $\dfrac{8}{-2}$
68. $\dfrac{-24}{6}$
69. $-28 \div 7$
70. $\dfrac{-32}{-4}$

71. $\dfrac{-50}{-10}$
72. $\dfrac{36}{-6}$
73. $\dfrac{-48}{-6}$
74. $\dfrac{-25}{-5}$

**75. BUS** You have decreased your house loan balance by $1,800 over the past 6 months. Show this figure as a signed number, and find the average monthly decrease.

**76.** A temperature of 10°C fell to −8°C in 3 h. Find the average hourly change expressed as a signed number.

**77.** $\dfrac{12}{0}$

**78.** $\dfrac{-15}{0}$

**79.** $\dfrac{0}{+3}$

**80.** $\dfrac{0}{-12}$

**81.** $\dfrac{0}{-8}$

**82.** $\dfrac{-20}{0}$

**83.** $\dfrac{-100}{0}$

**84.** $\dfrac{1}{0}$

**85.** $\dfrac{0}{8}$

**86.** $\dfrac{-350}{0}$

**87.** Which two operations have the same rules for handling signs when working with signed numbers?

**88.** A nor'easter storm blew into Green Bay, Wisconsin, and the temperature changed from 38°F to −22°F between 1 P.M. and 7 P.M. Represent the average hourly change in temperature with a signed number.

**89. BUS** A business started the year with a net worth of −$4,852. During the first six months of the year, the business recovered and increased its net worth to a value of $15,983. Find the average monthly increase in net worth for the 6-month period.

**90.** What value or values must be in the numerator and denominator of a division in order for the result to be zero?

---

## 4–4 | *Signed Fractions and Decimals*

**Learning Outcomes**

**1** Change a signed fraction to an equivalent signed fraction.

**2** Perform basic operations with signed fractions and decimals.

**3** Apply the order of operations with signed numbers.

**1 Change a Signed Fraction to an Equivalent Signed Fraction.**

A fraction has three signs: the sign of the fraction, the sign of the numerator, and the sign of the denominator. The fraction $\frac{2}{3}$ expressed as a *signed fraction* is $+\frac{+2}{+3}$. When a signed fraction has negative signs, we sometimes change the signed fraction to an equivalent signed fraction. The rules for operating with integers can be extended to apply to signed fractions.

**To find an equivalent signed fraction:**

1. Identify the three signs of the fraction (sign of the fraction, sign of the numerator, and sign of the denominator).
2. Change any two of the three signs to the opposite sign.

---

**EXAMPLE** Change $-\dfrac{-2}{-3}$ to three equivalent signed fractions.

$$-\dfrac{-2}{-3} = +\dfrac{+2}{-3} = \dfrac{2}{-3}$$
Change the signs of the fraction and the numerator.

$$-\dfrac{-2}{-3} = +\dfrac{-2}{+3} = \dfrac{-2}{3}$$
Change the signs of the fraction and the denominator.

$$-\dfrac{-2}{-3} = -\dfrac{+2}{+3} = -\dfrac{2}{3}$$
Change the signs of the numerator and the denominator.

Changing the signs of a fraction
perform basic operations with signed

...tool that simplifies our work when we

## Why Would We Want to Change the Signs of a Fraction?

When changing any two signs of a fraction, we can accomplish the

- Avoid dealing with negatives.

$$-\frac{-3}{+4} = +\frac{+3}{+4} \quad \text{or} \quad \frac{3}{4}$$

$$+\frac{-6}{-7} = +\frac{+6}{+7} \quad \text{or} \quad \frac{6}{7}$$

$$-\frac{+5}{-9} = +\frac{+5}{+9} \quad \text{or} \quad \frac{5}{9}$$

- Change subtraction to addition.

$$-\frac{+5}{+6} = +\frac{-5}{+6} \quad \text{or} \quad \frac{-5}{6}$$

- Avoid negative ...comes.

$$-\frac{+3}{-4} = +\frac{+3}{+4} \quad \text{...rs.}$$

$$+\frac{+2}{-5} = +\frac{-2}{+5} \quad \text{or}$$

- Deal with fewer negatives.

$$-\frac{-5}{-8} = +\frac{-5}{+8} \quad \text{or} \quad \frac{-5}{8}$$

**2** **Perform Basic Operations with Signed Fractions and Decimals.**

### To add and subtract signed fractions:

1. Change the signs of the fractions so the denominators are positive and subtractions are expressed as addition.
2. Apply the appropriate rules for adding integers (signed numbers) and for adding fractions.

---

**EXAMPLE** Add $\dfrac{-3}{4} + \dfrac{5}{-8}$.

$$\frac{-3}{4} + \frac{-5}{8}$$
Change the signs of the numerator and denominator in the second fraction so that both denominators are positive.

$$\frac{-6}{8} + \frac{-5}{8}$$
Change to equivalent fractions with a common denominator.

$$\frac{-11}{8}$$
Add numerators, applying the rule for adding numbers with like signs.

$$-1\frac{3}{8}$$
Change to a mixed number. The sign of the mixed number is determined by the rule for dividing numbers with unlike signs.

---

**EXAMPLE** Subtract $\dfrac{-3}{7} - \dfrac{-5}{7}$.

$$\frac{-3}{7} + \frac{5}{7}$$
Change subtraction to addition by changing the signs of the second fraction and the numerator.

$$\frac{2}{7}$$
Apply the rule for adding numbers with unlike signs.

---

**LE**    Multiply $\left(\dfrac{-4}{5}\right)\left(\dfrac{3}{-7}\right)$.

$$\dfrac{-4}{5} \cdot \dfrac{3}{-7}$$        Multiply the numerators and denominators.

$$\dfrac{-12}{-35}$$        Apply rule for dividing numbers with like signs.

$$\dfrac{12}{35}$$

**EXAMPLE**    Simplify $\left(\dfrac{-2}{3}\right)^3$.

$$\left(\dfrac{-2}{3}\right)^3$$        Cube the numerator and cube the denominator.

$$\dfrac{(-2)^3}{3^3}$$        $(-2)(-2)(-2) = -8$; $(3)(3)(3) = 27$

$$\dfrac{-8}{27} \quad \text{or} \quad -\dfrac{8}{27}$$        Manipulate signs if desired.

## To perform basic operations with signed decimals:

1. Determine the indicated operations.
2. Apply the appropriate rules for signed numbers and for decimals.

**EXAMPLE**    Add $-5.32$ and $-3.24$.

$$\begin{array}{r} -5.32 \\ -3.24 \\ \hline -8.56 \end{array}$$        Align decimals and use the rule for adding numbers with like signs.

**EXAMPLE**  Subtract $-3.7$ from 8.5.

$$8.5 - (-3.7) = 8.5 + 3.7 = \mathbf{12.2}$$

Change subtraction to an equivalent addition and add numbers with like signs.

**EXAMPLE**  Multiply 3.91 and $-7.1$.

$$
\begin{array}{r}
3.91 \\
\times \ -7.1 \\
\hline
391 \\
27\,37 \\
\hline
\mathbf{-27.761}
\end{array}
$$

Apply rule for multiplying numbers with unlike signs.

**EXAMPLE**  Divide: $(-1.2) \div (-0.4)$.

$$-0.4\overline{)\ -1.2\ }^{\,+3.}$$

Shift decimal points and divide. Use rule for dividing numbers with like signs.

**TIP!**

## The Rules of Signed Numbers Apply to All Types of Numbers

The rules of signed numbers apply to all types of numbers. First apply the appropriate rule for signed numbers, then perform the necessary calculations.

**3** **Apply the Order of Operations with Signed Numbers.**

Signed numbers follow the same order of operations as whole numbers.

### Perform operations in the following order as they appear from left to right:

1. Parentheses used as groupings and other grouping symbols.    P
2. Exponents (powers and roots).    E
3. Multiplications and divisions.    MD
4. Additions and subtractions.    AS

**EXAMPLE**  Evaluate $8(4 - 6)$.

Work within the grouping symbols first:

$$8\ (4 - 6) =$$    Add or subtract inside parentheses. $4 - 6 = -2$.    P

$$8\ (-2) = \mathbf{-16}$$    Multiply 8 by $-2$.    M D

Or use the distributive principle:

$$8(4 - 6) =$$  Distribute. Multiply each term in the parentheses by the factor 8.

$$8(4) + 8(-6) =$$

$$32 + -48 = -16$$  Add 32 and $-48$.

Symbols for division are also grouping symbols.

**EXAMPLE**  Evaluate $\dfrac{3-5}{2} - \dfrac{9}{2+1} + 4(5)$.

$$\dfrac{3-5}{2} - \dfrac{9}{2+1} + 4(5) =$$  Perform operations grouped by the fraction bar.

$$\dfrac{-2}{2} - \dfrac{9}{3} + 4(5) =$$  Multiply and divide.

$$-1 - 3 + 20 =$$  Add and subtract last.

$$-4 + 20 = 16$$

**EXAMPLE**  Evaluate $-12 \div 3 - (2)(-5)$.

$$-12 \div 3 - (2)(-5) =$$

$$-12 \div 3 - (2)(-5) =$$  Multiply and divide first.

$$-4 - (-10) =$$  Change to a single sign between the numbers.

$$-4 + 10 = 6$$  Add integers with unlike signs.

The problem in the preceding example can also be written without parentheses around the 2. The problem is then expressed as

$$-12 \div 3 - 2(-5)$$

We multiply and divide first, but notice how we perform the multiplication.

$$-12 \div 3 - 2(-5) =$$  Consider the minus sign as the sign of the 2. $-12 \div 3 = -4$ and $-2(-5) = 10$.

$$-4 + 10 =$$  Add the results of the division and the multiplication.

$$-4 + 10 = 6$$  Add integers with unlike signs.

**EXAMPLE**  Evaluate $10 - 3(-2)$.

$$10 - 3(-2) =$$  Think of the multiplication as $-3$ times $-2$.

$$10 + 6 =$$  Interpret as addition.

$$10 + 6 = 16$$  Add integers with like signs.

Only after parentheses and all multiplications and/or divisions are taken care of do we perform the final additions and/or subtractions from left to right.

Chapter 4 / Signed Numbers and Powers of 10

**EXAMPLE** Evaluate $4 + 5(2 - 8)$.

$$4 + 5 \boxed{(2 - 8)} = \qquad \text{Perform operations in parentheses.}$$
$$4 + 5 \boxed{(-6)} = \qquad \text{Multiply.}$$
$$4 - 30 = \qquad \text{Add integers with unlike signs.}$$
$$4 - 30 = \mathbf{-26}$$

TIP!

## The Order of Operations Is Important

Note what happens if we proceed *out of order* in the preceding example!

**Incorrectly Worked**

$$4 + 5(2 - 8) \qquad \text{\textit{Incorrectly} add first instead of last.}$$
$$9(2 - 8) \qquad \text{Perform operation in parentheses second instead of first.}$$
$$9(- 6) \qquad \text{Multiply last instead of second.}$$
$$9(- 6) = -54 \qquad \text{Incorrect solution.}$$

We get an incorrect answer. *The order of operations must be followed to arrive at a correct solution.*

---

**EXAMPLE** Evaluate $5 + (-2)^3 - 3(4 + 1)$.

$$5 + (-2)^3 - 3 \boxed{(4 + 1)} = \qquad \text{Perform operation in parentheses.}$$
$$5 + \boxed{(-2)^3} - 3 \boxed{(5)} = \qquad \text{Raise to a power.}$$
$$5 + \boxed{(-8)} - \boxed{3(5)} = \qquad \text{Multiply.}$$
$$5 + (-8) - \boxed{15} = \qquad \text{Change to one sign only between numbers.}$$
$$5 - 8 - 15 = \qquad \text{First addition of integers.}$$
$$- 3 - 15 = \qquad \text{Remaining addition of integers.}$$
$$\mathbf{-18}$$

---

Since negative numbers are such an integral part of our everyday lives, practically all types of calculators, even the basic calculator, can deal with negative values. However, the notation for negatives and the process for entering negatives varies widely from calculator to calculator. We will look at some of the most common options.

Some calculators use the subtraction key for both subtracting and entering negative numbers. Others have a special key for entering a negative sign that is labeled as a negative sign enclosed in parentheses $\boxed{(-)}$. A common option with basic calculators is the *sign-change key* $\boxed{+/-}$. This key is a *toggle key* and changes the sign of the number in the display to the opposite sign.

---

**EXAMPLE** Use a calculator to evaluate.

**(a)** $2 - 7$ **(b)** $(-7)(2)$ **(c)** $\dfrac{-4}{2} + 14$

The most common options:

**(a)** $2 - 7$ $\qquad 2 \boxed{-} 7 \boxed{=} \Rightarrow \mathbf{-5}$ $\qquad\qquad \boxed{=}$ may be labeled $\boxed{\text{ENTER}}$ or $\boxed{\text{EXE}}$

**(b)** $(-7)(2)$    $\boxed{(-)}\ 7\ \boxed{\times}\ 2\ \boxed{=}\ \Rightarrow\ \mathbf{-14}$

or

$\boxed{(-)}\ 7\ \boxed{(}\ 2\ \boxed{)}\ \boxed{=}\ \Rightarrow\ \mathbf{-14}$    Some calculators interpret parentheses with no operational symbol preceding the parentheses as multiplication.

or

$7\ \boxed{+/-}\ \boxed{\times}\ 2\ \boxed{=}\ \Rightarrow\ \mathbf{-14}$    The sign-change key is entered after the absolute value of the number.

**(c)** $\dfrac{-4}{2} + 14$    $\boxed{(-)}\ 4\ \boxed{\div}\ 2\ \boxed{=}\ \boxed{+}\ 14\ \boxed{=}\ \Rightarrow\ \mathbf{12}$    The first $\boxed{=}$ may not be required if your calculator applies the order of operations. A slash $\boxed{/}$ may be used for division on some calculators.

You may impose the appropriate order of operations by using the $\boxed{=}$ key when you perform a series of calculations. This is necessary with a basic calculator when no parentheses keys are available.

When a bar is used as both a grouping and a division symbol on a calculator, we must instruct the calculator to work the grouping first by enclosing the grouping in parentheses.

---

**EXAMPLE**   Evaluate using a calculator:

$$\frac{5+1}{3} + 2$$

$\dfrac{5+1}{3} + 2$    $\boxed{(}\ 5\ \boxed{+}\ 1\ \boxed{)}\ \boxed{\div}\ 3\ \boxed{+}\ 2\ \boxed{=}\ \Rightarrow\ \mathbf{4}$    Using parentheses for grouping.

or

$5\ \boxed{+}\ 1\ \boxed{=}\ \boxed{\div}\ 3\ \boxed{+}\ 2\ \boxed{=}\ \Rightarrow\ \mathbf{4}$    Using equal for grouping.

---

**EXAMPLE**   Evaluate $-\frac{1}{2} + 4\left(\frac{3}{8} - \frac{5}{8}\right)$.

$-\dfrac{1}{2} + 4\left(\dfrac{3}{8} - \dfrac{5}{8}\right) =$    Perform operation inside grouping. $\dfrac{3}{8} - \dfrac{5}{8} = \dfrac{-2}{8} = -\dfrac{1}{4}$.

$-\dfrac{1}{2} + 4\left(-\dfrac{1}{4}\right) =$    Multiply. $4\left(-\dfrac{1}{4}\right) = -1$.

$-\dfrac{1}{2} - 1 =$    Change $-1$ to an equivalent fraction with a denominator of 2.

$-\dfrac{1}{2} - \dfrac{2}{2} =$    Add like fractions with like signs.

$-\dfrac{3}{2}$ or $-1\dfrac{1}{2}$    Change to signed mixed number if desired.

---

**EXAMPLE**   Evaluate $3.2 - 1.7\,(0.2)^3$.

$3.2 - 1.7(0.2)^3 =$    Raise 0.2 to the third power. $(0.2)(0.2)(0.2) = 0.008$.

$3.2 - 1.7(0.008) =$    Multiply. $-1.7(0.008) = -0.0136$.

$3.2 - 0.0136 =$    Subtract (add opposite).

**3.1864**

**1** Change each fraction to three equivalent signed fractions.

1. $+\dfrac{+5}{+8}$

2. $-\dfrac{3}{4}$

3. $\dfrac{-2}{-5}$

4. $-\dfrac{-7}{-8}$

5. $\dfrac{7}{8}$

**2** Perform the operations.

6. $\dfrac{-7}{8} + \dfrac{5}{8}$

7. $\dfrac{-4}{5} + \left(\dfrac{-3}{10}\right)$

8. $\dfrac{3}{8} + \left(-\dfrac{7}{16}\right)$

9. $\left(-\dfrac{5}{16}\right) + \dfrac{1}{2}$

10. $\dfrac{1}{3} + \left(-\dfrac{5}{9}\right)$

11. $\left(-3\dfrac{1}{4}\right) + \left(-7\dfrac{3}{8}\right)$

12. $1\dfrac{7}{8} + \left(-4\dfrac{5}{12}\right)$

13. $-7\dfrac{1}{2} + \left(-1\dfrac{3}{4}\right) + \left(5\dfrac{3}{8}\right)$

14. $\dfrac{1}{2} - \left(\dfrac{-3}{5}\right)$

15. $\left(-\dfrac{3}{4}\right) + \left(-\dfrac{5}{8}\right)$

16. $-\dfrac{3}{8} + \left(+\dfrac{5}{16}\right)$

17. $1\dfrac{7}{8} - \left(-\dfrac{5}{6}\right)$

18. $3\dfrac{3}{4} - \left(+4\dfrac{1}{4}\right)$

19. $\dfrac{-3}{5} \times \left(\dfrac{10}{-11}\right)$

20. $\left(-\dfrac{4}{9}\right) \times \left(-\dfrac{5}{6}\right)$

21. $\left(-7\dfrac{1}{3}\right)\left(-4\dfrac{1}{11}\right)$

22. $\dfrac{1}{4} \div (-8)$

23. $-5\left(\dfrac{-2}{-5}\right)$

24. $\left(-\dfrac{2}{7}\right)\left(-\dfrac{14}{15}\right)\left(-\dfrac{3}{8}\right)$

25. $\left(-\dfrac{3}{5}\right)\left(\dfrac{-1}{-2}\right)\left(\dfrac{10}{21}\right)$

26. $-\dfrac{5}{8} \div \dfrac{4}{5}$

Round to hundredths if necessary.

27. $5.823 - 32.12$

28. $-8.32 + 7.21$

29. $-84.23 - 7.21$

30. $34.6(-3.2)$

31. $-7.2(8.2)$

32. $-83.1(-4.1)$

33. $83.2 \div (-3)$

34. $-0.826 \div (-2)$

35. $-3.2 + 7.8$

36. $4.23 - 4.2$

37. $4.6 \div (-2)$

38. $-3.8 + (-1.7)$

**3** Evaluate.

39. $\dfrac{4}{5} + 3\left(-\dfrac{2}{3} + \dfrac{1}{3}\right)$

40. $\dfrac{-5}{8} + 7\left(\dfrac{1}{8} - \dfrac{7}{8}\right)$

41. $\dfrac{1}{2} - 3\left(2\dfrac{3}{4} - \dfrac{1}{4}\right)$

42. $-\dfrac{3}{5} + \dfrac{1}{2}(2 - 8)$

43. $\left(-\dfrac{2}{3}\right)^2 - \dfrac{1}{2}\left(\dfrac{3}{8}\right)$

44. $-\dfrac{3}{5} + \left(\dfrac{2}{7}\right)^2$

45. $5.2 + 3.8(-4.1)$

46. $1.3 - (2.1)^2(1.2)$

47. $(0.3)^2 + 5.7(-2.1)$

---

## 4–5 | Powers of 10

*Learning Outcomes*

**1** Multiply and divide by powers of 10.

**2** Raise a power of 10 to a power.

### **1** Multiply and Divide by Powers of 10.

We learned that our number system is based on the number 10; that is, each place value is a *power of 10*. To show how each place value is related to the base 10, we need to define two special exponents. A number with an **exponent of zero** is defined to have a value of 1. On the place-value chart, the ones place is shown to be $10^0$. The decimal places have the power of 10 in the denominator of the fraction representing the place value. That is, the place value is the *reciprocal* of a power of 10 with a positive exponent. To illustrate these decimal places as powers of 10, we use **negative exponents.** Look at the place-value chart in Figure 4–7 to see how our number system relates to powers of 10.

TIP!

### Reciprocals and Negative Exponents

Any nonzero number raised to the zero power is equal to 1.

$$n^0 = 1 \quad \text{if} \quad n \neq 0$$

An expression with a *negative exponent* can be written as an equivalent expression with a positive exponent.

$$n^{-1} = \frac{1}{n} \qquad \frac{1}{n^{-1}} = n \qquad \text{if } n \neq 0$$

A nonzero number times its reciprocal equals 1.

$$n \cdot \frac{1}{n} = 1 \qquad n^1 \cdot n^{-1} = n^0 = 1$$

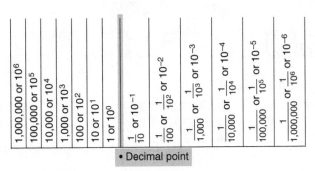

**Figure 4–7**   Base-10 place-value chart.

| Whole-Number Part | | Fractional or Decimal Part | |
|---|---|---|---|
| Millions | $1{,}000{,}000 = 10^6$ | $\frac{1}{10} = 0.1 = 10^{-1}$ | Tenths |
| Hundred thousands | $100{,}000 = 10^5$ | $\frac{1}{100} = 0.01 = 10^{-2}$ | Hundredths |
| Ten thousands | $10{,}000 = 10^4$ | $\frac{1}{1{,}000} = 0.001 = 10^{-3}$ | Thousandths |
| Thousands | $1{,}000 = 10^3$ | $\frac{1}{10{,}000} = 0.0001 = 10^{-4}$ | Ten-thousandths |
| Hundreds | $100 = 10^2$ | $\frac{1}{100{,}000} = 0.00001 = 10^{-5}$ | Hundred-thousandths |
| Tens | $10 = 10^1$ | $\frac{1}{1{,}000{,}000} = 0.000001 = 10^{-6}$ | Millionths |
| Ones | $1 = 10^0$ | | |

### What Does the Exponent in a Power of 10 Tell Us?

- When the exponent of 10 is positive, the exponent is the same as the number of zeros in the equivalent whole number in standard notation.
- When the exponent of 10 is negative, the exponent is the same as the number of zeros in the denominator of the equivalent fraction.
- When the exponent of 10 is negative, the exponent is the same as the number of decimal digits in the decimal equivalent.

## To multiply by a power of 10:

1. If the exponent is positive, shift the decimal point to the *right* the number of places indicated by the *positive* exponent. Attach zeros as necessary.
2. If the exponent is negative, shift the decimal point to the *left* the number of places indicated by the *negative* exponent. Insert zeros as necessary.

---

**EXAMPLE** Perform the multiplications using powers of 10.

    **(a)** $275 \times 10$     **(b)** $0.18 \times 100$     **(c)** $2.4 \times 1{,}000$     **(d)** $43 \times 0.1$

**(a)** $275 \times 10 = 275 \times 10^1 = \mathbf{2{,}750}$     The exponent is +1, so move the decimal one place to the right. Attach one zero.

**(b)** $0.18 \times 100 = 0.18 \times 10^2 = \mathbf{18}$     The exponent is +2, so move the decimal two places to the right.

**(c)** $2.4 \times 1{,}000 = 2.4 \times 10^3 = \mathbf{2{,}400}$     The exponent is +3, so move the decimal three places to the right. Attach two zeros.

**(d)** $43 \times 0.1 = 43 \times 10^{-1} = \mathbf{4.3}$     The exponent is −1, so move the decimal one place to the left.

## To divide by a power of 10:

1. Change the division to an equivalent multiplication.
2. Use the rule for multiplying by a power of 10.

---

**EXAMPLE** Perform the divisions using powers of 10.

    **(a)** $3.14 \div 10$     **(b)** $0.48 \div 100$     **(c)** $20.1 \div 1{,}000$

**(a)** $3.14 \div 10 = 3.14 \times \frac{1}{10}$     Change the division to an equivalent multiplication.

$\qquad\qquad = 3.14 \times 10^{-1}$     Express the fraction $\frac{1}{10}$ as a power of 10.

$\qquad\qquad = \mathbf{0.314}$     Because the exponent of 10 is −1, move the decimal one place to the *left*.

**(b)** $0.48 \div 100 = 0.48 \times \frac{1}{100} = 0.48 \times 10^{-2} = \mathbf{0.0048}$     Move the decimal two places to the left. Insert two zeros.

**(c)** $20.1 \div 1{,}000 = 20.1 \times \frac{1}{1{,}000} = 20.1 \times 10^{-3} = \mathbf{0.0201}$     Move decimal three places to the left. Insert one zero.

To multiply or divide a power of 10 by another power of 10, we can use special applications of the *laws of exponents*. These laws are introduced in Chapter 9, but we will look at these special cases now.

### To multiply powers of 10 written in exponential notation:

1. The product will be a power of 10.
2. Add the exponents for the power-of-10 factors.

Symbolically, $10^a \cdot 10^b = 10^{a+b}$, where $a$ and $b$ are signed numbers.

### To divide powers of 10 written in exponential notation:

1. The quotient will be a power of 10.
2. Subtract the exponents for the power-of-10 factors (numerator minus demoninator).
3. Powers of 10 with negative exponents can be written as reciprocal expressions with positive exponents, if desired.

Symbolically, $\dfrac{10^a}{10^b} = 10^{a-b}$, where $a$ and $b$ are signed numbers; $10^{-a} = \dfrac{1}{10^a}$.

**EXAMPLE**    Multiply or divide using the laws of exponents.

**(a)** $10^5(10^2)$     **(b)** $10^{-1}(10^2)$     **(c)** $10^0(10^3)$     **(d)** $\dfrac{10}{10^3}$

**(e)** $\dfrac{10^5}{10^4}$     **(f)** $\dfrac{10^2}{10^2}$     **(g)** $\dfrac{10^{-2}}{10^3}$

**(a)** $10^5(10^2) = 10^{5+2} = \mathbf{10^7}$     Add exponents.

**(b)** $10^{-1}(10^2) = 10^{(-1+2)} = \mathbf{10^1}$ **or** $\mathbf{10}$     Add exponents.

**(c)** $10^0(10^3) = 10^{0+3} = \mathbf{10^3}$     Add exponents.

**(d)** $\dfrac{10}{10^3} = 10^{1-3} = \mathbf{10^{-2}}$ **or** $\dfrac{\mathbf{1}}{\mathbf{10^2}}$     Subtract exponents.

**(e)** $\dfrac{10^5}{10^4} = 10^{5-4} = \mathbf{10^1}$ **or** $\mathbf{10}$     Subtract exponents.

**(f)** $\dfrac{10^2}{10^2} = 10^{2-2} = \mathbf{10^0}$ **or** $\mathbf{1}$     Subtract exponents.

**(g)** $\dfrac{10^{-2}}{10^3} = 10^{-2-3} = \mathbf{10^{-5}}$ **or** $\dfrac{\mathbf{1}}{\mathbf{10^5}}$     Subtract exponents.

**2** **Raise a Power of 10 to a Power.**

Recall that raising a number to a natural-number power is an extension of multiplication. Powers of 10 can be raised to a power by applying this same concept and thus establishing another law of exponents.

### To raise a power of 10 to a power:

1. The result will be a power of 10.
2. The exponent will be the product of the original exponent and the power.

Symbolically, $(10^a)^b = 10^{ab}$.

---

**EXAMPLE** Raise the powers of 10 to the indicated power.

**(a)** $(10^3)^2$     **(b)** $(10^{-3})^4$

**(a)** $(10^3)^2 = 10^{3\cdot2} = \mathbf{10^6}$
**(b)** $(10^{-3})^4 = 10^{-3(4)} = \mathbf{10^{-12}}$

---

### SECTION 4–5 SELF-STUDY EXERCISES

**1** Perform the operations.

1. $453 \times 100$
2. $0.27 \times 100$
3. $5.82 \times 1,000$
4. $8.97 \times 0.1$
5. $523 \times 0.01$
6. $8.06 \times 0.001$
7. $5.73 \div 10$
8. $0.293 \div 100$
9. $45.7 \div 1,000$
10. $85.79 \div \dfrac{1}{10}$
11. $43.7 \div \dfrac{1}{100}$
12. $8.37 \div \dfrac{1}{1,000}$

Multiply or divide as indicated.

13. $0.37 \times 10^2$
14. $1.82 \times 10^3$
15. $5.6 \times 10^{-1}$
16. $142 \times 10^{-2}$
17. $78 \times 10^4$
18. $62 \times 10^0$
19. $4.6 \div 10^4$
20. $6.1 \div 10$
21. $7.2 \div 10^1$
22. $42 \div 10^0$
23. $10^7 (10^5)$
24. $10^{-3} (10^4)$
25. $10^0 (10^2)$
26. $\dfrac{10^7}{10^3}$
27. $\dfrac{10^5}{10^5}$
28. $\dfrac{10^3}{10^5}$
29. $\dfrac{10^{-5}}{10^7}$
30. $\dfrac{10}{10^7}$
31. $10^4(10^6)$
32. $10^{-3}(10^{-4})$
33. $10^0(10^{-3})$
34. $10^{-3}(10^4)$
35. $10(10^2)$
36. $\dfrac{10^4}{10^2}$
37. $\dfrac{10}{10^4}$
38. $\dfrac{10^4}{10^4}$
39. $\dfrac{10^{-2}}{10^3}$
40. $\dfrac{10^0}{10^1}$

**2** Raise the powers of 10 to the indicated exponents.

41. $(10^2)^3$
42. $(10^3)^4$
43. $(10^{-2})^2$
44. $(10^{-5})^{-2}$
45. $(10^4)^{-2}$

---

*Learning Outcomes*

**1** Change a number from scientific notation to ordinary notation.

**2** Change a number from ordinary notation to scientific notation.

**3** Multiply and divide numbers in scientific notation.

**4** Raise a number in scientific notation to a power.

**1 Change a Number from Scientific Notation to Ordinary Notation.**

A number is expressed in **scientific notation** if it is the product of two factors. The absolute value of the first factor is a number greater than or equal to 1 but less than 10. The second factor is a power of 10.

A number that is written strictly according to place value is an **ordinary number**.

TIP!

**Characteristics of Scientific Notation**

- Numbers between 0 and 1 and between $-1$ and 0 require negative exponents when written in scientific notation.
- The first factor in scientific notation always has an absolute value greater than or equal to one but less than 10. There is only one digit to the left of the decimal and that digit cannot be a zero.
- The use of the times sign ($\times$) for multiplication is the most common representation for scientific notation.
- The second factor is a power of 10.

**EXAMPLE** Which of the terms is expressed in scientific notation?

(a) $4.7 \times 10^2$  (b) $0.2 \times 10^{-1}$  (c) $-3.4 \times 5^2$  (d) $2.7 \times 10^0$

(e) $34 \times 10^4$  (f) $8 \times 10^{-6}$  (g) $-2.8 \div 10^3$

(a) 4.7 is more than 1 but less than 10. $10^2$ is a power of 10. Multiplication is indicated. **Thus, the term is in scientific notation.**

(b) Even though $10^{-1}$ is a power of 10, **this term is not in scientific notation** because the first factor (0.2) is less than 1.

(c) The absolute value of $-3.4$ is more than 1 and less than 10, but $5^2$ is not a power of 10. **Thus, this term is not in scientific notation.**

(d) 2.7 is more than 1 and less than 10. $10^0$ is a power of 10. **Thus, this term is in scientific notation.**

(e) 34 is greater than 10. Even though $10^4$ is a power of 10, **this term is not in scientific notation** because the first factor (34) is 10 or more.

(f) 8 is more than 1 and less than 10. $10^{-6}$ is a power of 10. **Thus, this term is in scientific notation.**

(g) The absolute value of $-2.8$ is more than 1 and less than 10, but division rather than multiplication is the indicated operation. **This term is not in scientific notation.**

**To change from a number written in scientific notation to an ordinary number:**

1. Perform the indicated multiplication by moving the decimal point in the first factor the appropriate number of places. Affix or insert zeros as necessary.
2. Omit the power-of-10 factor.

When we multiply by a power of 10, the exponent of 10 tells us how many places and in which direction to move the decimal.

EXAMPLE    Change to ordinary numbers.

(a) $3.6 \times 10^4$       (b) $2.8 \times 10^{-2}$       (c) $1.1 \times 10^0$       (d) $-6.9 \times 10^{-5}$
(e) $-9.7 \times 10^6$

(a)  $3.6 \times 10^4 = 36000. = \mathbf{36,000}$          Move the decimal four places to the right.

(b)  $2.8 \times 10^{-2} = .028 = \mathbf{0.028}$          Move the decimal two places to the left.

(c)  $1.1 \times 10^0 = \mathbf{1.1}$          Move the decimal zero places.

(d)  $-6.9 \times 10^{-5} = -.000069 = \mathbf{-0.000069}$          Move the decimal five places to the left.

(e)  $-9.7 \times 10^6 = 9700000. = \mathbf{-9,700,000}$          Move the decimal six places to the right.

### 2  Change a Number from Ordinary Notation to Scientific Notation.

To express an ordinary number in scientific notation, we reverse the procedures we used before. Shifting the decimal point changes the value of a number. The power-of-10 factor is used to offset or balance this change so the original value of the number is maintained.

### Change from a number written in ordinary notation to scientific notation:

1. Insert a caret ($\wedge$) in the proper place to indicate where the decimal should be positioned in the ordinary number so that the absolute value of the number is valued at 1 or between 1 and 10.
2. Determine the number of places and in which direction the decimal shifts *from* the new position (caret) *to* the old position (decimal point). This number is the exponent of the power of 10.

### TIP!

### A Balancing Act: Why Count from the New to the Old?

Moving the decimal in the ordinary number changes the value of the number unless you balance the effect of the move in the power-of-10 factor. When a decimal is moved in the first factor, the value is changed. To offset this change, an opposite change must be made in the power-of-10 factor. Counting from the new to the old position indicates the proper number of places and the direction (positive or negative) for balancing with the power-of-10 factor.

Remember the word **NO**. Count from **N**ew to **O**ld.

$3,800 = 3.8 \times 10^3$          $3_\wedge 800. \times 10^3$ N → O = +3.
$0.0045 = 4.5 \times 10^{-3}$          $0.004_\wedge 5 \times 10^{-3}$ N → O = −3.

EXAMPLE    Express in scientific notation.

(a) 285       (b) 0.007       (c) 9.1       (d) 85,000       (e) 0.00074

(a)  $285 \to 2_\wedge 85 = \mathbf{2.85 \times 10^2}$.
The unwritten decimal is after the 5. Place the caret between 2 and 8 so the number 2.85 is between 1 and 10. Count *from* the caret *to* the decimal to determine the exponent of 10. A move two places to the right represents the exponent +2.

**(b)** $0.007 \rightarrow 0.\,\boxed{007}\,_\wedge = \mathbf{7 \times 10^{-3}}$

7 is between 1 and 10. Count *from* the caret *to* the decimal. A move three places to the left represents the exponent $-3$.

**(c)** $9.1 = \mathbf{9.1 \times 10^0}$

9.1 is already between 1 and 10, so the decimal does not move; that is, the decimal moves zero places.

**(d)** $85{,}000 \rightarrow 8\,_\wedge\,5000 = \mathbf{8.5 \times 10^4}$

*From* the caret *to* the decimal is four places to the right.

**(e)** $0.00074 \rightarrow 0.\,\boxed{0007}\,_\wedge 4 = \mathbf{7.4 \times 10^{-4}}$

*From* the caret *to* the decimal is four places to the left.

**TIP!**

### Between 1 and 10: One Digit to the Left of the Decimal and that Digit Cannot be Zero

The expression "between 1 and 10" means any number that is more than 1 but less than 10. The first factor in scientific notation must be equal to 1 *or* between 1 and 10. That means **there will be one and only one digit to the left of the decimal point and that digit cannot be zero.**

Occasionally a number is written in power-of-10 notation but not in scientific notation because the first factor is not equal to 1 or is not between 1 and 10. Write the first factor in scientific notation and multiply the power-of-10 factors.

**EXAMPLE**   Express in scientific notation.

**(a)** $37 \times 10^5$        **(b)** $0.03 \times 10^3$

**(a)**   $3\,_\wedge 7 \times 10^5 = 3.7 \times 10^1 \times 10^5$          Write the first factor in scientific notation and multiply
$\qquad\qquad\quad = \mathbf{3.7 \times 10^6}$          the power-of-10 factors.

**(b)**   $0.03\,_\wedge \times 10^3 = 3 \times 10^{-2} \times 10^3$          Write the first factor in scientific notation and multiply
$\qquad\qquad\quad = \mathbf{3 \times 10^1}$          the power-of-10 factors.

### 3   Multiply and Divide Numbers in Scientific Notation.

We can multiply or divide numbers expressed in scientific notation without first having to convert them to ordinary numbers.

**Multiply numbers in scientific notation:**

1. Multiply the first factors using the rules of signed numbers.
2. Multiply the power-of-10 factors using the laws of exponents.
3. Examine the first factor of the product (Step 1) to see if its value is equal to 1 or between 1 and 10.
    **(a)** If so, write the results of Steps 1 and 2.
    **(b)** If not, write the first factor in scientific notation and multiply power-of-10 factors.

**EXAMPLE**  Multiply and express the product in scientific notation.

(a) $(4 \times 10^2)(2 \times 10^3)$  (b) $(3.7 \times 10^3)(2.5 \times 10^{-1})$  (c) $(8.4 \times 10^{-2})(5.2 \times 10^{-3})$

(a) $(4 \times 10^2)(2 \times 10^3) = \mathbf{8 \times 10^5}$
The product is in scientific notation.

(b) $(3.7 \times 10^3)(2.5 \times 10^{-1}) = \mathbf{9.25 \times 10^2}$
The product is in scientific notation.

(c) $(8.4 \times 10^{-2})(5.2 \times 10^{-3}) = 43.68 \times 10^{-5}$
$43.68 \times 10^{-5}$ is not in scientific notation.

$43.68 \rightarrow 4_{\wedge}3.68$  or  $4.368 \times 10^1$       Write first factor in scientific notation.
$4.368 \times 10^1 \times 10^{-5} = 4.368 \times 10^{1-5}$       Multiply the powers-of-10 factors.

$\qquad\qquad = \mathbf{4.368 \times 10^{-4}}$

Division involving numbers in scientific notation is similar to multiplication.

## To divide numbers in scientific notation:

1. Divide the first factors using the rules of signed numbers.
2. Divide the power-of-10 factors using the laws of exponents.
3. Examine the first factor of the quotient (Step 1) to see if its value is equal to 1 or between 1 and 10.
   (a) If so, write the results of Steps 1 and 2.
   (b) If not, write the first factor in scientific notation and multiply the power-of-10 factors.

**EXAMPLE**  Divide and express the quotient in scientific notation.

(a) $\dfrac{3 \times 10^5}{2 \times 10^2}$  (b) $\dfrac{1.44 \times 10^{-3}}{6 \times 10^{-5}}$  (c) $\dfrac{9.6 \times 10^{29}}{3.2 \times 10^{111}}$  (d) $\dfrac{1.25 \times 10^3}{5}$

(a) $\dfrac{3 \times 10^5}{2 \times 10^2} = \dfrac{3}{2} \times 10^{5-2} = \mathbf{1.5 \times 10^3}$
The first factor is usually written in decimal notation.

(b) $\dfrac{1.44 \times 10^{-3}}{6 \times 10^{-5}} = \dfrac{1.44}{6} \times 10^{-3-(-5)} = 0.24 \times 10^{-3+5} = 0.24 \times 10^2$
$0.24$ is less than 1, so adjustments are necessary.
$0.24 \rightarrow 0.2_{\wedge}4 = 2.4 \times 10^{-1}$       Write the first factor in scientific notation.
Then $2.4 \times 10^{-1} \times 10^2 = \mathbf{2.4 \times 10^1}$.       Multiply the powers-of-10 factors.

(c) $\dfrac{9.6 \times 10^{29}}{3.2 \times 10^{111}} = \dfrac{9.6}{3.2} \times 10^{29-111} = \mathbf{3 \times 10^{-82}}$

(d) $\dfrac{1.25 \times 10^3}{5}$       5 is the same as $5 \times 10^0$.

$\dfrac{1.25 \times 10^3}{5 \times 10^0} = \dfrac{1.25}{5} \times 10^{3-0} = 0.25 \times 10^3$       0.25 is less than 1.

$0.25 \rightarrow 0.2_{\wedge}5 = 2.5 \times 10^{-1}$       Write the first factor in scientific notation and multiply the power-of-10 factors.

Then, $2.5 \times 10^{-1} \times 10^3 = 2.5 \times 10^{-1+3} = \mathbf{2.5 \times 10^2}$.

## Negative Exponents and Mental Adjustment of Exponents

In scientific notation, we **do** leave exponents as negative numbers when appropriate.

Once we understand the concept of balancing the effect of moving a decimal by adjusting the exponent of the power-of-10 factor, we can perform this adjustment mentally.

$$43.68 \times 10^{-5} = 4_{\wedge}3.68 \times 10^{-5} \qquad N \to O = +1.$$
$$= 4.368 \times 10^{-5+1} \qquad \text{Adjust mentally.}$$
$$= 4.368 \times 10^{-4}$$
$$0.25 \times 10^{3} = 0.2_{\wedge}5 \times 10^{3} \qquad N \to O = -1.$$
$$= 2.5 \times 10^{3-1} \qquad \text{Adjust mentally.}$$
$$= 2.5 \times 10^{2}$$

## Scientific Notation and the Calculator

### Power-of-10 Key

The power-of-10 key, labeled $\boxed{10^x}$, $\boxed{EXP}$ or $\boxed{EE}$ on most calculators, is a shortcut key for entering the following keys:

$$\boxed{\times}\ 10\ \boxed{x^y}$$

The shortcut key is used only for power-of-10 factors, and only the *exponent* of 10 is entered. If you enter $\times 10$, then $\boxed{EXP}$, your answer will have one extra factor of 10.

Look at $\dfrac{(3 \times 10^5)}{(2 \times 10^2)}$ on the calculator. *Steps will vary on various calculators.*

$$3\ \boxed{10^x}\ 5\ \boxed{)}\ \boxed{\div}\ \boxed{(}\ 2\ \boxed{10^x}\ 2\ \boxed{=} \Rightarrow 1500 \quad \text{or} \quad 3\ \boxed{EXP}\ 5\ \boxed{\div}\ 2\ \boxed{EXP}\ 2\ \boxed{=} \Rightarrow 1500$$

This result may be expressed in scientific notation if desired: $1{,}500 = 1.5 \times 10^3$.

The internal program of a calculator has predetermined how the output of a calculator is displayed. For example, even if you would like an answer displayed in scientific notation, it may fall within the guidelines for display as an ordinary number. You must make the conversion to scientific notation yourself. The reverse may also be true.

Experiment using problems for which you already know the answer to determine the limitations and requirements of your calculator. Some calcualtors require the parentheses and some do not.

| | |
|---|---|
| **EXAMPLE** | A star is 4.2 light-years from Earth. If 1 light-year is $5.87 \times 10^{12}$ mi, how many miles from Earth is the star? |
| **AVIA** | To solve this problem, multiply the number of light years times the distance of 1 light year. |
| **Estimation** | By rounding we estimate that the star will be approximately 4 times $6 \times 10^{12}$ mi away. |

$$4 \times 6 \times 10^{12} = 24 \times 10^{12} = 2.4 \times 10^{13}$$

$$4.2 \times 5.87 \times 10^{12} =$$

$$24.654 \times 10^{12} =$$

$$2.4654 \times 10^{12+1} =$$

$$2.4654 \times 10^{13}$$

or     $2.5 \times 10^{13}$

$4.2 = 4.2 \times 10^0$.

Perform scientific notation adjustment.

$N \rightarrow O = +1$.

Round the first factor to tenths.

**Interpretation**    **The star is $2.5 \times 10^{13}$ mi from Earth.**

---

**EXAMPLE**    An angstrom (Å) is $1 \times 10^{-7}$ mm. What is the length in millimeters of 14.82 Å?

**ELEC**    Solve the problem by mulitplying the number of angstrom units times the length of 1 Å.

**Estimation**    14.82 angstrom units are more than 10 times longer than $1 \times 10^{-7}$ mm, or more than $1 \times 10^{-6}$.

$$14.82 \times 1 \times 10^{-7} =$$

$$14.82 \times 10^{-7} =$$

$$1.482 \times 10^{-7+1} =$$

$$1.482 \times 10^{-6} \text{ mm}$$

Adjust the first factor and exponent.

$N \rightarrow O = +1$.

**Interpretation**    **The length of 14.82 Å is $1.482 \times 10^{-6}$ mm.**

---

**EXAMPLE**    One coulomb (C) is approximately $6.28 \times 10^{18}$ electrons. How many coulombs do $2.512 \times 10^{21}$ electrons represent?

**ELEC**    Divide the total number of electrons by the number of electrons in one coulomb.

$$\frac{2.512 \times 10^{21}}{6.28 \times 10^{18}}$$

**Estimation**    By rounding the first factors we have

$$\frac{3 \times 10^{21}}{6 \times 10^{18}} = 0.5 \times 10^{21-18} = 0.5 \times 10^3 = 500\,\text{C}$$

$$\frac{2.512 \times 10^{21}}{6.28 \times 10^{18}} =$$

$$0.4 \times 10^3 =$$

$$4 \times 10^{3-1} =$$

$$4 \times 10^2 \text{ or}$$

$$400\ \text{C}$$

Divide coefficients, subtract exponents.

Adjust the first factor.

$N \rightarrow O = -1$.

Write as an ordinary number.

**Interpretation**    **$2.512 \times 10^{21}$ electrons represent 400 C.**

### 4 Raise a Number in Scientific Notation to a Power.

Two more of the laws of exponents are applied when raising a number in scientific notation to a power. These laws address raising more than one factor to a power and raising a power to a power. Again, we will examine these laws in Chapter 9; but, for now we will examine the laws as they apply to scientific notation.

#### To raise a number in scientific notation to a power:

1. Raise the first factor to the power.
2. Raise the power of 10 to the power by multiplying exponents.
3. Adjust the first factor and power of 10 so that the first factor is greater than or equal to 1 or less than 10.

Symbolically, $(n \times 10^a)^b = n^b \times 10^{ab}$. Adjust so that $n \geq 1$ and $n < 10$.

---

**EXAMPLE**    Raise the following to the indicated powers: (a) $(2 \times 10^3)^2$    (b) $(4.2 \times 10^{-4})^3$

(a) $(2 \times 10^3)^2 = 2^2 \times 10^{3(2)}$        Square each factor.
$$= 4 \times 10^6$$

(b) $(4.2 \times 10^{-4})^3 = 4.2^3 \times 10^{-4(3)}$        Cube each factor.
$$= 74.088 \times 10^{-12}$$        Compute the first factor and power-of-10 exponent.
$$= 7.4088 \times 10^{-12+1}$$
$$= 7.4088 \times 10^{-11}$$

---

### SECTION 4–6 SELF-STUDY EXERCISES

**1**  Write as ordinary numbers.

| | | | |
|---|---|---|---|
| **1.** $4.3 \times 10^2$ | **2.** $6.5 \times 10^{-3}$ | **3.** $2.2 \times 10^0$ | **4.** $7.3 \times 10$ |
| **5.** $9.3 \times 10^{-2}$ | **6.** $8.3 \times 10^4$ | **7.** $5.8 \times 10^{-3}$ | **8.** $8 \times 10^4$ |
| **9.** $6.732 \times 10^0$ | **10.** $5.89 \times 10^{-3}$ | **11.** $7.83 \times 10$ | **12.** $1.59 \times 10^3$ |
| **13.** $3.97 \times 10^5$ | **14.** $4.723 \times 10^{-4}$ | **15.** $9.91 \times 10^{-6}$ | **16.** $1.03 \times 10^6$ |

**2**  Express in scientific notation.

| | | | |
|---|---|---|---|
| **17.** 392 | **18.** 0.02 | **19.** 7.03 | **20.** 42,000 |
| **21.** 0.081 | **22.** 0.0021 | **23.** 23.92 | **24.** 0.101 |
| **25.** 1.002 | **26.** 721 | **27.** $42 \times 10^4$ | **28.** $32.6 \times 10^3$ |
| **29.** $0.213 \times 10^2$ | **30.** $0.0062 \times 10^{-3}$ | **31.** $56,000 \times 10^{-3}$ | **32.** $0.197 \times 10^{-5}$ |
| **33.** $745 \times 10^{-1}$ | **34.** $18 \times 10^3$ | **35.** $0.701 \times 10^2$ | **36.** $72,500 \times 10^{-5}$ |

**3**  Perform the indicated operations. Express answers in scientific notation.

**37.** $(6.7 \times 10^4)(3.2 \times 10^2)$    **38.** $(1.6 \times 10^{-1})(3.5 \times 10^4)$    **39.** $(5.0 \times 10^{-3})(4.72 \times 10^0)$

**40.** $(8.6 \times 10^{-3})(5.5 \times 10^{-1})$    **41.** $\dfrac{3.15 \times 10^5}{4.5 \times 10^2}$    **42.** $\dfrac{4.68 \times 10^3}{7.2 \times 10^7}$

**43.** $\dfrac{4.55 \times 10^{-1}}{6.5 \times 10^{-4}}$    **44.** $\dfrac{7.84 \times 10^{-2}}{9.8 \times 10^0}$    **45.** $\dfrac{1.96 \times 10^{-3}}{8.0 \times 10^{-5}}$

**46.** A star is 5.5 light years from Earth. If one light-year is $5.87 \times 10^{12}$ miles, how many miles from Earth is the star?

**47.** An angstrom (Å) is $1 \times 10^{-7}$ mm. How many angstroms are in $4.2 \times 10^{-5}$ mm?

**4** Perform the indicated operation and express the result in scientific notation. Round appropriately.

**48.** $(8.3 \times 10^2)^3$

**49.** $(7 \times 10^{-2})^4$

**50.** $(1.93 \times 10^{-4})^2$

**51.** $(7.2 \times 10^2)^2 (3.1 \times 10^1)^2$

**52.** $\dfrac{(8.2 \times 10^{-3})^2}{(5.73 \times 10^4)^3}$

**53.** $63,000 \times 7,000$

**54.** $(2,500)(61,000)(300)$

**55.** $\dfrac{42,000}{60,000}$

**56.** $(52,000)^2$

**57.** $(0.00013)^{-2}$

**58.** $(17,000)^2$

**59.** $(83,000)^2(5,200)^2$

**60.** $\dfrac{(0.0071)^3}{(0.02)^4}$

**61.** $\left(\dfrac{210 \times 300}{510}\right)^2$

**62.** $\left(\dfrac{5,000 \times 820}{4,000}\right)^3$

**63.** $\left(\dfrac{16.7 \times 5,200}{0.16 \times 820}\right)^3$

**64.** $\left(\dfrac{0.0061 \times 2,300}{0.23 \times 46,000}\right)^3$

---

## CHAPTER REVIEW OF KEY CONCEPTS

| Learning Outcomes | What to Remember with Examples |
|---|---|

### Section 4–1

**1** Compare signed numbers (pp. 199–201).

Positive numbers are to the right of zero and negative numbers are to the left of zero on the number line.

> Arrange from smallest to largest: $5, -3, 0, 8, -5$
>
> $$-5, -3, 0, 5, 8$$

The "greater than" symbol is $>$.
The "less than" symbol is $<$.

> Use $>$ or $<$ to make a true statement: $5 \;?\; -3; \; 5 > -3 \quad$ or $\quad -3 < 5$

The absolute value of a number is its *distance* from zero without regard to direction.

> Evaluate the following absolute values: $|-3|, |5|$
>
> $$|-3| = 3, \qquad |5| = 5$$

Opposites are numbers that have the same absolute value but opposite signs.

> Give the opposite of the following: $8, -4, -2, +4$
>
> $$-8, \quad 4, \quad 2, -4$$

**2** Add signed numbers with like signs (p. 202).

To add signed numbers with like signs: **1.** Add the absolute values. **2.** Give the sum the common or like sign.

> Add: $13 + 7 = 20 \qquad -35 + (-13) = -48$

**3** Add signed numbers with unlike signs (pp. 203–204).

To add signed numbers with unlike signs: **1.** Subtract the smaller absolute value from the larger absolute value. **2.** Give the sum the sign of the number with the larger absolute value.

> Add: $-12 + 7$
>
> Subtract absolute values: $12 - 7 = 5$. Give the 5 the negative sign because $-12$ has the larger absolute value.
>
> $$-12 + 7 = -5$$

Chapter Review of Key Concepts

When zero is added to a signed number, the result is unchanged: $0 + a = a + 0 = a$.

> Add: $-42 + 0 = -42$      $0 + 38 = 38$

### Section 4–2

**1** Subtract signed numbers (pp. 205–206).

To subtract signed numbers: **1.** Change subtraction to addition. **2.** Change the sign of the second number, the subtrahend. **3.** Use the appropriate rule for adding signed numbers.

> Subtract: $-32 - (-28)$
>
> $$-32 - (-28) = -32 + (+28)$$
> $$= -4$$

To subtract with zero: **1.** Change the subtraction to addition. **2.** Use the appropriate rule for addition.

> $-8 - 0 = -8 + (0) = -8$      $0 - (-4) = 0 + (+4) = 4$

Subtracting an opposite from a number is the same as adding a number to itself.

> $8 - (-8) = 8 + 8 = 16$      $-12 - 12 = -24$

**2** Combine addition and subtraction (pp. 206–208).

**1.** Change all subtractions to additions. **2.** Add the series of signed numbers from left to right.

> Simplify: $5 - 3 + 2 - (-4)$
>
> $$5 + (-3) + 2 + 4 =$$
> $$2 + 2 + 4 = 4 + 4 = 8$$

### Section 4–3

**1** Multiply signed numbers (pp. 209–211).

To multiply signed numbers with like signs: **1.** Multiply the absolute values. **2.** Make the sign of the product positive.

> $-6(-7) = +42$      $8 \times 6 = 48$

To multiply signed numbers with unlike signs: **1.** Multiply the absolute values. **2.** Make the sign of the product negative.

> $-7(2) = -14$      $8(-3) = -24$

To multiply several signed numbers: **1.** The sign of the product is *positive* if the number of negative factors is even. **2.** The sign of the product is *negative* if the number of negative factors is odd.

> $5(-2)(3) = -10(3) = -30$

Any number (including signed numbers) multiplied by zero results in zero: $a \times 0 = 0 \times a = 0$

> $0 \times 3 = 0$      $-5 \times 0 = 0$      $0(-7) = 0$

**2** Evaluate powers of signed numbers (pp. 211–212).

A positive number raised to a natural-number power results in a positive number. Zero raised to any natural-number power is zero. To raise a negative number to a power requires parentheses. A negative number raised to an even natural-number power results in a positive number. A negative number raised to an odd natural-number power is a negative number.

> $(3)^3 = 27$      $(-2)^4 = 16$      $-2^4 = -16$
> $0^5 = 0$      $(-2)^3 = -8$      $-2^3 = -8$

**3** Divide signed numbers (pp. 212–213).

To divide signed numbers: **1.** Divide the absolute values of the dividend and the divisor. **2.** If the dividend and divisor have like signs, the sign of the quotient is positive. **3.** If the dividend and divisor have unlike signs, the sign of the quotient is negative.

$$-12 \div (-4) = 3 \qquad 15 \div (-3) = -5$$

Zero divided by any nonzero signed number is zero. $0 \div a = 0$ (if $a$ is not equal to zero). A nonzero number *cannot* be divided by zero. $a \div 0$ is undefined; $0 \div 0$ is indeterminate.

$$0 \div 12 = 0 \qquad -7 \div 0 \text{ is undefined.}$$
$$0 \div 0 \text{ is indeterminate.}$$

### Section 4–4

**1** Change a signed fraction to an equivalent signed fraction (pp. 214–215).

If any two of the three signs of a signed fraction are changed, the fraction's value does not change.

Write three equivalent signed fractions for $-\frac{7}{8}$.

$$-\frac{7}{8} = -\frac{+7}{+8} \quad \text{Equivalent fractions are } +\frac{+7}{-8} \text{ or } -\frac{-7}{-8} \text{ or } +\frac{-7}{+8}$$

**2** Perform basic operations with signed fractions and decimals (pp. 215–217).

To add or subtract signed fractions: **1.** Write equivalent fractions that have positive integers as denominators and have common denominators. **2.** Add or subtract the numerators using the rules for adding signed numbers.

To multiply or divide signed fractions: **1.** Multiply or divide the fractions. **2.** Apply the rules for multiplying or dividing signed numbers.

Add $\frac{-5}{8} + \frac{7}{8}$.

$$\frac{-5}{8} + \frac{7}{8} = \frac{2}{8} = \frac{1}{4}$$

To perform basic operations with signed decimals: **1.** Determine the indicated operations. **2.** Apply the appropriate rules for signed numbers and for decimals.

Add $4.37 + (-2.91)$

$$4.37 - 2.91 = 1.46 \qquad \text{Add decimals with unlike signs.}$$

**3** Apply the order of operations with signed numbers (pp. 217–220).

Expressions are evaluated in the following order from left to right: **1.** Parentheses. **2.** Exponents (powers and roots). **3.** Multiplication and division. **4.** Addition and subtraction.

**Evaluate:**                    **Work:**

$$5 - 4(7 + 2)^2 - 15 \div 5 =$$   Inside parentheses.
$$5 - 4(9)^2 - 15 \div 5 =$$       Raise to power.
$$5 - 4(81) - 15 \div 5 =$$        Multiply.
$$5 - 324 - 15 \div 5 =$$          Divide.
$$5 - 324 - 3 =$$                  Add or subtract.
$$-319 - 3 =$$
$$-322$$

Evaluate $-\frac{1}{4} - \frac{2}{3}\left(-\frac{1}{2}\right)$

$$-\frac{1}{4} - \frac{2}{3}\left(-\frac{1}{2}\right) \qquad \text{Multiply using rules for like signs and multiplying fractions.}$$

$$-\frac{1}{4} + \frac{1}{3} \qquad \text{Use the rule for adding numbers with unlike signs.}$$

Chapter Review of Key Concepts

$$-\frac{3}{12} + \frac{4}{12}$$

Change to equivalent fractions with the LCD.

$$\frac{1}{12}$$

Evaluate $(-0.3)^2 - 2.8$

$(-0.3)^2 - 2.8$       Square signed decimal.

$0.09 - 2.8$       Add decimals with unlike signs.

$-2.71$

## Section 4–5

**1** Multiply and divide by powers of 10 (pp. 221–224).

To multiply by powers of 10, add the exponents and keep the base of 10. To divide by powers of 10, subtract the exponents and keep the base of 10.

Multiply. $10^6(10^7) = 10^{13}$      Divide. $10^3 \div 10^5 = 10^{-2}$

**2** Raise a power of 10 to a power (p. 225).

**1.** The result will be a power of 10. **2.** The exponent will be the product of the original exponent and the power.

Symbolically, $(10^a)^b = 10^{ab}$.

Raise the powers of 10 to the indicated exponents.

$(10^4)^2$                                             $(10^{-3})^2$

$(10^4)^2 = 10^{4 \cdot 2} = 10^8$            $(10^{-3})^2 = 10^{-3 \cdot 2} = 10^{-6}$

## Section 4–6

**1** Change a number from scientific notation to ordinary notation (pp. 226–227).

To change a number from scientific notation to ordinary notation, perform the indicated multiplication by moving the decimal point in the first factor the appropriate number of places. Insert zeros as necessary. Omit the power-of-10 factor.

Write $3.27 \times 10^{-4}$ as an ordinary number.

$00003.27 = 0.000327.$      Shift the decimal 4 places to the *left*.

**2** Change a number from ordinary notation to scientific notation (pp. 227–228).

To change a number written in ordinary notation to scientific notation, insert a caret in the proper place to indicate where the decimal should be positioned so that the absolute value of the number is valued at 1 or between 1 and 10. Then determine how many places and in which direction the decimal shifts from the new position (caret) to the old position (decimal point). This number is the exponent of the power of 10.

Write 54,000 in scientific notation.

$5_\wedge 4000. = 5.4 \times 10^4$      From New to Old is 4 places to the right so the exponent is +4.

**3** Multiply and divide numbers in scientific notation (pp. 228–231).

To multiply numbers in scientific notation, multiply the first factors, then multiply the powers of 10 by adding exponents. Next, examine the first factor of the product to see if its absolute value is equal to 1 or is between 1 and 10. If the absolute value of the factor is 1 or is between 1 and 10, the process is complete. If the absolute value of the factor is not 1 or not between 1 and 10, shift the decimal so that the factor is equal to 1 or is between 1 and 10, and adjust the exponent of the power of 10 accordingly.

Multiply.

$(4.5 \times 10^{89})(7.5 \times 10^{36}) =$      Multiply first factors then power-of-10 factors.

$33.75 \times 10^{125} =$      Write first factor in scientific notation.

$3_\wedge 375 \times 10^1 \times 10^{125} =$      Multiply powers of 10.

$3.375 \times 10^{126}$

Chapter 4 / Signed Numbers and Powers of 10

To divide numbers in scientific notation, use steps similar to multiplication, but apply the rule for the division of signed numbers and subtract exponents.

Divide.

$$(3 \times 10^{-3}) \div (4 \times 10^2) =$$    Divide first factors, then power-of-10 factors.
$$0.75 \times 10^{-5} =$$    Write first factor in scientific notation.
$$07_\wedge 5 \times 10^{-1} \times 10^{-5} =$$    Multiply powers of 10.
$$7.5 \times 10^{-6}$$

**4** Raise a number in scientific notation to a power (p. 232).

**1.** Raise the first factor to the power. **2.** Raise the power of 10 to the power by multiplying exponents. **3.** Adjust the first factor and power of 10 so that the first factor is greater than or equal to 1 or less than 10.

Symbolically, $(n \times 10^a)^b = n^b \times 10^{ab}$. Adjust so that $n \geq 1$ and $n < 10$.

Raise the following to the indicated powers:

(a) $(3 \times 10^4)^2$    (b) $(2.3 \times 10^{-3})^3$

(a) $(3 \times 10^4)^2 = 3^2 \times 10^{4 \cdot 2} = 9 \times 10^8$    Square each factor.
(b) $(2.3 \times 10^{-3})^3 = 2.3^3 \times 10^{-3 \cdot 3} = 12.167 \times 10^{-9}$    Cube each factor.
$$= 1.2167 \times 10^{-9+1}$$    Adjust the first factor and the power-of-10 factor.

$$= 1.2167 \times 10^{-8}$$

## CHAPTER REVIEW EXERCISES

### Section 4–1

Give the value of each number:
**1.** $|5|$      **2.** $|-8|$      **3.** $|+7|$      **4.** $|-52|$

Give the opposite of each number:
**5.** $-12$      **6.** 8      **7.** $-2$      **8.** $-13$      **9.** 87

Add.
**10.** $-3 + (-8)$      **11.** $(-15) + 8$      **12.** $7 + (-11)$      **13.** $-25 + 0 + 12 + 7$

**14. BUS** A publicly traded company has a profit of $256,872 for one year and a loss of $38,956 for the following year. What is the net profit over the two-year period?

**15.** A football team gained and lost the following yardage during a series of plays beginning with first down: $+4$, $-5$, $+9$. What is the net yardage for the three plays?

**16. AVIA** Because of stormy weather, a pilot flying at 35,000 ft descends 8,000 ft. What is his new altitude?

**17. BUS** Agnes opens a checking account by depositing $500. She then writes checks for $42, $18, and $21. What is her balance after depositing another $150?

### Section 4–2

Evaluate.
**18.** $8 - 5$      **19.** $-9 - 4$      **20.** $-7 - (-2)$
**21.** $11 - (-3)$      **22.** $12 + 3 + (-8) - 5$      **23.** $-6 + 3 - 5 - 7$

**24.** Temperatures in northern Canada ranged as high as 37°F one summer. That same year the lowest temperature was $-28$°F. What was the range of temperatures for the year?

**25.** What is the difference (or range) in temperatures of 43° above zero and 27° below zero?

**26.** What is the difference in temperatures of 47° below zero and 28° below zero?

**27. HLTH/N** Two successive recordings for a surgery patient's temperature were 103.2°F and 97.8°F. Express the temperature change with a signed number.

**28. CON** A 10-ft-long fence post is placed in a hole that is 3 ft deep. How much of the post is above the ground?

**29.** If the temperature changes from $-22$°C to 14°C, what is the change?

## Section 4–3

Evaluate.

**30.** $-3(-7)$  **31.** $7(-2)$  **32.** $-7(3)$  **33.** $2(3)(-7)(0)$

**34.** $5(-2)(-1)(-3)$  **35.** $4(3)(-2)(7)$  **36.** $(-3)^2$  **37.** $(7)^3$

**38.** $(-4)^3$  **39.** $-4^2$  **40.** $-2^3$  **41.** $5^2$

**42. BUS** A stock dropped $2.00 for each of seven straight weeks. What was the total change in the stock value?

**43.** On one winter day, the temperature dropped $2°$ each hour for 5 h. What was the total drop in temperature?

**44.** If the temperature in Exercise 43 was $8°$ originally, what was the temperature at the end of the 5-h period?

**45. BUS** An article states, "XYZ stock has dropped 4 points each week for the past 5 weeks." Was the stock price higher or lower 5 weeks ago than it is today? How much higher or lower? Use negative numbers to express drops in prices. Use a signed number to express the amount the stock price changed.

Divide.

**46.** $-8 \div (-4)$  **47.** $12 \div 3$  **48.** $\dfrac{14}{-7}$  **49.** $\dfrac{-20}{-5}$  **50.** $\dfrac{16}{-4}$

**51.** $\dfrac{-51}{-3}$  **52.** $\dfrac{0}{-8}$  **53.** $\dfrac{-7}{0}$  **54.** $\dfrac{-51}{3}$  **55.** $\dfrac{51}{-17}$

**56. BUS** A company records the following gains and losses in net profit for a six-month period: $22,973; $-$12,357; $-$2,791; $32,872; $18,930; and $2,093. Find the net gain or loss for the six-month period.

## Section 4–4

Use the order of operations to evaluate the following. Verify the results with a calculator.

**57.** $7(3 + 5)$  **58.** $-2(3 - 1)$  **59.** $\dfrac{15 - 7}{8}$

**60.** $-20 \div 4 - 3(-2)$  **61.** $4 + (-3)^4 - 2(5 + 1)$  **62.** $(-3)^3 + 1 - 8$

Perform the indicated operations.

**63.** $\dfrac{-4}{5} \div \dfrac{7}{15}$  **64.** $3.23 + (-4.61)$  **65.** $-12.4 \div 0.2$  **66.** $\dfrac{-11}{12} - \left(\dfrac{-7}{8}\right)$  **67.** $-\dfrac{7}{8} + \left(-\dfrac{5}{12}\right)$

**68.** $1\dfrac{3}{5} \div \left(-7\dfrac{5}{8}\right)$  **69.** $-2\dfrac{5}{8} \times 4\dfrac{1}{2}$  **70.** $\dfrac{3}{4} - \left(-\dfrac{1}{2}\right)^2$  **71.** $0.2 - 3.1(-7.6)$  **72.** $-0.7 - (-7.2 + 5)$

## Section 4–5

Perform the indicated operations. Express as ordinary numbers.

**73.** $10^5 \cdot 10^7$  **74.** $10^{-2} \cdot 10^8$  **75.** $10^7 \cdot 10^{-10}$

**76.** $4.2 \times 10^5$  **77.** $8.73 \div 10^{-3}$  **78.** $5.6 \div 10^{-2}$

Write as ordinary numbers.

**79.** $3.75 \times 10^5$  **80.** $4.23 \times 10^4$  **81.** $3.87 \times 10^{-5}$  **82.** $7.37 \times 10^{-9}$

Write in scientific notation.

**83.** 52,000  **84.** 4,500  **85.** 0.00017  **86.** 3,800,000  **87.** 0.000000008

Perform the indicated operation and write the result in scientific notation.

**88.** $(4.2 \times 10^5)(3.9 \times 10^{-2})$  **89.** $\dfrac{1.25 \times 10^3}{3.7 \times 10^{-8}}$

Raise to the indicated power and express in scientific notation.

**90.** $(5.2 \times 10^3)^4$  **91.** $(8.3 \times 10^{-2})^3$  **92.** $(4.5 \times 10^{-2})^{-2}$

**93.** The United States population is approximately 250 million. Write the number in scientific notation.

**94.** One coulomb (C) is approximately $6.28 \times 10^{18}$ electrons. How many coulombs do $4.87 \times 10^{15}$ electrons represent?

1. Your team is examining the patterns that could be used for automobile license plates. Determine how many license plates can be made using the following patterns. Assume that no plate will be discarded from these patterns.
   (a) 3 digits followed by 3 letters and both digits and letters can be repeated
   (b) the same option from part (a) with either the group of letters or numbers coming first
   (c) 6 letters and all letters can be repeated

2. The national debt at one point was approximately $7.5 trillion. Assuming that a dollar bill is approximately 0.2 mm thick, answer the following questions.
   (a) Suppose the national debt is represented with dollar bills stacked on top of each other. How many kilometers high will the stack reach?
   (b) The distance from Earth to the moon is 380,000 km. How many times will the stack of bills represented by the national debt reach from Earth to the moon?
   (c) If there are approximately 300 million people in the United States, how much would it take per person to pay off the national debt?

# PRACTICE TEST

Use the symbols $>$ and $<$ to write the following as *true* statements.

1. $-8$ is less than 0
2. 2 is less than 3
3. $-5$ is more than $-10$

Answer the questions.

4. What is the value of $|-12|$?
5. What is the opposite of 8?

Perform the operations.

6. $-8 - 2$

7. $-3 + 7$

8. $\dfrac{8}{-2}$

9. $2(6)(-4)$

10. $-\dfrac{2}{5} + \dfrac{1}{10}$

11. $-1.3 - 2.4 + 5.8$

12. $-7 + (-3)$

13. $(-8)(3)(0)(-1)$

14. $-3 + 5 + 0 + 2 + (-5)$

15. $\dfrac{-7}{0}$

16. $4(-13)$

17. $\dfrac{4}{-2}$

18. $2 + 10(2) + 6(7)$

19. $2(3 - 9) \div 2^2 + 7$

20. $5(2 - 3) + 16 \div 4$

21. $(10^3)^2$

22. $\dfrac{10^{-5}}{10^3}$

Write as ordinary numbers.

23. $42 \times 10^3$

24. $0.83 \times 10^2$

25. $5.9 \times 10^{-2}$

Write in scientific notation.

26. $5.2301$

27. $0.021$

28. $52.3 \times 10^2$

29. $783 \times 10^{-5}$

Perform the indicated operations. Express the answers in scientific notation.

30. $(5.9 \times 10^5)(3.1 \times 10^4)$

31. $\dfrac{5.25 \times 10^4}{1.5 \times 10^2}$

32. A star is 3.4 light-years from Earth. If 1 light-year is $5.87 \times 10^{12}$ mi, how many miles from Earth is the star?

33. The total resistance (in ohms) of a dc series circuit equals the total voltage divided by the total amperage. If the total voltage is $3 \times 10^3$ V and the total amperage is $2 \times 10^{-3}$ A, find the total resistance (in ohms) expressed as an ordinary number.

34. In Grand Rapids, Minnesota, the temperature ranged from 42°F at 12 noon to $-14$°F at 7 P.M. Represent the change in temperature with a signed number.

35. Temperatures around the world may range from 135°F in Seville, Spain, to $-40$°F in Fairbanks, Alaska. What is the range (difference) of temperatures?

In Chapter 3 we examined some metric prefixes for very large and very small units of measure. In this chapter we examined power-of-10 exponents and scientific notation. We now examine power-of-10 exponents that are multiples of 3.

**Metric Prefixes**

| Prefix | Abbreviation | Meaning | Power of 10 |
|--------|--------------|---------|-------------|
| Yotta- | Y | one septillion times | $\times 10^{24}$ |
| Zetta- | X | one sextillion times | $\times 10^{21}$ |
| Exa- | E | one quintillion times | $\times 10^{18}$ |
| Peta- | P | one quadrillion times | $\times 10^{15}$ |
| Tera- | T | one trillion times | $\times 10^{12}$ |
| Giga- | G | one billion times | $\times 10^{9}$ |
| Mega- | M | one million times | $\times 10^{6}$ |
| kilo | k | one thousand times | $\times 10^{3}$ |
| **Standard Unit** | | | $\mathbf{\times 10^{0}}$ |
| milli- | m | one thousandth of | $\times 10^{-3}$ |
| micro- | $\mu$ | one millionth of | $\times 10^{-6}$ |
| nano- | n | one billionth of | $\times 10^{-9}$ |
| pico- | p | one trillionth of | $\times 10^{-12}$ |
| femto- | f | one quadrillionth of | $\times 10^{-15}$ |
| atto- | a | one quintillionth of | $\times 10^{-18}$ |
| zepto- | z | one sextillionth of | $\times 10^{-21}$ |
| yocto- | y | one septillionth of | $\times 10^{-24}$ |

These prefixes correlate with a modification of scientific notation called **engineering notation.** Engineering notation, like scientific notation, has a first factor and a power-of-10 factor. The first factor is greater than or equal to 1 and less than 1,000.

To change an ordinary number to engineering notation:

1. Indicate with a caret ( ^ ) where the decimal should be positioned.
   (a) If the number is greater than or equal to one and less than 1,000, the decimal will not shift and the power-of-10 factor will be $10^0$.
   (b) If the number is greater than or equal to 1,000, insert commas as appropriate to separate the place-value periods, and place the caret (the *new* position of the decimal) at the leftmost comma.
   (c) If there are no nonzero digits to the left of the decimal, count from the decimal to the right in groups of three places until you have at least one, but no more than three, significant digits to the left of the caret (*new* position of the decimal).
2. Determine the exponent of the power-of-10 factor and its sign by counting the number of places from the *new* position of the decimal to the *old* position. The resulting exponent will be a multiple of 3.

**EXAMPLE**    Change to engineering notation:
(a) 2,400,000    (b) 2,400    (c) 24
(d) 0.24    (e) 0.024    (f) 0.0000024

(a) 2,400,000    $2_\wedge 400{,}000$    Place the caret at the leftmost comma.
                $2.4 \times 10^6$    $N \rightarrow O = +6$

(b) 2,400    $2_\wedge 400$    Place the caret at the leftmost comma.
            $2.4 \times 10^3$    $N \rightarrow O = +3$

(c) 24    $24_\wedge$    Place the caret at the decimal point. The power-of-10 exponent can be only 0 or a multiple of 3.
       $24 \times 10^0$    $N \rightarrow O = 0$

(d) 0.24    $0.240_\wedge$    Place the caret after the third decimal place value. There can be no more than 3 significant digits on the left of the caret.
       $240 \times 10^{-3}$    $N \rightarrow O = -3$

| | | |
|---|---|---|
| (e) 0.024 | $0.024_\wedge$ | Place the caret after the third decimal value. There can be no more than 3 significant digits on the left of the caret. |
| | $24 \times 10^{-3}$ | $N \rightarrow O = -3$ |
| (f) 0.0000024 | $0.000002_\wedge 4$ | Place the caret after the sixth decimal value. There can be no more than 3 significant digits on the left of the caret. |
| | $2.4 \times 10^{-6}$ | $N \rightarrow O = -6$ |

Measures that are expressed in metric units can be written in engineering notation by showing the power-of-10 factor or by translating the power-of-10 factor to the appropriate metric prefix.

EXAMPLE    Write the engineering notation using metric prefixes.

(a) 4,382,000 Ω    (b) 0.00001 Å    (c) 5,870 µW    (d) 3,500 MHz

(a) $4,382,000\ \Omega = 4.382 \times 10^6\ \Omega$     $10^6 = M$
               $= 4.382\ M\Omega$

(b) $0.00001\ \text{Å} = 10 \times 10^{-6}\ \text{Å}$     $10^{-6} = \mu$
              $= 10\ \mu\text{Å}$

(c) $5,870\ \mu W = 5.87 \times 10^3\ \mu W$     $\mu = 10^{-6}$
             $= 5.87 \times 10^3 \times 10^{-6}\ W$     Multiply power-of-10 factors.
             $= 5.87 \times 10^{-3}\ W$     $10^{-3} = m$
             $= 5.87\ mW$

(d) $3,500\ MHz = 3.5 \times 10^3\ MHz$     $M = 10^6$
               $= 3.5 \times 10^3 \times 10^6\ Hz$     Multiply power-of-10 factors.
               $= 3.5 \times 10^9\ Hz$     $10^9 = G$
               $= 3.5\ GHz$

## Exercises

Write each number in engineering notation.

| | | | |
|---|---|---|---|
| **1.** 3,800,000 | **2.** 5,600 | **3.** 78 | **4.** 52,000 |
| **5.** 80,000,000 | **6.** 5,830,000 | **7.** 1,736,500,000 | **8.** 41,980,000 |
| **9.** 0.78 | **10.** 0.33 | **11.** 0.0000011 | **12.** 0.000000008 |
| **13.** 0.0009832 | **14.** 0.0000719 | **15.** 0.0120307 | **16.** 0.000675 |
| **17.** 0.00000092 | **18.** 0.004 | **19.** 8,400,000 | **20.** 5,900 |
| **21.** 41 | **22.** 31,000 | **23.** 17,000,000 | **24.** 129,000,000 |
| **25.** 3,084,000,000 | **26.** 0.0982 | **27.** 0.0000018 | **28.** 0.000989 |
| **29.** 0.007 | **30.** 0.12 | **31.** 0.00000000035 | **32.** 5,200,000,000 |

Change each unit to engineering notation using metric prefixes.

| | | | |
|---|---|---|---|
| **33.** 428,000 Ω | **34.** 5,700,000 V | **35.** 3,520,000,000 W | **36.** 79,000,000 Hz |
| **37.** 0.000081 s | **38.** 0.0973 Å | **39.** 0.00000000541 s | **40.** 0.89 s |
| **41.** 0.00058 MΩ | **42.** 0.00077 µs | **43.** 2,980,000 ps | **44.** 7,810,000 mW |
| **45.** 42,300 kV | **46.** 1,572,000 kW | **47.** 5,096,000 Ω | **48.** 0.000000008 Å |
| **49.** 182,000 Hz | **50.** 1,600 Ω | **51.** 5,200,000 V | **52.** 97,000,000 W |
| **53.** 0.00049 s | **54.** 0.00052 Å | **55.** 0.588 Å | **56.** 10.53 s |
| **57.** 246.7 V | **58.** 5,082 W | **59.** 42,000 mW | **60.** 7,800 kΩ |
| **61.** 5,729 µW | **62.** 25,000 ps | **63.** 4,800 GHz | **64.** 4,000 ns |
| **65.** 0.000001 Ms | **66.** 0.0047 GΩ | **67.** 0.0000829 W | **68.** 0.000000023 fs |

## Answers

| | | | |
|---|---|---|---|
| **1.** $3.8 \times 10^6$ | **2.** $5.6 \times 10^3$ | **3.** $78 \times 10^0$ | **4.** $52 \times 10^3$ |
| **5.** $80 \times 10^6$ | **6.** $5.83 \times 10^6$ | **7.** $1.7365 \times 10^9$ | **8.** $41.98 \times 10^6$ |
| **9.** $780 \times 10^{-3}$ | **10.** $330 \times 10^{-3}$ | **11.** $1.1 \times 10^{-6}$ | **12.** $8 \times 10^{-9}$ |
| **13.** $983.2 \times 10^{-6}$ | **14.** $71.9 \times 10^{-6}$ | **15.** $12.0307 \times 10^{-3}$ | **16.** $675 \times 10^{-6}$ |

| | | | |
|---|---|---|---|
| **17.** $920 \times 10^{-9}$ | **18.** $4 \times 10^{-3}$ | **19.** $8.4 \times 10^{6}$ | **20.** $5.9 \times 10^{3}$ |
| **21.** $41 \times 10^{0}$ | **22.** $31 \times 10^{3}$ | **23.** $17 \times 10^{6}$ | **24.** $129 \times 10^{6}$ |
| **25.** $3.084 \times 10^{9}$ | **26.** $98.2 \times 10^{-3}$ | **27.** $1.8 \times 10^{-6}$ | **28.** $989 \times 10^{-6}$ |
| **29.** $7 \times 10^{-3}$ | **30.** $120 \times 10^{-3}$ | **31.** $350 \times 10^{-12}$ | **32.** $5.2 \times 10^{9}$ |
| **33.** 428 kΩ | **34.** 5.7 MV | **35.** 3.52 GW | **36.** 79 MHz |
| **37.** 81 μs | **38.** 97.3 mA | **39.** 5.41 ns | **40.** 890 ms |
| **41.** 580 Ω | **42.** 770 ps | **43.** 2.98 μs | **44.** 7.81 kW |
| **45.** 42.3 MV | **46.** 1.572 GW | **47.** 5.096 MΩ | **48.** 8 nA |
| **49.** 182 kHz | **50.** 1.6 kΩ | **51.** 5.2 MV | **52.** 97 MW |
| **53.** 490 μs | **54.** 520 μA | **55.** 588 mA | **56.** 10.53 s |
| **57.** 246.7 V | **58.** 5.82 kW | **59.** 42 W | **60.** 7.8 MΩ |
| **61.** 5.729 mW | **62.** 25 ns | **63.** 4.8 THz | **64.** 4 μs |
| **65.** 1 s | **66.** 4.7 MΩ | **67.** 82.9 μW | **68.** 23 ys |

## CUMULATIVE PRACTICE TEST FOR CHAPTERS 1–4

**1.** Evaluate: $12 + 5(4) \div 2 - 8$

**2.** Evaluate: $6.7^2$

**3.** Find the perimeter of a rectangle that has a width of 24 cm and length of 40 cm.

**4.** Find the area of the rectangle in Exercise 3.

**5.** Round 42.8196 to the nearest hundredth.

**6.** Evaluate: $5^3 - \sqrt{49}(3 - 1)$

**7.** Write in factored form and using exponential notation the prime factorization of 420.

Evaluate and simplify.

**8.** $5\frac{3}{8} + 4 + 12\frac{1}{8}$

**9.** $3\frac{7}{8} + 9\frac{3}{4} + 5\frac{1}{2}$

**10.** $15\frac{3}{5} - 9\frac{3}{8}$

**11.** $4\frac{1}{2} \times \frac{4}{9}$

**12.** $\frac{3}{5} \div \frac{7}{10}$

**13.** $1\frac{3}{4} \div 1\frac{5}{12}$

**14.** 28% of 15 is what number?

**15.** What percent of 24 is 20? Round to the nearest percent.

**16.** 30% of what number is 12?

**17.** A salary of \$42,000 is increased to \$45,000. What is the percent of increase? Round to the nearest hundredth percent.

Change to the indicated unit.

**18.** 12 m = _____ cm

**19.** 14.3 dm = _____ m

**20.** 0.159 dkm = _____ dm

**21.** 86°F = _____ °C

**22.** How many significant digits are in 40.240?

**23.** What is the greatest possible error of the measurement $3\frac{1}{4}$ cm?

**24.** A scale calibrated to 10 grams actually reads 10.05 grams. What is the percent error?

**25.** Add and write the sum using appropriate precision: 5.7 cm + 3.05 cm + 21.46 cm

Perform the indicated operations and simplify.

**26.** $5 + (-2) + (-8)$

**27.** $-3 - 8 - 5 - (-2)$

**28.** $(-8)(2)(-4)$

**29.** $-24 \div 4$

**30.** $(-2)^3 - 5 + (-6) - 2^2$

**31.** $-5\frac{1}{2} + 4\frac{1}{4}$

**32.** Write 58,000 in scientific notation.

**33.** Write $5.3 \times 10^{-2}$ as an ordinary number.

Simplify and write the result in scientific notation.

**34.** $(5.1 \times 10^2)(4.1 \times 10^3)$

**35.** $\dfrac{4.6 \times 10^4}{2.3 \times 10^{-2}}$

# 5

# Linear Equations

## Focus on Careers

The hospitality industry provided more than 1.8 million jobs in 2002. Some of these were self-employed workers who run bed-and-breakfast inns and provide other services. Most employment is available in densely populated cities or resort areas and most employers have more than 100 employees.

There are many careers in this service sector. Hotel manager, lodging manager, food service manager, sales manager, purchasing manager, and executive chef are just a few careers found in the hospitality industry.

Skills and experience needed by workers vary widely and almost all workers undergo on-the-job training. About 200 community colleges offer programs in hotel and restaurant management. Graduates of these programs, especially those with good communication skills, have a better-than-average opportunity for entry and advancement in these careers.

*(continued)*

There is a wide range of pay in the industry. Nonsupervisory workers earned $10.01 an hour on average, but hourly earnings for restaurant cooks and maintenance and repair workers were $10.48 on average. Salaries of managers are usually higher, and managers may earn bonuses ranging up to 20% of their basic salaries.

Employment in this industry is expected to increase by about 17% for the 2002–2012 period compared with a 16% growth rate projected for all industries combined.

Source: *Occupational Outlook Handbook*, 2004–2005 Edition, U.S. Department of Labor, Bureau of Labor Statistics

# 5–1 | *Variable Notation*

*Learning Outcomes*

**1** Identify equations, terms, factors, constants, variables, and coefficients.
**2** Write verbal interpretations of symbolic statements.
**3** Translate verbal statements into symbolic statements using variables.
**4** Simplify variable expressions.

**1 Identify Equations, Terms, Factors, Constants, Variables, and Coefficients.**

Before we work with equations to solve problems, we need to understand the basic concepts and terminology of equations. In mathematics, we use the symbol "=" to show that quantities are "equal to" each other. We write the statement "5 is equal to 2 plus 3" as $5 = 2 + 3$. This symbolic statement is called an **equation.**

An *equation* is a symbolic statement that two expressions or quantities are equal in value. An equation may be true or false. However, our purpose is to find values of the missing numbers that make the equation *true*.

**To verify that an equation is true:**

1. Find the value of the expression on each side of the equal sign.
2. Compare the values for each side.
   (a) If the values for each side are equal, the equation is true.
   (b) If the values for each side are not equal, the equation is not true.

**EXAMPLE** Verify that the statements are true equations.

(a) $3(8) = 24$     (b) $12 - 3 = 2 + 7$

(a) $3(8) = 24$            3 times 8 is 24.
       $24 = 24$           The equation is true.

(b) $12 - 3 = 2 + 7$     12 minus 3 is 9. 2 plus 7 is 9.
         $9 = 9$           The equation is true.

Some types of equations contain a *missing* or *unknown* number. We **solve** an equation by finding the missing number that makes the equation *true*. We use letters such as *x, a,* or *z* to represent the missing number in the equation. These letters are called **variables.** In each equation, the letter has a certain but unknown value that depends on the other numbers and relationships in the equation.

**EXAMPLE** Find the value of the variable that makes the equation true.

(a) $x = 3 + 8$    (b) $\dfrac{18}{3} = n$    (c) $y = 3(4) - 5$

(a)  $x = 3 + 8$          8 added to 3 is 11.
     $x = \mathbf{11}$

(b)  $\dfrac{18}{3} = n$          18 divided by 3 is 6.
     $\mathbf{6 = n}$

(c)  $y = 3(4) - 5$        Multiply first. 3 times 4 is 12.
     $y = 12 - 5$          Add 12 and $-5$. 12 plus $-5 = 7$.
     $y = \mathbf{7}$

Because both sides of an equation are equivalent, it does not matter which side comes first in the equation. In these examples, for instance, $x = 3 + 8$ means the same as $3 + 8 = x$. Also $n = 6$ is equivalent to $6 = n$, and so on. The unknown, or variable, may appear on either the left or right side of an equation.

**Symmetric property of equality:**

If the sides of an equation are interchanged, equality is maintained.

Symbolically, if $a = b$, then $b = a$.

In algebra we carefully distinguish between terms and factors. **Factors** are numbers or variables that are multiplied. **Terms** are algebraic expressions that are single quantities or products or quotients of quantities.

A term can be a single letter; a single number; the product of numbers and letters; the product of numbers, letters, and/or groupings; or the quotient of numbers, letters, and/or groupings. The fraction line implies that the numerator and/or the denominator is grouped.

**TIP!**

**Multiplication Notation Conventions**

When a term is the product of letters alone (such as $ab$ or $xyz$) or the product of a number and one or more letters (such as $3x$ or $2ab$), parentheses or other symbols of multiplication are usually omitted. Thus, $ab$ means $a$ times $b$, $3x$ means 3 times $x$, and so on.

**To identify terms:**

1. Separate an expression into terms by identifying addition or subtraction signs that are not within a grouping.
2. The sign of each term is the sign that precedes the term.

$$3a + b \qquad 3(a + b) \qquad \frac{3}{a + b}$$

two terms          one term          one term

**EXAMPLE**  Identify the terms in each expression, and draw a box around each term.

(a) $3x + 5$    (b) $2ab + 4a - b + 2(a + b)$

(c) $\dfrac{2a + 1}{3}$    (d) $\dfrac{2a}{3} + 1$

(a)  $\boxed{3x} + \boxed{5}$

Terms are separated by "+" and "−" signs that are not within a grouping. The expression has two terms.

(b)  $\boxed{2ab} + \boxed{4a} - \boxed{b} + \boxed{2(a + b)}$

The plus sign within the grouping does not separate terms. The expression has four terms.

(c)  $\dfrac{2a + 1}{3}$

The fraction bar is a grouping symbol. The numerator is a grouping, $2a + 1$. The expression has one term.

(d)  $\dfrac{2a}{3} + \boxed{1}$

The expression has two terms.

A term that contains only numbers is a **number term** or **constant**. A term that contains only one letter, several letters used as factors, or a combination of letters and numbers used as factors is a **letter term** or **variable term.**

The numerical factor of a term is the **numerical coefficient.** Unless otherwise specified, we will use **coefficient** to mean the *numerical coefficient.* The numerical factor should be written *in front* of the variable.

### To identify the numerical coefficient of a term:

1. Since a term contains only factors, the coefficient is the number factor or the product of all number factors.
2. If the term is a fraction (an indicated division),
   (a) Write the number in the denominator (indicated division) as an equivalent multiplication.
   (b) Write the product of all numerical factors.
   (c) The numerical coefficient is the product from Step 2b.

**EXAMPLE**  Identify the numerical coefficient in each term.

(a) $2x$    (b) $-3ab$    (c) $4(x + 3)$    (d) $-\dfrac{n}{3}$    (e) $\dfrac{2b}{5}$

(a)  The coefficient of $2x$ is **2.**
(b)  The coefficient of $-3ab$ is **−3.**
(c)  The coefficient of $4(x + 3)$ is **4.**

(d)  $-\dfrac{n}{3}$ is the same as $-\dfrac{1}{3}n$, so the coefficient is $-\dfrac{1}{3}.$

(e)  $\dfrac{2b}{5} = \dfrac{1}{5}(2b) = \dfrac{1}{5}(2)b = \dfrac{2}{5}b$. The coefficient is $\dfrac{2}{5}.$

**TIP!**

### Coefficient of 1 or −1

When a letter term has no written numerical coefficient, the coefficient is understood to be 1, so $x = 1x.$ Similarly, the numerical coefficient of $-x$ is $-1$, so $-x = -1x.$

## 2 Write Verbal Interpretations of Symbolic Statements.

In early studies of mathematics, unknown values were represented by boxes, circles, underlining, or other symbols. That is, finding the missing value was written as $5 + \boxed{\phantom{x}} = 8$. The box identifies the missing value and the other numbers and symbols give the conditions of the problem.

We commonly use letters to represent the missing value. That is, we write $5 + x = 8$ or $5 + y = 8$ or $5 + a = 8$. The choice of letters is not significant. The position of the letter and the other conditions of the statement are important. We can phrase a mathematical statement several ways: $5 + x = 8$ can be stated

"What number added to 5 gives 8?"

or

"5 plus a number has a result of 8. What is the missing number?"

### To write verbal interpretations of symbolic statements:

1. Locate the variable or variable terms.
2. Examine the operations that link the factors and terms.
3. Write a verbal statement or statements that describe all the conditions of the symbolic statement.

---

**EXAMPLE**   State the equations in words.

   **(a)** $x - 7 = 4$   **(b)** $\dfrac{x}{5} = 3$   **(c)** $2x + 3 = 15$   **(d)** $2(x + 3) = 14$

Each equation can be stated several ways. Here is one choice for each equation.

   **(a) When 7 is subtracted from a number, the answer is 4.**
   **(b) A number divided by 5 is 3.**
   **(c) 15 is the result when 3 is added to 2 times a number.**
   **(d) If the sum of a number and 3 is doubled, the result is 14.**

---

## 3 Translate Verbal Statements into Symbolic Statements Using Variables.

Symbolic representations of written statements or real-life situations allow us to determine the value of missing amounts more systematically.

### To translate verbal statements into symbolic statements:

1. Assign a letter to represent the missing number.
2. Identify key words or phrases that imply or suggest specific operations.
3. Translate words into symbols.

---

**EXAMPLE**   Translate the statements into symbols.

   **(a)** The sum of 12, 23, and a third number is 52.
   The third number is missing.
   Let $x$ represent the third number.

   $$12 + 23 + x = 52$$   Translate the entire statement. *Sum* indicates addition. *Is* translates to *equals*.

**(b)** When a number is subtracted from 45, the result is 17.
The number being subtracted (the subtrahend or second number) is missing.
Let $x$ represent the missing number.

$$45 - x = 17 \qquad \text{45 is the minuend or first number.}$$

Key words can be examined to identify operations.

| | |
|---|---|
| **Addition:** | the sum of, plus, increased by, more than, added to, exceeds, longer, total, heavier, older, wider, taller, gain, greater than, more, expands |
| **Subtraction:** | less than, decreased by, subtracted from, the difference between, diminished by, take away, reduced by, less, minus, shrinks, younger, lower, shorter, narrower, slower, loss |
| **Multiplication:** | times, multiply, of, the product of, multiplied by |
| **Division:** | divide, divided by, divided into, how big is each part, how many parts can be made from |

Some words that imply multiplication or division may indicate a specific number in the multiplication or division. Examples include twice (2 times), double (2 times), triple (3 times), and half of (1/2 times or divided by 2).

**EXAMPLE** Translate the statement into symbols.

How many shelves that are each 3 feet long can be made from a board that is 12 feet long?

The missing number is the number of shelves that can be made from one 12-ft board. Let $x$ represent the number of shelves. The number of shelves, $x$, times the length of each shelf, 3, equals 12.

$$3x = 12.$$

**EXAMPLE** Explain the difference between the two statements.

4 times the difference between a number and 8 is 24.
The difference between 4 times a number and 8 is 24.

4 times the difference between a number and 8 is 24. This statement is written symbolically as $4(x - 8) = 24$. It shows that the difference is taken first and the result is multiplied by 4.

The difference between 4 times a number and 8 is 24. This statement is written symbolically as $4x - 8 = 24$. A number is multiplied by 4 and then 8 is subtracted from the result.

**4  Simplify Variable Expressions.**

Terms are **like terms** if they are numbers or if they are letter terms with exactly the same letter factors.

The terms $4y$ and $2y$ are like terms because both contain the same letter ($y$). Similarly, 3 and 1 are like terms because both are numbers. We *cannot* combine 3 and $4y$ because they are unlike terms (number term and letter term), and we *cannot* combine $4y$ and $2x$ because they are also unlike terms (different letters).

**To simplify variable expressions:**

1. Change subtraction to addition if appropriate.
2. Add the numbers using the appropriate rule for adding signed numbers. The sum is a signed number.
3. Add the numerical coefficients of the like variables using the appropriate rule for adding signed numbers. The sum has the same letter factor or factors as the like terms being added.

**EXAMPLE** Simplify the expressions by combining like terms.

(a) $5a + 2a - a$    (b) $3x + 5y + 8 - 2x + y - 12$

(a) $5a + 2a - a = 6a$          All terms are like terms. Add coefficients of $a$: $5 + 2 - 1 = 6$.

(b) $3x + 5y + 8 - 2x + y - 12$   Add like terms.

$\quad = x + 6y - 4$

$\qquad 3x - 2x = x$

$\qquad 5y + y = 6y$

$\qquad 8 - 12 = -4$

The **distributive principle** can be extended to include multiplying numbers and variables. When an expression has an instance of the distributive principle, first remove the grouping symbol by multiplying by the factor or factors in front of the grouping.

**EXAMPLE** Simplify by applying the distributive principle.

(a) $3(5x - 2)$    (b) $-7(x + y - 3z)$    (c) $-(-3x + 4)$

(a) $3(5x - 2) = 3(5x) - 3(2)$          Apply the distributive principle by mul-
$\qquad = 15x - 6$                        tiplying $(5x - 2)$ by 3.

(b) $-7(x + y - 3z) = -7(x) - 7(y) - 7(-3z)$   Apply the distributive principle by multi-
$\qquad = -7x - 7y + 21z$                   plying each term of the grouping by $-7$.

(c) $-(-3x + 4) = -1(-3x) - 1(+4)$        Apply the distributive principle by multi-
$\qquad = 3x - 4$                          plying each term of the grouping by $-1$.

**SECTION 5–1 SELF-STUDY EXERCISES**

**1** Verify that the statements are true.

**1.** $5(-3) = -15$

**2.** $17 - 6 = 11$

**3.** $12 - 5(2) = -7 + 9$

**4.** $8 - 3(5) = 2(-1 - 3) + 1$

Find the value of the variable that makes the equation true.

**5.** $n = 4 + 7$

**6.** $m = 8 - 2$

**7.** $5 - 9 = y$

**8.** $\dfrac{12}{2} = x$

**9.** $p = 2(5) - 1$

**10.** $b = 6 - 3(5)$

Identify the terms in each expression by drawing a box around each term.

**11.** $7 + c$

**12.** $4a - 7$

**13.** $3x - 2(x + 3)$

**14.** $\dfrac{a}{3}$

**15.** $7xy + 3x - 4 + 2(x + y)$

**16.** $14x + 3$

**17.** $\dfrac{7}{(a + 5)}$

**18.** $\dfrac{4x}{7} + 5$

**19.** Write an algebraic expression that contains three terms.

Identify the numerical coefficient of each term.

**20.** $5x$

**21.** $-4xy$

**22.** $\dfrac{n}{5}$

**23.** $\dfrac{2a}{7}$

**24.** $6(x + y)$

**25.** $\dfrac{-4}{5x}$

**26.** Write an expression that has one term and a numerical coefficient of $-15$.

5–1 Variable Notation

State the equations in words.

**27.** $x + 4 = 7$          **28.** $x - 5 = 2$          **29.** $3x = 15$          **30.** $3x + 1 = 7$

**3** Write the statements in symbols.

**31.** 5 more than a number is 12.          **32.** A number divided by 6 is 9.

**33.** 4 times the difference of a number and 3 is 12.          **34.** 3 less than 4 times a number is 12.

**35.** The sum of 12, 7, and a third number is 17.          **36.** 7 more than twice a number is 21.

**37.** **INDTR** If the temperature rises 15°, it will be 48°. What is the temperature?          **38.** **HLTH/N** If 15 mL of water are added to a medicine, there are 45 mL in all. What is the volume of the medicine?

**39.** **INDTEC** How many 5-ft pieces of I-beam can be cut from a piece that is 45 ft long?          **40.** **CON** A piece of oak flooring that is 18 ft long has an unacceptable flaw that requires 3 ft to be trimmed from one end. How many 5-ft boards can be cut from the remaining length of flooring?

**4** Simplify by combining like terms.

**41.** $3a - 7a + a$          **42.** $-8x + y - 3y$          **43.** $5x - 3y + 2x + y$

**44.** $-4a + b + 9 + a - b - 3$          **45.** $2x + 8 - x + 4 - x - 2$          **46.** $3a + 5b + 8c + 1 + b$

Simplify by applying the distributive principle and combining like terms if appropriate.

**47.** $5(2x - 4)$          **48.** $4 + 3(2x + 3)$          **49.** $3 - (4x - 2)$          **50.** $5 - 2(6a + 1)$

---

## 5–2 | Solving Basic Linear Equations

*Learning Outcomes*

**1** Solve linear equations using the addition axiom.

**2** Solve linear equations using the multiplication axiom.

**3** Solve linear equations with like terms on the same side of the equation.

**4** Solve linear equations with like terms on opposite sides of the equation.

A **linear equation in one variable** is an equation in which the same letter is used in all variable terms and the exponent of the variable is 1.

**1 Solve Linear Equations Using the Addition Axiom.**

An equation is **solved** when the letter or variable is alone on one side of the equal sign. That is, the coefficient of the variable is +1 and no other terms are on the same side as the variable. The number on the side opposite the letter or variable is called the **solution** or **root** of the equation. The solution is the number that makes the equation true.

The underlying principle for solving equations is the **basic principle of equality.**

**Basic principle of equality:**

To preserve equality, if an operation is performed on one side of an equation, the *same* operation must be performed on the other side.

We can illustrate this principle by visualizing a balanced scale. If we have a scale with 1 oz on one side and 1 oz on the other side, the scale is balanced. If we increase or decrease the weight on one side, we need to do the same on the other side or one side would be heavier than the other and the equality of both sides would be lost (see Figure 5–1).

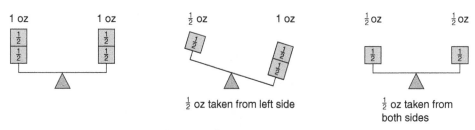

| 1 oz | 1 oz | ½ oz | 1 oz | ½ oz | ½ oz |
| --- | --- | --- | --- | --- | --- |

½ oz taken from left side

½ oz taken from both sides

**Figure 5–1**

If $\frac{1}{2}$ oz is taken from both sides, the scale is still balanced. The purpose in solving an equation is to **isolate** the variable or letter term on one side of the equation. Terms can be moved from one side of an equation to the other using the **addition property** or **addition axiom.** The addition axiom states that the same quantity can be added to both sides of an equation without changing the equality of the two sides. Symbolically, if $a = b$, then $a + c = b + c$.

Since subtraction is the same as adding the opposite, the addition axiom applies to both addition and subtraction.

**To solve a linear equation using the addition axiom:**

1. Locate the variable in the equation.
2. Identify the constant that is associated with the variable by addition (or subtraction).
3. Add the opposite of the constant to both sides of the equation.

**EXAMPLE**  Solve the equation $x + 4 = 8$.

$$x + 4 = 8 \qquad \text{4 is added to } x. \text{ Add } -4, \text{ the opposite of 4, to both sides (addition axiom).}$$
$$x + 4 - 4 = 8 - 4 \qquad 4 - 4 = 0, 8 - 4 = 4.$$
$$x + 0 = 4 \qquad x + 0 = x.$$
$$x = \mathbf{4}$$

TIP!

**Adding Opposites and Zero**

A number plus its opposite always equals zero.

$$n + (-n) = 0$$

Zero plus a number (or variable) leaves the number (or variable) unchanged.

$$n + 0 = n$$

These two properties and the addition axiom allow us to mentally move a term to the other side of the equation as its opposite.

| | | | |
| --- | --- | --- | --- |
| $x + 3 = 8$ | becomes | $x = 8 - 3$ | Mentally add $-3$ to both sides. |
| $x - 2 = 4$ | becomes | $x = 4 + 2$ | Mentally add $+2$ to both sides. |
| $5 + x = 2$ | becomes | $x = 2 - 5$ | Mentally add $-5$ to both sides. |
| $10 = x - 1$ | becomes | $10 + 1 = x$ | Mentally add $+1$ to both sides. |

**EXAMPLE** Solve the equations.

    **(a)** $x + 3 = 8$    **(b)** $x - 2 = 4$    **(c)** $5 + x = 2$    **(d)** $10 = x - 1$

    **(a)**  $x + 3 = 8$           To isolate $x$ mentally add $-3$ to both sides.
           $x = 8 - 3$         Combine like terms.
           **$x = 5$**

    **(b)**  $x - 2 = 4$           To isolate $x$ mentally add $+2$ to both sides.
           $x = 4 + 2$         Combine like terms.
           **$x = 6$**

    **(c)**  $5 + x = 2$           To isolate $x$ mentally add $-5$ to both sides.
           $x = 2 - 5$         Combine like terms.
           **$x = -3$**

    **(d)**       $10 = x - 1$      To isolate $x$ mentally add $+1$ to both sides.
        $10 + 1 = x$        Combine like terms.
          **$11 = x$**        Apply the symmetric property of equality.
           or
          **$x = 11$**

**TIP!**

### Use Simple Cases to Learn and Understand a Process or Procedure

Many equations can be solved mentally, or by using basic arithmetic and our knowledge of the relationships among the operations of addition, subtraction, multiplication, and division. In the example of $x + 3 = 8$, we can determine mentally that the missing number is 5. Our goal is to learn a procedure for solving simple equations, so that we can apply the same procedure to more complex equations.

**2** **Solve Linear Equations Using the Multiplication Axiom.**

The basic principle of equality can be extended to multiplication and division. The **multiplication property of equality** or the **multiplication axiom** states that both sides of an equation may be multiplied (or divided) by the same *nonzero* quantity without changing the equality of the two sides. Symbolically, if $a = b$ and $c \neq 0$, then $ac = bc$. Similarly, if $a = b$ and $c \neq 0$, then $\dfrac{a}{c} = \dfrac{b}{c}$.

### To solve a linear equation using the multiplication axiom:

1. Divide both sides of the equation by the numerical coefficient of the letter term; or
2. Multiply both sides of the equation by the reciprocal of the coefficient of the letter term.

**EXAMPLE** Solve the equation $4x = 48$.

Multiplication and division are inverse operations, so dividing by a number is the same as multiplying by its multiplicative inverse—its reciprocal.

$$4x = 48$$
$$\frac{1}{4}(4x) = \frac{1}{4}(48)$$
$$x = 12$$

Multiply by the reciprocal of the coefficient of the letter term.

$$4x = 48$$
$$\frac{4x}{4} = \frac{48}{4}$$
$$x = 12$$

Divide by the coefficient of the letter term.

**EXAMPLE**  Solve the equation $45 = 0.5x$.

$$\frac{45}{0.5} = \frac{0.5x}{0.5}$$

Divide both sides by the coefficient of the letter term.

$$90 = x$$

Apply the symmetric property of equality.

or

$$x = 90$$

**TIP!**

### On Which Side Should the Letter Be?

Multiplication and division are effective regardless of which side of the equation contains the letter term. Once the equation is solved, however, many people prefer to put the letter on the left ($x = 90$ instead of $90 = x$). Either way is correct.

When the equation contains only whole numbers or decimals, it is generally more convenient to divide by the coefficient of the variable or letter term than to multiply by the reciprocal of the coefficient of that variable.

When the equation contains a fraction, it is usually preferable to multiply by the reciprocal of the coefficient of the variable.

**EXAMPLE**  Solve the equation $\frac{1}{4}n = 7$.

Using multiplication:

$$\frac{1}{4}n = 7$$

Multiply both sides by the reciprocal of the coefficient of the variable term.

$$\left(\frac{4}{1}\right)\frac{1}{4}n = 7\left(\frac{4}{1}\right)$$

The reciprocal of $\frac{1}{4}$ is $\frac{4}{1}$.

$$n = 28$$

Using division:

$$\frac{1}{4}n = 7$$

Divide both sides by the coefficient of the variable term.

$$\frac{\frac{1}{4}n}{\frac{1}{4}} = \frac{7}{\frac{1}{4}}$$

$$7 \div \frac{1}{4} = \frac{7}{1} \cdot \frac{4}{1} = 28.$$

$$n = 28$$

An equation is *solved* when the coefficient of the letter term is $+1$. If the coefficient is $-1$, we apply the multiplication axiom to complete the solution.

**EXAMPLE**   Solve the equation $-n = 25$.

The coefficient is $-1$, so the equation is not solved. The coefficient must be $+1$ for the equation to be solved.

$$\frac{-1n}{-1} = \frac{25}{-1}$$   Divide by the coefficient of the letter term.

$$n = -25$$

When an equation has been solved, the *solution,* or *root,* can be checked to prove it is the solution.

**To check or verify the solution or root of an equation:**

1. Substitute the solution in place of the letter in the original equation.
2. Perform all indicated operations.
3. If the solution is correct, the value of the left side of the equation should equal the value of the right side.

**EXAMPLE**   In the example $4x = 48$ the solution was found to be 12. Check or verify this root.

$$4\ x\ = 48$$   Substitute 12 for *x*.
$$4(\ 12\ ) = 48$$   Perform the multiplication.
$$48 = 48$$   The root is verified if the final equality is true.

**3**   **Solve Linear Equations with Like Terms on the Same Side of the Equation.**

We can combine only *like* terms and only if they are on the *same side* of the equation. That means number terms can be added only to other number terms. Letter terms can be added only to other like letter terms. By **combine** we mean add like terms according to the appropriate signed number rule.

**To solve linear equations with like terms on the same side of the equation:**

1. Combine (add or subtract) like terms that are on the same side of the equal sign.
2. *Multiply* both sides of the equation by the reciprocal of the coefficient of the variable or *divide* both sides of the equation by the coefficient of the variable.

**EXAMPLE**   Solve the equation $4x - x = 7 + 8$.

$$4x - x = 7 + 8$$   The coefficient of *x* is $-1$.
$$4x - x\ =\ 7 + 8$$   Combine like terms on each side of the equal sign.
$$3x\ =\ 15$$   Divide both sides of the equation by 3, the coefficient of *x*.

$$\frac{3x}{3} = \frac{15}{3}$$

$$x = 5$$

**4** **Solve Linear Equations with Like Terms on Opposite Sides of the Equation.**

Like terms do not always appear on the same side of the equal sign. In such cases, the like terms must be manipulated so that they appear on the same side of the equal sign *before* they can be combined. The basic strategy for solving an equation is to *isolate* the variable or letter terms on one side of the equation.

**To solve linear equations with like terms on opposite sides of the equation:**

1. Identify the like terms and determine which term or terms should be moved to create an equation with like terms on each side.
2. Add to both sides of the equation the opposite of each term that is to be moved (addition axiom).
3. Combine like terms on each side of the equation.
4. Solve the resulting equation by using the multiplication axiom.

**EXAMPLE**  Solve the equation $2x + 4 = 8$.

$$2x + 4 = 8$$  4 and 8 are like terms on opposite sides of the equal sign. Add $-4$, the opposite of 4, to both sides (addition axiom).

$$2x \;\boxed{+ 4 - 4} = \boxed{8 - 4}$$  $4 - 4 = 0$, $8 - 4 = 4$.

$$2x \;\boxed{+ 0} = \boxed{4}$$  $2x + 0 = 2x$.

$$2x = 4$$  Divide both sides by the coefficient of the letter term, $2x$ (multiplication axiom).

$$\frac{2x}{2} = \frac{4}{2}$$

$$x = 2$$

**EXAMPLE**  Solve the equation $9 - 4x = 8x$.

Manipulate the terms in this equation so that both letter terms are isolated on one side of the equation and the number term is on the other side.

$$9 - 4x = 8x$$  $-4x$ and $8x$ are like terms on opposite sides of the equal sign. Add $4x$, the opposite of $-4x$, to both sides (addition axiom).

$$9 \;\boxed{-4x + 4x} = \boxed{8x + 4x}$$  $-4x + 4x = 0$, $8x + 4x = 12x$.

$$9 \;\boxed{+ 0} = \boxed{12x}$$  $9 + 0 = 9$.

$$9 = 12x$$  Divide both sides by the coefficient of $x$, which is 12 (multiplication axiom).

$$\frac{9}{12} = \frac{12x}{12}$$

$$\frac{9}{12} = x$$  Reduce to lowest terms.

$$\frac{3}{4} = x \quad \textbf{or} \quad x = \frac{3}{4}$$  Apply the symmetric property of equality if desired.

In the preceding examples each time we add the opposite of a term to both sides, we eliminate it from one side of the equation, and it appears as the *opposite* on the other side. Knowing this, we can *transpose* as a shortcut for adding opposites to both sides. *Transpose* means "move across as the opposite of."

**TIP!**

### Sorting or Transposing Terms: A Shortcut for the Addition Axiom

**Transposing** is a mental shortcut for applying the addition axiom. To do this, we add the opposite of a term to both sides of an equation. In other words, we omit the term from one side of the equation and write its opposite on the other side. This process is better described as *sorting* the terms so that like terms are on the same side of the equation.

What we are doing is *mentally* applying the addition axiom (adding the opposite term to both sides) but showing only part of the procedure.

Let's rework the previous two examples transposing like terms to isolate the variable.

$$2x \boxed{+4} = 8 \qquad\qquad 9 \boxed{-4x} = 8x$$
$$2x = 8 \boxed{-4} \qquad\qquad 9 = 8x \boxed{+4x}$$
$$2x = 4 \qquad\qquad 9 = 12x$$
$$\frac{2x}{2} = \frac{4}{2} \qquad\qquad \frac{9}{12} = \frac{12x}{12}$$
$$x = 2 \qquad\qquad \frac{3}{4} = x$$

---

**EXAMPLE**  Solve the equation $4x - 3 = 3x - 3 + x$.

| | |
|---|---|
| $4x - 3 = 3x - 3 + x$ | Combine like terms. $3x + x = 4x$ |
| $4x - 3 = 4x - 3$ | Mentally add 3 to both sides. |
| $4x = 4x$ | Mentally add $-4x$ to both sides. |
| $4x - 4x = 4x - 4x$ | |
| $0 = 0$ | Since $0 = 0$ is a *true* statement, the equation has many solutions. |

**The solution is all real numbers.**

---

**Real numbers** include all *rational numbers* and *irrational numbers*. Irrational numbers will be introduced in Chapter 11. Equations like $4x - 3 = 3x - 3 + 2$ that have many solutions are called **identities.**

---

**EXAMPLE**  Solve the equation $3x - 7 = x + 2x - 11$.

| | |
|---|---|
| $3x - 7 = x + 2x - 11$ | Combine like terms. |
| $3x - 7 = 3x - 11$ | Sort terms (addition axiom). |
| $3x - 3x = -11 + 7$ | Combine like terms. |
| $0 = -4$ | Zero and $-4$ are not equal. |

**No solution**

---

Chapter 5 / Linear Equations

When the variable terms add to zero, we examine the resulting statement. A true statement indicates the solution set is the *set of real numbers*. A false statement indicates the equation has *no solution*.

**Align Equal Signs! One Equal Sign per Line!**

Organizing the steps of the solution to an equation reduces errors.

- Arrange steps under each other and align the equal signs in a vertical line.
  This helps you determine if a term has moved from one side to the other.
  As a term is eliminated from one side move it to the other side as its opposite.
- Have only one equal sign per line.

  Each line should be one complete equation or statement.

| **Good Form** | **Poor Form** |
|---|---|
| $x = \dfrac{2}{4}$ | $x = \dfrac{2}{4} = \dfrac{1}{2}$ |
| $x = \dfrac{1}{2}$ | |

An equation can be solved two ways: (1) by isolating the letter terms on the left and (2) by isolating the letter terms on the right. No matter which side we choose for the letter terms, the number terms must be on the opposite side.

**EXAMPLE**   Solve the equation $9x + 2 = 6x - 10$ two ways.

**Letter Terms Isolated on Left**

$$9x + 2 = 6x - 10$$
$$9x - 6x = -10 - 2$$
$$3x = -12$$
$$\frac{3x}{3} = \frac{-12}{3}$$
$$x = -4$$

Sort or transpose.

Combine like terms.

Divide.

**Letter Terms Isolated on Right**

$$9x + 2 = 6x - 10$$
$$2 + 10 = 6x - 9x$$
$$12 = -3x$$
$$\frac{12}{-3} = \frac{-3x}{-3}$$
$$-4 = x$$

**EXAMPLE**   Solve the equation $2x - 3 + 5x = 8 - 6x$.

$$2x - 3 + 5x = 8 - 6x$$
$$7x - 3 = 8 - 6x$$
$$7x + 6x = 8 + 3$$
$$13x = 11$$
$$\frac{13x}{13} = \frac{11}{13}$$
$$x = \frac{11}{13}$$

Combine the like terms $2x$ and $5x$ on the left side.

Sort to collect like terms on the same side.

Combine the like terms on each side of the equation.

Divide by the coefficient of $x$.

**1** Solve the equations. Use the addition axiom.

1. $x - 3 = 5$    2. $x - 7 = 8$    3. $x - 9 = -12$    4. $x - 11 = -19$
5. $x + 5 = 9$    6. $x + 8 = -3$    7. $x + 11 = -15$    8. $x + 14 = 17$
9. $12 = x + 7$    10. $-15 = x - 8$    11. $14 = x - 2$    12. $18 = x + 9$
13. $7 = 5 - x$    14. $5 = 9 - x$

**2** Solve the equations. Use the multiplication axiom.

15. $7x = 56$    16. $2x = 18$    17. $15 = 3x$    18. $36 = 1.2n$
19. $-3a = 27$    20. $-18 = 9b$    21. $-7c = -49$    22. $-72 = 8x$
23. $5x = 32$    24. $36 = 8y$    25. $-6n = 15$    26. $-28 = -3b$
27. $-x = 7$    28. $-2 = -y$    29. $\frac{1}{3}x = 5$    30. $4 = \frac{1}{2}y$
31. $\frac{3}{5}x = 2$    32. $-\frac{3}{4}y = 5$    33. $-\frac{1}{2}x = -6$    34. $21 = \frac{3}{8}n$

**3** Solve.

35. $3x + 7x = 60$    36. $42 = 8m - 2m$    37. $5a - 6a = 3$
38. $3m - 9m = 3$    39. $y + 3y = 32$    40. $0 = 2x - x$
41. $8y - 6y = -14$    42. $5y - y = -16$

**4** Solve.

43. $-36 = 9x + 18$    44. $b + 6 = 5$    45. $1 = x - 7$
46. $5t - 18 = 12$    47. $10 - 2x = 4$    48. $2y + 7 = 17$
49. $3x + 7 = x$    50. $3a - 8 = 7a$    51. $10x + 18 = 8x + 18 + 2x$
52. $4x + 7 = 8 + 4x$    53. $4x = 5x + 8$    54. $2t + 6 = t + 13$
55. $12 + 5x = 6 - x$    56. $4y - 8 = 2y + 14$    57. $8 - 7y = y + 24$
58. $\frac{2x}{3} = 18$    59. $\frac{y + 1}{2} = 7$    60. $\frac{8 - R}{76} = 1$
61. $P = \frac{1}{2} + \frac{1}{3}$    62. $x + \frac{1}{7}x = 16$    63. $\frac{2}{5} - x = \frac{1}{2}x + \frac{4}{5}$
64. $\frac{3}{7}m - \frac{1}{2} = \frac{2}{3}$    65. $\frac{1}{4}s = \frac{1}{4} + \frac{1}{10} + \frac{1}{20}$    66. $m = 2 + \frac{1}{4}m$
67. $\frac{7}{9} + 3 = \frac{1}{2}T$    68. $2.3x = 4.6$

69. $0.8R = 0.6$ (round to nearest tenth)    70. $0.33x + 0.25x = 3.5$ (round to nearest hundredth)
71. $0.04x = 0.08 - x$ (round to nearest hundredth)    72. $0.47 = R + 0.4R$ (round to nearest hundredth)

---

## 5–3 | Applying the Distributive Property in Solving Equations

**Learning Outcome**   **1** Solve linear equations that contain parentheses.

When an equation contains an addition or subtraction in parentheses and that quantity in parentheses is multiplied by another factor, we have an example of the *distributive property.* (See Chapter 1, Section 1, Outcome 6.)

**1** Solve Linear Equations That Contain Parentheses.

## To solve linear equations that contain parentheses:

1. Apply the distributive property to *remove parentheses.*
2. *Combine like terms* on each side of the equation.
3. *Sort terms* to collect the variable or letter terms on one side and constant or number terms on the other (addition axiom).
4. *Combine like terms* on each side of the equation.
5. Multiply by the reciprocal of the coefficient of the variable term or divide by the coefficient of the variable term (multiplication axiom).

---

**EXAMPLE**    Solve the equation $28 = 7x - 3(x - 4)$.

| | |
|---|---|
| $28 = 7x - 3(x - 4)$ | Each term in the parentheses is multiplied by $-3$ using the distributive property. |
| $28 = \boxed{7x - 3x} + 12$ | Combine like terms on the right. |
| $28 = \boxed{4x} + \boxed{12}$ | Sort terms. |
| $28 \boxed{- 12} = 4x$ | Combine like terms on the left. |
| $16 = 4x$ | Divide. |
| $\dfrac{16}{4} = \dfrac{4x}{4}$ | |
| $\mathbf{4 = x}$ | |

---

As equations get more involved, the importance of checking becomes more apparent. To check the root 4, we substitute 4 for *x* in the equation.

| | |
|---|---|
| $28 = 7x - 3(x - 4)$ | Substitute 4 for *x*. |
| $28 = 7(4) - 3(4 - 4)$ | Simplify the grouping $4 - 4 = 0$. |
| $28 = 7(4) - 3(0)$ | Multiply. |
| $28 = 28 - 0$ | Subtract. |
| $\mathbf{28 = 28}$ | Solution checks. |

To solve any equation, it is very important to be careful with the sign of each term. The algebraic sign of any term is the sign that comes *before* the term.

---

**EXAMPLE**    Solve the equation $6 - (x + 3) = 2x$ for *x*.

Since $(x + 3)$ is in parentheses, it is a grouping and we handle it as one term. The sign of the term is negative and the numerical coefficient is understood to be $-1$.

The first step in solving this equation is to apply the distributive property by multiplying $(x + 3)$ by its understood coefficient, $-1$.

| | |
|---|---|
| $6 - (x + 3) = 2x$ | $-1$ is the understood coefficient of $(x + 3)$. |
| $6 - 1(x + 3) = 2x$ | Distribute. |
| $6 \boxed{-x - 3} = 2x$ | Combine. |
| $3 - x = 2x$ | Sort. |
| $3 = 2x + x$ | Combine. |
| $3 = 3x$ | Divide. |
| $\dfrac{3}{3} = \dfrac{3x}{3}$ | |
| $\mathbf{1 = x}$ | |

---

The usefulness of solving linear equations is in solving applied problems. We will continue to use the Six-Step Problem-Solving Plan as our guide for solving applied problems.

EXAMPLE

AG/H

A horticulturist marks off a 32-m-wide rectangular nursery plot with 158 m of fencing. Because the plants must be properly spaced, she needs to know the length of the plot. Find the length in meters.

**Unknown facts**

The length of the nursery plot in meters.

**Known facts**

The nursery plot is rectangular. The width is 32 m. The fencing gives us the perimeter of 158 m.

**Relationships**

Using the formula for the perimeter of a rectangle, we know that the perimeter is twice the sum of the length and width, or $p = 2(l + w)$.

**Estimation**

Two widths $(2 \cdot 32$ m) will use 64 m of fencing. There will be less than 100 m of fencing for the two lengths. The length will be less than 50 m.

**Calculations**

$$p = 2(l + w) \qquad \text{Substitute in the formula.}$$
$$158 = 2(l + 32) \qquad \text{Distribute.}$$
$$158 = 2(l) + 2(32)$$
$$158 = 2l + 64 \qquad \text{Sort.}$$
$$158 - 64 = 2l \qquad \text{Combine.}$$
$$94 = 2l \qquad \text{Divide.}$$
$$\frac{94}{2} = \frac{2l}{2}$$
$$47 = l$$

**Interpretation**

**Thus, the length of the nursery plot is 47 m.**

Check:
$$p = 2(l + w) \qquad \text{Substitute.}$$
$$158 = 2(47 + 32) \qquad \text{Distribute.}$$
$$158 = 94 + 64 \qquad \text{Combine.}$$
$$158 = 158$$

EXAMPLE

INDTEC

A machine part weighs 2.7 kg and is to be shipped in a carton weighing $x$ kg. If the total weight of three packaged machine parts is 9.3 kg, how much does each carton weigh? (See Figure 5–2.)

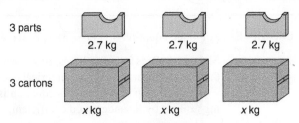

3 parts    2.7 kg    2.7 kg    2.7 kg

3 cartons    $x$ kg    $x$ kg    $x$ kg

**Figure 5–2**

**Unknown facts**

The weight of each empty carton.

**Known facts**

There are three parts and three cartons. Each part weighs 2.7 kg. The three packaged parts (in cartons) weigh a total of 9.3 kg.

**Relationships**

Let $x$ equal the weight of one empty carton. Each packaged part weighs the sum of the part (2.7 kg) plus the weight of a carton ($x$ kg). The sum of the three packaged parts equals 9.3 kg; that is, $3(2.7 + x) = 9.3$ or $(2.7 + x) + (2.7 + x) + (2.7 + x) = 9.3$.

**Estimation**

Three packaged parts weigh 9.3 kg, and $\frac{9.3}{3} = 3.1$ kg per packaged carton. Since one part weighs 2.7 kg, then the carton weighs well under 1 kg.

| Calculations | $3(2.7 + x) = 9.3$ | Distribute. |
| | $3(2.7) + 3(x) = 9.3$ | Make calculations. |
| | $8.1 + 3x = 9.3$ | Sort. |
| | $3x = 9.3 - 8.1$ | Combine. |
| | $3x = 1.2$ | Divide. |
| | $\dfrac{3x}{3} = \dfrac{1.2}{3}$ | |
| | $x = 0.4$ | |

**Interpretation**    **Each carton weighs 0.4 kg.**

Check:

$$3(2.7 + \boxed{x}\,) = 9.3 \qquad \text{Substitute.}$$
$$3(2.7 + \boxed{0.4}\,) = 9.3 \qquad \text{Make calculations.}$$
$$3(3.1) = 9.3$$
$$9.3 = 9.3$$

**TIP!**

**Additional Problem-Solving Strategies**

- Read the problem carefully. Read it several times and read it phrase by phrase.
- Understand all the words in the problem.
- Analyze the problem:
  What are you asked to find?
  What facts are given?
  What facts are implied?
- Visualize the problem.
- State the conditions or relationships of the problem "symbolically."
- Examine the options.
- Develop a *plan* for solving the problem.
- Write your *plan* symbolically; that is, write an equation.
- Anticipate the characteristics of a reasonable solution.
- Solve the equation.
- Verify your answer with the conditions of the problem.

There are many different problem-solving plans. Additional strategies can be considered.

SECTION 5–3 SELF-STUDY EXERCISES

**1**   Solve.

1. $4(x - 5) = 8$
2. $3(y - 7) = -15$
3. $6(x + 3) = -12$
4. $5(x + 7) = 25$
5. $-3(x + 8) = -39$
6. $-2(x - 7) = 26$
7. $3(2x - 1) = 3$
8. $4(x - 1) = 28$
9. $-4(-3x + 1) = 8$
10. $-9(2x - 3) = 27$
11. $7(x - 3) = 7x + 21$
12. $5x + 15 = 5(x + 3)$
13. $8(x - 5) = 8x + 12$
14. $6x + 7 = 3(2x + 4)$
15. $2(3x - 5) = 7x - 12$
16. $5(3x - 7) + 2 = 7x + 7$
17. $2x - 3 + 6x = 4(3x - 2) + 3$
18. $7(x - 1) = 2(x + 3) + 4(x - 7)$
19. $3(x - 4) - 2(3x - 1) = 17$
20. $6x - 4 + 2x = -(x - 5)$
21. $8x - (2x - 9) = -3$
22. $6 - 2(3x - 1) = 14$
23. $12 - 5(2x - 3) = -5$
24. $8 - 3(x - 2) = x - 6$

Write the statements as equations and solve.

**25.** The sum of $x$ and 4 equals 12. Find $x$.

**26.** 4 less than 2 times a number is 6. Find the number.

**27. INDTEC** Three parts totaling 27 lb are packaged for shipping. Two parts weigh the same. The third part weighs 3 lb less than each of the two equal parts. Find the weight of each part.

**28. HLTH/N** How many gallons of water must be added to 24 gal of pure disinfectant to make 60 gal of diluted disinfectant?

**29. INDTEC** A wet casting weighing 4.03 kg weighs 3.97 kg after drying. Write and solve an algebraic equation to find the weight loss due to drying.

**30. CON** A plumber needs 3 times as much perforated pipe as solid pipe to lay a drain line 400 ft long. How much of each type of pipe is needed?

**31. ELEC** An engineering student purchased a new circuits text, a used Spanish text, and a graphing calculator. She remembered that the calculator cost twice as much as the Spanish book and that the circuits text cost $70.00. The total before tax was $235.00. What was the cost of the calculator and the Spanish text?

**32. CON** Mary Jefferson purchased a home on a square lot 150 ft on each side. She wants to enclose the entire lot with a cedar fence. One estimate for the job was for $14.00 per linear foot. How much will the fence cost?

**33. AG/H** Phil Chu, owner of Chu's Landscape Service, knows that one of his fertilizer tanks holds twice as many gallons of liquid fertilizer as a second tank. The two tanks together hold 325 gal. If both tanks are filled to capacity, how many gallons of fertilizer does each tank hold?

**34. AG/H** Carolina Villa hired a stone mason to build a rectangular flower bed at one end of her patio. She needs enough mulch to cover 60 ft$^2$, the area of the flower bed. If the flower bed has a length of 10 ft, how wide is the bed?

---

## 5–4 | *Solving Linear Equations with Fractions by Clearing the Denominators*

*Learning Outcomes*

**1** Solve fractional equations by clearing the denominators.

**2** Solve applied problems involving rate, time, and work.

**3** Solve decimal equations by the clearing decimals.

### **1** Solve Fractional Equations by Clearing the Denominators.

Thus far, we have used mostly integers in equations. However, real-world situations require us to deal with various fraction and decimal quantities when we solve equations.

The techniques we used in Sections 5–2 and 5–3 will also help us solve equations with fractions and decimals. Other techniques minimize the calculations with fractions and decimals and allow more steps to be performed mentally. One of these techniques involves *clearing* the equation of all denominators in the first step. The resulting equation, which contains no fractions, is then solved using the procedures we learned earlier.

This process, called **clearing the fractions,** is another application of the multiplication axiom.

**To clear an equation of a single fraction:**

Apply the multiplication axiom by multiplying the entire equation by the fraction's denominator.

**EXAMPLE** Solve $\dfrac{4d}{3} = 12$.

$$\frac{4d}{3} = 12$$      Multiply both sides by the denominator 3. Reduce where possible.

$$(\overset{1}{\cancel{3}})\frac{4d}{\cancel{3}}_{1} = (3)(12)$$

$$4d = 36 \qquad \text{Divide.}$$

$$\frac{4d}{4} = \frac{36}{4}$$

$$d = \boxed{9}$$

Check: $\dfrac{4(\overset{3}{\cancel{9}})}{\underset{1}{\cancel{3}}} = 12$

$$12 = 12$$

**EXAMPLE** Solve $2x + \dfrac{3}{4} = 1$ by clearing the fraction.

$$2x + \frac{3}{4} = 1 \qquad \text{Multiply each term by the denominator 4.}$$

$$\boxed{4}\,(2x) + \boxed{4}\left(\frac{3}{4}\right) = \boxed{4}\,(1) \qquad \text{Reduce and multiply.}$$

$$4(2x) + \overset{1}{\cancel{4}}\left(\frac{3}{\cancel{4}\underset{1}{}}\right) = 4(1)$$

$$8x + 3 = 4 \qquad \text{Sort.}$$

$$8x = 4 - 3 \qquad \text{Combine.}$$

$$8x = 1 \qquad \text{Divide.}$$

$$\frac{8x}{8} = \frac{1}{8}$$

$$\boldsymbol{x = \frac{1}{8}}$$

Check: $2x + \dfrac{3}{4} = 1 \qquad$ Substitute $\dfrac{1}{8}$ in place of $x$.

$$2\overset{1}{}\left(\frac{1}{\underset{4}{\cancel{8}}}\right) + \frac{3}{4} = 1 \qquad \text{Multiply.}$$

$$\frac{1}{4} + \frac{3}{4} = 1 \qquad \text{Add.}$$

$$1 = 1$$

One advantage of clearing fractions is that it can be applied to equations in which the variable is in the *denominator* of the fraction. Let's use this procedure to solve $\dfrac{10}{x} = 2$. Note that the term $\dfrac{10}{x}$ is *not* the product of 10 and $x$. Therefore, the coefficient of $x$ is *not* 10.

EXAMPLE    Solve $\dfrac{10}{x} = 2$.

$$\dfrac{10}{x} = 2$$    Multiply both sides by denominator $x$. Reduce where possible.

$$(\overset{1}{\cancel{x}})\dfrac{10}{\underset{1}{\cancel{x}}} = (x)2$$

$$10 = 2x$$    Divide.

$$\dfrac{10}{2} = \dfrac{2x}{2}$$

$$\mathbf{5 = x}$$

Check:    $$\dfrac{\overset{2}{\cancel{10}}}{\underset{1}{\cancel{5}}} = 2$$

$$2 = 2$$

TIP!

## Check for Extraneous Roots

When you solve equations with a variable in the fraction's denominator, you may get a "root" that does not make a true statement when substituted in the original equation. Solutions or roots that do not make a true statement in the original equation are called **extraneous roots.** When solving an equation with the variable in the denominator, you *must* check the root to see if it makes a true statement in the original equation.

One situation that produces an extraneous root is a value that causes the denominator of any fraction to be zero. Such values are called **excluded values.**

### To identify an excluded value that causes a denominator of zero:

1. Write an equation for each fraction that has a variable in the denominator by setting the denominator equal to zero.
2. Solve each resulting equation.
3. The solution for each equation from Step 1 is an excluded value.

EXAMPLE    Solve $\dfrac{0}{x} = 5$.

Excluded value:

$$x = 0$$    Set the denominator containing a variable equal to zero.

Solve the equation:

$$\dfrac{0}{x} = 5$$    Multiply both sides by $x$.

$$x\left(\dfrac{0}{x}\right) = 5(x)$$

$$\frac{0}{5} = \frac{5x}{5}$$

$0 = x$        Possible solution.

Check:   $\dfrac{0}{0} \neq 5$        Does not check. Zero is an excluded value.

**The equation $\dfrac{0}{x} = 5$ has no solution.**

Equations in which the fraction contains more than one term in its numerator or denominator can also be cleared of fractions.

**EXAMPLE**   Solve $\dfrac{12}{Q + 6} = 1$. Identify any excluded values.

Excluded value:

$Q + 6 = 0$        Set denominator equal to zero and solve for $Q$.

$Q = 0 - 6$

$Q = -6$        Excluded value.

$\dfrac{12}{Q + 6} = 1$        Multiply both sides of the equation by the denominator, the quantity $Q + 6$. Reduce where possible.

$\overset{1}{(Q + 6)} \dfrac{12}{\underset{1}{Q + 6}} = (Q + 6)\, 1$        Distribute.

$12 = Q + 6$        Sort.

$12 - 6 = Q$        Combine like terms.

$6 = Q$        Because the numerical coefficient of $Q$ is already 1, the equation is solved.

Check:   $\dfrac{12}{6 + 6} = 1$        Substitute 6 for $Q$ and evaluate.

$\dfrac{12}{12} = 1$

$1 = 1$

**The solution of the equation is $Q = 6$.**

When an equation has more than one fractional term, we expand our process.

**To solve an equation by clearing all fractions:**

1. Multiply each term of the *entire* equation by the least common multiple (LCM) of the denominators of the equation.
2. Apply the distributive property to remove parentheses.
3. Combine like terms on each side of the equation.
4. Sort terms to collect the variable or letter terms on one side and number terms on the other (addition axiom).
5. Combine like terms on each side of the equation.
6. Solve the resulting equation by multiplying by the reciprocal of the coefficient of the variable term or dividing by the coefficient of the variable term (multiplying axiom).

**EXAMPLE** Solve $-\frac{1}{4}x = 9 - \frac{2}{3}x$ by clearing all fractions first.

There are no excluded values since there are no variables in a denominator.

$$-\frac{1}{4}x = 9 - \frac{2}{3}x$$

Multiply each term by the LCM of the denominators. LCM = 12 or 4(3).

$$(4)(3)\left(-\frac{1}{4}x\right) = (4)(3)(9) - (4)(3)\left(\frac{2}{3}x\right)$$

Reduce.

$$(\overset{1}{4})(3)\left(-\frac{1}{\underset{1}{4}}x\right) = (4)(3)(9) - (4)(\overset{1}{3})\left(\frac{2}{\underset{1}{3}}x\right)$$

Multiply the remaining factors.

$$-3x = 108 - 8x$$

This equation contains no fractions. Sort terms.

$$-3x + 8x = 108$$

Combine like terms.

$$5x = 108$$

Divide by the coefficient of $x$.

$$\frac{5x}{5} = \frac{108}{5}$$

$$x = \frac{108}{5}$$

Check using a calculator:

$$-\frac{1}{4}x = 9 - \frac{2}{3}x$$

Substitute $\frac{108}{5}$ for $x$.

$$-\frac{1}{4}\left(\frac{108}{5}\right) = 9 - \frac{2}{3}\left(\frac{108}{5}\right)$$

**Left Side:** $\boxed{(}\boxed{(-)}\,\boxed{1}\,\boxed{\div}\,\boxed{4}\,\boxed{)}\,\boxed{\times}\,\boxed{(}\,\boxed{108}\,\boxed{\div}\,\boxed{5}\,\boxed{)}\,\boxed{=} \Rightarrow -5.4$

**Right Side:** $9\,\boxed{-}\,\boxed{(}\,\boxed{2}\,\boxed{\div}\,\boxed{3}\,\boxed{)}\,\boxed{\times}\,\boxed{(}\,\boxed{108}\,\boxed{\div}\,\boxed{5}\,\boxed{)} = \Rightarrow -5.4$

Solutions of equations are not always whole numbers. Such solutions are usually written as *proper* or *improper fractions* in lowest terms. Solutions of applied problems are usually written in decimal or mixed-number form.

**EXAMPLE** Find the total resistance in a parallel dc circuit with three branches rated at 4 Ω, 10 Ω, and 20 Ω, respectively. Solve $\frac{1}{R} = \frac{1}{4} + \frac{1}{10} + \frac{1}{20}$ by clearing fractions first.

**ELEC** $R = 0$ is the excluded value.

$$\frac{1}{R} = \frac{1}{4} + \frac{1}{10} + \frac{1}{20}$$

The LCM is 20$R$ because $R$, 4, 10, and 20 all divide evenly into $20R$.

$$20R\left(\frac{1}{R}\right) = \overset{5}{20R}\left(\frac{1}{\underset{1}{4}}\right) + \overset{2}{20R}\left(\frac{1}{\underset{1}{10}}\right) + \overset{1}{20R}\left(\frac{1}{\underset{1}{20}}\right)$$

Multiply each term in the *entire* equation by 20$R$ and reduce.

$$20 = 5R + 2R + R$$

Combine like terms.

$$20 = 8R$$

Divide by the coefficient of $R$.

$$\frac{20}{8} = \frac{8R}{8}$$

Reduce.

$$\frac{5}{2} = R$$

$\frac{5}{2} = 2.5$

| Interpretation | The resistance is 2.5 Ω. |
|---|---|

Check: $\dfrac{1}{R} = \dfrac{1}{4} + \dfrac{1}{10} + \dfrac{1}{20}$      Substitute $\dfrac{5}{2}$ for $R$.

$$\dfrac{1}{\frac{5}{2}} = \dfrac{1}{4} + \dfrac{1}{10} + \dfrac{1}{20}$$

$$\dfrac{1}{\frac{5}{2}} = \dfrac{5}{20} + \dfrac{3}{20} + \dfrac{1}{20}$$

$$1 \cdot \dfrac{3}{5} = \dfrac{8}{20}$$

$$\dfrac{2}{5} = \dfrac{2}{5}$$

**TIP!**

### Fractions versus Decimals

Sometimes applied problems that require fractions in their equations require that their solutions be expressed as decimal numbers. In these cases, we perform the division indicated by the fraction.

The equation in the preceding example is derived from the formula for finding total resistance in a parallel dc circuit with three branches rated at 4, 10, and 20 ohms. Ohms are expressed in decimal numbers, so in an application the solution should be converted to a decimal equivalent.

$$R = \dfrac{5}{2}\,\Omega \qquad \text{or} \qquad 2.5\,\Omega$$

**2**   **Solve Applied Problems Involving Rate, Time, and Work.**

A **rate measure** often involves a unit of time. If car A travels 50 mi in 1 h, then car A's **rate of work** (travel) is 50 mi per 1 h, or $\dfrac{50 \text{ mi}}{1 \text{ h}}$ expressed as a fraction. If car A travels for 3 h, then the **amount of work** is $\dfrac{50 \text{ mi}}{\text{h}} \times 3 \text{ h} = 150 \text{ mi}$.

### To find the amount of work produced by one individual or machine:

1. Identify the rate of work and the time worked.
2. Use the formula for amount of work.

### Formula for amount of work:

amount of work completed = rate of work × time worked

A carpenter can install 1 door in 3 h. Find the number of doors the carpenter can install in 30 h.

| | |
|---|---|
| **Known facts** | Rate of work = 1 door per 3 h, $\frac{1}{3}$ door per hour, or $\frac{1 \text{ door}}{3 \text{ h}}$ |
| | Time worked = 30 h |
| **Unknown facts** | $W$ = amount of work or number of doors installed |
| **Relationships** | Amount of work = rate of work × time worked |
| **Estimation** | Several doors can be installed in 30 h, but less than 30 since 1 door would have to be installed every hour to install 30 doors. |
| **Calculations** | $W = \dfrac{1 \text{ door}}{\cancel{3} \text{ h}} \times \cancel{30}^{10} \text{ h}$   Reduce and multiply. |
| | $W = 10$ doors |
| **Interpretation** | **Thus, 10 doors can be installed in 30 h.** |

If two workers or machines do a job together, we can find the amount of work done by each worker or machine. Combined, the amounts equal 1 total job.

**To find the amount of work each individual or machine produces when working together:**

1. Identify the rate of work for each individual or machine. If unknown, assign a letter to represent the unknown.
2. Identify the time worked for each individual or machine. If unknown, assign a letter to represent the unknown. *Note:* Only one letter should be used and other unknowns should be written in relationship to the one letter.
3. Use the formula for completing one job.

**Formula for completing one job when A and B are working together:**

$$\begin{pmatrix} \text{A's} \\ \text{amount of} \\ \text{work} \end{pmatrix} + \begin{pmatrix} \text{B's} \\ \text{amount of} \\ \text{work} \end{pmatrix} = 1 \text{ completed job}$$

or

$$\begin{pmatrix} \text{A's} \\ \text{rate of} \times \text{time} \\ \text{work} \quad \text{worked} \end{pmatrix} + \begin{pmatrix} \text{B's} \\ \text{rate of} \times \text{time} \\ \text{work} \quad \text{worked} \end{pmatrix} = 1 \text{ completed job}$$

**EXAMPLE** Pipe 1 fills a tank in 6 min and pipe 2 fills the same tank in 8 min (Figure 5–3). How long does it take for both pipes together to fill the tank?

AG/H

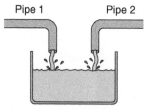

**Figure 5–3**

| | |
|---|---|
| **Known facts** | Pipe 1 fills the tank at a rate of 1 tank per 6 min, $\frac{1}{6}$ tank per minute, or $\dfrac{1 \text{ tank}}{6 \text{ min}}$.<br><br>Pipe 2 fills the tank at a rate of 1 tank per 8 min, $\frac{1}{8}$ tank per minute, or $\dfrac{1 \text{ tank}}{8 \text{ min}}$. |
| **Unknown facts** | $T$ = time (in minutes) for both pipes together to fill the tank. |
| **Relationships** | Amount of work of pipe 1 = $\dfrac{1 \text{ tank}}{6 \text{ min}}(T)$. Amount of work of pipe 2 = $\dfrac{1 \text{ tank}}{8 \text{ min}}(T)$.<br>Amount of work together = pipe 1's work + pipe 2's work. |
| **Estimation** | Both pipes together should fill the tank more quickly than the faster rate, or in less than 6 min. |

**Calculations**

$$\frac{1 \text{ tank}}{6 \text{ min}}(T \text{ min}) + \frac{1 \text{ tank}}{8 \text{ min}}(T \text{ min}) = 1 \text{ tank} \qquad \text{The LCM is } \boxed{24}.$$

$$\boxed{(24)}\left(\frac{1}{6}T\right) + \boxed{(24)}\left(\frac{1}{8}T\right) = \boxed{(24)}(1) \qquad \text{Clear fractions.}$$

$$(\overset{4}{24})\left(\frac{1}{\underset{1}{6}}T\right) + (\overset{3}{24})\left(\frac{1}{\underset{1}{8}}T\right) = (24)(1) \qquad \text{Reduce and multiply.}$$

$$4T + 3T = 24 \qquad \text{Combine.}$$

$$7T = 24 \qquad \text{Divide.}$$

$$\frac{7T}{7} = \frac{24}{7}$$

$$T = \frac{24}{7}\left(\text{or } 3\frac{3}{7}\right) \text{ min}$$

**Interpretation** **Both pipes together fill the tank in $3\frac{3}{7}$ min.**

---

**TIP!**

**Interpretation of Improper Fractions as Mixed Numbers, Decimal Equivalents, or Mixed Measurements**

In applied problems, it is desirable to change improper fractions like $\frac{24}{7}$ into mixed numbers or decimal equivalents. The problem statement generally dictates the interpretation.

In the preceding example, $\frac{24}{7}$ min can be interpreted as $3\frac{3}{7}$ min or 3.4 min (rounded). In some instances, you may need to change $3\frac{3}{7}$ min to seconds ($\frac{3}{7}$ min $\times \frac{60 \text{ s}}{1 \text{ min}} = 25.7$ s). Then, $\frac{3}{7}$ min becomes 3 min 26 s (rounded).

Two pipes, one a faucet and the other a drain, have opposite functions. Pipe 1 fills the tank. Pipe 2 is a drain and empties the tank. In this case, we subtract the work done by the drain from the work done by the faucet. This combined action, if the faucet fills at a faster rate than the drain empties, results in a full tank.

**Formula for completing one job when A and B are working in opposition:**

$$\left(\begin{array}{c} \text{A's} \\ \text{amount of} \\ \text{work} \end{array}\right) - \left(\begin{array}{c} \text{B's} \\ \text{amount of} \\ \text{work} \end{array}\right) = 1 \text{ completed job}$$

or

$$\left(\begin{array}{c} \text{A's} \\ \text{rate of} \times \text{time} \\ \text{work} \quad \text{worked} \end{array}\right) - \left(\begin{array}{c} \text{B's} \\ \text{rate of} \times \text{time} \\ \text{work} \quad \text{worked} \end{array}\right) = 1 \text{ completed job}$$

based on A's amount of work being greater than B's amount of work.

**EXAMPLE**

**AG/H**

A faucet fills a tank in 6 min. A drain empties the tank in 8 min (Figure 5–4). If both the faucet and drain are open, in how many minutes will the tank start to overflow if the faucet is not turned off?

**Known facts**

Faucet's rate of work = 1 tank filled per 6 min, $\frac{1}{6}$ tank per min, or $\dfrac{1 \text{ tank}}{6 \text{ min}}$.

Drain's rate of work = 1 tank emptied per 8 min, $\frac{1}{8}$ tank per min, or $\dfrac{1 \text{ tank}}{8 \text{ min}}$.

**Unknown facts**

$T$ = time (min) until tank overflows with faucet and drain both working.

**Relationships**

amount of work of faucet = $\dfrac{1 \text{ tank}}{6 \text{ min}}(T)$

amount of work of drain = $\dfrac{1 \text{ tank}}{8 \text{ min}}(T)$

amount of work when both faucet and drain are open = faucet's work − drain's work

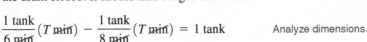

Faucet

Drain

**Figure 5–4**

**Estimation**

With both the faucet and drain open, it should take longer to fill the tank than if the faucet was open and the drain closed. It should take longer than 6 min.

**Calculations**

$\dfrac{1 \text{ tank}}{6 \text{ min}}(T \text{ min}) - \dfrac{1 \text{ tank}}{8 \text{ min}}(T \text{ min}) = 1 \text{ tank}$   Analyze dimensions.

$\dfrac{1}{6}T - \dfrac{1}{8}T = 1$   Clear fractions. LCM = 24.

$(24)\left(\dfrac{1}{6}T\right) - (24)\left(\dfrac{1}{8}T\right) = (24)(1)$   Reduce.

$(\overset{4}{24})\left(\dfrac{1}{\underset{1}{6}}T\right) - (\overset{3}{24})\left(\dfrac{1}{\underset{1}{8}}T\right) = (24)(1)$   Multiply remaining factors.

$4T - 3T = 24$   Combine like terms.

$T = 24 \text{ min}$

**Interpretation**

**With the faucet and drain both open, the tank will be full in 24 min.**

### 3  Solve Decimal Equations by Clearing the Decimals.

Since a decimal is a type of fraction, we can also solve equations containing decimals by clearing decimals. The place value of the last digit of the decimal determines the denominator of the fraction.

**To solve a decimal equation by clearing decimals:**

1. Multiply each term of the *entire* equation by the least common denominator (LCD) of the fractional amounts represented by the decimals.
2. Follow the same steps used in solving a linear equation.

**TIP!**

**LCD for Decimals**

Digits to the right of the decimal point represent fractions whose denominators are determined by the place value. To find the LCD for all the decimal numbers in an equation, find the decimal with the most digits after the decimal point. Use its denominator to clear the decimals.

This procedure allows you to avoid dividing by a decimal, which can be a common source of error when making calculations by hand.

**EXAMPLE**  Solve $0.38 + 1.1y = 0.6$ by first clearing the equation of decimals.

$$0.38 + 1.1y = 0.6 \qquad \text{The LCD is } 100.$$

$$100\,(0.38) + 100\,(1.1y) = 100\,(0.6) \qquad \text{Multiply by 100.}$$

$$38 + 110y = 60 \qquad \text{Sort.}$$

$$110y = 60 - 38 \qquad \text{Combine.}$$

$$\frac{110y}{110} = \frac{22}{110} \qquad \text{Divide.}$$

$$y = \mathbf{0.2}$$

The interest formula resembles the percentage formula, $P = RB$, but it includes the time period over which the money was borrowed or invested. If we know three of the four elements, we can find the fourth.

**Formula for simple interest:**

$$I = PRT$$

where $I = $ **interest,** $P = $ **principal,** $R = $ **rate** or percent in decimal form, and $T = $ **time.**

For comparison with the percentage formula, interest is the part or portion and the principal is the base.

**EXAMPLE**

**BUS**

A $1,000 investment is made for $2\frac{1}{2}$ years at 6.25%. Find the amount of interest.

When we use the interest formula, we change the percent to a decimal equivalent. When no time period is given for the rate, we assume the rate to be per year.

$$I = PRT$$  Substitute the values in the formula.

principal    rate        time

$$I = \$1,000 \quad (0.0625) \quad (2.5 \text{ years})$$  $6.25\% = 0.0625; 2\frac{1}{2}$ years $= 2.5$ years.

$$I = \$156.25$$

**The interest for $2\frac{1}{2}$ years is $156.25.**

TIP!

## Hand-Clearing Decimals Versus Using a Calculator

The preceding example illustrates that in some cases clearing decimals is *not* the most efficient way to solve an equation with decimals. The numbers produced may be extremely large and cumbersome.

$$I = \$1,000 \cdot 0.0625 \cdot 2.5$$  LCM is 10,000.

$$10,000I = \overset{1}{\cancel{10,000}} \left( \$1{,}\overset{100}{\cancel{000}} \cdot \frac{625}{\underset{1}{\cancel{10{,}000}}} \cdot \frac{25}{\underset{1}{\cancel{10}}} \right)$$  Multiply both sides of the equation by 10,000.

$$10,000I = 100(625)(25)$$  Multiply.

$$10,000I = 1,562,500$$  Divide.

$$I = \frac{1,562,500}{10,000}$$

$$I = \$156.25$$

A calculator gives the solution more quickly and more efficiently if we proceed using the decimals.

Let's use the interest formula to solve problems when facts other than the interest are missing.

**EXAMPLE**
**BUS**

Aetna Photo Studio borrowed $3,500 for some darkroom equipment and has to pay $1,890 in interest over a 3-year period. What is the interest rate?

**Known facts**    Principal = $3,500. Interest = $1,890. Time = 3 years.

**Unknown facts**    $R$ = rate.

**Relationships**    $I = PRT$.

**Estimation**    One year's interest is about $\frac{1}{3}$ of the total interest, or about $600.

$$10\% \text{ of } \$3,500 = \$350$$

$$20\% \text{ of } \$3,500 = \$350 \cdot 2 = \$700$$

Therefore, $R$ is more than 10% and less than 20%.

**Calculations**    $$I = PRT$$  Substitute known values.

$$1,890 = 3,500 \cdot R \cdot 3$$  Multiply.

$$1,890 = 10,500R \qquad \text{Solve for } R.$$

$$\frac{1,890}{10,500} = \frac{10,500R}{10,500}$$

$$0.18 = R \qquad \text{Translate as a percent. } 0.18 = 18\%.$$

**Interpretation**    **The interest rate is 18%.**

---

**EXAMPLE**
**AUTO**

The distance formula is distance = rate × time. If a truck was driven 82.5 mi at an average of 55 mi per h, how long did the trip take?

**Known facts**    Distance = 82.5 mi. Rate $= \dfrac{55 \text{ mi}}{\text{h}}$.

**Unknown facts**    $T$ = time in hours.

**Relationships**    Distance = rate × time.

**Estimation**    At 55 mi per h, a truck would travel 110 mi in 2 h. It would take less than 2 h to travel 82.5 mi.

**Calculations**    $82.5 = 55 \cdot T \qquad \text{Substitute known values.}$

$82.5 = 55T \qquad \text{Divide.}$

$\dfrac{82.5}{55} = \dfrac{55T}{55}$

$1.5 = T$

**Interpretation**    **The trip took 1.5 h.**

---

**EXAMPLE**
**ELEC**

The formula for voltage $V$ is wattage $W$ divided by amperage $A$: $V = \frac{W}{A}$. Find the voltage to the nearest hundredth needed for a circuit of 1,280 W with a current of 12.23 A.

**Estimation**    $1,200 \div 12 = 100$

$$V = \frac{W}{A} \qquad \text{Substitute values.}$$

$$V = \frac{1,280}{12.23} \qquad \text{Divide.}$$

$$V = 104.6606705 \qquad \text{or} \qquad 104.66 \ V$$

**Interpretation**    **The voltage is 104.66 V.**

---

**SECTION 5–4 SELF-STUDY EXERCISES**

**1**   Solve the equations. Identify excluded values as appropriate.

**1.** $\dfrac{2}{7}x = 8$

**2.** $-7 = \dfrac{21}{33}p$

**3.** $\dfrac{1}{3}r = \dfrac{6}{7}$

**4.** $0 = -\dfrac{2}{5}c$

**5.** $-\dfrac{5}{3}m = 9$

**6.** $-9m = \dfrac{5}{3}$

**7.** $\dfrac{5}{8}t = 1$

**8.** $-\dfrac{5}{7}p = -\dfrac{11}{21}$

**9.** $10 = -\dfrac{1}{35}t$

**10.** $\dfrac{5}{12}z = 20$

**11.** $\dfrac{2x}{3} = 18$

**12.** $\dfrac{7}{Q} = 21$

**13.** $\dfrac{y+1}{2} = 7$

**14.** $\dfrac{7}{p-4} = -8$

**15.** $0 = \dfrac{x}{4}$

**16.** $-\dfrac{8}{P} = -72$

**17.** $\dfrac{P}{-8} = -72$

**18.** $-8 = \dfrac{4B}{B-6}$

**19.** $\dfrac{3P}{7} = 12$

**20.** $\dfrac{8-R}{76} = 1$

**21.** $\dfrac{2}{9}c + \dfrac{1}{3}c = \dfrac{3}{7}$

**22.** $-\dfrac{1}{4}x = 9 - \dfrac{2}{3}x$

**23.** $\dfrac{2}{7}y + \dfrac{3}{8} = \dfrac{1}{7}y + \dfrac{5}{3}$

**24.** $\dfrac{1}{3}x + \dfrac{1}{2}x = \dfrac{20}{3}$

**25.** $\dfrac{7}{R} - \dfrac{2}{R} = -1$

**26.** $S = \dfrac{1}{15} + \dfrac{1}{5} + \dfrac{1}{30}$

**27.** $18 - \dfrac{1}{4}x = \dfrac{1}{2}$

**28.** $\dfrac{1}{7}H - \dfrac{1}{3}H = 0$

**29.** $\dfrac{7}{16}h + \dfrac{1}{9} = \dfrac{1}{3}$

**30.** $x + \dfrac{1}{4}x = 8$

**31.** $3y + 9 = \dfrac{1}{4}y$

**32.** $18 = \dfrac{4}{3x} - \dfrac{3}{2x}$

**33.** $\dfrac{2}{7}p + 1 = \dfrac{1}{3}p$

**34.** $S = \dfrac{1}{10} + \dfrac{1}{25} + \dfrac{1}{50}$

**35.** $-x + \dfrac{1}{7} = \dfrac{1}{2}x$

**36.** $0 = 1 + \dfrac{2}{9}c - c$

**2** Set up an equation and solve.

**37.** **INDTEC** A kiln fires 8 large vases in 2 h. How many vases can be fired in 20 h?

**38.** **INDTR** A printing press produces 1 day's newspaper in 4 h. A higher-speed press does 1 day's newspaper in 2 h. How much time does it take both presses to produce 1 day's newspaper?

**39.** **INDTEC** One machine packs 1 day's salmon catch in 8 h. A second machine packs 1 day's catch in 5 h. How much time does it require for 1 day's catch to be packed if both machines are used?

**40.** **CON** A painter can paint a house in 6 days. Another painter takes 8 days to paint the same house. If they work together, how much time will it take them to paint the house?

**41.** **INDTEC** One bottling machine can fill 400 bottles of water in 1 h and another machine can fill 400 bottles of water in $1\frac{1}{2}$ h. If both machines are working, how much time does it take to fill 400 bottles of water?

**42.** **AG/H** A tank has two pipes entering. Pipe 1 alone fills the tank in 4 min and pipe 2 takes 12 min to fill the tank. How much time does it take to fill the tank if both pipes are operating at the same time?

**43.** **AG/H** A tank has two pipes entering. Pipe 1 alone fills the tank in 30 min and together the pipes take 10 min to fill the tank. How much time does pipe 2 need to fill the tank alone?

**44.** **AG/H** A tank has two pipes entering it and one leaving it. Pipe 1 fills the tank in 3 min. Pipe 2 takes 7 min to fill the same tank. Pipe 3, however, empties the tank in 21 min. How much time does it take to fill the tank with all three pipes operating at the same time?

For Exercises 45–48, see example on page 266. Round to hundredths.

**45.** **ELEC** Find the total resistance in a parallel dc circuit with two branches rated at 12 Ω and 30 Ω, respectively.

**46.** **ELEC** A parallel dc circuit has three branches rated at 12 Ω, 15 Ω, and 20 Ω. Find the total resistance.

**47.** **ELEC** Three branches of a parallel dc circuit are rated at 16 Ω, 24 Ω, and 32 Ω. Find the total resistance.

**48.** **ELEC** A parallel dc circuit has a total resistance of 3 Ω. One branch alone produces 18 Ω resistance. What is the resistance of the other branch?

**3** Solve the equations.

**49.** $2.3x = 4.6$

**50.** $0.8R = 0.6$ (round to nearest tenth)

**51.** $0.33x + 0.25x = 3.5$ (round to nearest hundredth)

**52.** $0.3a = 4.8$

**53.** $1.5p = 7$ (round to nearest tenth)

**54.** $0.04x = 0.08 - x$ (round to nearest hundredth)

**55.** $0.4p = 0.014$

**56.** $0.47 = R + 0.4R$ (round to nearest hundredth)

**57.** $2.3 = 5.6 + y$

**58.** $4.3 = 0.3x - 7.34$

**59.** $2x + 3.7 = 10.3$

**60.** $0.16 + 2.3x = -0.3$

**61.** $1.5x + 2.1 = 3$

**62.** $3.82 - 2.5y = 1$

**63.** $0.15p = 2.4$

Solve the problems using decimal equations.

**64. AUTO** If the formula for force is force = pressure × area, how many pounds of force are produced by a pressure of 35 pounds per square inch on a piston whose surface area is 2.5 in²? Express the pounds of force as a decimal number.

**65. INDTR** The circumference of a circle equals $\pi$ times the diameter. If a steel rod has a diameter of 1.5 in., what is the circumference of the rod to the nearest hundredth? (Circumference is the distance around a circle.)

**66. AUTO** The distance formula is distance = rate × time. If a trucker drove 682.5 mi at 55 mi per h (mi/h), how long did she drive? (Answer to the nearest whole number.)

**67. AUTO** The distance formula is distance = rate × time. If a tractor-trailer rig is driven 422.5 mi at 65 mi/h on interstate highways, how long does the trip take?

**68. ELEC** Electrical resistance in ohms ($\Omega$) is voltage $V$ divided by amperage $A$. Find the resistance to the nearest tenth for a motor with a voltage of 12.4 V requiring 1.5 A.

**69. BUS** Rita earned $84.75 working 7.5 h at a college bookstore. What was her hourly wage?

**70. ELEC** The formula for electrical power is $V = \frac{W}{A}$: voltage ($V$) equals wattage ($W$) divided by amperage ($A$). Find the voltage to the nearest hundredth needed for a circuit of 500 W with a current of 3.2 A.

**71. BUS** Find the interest paid on a loan of $2,400 for 1 year at an interest rate of 11%.

**72. BUS** Find the interest paid on a loan of $800 at $8\frac{1}{2}\%$ interest for 2 years.

**73. BUS** Find the total amount of money (maturity value) that the borrower will pay back on a loan of $1,400 at $12\frac{1}{2}\%$ simple interest for 11 years.

**74. BUS** Find the rate of interest on an investment of $2,500 made by Nurse Honda for a period of 2 years if she received $612.50 in interest.

**75. AG/H** Maddy Brown needed start-up money for her landscape service. She borrowed $12,000 for 30 months and paid $360 interest on the loan. What interest rate did she pay?

---

## 5–5 | Formulas

*Learning Outcomes*  **1** Evaluate formulas.

**2** Rearrange formulas to solve for a specified variable.

The most common use of formulas is for finding missing values. If we know values for all but one variable of a formula, we can find the unknown value.

**1 Evaluate Formulas.**

To **evaluate** a formula is to substitute known values for some variables and perform the indicated operations to find the value of the variable in question. In so doing, we may need to use any or all of the steps and procedures for solving equations.

**To evaluate a formula:**

1. Write the formula.
2. Rewrite the formula substituting known values for variables of the formula.
3. Solve the equation from Step 2 for the missing variable.
4. Interpret the solution within the context of the formula.

We can solve for a variable that is not isolated. In the next example the formula for the perimeter of a rectangle is used. The variable $w$ is within parentheses and must be isolated.

**EXAMPLE**    Solve the formula $P = 2(l + w)$ for $w$ if $P = 12$ ft and $l = 4$ ft.

$P = 2(l + w)$      Perimeter of a rectangle = 2 times the sum of the length and the width.
                     Substitute values.

$12 = 2(4 + w)$      Apply the distributive property.

$12 = 8 + 2w$      Isolate the term with the variable (addition axiom).

$12 - 8 = 2w$      Combine like terms.

$4 = 2w$      Divide.

$$\frac{4}{2} = \frac{2w}{2}$$

$2 = w$      Interpret solution.

**The width is 2 ft.**

---

The interest formula may be used to illustrate a variable that is one of several factors.

---

**EXAMPLE**    Evaluate the formula $I = PRT$ for principal ($P$) if interest ($I$) = \$94.50, rate ($R$) = 21%, and time ($T$) = $\frac{1}{2}$ year.

**BUS**    For convenience in using a calculator, convert $\frac{1}{2}$ year to 0.5 year. In this formula, the rate should be expressed as a decimal equivalent, 21% = 0.21.

$I = PRT$      Substitute values and solve for $P$.

$94.50 = P(0.21)(0.5)$      Multiply 0.21 and 0.5.

$94.50 = 0.105P$      Divide.

$$\frac{94.50}{0.105} = \frac{0.105P}{0.105}$$

$900 = P$      Interpret solution.

**The principal is \$900.**

---

**EXAMPLE**    Evaluate the formula $E = \dfrac{I - P}{I}$ to the nearest thousandth if $I = 24{,}000$ calories (cal) and

**AUTO**    $P = 8{,}600$ cal. Round to thousandths.

$E = \dfrac{I - P}{I}$      Engine efficiency = difference between heat input and heat output divided by heat input.

$E = \dfrac{24{,}000 - 8{,}600}{24{,}000}$      Substitute given values. Perform calculations in numerator grouping.

$E = \dfrac{15{,}400}{24{,}000}$      Divide.

$E = 0.642$

**The engine efficiency is 0.642 (64.2% efficient).**

**EXAMPLE** Evaluate the formula $R_T = \dfrac{R_1 R_2}{R_1 + R_2}$ if $R_1 = 10\ \Omega$ and $R_2 = 6\ \Omega$.

ELEC $\quad R_T = \dfrac{R_1 R_2}{R_1 + R_2}$ $\qquad$ Total resistance = product of first resistance and second resistance divided by sum of first and second resistances.

$\qquad R_T = \dfrac{10(6)}{10 + 6}$ $\qquad$ Substitute given values. Perform calculations in numerator and denominator groupings.

$\qquad R_T = \dfrac{60}{16}$ $\qquad$ Divide.

$\qquad R_T = 3.75$ $\qquad$ Dimension analysis: $\dfrac{\text{ohms (}\cancel{\text{ohms}}\text{)}}{\cancel{\text{ohms}}} = \text{ohms}$

**The total resistance in the circuit is 3.75 $\Omega$.**

The next example has many steps and it is important to apply the order of operations. The formula is used to find the length of a belt connecting two pulleys.

**EXAMPLE** Evaluate $L = 2C + 1.57(D + d) + \dfrac{D + d}{4C}$ if $C = 24$ in., $D = 16$ in., and $d = 4$ in. Round to hundredths. See Figure 5–5.

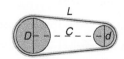

$L$ = length of belt joining two pulleys
$C$ = distance between centers of pulleys
$D$ = diameter of large pulley
$d$ = diameter of small pulley

**Figure 5–5**

INDTEC $\quad L = 2C + 1.57(D + d) + \dfrac{D + d}{4C}$ $\qquad$ Substitute the given values.

$\qquad L = 2(24) + 1.57(16 + 4) + \dfrac{16 + 4}{4(24)}$ $\qquad$ Work groupings in parentheses, numerator, and denominator.

$\qquad L = 2(24) + 1.57(20) + \dfrac{20}{96}$ $\qquad$ Work multiplications and division.

$\qquad L = 48 + 31.4 + 0.20833333$ $\qquad$ Add.

$\qquad L = 79.60833333$

**The pulley belt is 79.61 in. long.**

**2** **Rearrange Formulas to Solve for a Specified Variable.**

**Formula rearrangement** generally refers to isolating a letter term other than the one already isolated in the formula. Solving formulas in this manner shortens our work when doing repeated formula evaluations. After we solve the formula for the desired variable, we rewrite the formula with the variable on the left side for convenience and for use in electronic spreadsheets.

The following formulas require applying the addition axiom.

**EXAMPLE** Solve the markup formula, $M = S - C$, for $S$ (selling price).

BUS $\qquad M = S - C$ $\qquad$ Markup = selling price − cost. Isolate $S$.

$\qquad M + C = S$ $\qquad$ The coefficient of $S$ is positive, so the formula is solved.

$\qquad S = M + C$ $\qquad$ Rewrite $S$ on the left for convenience.

**Where Is the Variable or Unknown in a Formula?**

It may help to think of the one letter we are solving for as the unknown or variable, and to think of the other letters as if they were *coefficients* or *constants* in an ordinary equation.

**To rearrange a formula:**

1. Determine which variable of the formula will be isolated (solved for).
2. Highlight or mentally locate all instances of the variable to be isolated.
3. Treat all other variables of the formula as you would treat numbers in an equation, and perform normal steps for solving an equation.
4. If the isolated variable is on the right side of the equation, interchange the sides so that it appears on the left side.

**EXAMPLE**  Solve the formula $M = S - C$ for $C$ (cost).

**BUS**

$$M = S - C$$  Markup = selling price – cost. Isolate $C$.

$$M - S = -C$$  The coefficient of $C$ is negative, so divide both sides by $-1$.

$$\frac{M - S}{-1} = \frac{-C}{-1}$$  Note effect on signs after division by $-1$.

$$-M + S = C$$  Write the positive term first.

$$S - M = C$$  Rewrite with $C$ on the left for convenience.

$$C = S - M$$

When a formula contains a term of several factors and we need to solve for one of those factors, we treat the factor being solved for as the variable and the other factors as its coefficient.

**EXAMPLE**  Solve for $R$ in the formula $I = PRT$.

**BUS**

$$I = P\ R\ T$$  Interest = Principal × Rate × Time. Since we are solving for $R$, $PT$ is its coefficient.
Divide both sides by the coefficient of the variable.

$$\frac{I}{PT} = \frac{PRT}{PT}$$  Divide both sides of the equation by $PT$. Reduce.

$$\frac{I}{PT} = R$$  Rewrite with $R$ on the left.

$$R = \frac{I}{PT}$$

Sometimes formulas contain addition (or subtraction) and multiplication in which we must use the distributive property to solve for a particular letter. In the formula $A = P(1 + ni)$, we must use the distributive property to solve for either $n$ or $i$ because each appears inside the grouping.

**EXAMPLE**    Solve for $n$ in the formula for compound amount, $A = P(1 + ni)$.

**BUS**

$A = P(1 + \boxed{n}\,i)$    Identify the variable to be isolated. Use the distributive property to remove $n$ from parentheses.

$A = P + P\,\boxed{n}\,i$    Use the addition axiom to isolate the term containing $n$.

$A - P = P\,\boxed{n}\,i$    Divide both sides by the coefficient of $n$, $Pi$.

$$\frac{A - P}{Pi} = \frac{P\,\boxed{n}\,i}{Pi}$$    Reduce.

$$\frac{A - P}{Pi} = \boxed{n}$$    Rewrite with $n$ on the left.

$$\boxed{n} = \frac{A - P}{Pi}$$

When the formula contains division, we clear the denominator before taking further steps.

**EXAMPLE**    The formula $V = V = \dfrac{P}{I}$ represents the relationship among the voltage drop ($V$), the electrical power ($P$), and the current ($I$). Rearrange the formula to solve for $P$.

**ELEC**

$V = \dfrac{P}{I}$    Identify the variable to be isolated. Multiply both sides by the denominator $I$ to clear it.

$(I)V = \dfrac{P}{I}(I)$    Reduce.

$IV = \boxed{P}$    Rewrite with $P$ on the left.

$P = IV$

**SECTION 5–5 SELF-STUDY EXERCISES**

 Evaluate the formulas.

1. $D = RT$ if $R = 40$ mi/h and $T = 7$ h
2. $S = C + M$ if $C = \$40$ and $M = \$80$
3. $A = 4\pi r^2$ if $\pi = 3.14$ and $r = 15$ in.
4. $A = bh$ if $b = 12$ m and $h = 9.8$ m.
5. $P = a + b + c$ if $a = 50$ cm, $b = 43$ cm, and $c = 45$ cm

Evaluate the interest formula $I = PRT$ using the following values.

6. **BUS**    Find the interest if $P = \$800$, $R = 15.5\%$, and $T = 2\frac{1}{2}$ years.

7. **BUS**    Find the rate if $I = \$427.50$, $P = \$1,500$, and $T = 2$ years.

8. **BUS**    Find the time if $I = \$236.25$, $P = \$750$, and $R = 10.5\%$.

9. **BUS**    Find the principal if $I = \$838.50$, $R = 21\frac{1}{2}\%$, and $T = 1\frac{1}{2}$ years.

Evaluate Exercises 10-12 using the percentage formula, $P = RB$.

**10. AG/H** Find the portion if $R = 15\%$ and $B = 600$ lb.

**11.** Find the rate if $P = 24$ kg and $B = 300$ kg.

**12. BUS** Find the base if $P = \$250$ and $R = 7.4\%$. Round to hundredths.

**13.** Evaluate the rate formula, $R = \dfrac{P}{B}$, for $P$ if $R = 16\%$ and $B = 85$.

**14.** Evaluate the base formula, $B = \dfrac{P}{R}$, for $R$ if $B = \$2,200$ and $P = \$374$.

**15. CAD/ARC** Find the length of a rectangular work area if the perimeter is 180 in. and the width is 24 in. Use the formula $P = 2(l + w)$.

**16.** Find the area of a circle whose radius is 54.5 cm using the formula $A = \pi r^2$ and using 3.14 for $\pi$. Round to the nearest tenth.

**17. BUS** Find the cost ($C$) if the markup ($M$) on an item is \$5.25 and the selling price ($S$) is \$15.75. Use the formula $M = S - C$.

**18.** Using the formula for the side of a square, $s = \sqrt{A}$, find the length of a side of a square field whose area is 16 mi$^2$.

**19. CON** Evaluate the formula for the area of a circle, $A = \pi r^2$, if $r = 7$ in. and $\pi = 3.16$. Round to the nearest tenth.

**20.** Evaluate the formula for the area of a square, $A = s^2$, if $s = 2.5$ km.

**21. BUS** Using the markup formula, $M = S - C$, find the selling price ($S$) if the markup ($M$) is \$12.75 and the cost ($C$) is \$36.

**22. ELEC** Use the formula $R_T = \dfrac{R_1 R_2}{R_1 + R_2}$ to find the total resistance ($R_T$) if one resistance ($R_1$) is 12 $\Omega$ and the second resistance ($R_2$) is 8 $\Omega$.

**23. AUTO** What is the percent efficiency ($E$) of an engine if the input ($I$) is 25,000 calories and the output ($P$) is 9,600 calories? Use the formula $E = \dfrac{I - P}{I}$.

**24. AUTO** Distance is rate times time, or $D = RT$. Find the rate if the distance traveled is 140 mi and the time traveled is 4 h.

**25. ELEC** The formula for voltage (Ohm's law) is $E = IR$. Find the amperes of current ($I$) if the voltage ($E$) is 120 V and the resistance ($R$) is 80 $\Omega$.

**26. INDTEC** According to Boyle's law, if temperature is constant, the volume of a gas is inversely proportional to the pressure on it. Find the final volume ($V_2$) of a gas using the formula $\dfrac{V_1}{V_2} = \dfrac{P_2}{P_1}$ if the original volume ($V_1$) is 15 ft$^3$, the original pressure ($P_1$) is 60 lb per square inch (psi), and the final pressure ($P_2$) is 150 psi.

**27. ELEC** The formula for power ($P$) in watts ($W$) is $P = I^2 R$. Find the current ($I$) in amperes if a device draws 63 W and the resistance ($R$) is 7 $\Omega$.

**28. AUTO** Use the formula $H = \dfrac{D^2 N}{2.5}$ to find the number of cylinders ($N$) required in an engine of 3.2 hp ($H$) if the cylinder diameter ($D$) is 2 in.

**29. INDTEC** The formula for the speed ($s$) of a driven pulley in revolutions per minute (rpm) is $s = \dfrac{DS}{d}$. Find the speed of a driven pulley with a diameter ($d$) of 5 in. if the diameter ($D$) of the driving pulley is 10 in. and its speed ($S$) is 800 rpm.

**30. INDTEC** If the distance ($C$) between the centers of the pulleys in Exercise 29 is 24 in., find the length ($L$) to the nearest hundredth of the belt connecting them using the formula $L = 2C + 1.57(D + d) + \dfrac{D + d}{4C}$

**31. ELEC** Find the impedance ($Z$) in ohms using the formula $Z = \sqrt{R^2 + X^2}$ if the reactance ($X$) is 15 $\Omega$ and the resistance ($R$) is 8 $\Omega$. Round to tenths.

**2** Rearrange the formulas.

**32. BUS** Solve $I_n = I - S$ for $S$, where $I_n = $ new inventory, $I = $ current inventory, and $S = $ sales.

**33.** Solve $S = 2\pi rh$ for $r$.

**34.** Solve $y = mx + b$ for $b$.

**35.** Solve $V = \pi r^2 h$ for $h$.

**36.** Solve $S = C + M$ for $C$.

**37.** Solve $\dfrac{R}{100} = \dfrac{P}{B}$ for $R$.

**38.** Solve $P = 2(b + s)$ for $b$.

**39.** Solve $C = 2\pi r$ for $r$.

**40.** Solve $A = lw$ for $l$.

**41.** Solve $R = AC - B$ for $C$.

**42.** Solve $E = IR$ for $R$.

**43.** Solve $D = RT$ for $R$.

**44.** Solve $S = P - D$ for $D$.

**45.** Solve $C = \pi d$ for $d$.

**46.** Solve $C = 2\pi r$ for $r$.

**47. BUS** The formula for finding the amount of a repayment on a loan is $A = I + P$, where $A$ is the amount of the repayment, $I$ is the interest, and $P$ is the principal. Solve the formula for interest.

**48. BUS** The formula for finding interest is $I = PRT$, where $I$ represents interest, $P$ represents principal, $R$ represents rate, and $T$ represents time. Rearrange the formula to find the time.

---

## CHAPTER REVIEW OF KEY CONCEPTS

**Learning Outcomes**
**Section 5–1**

**1** Identify equations, terms, factors, constants, variables, and coefficients (pp. 244–246).

**What to Remember with Examples**

An *equation* is a statement that two quantities are equal. A *variable* is a letter that represents an unknown value. A *root* or *solution* of an equation is the value of the variable that makes the equation a true statement.

> $x = 5 + 2$ is an *equation*. $x$ is the *variable*. 7 is the *root* or *solution*.

*Factors* are expressions of multiplication.

> $5a$ means 5 *times a*. 5 is a *factor* of $5a$. $a$ is a *factor* of $5a$.

*Terms* are algebraic expressions that are added or subtracted.

> In the expression $5a + 3b + 7$, $5a$ is a *term* and $3b$ is a *term*. 7 is a *term*.

*Constants* are terms that contain only numbers.

> In $5a + 3b + 7$, the *constant* term is 7.

*Variable* terms are terms that have at least one letter.

> In $5a + 3b + 7$, $5a$ and $3b$ are *variable* terms.

A *coefficient* is one factor as it relates to the remaining factors of a term.

> In $5a + 7$, 5 is the *coefficient* of $a$ and $a$ is the *coefficient* of 5. 7 has no coefficient. The coefficient 5 is also called the *numerical coefficient*.

**2** Write verbal interpretations of symbolic statements (p. 247).

Mathematical symbols can be translated into phrases and statements.

> $2x - 3 = 5$ can be translated into "3 less than twice a number is five."

**3** Translate verbal statements into symbolic statements using variables (pp. 247–248).

Statements can be translated into mathematical symbols.

> The statement "A number increased by 13 results in 52" is translated into $x + 13 = 52$.

**4** Simplify variable expressions (pp. 248–249).

Combine like terms by adding or subtracting the coefficients of the terms and by using the same letter or letters for the sum or difference.

> $5m + 4m = (5 + 4)m = 9m$

Apply the distributive principle by multiplying the factor outside the parentheses by each term inside the parentheses.

> $-4(x + 2y - 5) = -4x - 8y + 20$

Distributing $-1$ changes only the sign of each term inside the parentheses.

$$-(2x - 3) = -1(2x - 3) = -2x + 3$$

## Section 5–2

**1** Solve linear equations using the addition axiom (pp. 250–252).

**1.** Locate the variable in the equation. **2.** Identify the constant that is associated with the variable by addition (or subtraction). **3.** Add the opposite of the constant to both sides.

| | |
|---|---|
| $x + 3 = 2$ | Add $-3$ to both sides. |
| $x + 3\ \boxed{-3} = 2\ \boxed{-3}$ | Combine like terms. |
| $x + 0 = -1$ | $x + 0 = x$ |
| $x = -1$ | |

| | |
|---|---|
| $9 = x - 6$ | Add 6 to both sides. |
| $9 + 6 = x - 6 + 6$ | Combine like terms. |
| $15 = x + 0$ | $x + 0 = x$ |
| $15 = x$ | |

**2** Solve linear equations using the multiplication axiom (pp. 252–254).

Multiply both sides of the equation by the reciprocal of the coefficient of the letter term *or* divide both sides of the equation by the coefficient of the letter term.

Solve:

$$\frac{x}{5} = 8 \qquad\qquad -8x = 24 \qquad\qquad \frac{1}{3}x = \frac{4}{9}$$

$$\frac{5}{1}\left(\frac{x}{5}\right) = 8\left(\frac{5}{1}\right) \qquad \frac{-8x}{-8} = \frac{24}{-8} \qquad \left(\frac{\overset{1}{\cancel{3}}}{1}\right)\left(\frac{1}{\cancel{3}}x\right) = \left(\frac{4}{\cancel{9}}\right)\left(\frac{\cancel{3}}{1}\right)$$

$$x = 40 \qquad\qquad x = -3$$

$$x = \frac{4}{3}$$

$$7.5 = 2.5x \qquad\qquad \frac{x}{0.6} = 2.9$$

$$\frac{7.5}{2.5} = \frac{2.5x}{2.5} \qquad\qquad 0.6\left(\frac{x}{0.6}\right) = 0.6(2.9)$$

$$3 = x \qquad\qquad x = 1.74$$

To check the solution of an equation, substitute the value of the variable in each place it appears in the equation. Perform operations on both sides of the equation. The two sides of the equation should be equal.

Verify that $x = 3$ is the solution for the equation $2x - 1 = 5$.

$$2(3) - 1 = 5$$
$$6 - 1 = 5$$
$$5 = 5$$

**3** Solve linear equations with like terms on the same side of the equation (pp. 254–255).

Combine the like terms on the same side of the equation. Solve the remaining equation using the multiplication axiom.

Solve $3x - 5x = 12$ for $x$.

$$3x - 5x = 12$$
$$-2x = 12$$
$$\frac{-2x}{-2} = \frac{12}{-2}$$
$$x = -6$$

Galenda assembles a product in 10 min and Marcus assembles the product in 15 min. How long will it take them together to assemble one product? Galenda's rate × time: $\frac{1}{10}T$. Marcus's rate × time: $\frac{1}{15}T$.

$$\frac{1}{10}T + \frac{1}{15}T = 1$$

$$(30)\left(\frac{1}{10}T\right) + (30)\left(\frac{1}{15}T\right) = (30)(1)$$

$$(\overset{3}{\cancel{30}})\left(\frac{1}{\cancel{10}}T\right) + (\overset{2}{\cancel{30}})\left(\frac{1}{\cancel{15}}T\right) = (30)(1)$$

$$3T + 2T = 30$$
$$5T = 30$$
$$\frac{5T}{5} = \frac{30}{5}$$
$$T = 6 \text{ min}$$

Estimation: less than 10 min but more than half of 10, or 5 min. for both to do one job.
Galenda's rate: $\frac{1}{10}$ job per min.
Marcus's rate: $\frac{1}{15}$ job per min.
Time in minutes: T. LCM = 30.
Multiply by 30.
Reduce.

Combine like terms.
Divide.

Completing one job when working in opposition:
(A's rate × time) − (B's rate × time) = 1 job

Solve as above but *subtract* the two amounts of work.

An inlet valve fills a vat in 2 hr. A drain valve empties the vat in 5 hr. With both valves open, how long does it take for the vat to fill? Inlet valve's rate × time: $\frac{1}{2}T$. Drain valve's rate × time: $\frac{1}{5}T$.

$$\frac{1}{2}T - \frac{1}{5}T = 1$$

$$(10)\left(\frac{1}{2}T\right) - (10)\left(\frac{1}{5}T\right) = (10)1$$

$$(\overset{5}{\cancel{10}})\left(\frac{1}{\cancel{2}}T\right) - (\overset{2}{\cancel{10}})\left(\frac{1}{\cancel{5}}T\right) = (10)1$$

$$5T - 2T = 10$$
$$3T = 10$$
$$\frac{3T}{3} = \frac{10}{3}$$
$$T = 3\frac{1}{3} \text{ hr}$$

Estimation: more than 2 hr.

LCM = 10.
Multiply by 10.

Reduce.

**3** Solve decimal equations by clearing the decimals (pp. 271–273).

To clear the equation of decimals: Multiply the *entire* equation by the least common denominator (LCD) of the fractional amounts represented by the decimals. (The place value of the decimal amount with the most places after the decimal point will be the LCD.)

$$3.5x + 2.75 = 10$$
$$(100)(3.5x) + (100)(2.75) = (100)(10)$$
$$350x + 275 = 1{,}000$$
$$350x = 1{,}000 - 275$$
$$350x = 725$$
$$\frac{350x}{350} = \frac{725}{350}$$
$$x = 2.071428571$$
$$x = 2.07 \quad \text{(rounded)}$$

Place value of the LCM = 100.
Multiply by 100.
Sort.
Combine like terms.
Divide.

Chapter Review of Key Concepts

## Section 5–5

**1** Evaluate formulas (pp. 275–277).

Substitute values. Solve using rules for solving equations and/or the order of operations.

---

Find the area ($A$) if the length ($l$) is 4 ft and the width ($w$) is 2 ft.

| | |
|---|---|
| $A = lw$ | Substitute values. |
| $A = (4)(2)$ | Multiply. |
| $A = 8 \text{ ft}^2$ | ft (ft) = ft² |

---

Sometimes the letter that is to be solved for must be isolated.

---

If the perimeter ($P$) of a square is 12 in., find the length of a side.

| | |
|---|---|
| $P = 4s$ | Substitute value for $P$. |
| $12 = 4s.$ | Divide. |
| $3 = s$ | Interchange sides. |
| $s = 3 \text{ in.}$ | |

---

**2** Rearrange formulas to solve for a specified variable (pp. 277–279).

Isolate the desired variable so that it appears on the left. This can make evaluation or using spreadsheets more efficient. Apply appropriate rules for solving equations.

---

Solve the formula $S = \dfrac{R + P}{2}$ for $R$.

| | |
|---|---|
| $(2)S = \dfrac{R + P}{2}(2)$ | Clear the denominator. |
| $2S = R + P$ | Isolate $R$ (sort terms). |
| $2S - P = R$ | Interchange sides so the variable appears on the left. |
| $R = 2S - P$ | |

---

## CHAPTER REVIEW EXERCISES

### Section 5–1

Verify that the statements are true equations.

1. $5 + 7 = 18 - 6$
2. $9(8) = 12 + 3(4) + 7^2 - 1$
3. $8 - 9 + 3^2 = -2^2 + 12$
4. $7 + 3 = 8 \cdot 1 - 5 + 4 + 3$

Find the value of the variable that makes the equation true.

5. $x = 5 + 9$
6. $8 - 5 = x$
7. $y = \dfrac{48}{-6}$
8. $7 + 3(5-2) = x$

Identify the terms in the expressions by drawing a box around each term.

9. $15x - \dfrac{3a}{7} + \dfrac{(x - 7)}{5}$
10. $5x - 8 + \dfrac{3}{y}$

State the equations in words.

11. $x + 5 = 2$
12. $x - 7 = 11$
13. $\dfrac{x}{8} = 7$
14. $3(x + 7) = -3$

Write the following statements in symbols.

15. 7 more than twice a number is 11.
16. A certain stock listed on the New York Stock Exchange closed at 42.375, a decrease of 3.125 points from the opening price.

**17.** Twice the sum of a number and 8 is 40.

**18.** The print shop used 31 cases of copy paper during one month. End-of-the-month inventory indicated 172 cases on hand. Write an equation to find the number of cases on hand at the beginning of the month. Then, solve the equation.

Simplify.

**19.** $3a + 2a$

**20.** $7a - 3b + 5 + 7a - 9b$

**21.** $3(2y - 4) + y$

**22.** $-(4y - 7)$

**23.** $-3(a + 2) - 5$

**24.** $5 - (a - 3)$

## Section 5–2

Solve.

**25.** $x - 5 = 8$

**26.** $y + 7 = 3$

**27.** $x - 8 = -10$

**28.** $x - 15 = -7$

**29.** $x - 5 = 14$

**30.** $-2 = 8 - x$

**31.** $3x = 21$

**32.** $4x = -28$

**33.** $-15 = 2b$

**34.** $-5 = -m$

**35.** $3 = \dfrac{1}{5}x$

**36.** $-\dfrac{2}{7}x = 8$

**37.** $4x + x = 25$

**38.** $36 = 9a - 5a$

**39.** $20 - 4 = 2x - 6x$

**40.** $13 - 27 = 3x - 10x$

**41.** $-12 = -8 - 2x$

**42.** $12x + 27 = 3x$

**43.** $3x + 9 = 10 + 3x + 1$

**44.** $10 + 4x = 5 - x$

**45.** $7 - 4y = y + 22$

**46.** $7x - 1 = 4x + 17 + 7x$

**47.** $7x - 5 + 2x = 3 - 4x + 12$

**48.** $2x - 3 + 15 = 7x - 8 - 6x$

## Section 5–3

Solve the equations.

**49.** $3x = 3(9 + 2x)$

**50.** $4a = 8 - (a + 7)$

**51.** $5x = 7 + (x + 5)$

**52.** $3(x + 2) - 5 = 2x + 7$

**53.** $4(3 - x) = 2x$

**54.** $3(2x - 4) = 4x - 6$

**55.** $-2(4 - 2x) = -16 + 2x$

**56.** $-16 = -2(-2x + 4)$

**57.** $8 = 6 - 2(3x - 1)$

**58.** $4x - (x + 3) = 3x - 3$

**59.** $3(x - 1) = 18 - 2(x + 3)$

**60.** $-(x - 1) = 2(x + 7)$

**61.** $-(2x + 1) = -7$

**62.** $2 + 3(x - 4) = 2x - 5$

**63.** $7 = 3 + 4(x + 2)$

**64.** $7(x + 2) = -6 + 2x$

**65.** $3(4x + 3) = 3 - 4(x - 1)$

**66.** $3(2 - x) - 1 = 4(3 - x)$

Write the statements as equations and solve.

**67.** 5 times the sum of $x$ and 6 is 42 more than $x$. Find $x$.

**68.** How many gallons of water must be added to 46 gal of pure alcohol to make 100 gal of alcohol solution?

**69.** **BUS** If one technician works 3 hr less than another and their total hours worked are 51, how many hours has each technician worked?

**70.** **CON** The shorter side of an L-shaped carpenter's square is 6 in. shorter than the longer side. If the total length of the carpenter's square is 24 in., what is the measure of each side?

**71.** **AG/H** Ms. Galendez's backyard is a rectangle whose length is twice the width. If the perimeter of the yard is 720 ft, what are the dimensions of her yard?

**72.** **BUS** Brubakers catered 32 chicken dinners for $409. This included a $25 delivery charge. Find the cost of each dinner excluding the delivery charge.

## Section 5–4

Solve the equations.

**73.** $m + \dfrac{1}{4} = \dfrac{3}{4}$

**74.** $\dfrac{3}{5}y = 12$

**75.** $p = \dfrac{1}{2} + \dfrac{1}{3}$

**76.** $x + \dfrac{1}{7}x = 16$

**77.** $\dfrac{2}{5} - x = \dfrac{1}{2}x + \dfrac{4}{5}$

**78.** $\dfrac{R}{7} - 6 = -R$

**79.** $\dfrac{3}{7}m - \dfrac{1}{2} = \dfrac{2}{3}$

**80.** $\dfrac{1}{4}S = \dfrac{1}{4} + \dfrac{1}{10} + \dfrac{1}{20}$

**81.** $m = 2 + \dfrac{1}{4}m$

Solve the equations. Identify extraneous values.

**82.** $\dfrac{1}{R} = \dfrac{1}{10} + \dfrac{1}{3} + \dfrac{1}{6}$

**83.** $\dfrac{2}{P} = \dfrac{1}{2} + \dfrac{1}{4} - \dfrac{5}{12}$

**84.** $\dfrac{3}{x} + 4 = \dfrac{1}{5} - 7$

Set up an equation with fractions and solve.

**85. AG/H** Melissa can complete a landscape project in 3 h. Henry can complete the same project in 7 h. How long would it take Melissa and Henry working together to complete the landscape project?

**86. COMP** One optical scanner reads a stack of sheets in 20 min. A second scanner reads the same stack in 12 min. How long does it take for both scanners together to process the one stack of sheets?

**87. ELEC** An apprentice electrician can install 5 light fixtures in 2 h. How many light fixtures can be installed in 10 h?

**88. CON** A brick mason can erect a retaining wall in 6 h. The brick mason's apprentice can do the same job in 10 h. How much time does it take both of them working together to erect the retaining wall?

**89. ELEC** A parallel dc circuit has 3 branches rated at 2 Ω, 6 Ω, and 12 Ω. Find the total resistance. See example on p. 266.

**90. ELEC** Find the total resistance of a parallel circuit with 2 branches rated at 12 Ω and 16 Ω. Round to hundredths.

Solve the equations. Round to tenths if necessary.

**91.** $2.3x - 4.1 = 0.5$

**92.** $0.22 + 1.6x = -0.9$

**93.** $0.3x - 2.15 = 0.8x + 3.75$

**94. AUTO** The distance formula is distance = rate × time. If a portable MRI unit traveled 350.8 mi to and from a rural hospital at 50 mi/h, how long to the nearest hour did the trip to and from the hospital take?

**95. ELEC** Electrical resistance in ohms (Ω) is voltage ($V$) divided by amperage ($A$). Find the resistance of a small motor with a voltage of 8.5 V requiring 0.5 A.

## Section 5–5

Evaluate the interest formula, $I = PRT$, using the given values.

**96.** Find the rate if $I = \$2{,}484$, $P = \$4{,}600$, and $T = 3$ years.

**97. BUS** Find the time if $I = \$387.50$, $P = \$1{,}550$, and $R = 12.5\%$.

**98.** Evaluate the formula for the area of a square, $A = s^2$, if $s = 3.25$ km.

**99. ELEC** Ohm's law is $E = IR$. Find the amperes of current ($I$) if the voltage ($E$) is 220 V and the resistance ($R$) is 80 Ω.

**100. ELEC** Use the formula $R_t = \dfrac{R_1 R_2}{R_1 + R_2}$ to find the total resistance ($R_t$) if one resistance ($R_1$) is 10 Ω and the second resistance ($R_2$) is 9 Ω. Round to tenths.

**101. ELEC** The formula for power ($P$) in watts (W) is $P = I^2 R$. Find the current ($I$) in amperes if a device draws 392 W and the resistance ($R$) is 8 Ω.

**102. AUTO** Use the formula $E = \dfrac{I - P}{I}$ to find the percent efficiency ($E$) of an engine if the input ($I$) is 22,600 cal and the output ($P$) is 5,600 cal. Round to the nearest tenth of a percent.

Solve the formulas for the indicated variable.

**103.** $V = lwh$ for $w$

**104.** $s = c + m$ for $c$

**105.** $s = r - d$ for $r$

**106.** $s = r - d$ for $d$

**107.** $v = v_0 - 32t$ for $t$

**108.** $V = \frac{1}{3}Bh$ for $h$

**109. BUS** The formula for finding the sale price on an item is $S = P - D$, where $S$ is the sale price, $P$ is the original price, and $D$ is the discount. Solve the formula for the original price.

**110. BUS** The formula for finding tax is $T = RM$, where $T$ represents tax, $R$ represents the tax rate, and $M$ represents the marked price. Rearrange the formula to find the marked price.

1. A formula is an equation that gives a model for solving a certain type of application. Devise formulas for the following relationships.
   (a) An electrical power company computes the monthly charges by multiplying the kilowatts of power used times the cost per kilowatt and adds to that a fixed monthly fee.
   (b) A store calculates the ending balance on a charge account by multiplying the interest rate times the previous unpaid balance and then adding the previous balance, the interest, and purchases and subtracting payments.
   (c) Profit on the sale of a particular item is the product of the number of items sold and the difference between the selling price of the item and its cost to the seller.

2. Formula rearrangement is a way of devising variations of formulas.
   (a) Explain the usefulness of formula rearrangement.
   (b) Select a formula that has at least three variables. Find a variation of the formula for each variable of the formula.
   (c) Describe at least two occasions when it is desirable to rearrange a formula.

## PRACTICE TEST

Solve. Round to hundredths if necessary.

1. $x + 5 = 19$

2. $-8y = 72$

3. $\dfrac{x}{2} = 5$

4. $5x + 2x = 49$

5. $5 - 2x = 3x - 10$

6. $5x + 3 - 7x = 2x + 4x - 11$

7. $3(x + 4) = 18$

8. $3x + 2 = 4(x - 1) - 1$

9. $\dfrac{8}{y + 2} = -7$

10. $\dfrac{4}{5}z + z = 8$

11. $5x + \dfrac{3}{5} = 2$

12. $3x + 2 = \dfrac{2}{3}$

13. $\dfrac{3}{5}x + \dfrac{1}{10}x = \dfrac{1}{3}$

14. $\dfrac{1}{x} = \dfrac{1}{3} + \dfrac{5}{6}$

Solve the equations. Round to hundredths when necessary.

15. $1.3x = 8.02$

16. $4.5y + 1.1 = 3.6$

17. $0.18x = 300 - x$

18. $7.9 = 0.5x - 8.35$

19. $0.23 + 7.1x = -0.8$

Solve the problems involving fractions and decimal numbers.

20. A pipe fills 1 tank in 4 h. If a second pipe empties 1 tank in 6 h, how long does it take for the tank to fill with both pipes operating?

21. Ohm's law is $E = IR$. Find the amperes of current ($I$) if the voltage ($E$) is 110 V and the resistance ($R$) is 50 $\Omega$?

22. The formula for the volume ($V$) of a solid rectangular figure is $V = lwh$ (length × width × height). If the volume of a mailing container is 7.5 m³, its length is 1.5 m, and its width is 0.5 m, what is its height?

23. The electrical resistance of a wire is found from the formula $R = \frac{PL}{A}$. Rearrange the formula to find the length $L$ of the wire.

24. A pipe fills 1 tank in 8 h. A second pipe fills the tank in 12 h. How long does it take to fill the tank with both pipes filling the tank?

25. Engine displacement $d$ is found using the formula $d = \pi r^2 sn$. Solve to find $s$.

26. Use the formula $R_t = \dfrac{R_1 R_2}{R_1 + R_2}$ to find the total resistance ($R_t$) if one resistance ($R_1$) is 9 $\Omega$ and the second resistance ($R_2$) is 8 $\Omega$. Round to tenths.

27. If the efficiency ($E$) of an engine is 70% and the input ($I$) is 40,000 cal, find the output ($P$) in calories. Use the formula $E = \dfrac{I - P}{I}$.

Kirchhoff's current law (KCL) and Kirchhoff's voltage law (KVL) form the basis of all electronics. Kirchhoff's current law (KCL) says that *the sum of all currents at a node equals zero.* A node is like a street intersection. Current entering the node counts as positive and current exiting the node counts as negative. Then the sum of all the currents will equal zero. Current is measured in amperes (A). Because an ampere is a large measuring unit, measurements are often in milliamperes (mA), or thousandths of an ampere. Figure 5–6 illustrates a node.

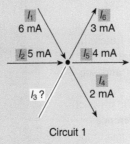

Circuit 1

**Figure 5–6**

Arrows are often used to indicate current. *Let each entering arrow have a + sign and each exiting arrow have a – sign.* Assume that there is a node with two currents entering the node ($I_1 = 6$ mA, $I_2 = 5$ mA), an unknown current $I_3$, and three currents exiting the node ($I_4 = 2$ mA, $I_5 = 4$ mA, and $I_6 = 3$ mA). See Figure 5–6, circuit 1. As an equation this is

$$I_1 \;+\; I_2 \;+ I_3 +\; I_4 \;+\; I_5 \;+\; I_6 \;= 0$$
$$6\,\text{mA} + 5\,\text{mA} + I_3 - 2\,\text{mA} - 4\,\text{mA} - 3\,\text{mA} = 0 \qquad \text{Insert appropriate numbers and signs.}$$
$$I_3 = 0 - 6\,\text{mA} - 5\,\text{mA} + 2\,\text{mA} + 4\,\text{mA} + 3\,\text{mA} \qquad \text{Solve for } I_3.$$
$$I_3 = -2\,\text{mA} \qquad \text{Combine like terms.}$$

This means that $I_3$ *is exiting the node and equals 2 mA,* as shown in Figure 5–7. To verify this, substitute $-2$ mA into the original equation.

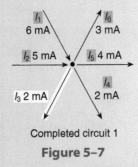

Completed circuit 1

**Figure 5–7**

$$I_1 \;+\; I_2 \;+\; I_3 \;+\; I_4 \;+\; I_5 \;+\; I_6 \;= 0$$
$$6\,\text{mA} + 5\,\text{mA} \; -2\text{mA} \; - 2\,\text{mA} - 4\,\text{mA} - 3\,\text{mA} = 0 \qquad \text{It works!}$$

Kirchhoff's voltage law (KVL) says that *the sum of all voltages in any loop equals zero.* Visualize current moving around a loop in a circuit. The circuit diagram includes points of change in voltage that may represent sources of current indicated by a positive sign as the first sign encountered or that may represent a use of current (drop) indicated by a negative sign as the first sign encountered. The equations for KVL look like those for KCL. The only difference is that a closed loop, rather than a node, is used.

Look at circuit 2 in Figure 5–8, which has two voltage sources ($V_1 = 8$ V, $V_2 = 7$ V), an unknown voltage called $V_3$, and three voltage drops ($V_4 = 3$ V, $V_5 = 4$ V, $V_6 = 2$ V). Let's start at the top of the circuit and move down the left side to the bottom and then up the right side.

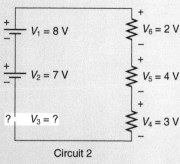

Circuit 2

**Figure 5–8**

As an equation,

$$V_1 + V_2 + \boxed{V_3} + V_4 + V_5 + V_6 = 0$$
$$8\,\text{V} + 7\,\text{V} + \boxed{V_3} - 3\,\text{V} - 4\,\text{V} - 2\,\text{V} = 0 \qquad \text{Insert appropriate numbers and signs.}$$
$$\boxed{V_3} = 0 - 8\,\text{V} - 7\,\text{V} + 3\,\text{V} + 4\,\text{V} + 2\,\text{V} \qquad \text{Solve for } V_3.$$
$$\boxed{V_3} = -6\,\text{V}$$

This means that $V_3$ *is a voltage drop of 6 V* as shown in Figure 5–9. To verify this, substitute $-6$ V into the original equation.

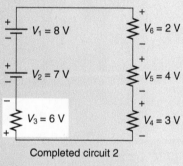

Completed circuit 2

**Figure 5–9**

$$V_1 + V_2 + \boxed{V_3} + V_4 + V_5 + V_6 = 0$$
$$8\,\text{V} + 7\,\text{V} \;\boxed{-6\,\text{V}}\; - 3\,\text{V} - 4\,\text{V} - 2\,\text{V} = 0 \qquad \text{It works!}$$

### Exercises

Redraw each circuit in Figs. 5–10 to 5–15. Find the missing currents and voltages in all the circuits and draw the correct component on the circuit. For each circuit, write a complete equation using all the values to show that the algebraic sum of the currents at a node or of the voltages in a loop equals zero.

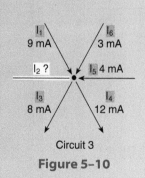

Circuit 3

**Figure 5–10**

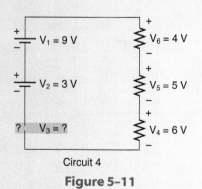

Circuit 4

**Figure 5–11**

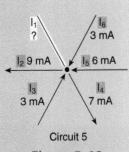

Circuit 5

**Figure 5–12**

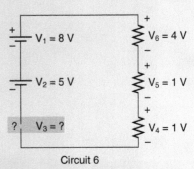

Circuit 6

**Figure 5–13**

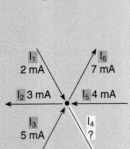

Circuit 7

**Figure 5–14**

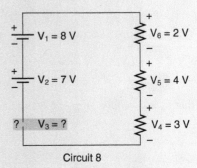

Circuit 8

**Figure 5–15**

## Answers

Circuit 3:  $I_1 + \boxed{I_3} + I_3 + I_4 + I_5 + I_6 = 0$

$9\text{ mA} \underline{+4\text{ mA}} - 8\text{ mA} - 12\text{ mA} + 4\text{ mA} + 3\text{ mA} = 0$

Circuit 4:  $V_1 + V_2 + \boxed{I_3} + V_4 + V_5 + V_6 = 0$

$9\text{ V} + 3\text{ V} \underline{+3\text{ V}} - 6\text{ V} - 5\text{ V} - 4\text{ V} = 0$

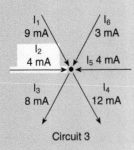

Circuit 3

**Figure 5–16**

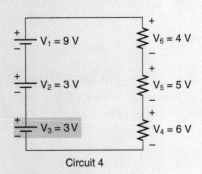

Circuit 4

**Figure 5–17**

Circuit 5:  $\boxed{I_1} + I_2 + I_3 + I_4 + I_5 + I_6 = 0$

$\underline{4\text{ mA}} - 9\text{ mA} + 3\text{ mA} - 7\text{ mA} + 6\text{ mA} + 3\text{ mA} = 0$

Circuit 6:  $V_1 + V_2 + \boxed{V_3} + V_4 + V_5 + V_6 = 0$

$8\text{ V} + 5\text{ V} \underline{-7\text{ V}} - 1\text{ V} - 1\text{ V} - 4\text{ V} = 0$

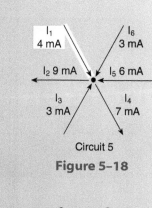

Circuit 5

**Figure 5–18**

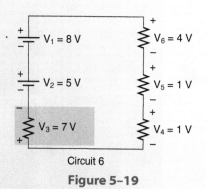

Circuit 6

**Figure 5–19**

Circuit 7:   $I_1 + I_2 + I_3 + \boxed{I_4} + I_5 + I_6 = 0$

$2\,\text{mA} - 3\,\text{mA} + 5\,\text{mA} \ \underline{-1\,\text{mA}} + 4\,\text{mA} - 7\,\text{mA} = 0$

Circuit 8:   $V_1 + V_2 + \boxed{V_3} + V_4 + V_5 + V_6 = 0$

$8\,\text{V} + 7\,\text{V} \ \underline{-6\,\text{V}} - 3\,\text{V} - 4\,\text{V} - 2\,\text{V} = 0$

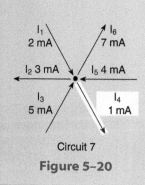

Circuit 7

**Figure 5–20**

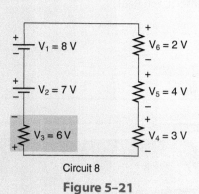

Circuit 8

**Figure 5–21**

# 6

# Ratio and Proportion

## Focus on Careers

Machinists use tools such as lathes and milling machines to produce precision metal parts. Machinists produce large quantities of the same part and also produce small batches of one-of-a-kind parts. They must be able to read blueprints, work with machines, and know the properties of metals.

Machinists learn the knowledge and skills required for their career at community or technical colleges or vocational schools, and some learn through a combination of apprenticeship programs and schooling. Machinists should be mechanically inclined, have good problem-solving abilities, and be able to do highly accurate work. In some highly precise parts, tolerances may reach 0.00001 in.! Mathematics, blueprint reading, metalworking, and drafting courses are musts for machinists.

Most machine shops are well ventilated, well lit, and relatively clean. Although there are some dangers in working with any machinery, machinists are expected to follow safety pre-

*(continued)*

cautions and wear protective equipment such as safety glasses, ear plugs, and gloves. Most machinists are expected to stand most of the day and may also do moderate to heavy lifting.

Although job growth is projected to be slower than average, job opportunities for machinists are expected to continue to be excellent because the number of workers entering the field is expected to be less than the number of job openings that arise from employment growth and from replacing machinists who retire.

The median hourly wage for machinists in 2002 was $15.66, but the top 10% of machinists earned more than $23.17 per hour. Metalworking machinery manufacturing jobs paid the highest hourly wage, and motor vehicle parts manufacturing paid slightly lower wages.

Source: *Occupational Outlook Handbook,* 2004–2005 Edition, U.S. Department of Labor, Bureau of Labor Statistics.

## 6-1 | Ratio and Proportion

*Learning Outcomes*  **1**  Solve equations that are proportions.

A common type of equation that contains fractions is a *proportion*. In a proportion, *each side of the equation is a fraction or ratio.*

### 1 Solve Equations That Are Proportions.

The equivalent fractions $\frac{5}{10}$ and $\frac{1}{2}$ can be written as the proportion $\frac{5}{10} = \frac{1}{2}$. Each fraction is also called a **ratio,** and two ratios that are equal or equivalent form a **proportion.**

To solve a proportion with a missing element, we use the property that the *cross products* in a proportion are equal. In a proportion, the **cross products** are the product of the numerator of the first fraction times the denominator of the second, and the product of the denominator of the first fraction times the numerator of the second. In the proportion $\frac{a}{b} = \frac{c}{d}$, the cross products are $a \cdot d$ and $b \cdot c$.

> **Property of proportions:**
>
> The cross products in a proportion are equal. Symbolically, if $\frac{a}{b} = \frac{c}{d}$, then $a \cdot d = b \cdot c$, provided that $b$ and $d$ are not equal to zero. Also, if $a \cdot d = b \cdot c$, then $\frac{a}{b} = \frac{c}{d}$.

This property can also help us find a missing element when three of the four elements of a proportion are known.

**EXAMPLE** Solve the proportions.

(a) $\dfrac{x}{4} = \dfrac{9}{6}$   (b) $\dfrac{4x}{5} = \dfrac{17}{20}$   (c) $\dfrac{x-2}{x+8} = \dfrac{3}{5}$   (d) $\dfrac{3}{x} = 7$

(a)   $\dfrac{x}{4} = \dfrac{9}{6}$         Cross multiply. $x(6) = 6x$ ; $4(9) = 36$

$6x = 36$         Solve for $x$.

$\dfrac{6x}{6} = \dfrac{36}{6}$

$x = 6$

**(b)** $\dfrac{4x}{5} = \dfrac{17}{20}$     Cross multiply. $4x(20) =$ 80*x* ; $5(17) =$ 85

85*x* = 85     Solve for *x*.

$\dfrac{80x}{80} = \dfrac{85}{80}$

$x = \dfrac{85}{80}$     Reduce.

$x = \dfrac{17}{16}$

**(c)** $\dfrac{x-2}{x+8} = \dfrac{3}{5}$     Cross multiply.

5(*x* − 2) = 3(*x* + 8)     Distribute.

$5x - 10 = 3x + 24$     Sort.

$5x - 3x = 24 + 10$     Combine terms.

$2x = 34$     Solve for *x*.

$\dfrac{2x}{2} = \dfrac{34}{2}$

$x = 17$

**(d)** $\dfrac{3}{x} = 7$     Write 7 as a fraction. $7 = \dfrac{7}{1}$

$\dfrac{3}{x} = \dfrac{7}{1}$     Cross multiply.

3 = 7*x*     Solve for *x*.

$\dfrac{3}{7} = \dfrac{7x}{7}$

$\dfrac{3}{7} = x$

---

**SECTION 6–1 SELF-STUDY EXERCISES**

**1** Solve the proportions. Round to four significant digits if necessary.

1. $\dfrac{x}{5} = \dfrac{9}{15}$

2. $\dfrac{3x}{16} = \dfrac{3}{8}$

3. $\dfrac{x-1}{x+6} = \dfrac{4}{5}$

4. $\dfrac{5}{x} = 8$

5. $\dfrac{2}{7} = \dfrac{x-4}{x+3}$

6. $\dfrac{2x+1}{8} = \dfrac{3}{7}$

7. $\dfrac{3x-2}{3} = \dfrac{2x+1}{3}$

8. $\dfrac{5}{2x-2} = \dfrac{1}{8}$

9. $\dfrac{8}{3x+2} = \dfrac{8}{14}$

10. $\dfrac{2x}{8} = \dfrac{3x+1}{7}$

11. $\dfrac{-5}{x-2} = \dfrac{5}{x}$

12. $\dfrac{7}{x-6} = \dfrac{-3}{x}$

13. $\dfrac{4}{8} = \dfrac{x}{60}$

14. $\dfrac{5.2}{16} = \dfrac{x}{8.3}$

15. $\dfrac{5.2}{340} = \dfrac{12.8}{x}$

16. $\dfrac{3x + 1}{4} = \dfrac{5x}{8}$

17. $\dfrac{7.35}{4} = \dfrac{5.21}{x}$

18. $\dfrac{9.12}{8} = \dfrac{x}{1.03}$

19. $\dfrac{2x}{7} = \dfrac{5.3}{6.7}$

20. $\dfrac{7x}{5} = \dfrac{91}{20}$

21. $\dfrac{0.107}{4x} = \dfrac{0.04}{321}$

22. $\dfrac{0.7x}{2.3} = \dfrac{5.6}{12.8}$

23. $\dfrac{0.04x}{5.2} = \dfrac{3.16}{14.08}$

24. $\dfrac{10^4}{x} = \dfrac{10^7}{10^5}$

25. $\dfrac{10^8}{10^5} = \dfrac{x}{10^9}$

26. $\dfrac{2x - 3}{7} = \dfrac{3}{28}$

27. $\dfrac{32}{21} = \dfrac{8}{5x - 1}$

28. $\dfrac{4}{3x - 5} = \dfrac{18}{12}$

29. $\dfrac{12 \text{ in.}}{15 \text{ in.}} = \dfrac{8 \text{ in.}}{x}$

30. $\dfrac{15 \text{ ft}}{38 \text{ ft}} = \dfrac{x}{57 \text{ ft}}$

31. $\dfrac{12 \text{ k}\Omega}{8 \text{ k}\Omega} = \dfrac{x}{6 \text{ k}\Omega}$

32. $\dfrac{x}{15 \text{ W}} = \dfrac{60 \text{ W}}{3.5 \text{ W}}$

33. $\dfrac{150 \text{ V}}{0.6 \text{ V}} = \dfrac{65 \text{ V}}{x}$

34. $\dfrac{12 \text{ mA}}{8.2 \text{ mA}} = \dfrac{x}{0.5 \text{ mA}}$

35. $\dfrac{5.12}{14.87} = \dfrac{x}{3.91}$

36. $\dfrac{21.25}{3.2x} = \dfrac{212.5}{32}$

37. $\dfrac{\frac{3}{5}}{\frac{5}{8}} = \dfrac{\frac{4}{5}}{x}$

38. $\dfrac{x}{\frac{5}{9}} = \dfrac{\frac{3}{10}}{\frac{4}{5}}$

39. $\dfrac{2\frac{1}{4}}{x} = \dfrac{\frac{7}{10}}{\frac{4}{9}}$

40. $\dfrac{4\frac{3}{8}}{\frac{5}{8}} = \dfrac{x}{\frac{4}{5}}$

41. $\dfrac{50 \text{ mi}}{1 \text{ h}} = \dfrac{400 \text{ mi}}{x \text{ h}}$

42. $\dfrac{2{,}500 \text{ mi}}{3.8 \text{ h}} = \dfrac{500 \text{ mi}}{x \text{ h}}$

43. $\dfrac{\$245{,}000}{2{,}500 \text{ ft}^2} = \dfrac{\$x}{1 \text{ ft}^2}$

44. $\dfrac{\$495{,}000}{5{,}200 \text{ ft}^2} = \dfrac{\$x}{1 \text{ ft}^2}$

45. $\dfrac{\$4.12}{20 \text{ oz}} = \dfrac{\$x}{1 \text{ oz}}$

46. $\dfrac{15 \text{ gal}}{40 \text{ acres}} = \dfrac{x \text{ gal}}{1 \text{ acres}}$

47. $\dfrac{350 \text{ mg}}{7.8 \text{ cm}^3} = \dfrac{x \text{ mg}}{1 \text{ cm}^3}$

48. $\dfrac{350 \text{ mg}}{60 \text{ cm}^3} = \dfrac{x \text{ mg}}{1 \text{ cm}^3}$

49. $\dfrac{5 \text{ gal}}{20 \text{ acres}} = \dfrac{x \text{ gal}}{1 \text{ acre}}$

50. $\dfrac{44 \text{ gal}}{1 \text{ min}} = \dfrac{x \text{ gal}}{3 \text{ min}}$

51. $\dfrac{1{,}500 \text{ mi}}{2.7 \text{ h}} = \dfrac{x \text{ mi}}{1 \text{ h}}$

52. $\dfrac{400 \text{ mg}}{20 \text{ cm}^3} = \dfrac{x \text{ mg}}{1 \text{ cm}^3}$

53. $\dfrac{1{,}000 \text{ mg}}{120 \text{ cm}^3} = \dfrac{x \text{ mg}}{1 \text{ cm}^3}$

54. $\dfrac{4{,}500 \text{ mg}}{560 \text{ cm}^3} = \dfrac{x \text{ mg}}{1 \text{ cm}^3}$

---

## 6–2  Direct Variation

*Learning Outcomes*

1  Solve problems with direct variation.

2  Solve problems that involve similar triangles.

### 1  Solve Problems with Direct Variation.

| Apples | Cost |
|--------|------|
| 4 | $1 |
| 8 | $2 |
| 12 | $3 |
| 16 | $4 |

Many problems in the workplace can be solved using proportions. The details of the problem can be grouped into two pairs of data that can be *directly* related.

A **direct variation** is one in which the quantities being compared are directly related, so that as one quantity increases (or decreases), the other quantity also increases (or decreases).

Data are often arranged in tables so these relationships can be examined. If 4 apples cost $1, set up a table to examine related costs of other amounts of apples.

In a pair of data, one item is identified as the **independent variable,** such as the *number of apples* purchased. The other item depends on the first item and it is identified as the **dependent variable,** such as the *total cost* of the apples.

**To set up a direct variation:**

1. Establish two pairs of related data.
2. Write one pair of data in the numerators of two ratios.
3. Write the other pair of data in the denominators of two ratios.
4. Form a proportion using the two ratios.

EXAMPLE   (a) Find the cost of 10 apples if 4 apples cost \$1. (b) How many apples can be purchased for \$10?

BUS   **(a)** Pair 1:  4 apples  cost  \$1 .

Pair 2:  10 apples  cost  $c$ dollars .

Estimation   8 apples would cost \$2. Therefore, 10 apples will cost more than \$2.

$$\frac{4 \text{ apples}}{10 \text{ apples}} = \frac{\$1}{\$c}$$   Pair 1 is the numerator of each ratio.
  Pair 2 is the denominator of each ratio. Cross multiply.

$$4c = 10$$   Divide.

$$\frac{4c}{4} = \frac{10}{4}$$

$$c = 2.50$$   To the nearest cent.

Interpretation   **10 apples cost \$2.50.**

**(b)** Pair 1:  4 apples  cost  \$1 .

Pair 2:  $a$ apples  cost  \$10 .

Estimation   If 10 apples cost \$2.50, 4 times as many apples can be bought for \$10. That is, 40 apples can be bought.

$$\frac{4 \text{ apples}}{a \text{ apples}} = \frac{\$1}{\$10}$$   Pair 1
  Pair 2 Cross multiply.

$$40 = a$$

Interpretation   **40 apples can be bought for \$10.**

Directly related data pairs can also be set up by making each pair a ratio. In the preceding example, we would write the ratio as

Pair 1   $$\frac{4 \text{ apples}}{\$1} = \frac{10 \text{ apples}}{\$c}$$   Pair 2

A *third way* to use directly related data is to identify a data pair in which both values are known and then use that relationship to find a conversion factor. For example, if 4 apples cost \$1, how much does 1 apple cost?

$$\$1 \div 4 \text{ apples} = \$0.25 \text{ per apple}$$

A conversion factor is used to multiply by the number of items. If each apple costs \$0.25, then 10 apples cost $10 \times \$0.25$, or \$2.50. A conversion factor is also called a **constant of direct variation.** The direct variation formula is $y = kx$, where $x$ is the **independent variable,** $y$ is the **dependent variable,** and $k$ is the *constant of direct variation.* Another way to express this is $k = \dfrac{y}{x}$.

## To find the constant of direct variation:

1. Identify a pair of related data in which both values are known.
2. Write the data pair as a ratio. The units in the denominator of the ratio should match the units of the independent variable. $k = \dfrac{y}{x}$
3. Leave the ratio as a fraction or change it to a decimal equivalent.

---

**EXAMPLE**   A 15-oz box of cereal costs $2.29. Find the constant of direct variation for the cost per ounce. (This is also referred to as the **unit cost.**)

**BUS**   $k = \dfrac{y}{x}$

$$\begin{array}{c}\text{Unit cost}\\\text{or}\\\text{Constant of direct variation}\end{array} = \dfrac{\text{cost}}{\text{oz}} = \dfrac{\$}{\text{oz}} \qquad \dfrac{\text{dependent variable}}{\text{independent variable}}$$

$k = \dfrac{\$2.29}{15 \text{ oz}}$      Change to a decimal equivalent.

$k = 0.1527$      Round.

**The unit cost or constant of direct variation is 0.1527 $\dfrac{\$}{\text{oz}}$.**

---

**EXAMPLE**   A 15-oz box of cereal costs $2.29 or a 36-oz box of cereal costs $4.89. Which is the better buy? Round to the nearest ten-thousandth.

**BUS**   In the preceding example we found the cost per ounce (unit cost) of the 15-oz box of cereal to be $0.1527.

Find the unit cost of the 36-oz box of cereal.

$k = \dfrac{\$4.89}{36} = \$0.1358$      Ratio of $\dfrac{\$}{\text{oz}}$

Compare the unit costs.

Unit cost of 15-oz box = $0.1527 per oz; unit cost of 36-oz box = $0.1358 per oz

**The 36-oz box of cereal costs less per ounce and is the better buy.**

---

In the preceding example, we make a judgment solely on the basis of the mathematical facts. In reality, other factors are considered. Do you have the money to buy the larger box? Will you be able to use the larger amount before it gets stale?

**TIP!**

### Which Method for Direct Variation Is Preferred?

The proportion method is the most versatile for a variety of situations. Using the constant of direct variation is useful in computer programs and electronic spreadsheets.

| | |
|---|---|
| **AUTO**     **EXAMPLE** | A truck travels 102 mi on 6 gal of gasoline. How far will it travel on 30 gal of gasoline? |
| **Known facts** | Pair 1:   102 mi   uses   6 gal   of gasoline. |
| **Unknown facts** | Pair 2:   $m$ mi   uses   30 gal   of gasoline. |
| **Estimation** | 30 gal ÷ 6 gal = 5. Then, approximately 5 times 100 miles or 500 miles can be driven on 30 gal. |

**Dimension Analysis**

**Calculations**

$$\frac{102 \text{ mi}}{m \text{ mi}} = \frac{6 \text{ gal}}{30 \text{ gal}} \qquad \frac{\text{distance}_1}{\text{distance}_2} = \frac{\text{gasoline}_1}{\text{gasoline}_2} \qquad \text{Pair 1} \atop \text{Pair 2}$$

$$\frac{102}{m} = \frac{6}{30} \qquad \frac{\text{mi}}{\text{mi}} = \frac{\text{gal}}{\text{gal}}$$

$$102(30) = 6m \qquad \text{Cross multiply. mi(gal)} = \text{gal(mi)}$$

$$\frac{3{,}060}{6} = \frac{6m}{6} \qquad \text{Divide by gal. Reduce. } \frac{\text{mi}(\cancel{\text{gal}})}{\cancel{\text{gal}}} = \frac{\cancel{\text{gal}}(\text{mi})}{\cancel{\text{gal}}}$$

$$510 = m \qquad m \text{ is expressed in miles.}$$

**Interpretation**    **The truck will travel 510 mi on 30 gal of gasoline.**

---

**TIP!**

### Analyze Dimensions

Even though we often remove the written dimensions from an equation, we should analyze the dimensions to be sure we use the correct units in the solution.

---

**EXAMPLE**    If a metal rod tapers 1 in. for every 24 in. of length, what is the amount of taper of a 30-in. piece of rod? (See Fig. 6–1.)

**INDTEC**

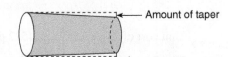
    — Amount of taper

**Figure 6–1**

| | |
|---|---|
| **Known facts** | Pair 1:   1-in. taper   for   24-in. length |
| **Unknown facts** | Pair 2:   $x$-in. taper   for   30-in length |
| **Estimation** | A 30-in. rod will taper more than 1 in. |

**Calculations**

$$\frac{1\text{-in. taper}}{x\text{-in. taper}} = \frac{24\text{-in. length}}{30\text{-in. length}} \qquad \text{Pair 1} \atop \text{Pair 2}$$

$$\frac{1}{x} = \frac{24}{30} \qquad \text{Cross multiply.}$$

$$1(30) = 24x$$

$$30 = 24x \qquad \text{Divide.}$$

$$\frac{30}{24} = \frac{24x}{24}$$

$$\frac{5}{4} = x \quad \text{or} \quad 1\frac{1}{4} = x$$

**Interpretation**     The amount of taper for a 30-in. length of rod is $1\frac{1}{4}$ in.

### 2   Solve Problems That Involve Similar Triangles.

Many applications involving proportions are based on the properties of similar triangles. Before we solve applied problems using similar triangles, let's examine the definitions of similar and congruent triangles. **Congruent triangles** have the same size and shape (Fig. 6–2). **Similar triangles** have the same shape but not the same size (Fig. 6–3).

Every triangle has six parts: three angles and three sides. Each angle or side of one similar or congruent triangle has a corresponding angle or side in the other similar or congruent triangle. The symbol for showing congruency is ≅ and is read "is congruent to." The symbol for a triangle is Δ.

**Properties of congruent triangles:**

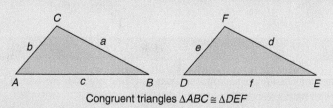

Congruent triangles $\triangle ABC \cong \triangle DEF$

**Figure 6–2**

Corresponding angles of congruent triangles are the same size; that is, they are equal in measure.

Angle $A$ = angle $D$
Angle $B$ = angle $E$
Angle $C$ = angle $F$

Corresponding sides of congruent triangles are the same size; that is, they are equal in measure.

Side $a$ = side $d$
Side $b$ = side $e$
Side $c$ = side $f$

The symbol for showing similarity is ~ and is read "is similar to."

**Properties of similar triangles:**

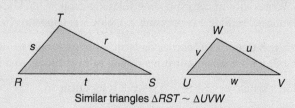

Similar triangles $\triangle RST \sim \triangle UVW$

**Figure 6–3**

Corresponding angles of similar triangles are equal in size.

Angle $R$ = angle $U$
Angle $S$ = angle $V$
Angle $T$ = angle $W$

Corresponding sides of similar triangles are directly proportional.

Side $r$ corresponds to side $u$
Side $s$ corresponds to side $v$
Side $t$ corresponds to side $w$

Examine the corresponding angles (Fig. 6–4). The symbol for angle is ∠.

∠ A corresponds to and is equal in measure to ∠ M.
∠ B corresponds to and is equal in measure to ∠ N.
∠ C corresponds to and is equal in measure to ∠ Q.

Examine the corresponding sides. Sides are named with the capital letters that identify the endpoints or with a lowercase letter that represents the angle opposite the side.

Side $BC$ corresponds to side $NQ$                $a$ corresponds to $m$
Side $AC$ corresponds to side $MQ$    or    $b$ corresponds to $n$
Side $AB$ corresponds to side $MN$                $c$ corresponds to $q$

The corresponding sides of similar triangles are proportional.

$$\frac{BC}{NQ} = \frac{AC}{MQ} = \frac{AB}{MN} \quad \text{or} \quad \frac{a}{m} = \frac{b}{n} = \frac{c}{q}$$

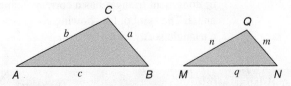

**Figure 6–4**

## To find a missing side of a similar triangle:

1. Examine the given dimensions of the corresponding sides and find one pair for which both measures are known.
2. Pair the unknown side with its corresponding known side.
3. Set up a direct proportion with the two pairs.
4. Solve the proportion.

**EXAMPLE**    Find the missing side in Fig. 6–5 if $\triangle ABC \sim \triangle DEF$.

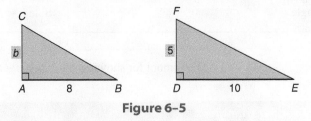

**Figure 6–5**

Write ratios of corresponding sides in a proportion. Use single lowercase letters to identify the sides. Side $a$ is opposite ∠$A$, side $b$ is opposite ∠$B$, and so on.

Pair 1: 8 from $\triangle ABC$ corresponds and is proportional to 10 from $\triangle DEF$.
Pair 2: $b$ from $\triangle ABC$ corresponds and is proportional to 5 from $\triangle DEF$.

**Estimation**    Side $b$ should be less than 5 since 8 is less than 10.

$$\frac{\text{Pair 1}\ 8}{\text{Pair 2}\ b} = \frac{10}{5}$$

Pair with both measures known.
Pair with one measure known.

$10b = 8(5)$      Cross multiply.

$10b = 40$      Divide.

$b = 4$

**Interpretation**    **Side $b$ is 4 units long.**

EXAMPLE

AG/H

A tree surgeon must know the height of a tree to determine which way to fell it so that it does not endanger lives, traffic, or property. A 6-ft pole casts a 4-ft shadow when the tree casts a 20-ft shadow (Fig. 6–6). What is the height of the tree?

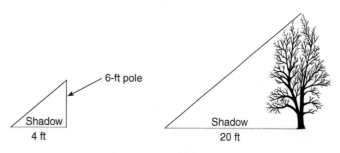

**Figure 6–6**

The triangles formed are similar. Let $x$ = the height of the tree.

Pair 1: 6-ft pole casts a 4-ft shadow.  Both measures known.

Pair 2: $x$-ft tree casts a 20-ft shadow.  One measure known.

**Estimation**  The tree is more than 20 ft tall.

Since the pole is taller than its shadow, the tree will be taller than its shadow of 20 ft.

$$\frac{6 \text{ (height of pole)}}{x \text{ (height of tree)}} = \frac{4 \text{ (shadow of pole)}}{20 \text{ (shadow of tree)}}$$  Pair 1

Pair 2

$6(20) = 4x$  Cross multiply.

$120 = 4x$  Divide.

$30 = x$

**Interpretation**  **The tree is 30 ft tall.**

---

EXAMPLE

CAD/ARC

A building lies between points $A$ and $B$, so the distance between these points cannot be measured directly by a surveyor. Find the distance using the similar triangles shown in Fig. 6–7.

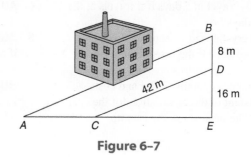

**Figure 6–7**

$\triangle ABE \sim \triangle CDE$. Note that $CD$ must be made parallel to $AB$ for the triangles to be similar. Parallel means the lines are the same distance apart from end to end.

Visualize the triangles as separate triangles (Fig. 6–8). Using the lowercase letter $e$ for the missing $AB$ measure is confusing because in $\triangle CDE$, side $CD$ can also be thought of as side $e$.

6–2 Direct Variation

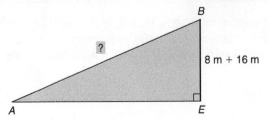

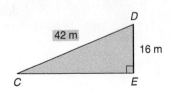

**Figure 6–8**

Pair 1: $AB$ from $\triangle ABE$ corresponds to $CD$ from $\triangle CDE$.

Pair 2: $BE$ from $\triangle ABE$ corresponds to $DE$ from $\triangle CDE$.

**Estimation**    $AB$ is more than 42 m.

$$\frac{AB}{BE} = \frac{CD}{DE}$$     Substitute values. $CD = 42$; $DE = 16$
$BE = BD + DE = 8 + 16 = 24$

$$\frac{AB}{24} = \frac{42}{16}$$     Cross multiply.

$16AB = (24)42$     $AB$ is interpreted as a single variable.

$16AB = 1{,}008$     Divide.

$AB = 63$ m

**Interpretation**    **Thus, the distance from $A$ to $B$ is 63 m.**

SECTION 6–2 SELF-STUDY EXERCISES

**1**  Solve using direct variation.

1. **BUS**  If 7 cans of dog food sell for $4.13, how much will 10 cans sell for?

2. **BUS**  If 6 cans of coffee sell for $22.24, how much will 20 cans sell for?

3. **AUTO**  A mechanic took 7 h to tune up 9 fuel-injected engines. At this rate, how many fuel-injected engines can be tuned up in 37.5 h? Round to the nearest whole number.

4. **BUS**  A costume maker took 9 h to make 4 headpieces for a Mardi Gras ball. At this rate, how many complete headpieces can be made in 35 h?

5. How far can a family travel in 5 days if it travels at the rate of 855 mi in 3 days?

6. **AUTO**  How far can a tractor-trailer rig travel in 8 days if it travels at the rate of 1,680 mi in 4 days?

7. **AG/H**  How much crystallized insecticide does 275 acres of farmland need if the insecticide treats 50 acres per 100 lb?

8. **AG/H**  How much fertilizer does 2,625 ft$^2$ of lawn need if the fertilizer treats 1,575 ft$^2$ per gallon? Express the answer to the nearest tenth of a gallon.

9. **BUS**  A can of tomatoes contains 793 g and costs $1.49. Find the constant of direct variation for the cost per gram. Round to the nearest ten-thousandth.

10. **BUS**  A can of tomatoes contains 411 g and costs $0.89. Find the constant of direct variation for the cost per gram. Round to the nearest ten-thousandth.

11. From Exercises 9 and 10 compare the unit cost of the small can of tomatoes with the unit cost of the large can to determine which size can is more economical.

12. **INDTEC** A bottling machine can fill 800 bottles in 2.5 h. Find the constant of direct variation for the number of bottles filled per hour.

13. **INDTEC** Two gears have a ratio of 8 to 2. If the larger gear has 48 teeth, how many teeth does the smaller gear have?

14. **INDTEC** Two gears have a ratio of 9 to 4. If the larger gear has 72 teeth, how many teeth does the smaller gear have?

15. **INDTR** The diameter of the larger of two pulleys connected with a belt is 36 cm. If the ratio is 9:1, what is the diameter of the smaller pulley?

16. **INDTR** The diameter of the smaller of two pulleys connected with a belt is 48 cm. If the ratio is 8:1, what is the diameter of the larger pulley?

17. **CON** The slope of a roof is the <u>ratio</u> of the rise to the run $\left(\dfrac{\text{rise}}{\text{run}}\right)$ (Fig. 6–9). A roof has a slope of $\dfrac{3}{12}$. What is the rise for this roof if it has a run of 28 ft?

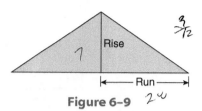

Figure 6–9

18. **CON** A roof has a slope of $\dfrac{1}{6}$. What is the rise for this roof if it has a run of 36 ft?

19. **CON** The *pitch* of a roof is the ratio of the rise of a roof to its span (Fig. 6–10). A roof has a pitch of 1:4. If the span is 40 ft, what is the rise?

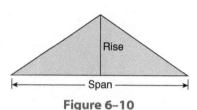

Figure 6–10

20. **CON** A roof has a pitch of 1:3. If the span is 42 ft, what is the rise?

21. **HLTH/N** A person who weighs 142 lb should be given how many milligrams of medication if the dosage is 25 mg for every 10 lb?

22. **HLTH/N** A person who weighs 185 lb should be given how many milligrams of medication if the dosage is 15 mg for every 10 lb?

23. **HLTH/N** The pediatric dosage for chlorpromazine hydrochloride is 0.25 mg/lb. What is the dosage for a child that weighs 40 lb?

24. **HLTH/N** Chlorpromazine injection strength contains 50 mg per 2 mL. A patient is prescribed 0.5 g. How many milliliters should be administered (1 g = 1,000 mg)?

25. **CAD/ARC** A blueprint has a scale of $\dfrac{1}{2}$ in. = 1 ft. On a blueprint a wall is drawn $7\dfrac{1}{2}$ in. long. What is the actual measure of the wall?

26. **CAD/ARC** A blueprint has a scale of $\dfrac{3}{4}$ in. = 1 ft. On a blueprint a wall is drawn $8\dfrac{3}{4}$ in. long. What is the actual measure of the wall?

27. **CAD/ARC** A building that is 120 ft long would be shown as what length on a blueprint if the scale is $\dfrac{1}{4}$ in. = 1 ft?

28. **CAD/ARC** A building that is 320 ft long would be shown as what length on a blueprint if the scale is $\dfrac{1}{8}$ in. = 1 ft?

29. **ELEC** The length of a wire is proportional to its resistance. The resistance of 200 ft of a certain wire is 0.0062 Ω. What is the resistance of 750 ft of the same wire?

30. **ELEC** The length of a wire is proportional to its resistance. The resistance of 100 ft of a certain wire is 0.048 Ω. What is the resistance of 1,200 ft of the same wire?

**31. AVIA** Use the map in Fig. 6–11 to find the air distance to the nearest mile from Upper Sandusky to Lima.

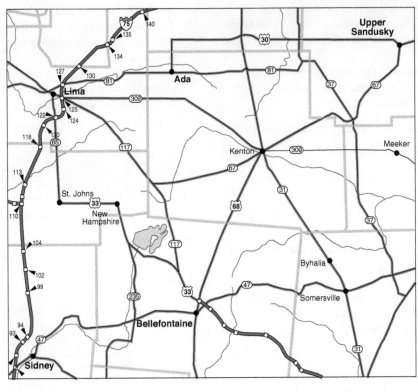

$\frac{13}{16}$ inch = 10 miles

**Figure 6–11**

**32. AVIA** Use the map in Figure 6–11 to find the air distance to the nearest mile from Lima to Bellefontaine.

**33. AVIA** Use the map in Figure 6–11 to find the air distance to the nearest mile from Sydney to Bellefontaine.

**34. INDTR** A machinist is to make the metal plate represented in Figure 6–12. What is the overall length of the plate?

**35. INDTR** What is the overall width of the actual plate represented in Figure 6–12?

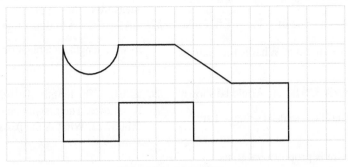

Side of 1 square = 2.5 in.

**Figure 6–12**

**36. INDTR** What is the diameter of the semicircle in Figure 6–12? The diameter is the distance across the center of a circle.

Use square-ruled paper to make a line drawing of objects described in Exercises 37–41.

**37.** A square 10 cm on a side, using a scale of 1 square = 1 cm

**38.** A rectangle 10 ft by 7.5 ft, using a scale of 1 square = 2.5 ft

**39.** A rectangle 15 m by 12 m, using a scale of 1 square = 3 m

**40.** A circle with a 24-in. diameter, using a scale of 1 square = 4 in.

**41.** a rectangle 4.5 cm by 3.5 cm, using a scale of 1 square = 0.5 cm

**43.** **HLTH/N** A patient is prescribed 50 mg of Librium® IM. The medication is available as 0.4 mg per 2 mL. How many milliliters should be injected?

**45.** **AG/H** Instructions for mixing a chemical pesticide state that 6 oz of chemical should be mixed with 64 oz of water. How many ounces of chemical should be used with 160 oz of water?

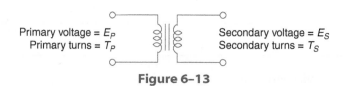

Primary voltage = $E_P$
Primary turns = $T_P$

Secondary voltage = $E_S$
Secondary turns = $T_S$

**Figure 6–13**

**42.** **HLTH/N** A doctor ordered streptomycin 250 mg IM for a patient. The dosage available for use contains 1 g per 4 mL. How many milliliters should be injected (1 g = 1,000 mg)?

**44.** **HLTH/N** A patient is prescribed an injection of 0.2 mg of Atropine® IM. The drug is available as 0.5 mg per milliliter. How many milliliters should be injected?

**46.** **ELEC** An electrical transformer shown in Fig. 6–13 is constructed by winding a number of turns of wire into a primary. Another set of turns is constructed by winding to form one or more secondaries. The ratio of the primary voltage ($E_p$) to the secondary voltage ($E_s$) is the same as the ratio of the number of primary turns ($T_p$) to the number of secondary turns ($E_s$).

$$\frac{E_p}{E_s} = \frac{T_p}{T_s}$$

Find the secondary voltage if the transformer has a 360-turn primary and a 30-turn secondary and 150 V are applied.

---

**2** Write the corresponding parts not given for the congruent triangles in Exercises 47–49 (Figure 6–14 to 6–16).

**47.** $AB = FE$
$BC = DE$
$AC = DF$

**48.** $\angle R$ and $\angle U = 90°$
$QR = TU$
$RP = US$

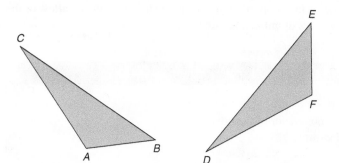

**Figure 6–14**

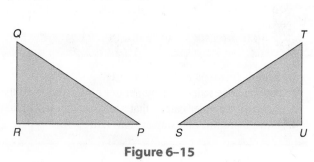

**Figure 6–15**

**49.** $JK = MN$
$\angle J = \angle M$
$\angle K = \angle N$

**50.** $\triangle ABC \sim \triangle EDF$. $\angle A = \angle E$, $\angle C = \angle F$. Find $a$ and $d$. Use Figure 6–17.

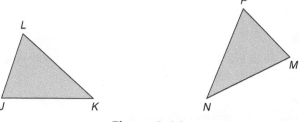

**Figure 6–16**

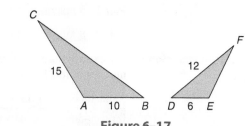

**Figure 6–17**

**51.** $\triangle ABC \sim \triangle DEC$. $\angle 1$ and $\angle 2$ have the same measure. Find $DC$ and $DE$. (*Hint:* Let $DC = x$ and $AC = x + 3$. Use Fig. 6–18.)

**52.** **AG/H**   Find the height of a tree that casts a 30-ft shadow when a 6-ft 6-in. pole casts a shadow of 3 ft.

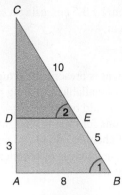

**Figure 6–18**

---

# 6–3 | *Inverse Variation*

*Learning Outcome*   **1**   Solve problems with inverse variation.

### **1**   Solve Problems with Inverse Variation.

An **inverse variation** is one in which the quantities being compared are inversely related. That is, as one quantity increases, the other decreases, or as one quantity decreases, the other increases.

For example, as we *increase* pressure on a gas, the gas compresses and so *decreases* in volume. Or, as we *decrease* pressure on the gas, it *increases* in volume as it expands. This relationship is the opposite, or inverse, of direct variation.

If 3 workers frame a house in 2 weeks and the contractor *increases* the number of workers to 6, the framing time *decreases* to 1 week, assuming the workers work at the same rate. In other words, the framing time is *inversely proportional* to the number of workers on the job.

Unlike the directly related ratios in a proportion, inversely related ratios do not allow us the flexibility we had in setting up ratios of unlike measures.

## To set up an inverse proportion:

1. Establish two pairs of related data.
2. Arrange one pair as the numerator of one ratio and the denominator of the other ratio.
3. Arrange the other pair so that each ratio contains like measures.
4. Form a proportion using the two ratios.

---

**EXAMPLE**   If 5 machines take 12 days to complete a job, how long will it take for 8 machines to do the job?

**INDTEC**   As the number of machines *increases,* the amount of time required to do the job *decreases.* Thus, the quantities are *inversely proportional.*

Pair 1: 5 machines finish in 12 days .

Pair 2: 8 machines finish in $x$ days .

**Estimation**   We expect more machines to do the job in less than 12 days. Also, we did not double the number of machines, so it will take more than half of the time (6 days).

$$\frac{5 \text{ machines}}{8 \text{ machines}} = \frac{x \text{ days for 8 machines}}{12 \text{ days for 5 machines}}$$

Each ratio uses like measures. The pairs are arranged inversely.

$$\frac{5}{8} = \frac{x}{12}$$

**Dimension Analysis**

$$\frac{\text{machines}}{\text{machines}} = \frac{\text{days}}{\text{days}}$$

$$5(12) = 8x$$

Cross multiply. machines(days) = machines(days)

$$60 = 8x$$

Divide by machines.

$$\frac{60}{8} = x$$

$$\frac{\cancel{\text{machines}} \ (\text{days})}{\cancel{\text{machines}}} = \text{days}$$

$$x = \frac{15}{2} \quad \text{or} \quad 7\frac{1}{2} \text{ days}$$

**Interpretation**   It will take $7\frac{1}{2}$ days for 8 machines to do the job.

**TIP!**

### Why Is Estimation So Important?

Suppose we arrange the data so that each pair forms a ratio of unlike measures and then we invert one of the ratios. Look at the value to see if the answer conforms to what we expect the answer to be.

Pair 1        Reciprocal of Pair 2

$$\frac{5 \text{ machines}}{12 \text{ days}} = \frac{x \text{ days}}{8 \text{ machines}}$$

$$12x = 40$$

$$x = 3.33 \text{ days}$$

How does this compare to our estimate? We expected our answer to be less than 12 days but more than 6 days.

$$\frac{\text{machines}}{\text{days}} = \frac{\text{days}}{\text{machines}}$$

Analyzing the dimensions we produce an incorrect statement.

$$\text{machines}(\text{machines}) = \text{days}(\text{days})$$

These measures are not equal. Estimation can identify proportions that are set up incorrectly!

Inverse variation has a conversion factor similar to the constant of direct variation. The inverse variation formula is $y = \frac{k}{x}$, where $x$ is the independent variable, $y$ is the dependent variable, and $k$ is the **constant of inverse variation.** This formula can also be written as $k = xy$.

### To find the constant of inverse variation:

1. Identify a pair of related data in which both values are known.
2. Write the related pair as a product. $k = xy$.
3. Multiply to find the constant of inverse variation.

The constant of inverse variation and the known data pair of 5 machines and 12 days can be used to solve the preceding example.

$k = 12(5)$      Using $k = xy$, find $k$ for $y = 5$ and $x = 12$.

$k = 60$      Constant of inverse variation.

$y = \dfrac{k}{x}$      Substitute $k = 60$ and $y = 8$.

$8 = \dfrac{60}{x}$      Solve for $x$.

$8x = 60$      Divide.

$x = \dfrac{60}{8}$

$x = 7\dfrac{1}{2}$      Days for 8 machines to complete the job.

The speed and size of gears and pulleys involve *inverse* relationships. Suppose a large gear and a small gear are in mesh, or a large pulley is connected by a belt to a smaller pulley. The larger gear or pulley has the *slower* speed, and the smaller gear or pulley has the *faster* speed.

**EXAMPLE**

**INDTEC**

A 10-in.-diameter gear is in mesh with a 5-in.-diameter gear (Figure 6–19). If the larger gear has a speed of 25 rpm, at how many rpm does the smaller gear turn?

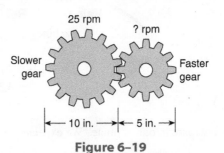

**Figure 6–19**

Because gears in mesh are *inversely* related, we set up an inverse proportion. Each ratio uses like measures and the ratios are in inverse order.

Pair 1: 25 rpm of larger gear for 10-in. size of larger gear.

Pair 2: x rpm of smaller gear for 5-in. size of smaller gear.

**Estimation**

We expect the speed of the smaller gear to be faster than the 25-rpm speed of the larger gear.

$$\frac{25 \text{ rpm of larger gear}}{x \text{ rpm of smaller gear}} = \frac{5 \text{ in. of smaller gear}}{10 \text{ in. of larger gear}} \quad \frac{\text{Pair 2}}{\text{Pair 1}}$$    Set up ratios in inverse order.

$$\frac{25}{x} = \frac{5}{10}$$    Cross multiply.

$$25(10) = 5x$$

$$250 = 5x$$    Divide.

$$\frac{250}{5} = x$$

$$50 = x$$

**Interpretation**    **Thus, the smaller gear turns at the faster speed of 50 rpm.**

### How to Distinguish Between Direct and Inverse Variation

We can distinguish between direct and inverse variation by anticipating cause-and-effect situations.

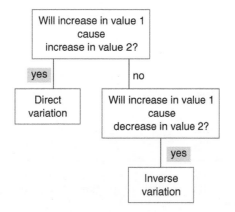

Similarly,

$$\text{decrease causes decrease} = \text{direct variation}$$
$$\text{decrease causes increase} = \text{inverse variation}$$

$X =$ INDEPENDENT VARIABLE

$y =$ DEPENDENT

$K =$ CONSTANT OF INVERSE VARIATION

$y = \dfrac{K}{X}$

### SECTION 6–3 SELF-STUDY EXERCISES

**1** Solve using inverse variation.

**1. INDTR** A 12-in. pulley turns at 30 rpm. Find the constant of inverse variation.

**2. INDTR** A 16-in. gear makes 60 rpm. Find the constant of inverse variation.

**3. INDTR** The fan pulley and alternator pulley are connected by a fan belt on an automobile engine. The fan pulley is 225 cm in diameter and the alternator pulley is 125 cm in diameter. If the fan pulley turns at 500 rpm, how many revolutions per minute does the faster alternator pulley turn?

**4. INDTR** The volume of a certain gas is inversely proportional to the pressure on it. If the gas has a volume of 160 in$^3$ under a pressure of 20 lb per square inch (psi), what is the volume if the pressure is decreased to 16 psi?

**5. INDTR** A pulley that measures 15 in. across (diameter) turns at 1,600 rpm and drives a larger pulley at the rate of 1,200 rpm. What is the diameter of the larger pulley in this inverse relationship?

**6. CON** Six painters can trim the exterior of all the new brick homes in a subdivision in 9 weeks. The contractor wants to have the homes ready in just 3 weeks and so needs more painters. How many painters are needed for the job? Assume that the painters all work at the same rate.

**7. AG/H** Two groundskeepers take 25 h to prepare a golf course for a tournament. How long would it take five groundskeepers to prepare the golf course?

**8. INDTEC** A gear measures 5 in. across. It turns another gear 2.5 in. across. If the larger gear has a speed of 25 rpm, what is the rpm of the smaller gear?

**9. INDTR** A small pulley 3 in. in diameter turns 250 rpm and drives a larger pulley at 150 rpm. What is the diameter (distance across) of the larger pulley?

**10. CON** Two painters working at the same speed can paint 800 ft$^2$ of wall space in 6 h. If a third painter paints at the same speed, how long will it take the three of them to paint the same wall space?

**11. INDTEC** Three machines complete a printing project in 5 h. How many machines are needed to finish the same project in 3 h?

**12. INDTEC** A gear measures 6 in. across. It is in mesh with another gear with a diameter of 3 in. If the larger gear has a speed of 60 rpm, what is the rpm of the smaller gear?

**13.** **HLTH/N** Nurse Lee prepares dosages for her patients in 30 min. If she gets help from assistants, who also work at her rate, and together they complete the preparation in 6 min, how many helpers did she get?

**14.** **INDTR** A 4.5-in. pulley turning at 1,000 rpm is belted to a larger pulley turning at 500 rpm. What is the size of the larger pulley?

Complete the table for two gears in mesh.

| | LARGE GEAR | | SMALL GEAR | |
|---|---|---|---|---|
| | DIAMETER | RPM | DIAMETER | RPM |
| **15.** | 30 in. | 120 | 15 in. | |
| **16.** | 24 in. | | 18 in. | 300 |
| **17.** | | 70 | 14 cm | 3,500 |
| **18.** | 20 cm | 100 | | 500 |
| **19.** | | 180 | 12 cm | 540 |
| **20.** | 32 in. | 225 | 12 in. | |

The number of teeth of two gears in mesh is inversely proportional to the number of revolutions per minute each turns. Complete the table for two gears in mesh.

| | LARGE GEAR | | SMALL GEAR | |
|---|---|---|---|---|
| | NUMBER OF TEETH | RPM | NUMBER OF TEETH | RPM |
| **21.** | 60 | 300 | 45 | |
| **22.** | 250 | 28 | 50 | |
| **23.** | | 1.7 | 60 | 8.5 |
| **24.** | 300 | 70 | | 600 |
| **25.** | 448 | | 56 | 300 |
| **26.** | 84 | 24 | | 96 |

**27.** **INDTEC** A gear with 40 teeth turns 30 rpm on another gear with 120 teeth. At how many rpm does the second gear turn?

**28.** **INDTEC** A gear turning at 19 rpm with 150 teeth turns a smaller gear with 75 teeth. At how many rpm does the smaller gear turn?

**29.** **INDTEC** A large gear turns at 40 rpm in mesh with a gear that has 60 teeth and turns at 120 rpm. How many teeth does the larger gear have?

**30.** **INDTEC** A small gear turns at 150 rpm in mesh with a large gear that has 180 teeth and turns at 90 rpm. How many teeth has the small gear?

---

## CHAPTER REVIEW OF KEY CONCEPTS

### Section 6-1

**1** Solve equations that are proportions (pp. 295–296).

Property of proportions: If $\dfrac{a}{b} = \dfrac{c}{d}$, then $ad = bc$ ($b, d \neq 0$).

To solve: **1.** Find the cross products. **2.** Divide both sides by the coefficient.

$$\frac{4}{2} = \frac{x}{6}$$

$(2)(x) = (4)(6)$  Cross products.

$2x = 24$  Divide by coefficient.

$$\frac{2x}{2} = \frac{24}{2}$$

$x = 12$

## Section 6-2

**1** Solve problems with direct variation (pp. 297–301).

Verify that the problem involves direct proportion: As one amount increases (decreases), the other amount increases (decreases).

Set up a direct proportion: **1.** Establish two pairs of related data. **2.** Write one pair of data in the numerators of two ratios. **3.** Write the other pair of data in the denominators of the two ratios. **4.** Form a proportion using the two ratios.

---

Kinta makes $72 for 8 h of work in a hospital business office. How many hours must he work to make $150?

Proportion is direct: As hours increase, pay increases.

$$\frac{\$72}{\$150} = \frac{8\text{ h}}{x\text{ h}} \qquad \text{Pair 1. Estimation: } h > 8\text{ h.}$$
$$\phantom{\frac{\$72}{\$150} = \frac{8\text{ h}}{x\text{ h}}} \qquad \text{Pair 2.}$$

$$\frac{72}{150} = \frac{8}{x} \qquad \text{Cross multiply.}$$

$$(72)(x) = (8)(150)$$

$$72x = 1{,}200 \qquad \text{Divide.}$$

$$\frac{72x}{72} = \frac{1{,}200}{72}$$

$$x = 16\frac{2}{3}\text{ h}$$

---

## Section 6-3

**1** Solve problems that involve similar triangles (pp. 301–304).

Each angle of one triangle is equal to its corresponding angle in the other similar triangle. Each side of one triangle is proportional to its corresponding side in the other triangle.

---

Find the height of a building that makes a shadow of 25 m when a meter stick makes a shadow of 1.5 m.

Direct proportion: As shadow increases, height increases.

$$\frac{x\text{-m building}}{1\text{-m stick}} = \frac{25\text{-m building shadow}}{1.5\text{-m stick shadow}} \qquad \begin{array}{l}\text{Pair 1} \quad \text{Estimation: meter stick}\\ \text{Pair 2} \quad < \text{shadow, so building} < 25\text{ m.}\end{array}$$

$$\frac{x}{1} = \frac{25}{1.5} \qquad \text{Cross multiply.}$$

$$1.5x = 25 \qquad \text{Divide.}$$

$$x = \frac{25}{1.5}$$

$$x = 16.67\text{ m}$$

---

**2** Solve problems with inverse variation (pp. 308–311).

Verify that the problem involves inverse proportion: As one amount increases (decreases), the other amount decreases (increases).

Set up an inverse proportion: **1.** Establish two pairs of related data. **2.** Arrange one pair as the numerator of one ratio and the denominator of the other. **3.** Arrange the other pair so that each ratio contains like measures. **4.** Form a proportion using the two ratios.

---

Jane can paint a room in 4 h. If she has two helpers who also can paint a room in 4 h, how long will it take all three to paint the same room?

Proportion is inverse: As number of painters increases, time decreases.

$$\frac{1\text{ painter}}{3\text{ painters}} = \frac{x\text{ h}}{4\text{ h}} \qquad \begin{array}{ll}\text{Pair 1: 1 painter, 4 h.} & \text{Estimation: } h < 4\text{ h.}\\ \text{Pair 2: 3 painters, } x\text{ h.}\end{array}$$

$$\frac{1}{3} = \frac{x}{4}$$

$$(3)(x) = (1)(4)$$

---

$$3x = 4$$

$$\frac{3x}{3} = \frac{4}{3}$$

$x = 1\frac{1}{3}$ h for 3 painters to paint the room.

## CHAPTER REVIEW EXERCISES

### Section 6–1

Solve the proportions.

**1.** $\dfrac{7}{x} = 6$

**2.** $\dfrac{2}{3} = \dfrac{x + 3}{x - 7}$

**3.** $\dfrac{4x + 3}{15} = \dfrac{1}{3}$

**4.** $\dfrac{3x - 2}{3} = \dfrac{2x + 1}{4}$

**5.** $\dfrac{5}{4x - 3} = \dfrac{3}{8}$

**6.** $\dfrac{8}{3x - 2} = \dfrac{2}{3}$

**7.** $\dfrac{4x}{7} = \dfrac{2x + 3}{3}$

**8.** $\dfrac{2x}{3x - 2} = \dfrac{5}{8}$

**9.** $\dfrac{5x}{3} = \dfrac{2x + 1}{4}$

**10.** $\dfrac{5}{9} = \dfrac{x}{2x - 1}$

**11.** $\dfrac{7}{x} = \dfrac{5}{4x + 3}$

**12.** $\dfrac{3}{5} = \dfrac{2x - 3}{7x + 4}$

### Section 6–2

Solve the problems using proportions.

**13.** There are 25 women in a class of 35 students. If this is typical of all classes in the college, how many women are enrolled in the college if it has 6,300 students altogether?

**14.** **HOSP**  In preparing a banquet for 30 people, Cedric Henderson uses 9 lb of potatoes. How many pounds of potatoes will be needed for a banquet for 175 people?

**15.** **CAD/ARC**  A CAD program scales a blueprint so that $\frac{5}{8}$ in. = 2 ft. What is the actual measure of a wall that is shown as $1\frac{5}{16}$ in. on the blueprint?

**16.** **AG/H**  A 6-ft landscape engineer casts a 5-ft shadow on the ground. How tall is a nearby tree that casts a 30-ft shadow?

**17.** **HLTH/N**  On recent trips to rural health centers, a portable mammography unit used 81.2 gal of unleaded gasoline. If the travel involved a total of 845 mi, how many gallons of gasoline would be used for travel of 1,350 mi? Round to tenths.

**18.** **BUS**  A certain fabric sells at a rate of 3 yd for $7.00. How many yards can Clemetee Whaley buy for $35.00?

**19.** **CON**  A contractor estimates that a painter can paint 300 ft$^2$ of wall space in 3.5 h. How many hours should the contractor estimate for the painter to paint 425 ft$^2$? Express your answer to the nearest tenth.

**20.** **CAD/ARC**  An architect's drawing is scaled at $\frac{3}{4}$ in. = 6 ft. What is the actual height of a door that measures $\frac{7}{8}$ in. on the drawing?

**21.** **ELEC**  A wire 825 ft long has a resistance of 1.983 Ω. How long is a wire of the same diameter if the resistance is 3.247 Ω? Round to the nearest whole foot.

**22.** **BUS**  A coffee company mixes 1.6 lb of chicory with every 3.5 lb of coffee. At this ratio, how many pounds of chicory are needed to mix with 2,500 lb of coffee? Round to the nearest whole number.

**23.** Write proportions for the similar triangles in Fig. 6–20.

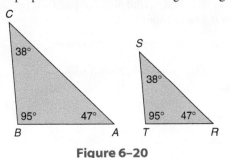

**Figure 6–20**

**24.** $\triangle LMN \sim \triangle XYZ$, $\angle L = \angle X$, $\angle M = \angle Y$. Find $m$ and $x$ (see Fig. 6–21).

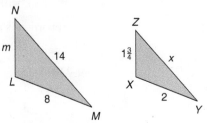

**Figure 6–21**

Chapter 6 / Ratio and Proportion

**25.** Find *AB* if $\triangle DEC \sim \triangle AEB$ (see Fig. 6–22).

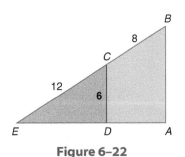

**Figure 6–22**

## Section 6–3

Solve the problems using proportions.

**26. AG/H** It takes five people 7 days to clear an acre of land of debris left by a tornado; inversely, more people can do the job in less time. How long would it take seven people all working at the same rate?

**27. INDTEC** If 15 machines can complete a job in 6 weeks, how many machines are needed to complete the job in 4 weeks?

**28. INDTR** A 10-in. pulley makes 900 revolutions every minute. It drives a larger pulley at 500 rpm. What is the diameter of the larger pulley in this inverse relationship?

**29.** A car with a speed control device travels 100 mi at 50 mi/h. The trip takes 2 h. If the car traveled at 40 mi/h, how much time would the driver need to reach the same destination?

**30. INDTR** A pulley whose diameter is 3.5 in. is belted to a pulley whose diameter is 8.5 in. In this inverse relationship, if the smaller, faster pulley turns at the rate of 1,200 rpm, what is the rpm of the slower pulley? Round to the nearest whole number.

**31. INDTR** A gear with a diameter of 45 cm is in mesh with a gear that has a diameter of 30 cm. If the larger gear turns at 1,000 rpm, how many revolutions per minute does the smaller gear make?

**32. INDTR** A small pulley with a 6-in. diameter turns 350 rpm and drives a larger pulley at 150 rpm. What is the diameter (distance across) of the larger pulley?

**33. INDTEC** Three workers take 5 days to assemble a shipment of microwave ovens; inversely, more workers can do the job in less time. How long will it take five workers to do the same job?

**34. INDTEC** Waylon can install a hard drive in 10 PCs in 6 h. He gets help from assistants who work at his rate and together they complete the installations in 2 h. How many helpers did he get?

**35. INDTR** A gear is 4 in. across. It turns a smaller gear 2 in. across. If the larger gear has a speed of 30 rpm, what is the rpm of the smaller gear?

## TEAM PROBLEM-SOLVING EXERCISES

**1.** An effective measure of your understanding of a concept is your ability to apply the concept to real-world situations.
   **(a)** Develop and solve a word problem that can be solved using a direct proportion. Include in your storyline a lawn mower, tanks of gasoline, and acres to be mowed.
   **(b)** Develop and solve a word problem that can be solved using a direct proportion and similar triangles.

**2.** Inverse proportions can be used to solve certain types of real-world applications.
   **(a)** Develop and solve a word problem that can be solved using an inverse proportion. Include in your storyline a belt, pulleys, rpm's and diameters of pulleys.
   **(b)** Develop and solve a word problem that can be solved using an inverse proportion.

## PRACTICE TEST

Solve the proportions.

**1.** $\dfrac{x}{12} = \dfrac{5}{8}$

**2.** $\dfrac{x}{6} = \dfrac{12}{8}$

**3.** $\dfrac{640}{24} = \dfrac{x}{360}$

**4.** $\dfrac{45 \text{ cm}}{18 \text{ cm}} = \dfrac{9 \text{ cm}}{x}$

**5.** $\dfrac{2\frac{1}{2}}{x} = \dfrac{1\frac{1}{4}}{3\frac{1}{5}}$

**6.** $\dfrac{3.9}{5.4} = \dfrac{x}{8.1}$

**7.** A large gear with 300 teeth turns at 40 rpm. Find the rpm of a small gear that has 60 teeth.

Solve the problems involving fractions, decimal numbers, and proportions.

**8.** A pipe fills a 120-gal tank in 4 h. How long does it take for the pipe to fill a 420-gal tank??

**9.** A 9-in. gear is in mesh with a 4-in. gear. If the larger gear makes 75 rpm, how many revolutions per minute does the smaller gear make in this inverse relationship?

**10.** One employee can wallpaper a room in 12 h. Four employees can wallpaper the same room in how many hours?

**11.** A force of 32.75 lb exerts pressure on a surface of 24.65 in$^2$ in a hydraulic system? A force of 117.9 lb exerts the same pressure on what area?

**12.** A $34,475 loan for the purchase of an electronically controlled assembly machine in a factory cost the management $2,758 in simple interest. How much would a machine cost (if the interest was $1,879)? Round to the nearest dollar.

**13.** If a compact car used 62.5 L of unleaded gasoline to travel 400 mi, how many liters of gasoline would the driver use to travel 350 mi? Round to tenths.

**14.** If three workers take 8 days to complete a job, how many workers would be needed to finish the same job in only 6 days if each worked at the same rate? (More workers take fewer days.)

**15.** If voltage is 40 V and amperage is 3.5 A, find the equivalent amperage for a voltage of 100V. Express the answer as a decimal rounded to tenths.

**16.** The ratio of men to women in technical and trade occupations is estimated to be 3 to 1, that is, $\frac{3}{1}$. If 56,250 men are employed in such occupations in a certain city, how many employees are women?

**17.** If an ice maker produces 75 lb of ice in $3\frac{1}{2}$ h, how many pounds of ice would it produce in 5 h? Round to the nearest whole number.

**18.** Find *HI* if $\triangle ABC \sim \triangle GHI$ (Fig. 6–23).

**19.** Find *DB* if $\triangle ABE \sim \triangle CDE$, $CD = 9$, $AB = 12$, and $DE = 15$ (Fig. 6–24).

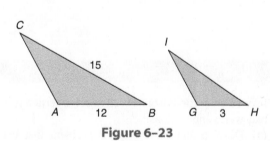

Figure 6–23

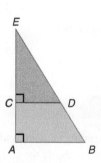

Figure 6–24

**20.** A model of a triangular part is shown in Fig. 6–25. Find side *x* if the part is to be similar to the model.

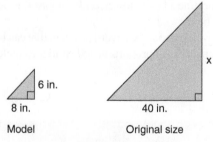

Model          Original size

Figure 6–25

In 1990 Congress passed the Nutrition Labeling and Education Act. This act requires that nutritional labels following the new standards appear on all processed food products. Nutrition information for fresh produce, meat, poultry, and fish must appear in the areas of the store where they are displayed.

To interpret the information given on the label it is important to determine the *base* for the percents given on the label. Regulations also require that *standard daily nutritional value* be used as the basis for all labels. The column heading "% Daily Value" has an asterisk that guides us to a footnote on the label that reads, "Percent Daily Values are based on a 2,000-calorie diet. Your daily values may be higher or lower depending on your calorie needs." The label also includes the recommended daily intakes that are used as a basis for a 2,000- and 2,500-cal diet.

**Nutrition Facts**
Serving Size 1/2 cup (114g)
Servings per Container 4

Amount Per Serving
Calories 260   Calories from Fat 120

| | % Daily Value* |
|---|---|
| Total Fat 13g | 20% |
| Saturated Fat 5g | 25% |
| Cholesterol 30mg | 10% |
| Sodium 660mg | 28% |
| Total Carbohydrate 31g | 11% |
| Dietary Fiber 0g | 0% |
| Sugar 5g | |
| Protein 5g | |

Vitamin A 4%    Vitamin C 2%
Calcium 15%    Iron 4%

*Percent Daily Values are based on a 2,000 calorie diet.
Your daily values may be higher or lower depending on your calorie needs:

| | Calories: | 2,000 | 2,500 |
|---|---|---|---|
| Total Fat | Less than | 65 g | 80 g |
| Sat Fat | Less than | 20 g | 25 g |
| Cholesterol | Less than | 300 mg | 300 mg |
| Sodium | Less than | 2,400 mg | 2,400 mg |
| Total Carbohydrate | | 300 g | 375 g |
| Dietary Fiber | | 25 g | 30 g |

Are the recommended daily intakes proportional to the daily calories?
First, examine the recommended daily intake values for total fat. If the values are proportional,

$$\frac{65 \text{ g}}{80 \text{ g}} = \frac{2,000 \text{ calories}}{2,500 \text{ calories}}$$

Before we continue we need to realize that the recommended values have been rounded, so if cross products are *reasonably* close, we will say the relationships are proportional. To verify that the total fat recommended is or is not proportional to the number of calories per day, check the cross products.

$$65 (2,500) = 2,000 (80)$$
$$162,500 = 160,000$$

If we round the values to the nearest ten thousand, we can say that the recommended daily intake of total fat is proportional.
If total fat is proportional to calories, then we can determine the recommended total fat intake for diets based on any number of calories.

**EXAMPLE**  Find the recommended intake value of total fat for a 1,600-cal diet. Round the value to the nearest multiple of 5.

$$\frac{x}{65 \text{ g}} = \frac{1,600}{2,000}$$  Find cross products.

$$2,000x = 1,600 \,(65)$$  Multiply 1,600 (65).

$$2,000x = 104,000$$

$$x = \frac{104,000}{2,000}$$  Divide by 2,000.

$$x = 52 \text{ g}$$

**Rounded to the nearest 10 g, the recommended total fat for a 1,600-cal diet is 50 g.**

## Exercises

Use the nutrition label on p. 317 for Exercises 1-5.

1. Determine if the following recommended daily intake values are proportional (based on rounded values).

   Saturated fat _____     Cholesterol _____
   Sodium _____     Total carbohydrate _____
   Dietary fiber _____

2. Find the recommended daily intake values for a 1,600-cal diet for the other recommended values that are proportional.

3. Find the recommended daily intake values for a 1,200-cal diet for the recommended values that are proportional.

4. Find the recommended daily intake values for a 1,000-cal diet for the recommended values that are proportional. Round to a whole number.

5. Calculate the percent daily value for each category of one serving of the food item for the given nutrition label based on a 2,500-cal diet. Round to a whole number.

## Answers

1. Saturated fat—proportional
   Cholesterol—not proportional
   Sodium—not proportional
   Total carbohydrate—proportional
   Dietary fiber—proportional
2. Saturated fat—16 g
   Total carbohydrate—240 g
   Dietary fiber—20 g
3. Total fat—39 g
   Saturated fat—12 g
   Total carbohydrate—180 g
   Dietary fiber—15 g
4. Total fat—33 g
   Saturated fat—10 g
   Total carbohydrate—150 g
   Dietary fiber—13 g
5. Total fat—16%
   Saturated fat—20%
   Cholesterol—10%
   Sodium—28%
   Total carbohydrate—8%
   Dietary fiber—0%

# 7

# Graphing Linear Equations

## Focus on Careers

Have you always wanted to be a fire-fighter? Every year fires cause thousands of deaths and injuries and destroy billions of dollars worth of property. Firefighters are frequently the first emergency personnel to arrive at the scene of a fire or other emergency. They are required to put out fires, treat injuries, and perform many other important duties.

Firefighting requires a high level of organization and teamwork, and firefighters are required to be able to follow directions at the scene of a fire. They may be asked to connect hose lines to hydrants, position ladders, control water hoses that deliver water to the fire, rescue victims, and provide emergency medical assistance.

About 90% of firefighters are employed by municipal (city) or county fire departments. Competition for these jobs is very high, and applicants usually are required to pass written, physical, and medical examinations.

Firefighting employees may also become involved in fire prevention, public fire education and safety, emergency medical services, and public relations. A high degree of skills and

*(continued)*

abilities is required of firefighters. Some firefighters attend the U.S. National Fire Academy and many colleges and universities offer 2-year and 4-year degrees in fire science.

Employment of firefighters is expected to grow about as fast as average through 2012, but competition for the available jobs is expected to be very high. The median hourly wage of firefighters was $17.42 in 2002. Local governments paid the highest median hourly earnings at $17.92, and state governments paid $13.58 per hour.

Source: *Occupational Outlook Handbook*, 2004–2005 Edition, U.S. Department of Labor, Bureau of Labor Statistics.

## 7–1 | *Graphical Representation of Linear Equations*

*Learning Outcomes*

**1** Locate points on a rectangular coordinate system.

**2** Graph linear equations using a table of solutions.

Linear equations in one variable have at most one solution. The solution can be represented graphically as a point on a number line. A linear equation in two variables has an unlimited number of solutions. The solutions of a linear equation in two variables can be represented graphically as a line on a rectangular coordinate system.

**1** **Locate Points on a Rectangular Coordinate System.**

The number line shows a value pictorially. This kind of visual representation is called a **one-dimensional graph**. The one dimension represents the distance and direction that a value is from zero.

The **rectangular coordinate system** gives a graphical representation of two-dimensional values. In the rectangular coordinate system, two number lines are positioned to form a right angle or square corner (see Figure 7–1). The number line that runs from left to right is the **horizontal axis** and is represented by the letter *x*. The number line that runs from top to bottom is the **vertical axis** and is represented by the letter *y*.

Zero on both number lines is located at the point where the two number lines cross. This point is called the **origin**. Other points are located by two dimensions, the horizontal distance and the vertical distance from the origin.

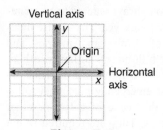

Vertical axis

Origin

Horizontal axis

**Figure 7–1**

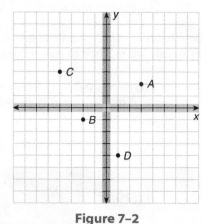

**Figure 7–2**

**EXAMPLE**  Describe the location of the points in Figure 7–2 by giving the amount of horizontal and vertical movement from the origin.

**Point A:** horizontal movement, +3 ; vertical movement, +2
**Point B:** horizontal movement, −2 ; vertical movement, −1
**Point C:** horizontal movement, −4 ; vertical movement, +3
**Point D:** horizontal movement, +1 ; vertical movement, −4

The location of points on a rectangular coordinate system can be written in an abbreviated, symbolic form. **Point notation** uses two signed numbers to show the horizontal and vertical movement from the origin to a given point. Horizontal movement is represented by the **x-coordinate** and is written first. Vertical movement is represented by the **y-coordinate** and is written second. These signed numbers are separated with a comma and enclosed in parentheses.

Symbolically, we represent the location of a point as an ordered pair of numbers

$$(x, y)$$

where $x$ is the horizontal movement from the origin, indicated by the $x$-coordinate, and $y$ is the vertical movement from the $x$-axis, indicated by the $y$-coordinate.

**EXAMPLE**    Write the points in the preceding example (Figure 7–2) using point notation.

Point $A$ = (horizontal movement of $+3$, vertical movement of $+2$) or $(\mathbf{+3, +2})$
Point $B$ = (horizontal movement of $-2$, vertical movement of $-1$) or $(\mathbf{-2, -1})$
Point $C$ = (horizontal movement of $-4$, vertical movement of $+3$) or $(\mathbf{-4, +3})$
Point $D$ = (horizontal movement of $+1$, vertical movement of $-4$) or $(\mathbf{+1, -4})$

To **plot** a point means to show its location on the rectangular coordinate system.

## To plot a point on the rectangular coordinate system:

1. Start at the origin.
2. Count to the left or right the number of units of the first signed number in the ordered pair.
3. From the ending point of Step 2, count up or down the number of units from the second signed number.

**EXAMPLE**    Plot these points: Point $A = (3, 1)$, point $B = (-2, 5)$, point $C = (-3, -2)$, point $D = (1, -3)$.

See Figure 7–3.

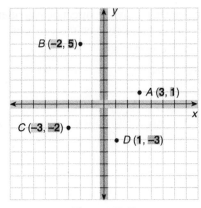

*A:* **right 3, up 1**
*B:* **left 2, up 5**
*C:* **left 3, down 2**
*D:* **right 1, down 3**

**Figure 7–3**

Moving in the wrong direction is a common mistake when plotting points. To develop your spatial sense, consider the signs of the coordinates of points being plotted. Then, visualize in which part of the graph the point will fall.

Refer to the points $A = (3, 1)$; $B = (-2, 5)$; $C = (-3, -2)$; and $D = (1, -3)$.

Point $A$ (3, 1) moves *right* and *up,* as shown in Figure 7–4.

Point $B$ ($-2$, 5) moves *left* and *up,* as shown in Figure 7–5.

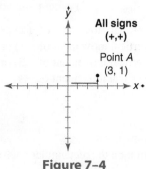

**Figure 7–4**

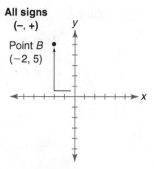

**Figure 7–5**

Point $C$ ($-3$, $-2$) moves *left* and *down,* as shown in Figure 7–6.

Point $D$ (1, $-3$) moves *right* and *down,* as shown in Figure 7–7.

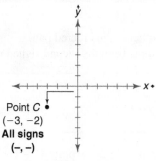

**Figure 7–6**

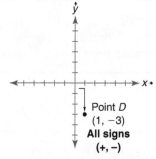

**Figure 7–7**

A point that shows movement in only one direction is written in point notation by using the number 0 to represent no movement. Points on the *x*-axis have no vertical movement. Points on the *y*-axis have no horizontal movement.

**EXAMPLE**   Plot the points: $A = (2, 0)$, $B = (0, 2)$, $C = (-2, 0)$, $D = (0, -2)$.

See Figure 7–8.

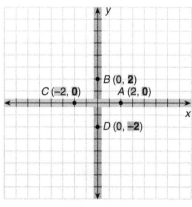

| | |
|---|---|
| Point $A = (2, 0)$ | **right 2, then stop** |
| Point $B = (0, 2)$ | **up 2** |
| Point $C = (-2, 0)$ | **left 2, then stop** |
| Point $D = (0, -2)$ | **down 2** |

**Figure 7–8**

### 2   Graph Linear Equations Using a Table of Solutions.

For every value of the independent variable $x$, there is a corresponding value of the dependent variable $y$. These two values form a pair of values called an **ordered pair**. A representative selection of these solutions can be written in a **table of solutions**. The **domain** of the equation is a set of suitable values determined by the context of the problem that can be selected for the independent variable of the equation. The set of corresponding values of the dependent variable that is formed by considering each value in the domain is called the **range**.

**To prepare a table of solutions for an equation with two variables:**

1. Select any appropriate value from the domain of the independent variable.
2. Substitute the selected value for the independent variable and solve the resulting equation for the dependent variable.
3. Write the solution as an ordered pair.
4. Repeat Steps 1 and 2 until the desired number of solutions is obtained.

The *table-of-solutions* or **table-of-values** procedure for graphing an equation works for *all* types of equations. Other methods for graphing equations focus on specific properties of each type of equation; these methods are generally less time-consuming and more efficient than the table-of-solutions procedure.

## To graph a linear equation using a table of solutions:

1. Prepare a table of solutions by evaluating the equation at different values (at least three) in the domain of the independent variable. Each pair of values is represented by a point on the graph.
2. Plot the points from the table of solutions on a rectangular coordinate system, with the independent variable on the x-axis and the dependent variable on the y-axis.
3. Connect the points with a straight line. Extend the graph beyond the three points and place an arrow on each end as appropriate to indicate that the line extends indefinitely.

In graphing linear equations in two variables, the independent variable is generally represented by the letter $x$ and the dependent variable is represented by the letter $y$. Then, solutions are represented by the ordered pair $(x, y)$ and are plotted on the rectangular coordinate system.

**EXAMPLE**    Make a table of solutions and graph the equation $y = -2x + 5$. Discuss the characteristics of the domain and range.

Select at least three values for the independent variable: $-1, 0, 1$.

When $x = -1$, $y = -2(-1) + 5$      Substitute $-1$ for $x$.
$\qquad\qquad = 2 + 5$      Evaluate.
$\qquad\qquad = 7$      Solution: $x = -1$ and $y = 7$ or $(-1, 7)$.

When $x = 0$, $y = -2(0) + 5$      Substitute.
$\qquad\qquad = 0 + 5$      Evaluate.
$\qquad\qquad = 5$      Solution: $(0, 5)$.

When $x = 1$, $y = -2(1) + 5$      Substitute.
$\qquad\qquad = -2 + 5$      Evaluate.
$\qquad\qquad = 3$      Solution: $(1, 3)$.

Make a table of the three solutions.

| $x$ | $y$ | Points on Graph |
|---|---|---|
| $-1$ | 7 | $(-1, 7)$ |
| 0 | 5 | $(0, 5)$ |
| 1 | 3 | $(1, 3)$ |

Plot the solutions in the table of solutions as points on a rectangular coordinate system (Figure 7–9).

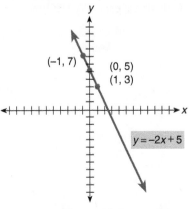

**The domain is all real numbers and the range is all real numbers. That is, there are no restrictions on the domain and range.**

**Figure 7–9**

Chapter 7 / Graphing Linear Equations

A **linear equation** is an equation for which the graph is a straight line. The solutions of a linear equation in two variables are **ordered pairs** of numbers. There is one number for each variable. The numbers are generally ordered in the alphabetical order of the two variables rather than in the order the letters appear in the equation. Thus, in the equation $2y + x = 8$, a solution of (4, 2) indicates that $x = 4$ and $y = 2$.

We can *check* each solution of an equation in two variables by substituting the values for the respective variables and performing the operations indicated.

### To check a solution of an equation with two variables:

1. Substitute the values for the respective variables into the equation.
2. Evaluate each side of the equation.
3. If the ordered pair is a solution of the equation, the two sides of the equation will be equal.

**EXAMPLE**   Check the solution $(-2, 16)$ for the equation $2x + y = 12$.

$$2x + y = 12 \qquad \text{Substitute } -2 \text{ for } x \text{ and } 16 \text{ for } y.$$
$$2(-2) + 16 = 12 \qquad \text{Evaluate.}$$
$$-4 + 16 = 12$$
$$12 = 12 \qquad \text{True.}$$

**The ordered pair makes the equation true. So $(-2, 16)$ is a solution.**

**EXAMPLE**   Check the ordered pair $(3, -2)$ for the equation $y = 2x - 5$.

$$y = 2x - 5 \qquad \text{Substitute 3 for } x \text{ and } -2 \text{ for } y.$$
$$-2 = 2(3) - 5 \qquad \text{Evaluate.}$$
$$-2 = 6 - 5$$
$$-2 = 1 \qquad \text{False.}$$

**The ordered pair does not make the equation true. So $(3, -2)$ is not a solution.**

If we are given the value of one variable for an equation with two variables, we can substitute the given value in the equation and solve the equation for the other variable.

### To find a specific solution of an equation with two variables when given the value of one variable:

1. Substitute the given value of the variable in each place that variable occurs in the equation.
2. Solve the equation for the other variable.

**EXAMPLE**   Solve the equation $2x - 4y = 14$, if $x = 3$.

$$2x - 4y = 14 \qquad \text{Substitute 3 for } x.$$
$$2(3) - 4y = 14$$
$$6 - 4y = 14 \qquad \text{Sort.}$$
$$-4y = 14 - 6$$
$$-4y = 8 \qquad \text{Divide both sides by } -4.$$
$$\frac{-4y}{-4} = \frac{8}{-4}$$
$$y = -2$$

**The solution is $(3, -2)$; $x = 3$ and $y = -2$.**

The **standard form of a linear equation in two variables** is $ax + by = c$. Both $a$ and $b$ are not zero; however, either $a$ or $b$ may be zero.

When $a$ or $b$ is zero, the variable for which zero is the coefficient is eliminated from the equation.

| | |
|---|---|
| $ax + by = 15$ | Let $a = 5$ and $b = 0$. |
| $5x + 0(y) = 15$ | $0(y) = 0$ |
| $5x + 0 = 15$ | Linear equation in one variable $n + 0 = n$ |
| $5x = 15$ | Divide both sides by 5. |
| $\dfrac{5x}{5} = \dfrac{15}{5}$ | |
| $x = 3$ | |

An equation like $x = 3$ is a linear equation, and we think of the coefficient of the missing variable term as zero:

$$x = 3 \quad \text{is considered as} \quad x + 0y = 3 \quad \text{in standard form}$$

This means that, for any value of $y$, $x$ is 3.

---

**EXAMPLE**  For the equation $x = 6$, complete the given ordered pairs: ( , 2); ( , −3); ( , 1).

**Equation**          **Ordered Pairs**

$x = 6$      ( , 2)      ( , −3)      ( , 1)      $x$ is 6 for *any* value of $y$.

**(6, 2)      (6, −3)      (6, 1)**

---

Numerous application problems can be solved with equations in two variables.

---

**EXAMPLE**

**BUS**

A small business photocopied a report that included 12 black-and-white pages and 25 color pages. The cost was $23.70. Letting $b =$ the cost for a black-and-white page and $c =$ the cost for a color page, write an equation with two variables.

| | |
|---|---|
| Cost of black-and-white copies: | $12b$ (12 times cost per page) |
| Cost of color copies: | $25c$ (25 times cost per page) |
| Total cost: | black-and-white plus color = $23.70 |

**Equation: $12b + 25c = 23.70$**

---

An equation containing two variables shows the relationship between the two unknown amounts. If a value is known for either amount, the other amount can be found by solving the equation for the missing amount.

---

**EXAMPLE**  Solve the equation from the preceding example to find the cost for each black-and-white copy if color copies cost $0.90 each.

| | |
|---|---|
| $12b + 25c = 23.70$ | Substitute 0.90 for $c$. |
| $12b + 25(0.90) = 23.70$ | |
| $12b + 22.50 = 23.70$ | Sort. |
| $12b = 23.70 - 22.50$ | |
| $12b = 1.20$ | Divide both sides by 12. |
| $\dfrac{12b}{12} = \dfrac{1.20}{12}$ | |
| $b = 0.10$ | |

**Each black-and-white copy costs $0.10.**

---

Chapter 7 / Graphing Linear Equations

**1** Locate the points in Figure 7–10 by giving the amount of horizontal and vertical movement from the origin.

**1.** $R$ ⁴,²     **2.** $S$ ⁻⁴,³     **3.** $T$ ⁵,⁻³     **4.** $U$ ⁻²,⁻⁵     **9.** Write the coordinates for the points $A$, $B$, $C$, and $D$ on
**5.** $V$ ⁷,⁰     **6.** $W$ ⁰,⁵     **7.** $X$ ⁰,⁰     **8.** $Y$ ⁻³,⁰      the graph in Figure 7–11.

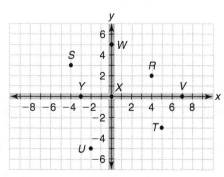

**Figure 7–10**

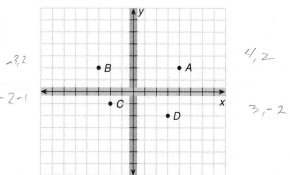

**Figure 7–11**

Draw a rectangular coordinate system, and locate the points given in Exercises 10–15.

**10.** $A = (-7, 4)$           **11.** $B = (0, -10)$           **12.** $C = (-2, 0)$
**13.** $D = (-7, -7)$         **14.** $E = (3, 0)$             **15.** $F = (4, -3)$

**16.** What common property describes the coordinates of the points on the $x$-axis?
**17.** What common property describes the coordinates of the points on the $y$-axis?
**18.** Describe the signs of the coordinates of points that lie in the upper-left quarter of a graph.
**19.** Describe the signs of the coordinates of points that lie in the lower-right quarter of a graph.

**2**

**20.** Make a table of at least three solutions for the equation $y = 3x - 7$.
**21.** Make a table of three solutions for the equation $y = -2x + 3$.

Prepare a table of three solutions for each of the equations.

**22.** $y = -4x + 5$        **23.** $y = \dfrac{1}{2}x + 3$        **24.** $y = -\dfrac{2}{3}x - 1$

**25.** $y = 5x + 2$          **26.** $y = -\dfrac{1}{2}x + 2$       **27.** $y = \dfrac{1}{3}x + 2$

**3** Use the table of solutions prepared in Exercises 20–27 to graph the following solutions.

**28.** $y = 3x - 7$ (See Exercise 20.)           **29.** $y = -2x + 3$ (See Exercise 21.)

**30.** $y = -4x + 5$ (See Exercise 22.)         **31.** $y = \dfrac{1}{2}x + 3$ (See Exercise 23.)

**32.** $y = -\dfrac{2}{3}x - 1$ (See Exercise 24.)      **33.** $y = 5x + 2$ (See Exercise 25.)

**34.** $y = -\dfrac{1}{2}x + 2$ (See Exercise 26.)      **35.** $y = \dfrac{1}{3}x + 2$ (See Exercise 27.)

Determine which of the ordered pairs are solutions for the equation $2x - 5y = 9$.

**36.** $(4, 5)$              **37.** $(17, 5)$            **38.** $(12, 3)$
**39.** $(6, 1)$              **40.** $(7, 1)$             **41.** $(4.5, 0)$

Determine which of the ordered pairs are solutions for the equation $3y = 2 - 4x$.

**42.** $(2, -2)$           **43.** $(-2, 2)$          **44.** $(5, -6)$
**45.** $(-6, 5)$          **46.** $\left(0, \frac{2}{3}\right)$         **47.** $\left(6, -7\frac{1}{3}\right)$

7–1 Graphical Representation of Linear Equations        

Write the specific solution for each equation in ordered pair form.

**48.** $x - y = 3$, if $x = 8$      **49.** $x + y = 7$, if $x = 2$      **50.** $x + 2y = -4$, if $y = 1$

**51.** $x - 3y = 12$, if $y = 3$      **52.** $5x - y = 10$, if $x = 5$      **53.** $4x + y = 8$, if $x = 2$

Complete the ordered pairs for each equation.

**54.** $x = 1$: ( , 3); ( , −2); ( , 0); ( , 1)      **55.** $y = 4$: (−1, ); (3, ); (0, ); (2, )

Solve using equations with two variables.

**56. BUS** Anne Richards purchased three shirts at one price and five belts at another price. Let $s$ = the price of each shirt and $b$ = the price of each belt. Each shirt cost $22. How much did each belt cost if the total purchase price was $126?

**57. AUTO** Allen and Cecile Bell paid $13 for 4 spark plugs and 5 quarts of motor oil. Let $p$ = the cost of each plug and $q$ = the cost of each quart of oil. If the oil cost $1 per quart, how much did each spark plug cost?

**58. BUS** Lola Bowers charged $160 for four pairs of pants and two shirts. If the shirts cost $12 each, how much did each pair of pants cost? Let $p$ = the cost of each pair of pants and $s$ = the cost of each shirt.

**59. AUTO** Coach Barton paid $15,000 for an automobile. The purchase price included three extended warranty payments of $400 each. How much did the car cost without the extended warranty? Let $c$ = the cost of the car and $w$ = the cost of each warranty payment.

---

## 7–2 | Graphing Linear Equations with Two Variables Using Alternative Methods

*Learning Outcomes*

**1** Graph linear equations using intercepts.

**2** Graph linear equations using the slope and $y$-intercept.

**3** Graph linear equations using a graphing calculator.

**4** Solve a linear equation using a graph.

We have graphed linear equations using a table of solutions. Now let's examine other procedures and specific properties of linear equations.

**1** **Graph Linear Equations Using Intercepts.**

Two important points on the graph of an equation are the points where the graph crosses the $x$- and $y$-axis. These points are called **intercepts**. The **$x$-intercept** is the point on the $x$-axis through which the line of the equation passes; that is, the $y$-value is zero $(x, 0)$. The **$y$-intercept** is the point on the $y$-axis through which the line of the equation passes; that is, the $x$-value is zero $(0, y)$.

**To find the intercepts of a linear equation:**

1. To find the $x$-intercept, let $y = 0$ and solve for $x$.
2. To find the $y$-intercept, let $x = 0$ and solve for $y$.

**EXAMPLE** Find the intercepts of the equation $3x - y = 5$.

$$3x - y = 5$$      For the $x$-intercept, let $y = 0$.

$$3x - 0 = 5$$      Solve for $x$.

$$3x = 5$$

$$x = \frac{5}{3}$$      A mixed number or decimal is easier to plot than an improper fraction.

$$x = 1\frac{2}{3} \quad \text{or} \quad 1.67$$

**The $x$-intercept is $(1\frac{2}{3}, 0)$.**

$$3x - y = 5$$ For the *y*-intercept, let $x = 0$.

$$3(0) - y = 5$$ Solve for *y*.

$$-y = 5$$

$$y = -5$$

**The *y*-intercept is (0, −5).**

## To graph linear equations by the intercepts method:

1. Find the *x*- and *y*-intercepts.
2. Plot the intercepts on a rectangular coordinate system.
3. Draw the line through the two points and extend it beyond each point.
4. Check by examining one additional solution of the equation.

**EXAMPLE** Graph the equation $3x - y = 5$ by using the intercepts of each axis.

Plot the two intercepts found in the preceding example, $(1\frac{2}{3}, 0)$ and $(0, -5)$. Draw the line connecting these points and extending beyond. See Figure 7–12. We can check by finding one other point in the usual way and plotting it. If it is on the line, the graph is correct. Check for $x = 2$:

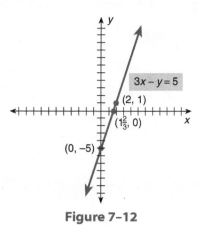

$$3x - y = 5$$ Substitute 2 for *x*.

$$3(2) - y = 5$$ Solve for *y*.

$$6 - 5 = y$$

$$1 = y$$ The point (2, 1) is on the graph.

**Figure 7–12**

**TIP!**

### Graphing Equations When Both Intercepts Are (0, 0)

If both intercepts are (0, 0), they coincide at the origin and form only *one point*. An additional point must be found by the table-of-solutions method so that you will have two distinct points for drawing the line. A third point is still useful to check your work.

**EXAMPLE** Graph $y = 7x$ using the intercepts method.

Find the intercepts.

$y = 7x$      Let $x = 0$.
$y = 7(0)$
$y = 0$      $y$-intercept

$y = 7x$      Let $y = 0$.
$0 = 7x$
$\dfrac{0}{7} = \dfrac{7x}{7}$
$0 = x$      $x$-intercept

Find another point.

$y = 7x$      Let $x = 1$.
$y = 7(1)$
$y = 7$

Additional point $= (1, 7)$

Plot the points $(1, 7)$ and $(0, 0)$. Draw the line through the two points and extend it beyond each point, as shown in Figure 7–13.

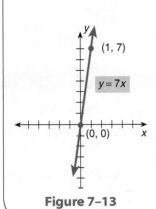

**Figure 7–13**

## 2 Graph Linear Equations Using the Slope and *y*-Intercept.

We can identify characteristics of the graph of an equation by inspection. To see these characteristics, we solve the equation for $y$. The equation will be in the form $y = mx + b$, where $m$ is the coefficient of $x$ and $b$ is a constant.

First, let's examine the characteristics identified by the constant. Let the coefficient of $x$ equal 1 and look at the graphs of the equations in Figure 7–14.

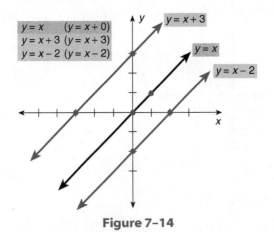

| $y = x$ | | $y = x + 3$ | | $y = x - 2$ | |
|---|---|---|---|---|---|
| $x$ | $y$ | $x$ | $y$ | $x$ | $y$ |
| 0 | 0 | 0 | 3 | 0 | $-2$ |
| 1 | 1 | $-3$ | 0 | 2 | 0 |

**Figure 7–14**

The three graphs have the same slope (steepness or slant), but they have different $y$-intercepts. That is, they cross the $y$-axis at different points.

$y = x + 0$    crosses the $y$-axis at 0;    $y$-intercept $= (0, 0)$
$y = x + 3$    crosses the $y$-axis at $+3$;    $y$-intercept $= (0, 3)$
$y = x - 2$    crosses the $y$-axis at $-2$;    $y$-intercept $= (0, -2)$

The $y$-coordinate of the point where the graph crosses the $y$-axis is the same as the constant in the equation. The constant ($b$) in an equation in the form $y = mx + b$ is the $y$-coordinate of the $y$-intercept.

Let's examine equations that have graphs with a different slant or steepness but the same $y$-intercept. The $x$- and $y$-intercepts in each of the three equations in Figure 7–15 are $(0, 0)$. We will find one additional point to graph.

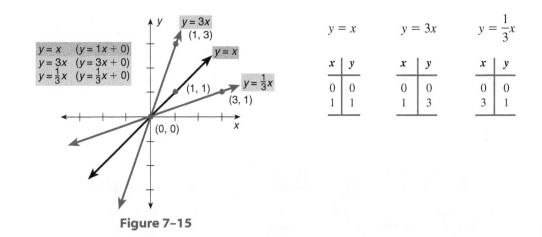

**Figure 7–15**

The three graphs have the same $y$-intercepts, but they have different slopes (steepness or slants). We define this slope or steepness so that a numerical value identifies it. The **slope** of a line is the ratio of the vertical change to the horizontal change.

$$\text{slope} = \frac{\text{vertical change}}{\text{horizontal change}}$$

A slope of 1 (written as $\frac{1}{1}$) means a change of 1 vertical unit for every 1 horizontal unit of change. A slope of 3 (written as the ratio $\frac{3}{1}$) means a change of 3 vertical units for every 1 horizontal unit. A slope of $\frac{1}{3}$ means a change of 1 vertical unit for every 3 horizontal units (see Figure 7–16).

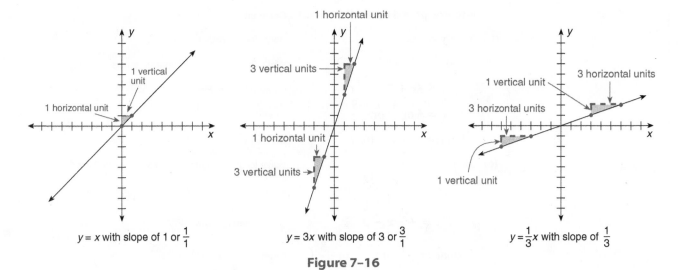

**Figure 7–16**

Notice that the slope of the line is the same as the fractional form of the coefficient of the $x$-term in the equation. We make one final observation. Look again at the equations in Figures 7–14 and 7–15.

| From Figure 7–14 | Slope-Intercept Form | From Figure 7–15 | Slope-Intercept Form |
|---|---|---|---|
| $y = x$ | $y = 1x + 0$ | $y = x$ | $y = 1x + 0$ |
| $y = x + 3$ | $y = 1x + 3$ | $y = 3x$ | $y = 3x + 0$ |
| $y = x - 2$ | $y = 1x - 2$ | $y = \frac{1}{3}x$ | $y = \frac{1}{3}x + 0$ |

In each case, the equation is solved for $y$. This form of equation is called the **slope-intercept form of a linear equation**, or $y = mx + b$, where $m$ is the slope and $b$ is the $y$-coordinate of the $y$-intercept. When equations are written in this form, the slope and $y$-intercept can be identified by inspection.

$$y = mx + b$$

where $m$ = slope and $b$ = $y$-coordinate of the $y$-intercept

### To determine the slope and $y$-intercept of a linear equation:

1. Rearrange the equation so that $y$ is isolated on the left side of the equation. That is, solve the equation for $y$.
2. The *slope* is $m$ which is the coefficient of $x$ when the equation is in the form $y = mx + b$.
3. The $y$-coordinate of the $y$-intercept is $b$ which is the constant when the equation is in the form $y = mx + b$. The $y$-intercept is $(0, b)$.

**EXAMPLE**  Write the equations in slope-intercept form and identify the slope and $y$-intercept.

(a) $2x + y = 4$      (b) $5x - y = -2$      (c) $3x + 4y = -12$

(a) $2x + y = 4$      Solve for $y$.

$\qquad y = -2x + 4$      Slope-intercept form.

$\qquad$ **slope** $= -2$ or $\dfrac{-2}{1}$ or $\dfrac{2}{-1}$      Coefficient of $x$.

$\qquad$ **$y$-intercept** $= 4$ or $(0, 4)$      Constant.

(b) $5x - y = -2$      Solve for $y$. Divide by $-1$.

$\qquad \dfrac{-y}{-1} = \dfrac{-5x - 2}{-1}$      Write the right side as separate terms and simplify.

$\qquad y = 5x + 2$      Slope-intercept form.

$\qquad$ **slope** $= 5$ or $\dfrac{5}{1}$ or $\dfrac{-5}{-1}$      Coefficient of $x$.

$\qquad$ **$y$-intercept** $= 2$ or $(0, 2)$      Constant.

(c) $3x + 4y = -12$      Solve for $y$. Sort terms to isolate $4y$. Divide by 4.

$\qquad \dfrac{4y}{4} = \dfrac{-3x - 12}{4}$      Write the right side as separate terms and reduce.

$\qquad y = -\dfrac{3}{4}x - 3$      Slope-intercept form.

$\qquad$ **slope** $= -\dfrac{3}{4}$ or $\dfrac{-3}{4}$ or $\dfrac{3}{-4}$      Coefficient of $x$.

$\qquad$ **$y$-intercept** $= -3$ or $(0, -3)$      Constant.

An equation in the slope-intercept form can be graphed using just the slope and $y$-intercept.

### To graph a linear equation in the form $y = mx + b$ using the slope and $y$-intercept method:

1. Locate the $y$-intercept on the $y$-axis.
2. Using the slope in fractional form, determine the amount of vertical and horizontal movement indicated.
3. From the $y$-intercept, locate additional points on the graph of the equation by counting the indicated vertical and horizontal movement.
4. Draw the line connecting the points and extending beyond the points.

**EXAMPLE** Graph the equations using the slope and *y*-intercept method.

(a) $2x + y = 4$      (b) $5x - y = -2$      (c) $3x + 4y = -12$

(a) $2x + y = 4$                Solve for *y*.

$\quad y = -2x + 4$

*y*-intercept $= (0, 4)$      Locate this point on the *y*-axis.

slope $= \dfrac{-2}{1}$ or $\dfrac{2}{-1}$      Write the slope in fractional form.

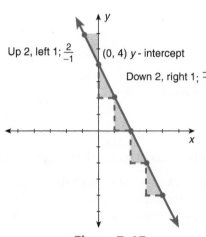

Up 2, left 1; $\frac{2}{-1}$      (0, 4) *y* - intercept

Down 2, right 1; $\frac{-2}{1}$

$\frac{-2}{1}$ indicates vertical movement of $-2$ and horizontal movement of $+1$ from the *y*-intercept.

$\frac{2}{-1}$ indicates vertical movement of $+2$ and horizontal movement of $-1$. See Figure 7–17.

**Figure 7–17**

(b) $5x - y = -2$              Solve for *y*.

$\quad y = 5x + 2$

*y*-intercept $= (0, 2)$      Locate this point on the *y*-axis.

slope $= \dfrac{5}{1}$ or $\dfrac{-5}{-1}$

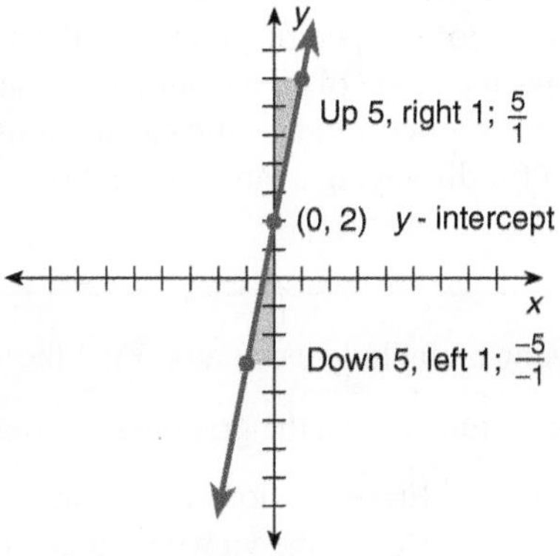

Up 5, right 1; $\frac{5}{1}$

(0, 2)   *y* - intercept

Down 5, left 1; $\frac{-5}{-1}$

$\frac{5}{1}$ indicates vertical movement of $+5$ and horizontal movement of $+1$ from the *y*-intercept.

$\frac{-5}{-1}$ indicates vertical movement of $-5$ and horizontal movement of $-1$. See Figure 7–18.

**Figure 7–18**

(c) $3x + 4y = -12$      Solve for *y*.

$\quad y = -\dfrac{3}{4}x - \dfrac{12}{4}$

$\quad y = -\dfrac{3}{4}x - 3$

*y*-intercept $= (0, -3)$      Locate this point on the *y*-axis.

slope $= \dfrac{-3}{4}$ or $\dfrac{3}{-4}$

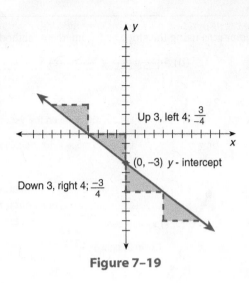

$\frac{-3}{4}$ indicates vertical movement of $-3$ and horizontal movement of $+4$ from the $y$-intercept.
$\frac{3}{-4}$ indicates vertical movement of $+3$ and horizontal movement of $-4$. See Figure 7–19.

Up 3, left 4; $\frac{3}{-4}$

$(0, -3)$ $y$ - intercept

Down 3, right 4; $\frac{-3}{4}$

**Figure 7–19**

### 3 Graph Linear Equations Using a Graphing Calculator.

A graphing calculator can be used to graph linear equations written in the form $y = mx + b$. Also, computer software can be used to graph linear equations. These tools free us from the tedious task of graphing equations and allow us to focus on the patterns and properties of graphs.

Every model of graphing calculator and every brand of computer software may have a different set of keystrokes for graphing equations. You will need to refer to the owner's manual to adapt to a particular model or brand. However, we will illustrate the usefulness of these tools by showing a few features common to most calculators.

| Feature | Purpose |
|---------|---------|
| Y = | Screen for entering equation that is solved for $y$ |
| Window | Setting the range (high and low values of each axis) for the viewing window |
| Zoom | Zooming to get a closer or wider view or a view to "fit" the full range of $y$-values for the selected $x$-values |
| Trace | Tracing to locate specific points and determine their coordinates |
| Graph | Displays the graph of all equations that have been entered |
| Calc | Menu for shortcuts to specific calculations |
| Table | Screen for displaying a table of solutions |

**EXAMPLE**    Graph $y = 5x + 2$ using a graphing calculator. Find the $y$-value for $x = -2$.

Perform the appropriate function on the graphing calculator of your choice.

| | |
|---|---|
| Clear graphing screen. | Erase previous equations from the y = screen. |
| Set or initialize range. | Define the viewing range of the graph window. One option is to select the standard option from the Zoom screen. That means the range will be reset at a predetermined, factory-set domain and range. |
| Enter equation. | Equation must be solved for $y$. On the y = screen enter the equation using the appropriate key for the variable. This key is often labeled $\boxed{\text{X, T, }\theta, n}$. |
| View graph. | Use the Graph key. Determine a different viewing window by setting the values on the Window screen or using the Zoom feature. |
| Determine key points on the graph. | Using the trace function, move the cursor using the left or right arrow keys. The $x$- and $y$-coordinates of the point at the cursor will appear at the bottom of the screen. |

Find the corresponding $y$-value for a given $x$-value by using the *value* function from the CALC screen. Enter $(-)2$ for $x$ at the prompt.

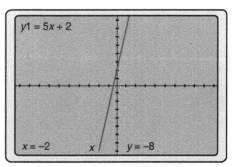

**Figure 7–20**

**When $x = -2, y = -8$.**

**4** **Solve a Linear Equation Using a Graph.**

A linear equation written in slope-intercept form is also referred to as a **function** of $x$, the independent variable. An equation in one variable, such as $2x + 5 = 3$, can be written in the form of a function by first rewriting all terms on either side of the equation.

$$2x + 5 = 3 \qquad \text{or} \qquad 2x + 5 = 3$$
$$2x + 5 - 3 = 0 \qquad\qquad 0 = 3 - 2x - 5$$
$$2x + 2 = 0 \qquad\qquad 0 = -2x - 2$$

Then, write the nonzero side of each equation as a function of $x$.

$$y = 2x + 2 \qquad \text{or} \qquad y = -2x - 2$$

**To write an equation in one variable as a function:**

1. Rewrite the equation so that all terms are on one side of the equation.
2. Write the equation in function notation using $y =$ the nonzero side of the equation from Step 1.

*The solution of the original equation is the value of* x *that makes the value of either function zero.*

If an equation in one variable is written in function notation, the solution of the equation is shown on the graph at the point where the graph crosses the $x$-axis. In Figure 7–21, both functions cross the $x$-axis at $(-1, 0)$ and the solution of the equation $2x + 5 = 3$ is $x = -1$. Thus, we can solve equations in one variable by using the graphing function on a calculator or computer.

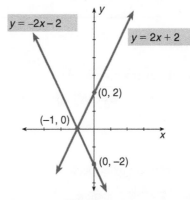

**Figure 7–21**

## To solve an equation in one variable using a graph:

1. Write the equation as a function.
2. Graph the function using a calculator or computer.
3. The solution is the $x$-intercept. This point is also referred to as the *zero point* ($y = 0$) of the function.

**TIP!**

### Find the Zero of a Function Using a Calculator

On most graphing calculators there is a feature to find the zero point or points of a function. On one popular model, the TI-84 Plus Silver Edition™, the feature is a selection on the CALC screen.

The option for finding the zero point requires several steps that are illustrated in the next example.

**EXAMPLE**   Solve the equation $2x + 5 = 3$ using the TI-84 Plus Silver Edition™.

$$2x + 5 = 3$$
$$2x + 5 - 3 = 0$$
$$2x + 2 = 0$$
$$y = 2x + 2$$   Function form of the equation.

$\boxed{Y=}$   Enter right side of equation: 2 $\boxed{X, T, \theta, n}$ + 2

CALC

CALC   $\boxed{2\text{nd}}$ $\boxed{\text{TRACE}}$

Zero   Option 2. Press $\boxed{\text{ENTER}}$.

Left bound?   Move cursor to any point to the left of where the graph crosses the $x$-axis. Press $\boxed{\text{ENTER}}$.

Right bound?   Move cursor to any point to the right of where the graph crosses the $x$-axis. Press $\boxed{\text{ENTER}}$.

Guess?   Move the cursor close to the $x$-intercept. Press $\boxed{\text{ENTER}}$.

$x = -1$   Read solution from $x$ value at the bottom left of the screen.

### SECTION 7–2 SELF-STUDY EXERCISES

**1**  Graph the equations using the intercepts procedure.

**1.** $x + y = 5$

**2.** $x + 3y = 5$

**3.** $\frac{1}{2}y = 4 + x$

**4.** $y = 3x - 1$

**5.** $5x = y + 2$

**6.** $x = -2y + 3$

**2**  Determine the slope and $y$-intercept by inspection.

**7.** $y = 4x + 3$

**8.** $y = -5x + 6$

**9.** $y = -\frac{7}{8}x - 3$

**10.** $y = 3$

Rewrite the following equations in slope-intercept form and determine the slope and $y$-intercept.

**11.** $4x - 2y = 10$

**12.** $2y = 5$

**13.** $\frac{1}{2}y + x = 3$

**14.** $2.1y - 4.2x = 10.5$

**15.** $2x = 8$

**16.** $x - 5 = 4$

Graph the equations using the slope-intercept procedure.

**17.** $y = 2x - 3$

**18.** $y = -\frac{1}{2}x - 2$

**19.** $y = -\frac{3}{5}x$

**20.** $x - 2y = 3$

**21.** $2x + y = 1$

**22.** $y = \frac{3}{4}x + 2$

**3** Graph the equations using a graphing calculator and find the value of $y$ as indicated.

**23.** $y = 0.3x + 0.5$

**24.** Find $y$ when $x = -0.4$ in $y = 0.3x + 0.5$.

**25.** $y = -2.6x - 1.7$

**26.** Find $y$ when $x = 1.3$ in $y = -2.6x - 1.7$.

**27.** $y = 212x + 757$

**28.** Find $y$ when $x = 5$ in $y = 212x + 757$.

**29.** $y = 48x - 200$

**30.** Find $y$ when $x = 7$ in $y = 48x - 200$.

**31.** The cost of printing a magazine is $5,000 to typeset and prepare copy for printing and $3 per copy to print. This is expressed as an equation $y = 3x + 5,000$, where $x$ is the number of copies printed and $y$ is the total cost of printing $x$ copies. Find the cost of printing 10,000 copies. Find the cost of printing 30,000 copies.

**32.** Income is expressed by the equation $y = \$8.30x$, where $x$ is the number of hours per week an employee works. Find the income for an employee who works 40 hours in a week. Find the annual income if a person works for an entire year (52 weeks).

**33.** Some researchers say a person's maximum heart rate in beats per minute can be found by subtracting age from 220. This can be expressed as an equation $y = 220 - x$. What is the maximum heart rate for a person 25 years old? What is the maximum heart rate for a person 65 years old?

**34.** An inventory shows 196 jigsaw puzzles in stock. These puzzles can be produced at a rate of 15 per hour. The number of puzzles on hand can be expressed as an equation $y = 15x + 196$, where $x$ represents the number of hours of production and $y$ represents the total inventory. How many puzzles are in inventory if they have been produced for 60 h?

**4** Solve the following equations using a graphing calculator.

**35.** $3x + 2 = 7$

**36.** $5x - 4 = 2$

**37.** $4 = 6x - 1$

**38.** $3x + 7 = 4x + 6$

**39.** $3x - 1 = 4(x + 2)$

**40.** $2x - 8 = 4(x + 2)$

---

# 7-3 | *Slope*

*Learning Outcomes*

**1** Calculate the slope of a line, given two points on the line.

**2** Determine the slope of a horizontal or vertical line.

The concept of the rate of change is important in real-world applications. This concept is often referred to as *slope* and is used when we consider the slant of a roof or the grade of a roadway. We use the rate of change in applications, like changes in temperature, in the quality of a product, or in sales.

**1** **Calculate the Slope of a Line, Given Two Points on the Line.**

The *slope* of a straight line is the rate of change of the line or the ratio of the vertical rise of a line to the horizontal run of the line (see Figure 7–22).

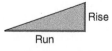

**Figure 7–22**

$$\text{slope} = \frac{\text{rise}}{\text{run}} \quad \text{or} \quad \frac{\text{vertical change}}{\text{horizontal change}}$$

**To find the slope of a line from two given points on the line:**

1. Designate one point as point 1 with coordinates $(x_1, y_1)$. Designate the other point as point 2 with coordinates $(x_2, y_2)$.
2. Calculate the change (*difference*) in the $y$-coordinates to find the vertical *rise* $(y_2 - y_1)$ and in the $x$-coordinates to find the horizontal *run* $(x_2 - x_1)$.
3. Write a ratio of the rise to the run and reduce the ratio to lowest terms.

Symbolically, if $P_1$ and $P_2$ are any two points on the line,

$$\text{slope} = \frac{\Delta y}{\Delta x} = \frac{y_2 - y_1}{x_2 - x_1}$$

where $P_1 = (x_1, y_1)$ and $P_2 = (x_2, y_2)$.
 The Greek capital letter delta ($\Delta$) is used to indicate a change and to write the slope definition symbolically.

---

**EXAMPLE**  Find the slope of a line if the points $(2, -1)$ and $(5, 3)$ are on the line (see Figure 7–23).

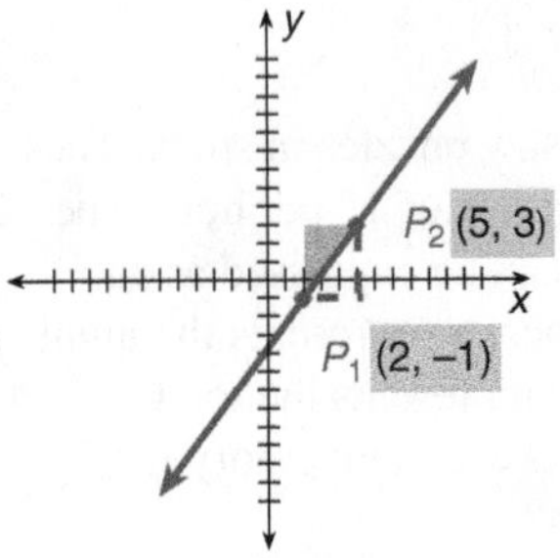

Identify $(2, -1)$ as point 1 $(P_1)$ and $(5, 3)$ as point 2 $(P_2)$. The coordinates of $P_1$ are $(x_1, y_1)$ and the coordinates of $P_2$ are $(x_2, y_2)$. Write the slope definition symbolically and substitute known values.

$$\text{change in } y = \Delta y = \text{rise} = y_2 - y_1 = 3 - (-1) = 4$$
$$\text{change in } x = \Delta x = \text{run} = x_2 - x_1 = 5 - 2 = 3$$

$$\text{slope} = \frac{\Delta y}{\Delta x} = \frac{\text{rise}}{\text{run}} = \frac{y_2 - y_1}{x_2 - x_1} = \frac{3 - (-1)}{5 - 2} = \frac{4}{3}$$

**The slope of the line through points $(2, -1)$ and $(5, 3)$ is $\dfrac{4}{3}$.**

**Figure 7–23**

Any line has an infinite number of points. *Any* two points on the line can be used to find the slope. Also, the designation of $P_1$ and $P_2$ is not critical. Let's find the slope of the line in the preceding example by designating $P_1$ as $(5, 3)$ and $P_2$ as $(2, -1)$.

$$\Delta y = \text{rise} = y_2 - y_1 = -1 - 3 = -4$$
$$\Delta y = \text{run} = x_2 - x_1 = 2 - 5 = -3$$

$$\frac{\Delta y}{\Delta x} = \frac{\text{rise}}{\text{run}} = \frac{-4}{-3} = \frac{4}{3}$$

---

**EXAMPLE**  Find the slope of the line passing through the pair of points.

$(-5, -1)$ and $(3, -2)$

Visualize the line on a coordinate system (Figure 7–24).

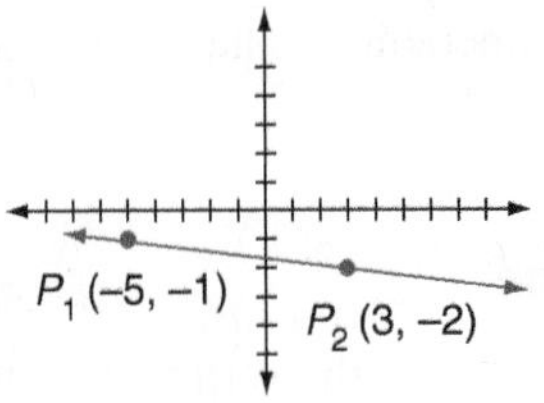

 Let $P_1 = (-5, -1)$ and $P_2 = (3, -2)$.

$$\text{slope} = \frac{\Delta y}{\Delta x} = \frac{-2 - (-1)}{3 - (-5)} = \frac{-2 + 1}{3 + 5} = \frac{-1}{8} = -\frac{1}{8}$$

**Figure 7–24**

**The slope is $-\dfrac{1}{8}$.**

Examine the slopes found in the two previous examples and the corresponding figures (Fig. 7–23 and Fig. 7–24). The first example has a positive slope of $\frac{4}{3}$. The graph of the line in Figure 7–23 rises from left to right. The second example has a negative slope of $-\frac{1}{8}$. The graph of the line in Figure 7–24 falls from left to right.

---

**Lines with positive or negative slope:**

- A line that has a positive slope will *rise* as the line moves from left to right.
- A line that has a negative slope will *fall* as the line moves from left to right.

---

**EXAMPLE** Table 7–1 lists the annual cost in dollars of tuition and fees at public 2-year colleges for selected years. Find the rate of change in tuition and fees from 1998–1999 and from 1999–2000.

| Table 7–1 Tuition and Fees at 2-Year Public Colleges | |
| --- | --- |
| Academic Year | Tuition and Fees |
| 94–95 | $1,310 |
| 95–96 | $1,330 |
| 96–97 | $1,465 |
| 97–98 | $1,567 |
| 98–99 | $1,554 |
| 99–00 | $1,649 |
| 00–01 | $1,642 |
| 01–02 | $1,608 |
| 02–03 | $1,674 |
| 03–04 | $1,909 |
| 04–05 | $2,076 |

Source: *Annual Survey of Colleges*, The College Board, New York, NY.

Find the rate of change of tuition and fees from 1998–1999 to 1999–2000.

(98–99, 1,554)     and     (99–00, 1,649)     Write the data as ordered pairs. Use the initial year of the academic year as the reference.

$$m = \frac{y_2 - y_1}{x_2 - x_1}$$     Use the slope formula to find the rate of change.

rate of change $= \dfrac{1,649 - 1,554}{1999 - 1998}$

rate of change $= \dfrac{95}{1} =$ **$95 per year**

Find the rate of change from 1999–2000 to 2004–2005.

(1999, 1,649)     and     (2004, 2,076)

$$m = \frac{y_2 - y_1}{x_2 - x_1}$$

rate of change $= \dfrac{2,076 - 1,649}{2004 - 1999}$

rate of change $= \dfrac{427}{5} =$ **$85.40 per year**

**The rate of change in tuition for one year from 1998–1999 to 1999–2000 was $95. But the rate of change from 1999–2000 to 2004–2005 was $85.40 per year, which represents an average change per year over the 5-year period.**

---

**2** **Determine the Slope of a Horizontal or Vertical Line.**

Let's examine the slopes of a horizontal line and a vertical line (Figures 7–25 and 7–26). First look at the horizontal line that passes through the points $(3, 2)$ and $(-3, 2)$.

7–3 Slope

**EXAMPLE** Find the slope of the horizontal line that passes through $(3, 2)$ and $(-3, 2)$.

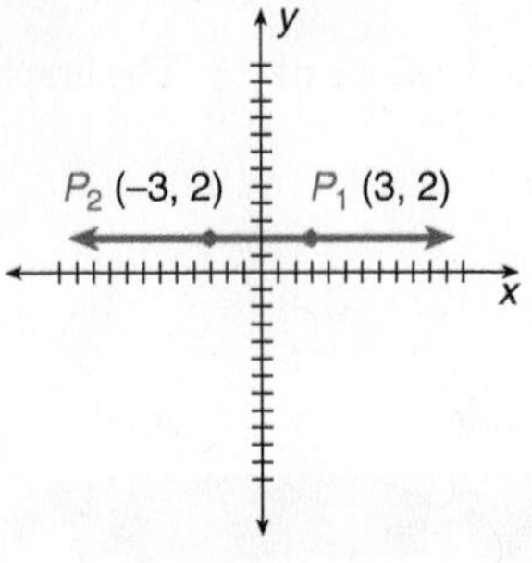

Let $P_1 = (3, 2)$ and $P_2 = (-3, 2)$. See Figure 7–25.

$$\text{slope} = \frac{\Delta y}{\Delta x} = \frac{y_2 - y_1}{x_2 - x_1} = \frac{2 - 2}{-3 - 3} = \frac{0}{-6} = 0$$

**The line has a slope of 0.**

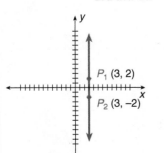

Figure 7–25

## Slope of horizontal lines:

The slope of a horizontal line is zero.

- Two points are on the same horizontal line if their $y$-coordinates are equal.
- A horizontal line has no rise or no change in $y$-values.

Next, let's look at the vertical line that passes through the points $(3, 2)$ and $(3, -2)$.

**EXAMPLE** Find the slope of the vertical line that passes through the points $(3, 2)$ and $(3, -2)$.

Let $P_1 = (3, 2)$ and $P_2 = (3, -2)$. See Figure 7–26.

$$\text{slope} = \frac{\Delta y}{\Delta x} = \frac{y_2 - y_1}{x_2 - x_1} = \frac{-2 - 2}{3 - 3} = \frac{-4}{0}$$

Division by zero is undefined; **therefore, the slope of the vertical line passing through $(3, 2)$ and $(3, -2)$ is undefined.**

**The line has undefined slope.**

Figure 7–26

## Slope of vertical lines:

The slope of a vertical line is undefined.

- Two points are on the same vertical line if their $x$-coordinates are equal.
- A vertical line has no run or no change in $x$-values.

## Zero Slope Versus Undefined Slope

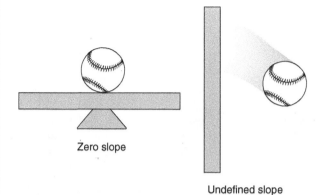

Zero slope

Undefined slope

**Figure 7–27**

A slope of zero and an undefined slope are *not* the same. Zero is a real number. A slope of zero is a real number. When zero is the denominator of a fraction, the number is undefined. As a memory aid, think of the number zero as a ball. It can be balanced on a horizontal line. It will not balance on a vertical line (Figure 7–27).

### To identify a horizontal or vertical line:

1. Examine the *y*-coordinates of two points on the line. If the *y*-coordinates are the same, the line is horizontal and the slope is 0.
2. Examine the *x*-coordinates of two points on the line. If the *x*-coordinates are the same, the line is vertical and the slope is undefined.

### SECTION 7–3 SELF-STUDY EXERCISES

**1** Find the slope of the line passing through the pairs of points.

**1.** $(-3, 3)$ and $(1, 5)$      **2.** $(4, -1)$ and $(1, 4)$      **3.** $(3, 1)$ and $(5, 7)$

**4.** $(-1, -1)$ and $(3, 3)$      **5.** $(4, 5)$ and $(-4, -1)$      **6.** $(7, 2)$ and $(-3, 2)$

**7.** $(4, -5)$ and $(0, 0)$      **8.** $(2, -2)$ and $(5, -5)$      **9.** $(-5, 2)$ and $(-5, -4)$

**10.** $(4, -4)$ and $(1, 5)$      **11.** $(3, 5)$ and $(-4, 5)$      **12.** $(-7, 3)$ and $(-7, 5)$

**13.** Don Lewis's salary was $22,000 in 2001 and increased to $24,800 in 2003. What was the annual rate of change in his salary?

**14.** If 100 backpacks cost $2,000 to produce and 800 backpacks cost $9,000 to produce, find the unit cost (slope) of producing the backpacks.

**15.** An airplane takes off from the ground (altitude is 0 ft and time is 0 min) and after 3 min it has an altitude of 6,000 ft. Find the rate of change in feet per minute.

Use Table 7–2 for Exercises 16–22.

**16.** Find the rate of change in tuition and fees at public 4-year colleges from 1993 to 2003.

**17.** Find the rate of change in tuition and fees at public 4-year colleges from 1998 to 2003.

**18.** Find the rate of change in tuition and fees at public 4-year colleges from 2002 to 2003.

**19.** Find the rate of change in tuition and fees at public 2-year colleges from 1993 to 2003.

**20.** Find the rate of change in tuition and fees at public 2-year colleges from 1998 to 2003.

**21.** The rate of change in tuition and fees for the 10-year period 1993–2003 is not the same as the rate of change from 1998 to 2003 for public 4-year colleges. Will the data graph to form a straight line? Explain.

**22.** Compare the rate of change in tuition and fees at public 4-year colleges with the rate of change in tuition and fees at public 2-year colleges for the 10-year period 1993–2003.

**Table 7–2  Average Tuition and Fees in Constant 2002 Dollars for U.S. Colleges**

| Year | Public Four-Year | Public Two-Year |
|------|-----------------|-----------------|
| 1992 | 2,746 | 1,526 |
| 1993 | 2,949 | 1,410 |
| 1994 | 3,122 | 1,534 |
| 1995 | 3,239 | 1,569 |
| 1996 | 3,277 | 1,550 |
| 1997 | 3,372 | 1,660 |
| 1998 | 3,464 | 1,745 |
| 1999 | 3,557 | 1,702 |
| 2000 | 3,581 | 1,756 |
| 2001 | 3,587 | 1,689 |
| 2002 | 3,765 | 1,625 |
| 2003 | 4,081 | 1,735 |

**2** Sketch the line passing through the pair of points and identify the line as having a positive slope, a negative slope, a zero slope, or an undefined slope.

**23.** $(-3, 3)$ and $(1, 5)$      **24.** $(4, -1)$ and $(1, 4)$      **25.** $(3, 1)$ and $(5, 7)$

**26.** $(-1, -1)$ and $(3, 3)$      **27.** $(4, 5)$ and $(-4, -1)$      **28.** $(7, 2)$ and $(-3, 2)$

**29.** $(4, -5)$ and $(0, 0)$      **30.** $(2, -2)$ and $(5, -5)$      **31.** $(-5, 2)$ and $(-5, -4)$

**32.** $(4, -4)$ and $(1, 5)$      **33.** $(3, 5)$ and $(-4, 5)$      **34.** $(-7, 3)$ and $(-7, 5)$

---

## 7-4 | Linear Equation of a Line

*Learning Outcomes*

**1** Find the equation of a line, given the slope and one point.

**2** Find the equation of a line, given two points on the line.

**3** Find the equation of a line, given the slope and *y*-intercept.

**4** Find the equation of a line, given a point on the line and an equation of a line parallel to that line.

**5** Find the equation of a line, given a point on the line and an equation of a line perpendicular to that line.

**1** **Find the Equation of a Line, Given the Slope and One Point.**

It is often desirable to know the equation of the line. The equation of the line can serve as a model for finding any point on the line and for solving various applications.

**To find the equation of a line if the slope and one point on the line are known:**

**1.** Use the following version of the *point-slope form of an equation* of a straight line.

$$y - y_1 = m(x - x_1)$$

where $x_1$ and $y_1$ are coordinates of the known point, $m$ is the slope of the line, and $x$ and $y$ are the variables of the equation.

**2.** Substitute known values for $x_1$, $y_1$ and $m$.

**3.** Rearrange the equation to be in standard form ($ax + by = c$) or slope-intercept form ($y = mx + b$).

**Standard form of an equation:**

$$ax + by = c$$

The characteristics of an equation in standard form are

- Both variable terms are on the left.
- The leading term has the $x$ variable and is positive.
- The equation contains no fractions.

---

**EXAMPLE** Find the equation of the line passing through the point $(3, -2)$ with a slope of $\frac{2}{3}$. Write the equation in slope-intercept form.

$$y - y_1 = m(x - x_1)$$  Substitute into the point-slope form. $m = \frac{2}{3}$; $x_1 = 3$; $y_1 = -2$.

$$y - (-2) = \frac{2}{3}(x - 3)$$  Simplify left side.

$$y + 2 = \frac{2}{3}(x - 3)$$  Distribute $\frac{2}{3} \cdot \frac{-\overset{-1}{\cancel{3}}}{1} = -2$.

$$y + 2 = \frac{2}{3}x - 2$$  Sort terms.

$$y = \frac{2}{3}x - 2 - 2$$  Combine like terms.

$$\mathbf{y = \frac{2}{3}x - 4}$$  Solved for $y$.

---

**EXAMPLE** Write $y = \frac{2}{3}x - 4$ in standard form.

$$y = \frac{2}{3}x - 4$$  Rearrange with variable terms on left with the $x$-variable as the first term.

$$-\frac{2}{3}x + y = -4$$  Clear fraction by multiplying by denominator 3.

$$3\left(-\frac{2}{3}x + y\right) = 3(-4)$$  Distribute.

$$-2x + 3y = -12$$  Multiply by $-1$ to make the leading term positive.

$$\mathbf{2x - 3y = 12}$$  Standard form.

---

**2**  **Find the Equation of a Line, Given Two Points on the Line.**

To find the equation of a line when two points on the line are known, we use both the slope and the point-slope formulas.

## To find the equation of a line if two points on the line are known:

1. Use the slope formula to find the slope, given two points.

$$m = \frac{y_2 - y_1}{x_2 - x_1}$$

2. Use the point-slope form of an equation of a straight line, the calculated slope from Step 1, and the coordinates of either one of the given points.

$$(y - y_1) = m(x - x_1)$$

3. Write the equation in slope-intercept or standard form.

---

**EXAMPLE**   Find the equation of the line that passes through the points $(0, 8)$ and $(5, 0)$. Write it in slope-intercept form.

$$\text{slope} = \frac{\Delta y}{\Delta x} = \frac{y_2 - y_1}{x_2 - x_1} = \frac{0 - 8}{5 - 0} = -\frac{8}{5}$$

Substitute known values into the slope formula. Let $P_1 = (0, 8)$ and $P_2 = (5, 0)$.

$$y - y_1 = m(x - x_1)$$

Substitute the slope and values for one point into the point-slope form. Use $P_1$.

$$y - 8 = -\frac{8}{5}(x - 0)$$

Simplify.

$$y - 8 = -\frac{8}{5}x$$

Solve for $y$.

$$y = -\frac{8}{5}x + 8$$

---

**EXAMPLE**   Find the equation of the line that passes through the points $(-3, 4)$ and $(7, 4)$.

We let $P_1 = (-3, \boxed{4})$ and $P_2 = (7, \boxed{4})$.

$$\text{slope} = \frac{\Delta y}{\Delta x} = \frac{y_2 - y_1}{x_2 - x_1} = \frac{4 - 4}{7 - (-3)} = \frac{0}{7 + 3} = \frac{0}{10} = 0$$

Substitute values.

Use the point-slope form of an equation and $P_1$.

$$y - y_1 = m(x - x_1)$$   Substitute values.

$$y - 4 = 0[x - (-3)]$$   Simplify.

$$y - 4 = 0$$   Solve for $y$.

$$y = 4$$

The situation in the preceding example is generalized in the equation of a horizontal line.

The equation of a line with a slope of zero (a horizontal line) is

$$y = k$$

where $k$ is the common $y$-coordinate for all points on the line.

There is one special situation in which the point-slope form of an equation *cannot* be used to determine the equation of a line, that is, when the slope is undefined.

The equation of a line with a slope that is undefined (a vertical line) is

$$x = k$$

where $k$ is the common $x$-coordinate for all points on the line.

**EXAMPLE**  Find the equation of the line that passes through the points (5, 3) and (5, 7).

Since the $x$-coordinate is the same in both points, the slope of the line is undefined and the equation of the line is $x = k$, where $k$ is the common $x$-coordinate. In this example, $k = 5$.

**The equation of the line through (5, 3) and (5, 7) is $x = 5$.**

### 3  Find the Equation of a Line, Given the Slope and $y$-Intercept.

So far we have written the equation of a line when we know either two points on the line or the slope and at least one point on the line. If the one point is the $y$-intercept $(0, b)$, then we use the slope-intercept form of an equation, $y = mx + b$, and we can write the equation by inspection.

**To find the equation of a line if the slope and $y$-intercept are known:**

**1.** Use the slope-intercept form of an equation of a straight line:

$$y = mx + b$$

where $m$ = slope and $b$ = the $y$-coordinate of the $y$-intercept.
**2.** Substitute known values for $m$ and $b$.
**3.** Write the equation in standard form or slope-intercept form.

**EXAMPLE**  Write the equation for a line with a slope of $-3$ and a $y$-intercept of 5.

Slope = $m = -3$;      $y$-intercept = $b = 5$

$y = mx + b$          Substitute values.

$\boldsymbol{y = -3x + 5}$          Slope-intercept form.

The necessary facts for writing the equation of a line are often obtained from the graph of the equation.

**EXAMPLE** Figure 7–28 shows the cost of producing picture frames.

(a) Write an equation that represents the graph.

(b) Using the equation, find the cost of producing 20 picture frames.

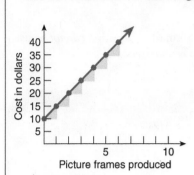

**Figure 7–28**

(a) The y-intercept represents the fixed cost of producing picture frames. $b = \$10$. The slope $m$ is 5 units vertical change for every 1 unit horizontal change, which is $\frac{5}{1} = 5$.

$$y = 5x + 10 \qquad \text{Equation of the line.}$$

This type of equation is a **cost function.**

(b) Use the equation of the line found in part (a) to find $y$ when $x = 20$.

| | |
|---|---|
| $y = 5x + 10$ | Substitute $x = 20$. |
| $y = 5(20) + 10$ | Evaluate. |
| $y = 100 + 10$ | |
| $y = \$110$ | Interpret. |

**The cost of producing 20 frames is $110.**

**4** **Find the Equation of a Line, Given a Point on the Line and an Equation of a Line Parallel to That Line.**

**Parallel lines** are two or more lines that are the same distance apart everywhere. They have no points in common. See Figure 7–29.

If we write $x + y = 5$ in slope-intercept form, we have $y = -x + 5$ and the slope is $-1$. *Any* equation that has a slope of $-1$ either is parallel to or coincides with the line formed by the equation $x + y = 5$. **Coincides** means one line lies on top of the other; that is, they are the same line.

While parallel lines have the same slopes they have different $x$- and $y$-intercepts.

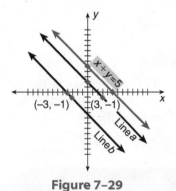

**Figure 7–29**

**Slope of parallel lines:**

The slopes of parallel lines are equal.

**EXAMPLE** Write the equations for lines $a$ and $b$ in Figure 7–29.

| | |
|---|---|
| $x + y = 5$ | Solve for $y$ to determine the slope of all lines parallel to this line. |
| $y = -x + 5$ | Slope $= -1$. |

The y-intercepts for lines $a$ and $b$ can be determined from the graph.

| | |
|---|---|
| **Line $a$:** $y = -x + 2$ | Substitute $m = -1$ and $b = 2$ |
| **Line $b$:** $y = -x - 4$ | Substitute $m = -1$ and $b = -4$ |

Chapter 7 / Graphing Linear Equations

**To write the equation of a line that is parallel to a given line when at least one point on the parallel line is known:**

1. Solve the equation of the given line for *y*.
2. Determine the slope *m* from the equation in Step 1.
3. The slope of the parallel line is the same as the slope of the given line.
4. Use the point-slope form of a straight line $y - y_1 = m(x - x_1)$ and substitute values for *m*, $x_1$, and $y_1$.
5. Rearrange the equation to be in standard form ($ax + by = c$) or slope-intercept form ($y = mx + b$).

**EXAMPLE**  Write the equation of a line that is parallel to $2y = 3x + 8$ and passes through the point (2, 1). Write the equation in slope-intercept form and in standard form.

$$2y = 3x + 8 \qquad \text{Solve for } y.$$

$$\frac{2y}{2} = \frac{3x + 8}{2}$$

$$y = \frac{3}{2}x + 4 \qquad \text{Identify the slope.}$$

The slope of $2y = 3x + 8$ is $\dfrac{3}{2}$.

$$y - y_1 = \boxed{m}\ (x - x_1) \qquad \begin{array}{l}\text{Substitute. The slope of a parallel line is equal to the slope of the given}\\ \text{line. } m = \frac{3}{2}; x_1 = 2; y_1 = 1.\end{array}$$

$$y - 1 = \boxed{\frac{3}{2}}\ (x - 2) \qquad \text{Distribute.}$$

$$y - 1 = \frac{3}{2}x - 3 \qquad \text{Solve for } y.$$

$$y = \frac{3}{2}x - 3 + 1 \qquad \text{Combine like terms.}$$

$$\mathbf{y = \frac{3}{2}x - 2} \qquad \text{Slope-intercept form.}$$

To write the equation in standard form, begin by clearing fractions.

$$2y = 2\left(\frac{3}{2}x - 2\right) \qquad \text{To clear fractions, multiply by 2.}$$

$$2y = 3x - 4 \qquad \text{Move variable term } 3x \text{ to the left side of the equation.}$$

$$-3x + 2y = -4 \qquad \text{Multiply each term by } -1 \text{ so that the coefficient of } x \text{ is positive.}$$

$$\mathbf{3x - 2y = 4} \qquad \text{Standard form.}$$

### 5  Find the Equation of a Line, Given a Point on the Line and an Equation of a Line Perpendicular to That Line.

**Perpendicular lines** are two lines that intersect to form right angles (90° angles). Another term for *perpendicular* is **normal**. In Figure 7–30, the perpendicular (or normal) line that passes through the given point is indicated by the symbol ⌐, which means that a right angle is formed by the lines.

7–4 Linear Equation of a Line          347

Any line can have several lines perpendicular to it, but only one line is perpendicular *and* passes through a *given* point. Figure 7–30 shows line *a*, which is perpendicular to the line $y = -2x + 7$ and passes through the point (6, 8). In some cases, the given point may lie on the given line.

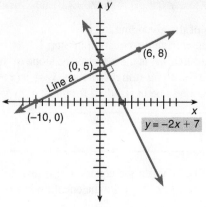

**Figure 7–30**

The slope of the equation $y = -2x + 7$ is $-2$. Two points on line *a* are the *x*-intercept $(-10, 0)$ and the *y*-intercept $(0, 5)$. The slope of line *a* is

$$m = \frac{5 - 0}{0 - (-10)} = \frac{5}{10} = \frac{1}{2} \qquad P_1 = (-10, 0) \text{ and } P_2 = (0, 5)$$

**EXAMPLE** Find the equation of line *a* in Figure 7–31. Write the equation in slope-intercept form.

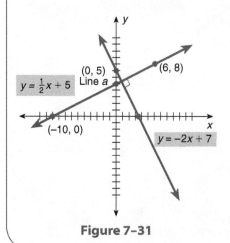

**Figure 7–31**

Line *a* passes through point (6, 8).

| | |
|---|---|
| $y - y_1 = m(x - x_1)$ | Substitute. $m = \dfrac{1}{2}$, $P_1 = (6, 8)$. |
| $y - 8 = \dfrac{1}{2}(x - 6)$ | Distribute. |
| $y - 8 = \dfrac{1}{2}x - 3$ | Solve for *y*. |
| $y = \dfrac{1}{2}x - 3 + 8$ | |
| $y = \dfrac{1}{2}x + 5$ | Slope-intercept form. |

To further explore the relationship of the slopes of perpendicular lines, examine the graphs and equations in Figure 7–32.

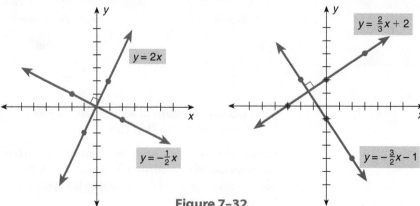

**Figure 7–32**

Chapter 7 / Graphing Linear Equations

The slope of *any* line that is perpendicular to a given line is the **negative reciprocal** of the slope of the given line.

**TIP!**

### Negative Reciprocals

The negative reciprocal of a number is not necessarily a negative value. It is the opposite of the given number. To find the negative reciprocal of a number:

1. Interchange the numerator and denominator to form the reciprocal.
2. Give the reciprocal the sign opposite the original number.

The negative reciprocal of $-5$ is $+\dfrac{1}{5}$. The negative reciprocal of $-\dfrac{3}{4}$ is $+\dfrac{4}{3}$.

The negative reciprocal of $\dfrac{4}{5}$ is $-\dfrac{5}{4}$. The negative reciprocal of $3$ is $-\dfrac{1}{3}$.

**To find the equation of a line that is perpendicular (normal) to a given line and passes through a given point:**

1. Determine the slope of the *given* line.
2. Find the *negative reciprocal* of this slope. The negative reciprocal is the slope of the perpendicular line.
3. Write the equation for the perpendicular line by substituting the coordinates of the *given* point for $x_1$ and $y_1$ and the slope of the *perpendicular* line for $m$ into the point-slope form of the equation $y - y_1 = m(x - x_1)$.
4. Write the equation in slope-intercept or standard form.

**TIP!**

### Words as Subscripts

In the next example, we clarify the notation with subscripts. The phrase *slope of a given line* is notated as "slope$_{\text{given}}$." The *slope of the perpendicular line* is notated as "slope$_{\text{perpendicular}}$."

---

**EXAMPLE**  Find the equation of the line perpendicular to $4x + y = -3$ and passing through $(0, -3)$.

$$4x + y = -3 \qquad \text{Given equation. Solve for } y.$$

$$y = -4x - 3 \qquad \text{Identify the slope.}$$

$$\text{slope}_{\text{given}} = -4$$

$$\text{slope}_{\text{perpendicular}} = +\frac{1}{4} \qquad \text{Negative reciprocal of } -4.$$

$$y - y_1 = m(x - x_1) \qquad \text{Point-slope form. Substitute } m = \tfrac{1}{4}, P_1 = (0, -3).$$

$$y - (-3) = \frac{1}{4}(x - 0) \qquad \text{Simplify.}$$

$$y + 3 = \frac{1}{4}x \qquad \text{Solve for } y.$$

$$y = \frac{1}{4}x - 3$$

Slope-intercept form. For standard form clear the fraction and rearrange.

or

$$4y = x - 12$$

Move the $x$-variable term to the left and make the leading coefficient positive.

$$x - 4y = 12$$

Standard form.

---

**EXAMPLE**  Which of the equations represents a line that is perpendicular to $2x - 3y = 1$ passing through $(2, -1)$?

(a) $y = \frac{2}{3}x - \frac{1}{3}$    (b) $y = \frac{2}{3}x - \frac{7}{3}$

(c) $y = -\frac{3}{2}x - \frac{1}{3}$    (d) $y = -\frac{3}{2}x + 2$

$$2x - 3y = 1$$    Rewrite the given equation in slope-intercept form. Solve for $y$.
$$-3y = -2x + 1$$

$$\frac{-3y}{-3} = \frac{-2x}{-3} + \frac{1}{-3}$$

$$y = \frac{2}{3}x - \frac{1}{3}$$    Slope-intercept form.

$$\text{slope}_{\text{given}} = \frac{2}{3}$$    From $y = \frac{2}{3}x - \frac{1}{3}$.

$$\text{slope}_{\text{perpendicular}} = -\frac{3}{2}$$    Negative reciprocal.

The choice is now limited to (c) or (d) since (a) and (b) have slopes of $\frac{2}{3}$.

$$y - y_1 = m(x - x_1)$$    Substitute $m = -\frac{3}{2}$, $x_1 = 2$, $y_1 = -1$.

$$y - (-1) = -\frac{3}{2}(x - 2)$$    Distribute and simplify.

$$y + 1 = -\frac{3}{2}x + 3$$    Solve for $y$.

$$y = -\frac{3}{2}x + 3 - 1$$

$$y = -\frac{3}{2}x + 2$$

**The correct equation is $y = -\frac{3}{2}x + 2$, or choice (d).**

**1** Find the equation of a line passing through the given point with the given slope. Solve the equation for $y$ when necessary.

**1.** $(-8, 3), m = \dfrac{2}{3}$      **2.** $(4, 1), m = -\dfrac{1}{2}$      **3.** $(-3, -5), m = 2$      **4.** $(0, -1), m = 1$

**2** Find the equation of a line passing through the given pairs of points. Solve the equation for $y$.

**5.** $(4, 6)$ and $(7, 1)$      **6.** $(-1, 6)$ and $(-1, 4)$      **7.** $(-1, -3)$ and $(3, -3)$
**8.** $(-4, 4)$ and $(-4, -2)$      **9.** $(-2, -4)$ and $(5, 10)$      **10.** $(-4, 0)$ and $(6, 0)$

**11.** If 50 DVDs cost $2,185 to produce and 2,000 DVDs cost $11,935 to produce, write a cost function for producing DVDs.

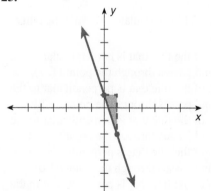

**12.** The first 5,000 leather bags can be produced at a cost of $160,000, and 50,000 bags can be produced at a cost of $1,150,000. Write a cost function for producing these bags.

**3** Write the equation of the line with the given slope and $y$-intercept.

**13.** $m = \dfrac{1}{4}, b = 7$      **14.** $m = -8, b = -4$      **15.** $m = -2, b = 3$      **16.** $m = \dfrac{3}{5}, b = -2$

**17.** $m = 1, b = 0$      **18.** $m = 5, b = -\dfrac{1}{5}$      **19.** $m = 2, b = -2$      **20.** $m = -\dfrac{3}{4}, b = 0$

**21.** A local business rents computer time for a $3 setup charge and $0.20 for every minute the computer is used. Write an equation to represent the cost $y$ of using the computer for $x$ minutes. How much does 45 min of computer time cost?

**22.** The cost of producing fine china plates is $8 per plate plus a one-time equipment charge of $12,000. Write an equation that represents the total cost $y$ of producing $x$ plates. What is the total cost of producing 8,000 plates?

Write the equations of the lines that are graphed in Figures 7–33 through 7–35.

**23.**

**24.**

**25.**

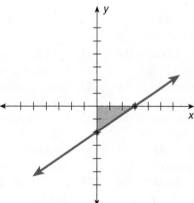

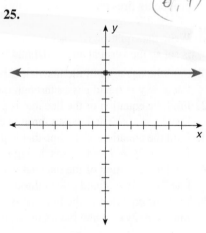

Figure 7–33        Figure 7–34        Figure 7–35

Solve Exercises 26 and 27 using the information given in Figure 7–36, which shows the cost of producing widgets in dollars.

**26.** Write an equation that represents the cost of producing widgets shown in the graph.

**27.** Use the equation to find the cost of producing 10 widgets.

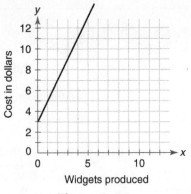

**Figure 7–36**

---

**4** Write the equations in standard form. Verify your answers using a graphing calculator.

**28.** Find the equation of the line that is parallel to the line $x + y = 6$ and passes through the point $(2, -3)$.

**29.** Find the equation of the line that is parallel to the line $2x + y = 5$ and passes through the point $(1, 7)$.

**30.** Find the equation of the line that is parallel to the line $3y = x - 2$ and passes through the point $(4, 0)$.

**31.** Find the equation of the line that is parallel to the line $3x - y = -2$ and passes through the point $(-3, -2)$.

**32.** Find the equation of the line that is parallel to the line $x + 3y = 7$ and passes through the point $(4, 1)$.

**33.** Find the equation of the line that is parallel to the line $3x - y = 4$ and passes through the point $(0, 3)$.

**34.** Find the equation of the line that is parallel to the line $2x + 3y = 5$ and passes through the point $(1, 1)$.

**35.** Find the equation of the line that is parallel to the line $3x + 2y = 1$ and passes through the point $(2, 0)$.

**36.** Find the equation of the line that is parallel to the line $2x - 5y = 0$ and passes through the point $(3, -1)$.

**37.** Find the equation of the line that is parallel to the line $-3x + 4y = -1$ and passes through the point $(\frac{1}{2}, 0)$.

**38.** The cost of producing tires on an assembly line includes a fixed cost of \$45,000 and a variable cost of \$12 per tire. This can be written as a cost function $y = 12x + 45,000$, where $x$ is the number of tires produced and $y$ is the total cost of $x$ tires. Research shows that 10,000 tires can be produced on a newly installed assembly line at a total cost of \$140,000 with the same variable cost per tire as the original assembly line. Write an equation that represents the total cost of producing tires on the new assembly line.

**39.** A car rental agency uses the function $y = 0.5x + 25$, where $x$ is the number of miles driven and $y$ is the total cost of car rental. Another car rental agency has the same variable cost of \$0.50 per mile and quotes the total cost of driving 500 mi as \$295. Write an equation to represent the second agency's rental charges.

---

**5** Write the equation for each exercise in standard form. Verify your results using a graphing calculator. Be sure the range is set so the vertical and horizontal increments are equal.

**40.** Find the equation of the line that is perpendicular to the line $x + y = 6$ and passes through the point $(2, 3)$.

**41.** Find the equation of the line that is perpendicular to the line $2x + y = 5$ and passes through the point $(1, 7)$.

**42.** Find the equation of the line that is perpendicular to the line $3y = x - 2$ and passes through the point $(4, 0)$.

**43.** Find the equation of the line that is perpendicular to the line $3x - y = 2$ and passes through the point $(-3, -2)$.

**44.** Find the equation of the line that is perpendicular to the line $x + 3y = 7$ and passes through the point $(4, 1)$.

**45.** Find the equation of the line that is perpendicular to the line $x + 2y = 7$ and passes through the point $(-2, 3)$.

**46.** Find the equation of the line that is perpendicular to the line $2x + 3y = 4$ and passes through the point $(3, -1)$.

**47.** Find the equation of the line that is perpendicular to the line $4x + y = 1$ and passes through the point $(0, 0)$.

**48.** Find the equation of the line that is perpendicular to the line $2x + 2y = 3$ and passes through the point $(\frac{1}{2}, 2)$.

**49.** Find the equation of the line that is perpendicular to the line $5x - y = 6$ and passes through the point $(5, -\frac{1}{5})$.

**Learning Outcomes**　　　**What to Remember with Examples**

## Section 7–1

**1** Locate points on a rectangular coordinate system (pp. 320–323).

To plot a point on the rectangular coordinate system: **1.** Start at the origin. **2.** Count to the left or right (horizontally) the number of units of the first signed number. **3.** Start at the ending point found in Step 2, and count up or down (vertically) the number of units of the second signed number.

Draw a coordinate system and locate the following points: $A\,(-3, -2)$, $B\,(4, -1)$, $C\,(0, -3)$, $D\,(2, 3)$. See Figure 7–37.

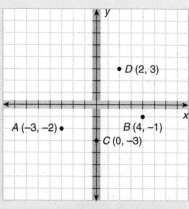

**Figure 7–37**

**2** Graph linear equations using a table of solutions (pp. 323–326).

To prepare a table of solutions: **1.** Select any appropriate value from the domain of the independent variable, **2.** Substitute the selected value for the independent variable and solve the resulting equation, **3.** Repeat Step 1 and Step 2 until the desired number of solutions are obtained.

Make a table of solutions for the equation $y = -x + 3$. The domain is all real numbers.

When $x = -2$　　　　　When $x = 0$　　　　　When $x = 2$
$y = -(-2) + 3$　　　　$y = -0 + 3$　　　　　$y = -2 + 3$
$y = 2 + 3$　　　　　　$\mathbf{y = 3}$　　　　　　$\mathbf{y = 1}$
$\mathbf{y = 5}$

Each solution is an ordered pair that can be written in point notation: $(-2, 5)$, $(0, 3)$, and $(2, 1)$. The range is all real numbers.

...ng a table of solutions: **1.** Prepare a table of solutions. **2.** Plot the points ...dinate system. **3.** Connect the points with a straight line.

Prepare a table of values and graph $y = -x + 3$.

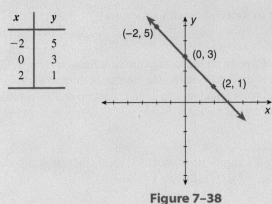

| $x$ | $y$ |
|-----|-----|
| $-2$ | 5 |
| 0 | 3 |
| 2 | 1 |

**Figure 7–38**

To check the solution of an equation in two variables: **1.** Substitute the value for $x$ in each place it occurs and substitute the value for $y$ in each place it occurs. **2.** Simplify each side of the equation using the order of operations. **3.** The solution makes a true statement; that is, both sides of the equation will be equal to the same number.

A solution for the equation $3x - y = 1$ is (1, 2). Check to verify the solution.

$$3x - y = 1 \qquad \text{Substitute values.}$$
$$3(1) - 2 = 1 \qquad \text{Evaluate.}$$
$$3 - 2 = 1$$
$$1 = 1 \qquad \text{True. Solution checks.}$$

To evaluate an equation when the value of one variable is given: **1.** Substitute the given value of the variable in each place that variable occurs in the equation. **2.** Solve the equation for the other variable.

Evaluate the equation $4x - y = 6$ for $x$, if $y = 14$.

$$4x - y = 6 \qquad \text{Substitute 14 for } y.$$
$$4x - 14 = 6 \qquad \text{Solve for } x.$$
$$4x = 20$$
$$x = 5$$

Solution: (5, 14)

**Section 7–2**

1 Graph linear equations using intercepts (pp. 328–330).

**1.** Find the $x$-intercept by letting $y = 0$ and solving for $x$. **2.** Find the $y$-intercept by letting $x = 0$ and solving for $y$. **3.** Plot the two points and graph the equation. **4.** Find a third point on the graph as a check point.

Graph the equation $y = 2x - 1$ by using the intercepts method.

$x$-intercept: $0 = 2x - 1$    Substitute $y = 0$    $y$-intercept: $y = 2(0) - 1$    Substitute $x = 0$.
$\qquad\qquad\quad 1 = 2x$    Solve for $x$.                 $y = 0 - 1$    Solve for $y$.
$\qquad\qquad\quad \dfrac{1}{2} = x$                                  $y = -1$

$\qquad\qquad\qquad\qquad\qquad\qquad\qquad\qquad\qquad\qquad\qquad (0, -1)$

$\left( \dfrac{1}{2}, 0 \right)$

Chapter 7 / Graphing Linear Equations

Plot the two points: then draw the graph.

Check point: For $x = 3$
$y = 2x - 1$
$y = 2(3) - 1$
$y = 6 - 1$
$y = 5$
$(3, 5)$

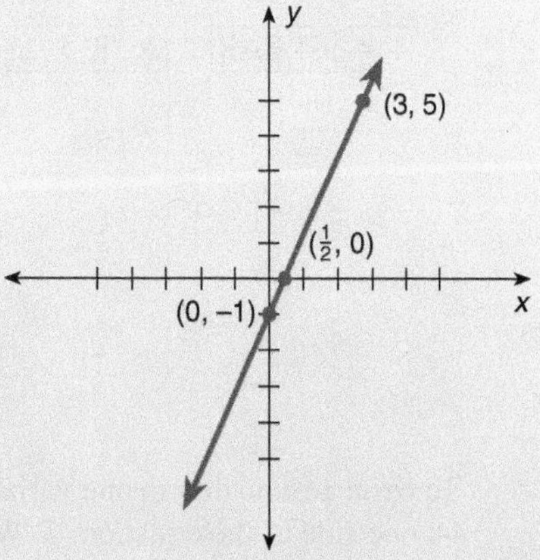

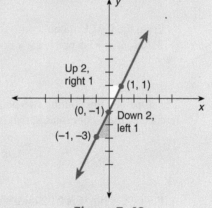

**Figure 7–39**

**2** Graph linear equations using the slope and $y$-intercept (pp. 330–334).

**1.** Write the equation in the form $y = mx + b$. **2.** The slope is $m$. In fraction form the numerator of $m$ is the vertical movement and the denominator of $m$ is the horizontal movement. **3.** The $y$-coordinate of the $y$-intercept is $b$.

To graph using the slope-intercept method: **1.** Locate the $y$-intercept or $b$ by counting vertically along the $y$-axis. **2.** From this point, count the slope (vertical, then horizontal) and locate the second point. **3.** Draw the line through the two points.

Use the slope-intercept method to graph $y = 2x - 1$. The $y$-intercept is $-1$, so count down 1 from the origin. The coordinates of this point are $(0, -1)$. From this point, move $+2$ vertically and $+1$ horizontally. The coordinates of the second point are $(1, 1)$. Connect the two points.

$y$-intercept $= (0, -1)$

$$m = 2 = \frac{2}{1} \text{ or } \frac{-2}{-1}$$

**Figure 7–40**

**3** Graph linear equations using a graphing calculator (pp. 334–335).

To graph a linear equation using a graphing caculator: **1.** Solve the equation for $y$. **2.** Press the function key $\boxed{Y=}$ ; then enter the right side of the equation. **3.** Press the graph key to show the graph on the screen. Adjust the view with the Window or Zoom key if appropriate.

Graph the equation $y = 2x - 1$.
Using a TI-84 Plus Silver Edition™:

$\boxed{Y=}$ $\boxed{\text{CLEAR}}$ 2 $\boxed{\text{X, T, }\theta\text{, }n}$ $\boxed{-}$ 1
$\boxed{\text{GRAPH}}$

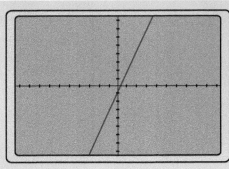

**Figure 7–41**

---

**4** Solve a linear equation using a graph (pp. 335–336).

To write an equation in one variable as a function: **1.** Rewrite the equation so that all terms are on one side of the equation. **2.** Write the equation in function notation using $y$ = the nonzero side of the equation. **3.** Evaluate the function for at least two values and graph. The solution is the $x$-coordinate of the point where the line crosses the $x$-axis.

---

Write $2x + 3 = 6$ as a function, graph the function, and find the solution from the graph.

$$2x + 3 = 6$$
$$2x + 3 - 6 = 0$$
$$2x - 3 = 0$$

$$y = 2x - 3 \qquad \text{At } x = 0$$
$$y = 2(0) - 3$$
$$y = -3$$

$$y = 2(1) - 3 \qquad \text{At } x = 1$$
$$y = -1$$

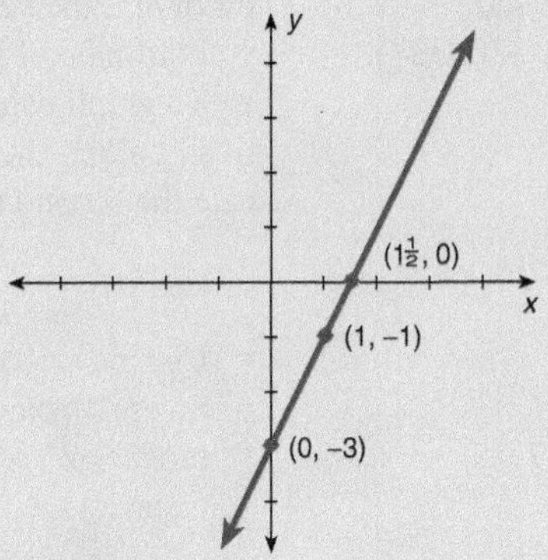

Graph points: $(1\frac{1}{2}, 0)$, $(1, -1)$, $(0, -3)$

**Figure 7–42**

From the CALC menu, chose the zero option.
Solution: $x = 1.5$ or $1\frac{1}{2}$, the $x$-coordinate of the point where the line crosses the $x$-axis.

---

## Section 7-3

**1** Calculate the slope of a line, given two points on the line (pp. 337–339).

The slope of a line joining two points is the difference of the $y$-coordinates divided by the difference of the $x$-coordinates, that is, $\frac{\text{rise}}{\text{run}}$.

$$m = \frac{\text{rise}}{\text{run}} = \frac{\Delta y}{\Delta x} = \frac{y_2 - y_1}{x_2 - x_1}$$

---

Find the slope of the line passing through the points $(3, 1)$ and $(-2, 5)$.

$$\frac{y_2 - y_1}{x_2 - x_1} \qquad \frac{5 - 1}{-2 - 3} = \frac{4}{-5} \quad \text{or} \quad -\frac{4}{5} \qquad p_1 = (3, 1), p_2 = (-2, 5)$$

---

**2** Determine the slope of a horizontal or vertical line (pp. 339–341).

The slope of a horizontal line is zero, and all points on the same horizontal line have the same $y$-coordinate. The slope of a vertical line is not defined, and all points on the same vertical line have the same $x$-coordinate.

---

Examine the coordinate pairs and indicate which points lie on the same horizontal line and which lie on the same vertical line: $A$ $(4, -3)$, $B$ $(2, 5)$, $C$ $(4, 5)$, $D$ $(2, -3)$.

Points $A$ and $D$ and points $B$ and $C$ lie on the same horizontal lines, respectively. Points $A$ and $C$ and points $B$ and $D$ lie on the same vertical lines, respectively.

---

Chapter 7 / Graphing Linear Equations

## Section 7-4

**1** Find the equation of a line, given the slope and one point (pp. 342–343).

Use the point-slope form of an equation to find the equation when given the slope and coordinates of a point on a line.

$$y - y_1 = m(x - x_1)$$

A line has slope 2 and passes through the point $(3, 4)$. Find the equation of the line.

| | |
|---|---|
| $y - y_1 = m(x - x_1)$ | Substitute values. |
| $y - 4 = 2(x - 3)$ | Distribute. |
| $y - 4 = 2x - 6$ | Solve for $y$. $-6 + 4 = -2$ |
| $y = 2x - 2$ | |

**2** Find the equation of a line, given two points on the line (pp. 343–345).

Find the slope, then use the point-slope form of the equation to find the equation when given coordinates of two points on a line.

Find the equation of a line that passes through the two points $(3, 1)$ and $(-3, 2)$. First, find the slope.

| | |
|---|---|
| $m = \dfrac{2 - 1}{-3 - 3} = \dfrac{1}{-6} \text{ or } -\dfrac{1}{6}$ | Find the slope. |
| $y - y_1 = m(x - x_1)$ | Substitute $m = -\dfrac{1}{6}$, $x_1 = 3$, and $y_1 = 1$. |
| $y - 1 = -\dfrac{1}{6}(x - 3)$ | Distribute. |
| $y - 1 = -\dfrac{1}{6}x + \dfrac{1}{2}$ | Solve for $y$. $\dfrac{1}{2} + 1 = \dfrac{1}{2} + \dfrac{2}{2} = \dfrac{3}{2}$ |
| $y = -\dfrac{1}{6}x + \dfrac{3}{2}$ | Slope-intercept form. |

or

| | |
|---|---|
| $6y = -1x + 9$ | Clear fractions and rearrange. |
| $x + 6y = 9$ | Standard form. |

**3** Find the equation of a line, given the slope and $y$-intercept (pp. 345–346).

Use the slope-intercept form of the equation, $y = mx + b$, and substitute values for $m$ and $b$, the slope and $y$-intercept, respectively, to find the equation of a line.

Write the equation of a line that has slope $\frac{2}{3}$ and the $y$-intercept $(0, -1)$. Use the slope-intercept form of the equation:

| | |
|---|---|
| $y = mx + b$ | Substitute values, $m = \dfrac{2}{3}$, $b = -1$ |
| $y = \dfrac{2}{3}x - 1$ | |

**4** Find the equation of a line given a point on the line and an equation of a line parallel to that line (pp. 346–347).

The slopes of parallel lines are equal; that is, lines that have the same slope are parallel lines. Equations that have equal slopes have graphs that are parallel lines.

Determine which two of the three given equations have graphs that are parallel lines:

(a) $y = 3x - 5$  (b) $3x - 2y = 10$  (c) $6x - 2y = 8$

Write each of the three equations in slope-intercept form and compare the slopes (coefficients of $x$).

(a) $y = 3x - 5$  (b) $y = \dfrac{3}{2}x - 5$  (c) $y = 3x - 4$

(a) and (c) have the same slope and their graphs are parallel.

To find the equation of a line parallel to a given line and passing through a given point, use the point-slope form of the linear equation. Substitute the value of the slope of the given equation and the coordinates of the given point into the point-slope form of an equation.

Find the equation of a line that passes through the point (2, 3) and is parallel to the line that has equation $y = 4x - 1$. Write the new equation in slope-intercept form.
Use the slope of the given equation, 4, for the slope of the new equation.

| | |
|---|---|
| $y - y_1 = m(x - x_1)$ | Substitute $m = 4$, $x_1 = 2$, and $y_1 = 3$. |
| $y - 3 = 4(x - 2)$ | Distribute. |
| $y - 3 = 4x - 8$ | Solve for $y$. $-8 + 3 = -5$ |
| $y = 4x - 5$ | Slope-intercept form. |

**5** Find the equation of a line, given a point on the line and an equation of a line perpendicular to that line (pp. 347–350).

The slopes of perpendicular lines are negative reciprocals. Equations with slopes that are negative reciprocals have graphs that are perpendicular lines.

Determine which two of the three given equations have graphs that are perpendicular lines:

(a) $y = 3x - 5$    (b) $3x + 9y = 10$    (c) $9x + 3y = 12$

Write each of the three equations in slope-intercept form and compare the slopes (coefficients of $x$).

(a) $y = 3x - 5$    (b) $y = -\dfrac{1}{3}x + \dfrac{10}{9}$    (c) $y = -3x + 4$

(a) and (b) have slopes that are negative reciprocals and their graphs are perpendicular lines. Note that (a) and (c) are not perpendicular. Their slopes are opposites but *not* reciprocals. The slopes of (b) and (c) are reciprocals but *not* opposites.

To write the equation of a line perpendicular to a given line, substitute the negative reciprocal of the slope of the given equation and the coordinates of the given point into the point-slope form of an equation.

Find the equation of a line that passes through the point $(-1, 2)$ and is perpendicular to the line represented by the equation $y = \frac{1}{3}x - 5$. Write the equation in standard form.

The slope for the new equation is the negative reciprocal of $\frac{1}{3}$, which is $-3$.

| | |
|---|---|
| $y - y_1 = m(x - x_1)$ | Substitute $m = -3$, $x_1 = -1$, and $y_1 = 2$. |
| $y - 2 = -3[x - (-1)]$ | Simplify and distribute. |
| $y - 2 = -3x - 3$ | Rearrange with variables on left. |
| $3x + y = -3 + 2$ | Combine like terms. |
| $3x + y = -1$ | Standard form. |

## Section 7–1

Draw a rectangular coordinate system and locate the points.

**1.** $A = (5, -2)$       **2.** $B = (-8, -3)$       **3.** $C = (0, -4)$
**4.** $D = (3, 7)$       **5.** $E = (-3, 2)$       **6.** $F = (-3, 0)$

**7.** What are the coordinates of the origin?
**8.** Which of the four quarters of the coordinate system is used to plot points with coordinates that are both negative?
**9.** Write the coordinates for points $A$ through $E$ on the graph in Figure 7–43.
**10.** Write the coordinates of a point that lies on the $y$-axis and is four units below the $x$-axis.

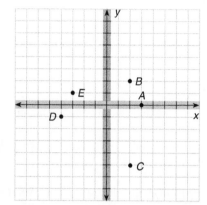

**Figure 7–43**

Represent the solutions of the equations in a table of solutions and on a graph.

**11.** $y = 2x - 3$      **12.** $y = -4x + 1$      **13.** $y = 3x$
**14.** $y = 4x$      **15.** $y = -3x$      **16.** $y = -4x$
**17.** $y = 2x + 1$      **18.** $y = 2x + 5$      **19.** $y = 4x - 2$
**20.** $y = -5x + 1$      **21.** $y = \frac{1}{2}x - 2$      **22.** $y = -\frac{1}{4}x + 1$

Which of the coordinate pairs are solutions for the equation $2x - 3y = 12$?

**23.** $(-2, -3)$      **24.** $(1, -3)$      **25.** $(3, -2)$
**26.** $(0, 4)$      **27.** $(6, 0)$      **28.** $(3, 4)$

**29.** In the equation $y = 3x - 1$, find $y$ when $x = -2$.
**30.** In the equation $2x + y = 8$, find $x$ when $y = 10$.
**31.** Find $y$ in the equation $x - 3y = 5$ when $x = 8$.
**32.** Find $y$ if $x = 7$ in the equation $3x - 4y = -2$.

## Section 7–2

Graph the equations using the intercepts procedure.

**33.** $x = -4y - 1$      **34.** $x + y = -4$      **35.** $3x - y = 1$
**36.** $x = -4y$      **37.** $\frac{1}{2}x + \frac{1}{3}y = 1$      **38.** $2x + 3y = 6$

Graph using the slope and $y$-intercept procedure.

**39.** $y = 5x - 2$      **40.** $y = -x$      **41.** $y = -3x - 1$
**42.** $y = \frac{1}{2}x + 3$      **43.** $x - y = 4$      **44.** $2y + 4 = -3$
**45.** $x - 2y = -1$      **46.** $x + y = 5$      **47.** $y - 2x = -2$
**48.** $y + 3x = 4$

Graph the equations using a graphing calculator.

**49.** $y = 0.5x - 3$      **50.** $y = 2.1x + 0.5$      **51.** $y = 20x - 15$      **52.** $y = -30x + 10$

**53.** The function $y = 5x + 8,000$ is used to express the cost of manufacturing a widget, where $8,000 is the fixed cost of constructing molds and $5 is the cost of materials to make each widget. What is the cost of manufacturing 10,000 widgets?

**54.** A salesperson earns a salary of $200 weekly plus $12.50 commission for each photo sitting. Weekly income is expressed by the function $y = 12.50x + 200$, where $x$ represents the number of sittings in a week. Find the weekly income if 28 sittings are sold.

Write each equation as a function and find the solution graphically.

**55.** $x + 2 = 8$      **56.** $3x - 7 = 1$      **57.** $2x + 1 = 5x + 7$
**58.** $-14 = 2(x - 7)$      **59.** $5(x + 2) = 3(x + 4)$      **60.** $2x + 1 = 7x + 2 - 8$

Determine the slope and $y$-intercept of the given equations by inspection.

**61.** $y = 3x + \dfrac{1}{4}$  **62.** $y = \dfrac{2}{3}x - \dfrac{3}{5}$  **63.** $y = -5x + 4$  **64.** $y = 7$

**65.** $x = 8$  **66.** $y = \dfrac{1}{3}x - \dfrac{5}{8}$  **67.** $y = \dfrac{x}{8} - 5$  **68.** $y = -\dfrac{x}{5} + 2$

Write the given equations in slope-intercept form and determine the slope and $y$-intercept.

**69.** $2x + y = 8$  **70.** $4x + y = 5$  **71.** $3x - 2y = 6$  **72.** $5x - 3y = 15$

**73.** $\dfrac{3}{5}x - y = 4$  **74.** $2.2y - 6.6x = 4.4$  **75.** $3y = 5$  **76.** $3x - 6y = 12$

## Section 7–3

Find the slope of the line passing through the given pairs of points.

**77.** $(-2, 2)$ and $(1, 3)$  **78.** $(3, -1)$ and $(1, 3)$  **79.** $(3, 2)$ and $(5, 6)$
**80.** $(-1, -1)$ and $(2, 2)$  **81.** $(4, 3)$ and $(-4, -2)$  **82.** $(6, 2)$ and $(-3, 2)$
**83.** $(3, -4)$ and $(0, 0)$  **84.** $(1, -1)$ and $(5, -5)$  **85.** $(-4, 1)$ and $(-4, 3)$
**86.** $(4, -4)$ and $(1, 3)$  **87.** $(5, 0)$ and $(-2, 4)$  **88.** $(-2, 1)$ and $(0, 3)$
**89.** $(-4, -8)$ and $(-2, -1)$  **90.** $(3, 3)$ and $(3, 0)$  **91.** $(5, -3)$ and $(-1, -3)$
**92.** $(-5, -1)$ and $(-7, -3)$  **93.** $(-7, 0)$ and $(-7, 5)$  **94.** $(3, 5)$ and $(2, 5)$

**95.** Write the coordinates of two points that lie on the same horizontal line.

**96.** Write the coordinates of two points that lie on the same vertical line.

Use Table 7–2 on page 342 for Exercises 97–102.

**97.** What is the rate of change in tuition and fees at public 2-year colleges from 1998 to 1999?

**98.** What is the rate of change in tuition and fees at public 2-year colleges from 2002 to 2003?

**99.** What is the rate of change in tuition and fees at public 2-year colleges from 1993 to 1994?

**100.** What is the rate of change in tuition and fees at public 2-year colleges from 1992 to 2003?

**101.** Explain why the rate of change found in Exercises 98, 99, and 100 are different.

**102.** If the data in Table 7–2 formed a straight line when graphed, what would you expect to find for the rate of change for the data in Exercises 98, 99, and 100?

## Section 7–4

Find the equation of a line passing through the given point with the given slope. Solve the equation for $y$ if necessary.

**103.** $(-6, 2)$, $m = \dfrac{1}{3}$  **104.** $(3, 2)$, $m = -\dfrac{2}{5}$  **105.** $(4, 0)$, $m = \dfrac{3}{4}$

**106.** $(0, -2)$, $m = 2$  **107.** $(2, 3)$, $m = 4$  **108.** $(6, 0)$, $m = -1$

Find the equation of a line passing through the given pairs of points. Solve the equation for $y$ if necessary.

**109.** $(-5, 2)$ and $(6, 1)$  **110.** $(1, 4)$ and $(-1, 3)$  **111.** $(-1, -3)$ and $(3, 4)$
**112.** $(-3, 0)$ and $(4, 0)$  **113.** $(-2, -3)$ and $(3, 6)$  **114.** $(2, -4)$ and $(3, -4)$
**115.** $(5, 2)$ and $(6, 3)$  **116.** $(4, 6)$ and $(1, -1)$  **117.** $(-1, -2)$ and $(-3, -4)$
**118.** $(4, 0)$ and $(4, -3)$  **119.** $(5, -2)$ and $(3, -2)$  **120.** $(5, 4)$ and $(0, 4)$

**121.** A salesperson sells 80 items and earns $3,800 in one month, and in another month 120 items are sold resulting in $4,200 earnings. If we assume this is a linear function, write a function that expresses the salesperson's monthly salary plus commission, $S$, as a function of the number of items sold, $x$. Salary is $y$-intercept.

**122.** Simple interest is a linear function. Write an equation to express interest earned if $500 is earned in 2 years and $2,000 is earned in 8 years.

Write the equations using the given slope and $y$-intercept.

**123.** $m = 3$, $b = -2$

**124.** Slope $= \dfrac{3}{5}$; $y$-intercept $= -7$

Write the equations using information from the graphs in Figures 7–44 and 7–45.

**125.**

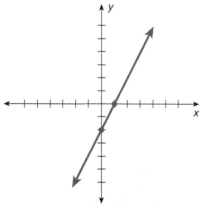

**Figure 7–44**

**126.**

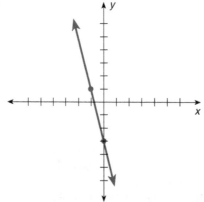

**Figure 7–45**

Write equations in standard form. Graph both equations for each exercise on a graphing calculator to verify parallelism.

**127.** Find the equation of the line that is parallel to the line $x + y = 4$ and passes through the point $(2, 5)$.

**128.** Find the equation of the line that is parallel to the line $3x + y = 6$ and passes through the point $(1, 0)$.

**129.** Find the equation of the line that is parallel to the line $2y = x - 3$ and passes through the point $(2, -3)$.

**130.** Find the equation of the line that is parallel to the line $4x - y = -1$ and passes through the point $(0, -2)$.

**131.** Find the equation of the line that is parallel to the line $x - 3y = 5$ and passes through the point $(5, -5)$.

**132.** Find the equation of the line that is parallel to the line $3x - 2y = 2$ and passes through the point $(0, -3)$.

**133.** Find the equation of the line that is parallel to the line $x + 3y = 6$ and passes through the point $(-4, -2)$.

**134.** Find the equation of the line that is parallel to the line $4x + 3y = 1$ and passes through the point $(3, -\frac{1}{2})$.

**135.** Find the equation of the line that is parallel to the line $3x - 4y = 0$ and passes through the point $(\frac{1}{3}, 2)$.

**136.** Find the equation of the line that is parallel to the line $-2x + 3y = 2$ and passes through the point $(-1, -1)$.

Write the equations in standard form. Use a graphing calculator to verify that the two lines in each exercise are perpendicular.

**137.** Find the equation of the line that is perpendicular to the line $x + y = 4$ and passes through the point $(-3, 1)$.

**138.** Find the equation of the line that is normal to the line $3x + y = 6$ and passes through the point $(1, 4)$.

**139.** Find the equation of the line that is perpendicular to the line $x + 2y = 5$ and passes through the point $(-2, 0)$.

**140.** Find the equation of the line that is normal to the line $2x + 2y = 4$ and passes through the point $(0, 0)$.

**141.** Find the equation of the line that is normal to the line $5x + y = 8$ and passes through the point $(-1, 2)$.

**142.** Find the equation of the line that is perpendicular to the line $3y = x - 4$ and passes through the point $(2, 3)$.

**143.** Find the equation of the line that is perpendicular to the line $5x - y = 10$ and passes through the point $(\frac{1}{2}, 3)$.

**144.** Find the equation of the line that is normal to the line $x - 3y = 6$ and passes through the point $(-2, 4)$.

**145.** Find the equation of the line that is perpendicular to the line $4x - y = 8$ and passes through the point $(4, -\frac{1}{2})$.

**146.** Find the equation of the line that is perpendicular to the line $4y = 2x + 1$ and passes through the point $(3, -1)$.

1. A 1-gal can of indoor house paint is advertised to cover 400 ft$^2$ of wall surface.
   (a) Make a table to show the amount of wall surface area that can be covered by 1, 2, 3, . . . , 10 gal of paint.
   (b) Graph these data and extend the line beyond the data to 15 gal of paint.
   (c) Use the graph to decide how many 1-gal cans of paint it would take to cover 5,500 ft$^2$ of wall surface.
   (d) The paint being used for this job can be purchased for $19.05 a gallon. Find the cost of the paint for the 5,500 ft$^2$ of wall surface.
   (e) The sales tax rate is 8.25%. Calculate the total cost of the paint.

2. Linda Kodama is introducing a new lipstick in Bright-glow's product line. As product development manager, she estimates the cost of the new lipstick to be $4.53 per item, plus an additional cost of $5,000 for product development.
   (a) Make a table of the estimated cost for producing 0 lipsticks, 100 lipsticks, 1,000 lipsticks, and 2,000 lipsticks.
   (b) Represent these costs as ordered pairs (number of lipsticks, cost).
   (c) Plot the ordered pairs and graph a line that fits the ordered pairs.
   (d) Write an equation that represents the cost of the new lipstick as a function of the number of lipsticks produced.
   (e) Linda projects the selling price of the new lipstick to be $8.99. How many lipsticks must the company sell to recover the cost of producing the new items?

## PRACTICE TEST

Make a table of values to represent the solutions to the following equations and show the solutions on a graph.

1. $y = \frac{1}{2}x$

2. $y = \frac{1}{2}x + 1$

3. $y = 2x - 4$

4. $y = 5 - x$

Write as functions and find the solution graphically.

5. $3x + 2 = 5$

6. $2(x + 1) = 3x$

Write the specific solution in ordered-pair form for each equation.

7. $x + y = 7$, if $x = 2$

8. $2x - y = 1$, if $y = -3$

9. A cake bakery estimates the profit function to be $y = 3x - 800$, where $x$ is the number of cakes produced in a month and $800 is cost of overhead such as utilities. Find the profit if 8,000 cakes are produced.

10. A cost function is known to be $y = 10x + 250$, where $x$ is the number of units produced and $250 is the fixed cost. Find the total cost of producing 5,000 units.

Graph using the intercepts method.

11. $x + y = -5$

12. $2x - 3y = 6$

13. $x + 2y = 8$

Graph using the slope-intercept method.

14. $y = -3x + 1$

15. $2x + y = -3$

16. $x + 2y = 1$

Find the slope of the line passing through the given pairs of points.

17. $(-3, 6)$ and $(3, 2)$

18. $(0, 4)$ and $(-1, 6)$

Write the equations in slope-intercept form and determine the slope and $y$-intercept.

19. $-2x + y = 34$

20. $x - y = 4$

21. $x = 4y$

22. $2y - x = 3$

Find the equation of the line passing through the given point with the given slope. Solve the equation for $y$.

23. $(3, -5)$, $m = \frac{2}{3}$

24. $(5, 1)$, $m = -2$

Find the equation of the line passing through the given pairs of points. Solve the equation for *y*.

**25.** (1, 3) and (4, 5)  **26.** (−1, 1) and (4, −4)  **27.** (5, 2) and (−1, 2)

**28.** Write the equation in slope-intercept form of the line that has a slope = −2 and *y*-intercept = −3.

**29.** Write the equation of the line shown in Figure 7–46 in slope-intercept form.

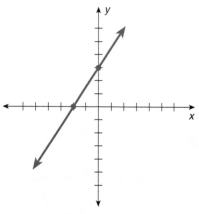

**Figure 7–46**

Write the equations in standard form.

**30.** Find the equation of the line that is parallel to $y - 2x = 3$ and passes through the point (2, 5).

**31.** Find the equation of the line that is parallel to $2x + y = 4$ and passes through the point (4, −3).

**32.** Find the equation of the line perpendicular to $y - 2x = 3$ and passing through the point (2, 5).

**33.** Find the equation of the line perpendicular to $2x + y = 4$ and passing through the point (4, −3).

---

## CAREER APPLICATION: ELECTRONICS – USING THE SLOPE TO FIND RESISTANCE

To find the slope of the line that passes through the two points $(x_1, y_1)$ and $(x_2, y_2)$, use the formula

$$\text{slope} = \frac{\text{rise}}{\text{run}} = \frac{y_2 - y_1}{x_2 - x_1} \quad \text{or} \quad \text{slope} = \frac{y_1 - y_2}{x_1 - x_2}$$

Assume that you have a graph with current on the horizontal axis and voltage on the vertical axis (Figure 7–47). Find the slope of the line through the points (1 mA, 3 V) and (4 mA, 6 V).

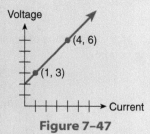

**Figure 7–47**

$$\text{slope} = \frac{3\text{ V} - 6\text{ V}}{1\text{ mA} - 4\text{ mA}} = \frac{-3\text{ V}}{-3\text{ mA}} = 1\text{ k}\Omega \quad \text{or} \quad \text{slope} = \frac{6\text{ V} - 3\text{ V}}{4\text{ mA} - 1\text{ mA}} = \frac{3\text{ V}}{3\text{ mA}} = 1\text{ k}\Omega$$

To analyze the dimensions, use the following relationships: volts divided by amperes gives ohms, $\frac{V}{A} = \Omega$. m = $\frac{1}{1,000}$; $1 \div \frac{1}{1,000} = 1,000$; $1,000 = k$; that is, the reciprocal of the unit *m* (milli-) is the unit *k* (kilo-).

Find the slope of the line through (4 V, 5 mA) and (6 V, 4 mA).

$$\text{slope} = \frac{5\text{ mA} - 4\text{ mA}}{4\text{ V} - 6\text{ V}} = \frac{1\text{ mA}}{-2\text{ V}} = -0.5\text{ mS}$$

To analyze the dimensions, amperes divided by volts gives siemens, $\frac{A}{V} = S$.

A graph like the one in Figure 7–48 that compares current and voltage for a **series circuit** is usually drawn with the current or the independent variable on the horizontal axis, because the current is the reference. The voltage or dependent variable is on the vertical axis. A line drawn on this graph that goes through the origin (vertical or $y$-intercept = 0) has

$$\text{slope} = \frac{\text{rise}}{\text{run}} = \frac{\Delta \text{ voltage } (E)}{\Delta \text{ current } (I)} = \text{resistance } (R \text{ or } Z)$$

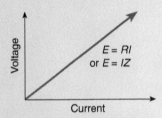

**Figure 7–48** Series circuit.

The equation of the line for a **series circuit** is

$$E = RI \text{ (for dc circuits)} \qquad \text{or} \qquad E = IZ \text{ (for ac circuits)}$$

A graph like the one in Figure 7–49 that compares current and voltage for a **parallel circuit** is usually drawn with the voltage on the horizontal axis because the voltage is the reference or independent variable, and the current is on the vertical axis as the dependent variable. A line drawn on this graph that passes through the origin has

$$\text{slope} = \frac{\text{rise}}{\text{run}} = \frac{\Delta \text{ current } (I)}{\Delta \text{ voltage } (E \text{ or } V)} = \text{conductance } (G \text{ or } Y)$$

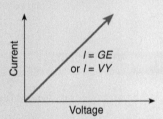

**Figure 7–49** Parallel circuit.

The equations of the line for a **parallel circuit** are

$$I = GE \text{ (for dc circuits)} \qquad \text{or} \qquad I = VY \text{ (for ac circuits)}$$

### Exercises

Find the slope of the line that passes through each of the given pairs of points. Be sure to include sign, prefix, and units as well as the number in the answer. Use engineering notation.

1. (4 mA, 12 V) and (7 mA, 15 V)
2. (4 mA, 12 V) and (9 mA, 8 V)
3. (15 V, 7 mA) and (20 V, 9 mA)
4. (25 V, 4 mA) and (37 V, 1 mA)
5. (0.65 V, 1 mA) and (0.75 V, 2 mA)
6. (9 μS, 11 V) and (12 μS, 17 V)
7. (9 μS, 11 V) and (12 μS, 5 V)
8. (4 hr, 200 mi) and (5 hr, 250 mi)
9. (7 hr, 700 km) and (9 hr, 900 km)
10. (2 class-hours, 4 h of homework) and (3 class-hours, 6 h of homework)

### Answers

1. $\dfrac{12 \text{ V} - 15 \text{ V}}{4 \text{ mA} - 7 \text{ mA}} = \dfrac{-3 \text{ V}}{-3 \text{ mA}} = +1 \text{ k}\Omega$

2. $\dfrac{12 \text{ V} - 8 \text{ V}}{4 \text{ mA} - 9 \text{ mA}} = \dfrac{4 \text{ V}}{-5 \text{ mA}} = -0.8 \text{ k}\Omega \text{ or } -800 \text{ }\Omega$

3. $\dfrac{7 \text{ mA} - 9 \text{ mA}}{15 \text{ V} - 20 \text{ V}} = \dfrac{-2 \text{ mA}}{-5 \text{ V}} = 0.4 \text{ mS or } 400 \text{ }\mu\text{S}$

4. $\dfrac{4 \text{ mA} - 1 \text{ mA}}{25 \text{ V} - 37 \text{ V}} = \dfrac{3 \text{ mA}}{-12 \text{ V}} = -0.25 \text{ mS or } -250 \text{ }\mu\text{S}$

5. $\dfrac{1 \text{ mA} - 2 \text{ mA}}{0.65 \text{ V} - 0.75 \text{ V}} = \dfrac{-1 \text{ mA}}{-0.10 \text{ V}} = 10 \text{ mS}$

6. $\dfrac{11 \text{ V} - 17 \text{ V}}{9 \text{ }\mu\text{S} - 12 \text{ }\mu\text{S}} = \dfrac{-6 \text{ V}}{-3 \text{ }\mu\text{S}} = 2 \text{ M}\Omega$

7. $\dfrac{11\text{ V} - 5\text{ V}}{9\ \mu\text{S} - 12\ \mu\text{S}} = \dfrac{+6\text{ V}}{-3\ \mu\text{S}} = -2\ \text{M}\Omega$

8. $\dfrac{200\text{ mi} - 250\text{ mi}}{4\text{ h} - 5\text{ h}} = \dfrac{-50\text{ mi}}{-1\text{ h}} = 50\text{ mi/h}$

9. $\dfrac{700\text{ km} - 900\text{ km}}{7\text{ h} - 9\text{ h}} = \dfrac{-200\text{ km}}{-2\text{ h}} = 100\text{ km/h}$

10. $\dfrac{4\text{ hours of homework} - 6\text{ hours of homework}}{2\text{ class-hours} - 3\text{ class-hours}} = \dfrac{-2\text{ h of homework}}{-1\text{ class-hour}} = 2\text{ h of homework per class-hour}$

# 8

# Systems of Linear Equations

## Focus on Careers

Careers in telecommunications include voice, video, and Internet communication services. Those persons who have up-to-date technical skills will have the best job opportunities. Maintaining up-to-date technical skills in this environment is a must.

Fiber-optic cables to transmit high-speed, high-capacity communications and radio towers to provide wireless telecommunications services are just two of the areas where highly trained technicians find employment. Wireless telephone communication is a primary sector of this industry. Emerging technology will allow higher-speed data transmission and better telephone and Internet access. Replacement of landlines with cellular phone service will become increasingly common as new technology is deployed.

Jobs in the telecommunications industry are steady, year-round jobs and overtime is sometimes required. More than 1.2 million people were employed in the telecommunications

*(continued)*

industry in 2002. More than 60% of those employed worked for companies with 100 or more employees. Most jobs are found in cities.

Although some jobs in this industry require a high school education, most companies choose persons that have training at 2-year or 4-year colleges. Computer literacy and keyboarding skills are important.

Average weekly earnings in the telecommunications industry were $761 in 2002. This is significantly higher than the average weekly earnings of $506 for all of private industry.

Job growth in the telecommunications industry is expected to be slower than overall job growth. However, rising demand for these services will eventually result in a higher job-growth rate.

Source: *Occupational Outlook Handbook,* 2004–2005 Edition. U.S. Department of Labor, Bureau of Labor Statistics.

Many real-world situations involve problems in which several conditions or constraints have to be considered. These conditions can be written in separate equations or inequalities that form a system of equations or inequalities. The solution of the system will be the value or values that satisfy all conditions.

## 8–1 | Solving Systems of Linear Equations and Inequalities Graphically

*Learning Outcome*  **1**  Solve a system of linear equations by graphing.

### **1**  Solve a System of Linear Equations by Graphing.

A **system of two linear equations,** each having two variables, is solved when we find the one ordered pair of solutions that satisfies *both* equations. One method of solving systems of two equations is to graph each equation and find the intersection of these graphs. The point where the two graphs intersect represents the ordered pair of solutions that the two graphs have in common.

### To solve a system of two linear equations with two variables by graphing:

1. Graph each equation on the same pair of axes.
2. The solution will be the common point or points.

**EXAMPLE**  A board is 20 ft long. It needs to be cut so that one piece is 2 ft longer than the other. What should be the length of each piece?

**CON**  Write two equations to describe all the conditions of the problem.

Since the board is not cut into equal pieces, we let the letter *l* represent the *longer* piece and the letter *s* represent the *shorter* piece.

**Condition 1:** The total length of the board is 20 ft. Thus, the two pieces (*l* and *s*) total 20 ft: $l + s = 20$.

**Condition 2:** One piece is 2 ft longer than the other. Thus, the shorter piece plus 2 ft equals the longer piece: $s + 2 = l$.

The two equations become a *system of equations.*

$l + s = 20$      Condition 1.
$s + 2 = l$      Condition 2.

Graph each equation on the same set of axes and examine the intersection of the graphs. Let *s* be the independent variable and *l*, the dependent variable.

| Condition 1 | | Condition 2 | |
|---|---|---|---|
| $l + s = 20$ | | $s + 2 = l$ | Intercepts: $(0, 2)$ and $(-2, 0)$. |
| or | | or | |
| $l = 20 - s$ | Domain: $[0, 20]$. | $l = s + 2$ | Domain: $[0, 20]$. |

Make a table of solutions for each equation.

| $s$ | $l$ | |
|---|---|---|
| 8 | 12 | Choose values that are near the |
| 10 | 10 | middle of the domain. |
| 12 | 8 | |

| $s$ | $l$ | |
|---|---|---|
| 0 | 2 | Choose only zero and positive values. |
| 5 | 7 | |
| 10 | 12 | |

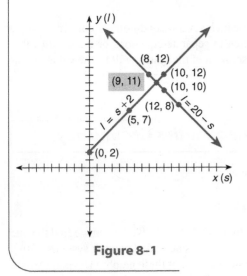

**Figure 8–1**

The point of intersection (Fig. 8–1) is $(9, 11)$, which means that $s = 9$ and $l = 11$. **The shorter length is 9 ft and the longer length is 11 ft.**
Check to see if the solution ($s = 9$, $l = 11$) satisfies both equations.

$$l + s = 20 \qquad s + 2 = l$$
$$11 + 9 = 20 \qquad 9 + 2 = 11$$
$$20 = 20 \qquad 11 = 11 \qquad \text{The ordered pair checks in both equations.}$$

## Line relationships:

When graphing two straight lines, three possibilities can occur.

1. The two lines can intersect at *just one point.* This means the system is **independent** and one pair of values satisfies both equations.
2. The two lines *do not intersect* at all. This means that the system is **inconsistent** and no pair of values satisfies both equations.
3. The two lines *coincide* or fall exactly in the same place. This means that the system is **dependent** and the equations are identical or are multiples, and any pair of values that satisfies one equation satisfies both equations.

## To solve a system of equations using a graphing calculator:

1. Graph the first equation.
2. Graph the second equation without clearing the graph screen.
3. Use the Trace feature or the CALC/Intersect feature to determine the approximate coordinates of the intersection of the graphs.
4. Use the Zoom or Box (Window) feature to get a closer view and thus a more accurate approximation of the intersection of the graphs.

Chapter 8 / Systems of Linear Equations

**1** Solve the systems of equations by graphing.

1. $x + y = 12$
   $x - y = 2$

2. $2x + y = 9$
   $3x - y = 6$

3. $2x - y = 5$
   $4x - 2y = 8$

4. $x - y = 9$
   $3x - 3y = 27$

5. $3x + 2y = 8$
   $x + y = 2$

6. $y = x - 2$
   $y = -2x + 7$

7. $y = -3$
   $y = x - 7$

8. $y = 5x$
   $y = -2x - 7$

9. $y = -2x$
   $y = x - 3$

10. $x - y = 3$
    $x + 2y = 6$

11. $5x + y = -2$
    $x - y = 8$

12. $2x - y = -3$
    $x + 2y = 6$

13. $4x + y = -1$
    $x + y = -4$

14. $x + y = 3$
    $2y = 3x - 4$

15. $y = 2x - 5$
    $y = x + 2$

16. $x + 2y = 6$
    $x + y = 2$

17. $2x - y = 7$
    $x + y = 5$

18. $3x + y = 5$
    $x - y = 7$

19. $2x - y = 6$
    $5x + y = 8$

20. $5x + y = 0$
    $x = y$

21. $-x + y = -2$
    $2x + y = 7$

---

## 8-2 | Solving Systems of Linear Equations Using the Addition Method

*Learning Outcomes*

**1** Use the addition method to solve a system of linear equations that contains opposite variable terms.

**2** Use the addition method to solve a system of linear equations that does not contain opposite variable terms.

**3** Apply the addition method to a system of linear equations with no solution or with many solutions.

We can see that solving systems of equations graphically is tedious and time-consuming. Also, many solutions to systems of equations are not whole numbers. Graphically, it is difficult to plot fractions or to read fractional intersection points. Therefore, we need a convenient algebraic procedure for solving systems of equations. We examine the two methods most commonly used to solve a system of equations with two variables, but there are other methods. The first method we introduce is the *addition method*.

**1** **Use the Addition Method to Solve a System of Linear Equations That Contains Opposite Variable Terms.**

The **addition method** incorporates the concept that *equals added to equals give equals*. Thus, when we add two equations, the result is still an equation. However, the addition method is effective *only* if one variable is *eliminated* in the addition process. This method is also called the **elimination method.**

**To solve a system of linear equations that contains opposite variable terms by the addition (elimination) method:**

1. Write each equation in standard form ($ax + by = c$).
2. Add the two equations.
3. Solve the equation from Step 2. This is one part of the ordered pair solution.
4. Substitute the solution from Step 3 in either equation and solve for the remaining variable. This completes the solution ordered pair.
5. Check the solution in each original equation.

**EXAMPLE**   Solve the systems of equations by the addition method:

(a) $x + y = 7$
$\phantom{(a)}\ x - y = 5$

(b)  $3x + 8y = 7$
$\phantom{(b)\ \ }-3x - 3y = 8$

(a)  $x + y =\ \ 7$       Equation 1. Add the equations to eliminate one variable.

$\phantom{(a)\ \ } x - y =\ \ 5$       Equation 2.

$\phantom{(a)\ \ } 2x\phantom{ - y}\ \ = 12$       Solve for $x$.

$\phantom{(a)\ \ \ \ \ \ } x = 6$       Substitute 6 in place of $x$ in *either* of the given equations.

$\phantom{(a)\ } x + y = 7$       Equation 1. Substitute $x = 6$.

$\phantom{(a)\ } 6 + y = 7$       Solve for $y$.

$\phantom{(a)\ \ \ \ \ } y = 7 - 6$       Simplify.

$\phantom{(a)\ \ \ \ \ } y = 1$

**The solution is $x = 6$ and $y = 1$, or (6, 1).**

Check in the *other* equation:

$x - y = 5$       Equation 2. Substitute values from the ordered-pair solution.

$6 - 1 = 5$       Simplify.

$\phantom{6 - }5 = 5$       The solution checks.

Checking in both equations helps ensure that no errors have been made in the addition process or in the substitution process.

(b)  $3x + 8y\ =\ \ 7$       Equation 1. Add the equations to eliminate one variable.

$\phantom{(b)\ }-3x - 3y\ =\ \ 8$       Equation 2.

$\phantom{(b)\ \ \ \ \ \ \ \ } 5y = 15$       Solve for $y$.

$\phantom{(b)\ \ \ \ \ \ \ \ \ \ } y = 3$       Substitute 3 for $y$ in Equation 1.

$\phantom{(b)\ } 3x + 8y\ = 7$       Equation 1. Substitute $y = 3$.

$\phantom{(b)\ } 3x + 8(3) = 7$       Solve for $x$.

$\phantom{(b)\ \ } 3x + 24 = 7$

$\phantom{(b)\ \ \ \ \ \ } 3x = 7 - 24$

$\phantom{(b)\ \ \ \ \ \ } 3x = -17$

$\phantom{(b)\ \ \ \ \ \ \ } x = \dfrac{-17}{3}$

**The solution is $\left(-\frac{17}{3}, 3\right)$.**

Check in equation 2:

$-3x - 3y = 8$       Equation 2. Substitute $x = -\dfrac{17}{3}$ and $y = 3$.

$-3\left(\dfrac{-17}{3}\right) - 3(3) = 8$       Multiply.

$\phantom{-3\left(\dfrac{-17}{3}\right)\ } 17 - 9 = 8$       Subtract.

$\phantom{-3\left(\dfrac{-17}{3}\right)\ \ \ \ \ \ } 8 = 8$       The solution checks.

**2**  **Use the Addition Method to Solve a System of Linear Equations That Does Not Contain Opposite Variable Terms.**

In the addition method one unknown must be eliminated; that is, the terms must add to 0. Thus, if neither pair of variable terms in a system of equations is opposite terms that will add to 0, we multiply one or both of the equations by numbers that *cause* the terms of one variable to add to 0. This is an application of the multiplication axiom (see Chapter 5, Section 2, Outcome 2).

**To solve a system of linear equations using the addition method (elimination):**

1. Write each equation in standard form ($ax + by = c$).
2. If necessary, multiply one or both equations by numbers that cause the terms of one variable to add to zero.
3. Add the two equations to eliminate a variable.
4. Solve the equation from Step 3 for the remaining variable.
5. Substitute the solution from Step 4 in either equation and solve for the remaining variable.
6. Check the solution in one or both original equations.

**EXAMPLE** Solve the system of equations: $2x + y = 7$ and $x + y = 3$.

Choice 1.
$$-2x - y = -7$$
$$x + y = \phantom{-}3$$
$$\overline{-x \phantom{-y} = -4}$$
$$\boxed{x = 4}$$

Multiply the first equation by $-1$ and add the equations to eliminate the $y$-terms.

Solve for $x$.

Choice 2.
$$2x + y = \phantom{-}7$$
$$-x - y = -3$$
$$\overline{\phantom{2}x \phantom{-y} = 4}$$

Multiply the second equation by $-1$ and add the equations to eliminate the $y$-terms.

Choice 3.
$$2x + \phantom{2}y = \phantom{-}7$$
$$-2x - 2y = -6$$
$$\overline{\phantom{2x +}-y = 1}$$
$$\boxed{y = -1}$$

Multiply the second equation by $-2$ and add the equations to eliminate the $x$-terms.

In choices 1 and 2, substitute $x = 4$:

$$-2\boxed{x} - y = -7$$
$$-2(\boxed{4}) - y = -7$$
$$-8 - y = -7$$
$$-y = -7 + 8$$
$$-y = 1$$
$$\boxed{y = -1}$$

In choice 3, substitute $y = -1$:

$$2x + \boxed{y} = 7$$
$$2x + (\boxed{-1}) = 7$$
$$2x - 1 = 7$$
$$2x = 7 + 1$$
$$2x = 8$$
$$\boxed{x = 4}$$

**The solution is $(4, -1)$.**

Check the solution in the first original equation:

$$2\boxed{x} + \boxed{y} = 7$$
$$2(\boxed{4}) + (\boxed{-1}) = 7$$
$$8 + (-1) = 7$$
$$7 = 7$$

Check the solution in the second original equation:

$$\boxed{x} + \boxed{y} = 3$$
$$\boxed{4} + (\boxed{-1}) = 3$$
$$3 = 3$$

When an original equation is altered, it is important to check your solution in *both* original equations. This enables you to identify mistakes such as forgetting to multiply *each* term in the equation by a number.

**EXAMPLE**   Solve the system of equations: $2x + 3y = 1$ and $3x + 4y = 2$.

In this system, no number can be multiplied by just one equation to eliminate a letter. Therefore, we need to multiply each equation by some number. There are several possibilities; we examine just two.

Choice 1.

$$-3(2x + 3y) = -3(1)$$ — Multiply Equation 1 by $-3$.
$$2(3x + 4y) = \phantom{-}2(2)$$ — Multiply Equation 2 by $+2$.
$$-6x - 9y = -3$$ — Add equations to eliminate $x$.
$$\underline{\phantom{-}6x + 8y = \phantom{-}4\phantom{-}}$$
$$-y = \phantom{-}1$$ — Solve for $y$.
$$\boldsymbol{y = -1}$$ — $y$-value of solution.
$$2x + 3y = \phantom{-}1$$ — Substitute $y = -1$.
$$2x + 3(-1) = \phantom{-}1$$ — Solve for $x$.
$$2x - 3 = \phantom{-}1$$
$$2x = \phantom{-}1 + 3$$
$$2x = \phantom{-}4$$
$$\boldsymbol{x = \phantom{-}2}$$ — $x$-value of solution.

Choice 2.

$$4(2x + 3y) = \phantom{-}4(1)$$ — Multiply Equation 1 by 4.
$$-3(3x + 4y) = -3(2)$$ — Multiply Equation 2 by $-3$.
$$8x + 12y = \phantom{-}4$$ — Add equations to eliminate $y$.
$$\underline{-9x - 12y = -6\phantom{-}}$$
$$-x = -2$$ — Solve for $x$.
$$\boldsymbol{x = \phantom{-}2}$$ — $x$-value of solution.
$$2x + 3y = \phantom{-}1$$ — Substitute $x = 2$.
$$2(2) + 3y = \phantom{-}1$$ — Solve for $y$.
$$3y = \phantom{-}1 - 4$$
$$3y = -3$$
$$\boldsymbol{y = -1}$$ — $y$-value of solution.

**The solution is $(2, -1)$.**

Check the solution in the original equations:

| | | |
|---|---|---|
| $2x + 3y = 1$ | $3x + 4y = 2$ | Substitute $x = 2$ and $y = -1$. |
| $2(2) + 3(-1) = 1$ | $3(2) + 4(-1) = 2$ | |
| $4 + (-3) = 1$ | $6 + (-4) = 2$ | |
| $1 = 1$ | $2 = 2$ | Solution checks. |

**3** **Apply the Addition Method to a System of Linear Equations with No Solution or with Many Solutions.**

Sometimes a system of equations has no solution; the graphs of the two equations do not intersect. There are also instances when a system of equations has all solutions in common; the graphs of the two equations coincide.

**EXAMPLE**   Solve the system $x + y = 7$ and $x + y = 5$.

$$x + y = \phantom{-}7$$ — Multiply Equation 2 by $-1$.
$$-1(x + y) = -1(5)$$
$$x + y = \phantom{-}7$$ — Add the equations to eliminate a variable.
$$\underline{-x - y = -5\phantom{-}}$$
$$0 = \phantom{-}2$$ — Both variables are eliminated.

Notice, both variables are eliminated and the resulting equation, $0 = 2$, is *false*.

**There are no solutions to this system.**

If both variables in a system of equations are eliminated and the resulting statement is false, the equations are **inconsistent** and have no solution. The graphs of the equations are parallel lines.

**EXAMPLE** Solve the system $2x - y = 7$ and $4x - 2y = 14$.

$$-2\,(2x - y) = -2(7) \qquad \text{Multiply Equation 1 by } -2.$$
$$\begin{array}{r} -4x + 2y = -14 \\ 4x - 2y = \phantom{-}14 \\ \hline 0 = \phantom{-}0 \end{array} \qquad \begin{array}{l} \text{Add the equations to eliminate a variable.} \\ \\ \text{Both variables are eliminated.} \end{array}$$

Both variables are eliminated; however, the result is a *true* statement ($0 = 0$). In this situation, **all solutions of one equation are also solutions of the other equation.** For example, $x = 4$ and $y = 1$ are a solution of both equations.

If both variables in a system of equations are eliminated and the resulting statement is true, then the equations are **dependent** and have many solutions. The graphs of the equations coincide.

SECTION 8–2 SELF-STUDY EXERCISES

**1** Solve the systems of equations using the addition method.

1. $a - 2b = 7$
   $3a + 2b = 13$

2. $3m + 4n = 8$
   $2m - 4n = 12$

3. $x - 4y = 5$
   $-x - 3y = 2$

4. $a - b = 6$
   $2a + b = 3$

5. $x + 2y = 5$
   $3x - 2y = 3$

6. $x - 5y = 7$
   $2x + 5y = 5$

7. $5x + 2y = -3$
   $-5x - 4y = -7$

8. $-8x - 3y = 7$
   $8x + 5y = 3$

9. $x + 3y = 5$
   $-x + 3y = 13$

**2** Solve the systems of equations using the addition method.

10. $3x + y = 9$
    $x + y = 3$

11. $7x + 2y = 17$
    $y = 3x + 2$

12. $a + 6b = 18$
    $4a - 3b = 0$

13. $3x + y = -1$
    $4x - 2y = -8$

14. $a = 6y$
    $2a - y = 11$

15. $5x - 3y = -4$
    $2x - 3y = -7$

16. $2x - 3y = 10$
    $3x + 2y = 2$

17. $4x - 3y = -19$
    $-2x - 4y = 4$

18. $2x + 5y = 0$
    $-3x - 2y = -11$

19. $x + 2y = 6$
    $2x + 5y = 15$

20. $5x - y = -8$
    $x - 2y = -7$

21. $-x + 2y = -7$
    $x - 6y = -5$

**3** Solve the systems of equations using the addition method.

22. $x + y = 8$
    $x + y = 3$

23. $x + y = 9$
    $x + y = 3$

24. $3x - 2y = 6$
    $9x - 6y = 18$

25. $2a + 4b = 10$
    $a + 2b = 5$

26. $3a - 2b = 14$
    $3a = 2b + 2$

27. $5x - 7y = 8$
    $10x - 14y = 16$

28. $-3x + 2y = 3$
    $9x - 6y = -8$

29. $2x + 5y = 3$
    $16x + 40y = 38$

30. $-x - 2y = 3$
    $4x + 8y = 15$

*Learning Outcome*     **1**  Use the substitution method to solve a system of linear equations.

**1**  **Use the Substitution Method to Solve a System of Linear Equations.**

Another method for solving systems of equations is by substitution. Recall that in formula re-arrangement (Chapter 5, Section 5, Outcome 2), whenever more than one variable is used in an equation or formula, we can rearrange the equation or solve for a particular variable. In the **substitution method** for solving systems of equations, we solve one equation for one variable and then substitute the equivalent expression in place of the variable in the other equation.

### To solve a system of equations by substitution:

1. Rearrange either equation to isolate one variable.
2. Substitute the equivalent expression from Step 1 into the *other* equation and solve for the remaining variable.
3. Substitute the portion of the solution from Step 2 into the equation from Step 1 to find the remaining value of the solution.
4. Check the solution in both original equations.

**EXAMPLE**  Solve the system of equations $x + y = 15$ and $y = 2x$ using the substitution method.

$$x + y = 15$$
$$y = 2x$$   Equation 2 is already solved for *y*.
$$x + \boxed{y} = 15$$   Substitute $2x$ for *y* in Equation 1 and solve.
$$x + \boxed{2x} = 15$$   Solve for *x*.
$$3x = 15$$
$$\boxed{x = 5}$$   *x*-value of solution.
$$y = 2\,\boxed{x}$$   Substitute the solution for *x* in Equation 2 to find *y*.
$$y = 2(\,\boxed{5}\,)$$   Simplify.
$$\boxed{y = 10}$$   *y*-value of solution.

**The solution is (5, 10).**

Check:

$$\boxed{x} + \boxed{y} = 15 \qquad \boxed{y} = 2\,\boxed{x} \qquad \text{Substitute } x = 5 \text{ and } y = 10 \text{ in both equations.}$$
$$\boxed{5} + \boxed{10} = 15 \qquad \boxed{10} = 2(\,\boxed{5}\,) \qquad \text{Simplify.}$$
$$15 = 15 \qquad\qquad 10 = 10 \qquad\quad \text{The solution checks in both equations.}$$

**EXAMPLE**  Solve the following system of equations using the substitution method.

$$2x - 3y = -14$$
$$x + 5y = 19$$

When using the substitution method, either equation can be solved for either unknown. In this example, the *x*-term in the second equation has a coefficient of 1, so the simplest choice would be to solve the second equation for *x*.

|                Step 1                |                Step 2                |                Step 3                |
|--------------------------------------|--------------------------------------|--------------------------------------|
| $x + 5y = 19$                        | $2x - 3y = -14$                      | $x = 19 - 5y$                        |
| $x = 19 - 5y$                        | $2(19 - 5y) - 3y = -14$              | $x = 19 - 5(4)$                      |
|                                      | $38 - 10y - 3y = -14$                | $x = 19 - 20$                        |
|                                      | $38 - 13y = -14$                     | $x = -1$                             |
|                                      | $-13y = -14 - 38$                    |                                      |
|                                      | $-13y = -52$                         |                                      |
|                                      | $\dfrac{-13y}{-13} = \dfrac{-52}{-13}$ |                                      |
|                                      | $y = 4$                              |                                      |

**The solution is $(-1, 4)$.**

Check the roots $x = -1$, $y = 4$ in both original equations:

**Step 4**

$$2x - 3y = -14 \qquad\qquad x + 5y = 19$$
$$2(-1) - 3(4) = -14 \qquad -1 + 5(4) = 19$$
$$-2 - 12 = -14 \qquad\qquad -1 + 20 = 19$$
$$-14 = -14 \qquad\qquad\qquad 19 = 19$$

### TIP!

### Long Problems Don't Have to Be Difficult

Sometimes we let ourselves become overwhelmed by the mere length of a problem. Look at the previous example. Each step of the solution involves skills we have previously used many times. Here are some tips to help you manage longer problems.

- Get a global or overall understanding of the problem you are solving.
- Make a prediction or estimate of the solution.
- Get a global or overall understanding of the process you are using to solve the problem.
- List in your own words (as briefly as possible) the steps of the process.
- Focus on one step at a time.
- Examine the solution to see if it matches your prediction or estimate. Check if appropriate.

SECTION 8–3 SELF-STUDY EXERCISES

**1** Solve the systems of equations using the substitution method.

**1.** $2a + 2b = 60$
$\quad a = 10 + b$

**2.** $7r + c = 42$
$\quad 3r - 8 = c$

**3.** $x - 35 = -2y$
$\quad 3x - 2y = 17$

**4.** $x + y = 12$
$\quad x = 2 + y$

**5.** $2p + 3k = 2$
$\quad 2p - 3k = 0$

**6.** $x + 2y = 5$
$\quad x = 3y$

**7.** $x - 3 = 2y$
$\quad x = 3y - 2$

**8.** $a = 3x - 1$
$\quad x = a + 5$

**9.** $5x + 2y = 7$
$\quad x = 2y - 1$

**10.** $x + y = 7$
$\quad x = y - 5$

**11.** $x - 2y = 6$
$\quad 4x + 3y = 35$

**12.** $y = 3x - 2$
$\quad y = x + 6$

**13.** $2x - 3y = 0$
$3x - y = 7$
**14.** $2x - 3y = -3$
$4x + 2y = 18$
**15.** $6x - 2y = 3$
$3x + 4y = 9$
**16.** $6x + 2y = 3$
$3x - 4y = -1$
**17.** $8x - 6y = -2$
$4x + 9y = 7$
**18.** $4x + 6y = 1$
$8x - 4y = -6$
**19.** $x - y = -4$
$5x + 9y = 8$
**20.** $x - 6y = 8$
$x + 3y = -1$
**21.** $x + 3y = 1$
$2x + 4y = -4$

---

## 8–4 | Problem Solving Using Systems of Linear Equations

*Learning Outcome*  **1**  Use a system of linear equations to solve application problems.

**1  Use a System of Linear Equations to Solve Application Problems.**

Many job-related problems can be solved by setting up and solving systems of equations.

---

**EXAMPLE**

**ELEC**

Two dry cells connected in series have a total internal resistance of 0.09 Ω. The difference between the internal resistances of the individual dry cells is 0.03 Ω. How much is each internal resistance?

**Known facts**

The total internal resistance of the two dry cells is 0.09 Ω.
The difference in the internal resistances of the two dry cells is 0.03 Ω.

**Unknown facts**

What is the resistance of dry cell 1 ($r_1$)?
What is the resistance of dry cell 2 ($r_2$)?

**Relationships**

$r_1 + r_2 = 0.09$      Equation 1.
$r_1 - r_2 = 0.03$      Equation 2.

**Estimation**

If resistances were the same, they would each be 0.045 Ω. Because they are not the same, one will be more than 0.045 and one will be less than 0.045.

**Calculations**

$r_1 + r_2 = 0.09$      Add the equations to eliminate $r_2$.

$\underline{r_1 - r_2 = 0.03}$

$2r_1 \quad\;\; = 0.12$      Solve for $r_1$.

$r_1 = \dfrac{0.12}{2}$

$r_1 = 0.06$      $r_1$-value of solution.

$r_1 + r_2 = 0.09$      Substitute for $r_1$ in Equation 1.

$0.06 + r_2 = 0.09$      Solve for $r_2$.

$r_2 = 0.09 - 0.06$

$r_2 = 0.03$      $r_2$-value of solution.

**Interpretation**

**The larger internal resistance is 0.06 Ω and the smaller internal resistance is 0.03 Ω.**

---

**EXAMPLE**

**AG/H**

A tank holds a solution that is 10% herbicide. Another tank holds a solution that is 50% herbicide. If a farmer wants to mix the two solutions to get 200 gal of a solution that is 25% herbicide, how many gallons of each solution should be mixed?

Chapter 8 / Systems of Linear Equations

| | |
|---|---|
| **Known facts** | There are two strengths of herbicide, 10% and 50%. 200 gal of 25% herbicide are needed. |
| **Unknown facts** | How many gallons of 10% herbicide ($h$) are needed? How many gallons of 50% herbicide ($H$) are needed? |
| **Relationships** | 200 gal of the new herbicide are needed: $h + H = 200$ Amount of pure herbicide in $h$ gal of 10% herbicide: $0.1h$ Amount of pure herbicide in $H$ gal of 50% herbicide: $0.5H$ Amount of pure herbicide in 200 gal of 25% herbicide: $0.25(200)$ |

$$h + H = 200 \qquad \text{Equation 1 (total gallons).}$$
$$0.1h + 0.5H = 0.25(200) \qquad \text{Equation 2 (gallons of pure herbicide).}$$

| | |
|---|---|
| **Estimation** | If equal amounts of herbicide were needed, we would need 100 gal of each solution. However, since the desired solution strength is not exactly halfway between the two original herbicide strengths, we will need unequal amounts of herbicide. One amount will be less than 100 gal and the other will be more than 100 gal. |
| **Calculations** | Solve by the substitution method. |

$$h + H = 200 \qquad \text{Solve Equation 1 for } h.$$
$$h = 200 - H \qquad \text{Equivalent expression for } h.$$
$$0.1\,h + 0.5H = 0.25(200) \qquad \text{Substitute } (200 - H) \text{ for } h \text{ in Equation 2.}$$
$$0.1\,(\,200 - H\,) + 0.5H = 0.25(200) \qquad \text{Solve for } H.$$
$$20 - 0.1H + 0.5H = 50$$
$$20 + 0.4H = 50$$
$$0.4H = 50 - 20$$
$$0.4H = 30$$
$$H = \frac{30}{0.4}$$
$$H = 75 \text{ gal} \qquad H\text{-value of solution.}$$
$$h + H = 200 \qquad \text{Substitute 75 for } H \text{ in Equation 1.}$$
$$h + 75 = 200 \qquad \text{Solve for } h.$$
$$h = 200 - 75$$
$$h = 125 \text{ gal} \qquad h\text{-value of solution.}$$

| | |
|---|---|
| **Interpretation** | **The farmer must mix 75 gal of the 50% herbicide and 125 gal of the 10% herbicide to make 200 gal of a 25% herbicide.** |

| | |
|---|---|
| **EXAMPLE** **BUS** | Rosita has $5,500 to invest and for tax purposes wants to earn exactly $500 interest for 1 year. She wants to invest part at 10% and the remainder at 5%. How much must she invest at each interest rate to earn exactly $500 interest in 1 year? |
| | Let $x$ = the amount invested at 10%. Let $y$ = the amount invested at 5%. Interest for 1 year = rate × amount invested. Convert percents to decimals. Using these relationships, we derive a system of equations. |
| **Known facts** | Total of $5,500 to be invested $500 interest to be earned in one year |
| **Unknown facts** | How much should be invested at 10%? How much should be invested at 5%? |
| **Relationships** | Amount invested at 10%: $x$ Interest earned at 10%: $0.1x$ |

Amount invested at 5%: $y$
Interest earned at 5%: $0.05y$

$$x + y = 5{,}500 \qquad \text{Equation 1 (total investment).}$$
$$0.1x + 0.05y = 500 \qquad \text{Equation 2 (total interest in 1 year).}$$

**Estimation**

If the total amount were invested at 10%, the interest (in 1 year) would be $550 ($0.1 \times$ $5,500). Since we want $500 in interest, most of the money will need to be invested at 10%.

**Calculations**

Solve by the substitution method.

| | |
|---|---|
| $x + y = 5{,}500$ | Solve Equation 1 for $x$. |
| $x = 5{,}500 - y$ | Equivalent expression for $x$. |
| $0.1\,x + 0.05y = 500$ | Substitute $(5{,}500 - y)$ for $x$ in Equation 2. |
| $0.1(\,5{,}500 - y\,) + 0.05y = 500$ | Solve for $x$. |
| $550 - 0.1y + 0.05y = 500$ | $-0.1y + 0.05y = -0.05y$. |
| $550 - 0.05y = 500$ | |
| $-0.05y = 500 - 550$ | |
| $-0.05y = -50$ | |
| $y = \dfrac{-50}{-0.05}$ | |
| $y = \$1{,}000$ | Amount invested at 5%. |
| $x + y = \$5{,}500$ | Substitute $1,000 for $y$ in Equation 1. |
| $x + \$1{,}000 = \$5{,}500$ | |
| $x = \$5{,}500 - \$1{,}000$ | |
| $x = \$4{,}500$ | Amount invested at 10%. |

**Interpretation**

**Rosita must invest $4,500 at 10% and $1,000 at 5% for 1 year to earn $500 interest.**

---

## SECTION 8–4 SELF-STUDY EXERCISES

**1** Solve the problems using systems of equations with two unknowns.

1. **CON** Two boards together are 48 in. If one board is 17 in. shorter than the other, find the length of each board.

2. **BUS** A broker invested $35,000 in two different stocks. One earned dividends at 4% and the other at 5%. If a $1,570 dividend was earned on both stocks together, how much was invested in each? (*Reminder:* Change 4% to 0.04 and 5% to 0.05.)

3. **BUS** A department store buyer ordered 12 shirts and 8 hats for $380 one month and 24 shirts and 10 hats for $664 the following month. What was the cost of each shirt and each hat?

4. **AUTO** A mechanic makes $105 on each 8-cylinder engine tune-up and $85 on each 4-cylinder engine tune-up. If the mechanic did 10 tune-ups and made a total of $990, how many 8-cylinder jobs and how many 4-cylinder jobs were completed?

5. **ELEC** 30 resistors and 15 capacitors cost $12. And 10 resistors and 20 capacitors cost $8.50. How much does each capacitor and resistor cost?

6. **AVIA** A private airplane flew 420 mi in 3 h with the wind. The return trip against the wind took 3.5 h. Find the rate of the plane in calm air and the rate of the wind.

7. **BUS** In 1 year, Dee Wallace earned $660 in interest on two investments totaling $8,000. If he received 7% and 9% rates of return, how much did he invest at each rate?

8. A motorboat went 40 mi with the current in 3 h. The return trip against the current took 4 h. How fast was the current? What would have been the speed of the boat in calm water?

9. **BUS** A visitor to south Louisiana purchased 3 lb of dark-roast pure coffee and 4 lb of coffee with chicory for $27.30 in a local supermarket. Another visitor at the same store purchased 2 lb of coffee with chicory and 5 lb of dark-roast pure coffee for $28. How much did each coffee cost per pound?

10. **AG/H** A lawn-care technician wants to spread a 200-lb seed mixture that is 50% bluegrass. If the technician has on hand a mixture that is 75% bluegrass and a mixture that is 10% bluegrass, how many pounds of each mixture are needed to make 200 lb of the 50% mixture? Round to the nearest whole pounds.

11. **BUS** A college bookstore received a partial shipment of 50 scientific calculators and 25 graphing calculators at a total cost of $2,200. Later the bookstore received the balance of the calculators: 25 scientific and 50 graphing at a cost of $3,800. Find the cost of each calculator.

12. **BUS** A consumer received two 1-yr loans totaling $10,000 at interest rates of 10% and 15%. If the consumer paid $1,300 interest, how much money was borrowed at each rate?

13. **HOSP** For the first performance at the Overton Park Shell, 40 reserved seats and 80 general admission seats were sold for $2,000. For the second performance, 50 reserved seats and 90 general admission seats were sold for $2,350. What was the cost for a reserved seat and for a general admission seat?

14. A photographer has a container with a solution of 75% developer and a container with a solution of 25% developer. If she wants to mix the solutions to get 8 pt of solution with 50% developer, how many pints of each solution does she need to mix?

15. **ELEC** Two resistances have a sum of 21 Ω. Their difference is 13 Ω. Write two equations using $R_1$ as the first resistance and $R_2$ as the second resistance. Find each resistance.

16. **AG/H** The total weight of a fertilizer composed of nitrogen and potassium is 480 lb. The fertilizer has three times as much nitrogen as potassium. How many pounds of each chemical are in the fertilizer?

17. **AG/H** A plant nursery purchased holly shrubs that cost $4 each and nandinas that cost $5 each. The total cost was $260 for 60 shrubs. How many of each type of shrub did the nursery purchase?

18. **CON** A mortar mix contains five times as much sand as water. The total volume is 12 ft³. How much of each ingredient is in the mix?

---

## CHAPTER REVIEW OF KEY CONCEPTS

**Learning Outcomes**  **What to Remember with Examples**

### Section 8–1

**1** Solve a system of linear equations by graphing (pp. 367–368).

Graph each equation. The intersection of the two lines is the solution to the system. (The table-of-solutions, intercepts, or slope-intercept method may also be used to graph each equation.)

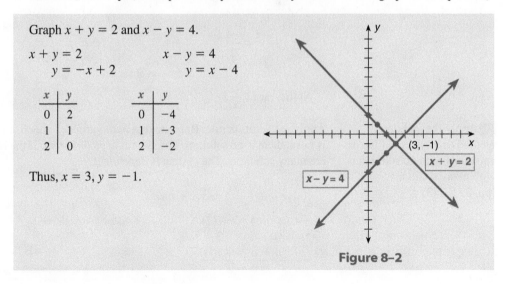

Graph $x + y = 2$ and $x - y = 4$.

$$x + y = 2 \qquad\qquad x - y = 4$$
$$y = -x + 2 \qquad\qquad y = x - 4$$

| x | y |
|---|---|
| 0 | 2 |
| 1 | 1 |
| 2 | 0 |

| x | y |
|---|---|
| 0 | −4 |
| 1 | −3 |
| 2 | −2 |

Thus, $x = 3$, $y = -1$.

**Figure 8–2**

## Section 8–2

**1** Use the addition method to solve a system of linear equations that contains opposite variable terms (pp. 369–370).

To solve a system of equations with two opposite variable terms, add the equations so that the opposite terms add to zero. Solve the new equation and substitute the root in an original equation to find the value of the remaining variable.

Solve $x - y = 6$ and $x + y = 4$ ($-y$ and $+y$ are opposites).

| | |
|---|---|
| $x - y = 6$ | Equation 1. Add the equations to eliminate $y$. |
| $x + y = 4$ | Equation 2. |
| $2x \quad = 10$ | Solve for $x$. |

$$\frac{2x}{2} = \frac{10}{2}$$

$x = 5$     $x$-value of solution.

$x - y = 6$     Equation 1. Substitute $x = 5$.

$5 - y = 6$     Solve for $y$.

$-y = 6 - 5$

$-y = 1$

$y = -1$     $y$-value of solution.

**The solution is $(5, -1)$**

Check:

$x - y = 6$

$5 - (-1) = 6$

$5 + 1 = 6$

$6 = 6$

$x + y = 4$

$5 + (-1) = 4$

$4 = 4$

Solution checks.

---

**2** Use the addition method to solve a system of linear equations that does not contain opposite variable terms (pp. 370–372).

To solve a system of equations that does not contain opposite variables: **1.** Multiply one or both equations by signed numbers that cause the terms of one variable to add to zero. **2.** Add the equations. **3.** Solve the equation from Step 2. **4.** Substitute the root in an original equation and solve for the remaining variable. **5.** Check solutions.

Solve $2x - y = 6$ and $4x + 2y = 4$.

$2(2x - y) = 2(6)$     Multiply Equation 1 by 2. Add equations to eliminate $y$.

$4x - 2y = 12$

$4x + 2y = 4$

$8x \quad = 16$     Solve for $x$.

$$\frac{8x}{8} = \frac{16}{8}$$

$x = 2$     $x$-value of solution.

$4x + 2y = 4$     Equation 2. Substitute $x = 2$.

$4(2) + 2y = 4$

$8 + 2y = 4$     Solve for $y$.

$2y = 4 - 8$

$2y = -4$

$$\frac{2y}{2} = \frac{-4}{2}$$

$y = -2$     $y$-value of solution.

**Solution: $(2, -2)$**

Check:

$2x - y = 6$

$2(2) - (-2) = 6$

$4 + 2 = 6$

$6 = 6$

$4x + 2y = 4$

$4(2) + 2(-2) = 4$

$8 - 4 = 4$

$4 = 4$

Solution checks.

---

**3** Apply the addition method to a system of linear equations with no solution or with many solutions (pp. 372–373).

Apply the addition rule. But note that *both* variables are eliminated. If the resulting statement is false, there is no solution. The system is *inconsistent*. If the resulting statement is true, there are many solutions. The system is *dependent*.

Solve $a + b = 2$ and $a + b = 4$.

$-1(a + b) = -1(2)$     Multiply the first equation by $-1$. Add to eliminate variables.

$-a - b = -2$

$a + b = 4$

$0 = 2$     False. No solution. Inconsistent system.

**Section 8–3**

**1** Use the substitution method to solve a system of linear equations (pp. 374–375).

To solve a system of equations by substitution: **1.** Rearrange either equation to isolate one of the variables. **2.** Substitute the equivalent expression from Step 1 in the *other* equation and solve. **3.** Substitute the root from Step 2 in one of the equations and solve for the other variable. **4.** Check.

Solve $2x + y = 6$ and $2x - 2y = 8$.

| | |
|---|---|
| $y = 6 - 2x$ | Isolate $y$ in the first equation. |
| $2x - 2y = 8$ | Substitute $6 - 2x$ for $y$ in second equation. |
| $2x - 2(6 - 2x) = 8$ | |
| $2x - 12 + 4x = 8$ | Solve for $x$. |
| $6x - 12 = 8$ | |
| $6x = 8 + 12$ | |
| $6x = 20$ | |
| $\dfrac{6x}{6} = \dfrac{20}{6}$ | |
| $x = \dfrac{10}{3}$ | $x$-value of solution. |

$$\text{Check: } 2x + y = 6 \qquad\qquad 2x - 2y = 8$$

$$y = 6 - 2\left(\frac{10}{3}\right) \qquad 2\left(\frac{10}{3}\right) + \left(\frac{-2}{3}\right) = 6 \qquad 2\left(\frac{10}{3}\right) - 2\left(-\frac{2}{3}\right) = 8$$

$$y = \frac{18}{3} - \frac{20}{3} \qquad \frac{20}{3} + \frac{-2}{3} = 6 \qquad \frac{20}{3} + \frac{4}{3} = 8$$

$$y = -\frac{2}{3} \quad y\text{-value of solution.} \qquad \frac{18}{3} = 6 \qquad\qquad \frac{24}{3} = 8$$

$$6 = 6 \qquad\qquad 8 = 8$$

**Solution:** $\left(\dfrac{10}{3}, -\dfrac{2}{3}\right)$     Solution checks.

**Section 8–4**

**1** Use a system of linear equations to solve application problems (pp. 376–378).

Let two variables represent the unknown values. Use numbers and the variables to represent the conditions of the problem. Set up a system of equations and solve.

A taxidermist bought two pairs of glass deer eyes and five pairs of glass duck eyes for $15. She later purchased three pairs of deer eyes and five pairs of duck eyes for $20. Find the price per pair of each type of glass eye.

Let $x$ = the price of a pair of deer eyes. Let $y$ = the price of a pair of duck eyes.
Price = the number of pairs of eyes times price of eye type:

| | |
|---|---|
| $2x + 5y = 15$ | 2 pairs of deer eyes plus 5 pairs of duck eyes cost $15. |
| $3x + 5y = 20$ | 3 pairs of deer eyes plus 5 pairs of duck eyes cost $20. |
| $-1(2x + 5y) = -1(15)$ | Multiply first equation by $-1$. Add equations to eliminate $y$. |
| $-2x - 5y = -15$ | |
| $\underline{\phantom{-}3x + 5y = 20}$ | |
| $x \qquad\quad = 5$ | $x$-value of solution. |
| $2x + 5y = 15$ | Use first equation to find $y$. Substitute $x = 5$. |
| $2(5) + 5y = 15$ | |
| $10 + 5y = 15$ | Solve for $y$. |
| $5y = 15 - 10$ | |
| $5y = 5$ | |
| $\dfrac{5y}{5} = \dfrac{5}{5}$ | |
| $y = 1$ | $y$-value of solution. |

**The deer eyes cost $5 a pair and the duck eyes cost $1 a pair.**

Chapter Review of Key Concepts

## Section 8–1

Solve the systems of equations by graphing.

**1.** $x + y = 8$
$x - y = 2$

**2.** $3x + 2y = 13$
$x - 2y = 7$

**3.** $2x + 2y = 10$
$3x + 3y = 15$

**4.** $x + y = 1$
$3x - 4y = 10$

## Section 8–2

Solve the systems of equations using the addition method.

**5.** $3x + y = 9$
$2x - y = 6$

**6.** $2a + 3b = 8$
$a - b = 4$

**7.** $Q = 2P + 8$
$2Q + 3P = 2$

**8.** $4j + k = 3$
$8j + 2k = 6$

**9.** $r = 2y + 6$
$2r + y = 2$

**10.** $3a + 3b = 3$
$2a - 2b = 6$

**11.** $c = 2y$
$2c + 3y = 21$

**12.** $2x + 4y = 9$
$x + 2y = 3$

**13.** $3R - 2S = 7$
$-14 = -6R + 4S$

**14.** $x - 3 = -y$
$2y = 9 - x$

**15.** $c = 2 + 3d$
$3c - 14 = d$

**16.** $Q - 10 = T$
$T = 2 - 2Q$

**17.** $x - 18 = -6y$
$4x - 0 = 3y$

**18.** $R + S = 3$
$S - 9 = -3R$

**19.** $3a - 2b = 6$
$6a - 12 = b$

**20.** $a - b = 2$
$a + b = 12$

**21.** $x + 2y = 7$
$x - y = 1$

**22.** $2c + 3b = 2$
$2c - 3b = 0$

## Section 8–3

Solve the systems of equations using either the addition or the substitution method.

**23.** $a + 7b = 32$
$3a - b = 8$

**24.** $x + y = 1$
$4x + 3y = 0$

**25.** $c - d = 2$
$c = 12 - d$

**26.** $3a + 4b = 0$
$a + 3b = 5$

**27.** $7x - 4 = -4y$
$3x + y = 6$

**28.** $5Q - 4R = -1$
$R + 3Q = -38$

**29.** $a = 2b + 11$
$3a + 11 = -5b$

**30.** $y = 5 - 2x$
$3x - 2y = 4$

**31.** $c = 2q$
$2c + q = 2$

**32.** $3x + 2y = 10$
$y = 6 - x$

**33.** $4x - 2.5y = 2$
$2x - 1.5y = -10$

**34.** $2a - c = 4$
$a = 2 + c$

**35.** $4d - 7 = -c$
$3c - 6 = -6d$

**36.** $x = 10 - y$
$5x + 2y = 11$

**37.** $3.5a + 2b = 2$
$0.5b = 3 - 1.5a$

**38.** $c + d = 12$
$c - d = 2$

**39.** $x + 4y = 20$
$4x + 5y = 58$

**40.** $a + 5y = 7$
$a + 4y = 8$

## Section 8–4

Solve the problems using systems of equations with two unknowns.

**41. ELEC** Three electricians and four apprentices earned a total of $365 on one job. At the same rate of pay, one electrician and two apprentices earned a total of $145. How much pay did each apprentice and electrician receive?

**42. AG/H** Six bushels of bran and 2 bushels of corn weigh 182 lb. If 2 bushels of bran and 4 bushels of corn weigh 154 lb, how much do 1 bushel of bran and 1 bushel of corn each weigh?

**43. CON** A painter paid $22.50 for 2 qt of white shellac and 5 qt of thinner. If 3 qt of shellac and 2 qt of thinner cost the painter's helper $14.50, what is the cost of each quart of shellac and thinner?

**44. ELEC** A main current of electricity is the sum of two smaller currents whose difference is 0.8 A. What are the two smaller currents if the main current is 10 A?

**45.** The sum of two angles is 175°. Their difference is 63°. What is the measure of each angle?

**46. HOSP** A tour boat traveled 20 mi in 2 h with the current. The return trip took 3 h. Find the rate of the boat in calm water and the rate of the current.

**47. BUS** In 1 year, Sholanda Brown earned $280 on two investments totaling $5,000. If she received 5% and 6% rates of return, how much did she invest at each rate?

**48. AVIA** A plane flew 300 km against the wind in 4 h. The return trip with the wind took 3 h. How fast was the wind? What would have been the speed of the plane in calm air?

**49.** **HOSP** A restaurant purchased 30 lb of Colombian coffee and 10 lb of blended coffee for $190. In a second purchase, the same restaurant paid $120 for 20 lb of Colombian coffee and 5 lb of blended coffee. How much did each coffee cost per lb?

**51.** **AUTO** An automotive service station purchased 25 maps of Ohio and 8 maps of Alaska at a total cost of $65.55. Later the station purchased 20 maps of Ohio and 5 maps of Alaska for $49.50. Find the cost of each map.

**53.** **BUS** At the first of the month, store buyer Selena Henson placed a $12,525 order for 20 name-brand suits and 35 suits with generic labels. At the end of the month, she placed a $15,725 order for 30 name-brand suits and 35 generic-label suits. How much did she pay for each type of suit?

**55.** **BUS** Jorge makes 5% commission on telephone sales and 6% commission on showroom sales. If his sales totaled $40,000 and his commission was $2,250, how much did he sell by telephone? How much did he sell on the showroom floor?

**50.** **AG/H** A rancher wants to spread a 300-lb grass seed mixture that is 50% tall fescue. If the rancher has on hand a seed mixture that is 80% tall fescue and a mixture that is 20% tall fescue, how many pounds of each mixture are needed to make 300 lb of the 50% mixture?

**52.** **BUS** Kristen Ammons made two 1-yr investments totaling $7,000 at interest rates of 4% and 7%. If she received $415 in return, how much money was invested at each rate?

**54.** **BUS** A taxidermist has a container with a solution of 10% tanning chemical and a container with a solution of 50% tanning chemical. If the taxidermist wants to mix the solutions to get 10 gal of solution with 25% tanning chemical, how many gallons of each solution should be mixed?

**56.** **BUS** Sing-Fong has 60 coins in nickels and quarters. The total value of the coins is $12. How many coins of each type does she have?

## TEAM PROBLEM-SOLVING EXERCISES

**1.** A true measure of your understanding of a concept is the ability to apply this understanding to real-world situations.
   **(a)** Write a story line for a mixture problem that can be solved using the system

$$0.5x + 0.3y = 8$$
$$x + y = 20$$

   **(b)** Solve the system and interpret the results within the context of your story line.

**2.** Select a career or business situation of interest to your team.
   **(a)** Write a story line for a mixture problem that can be solved using the system

$$3x + 5y = 28$$
$$x - y = 4$$

   **(b)** Solve the system and interpret the results within the context of your story line.

## PRACTICE TEST

Solve the systems of equations graphically.

**1.** $2a + b = 10$
$\phantom{2}a - b = 5$
**4.** $a - 3b = 7$
$\phantom{a}a - 5 = b$

**2.** $x + y = 8$
$\phantom{x}3x + 2y = 12$
**5.** $2c - 3d = 6$
$\phantom{2}c - 12 = 3d$

**3.** $3x + 4y = 6$
$\phantom{3}x + y = 5$

Solve the systems of equations using the addition method.

**6.** $x + y = 6$
$\phantom{x}x - y = 2$
**9.** $3x + y = 5$
$\phantom{3}2x - y = 0$

**7.** $p + 2m = 0$
$\phantom{p + 2}2p = -m$
**10.** $7c - 2b = -2$
$\phantom{7}c - 4b = -4$

**8.** $6p + 5t = -16$
$\phantom{6}3p - 3 = 3t$

Solve the systems of equations with two unknowns using the substitution method.

**11.** $4x + 3y = 14$
$\phantom{4}x - y = 0$

**12.** $a + 2y = 6$
$\phantom{a}a + 3y = 3$

**13.** $7p + r = -6$
$\phantom{7}3p + r = 6$

Solve the systems of equations with two unknowns using either the addition or the substitution method.

**14.** $4x + 4 = -4y$
$\phantom{4}6 + y = -6x$

**15.** $38 + d = -3a$
$\phantom{3}5a + 1 = 4d$

Solve the problems using systems of equations with two unknowns.

**16.** Two lengths of stereo speaker wire total 32.5 ft. One length is 2.9 ft longer than the other. How long is each length of speaker wire?

**17.** Two currents add to 35 A and their difference is 5 A. How many amperes are in each current?

**18.** Six packages of common nails and four packages of finishing nails weigh 6.5 lb. If two packages of common nails and three packages of finishing nails weigh 3.0 lb, how much does one package of each kind of nail weigh?

**19.** The length of a piece of sheet metal is $1\frac{1}{2}$ times the width. The difference between the length and the width is 17 in. Find the length and the width.

**20.** A mixture of fieldstone is needed for a construction job and will cost $954 for the 27 tons of stone. The stone is of two types, one costing $38 per ton and one costing $32 per ton. How many tons of each are required?

**21.** A broker invested $25,000 in two different stocks. One stock earned dividends at 3.5% and the other at 4%. If a dividend of $900 was earned on both stocks together, how much was invested in each stock?

**22.** A mason purchased 2-in. cold-rolled channels and $\frac{3}{4}$-in. cold-rolled channels whose total weight was 820 lb. The difference in weight between the heavier 2-in. and lighter $\frac{3}{4}$-in. channels was 280 lb. How many pounds of each type of channel did the mason purchase?

**23.** A total capacitance in parallel is the sum of two capacitances. If the capacitances total 0.00027 farad (F) and the difference between the two capacitances is 0.00016 F, what is the value of each capacitance in the system?

---

## CAREER APPLICATION: ELECTRONICS—MESH CURRENTS

Circuits are frequently solved using mesh currents to determine individual currents in a circuit. For instance, the analysis of a circuit with three currents might yield the following system of equations for $I_1$, $I_2$, and $I_3$.

$$I_1 - I_2 - I_3 = 0 \qquad \text{Equation 1.}$$
$$6I_1 + 4I_3 = 12 \text{ A} \qquad \text{Equation 2.}$$
$$3I_2 - 4I_3 = -3 \text{ A} \qquad \text{Equation 3.}$$

There are several ways to solve this system. One way is to solve the first equation for $I_1$ in terms of $I_2$ and $I_3$ and substitute that into the second equation. Then the second and third equations form a system of equations in two unknowns, which can be solved using several methods. The proof is to put all the final values back into the original equation to see if they work.

$$I_1 - I_2 - I_3 = 0 \qquad \text{Rearrange Equation 1 to give} \quad I_1 = I_2 + I_3$$

Substituting that expression for $I_1$ into the second equation gives

$$6\boxed{I_1} + 4I_3 = 12 \qquad \text{Equation 2.}$$
$$6(\boxed{I_2 + I_3}) + 4I_3 = 12 \qquad \text{Distribute and combine like terms.}$$
$$6I_2 + 10I_3 = 12 \qquad \text{Divide each term by 2.}$$
$$3I_2 + 5I_3 = 6$$

This gives the following system of equations from the modified Equation 2 and the original Equation 3. (Illustrated is the addition method of solving a system of equations.)

**Substituting into Equation 3:**

| | | | |
|---|---|---|---|
| **New Equation 2:** | $3I_2 + 5I_3 = 6 \text{ A}$ | $3I_2 + 5I_3 = 6$ | $3I_2 - 4\boxed{I_3} = -3 \text{ A}$ |
| **Equation 3:** | $3I_2 - 4I_3 = -3 \text{ A}$ | $\underline{-3I_2 + 4I_3 = 3}$ | $3I_2 - 4(\boxed{1 \text{ A}}) = -3 \text{ A}$ |
| | | $9I_3 = 9$ | $3I_2 = -3 \text{ A} + 4 \text{ A}$ |
| | | $\boxed{I_3 = 1 \text{ A}}$ | $3I_2 = 1 \text{ A}$ |
| | | | $I_2 = 0.333333 \text{ A}$ |

Then we go back to rearranged Equation 1 and substitute values for $I_2$ and $I_3$.

$$I_1 = I_2 + I_3 = 0.333333 \text{ A} + 1 \text{ A} = +1.333333 \text{ A}$$

**Thus, the solution is $I_1 = 1.333333 \text{ A}$, $I_2 = 0.333333 \text{ A}$, and $I_3 = 1 \text{ A}$.**

**Proof:**

**Equation 1**

$$I_1 - I_2 - I_3 = 0$$
$$1.333333\,\text{A} - 0.333333\,\text{A} - 1\,\text{A} = 0$$
$$0 = 0$$

**Equation 2**

$$6I_1 + 4I_3 = 12\,\text{A}$$
$$6(1.333333\,\text{A}) + 4(1\,\text{A}) = 12\,\text{A}$$
$$12\,\text{A} = 12\,\text{A}$$

**Equation 3**

$$3I_2 - 4I_3 = -3\,\text{A}$$
$$3(0.333333\,\text{A}) - 4(1\,\text{A}) = -3\,\text{A}$$
$$-3\,\text{A} = -3\,\text{A}$$

## Exercises

Solve each system of equations, which are derived from actual circuits. Prove your answers by going back to the original equations.

1. $I_1 - 2I_2 + 3I_3 = 4\,\text{A}$
   $2I_1 + I_2 - 4I_3 = 3\,\text{A}$
   $I_1 + 2I_3 = 8\,\text{A}$

2. $I_1 + I_2 + I_3 = -4\,\text{A}$
   $2I_1 + 3I_2 + 4I_3 = 0$
   $-I_1 - I_2 + 2I_3 = 8\,\text{A}$

3. $I_1 + I_2 + I_3 = 4\,\text{A}$
   $2I_1 + 3I_2 + 4I_3 = 0$
   $-I_1 - I_2 + 2I_3 = 8\,\text{A}$

4. $3I_1 + 2I_2 + 2I_3 = 3\,\text{A}$
   $2I_1 + 6I_2 + 3I_3 = 0$
   $I_1 + 2I_2 + I_3 = 1\,\text{A}$

5. $I_1 - I_2 + 3I_3 = 2\,\text{A}$
   $-I_1 + I_2 = 7\,\text{A}$
   $-I_1 + 2I_2 + 6I_3 = 4\,\text{A}$

## Answers

| System | Solutions | Proof |
|---|---|---|
| **1.** $I_1 - 2I_2 + 3I_3 = 4\,\text{A}$ <br> $2I_1 + I_2 - 4I_3 = 3\,\text{A}$ <br> $I_1 + 2I_3 = 8\,\text{A}$ | $I_1 = 4\,\text{A}$ <br> $I_2 = 3\,\text{A}$ <br> $I_3 = 2\,\text{A}$ | $4 - 6 + 6 = 4$ <br> $8 + 3 - 8 = 3$ <br> $4 + 4 = 8$ |
| **2.** $I_1 + I_2 + I_3 = -4\,\text{A}$ <br> $2I_1 + 3I_2 + 4I_3 = 0$ <br> $-I_1 - I_2 + 2I_3 = 8\,\text{A}$ | $I_1 = -10.667\,\text{A}$ <br> $I_2 = 5.333\,\text{A}$ <br> $I_3 = 1.333\,\text{A}$ | $-10.667 + 5.333 + 1.333 = -4$ <br> $-21.333 + 16 + 5.333 = 0$ <br> $10.667 - 5.333 + 2.667 = 8$ |
| **3.** $I_1 + I_2 + I_3 = 4\,\text{A}$ <br> $2I_1 + 3I_2 + 4I_3 = 0$ <br> $-I_1 - I_2 + 2I_3 = 8\,\text{A}$ | $I_1 = 16\,\text{A}$ <br> $I_2 = -16\,\text{A}$ <br> $I_3 = 4\,\text{A}$ | $16 - 16 + 4 = 4$ <br> $32 - 48 + 16 = 0$ <br> $-16 + 16 + 8 = 8$ |
| **4.** $3I_1 + 2I_2 + 2I_3 = 3\,\text{A}$ <br> $2I_1 + 6I_2 + 3I_3 = 0$ <br> $I_1 + 2I_2 + I_3 = 1\,\text{A}$ | $I_1 = 3\,\text{A}$ <br> $I_2 = 1\,\text{A}$ <br> $I_3 = -4\,\text{A}$ | $9 + 2 - 8 = 3$ <br> $6 + 6 - 12 = 0$ <br> $3 + 2 - 4 = 1$ |
| **5.** $I_1 - I_2 + 3I_3 = 2\,\text{A}$ <br> $-I_1 + I_2 = 7\,\text{A}$ <br> $-I_1 + 2I_2 + 6I_3 = 4\,\text{A}$ | $I_1 = -28\,\text{A}$ <br> $I_2 = -21\,\text{A}$ <br> $I_3 = 3\,\text{A}$ | $-28 + 21 + 9 = 2$ <br> $28 - 21 = 7$ <br> $28 - 42 + 18 = 4$ |

Simplify.

**1.** $4 - 3(2x - 5)$

**2.** $5x - (3x - 2)$

Solve.

**3.** $7x - 3(x - 8) = 28$

**4.** $18 - 6(2 - y) = 24$

**5.** $\dfrac{5}{12}x - \dfrac{3}{4} = \dfrac{1}{9} - \dfrac{2}{3}x$

**6.** $4x - 3.2 + x = 3.3 - 2.4x$

**7.** Find the impedance ($Z$) in ohms using the formula $Z = \sqrt{R^2 + X^2}$ if the resistance ($R$) is 4 $\Omega$ and the reactance ($X$) is 7 $\Omega$. Round to tenths.

Solve.

**8.** $\dfrac{3x}{8} = \dfrac{3}{4}$

**9.** $\dfrac{2x}{7} = \dfrac{3}{5}$

**10.** $\dfrac{2}{12} = \dfrac{7}{4x}$

**11.** A large gear has 400 teeth and turns at 30 rpm. Find the number of teeth of a small gear in mesh that turns at 75 rpm.

**12.** A blueprint has a scale of $\dfrac{1}{4}$ in. = 1 ft. What is the actual measurement of a wall that is five inches on the blueprint?

**13.** Graph using a table of solutions: $y = -2x + 4$

**14.** Graph using the intercepts procedure: $2x - y = 6$

**15.** Graph using the slope and $y$-intercept procedure:

$y - \dfrac{3}{4}x = 2$

Determine the slope and $y$-intercept.

**16.** $y = \dfrac{3}{4}x + 2$

**17.** $3x - 2y = 10$

**18.** Find the slope of the line passing through the points $(5, 2)$ and $(-2, 3)$.

Find the equation of the line passing through the given point with the given slope. Write the equation in slope $y$-intercept form.

**19.** $(5, -4), m = -\dfrac{2}{3}$

**20.** $(-1, -5), m = 3$

Find the equation of the line passing through the given pairs of points. Solve the equation for $y$.

**21.** $(3, 2)$ and $(5, 6)$

**22.** $(-2, 1)$ and $(4, -2)$

Write the equation with the given slope and $y$-intercept. Solve the equation for $y$.

**23.** $m = 4, b = 3$

**24.** slope $= -2$, $y$-intercept $= \dfrac{1}{3}$

**25.** Write the equation in standard form of the line that is parallel to the line $x - y = 3$ and contains the point $(5, 4)$.

**26.** Write the equation in standard form of the line that is perpendicular to the line $x - 2y = 6$ and passes through the point $(3, 0)$.

Solve the system of equations using the addition method.

**27.** $2x - y = -1$
$\phantom{2}x + y = 4$

**28.** $3x - 2y = -1$
$2x + 4y = 10$

**29.** Two resistances have a sum of 32 $\Omega$ and a difference of 8 $\Omega$. Find each resistance.

**30.** The total liquid in a herbicide mixture is 20 L. If the mixture has 4 times as much water as herbicide, how many liters of each substance is in the mixture?

# 9

# Powers and Polynomials

## Focus on Careers

If you have always liked to take apart and fix things, you may want to consider a career as a diesel service technician or mechanic. The nation's trucks and buses are powered by diesel engines because the diesel engine is more powerful and more durable than comparable gasoline-burning engines. Diesel technicians repair and maintain these trucks and buses. They also maintain locomotive engines and other heavy equipment such as bulldozers, cranes, road graders, fork lifts, farm tractors, and other farm equipment.

Diesel technicians most often work in a covered shop, but sometimes they repair vehicles on the road. They are required to lift heavy parts and handle greasy and dirty parts.

Most community colleges and trade and vocational schools offer programs in diesel repair. Employers prefer to hire graduates of formal training programs when they are hiring entry-level diesel mechanics. These training programs offer the most current diesel technology and

*(continued)*

instruction, and graduates also acquire increased skills needed to interpret technical manuals and communicate with coworkers and customers.

The National Institute for Automotive Service Excellence offers certification as master truck technicians or in specific areas of truck repair. Certifications give job seekers a clear advantage over those without certification.

Jobs in the diesel service technical field are expected to grow about as fast as average through 2012. Opportunities will be best for those who complete formal training at community and junior colleges or vocational and technical schools.

The middle 50% of diesel engine specialists earned between $13.13 and $24.54 per hour in 2002 and the highest 10% earned more than $24.61 per hour! Most service technicians work a standard 40-hr week and usually earn overtime for additional hours and a higher rate for evening, night, or weekend hours.

Source: *Occupational Outlook Handbook* 2004–2005 Edition. U.S. Department of Labor, Bureau of Labor Statistics.

## 9–1 | Laws of Exponents

*Learning Outcomes*

**1** Multiply powers with like bases.

**2** Divide powers with like bases.

**3** Find a power of a power.

We have seen in various equations and formulas that a variable can represent all types of numbers: natural numbers, whole numbers, integers, fractions, decimals, and signed numbers. When raising a variable to a power, we must consider all the types of numbers the variable can be.

The **laws of exponents** have evolved from the patterns that formed when interpreting powers as repeated multiplication. These laws allow us to shorten our processes and to take advantage of the accessibility of technology. Be sure to notice for what circumstances a law is applicable. For example, many of the laws apply to factors with *like* bases.

### **1** Multiply Powers with Like Bases.

To multiply $2^3$ by $2^2$ using repeated multiplication, we have $2(2)(2)$ times $2(2)$ or $2(2)(2)(2)(2)$. Another way of writing this is $2^3 \times 2^2 = 2^5$. We can multiply $x^4$ and $x^2$. Even though the value of $x$ is not known, $x^4(x^2)$ is $x(x)(x)(x)$ times $x(x)$ or $x(x)(x)(x)(x)(x)$.

$$x^4(x^2) = x^6$$

The product contains $4 + 2$ or 6 factors of $x$. The following law is a shortcut for using repeated multiplication.

### To multiply powers that have like bases:

1. Verify that the bases are the same. Use this base as the base of the product.
2. Add the exponents for the exponent of the product.

This rule can be stated symbolically as

$$a^m(a^n) = a^{m+n} \quad \text{where } a, m, \text{ and } n \text{ are real numbers.}$$

---

**EXAMPLE** Write the products.

**(a)** $y^4(y^3)$    **(b)** $a(a^2)$    **(c)** $b(b)$    **(d)** $x(x^3)(x^2)$    **(e)** $x^2(y^3)$    **(f)** $5x^3(2x^4)$

**(a)** $y^4(y^3) = y^{4+3} = \boldsymbol{y^7}$      The bases are the same, so add the exponents.

**(b)** $a(a^2) = a^{1+2} = \boldsymbol{a^3}$      When the exponent for a base is not written, it is 1.

**(c)** $b(b) = b^{1+1} = b^2$

**(d)** $x(x^3)(x^2) = x^{1+3+2} = x^6$

**(e)** $x^2(y^3) = x^2y^3$      Bases are unlike.

**(f)** $5x^3(2x^4) = 10x^7$      Multiply coefficients. Add exponents of like bases.

**TIP!**

## Do You Always Add Exponents When Multiplying?

The previous law applies *only* to expressions with *like* bases. Thus, $x^2(y^3)$ can only be written as $x^2y^3$. No other simplification can be made.

### 2   Divide Powers with Like Bases.

When reducing fractions, we can reduce to a factor of 1 any factors common to both the numerator and denominator. We use this concept when dividing powers that have like bases.

**EXAMPLE**   Reduce or simplify the fractions; that is, perform the division indicated by each fraction.

**(a)** $\dfrac{x^5}{x^2}$     **(b)** $\dfrac{a^2}{a}$     **(c)** $\dfrac{b^7}{c^5}$     **(d)** $\dfrac{m^3}{m^3}$     **(e)** $\dfrac{y^3}{y^4}$

**(a)** $\dfrac{x^5}{x^2} = \dfrac{x(x)(x)(\cancel{x})(\cancel{x})}{\cancel{x}(\cancel{x})} = x^3$      The quotient contains $5 - 2 = 3$ factors of $x$, or $x^3$.

**(b)** $\dfrac{a^2}{a} = \dfrac{a(\cancel{a})}{\cancel{a}} = a$  (**or** $a^1$)      The quotient contains $2 - 1 = 1$ factor of $a$.

**(c)** $\dfrac{b^7}{c^5} = \dfrac{b^7}{c^5}$      Bases are unlike, so no simplification can be made.

**(d)** $\dfrac{m^3}{m^3} = \dfrac{(\cancel{m})(\cancel{m})(\cancel{m})}{(\cancel{m})(\cancel{m})(\cancel{m})} = 1$      All factors of $m$ reduce to 1.

**(e)** $\dfrac{y^3}{y^4} = \dfrac{(\cancel{y})(\cancel{y})(\cancel{y})}{(y)(\cancel{y})(\cancel{y})(\cancel{y})} = \dfrac{1}{y}$      The denominator has more factors of $y$ than the numerator.

In the example above examine the exponents in the division and the exponent of the quotient. In parts a and b, the exponent of the quotient is the difference of the exponents of the dividend and the divisor.

**(a)** $\dfrac{x^5}{x^2} = x^{5-2} = x^3$      **(b)** $\dfrac{a^2}{a} = a^{2-1} = a^1 = a$

The results in parts d and e illustrate that the interpretation of exponents goes beyond just repeated multiplication; a new notation for 1 and for reciprocals is needed.

**(d)** $\dfrac{m^3}{m^3} = m^{3-3} = m^0 = 1$      Any nonzero base raised to the zero power equals 1.

**(e)** $\dfrac{y^3}{y^4} = y^{3-4} = y^{-1} = \dfrac{1}{y^1} = \dfrac{1}{y}$      An expression with a negative exponent equals the reciprocal of the expression written with a positive exponent having the same absolute value.

9–1 Laws of Exponents              

## Reciprocals and Negative Exponents

Any nonzero number raised to the zero power is equal to 1.

$$n^0 = 1 \quad \text{if} \quad n \neq 0.$$

An expression with a *negative exponent* can be written as an equivalent expression with a positive exponent.

$$n^{-1} = \frac{1}{n} \qquad \frac{1}{n^{-1}} = n \qquad \text{if } n \neq 0$$

A nonzero number times its reciprocal equals 1.

$$n \cdot \frac{1}{n} = 1 \qquad n^1 \cdot n^{-1} = n^0 = 1$$

### To divide powers that have like bases:

1. Verify that the bases are the same. Use this base as the base of the quotient.
2. Subtract the exponents for the exponent of the quotient.

This rule can be stated symbolically as

$$\frac{a^m}{a^n} = a^{m-n}$$

where *a*, *m*, and *n* are real numbers except that $a \neq 0$.

Because expressions with positive integral exponents are evaluated by using repeated multiplication, it is preferable to rewrite expressions with negative exponents as equivalent expressions with positive exponents. This manipulation is accomplished by applying the definition of negative exponents and exponents of zero.

---

**EXAMPLE**  Write the quotients using positive exponents.

(a) $\dfrac{x^5}{x^8}$  (b) $\dfrac{a}{a^4}$  (c) $\dfrac{y^{-3}}{y^2}$  (d) $\dfrac{x^3}{x^{-5}}$  (e) $\dfrac{12\,x^7}{8\,x^5}$

(a) $\dfrac{x^5}{x^8} = x^{5-8} = x^{-3} = \dfrac{1}{x^3}$
  (b) $\dfrac{a}{a^4} = a^{1-4} = a^{-3} = \dfrac{1}{a^3}$

(c) $\dfrac{y^{-3}}{y^2} = y^{-3-2} = y^{-5} = \dfrac{1}{y^5}$
  (d) $\dfrac{x^3}{x^{-5}} = x^{3-(-5)} = x^{3+5} = x^8$

(e) $\dfrac{12\,x^7}{8\,x^5} = \dfrac{3\,x^{7-5}}{2} = \dfrac{3\,x^2}{2}$ or $\dfrac{3}{2}x^2$

---

TIP!

The inverse relationship of multiplication and division allows us flexibility in applying the laws of exponents. Look at part a of the previous example.

$$\frac{x^5}{x^8} \quad \text{is} \quad x^5 \div x^8 \quad \text{or} \quad x^5 \cdot \frac{1}{x^8} \quad \text{or} \quad x^5 \cdot x^{-8}$$

The multiplication law of exponents can be used.

$$x^5 \cdot x^{-8} = x^{5+(-8)} = x^{-3}$$

**A factor can be moved from a numerator to a denominator (or vice versa) by changing the sign of its exponent.**

$$x^{-2} = \frac{x^{-2}}{1} = \frac{1}{x^2}, \qquad \frac{1}{x^{-3}} = \frac{x^3}{1} = x^3$$

This property *does not apply* to a term that is part of a numerator or denominator that has two or more terms.

$$\frac{x^{-2}+1}{3} \quad does\ not\ equal \quad \frac{1}{3x^2}.$$

In other words, *only factors* of the entire numerator or denominator can be moved. If more than one term is in the numerator, each term is divided by the denominator.

$$\frac{x^{-2}+1}{3} = \frac{x^{-2}}{3} + \frac{1}{3} = \frac{1}{3x^2} + \frac{1}{3}$$

---

**EXAMPLE** Simplify the expressions and make all exponents positive.

(a) $\dfrac{a^2 b^{-3}}{ab^{-1}}$  (b) $\dfrac{xy^{-1}}{xy^2}$  (c) $\dfrac{x^3 + y^2}{xy}$

There is more than one way to simplify the expressions.

**(a)**

**Option 1**

$$\frac{a^2 b^{-3}}{ab^{-1}} = a^{2-1} b^{-3-(-1)}$$

Apply the division law of exponents. $2 - 1 = 1$; $-3 - (-1) = -3 + 1 = -2$. Be careful with the signs.

$$= ab^{-2}$$

Make all exponents positive.

$$= \frac{a}{b^2}$$

**Option 2**

$$\frac{a^2 b^{-3}}{ab^{-1}} = \frac{a^2 a^{-1} b^{-3} b^1}{1}$$

Apply the property of negative exponents to write all factors in the numerator.

$$= ab^{-2} = \frac{a}{b^2}$$

Apply the multiplication law of exponents and make all exponents positive.

**(b)** $\dfrac{xy^{-1}}{xy^2} = x^{1-1} y^{-1-(2)}$      Apply the division law of exponents.

$= y^{-3}$      Make exponent positive.

$= \dfrac{1}{y^3}$

**(c)** $\dfrac{x^3 + y^2}{xy} = \dfrac{x^3}{xy} + \dfrac{y^2}{xy}$      Separate into two terms and apply division law to like bases.

$= \dfrac{x^{3-1}}{y} + \dfrac{y^{2-1}}{x}$

$= \dfrac{x^2}{y} + \dfrac{y}{x}$

### 3   Find a Power of a Power.

In the term $(2^3)^2$, we have a power raised to a power. $2^3$ is the base of the expression and 2 is the exponent.

$$(2^3)^2 \quad \text{is} \quad (2^3)(2^3) = 2^{3+3} = 2^6 \quad \text{or} \quad 64$$

Let's look at some examples of raising numerical and variable powers to a power.

**EXAMPLE**    Find the powers of powers by first expressing each expression as repeated multiplication and then applying the multiplication law of exponents.

**(a)** $(3^2)^3$      **(b)** $(x^3)^4$      **(c)** $(a^2)^2$      **(d)** $(n^3)^5$

**(a)** $(3^2)^3 = (3^2)(3^2)(3^2) = 3^{2+2+2} = 3^6 = \boldsymbol{729}$

**(b)** $(x^3)^4 = (x^3)(x^3)(x^3)(x^3) = x^{3+3+3+3} = \boldsymbol{x^{12}}$

**(c)** $(a^2)^2 = (a^2)(a^2) = a^{2+2} = \boldsymbol{a^4}$

**(d)** $(n^3)^5 = (n^3)(n^3)(n^3)(n^3)(n^3) = n^{3+3+3+3+3} = \boldsymbol{n^{15}}$

From the preceding example we can see a pattern developing. In each case, if we multiply the exponent, we get the exponent of the new power.

**To raise a power to a power:**

1. Multiply exponents.
2. Keep the same base.

$$(a^m)^n = a^{mn}$$

where $a$, $m$, and $n$ are real numbers.

Applying this rule to the problems in the preceding example, we have

(a) $(3^2)^3 = 3^{2(3)} = 3^6 = 729$        (b) $(x^3)^4 = x^{3(4)} = x^{12}$

(c) $(a^2)^2 = a^{2(2)} = a^4$        (d) $(n^3)^5 = n^{3(5)} = n^{15}$

     Other laws of exponents are extensions or combinations of the three laws we have already examined. Although these additional laws are not necessary, they are useful tools for simplifying expressions containing exponents.

     Chapter 9 / Powers and Polynomials

## To raise a fraction or quotient to a power:

1. Raise the numerator to the power.
2. Raise the denominator to the power.

$$\left(\frac{a}{b}\right)^n = \frac{a^n}{b^n} \qquad b \neq 0$$

**EXAMPLE** Raise the fractions to the indicated power.

(a) $\left(\dfrac{2}{3}\right)^2$      (b) $\left(\dfrac{-3}{4}\right)^2$      (c) $\left(\dfrac{-1}{3}\right)^3$      (d) $\left(\dfrac{x}{y^3}\right)^3$      (e) $\left(\dfrac{x^2}{y^3}\right)^4$

(a) $\left(\dfrac{2}{3}\right)^2 = \dfrac{2^2}{3^2} = \dfrac{4}{9}$

(b) $\left(\dfrac{-3}{4}\right)^2 = \dfrac{(-3)^2}{4^2} = \dfrac{9}{16}$      $(-3)(-3) = +9.$

(c) $\left(\dfrac{-1}{3}\right)^3 = \dfrac{(-1)^3}{3^3} = \dfrac{-1}{27}$      $(-1)(-1)(-1) = -1.$

(d) $\left(\dfrac{x}{y^3}\right)^3 = \dfrac{x^3}{(y^3)^3} = \dfrac{x^3}{y^9}$

(e) $\left(\dfrac{x^2}{y^3}\right)^4 = \dfrac{(x^2)^4}{(y^3)^4} = \dfrac{x^8}{y^{12}}$

## To raise a product to a power:

Raise each factor to the indicated power.

$$(ab)^n = a^n b^n$$

**EXAMPLE** Raise the products to the indicated powers.

(a) $(ab)^2$    (b) $(a^2b)^3$    (c) $(xy^2)^2$    (d) $(3x)^2$    (e) $(2x^2y)^3$    (f) $(-5xy)^2$

(a) $(ab)^2 = a^{1(2)}\, b^{1(2)} = a^2\, b^2$      Unwritten exponents are understood to be 1.

(b) $(a^2b)^3 = a^{2(3)}\, b^{1(3)} = a^6\, b^3$

(c) $(xy^2)^2 = x^{1(2)}\, y^{2(2)} = x^2\, y^4$

(d) $(3x)^2 = 3^{1(2)}\, x^{1(2)} = 3^2\, x^2 = 9x^2$      Evaluate numerical factors.

(e) $(2x^2y)^3 = 2^{1(3)}\, x^{2(3)}\, y^{1(3)} = 2^3\, x^6\, y^3 = 8x^6y^3$

(f) $(-5xy)^2 = (-5)^{1(2)}\, x^{1(2)}\, y^{1(2)} = (-5)^2\, x^2\, y^2 = 25x^2y^2$

## Limitations of the Laws of Exponents

It is very important to understand what the laws of exponents *do not* include.

- The product-raised-to-power law applies to factors, not terms.

  $(a + b)^3$ **does not equal** $a^3 + b^3$

- The multiplication-of-powers law applies to like bases.

  $a^2(b^3)$ **does not equal** $ab^5$ or $(ab)^5$

- An exponent affects only the one factor or grouping immediately to the left.

  $3x^2$ and $(3x)^2$ **are not equal**

  In the term $3x^2$, the numerical coefficient 3 is multiplied times the square of $x$. In the term $(3x)^2$, $3x$ is squared. Thus, $(3x)^2 = 3^2x^2 = 9x^2$.

- A negative coefficient of a base is not affected by the exponent.

  $-x^3$ **means** $-(x)(x)(x)$

  If $x = 4$, $-x^3 = -(4)^3$ or $-(64) = -64$. If $x = -4$, $-x^3 = -(-4)^3$ or $-(-64) = 64$.

### SECTION 9–1 SELF-STUDY EXERCISES

**1** Write the products.

1. $x^3(x^4)$
2. $m(m^3)$
3. $a(a)$
4. $x^2(x^3)(x^5)$
5. $y(y^2)(y^3)$
6. $a^2(b)$
7. $3a^5(4b^7)$
8. $2x^3(5x^8)$

9. $(5x^2yz)(-2xy^2z^3)$
10. $(-8x^2y)(-3xy^3)$
11. $\left(\dfrac{3}{5}a^2b\right)\left(\dfrac{10}{21}ab^4\right)$
12. $\left(\dfrac{2}{3}a^3b^2\right)\left(\dfrac{9}{16}ab\right)$

13. $(-1.2m^2n^7)(3.5m^{-4}n^3)$
14. $(-3a^2)(-4a^3)(-6a)$
15. $(32x^3)\left(\dfrac{3}{4}x^4\right)$
16. $(4a^2b)(-3ab)(-2a^{-2}b^2)$

**2** Write the quotients with positive exponents.

17. $\dfrac{y^7}{y^2}$
18. $\dfrac{x^5}{x}$
19. $\dfrac{a^3}{a^4}$
20. $\dfrac{b^6}{b^5}$

21. $\dfrac{m^2}{m^2}$
22. $\dfrac{x}{x^3}$
23. $\dfrac{y^5}{y}$
24. $\dfrac{n^2}{n^{-5}}$

25. $\dfrac{x^{-4}}{x^6}$
26. $\dfrac{8n^2}{4n^3}$
27. $\dfrac{18x^7}{12x^0}$
28. $\dfrac{x^{-3}}{x^2}$

29. $\dfrac{x^2y^{-2}}{xy^4}$
30. $\dfrac{ab^2}{a^{-2}b^4}$
31. $\dfrac{x^2y^4}{xy^2}$
32. $\dfrac{a^3b^2}{a^2b^3}$

33. $\dfrac{21a^3b^{-4}}{7a^2b}$
34. $\dfrac{39r^2s^7}{26rs^{-5}}$
35. $\dfrac{12x^3y^5}{18x^7y^3}$
36. $\dfrac{51m^2n^5}{34m^{-4}n^2}$

**3** Simplify the expressions.

37. $(4^2)^3$
38. $(x^4)^2$
39. $(y^7)^0$
40. $(x^{10})^4$

**41.** $(a^7)^3$      **42.** $\left(-\dfrac{1}{2}\right)^3$      **43.** $\left(-\dfrac{2}{7}\right)^2$      **44.** $\left(\dfrac{a}{b}\right)^4$

**45.** $(2m^2n)^3$      **46.** $\left(\dfrac{x^2}{y}\right)^3$      **47.** $(x^2y^4)^3$      **48.** $(-2a)^2$

**49.** $(x^2y)^3$      **50.** $(-3ab^2)^3$      **51.** $(-7x^2)^5$      **52.** $(4x^2y^8)^2$

**53.** $-7x^3(x^{-5})^4$      **54.** $-x^4$      **55.** $(-x)^4$      **56.** $-6(x^2)^4$

**57.** $-x^3y(x^2)^5$      **58.** $-3x^9(x^{-2})^3$      **59.** $-x^2y^{-7}$      **60.** $(-3a^2bc)^3$

---

## 9–2 | Polynomials

*Learning Outcomes*

**1**   Identify polynomials, monomials, binomials, and trinomials.

**2**   Identify the degree of terms and polynomials.

**3**   Arrange polynomials in descending order.

**1**   **Identify Polynomials, Monomials, Binomials, and Trinomials.**

A **polynomial** is an algebraic expression in which the exponents of the variables are nonnegative integers and there are no variables in a denominator.

> **EXAMPLE**   Identify which expressions are polynomials. If an expression is not a polynomial, explain why.
>
> **(a)** $5x^2 + 3x + 2$     **(b)** $5x - \dfrac{3}{x}$     **(c)** $9$     **(d)** $-\dfrac{1}{2}x + 3x^{-2}$
>
> **(a)**   $5x^2 + 3x + 2$ **is a polynomial.**
>
> **(b)**   $5x - \dfrac{3}{x}$ **has a variable in the denominator** because the term $-\dfrac{3}{x}$ is equivalent to $-3x^{-1}$.
> A polynomial cannot have a variable with a negative exponent.
>
> **(c)**   **9 is a polynomial** because it is equivalent to $9x^0$ and the exponent, zero, is a nonnegative integer.
>
> **(d)**   $-\dfrac{1}{2}x + 3x^{-2}$ **is not a polynomial** because $3x^{-2}$ has a negative exponent.

Some polynomials have special names depending on the number of terms contained in the polynomial.

A **monomial** is a polynomial containing one term. A term may have more than one factor.

$$3, -2x, 5ab, 7xy^2, \frac{3a^2}{4} \text{ are monomials.}$$

A **binomial** is a polynomial containing two terms.

$$x + 3, 2x^2 - 5x, x + \frac{y}{4} \text{ are binomials.}$$

A **trinomial** is a polynomial containing three terms.

$$a + b + c, x^2 - 3x + 4, x + \frac{2a}{7} - 5 \text{ are trinomials.}$$

1. Write all variables in the numerator if necessary.
2. Expressions that have a negative exponent in any term are not polynomials.
3. Identify the expression based on the number of terms it contains.

---

**EXAMPLE**   Identify which of the expressions are polynomials; then state whether each polynomial is a monomial, binomial, or trinomial.

**(a)** $x^2y - 1$                         **(b)** $4(x - 2)$

**(c)** $\dfrac{2x + 5}{2y}$               **(d)** $3x^2 - x + 1$

**(a)** $x^2y - 1$         **Binomial**

**(b)** $4(x - 2)$         **Monomial**

**(c)** $\dfrac{2x + 5}{2y}$    This one-term expression is **not a monomial because it is not a polynomial** (there is a variable in the denominator).

**(d)** $3x^2 - x + 1$      **Trinomial**

---

**2**  **Identify the Degree of Terms and Polynomials.**

The **degree of a term** that has only one variable with a nonnegative exponent is the same as the exponent of the variable.

$$3x^4, \quad \text{fourth degree} \qquad -5x, \quad \text{first degree}$$

The degree of a *constant* is 0. A variable to the zero power is implied.

$$5 = 5x^0, \quad \text{degree zero}$$

If a term has more than one variable, the degree of the term is the *sum* of the exponents of all the variable factors.

$$2xy = 2x^1y^1, \quad \text{second degree} \qquad -2ab^2 = -2a^1b^2, \quad \text{third degree}$$

**To identify the degree of a term:**

1. Exponents of variable factors must be integers greater than zero.
2. For a term that is a constant, the degree is zero.
3. For a term that has only one variable factor, the degree is the same as the exponent of the variable.
4. For a term that has more than one variable factor, the degree is the sum of the exponents of the variables.

---

**EXAMPLE**   Identify the degree of each term in the polynomial $5x^3 + 2x^2 - 3x + 3$.

$5x^3$,  degree 3 (or third degree)
$2x^2$,  degree 2 (or second degree)
$-3x$,  degree 1 (or first degree)
$3$,   degree 0

---

Special names are associated with terms of degree 0, 1, 2, and 3. A **constant term** has degree 0. A **linear term** has degree 1. A **quadratic term** has degree 2. A **cubic term** has degree 3.

The **degree of a polynomial** that has only one variable and only positive integral exponents is the degree of the term with the largest exponent.

A **linear polynomial** has degree 1. A **quadratic polynomial** has degree 2. A **cubic polynomial** has degree 3.

$5x^3 - 2$ has a degree of 3 and is a cubic polynomial.

$x + 7$ has a degree of 1 and is a linear polynomial.

$7x^2 - 4x + 5$ has a degree of 2 and is a quadratic polynomial.

### To identify the degree of a polynomial:

1. Identify the degree of each term of the polynomial.
2. Compare the degrees of each term of the polynomial and select the greatest degree as the degree of the polynomial.

**EXAMPLE**  Identify the degree of the polynomials.

(a) $5x^4 + 2x - 1$     (b) $3x^3 - 4x^2 + x - 5$     (c) 7     (d) $x - \dfrac{1}{2}$

(a) $5x^4 + 2x - 1$ has a **degree of 4.**
(b) $3x^3 - 4x^2 + x - 5$ has a **degree of 3 and is a cubic polynomial.**
(c) 7 has a **degree of 0 and is a constant.**
(d) $x - \dfrac{1}{2}$ has a **degree of 1 and is a linear polynomial.**

**3**  **Arrange Polynomials in Descending Order.**

The terms of a polynomial are customarily arranged in order based on the degree of each term of the polynomial. The terms can be arranged beginning with the term with the highest degree (*descending order*) or beginning with the term with the lowest degree (*ascending order*).

**TIP!**

### Most Common Arrangement of Polynomials

Polynomials are most often arranged in **descending order** so that the degree of the polynomial is the degree of the first term.

The first term of a polynomial arranged in descending order is called the **leading term** of the polynomial. The coefficient of the leading term of a polynomial is called the **leading coefficient**.

### To arrange polynomials in descending order of a variable:

1. Identify the variable on which the terms of the polynomial will be arranged if the polynomial has more than one variable.
2. Compare the degrees of the selected variable for each term.
3. List the term with highest degree of the specified variable first.
4. Continue to list the terms of the polynomial in descending order of the selected variable.

**EXAMPLE** Arrange each polynomial in descending order and identify the degree, the leading term, and the leading coefficient of the polynomial.

(a) $5x + 3x^3 - 7 + 6x^2$      (b) $x^4 - 2x + 3$      (c) $4 + x$      (d) $x^2 + 5$

(a) $5x + 3x^3 - 7 + 6x^2 = 3x^3 + 6x^2 + 5x - 7.$
**Third degree, leading term is $3x^3$, leading coefficient is 3.**

(b) $x^4 - 2x + 3$ is already in descending order.
**Fourth degree, leading term is $x^4$, leading coefficient is 1.**

(c) $4 + x = x + 4.$
**First degree or linear polynomial, leading term is $x$, leading coefficient is 1.**

(d) $x^2 + 5$ is already in descending order.
**Second degree or quadratic polynomial, leading term is $x^2$, leading coefficient is 1.**

A polynomial arranged in descending order can be written with every successive degree represented. Missing terms have a coefficient of 0.

$$x^4 - 2x + 3 \text{ is the same as } x^4 + 0x^3 + 0x^2 - 2x^1 + 3x^0.$$

**SECTION 9–2 SELF-STUDY EXERCISES**

**1** Identify each of the following expressions as a monomial, binomial, or trinomial.

1. $2x^3y - 7x$
2. $5xy^2 + 8y$
3. $3xy$
4. $7ab$
5. $5(3x - y)$
6. $(x - 5)(2x + 4)$
7. $5x^2 - 8x + 3$
8. $7y^2 + 5y - 1$
9. $\dfrac{4x - 1}{5}$
10. $\dfrac{x}{6} - 5$
11. $4(x^2 - 2) + x^3$
12. $3x^2 - 7(x - 8)$

**2** Identify the degree of each term in each polynomial.

13. $6x$
14. $8x^2$
15. $6x^2 - 8x + 12$
16. $7x^3 - 8x + 12$
17. $x - 12$
18. $3x^2 - 8$
19. $15$
20. $21$
21. $2x - \dfrac{1}{4}$
22. $8x^2 + \dfrac{5}{6}$
23. $5x^2y^3 + 7xy$
24. $8r^4s - 5r^3s^5$

Identify the degree of each polynomial.

25. $5x^2 + 8x - 14$
26. $x^3 - 8x^2 + 5$
27. $9 - x^3 + x^6$
28. $12 - 15x^2 - 7x^5$
29. $2x - \dfrac{4}{5}x^2$
30. $\dfrac{7}{8} - x$
31. $5x^3y + 3x^2y^2 + 5xy^5$
32. $7x^2y - 4xy^5$

**3** Arrange each polynomial in descending powers of $x$, and identify the degree, leading term, and leading coefficient of the polynomial.

33. $5x - 3x^2$
34. $7 - x^3$
35. $4x - 8 + 9x^2$
36. $5x^2 + 8 - 3x$
37. $7x^3 - x + 8x^2 - 12$
38. $7 - 15x^4 + 12x$
39. $-7x + 8x^6 - 7x^3$
40. $15 - 14x^8 + x$
41. $12x + 8x^4 + 15x^3$
42. $5x^3 - 7x^4 + 3x - 8$
43. $3x - 5x^4 + 2x^2 - 5$
44. $-3 + 7x - 8x^4 - x^3$

---

## 9–3   | *Basic Operations with Polynomials*

*Learning Outcomes*
**1** Add or subtract polynomials.
**2** Multiply polynomials.
**3** Divide polynomials.

### 1 Add or Subtract Polynomials.

Now that variables with exponents have been introduced, we can broaden our concept of **like terms**. For letter terms to be like terms, the letters as well as the exponents of the letters must be exactly the same; that is, $2x^2$ and $-4x^2$ are like terms because the $x$'s are both squared. But $2x^2$ and $4x$ are not like terms because the $x$'s do not have the same exponents.

---

**EXAMPLE**   Which pairs of terms are like terms?

(a) $3a^2b$ and $-\frac{2}{3}a^2b$      (b) $-8xy^2$ and $7x^2y$      (c) $10ab^2$ and $(-2ab)^2$
(d) $5x^4y^3$ and $-2y^3x^4$      (e) $2x^2$ and $3x^3$

(a) **$3a^2b$ and $-\frac{2}{3}a^2b$ are like terms.** All letters and their corresponding exponents are the same.

(b) **$-8xy^2$ and $7x^2y$ are *not* like terms.** In $-8xy^2$ the exponent of $x$ is 1, and in $7x^2y$ the exponent of $x$ is 2. Also, the exponents of $y$ are not the same.

(c) **$10ab^2$ and $(-2ab)^2$ are *not* like terms.** In the term $10ab^2$, only the $b$ is squared. In $(-2ab)^2$, the entire term is squared, and the result of the squaring is $4a^2b^2$.

(d) **$5x^4y^3$ and $-2y^3x^4$ are like terms.** Both $x$ factors have an exponent of 4, and both $y$ factors have an exponent of 3. Because multiplication is commutative, the order of factors does not matter.

(e) **$2x^2$ and $3x^3$ are *not* like terms.** The exponents of $x$ are not the same.

---

Algebraic expressions containing several terms are simplified as much as possible by combining like terms.

**To combine like terms:**

1. Combine the coefficients using the rules for adding or subtracting signed numbers.
2. The letter factors and exponents do not change.

**TIP!**

**Is Combining Like Terms the Same as Adding? And, What Does *Simplify* Mean?**

- Combining versus adding

   With the introduction of integers into our number system, we broaden our concept of addition to include both addition and subtraction. That is, when adding integers with like signs, we add absolute values, and when adding integers with unlike signs, we subtract absolute values. Also, we developed a strategy for interpreting any subtraction as an equivalent addition by changing the subtrahend to its opposite.

   It is common to use the word *combine* when referring to addition or to subtraction of signed numbers.

- Simplifying

   The instructions to "simplify the expression" are vague but often used in mathematics exercises. In general, to simplify an expression means to write the expression using fewer terms or reduced or with lower coefficients or exponents. When the instructions say "simplify," the intent is for you to examine the expression and see which laws or operations allow you to rewrite the expression in a simpler form.

**EXAMPLE** Simplify the algebraic expressions by combining like terms.

**(a)** $5x^3 + 2x^3$       **(b)** $x^5 - 4x^5$       **(c)** $a^3 + 4a^2 + 3a^3 - 6a^2$
**(d)** $3m + 5n - m$    **(e)** $y + y - 5y^2 + y^3$

**(a)**   $5\ x^3 + 2\ x^3 = (5 + 2)\ x^3 = \mathbf{7\ x^3}$      $5x^3$ and $2x^3$ are like terms; add the coefficients 5 and 2. The sum has the same letter factor and exponent as the like terms.

**(b)**   $x^5 - 4\ x^5 = (1 - 4)\ x^5 = \mathbf{-3\ x^5}$      $x^5$ and $-4x^5$ are like terms. $1 - 4 = -3$. The difference has the same letter factor and exponent as the like terms.

**(c)**   $a^3 + 4a^2 + 3a^3 - 6a^2 = \mathbf{4a^3 - 2a^2}$      $a^3$ and $3a^3$ are like terms.
$4a^2$ and $-6a^2$ are like terms.
Combine coefficients mentally.

**(d)**   $3m + 5n - m = \mathbf{2m} + \mathbf{5n}$      $3m$ and $-m$ are like terms.

**(e)**   $y + y - 5y^2 + y^3 = \mathbf{2y} - \mathbf{5y^2 + y^3}$      $y$ and $y$ are like terms. Coefficients are 1.

When an algebraic expression contains a grouping preceded immediately by a negative sign, we subtract the entire grouping. Another interpretation is that the grouping is multiplied by $-1$, the implied coefficient of the grouping. Either interpretation causes *each* sign within the grouping to be changed to its opposite and at the same time removes the parentheses. If the grouping is preceded by a positive sign or an unexpressed positive sign, parentheses are removed without changing signs, as if each term in the grouping were multiplied by $+1$, the implied coefficient.

**EXAMPLE** Simplify the expressions.

**(a)** $y^2 + 2y - (3y^2 + 5y)$
**(b)** $(m^3 - 3m^2 - 5m + 4) - (4m^3 - 2m^2 - 5m + 2)$

**(a)**   $y^2 + 2y - (3y^2 + 5y) =$

$y^2 + 2y -1\ (3y^2 + 5y) =$                Distribute the implied coefficient of $-1$.

$y^2 + 2y - 3y^2 - 5y =$                  Combine like terms.
$\mathbf{-2y^2 - 3y}$

**(b)**   $(m^3 - 3m^2 - 5m + 4) - (4m^3 - 2m^2 - 5m + 2) =$      Distribute the implied coefficient of $-1$.

$(m^3 - 3m^2 - 5m + 4) -1\ (4m^3 - 2m^2 - 5m + 2) =$    Combine like terms.

$m^3 - 3m^2 - 5m + 4 - 4m^3 + 2m^2 + 5m - 2 =$

$\mathbf{-3m^3 - m^2 + 2}$

**2 Multiply Polynomials.**

Simplifying algebraic expressions may also involve multiplication.

**To multiply by a monomial:**

1. Multiply the coefficients using the rules for signed numbers.
2. Multiply the letter factors using the laws of exponents for factors with like bases.
3. Distribute if multiplying a polynomial by a monomial.

**EXAMPLE**  Multiply.

    **(a)** $(4x)(3x^2)$    **(b)** $(-6y^2)(2y^3)$    **(c)** $(-a)(-3a)$

    **(a)** $(4x)(3x^2) = \;4(3)\;(x^{1+2})\; = \;\mathbf{12\;x^3}$    Multiply coefficients. Add exponents.

    **(b)** $(-6y^2)(2y^3) = \;-6(2)\;(y^{2+3})\; = \;\mathbf{-12\;y^5}$    Multiply coefficients. Add exponents.

    **(c)** $(-a)(-3a) = \;-1(-3)\;(a^{1+1})\; = \;\mathbf{3\;a^2}$    Multiply coefficients. Add exponents. The co-efficient of $-a$ is $-1$. The exponent of $-a$ is 1.

If factors are to be multiplied times more than one term, apply the distributive property.

**EXAMPLE**  Perform the multiplications.

    **(a)** $2x(x^2 - 4x)$    **(b)** $-2y^3(2y^2 + 5y - 6)$    **(c)** $4a(3a^3 - 2a^2 - a)$

    **(a)** $2x\;(x^2 - 4x) = \mathbf{2x^3 - 8x^2}$    Distribute.

    **(b)** $-2y^3\;(2y^2 + 5y - 6) = \mathbf{-4y^5 - 10y^4 + 12y^3}$    Distribute.

    **(c)** $4a\;(3a^3 - 2a^2 - a) = \mathbf{12a^4 - 8a^3 - 4a^2}$    Distribute.

**3**  **Divide Polynomials.**

Simplifying algebraic expressions may also involve division.

## To divide algebraic expressions:

1. Divide the coefficients using the rules for signed numbers.
2. Divide the letter factors using the laws of exponents for factors with like bases.

**EXAMPLE**  Divide. Express answers with positive exponents.

    **(a)** $\dfrac{2x^4}{x}$    **(b)** $\dfrac{-6y^5}{2y^3}$    **(c)** $\dfrac{-4x}{4x^2}$    **(d)** $\dfrac{-5x^4}{15x^2}$    **(e)** $\dfrac{3x}{12x^3}$

    **(a)** $\dfrac{2x^4}{x} = \dfrac{2}{1}\;(x^{4-1})\; = \;\mathbf{2\;x^3}$    Divide coefficients. Subtract exponents. The coefficient of $x$ in the denominator is 1.

    **(b)** $\dfrac{-6y^5}{2y^3} = \dfrac{-6}{2}\;(y^{5-3})\; = \;\mathbf{-3\;y^2}$    Divide coefficients. Subtract exponents.

    **(c)** $\dfrac{-4x}{4x^2} = \dfrac{-4}{4}\;(x^{1-2})\; = \;\mathbf{-1\;x^{-1}}\;$ **or** $\;-\dfrac{1}{x}$    Divide coefficients. Subtract exponents. Write factors with negative exponents as equivalent positive exponents.

    This can also be written as $\dfrac{1}{-x}$ or $\dfrac{-1}{x}$.

**(d)** $\dfrac{-5x^4}{15x^2} = \dfrac{-5}{15}\left(x^{4-2}\right) = -\dfrac{1}{3}x^2$ or $-\dfrac{x^2}{3}$

Reduce coefficients. Subtract exponents.

$\dfrac{-1}{3}x^2$ is the same as $-\dfrac{1}{3}\left(\dfrac{x^2}{1}\right)$ or $-\dfrac{x^2}{3}$.

**(e)** $\dfrac{3x}{12x^3} = \dfrac{3}{12}\left(x^{1-3}\right) = \dfrac{1}{4}x^{-2} = \dfrac{1}{4\,x^2}$

Reduce coefficients. Subtract exponents. Make exponent positive.

Since $\dfrac{1}{4}x^{-2} = \dfrac{1}{4}\left(\dfrac{1}{x^2}\right)$, this can be written as $\dfrac{1}{4x^2}$.

When a polynomial is divided by a monomial, *each* term in the dividend (numerator) is divided by the divisor (denominator).

**EXAMPLE** Perform the divisions.

**(a)** $\dfrac{18a^4 + 15a^3 - 9a^2 - 12a}{3a}$     **(b)** $\dfrac{3x^3 - x^2}{x^3}$     **(c)** $\dfrac{6x^3 + 2x^2}{2x^2}$

**(a)** $\dfrac{18a^4 + 15a^3 - 9a^2 - 12a}{3a} = \dfrac{18a^4}{3a} + \dfrac{15a^3}{3a} - \dfrac{9a^2}{3a} - \dfrac{12a}{3a}$

Distribute or write as separate terms.

$= 6a^3 + 5a^2 - 3a - 4$

Simplify each term.

**(b)** $\dfrac{3x^3 - x^2}{x^3} = \dfrac{3x^3}{x^3} - \dfrac{x^2}{x^3} = 3 - x^{-1}$ or $3 - \dfrac{1}{x}$

Write as separate terms and simplify each term.

**(c)** $\dfrac{6x^3 + 2x^2}{2x^2} = \dfrac{6x^3}{2x^2} + \dfrac{2x^2}{2x^2} = 3x + 1$

Write as separate terms and simplify each term.

**TIP!**

**Why Can't We Cancel Terms?**

**You reduce or cancel *factors*, but not *terms*.** Look at part c of the previous example, $\dfrac{6x^3 + 2x^2}{2x^2}$. A common *mistake* is to cancel the terms.

$\dfrac{6x^3 + 2x^2}{2x^2} = 6x^3$    Incorrect!

Why is this not correct? To check a division, multiply the quotient by the divisor (denominator). The result will be the dividend (numerator).

Does $6x^3(2x^2) = 6x^3 + 2x^2$? NO!

Expressions that have several operations can be simplified.

**EXAMPLE**    Simplify and write all exponents as positive exponents.

(a) $\dfrac{5x^5}{10x^2} + 3x(2x^2)$        (b) $\dfrac{3ab^2 - a^2b + 4}{ab}$        (c) $3x(4xy^2)^2$

(a)   $\dfrac{5x^5}{10x^2} + 3x(2x^2) =$        Simplify each term.

$\dfrac{1}{2}x^3 + 6x^3 =$        Combine terms. $\dfrac{1}{2} + 6 = \dfrac{1}{2} + \dfrac{12}{2} = \dfrac{13}{2}$

$\dfrac{13}{2}x^3$ or $\dfrac{13x^3}{2}$

(b)   $\dfrac{3ab^2 - a^2b + 4}{ab} =$        Write or mentally visualize as separate terms.

$\dfrac{3ab^2}{ab} - \dfrac{a^2b}{ab} + \dfrac{4}{ab} =$        Simplify each term.

$3b - a + \dfrac{4}{ab}$

(c)   $3x(4xy^2)^2 =$        Follow the order of operations. Raise to a power first.
$3x(16x^2y^4) =$        Multiply.

$48x^3y^4$

## SECTION 9–3 SELF-STUDY EXERCISES

**1**  Simplify.

**1.** $3a^2 + 4a^2$

**2.** $5x^3 - 2x^3$

**3.** $b^2 + 3a^2 + 2b^2 - 5a^2$

**4.** $3a - 2b - a$

**5.** $x - 3x - 2x^2 - 3x^2$

**6.** $3a^2 - 2a^2 + 4a^2$

**7.** $x^2 + 3y - (2x^2 + 5y)$

**8.** $4m^2 - 2n^2 - (2m^2 - 3n^2)$

**9.** $7a + 3b + 8c + 2a - (b - 2c)$

**10.** $5x + 3y - (7x - 2z)$

**11.** $(4x^2 - 3) - (3x^2 + 2) - (x^2 - 1)$

**12.** $5x - (3x + 7) - (-x - 8)$

**2**  Multiply.

**13.** $7x(2x^2)$

**14.** $(-2m)(-m^2)$

**15.** $(-3m)(7m)$

**16.** $(-y^3)(2y^3)$

**17.** $4x^2(2x - 7)$

**18.** $-3ab(2a^2b - 5ab^2)$

**19.** $7x(5x^3 - 3x^2 - 7)$

**20.** $2xy(3x^2y - 5xy^2)$

**21.** $3x(x - 6)$

**22.** $4x(3x^2 - 7x + 8)$

**23.** $-4x(2x - 3)$

**24.** $2x^2(5 + 2x)$

**3**  Divide.

**25.** $\dfrac{6x^4}{3x^2}$

**26.** $\dfrac{-5a^2}{10a}$

**27.** $\dfrac{-7x}{-14x^3}$

**28.** $\dfrac{-9x^5}{12x^2}$

**29.** $\dfrac{6x^2 - 4x}{2x}$

**30.** $\dfrac{12x^5 - 6x^3 - 3x^2}{3x^2}$

**31.** $\dfrac{7x^4 - x^2}{x^3}$

**32.** $\dfrac{8x^4 + 6x^3}{2x^2}$

**33.** $\dfrac{6x^4 + 8x^2}{18x^2}$

**34.** $\dfrac{5a^2b^3 - 3ab^2 - 7}{ab}$

**35.** $\dfrac{15a^2b^3 - 3ab^2}{6ab}$

**36.** $\dfrac{18x^2 - 12y^2 - 6xy}{6xy}$

**37.** $\dfrac{15x^2}{3x} - 2x(7x^3)$

**38.** $\dfrac{6x(2x^3y^2)^3 + 8x}{2x}$

**39.** $\dfrac{-3x - (5x + 8x^3)}{2x}$

**40.** $\dfrac{3x^2 - 5y^2}{15xy} - \dfrac{8x^3}{4x}$

| Learning Outcomes | What to Remember with Examples |
|---|---|

**Section 9–1**

**1** Multiply powers with like bases (pp. 388–389).

To multiply powers with like bases, add the exponents and keep the common base as the base of the product. $a^m(a^n) = a^{m+n}$, where $a$, $m$, and $n$ are real numbers.

> Multiply. $x^5(x^{-7}) =$
>
> $$x^{5+(-7)} = x^{-2} = \frac{1}{x^2}$$

**2** Divide powers with like bases (pp. 389–392).

To divide powers with like bases, subtract the exponents and keep the common base as the base of the quotient. $\frac{a^m}{a^n} = a^{m-n}$, where $a$, $m$, and $n$ are real numbers except that $a \neq 0$.

> Divide. $\frac{a^7}{a^4} =$
>
> $$a^{7-4} = a^3$$

**3** Find a power of a power (pp. 392–394).

To find a power of a power, multiply the exponents and keep the same base. $(a^m)^n = a^{mn}$, where $a$, $m$, and $n$ are real numbers.

> Simplify. $(y^5)^4 =$
>
> $$y^{5(4)} = y^{20}$$

To raise a fraction or quotient to a power, raise both the numerator and the denominator to the power. $\left(\frac{a}{b}\right)^n = \frac{a^n}{b^n}$ and $b \neq 0$.

> $$\left(\frac{-2}{x^2}\right)^3 = \frac{(-2)^3}{(x^2)^3} = \frac{-8}{x^6}$$

To raise a product to a power, raise each factor to the power. $(ab)^n = a^n b^n$.

> $$(5x^2y)^2 = 5^2 x^4 y^2 = 25x^4 y^2$$

**Section 9–2**

**1** Identify polynomials, monomials, binomials, and trinomials (pp. 395–396).

Polynomials are algebraic expressions in which the exponent of the variable is a nonnegative integer and there are no variables in a denominator. A monomial is a polynomial with a single term. A binomial is a polynomial containing two terms. A trinomial is a polynomial containing three terms.

> Give an example of a polynomial, a monomial, a binomial, and a trinomial.
>
> Polynomial: $4x^3 + 6x^2 - x + 3$  Monomial: $8x$
> Binomial: $6x + 3$  Trinomial: $8x^2 - 4x - 3$

**2** Identify the degree of terms and polynomials (pp. 396–397).

The degree of a term containing one variable is the exponent of the variable. The degree of a term of more than one variable is the sum of the exponents of all the variable factors. The degree of a constant term is zero. The degree of a polynomial that has only one variable is the degree of the term that has the largest exponent.

> Identify the degree of each term and the degree of the polynomial.
>
> $4x^3 \quad + \quad 6x^2 \quad - \quad x \quad + \quad 3$  The polynomial has a degree of 3.
>
> degree 3  degree 2  degree 1  degree 0

| **3** Arrange polynomials in descending order (pp. 397–398). | To arrange polynomials in descending order, list the term that has the highest degree first, the term that has the next highest degree second, and so on, until all terms have been listed. |
| --- | --- |

Arrange the polynomial in descending order.

$$4 - 2x + 7x^5 - 3x^2 + x^3$$

Descending order:

$$7x^5 + x^3 - 3x^2 - 2x + 4$$

## Section 9–3

| **1** Add or subtract polynomials (pp. 399–400). | Like terms are terms that not only have the same letter factors, but also have the same exponent. To add or subtract like terms, add or subtract the coefficients and keep the variable factors and their exponents exactly the same. |
| --- | --- |

Simplify. $3x^4 + 8x^2 - 7x^2 + 2x^4 = 5x^4 + x^2$

| **2** Multiply polynomials (pp. 400–401). | To multiply expressions containing powers, multiply the coefficients; then add the exponents of like bases. Write with positive exponents. |
| --- | --- |

Simplify. $(3x^4y^5)(7x^2yz) = 21x^6y^6z$

| **3** Divide polynomials (pp. 401–403). | To divide expressions containing powers, divide (or reduce) the coefficients; then subtract the exponents of the like bases. Write with positive exponents. |
| --- | --- |

Simplify. $\dfrac{10x^7y^4}{5x^8y^2} = 2x^{-1}y^2 = \dfrac{2y^2}{x}$

## CHAPTER REVIEW EXERCISES

### Section 9–1

Perform the indicated operations. Write the answers with positive exponents.

**1.** $x^5 \cdot x^5$

**2.** $x^2(x^4)$

**3.** $3x^4 \cdot 7x^5$

**4.** $5x^3 \cdot 8x$

**5.** $\dfrac{x^8}{x^5}$

**6.** $\dfrac{x^3}{x^5}$

**7.** $\dfrac{21x^4}{3x}$

**8.** $\dfrac{24y^7}{18y^{10}}$

**9.** $\dfrac{x^3y^{-1}}{x^2y^2}$

**10.** $\dfrac{x^3 + y^2}{x^2y}$

**11.** $(x^3)^4$

**12.** $(-x^3)^3$

**13.** $(x^{-3})^{-5}$

**14.** $(x^2y^5)^2$

**15.** $(-3x^2)^3$

**16.** $\dfrac{a^2b^7}{ab^2}$

**17.** $\dfrac{xy^3}{xy^5}$

**18.** $(-x^4)^3$

**19.** $5x^{-2} \cdot 8x^4$

**20.** $\dfrac{28x^3y^{-3}}{7xy^{-4}}$

### Section 9–2

Identify the degree of each polynomial.

**21.** $7m^2 - 8m + 12m^4$

**22.** $5a^4 - 7a^3 + 12a^2 - 38$

**23.** $2x^3y^2 - 15xy^3 + 21y^4$

**24.** Is the expression $5x^3 - 3x^{-2}$ a polynomial? Why or why not?

Arrange the following polynomials in descending order and identify the degree of each polynomial, the leading term, and the leading coefficient.

**25.** $5x + 3x^3 - 8 + x^2$

**26.** $3y^5 - 7y - 8y^4 + 12$

**27.** $5x^2 - 12x + 2x^4 - 32$

Simplify the following:

**28.** $4x^3 + 7x - 3x^3 - 5x$     **29.** $8x - 2x^4 - (3x^3 + 5x - x^3)$     **30.** $5x - 3x + (7x^2 - 8x)$

**31.** $4x^2 - (3y^2 + 7x^2 - 8y^2)$     **32.** $4x^3(-3x^4)$     **33.** $-7x^8(-3x^{-2})$

**34.** $5a(a^2 - 7)$     **35.** $2x(x^2 + 3x - 5)$     **36.** $-2y(3y^2 - 7y - 12)$

**37.** $-2x(x^3 - 7x^2 + 15)$     **38.** $\dfrac{12x^5}{6x^3}$     **39.** $\dfrac{12x^7}{-18x^4}$

**40.** $\dfrac{11x^4}{22x^7}$     **41.** $\dfrac{42x^3y}{-15x^3y^3}$     **42.** $\dfrac{-8x^3y^5}{20x^4y}$

**43.** $\dfrac{6x^3 - 12x^2 + 21x}{3x}$     **44.** $\dfrac{25y^5 - 85y^3 + 70y^2}{-5y}$     **45.** $\dfrac{4x^5}{8x^2} - 3x^2(2x^4)$

**46.** $2x(3x^2y)^3 + \dfrac{7x^3}{21x^2}$     **47.** $\dfrac{16x^3 + 12x^4 - 20x^5}{-4x^2}$     **48.** $\dfrac{9x^2y^5 - 15xy^3}{3xy^3}$

---

## TEAM PROBLEM-SOLVING EXERCISES

**1.** The symbolic representation of the laws of exponents illustrates many properties and restrictions. Explain these properties and restrictions in words.

  **(a)** $a^m \cdot a^n = a^{m+n}$

  **(b)** $\dfrac{a^m}{a^n} = a^{m-n}, a \neq 0$

  **(c)** $(a^m)^n = a^{mn}$

  **(d)** $\left(\dfrac{a}{b}\right)^n = \dfrac{a^n}{b^n}, b \neq 0$

  **(e)** $(ab)^n = a^n b^n$

**2.** Give an example to illustrate each of the laws of exponents.

  **(a)** $a^m \cdot a^n = a^{m+n}$

  **(b)** $\dfrac{a^m}{a^n} = a^{m-n}, a \neq 0$

  **(c)** $(a^m)^n = a^{mn}$

  **(d)** $\left(\dfrac{a}{b}\right)^n = \dfrac{a^n}{b^n}, b \neq 0$

  **(e)** $(ab)^n = a^n b^n$

---

## PRACTICE TEST

Perform the indicated operations. Write the answers with positive exponents.

**1.** $(x^4)(x)$     **2.** $\dfrac{x^0}{x^2}$     **3.** $\left(\dfrac{4}{7}\right)^2$

**4.** $\dfrac{24x^2y^{-1}}{16xy^3}$     **5.** $\dfrac{x^{-7}}{x^3}$     **6.** $(6a^2b)^2$

**7.** $\left(\dfrac{x^2}{y}\right)^2$     **8.** $\dfrac{12x^2}{4x^3}$     **9.** $4a(3a^2 - 2a + 5)$

**10.** $\dfrac{60x^3 - 45x^2 - 5x}{5x}$     **11.** $5x^2 - 3x - (2x + 4x^2)$     **12.** $4x^3 + 2x - 7 + (3x^3 - 7x - 5)$

**13.** $-7x^3(-8x^4)$     **14.** $-(2x - 8) + 12$     **15.** $4xy - (3x - 2) + 4$

What is the degree of the polynomial?

**16.** $5x^3 - 4x^3 + x^2$             **17.** $14x - 3x + 21x$

Arrange each polynomial in decreasing powers of $x$.

**18.** $4 - 12x + 15x^3$     **19.** $6 - 3x^4 - 2x^3$     **20.** $6x^4 - 2x^5 - 5$

Many formulas have terms that contain exponents. A commonly used formula that contains an exponent is the formula for calculating the compound amount for compound interest. When compound interest is applied to an investment, interest is calculated at the end of each period and then added to the principal for the next period. Then, the interest and the principal earn interest for the next period. The **compound amount** or the accumulated amount is the combined principal and interest accumulated during the entire time of the investment. This accumulated amount is also called the **future value,** or **maturity value.**

---

**Compound amount (accumulated amount):**

$$A = P\left(1 + \frac{r}{n}\right)^{nt}$$

where  $A$ = accumulated amount
$P$ = original principal
$t$ = time in years
$r$ = rate per year expressed as a decimal equivalent
$n$ = number of compounding periods per year

---

**EXAMPLE**  Using the formula $A = P\left(1 + \frac{r}{n}\right)^{nt}$ and a calculator, find the accumulated amount on an investment of \$1,500, invested at an interest rate of 9% for 3 years, if the interest is compounded quarterly.

**Estimation**  We expect to have more than \$1,500.

$A = P\left(1 + \dfrac{r}{n}\right)^{nt}$    $P = \$1,500; r = 9\%$ or 0.09; $n$ = quarterly or 4 times a year; $t = 3$ years.

$A = 1,500\left(1 + \dfrac{0.09}{4}\right)^{(4)(3)}$    Simplify exponent and division term in grouping.

$A = 1,500(1 + 0.0225)^{12}$    Combine terms in grouping.

$A = 1,500(1.0225)^{12}$    1.0225 $\boxed{\wedge}$ 12 $\boxed{=}$ ⇒ 1.30604999.

$A = 1,500(1.30604999)$    Multiply.

$A = 1,959.07$    Rounded.

**Interpretation**  **The accumulated amount of the \$1,500 investment after 3 years is \$1,959.07 to the nearest cent.**

---

The greater benefits of compound interest are realized over longer periods of time. Suppose you are 35 and invest \$25,000 at an annual compound interest rate of 8%. What will the investment be worth 30 years later when you are 65?

**EXAMPLE**  Invest \$25,000 at 8% for 30 years. Interest is compounded annually.

$A = P\left(1 + \dfrac{r}{n}\right)^{nt}$    Substitute $P = \$25,000$, $r = 0.08$, $n = 1$, and $t = 30$.

$A = \$25,000\left(1 + \dfrac{0.08}{1}\right)^{1(30)}$    Simplify grouping.

$A = \$25,000(1.08)^{30}$    Raise to the power; then multiply.

$A = \$251,566$    Round to the nearest dollar.

## Exercises

Find the compound amount to the nearest dollar that each investment will be at the retirement age of 65 if it is compounded annually.

1. $25,000 at 8% invested at age 45
2. $50,000 at 8% invested at age 45
3. $5,000 at 6% invested at age 25
4. $5,000 at 6% invested at age 40
5. $5,000 at 6% invested at age 50
6. $20,000 at 7% invested at age 30
7. $20,000 at 7% invested at age 40
8. $20,000 at 7% invested at age 60

## Answers

1. $116,524
2. $233,048
3. $51,429
4. $21,459
5. $11,983
6. $213,532
7. $108.549
8. $28,051

# 10

# Products and Factors

## Focus on Careers

Landscape and grounds maintenance workers create and maintain a pleasant and functional outdoor environment. Landscaping workers install and maintain landscaped areas. They slope land, install outdoor lighting and sprinkler systems, and build walkways, terraces, patios, and decks. They also transplant, mulch, fertilize, mow, and water lawns and trees for residential and commercial clients.

Supervisors of landscaping workers prepare cost estimates, schedule work for crews, perform quality checks, suggest and implement changes in work procedures to improve efficiencies, hire employees, and keep employees' time and work-performed records. Pesticide handlers mix and apply pesticides, herbicides, fungicides, or insecticides.

Work in these careers is almost always performed outdoors and in all kinds of weather. The work can be physically demanding and can involve bending and lifting. Safety precautions must be taken when using tools and applying chemicals.

*(continued)*

The Professional Grounds Management Society offers certification to grounds managers who have a combination of 8 years of experience and formal education beyond high school and who pass an examination.

Most states require certification for workers who apply pesticides. Certification usually includes passing a test on the proper and safe use and disposal of these chemicals and some formal education beyond high school. Many 2-year colleges have landscape maintenance programs.

Employment of landscape and grounds maintenance workers is expected to grow faster than the average for all occupations through 2012. Both new construction and renovation of existing landscaping and grounds will provide many jobs.

Supervisors and managers of landscaping and lawn service workers had median hourly earnings of $15.89 in 2002, and tree trimmers and pruners had median hourly earnings of $12.07 in 2002.

Source: *Occupational Outlook Handbook,* 2004–2005 Edition. U.S. Department of Labor, Bureau of Labor Statistics.

Throughout our study of mathematics, we have examined products and factors. To reduce fractions, we looked for factors common to both the numerator and the denominator. In this chapter, we again find it useful to examine products and factors.

## 10-1 | The Distributive Property and Common Factors

*Learning Outcome*    **1** Factor an expression containing a common factor.

We discussed the distributive property and finding common factors earlier in the text and applied them in different contexts. In this section, rather than use the distributive property to multiply and obtain a product, we will start with a product and regenerate the factors that produce the product. In other words, we want to undo the multiplication. Factoring resembles division, which is the inverse operation of multiplication.

### **1**  Factor an Expression Containing a Common Factor.

The multiplication problem $7a(3a + 2)$ is the indicated product of $7a$ and the grouped quantity $3a + 2$. This is the **factored form** of the expression. After the expression is multiplied, we have two terms written as the indicated sum $21a^2 + 14a$. This is the **expanded form.** To rewrite the expression $21a^2 + 14a$ as the indicated product $7a(3a + 2)$ is to **factor** it.

Let's look at a general example of the distributive property:

$$\underset{\text{factored form}}{a(x + y)} \quad = \quad \underset{\text{expanded form}}{ax + ay}$$

Notice that $a$ appears as a factor in both terms in the expanded form. When a factor appears in each of several terms, it is called a **common factor** of the terms. The distributive property in reverse can be used to write the addition as a multiplication. In other words, we can *factor* the expression.

$$ax + ay = a(x + y)$$

**To factor an expression containing a common factor:**

1. Find the *greatest* factor common to *each* term of the expression.
2. Divide each term by the common factor. Divide mentally if practical.
3. Rewrite the expression as the indicated product of the greatest common factor (GCF) and the quotients in Step 2.

**EXAMPLE**   Write $3a + 3b$ in factored form.

We can use the distributive property to factor the expression.

$\boxed{3}\,a + \boxed{3}\,b =$          Write 3 as a factor and divide each term by 3.

$3\left(\dfrac{3a}{3} + \dfrac{3b}{3}\right) =$          Simplify each fraction in parentheses.

$\mathbf{3(a + b)}$          Factored form.

The distributive property also applies if we have more than two terms.

**EXAMPLE**   Write $3ab + 9a + 12b$ in factored form.

$\boxed{3}\,ab + 9a\ +\ 12b$          3 is the common factor. Divide.

$\qquad\qquad \boxed{3} \cdot 3 \quad \boxed{3} \cdot 4$

$3\left(\dfrac{3ab}{3} + \dfrac{3 \cdot 3a}{3} + \dfrac{3 \cdot 4b}{3}\right)$          Simplify.

$\boxed{3}\,(ab + 3a + 4b)$          Factored form

When looking for a common factor, we always look for *all* common factors.

**EXAMPLE**   Factor $10a^2 + 6a$ completely.

$10a^2 + 6a =$          The GCF is 2*a*. Write 2*a* as a factor and divide each term by 2*a*.

$2a\left(\dfrac{10a^2}{2a} + \dfrac{6a}{2a}\right) =$          Simplify.

$\mathbf{2a(5a + 3)}$          Factored form

**EXAMPLE**   Factor $2x^2 + 4x^3$ completely.

$2x^2 + 4x^3 =$          The GCF is $2x^2$. Write $2x^2$ as a factor and divide each term by $2x^2$.

$2x^2\left(\dfrac{2x^2}{2x^2} + \dfrac{4x^3}{2x^2}\right) =$          Simplify.

$\mathbf{2x^2(1 + 2x)}$          Term of 1 must be written.

**EXAMPLE**   Write $2x + 3y$ in factored form.

$2x + 3y =$          The GCF is 1. The expression can only be written in factored form as $1(2x + 3y)$.
$\mathbf{1(2x + 3y)}$          Factored form

When an expression can only be written in factored form as 1 times the entire expression, the expression is a **prime polynomial.**

**When Is It Necessary to Write a 1?**

We have found that it is not always necessary to write the number 1. When is it necessary?

**When 1 is a term, it must be written.**

$$2x^2 - x = x\left(\frac{2x^2}{x} - \frac{x}{x}\right) = x(2x - \boxed{1})$$

**When 1 is a factor, writing the 1 is optional: $1 \cdot n = n$.**

$$2a + 2b = 2\left(\frac{2a}{2} + \frac{2b}{2}\right) = 2(1a + 1b) \qquad \text{or} \qquad 2(a + b)$$

**When 1 is an exponent, writing the 1 is optional: $n^1 = n$.**

$$2x^3 - 5x^2 = x^2\left(\frac{2x^3}{x^2} - \frac{5x^2}{x^2}\right) = x^2\left(2x^1 - 5x^0\right) = x^2(2x - 5)$$

Also, recall that $n^0 = 1$.

Sometimes a binomial factor or a grouping is the common factor.

**EXAMPLE**   Factor $7y(2y - 5) + 3(2y - 5)$.

$$7y(2y - 5) + 3(2y - 5) = \qquad \text{Common factor is } (2y - 5).$$

$$(2y - 5)\left[\frac{7y\cancel{(2y-5)}}{\cancel{(2y-5)}} + \frac{3\cancel{(2y-5)}}{\cancel{(2y-5)}}\right] =$$

$$\mathbf{(2y - 5)\,(7y + 3)}$$

If the leading coefficient of a polynomial is negative, it is often helpful to factor a common factor of $-1$.

**EXAMPLE**   Factor $-3x^2 + 2x - 5$.

$$-3x^2 + 2x - 5 = \qquad \text{Common factor is } -1.$$

$$-1\left(\frac{-3x^2}{-1} + \frac{2x}{-1} - \frac{5}{-1}\right) =$$

$$\mathbf{-1(3x^2 - 2x + 5)}$$

**1** Factor completely. Check.

**1.** $7a + 7b$

**2.** $12x + 12y$

**3.** $m^2 + 2m$

**4.** $5y^3 + 8y^2$

**5.** $6x^2 + 3x$

**6.** $12y^3 + 18y^4$

**7.** $12x^5 - 6x^4$

**8.** $5x - 15xy$

**9.** $5y + 3z$

**10.** $8x - 7y$

**11.** $5ab + 10a + 20b$

**12.** $4ax^2 + 6a^2x + 10a^2x^2$

**13.** $5a - 7ab + 35b$

**14.** $12a^2 - 15a + 6$

**15.** $3x^3 - 9x^2 - 6x$

**16.** $8a^2b + 14ab^3 + 28a^3b^3$

**17.** $3m^2 - 6m^3 + 12m^4$

**18.** $12x^2y - 18xy^3 + 24x^2y^2$

**19.** $15a^2bc + 18a^3b^2c^3 - 21a^4bc^5$

**20.** $20x^2y^3z - 35x^3y^2z - 40x^2y^2z$

**21.** $8x^4y^2z - 12x^2y^4 - 4x^2y^2$

Write in factored form so the leading coefficient of the polynomial factor is positive.

**22.** $-x - 7$

**23.** $-3x - 8$

**24.** $-5x + 2$

**25.** $-12x + 7$

**26.** $-x^2 + 3x - 8$

**27.** $-2x^2 - 7x - 11$

**28.** $-2x^2 + 6x - 8$

**29.** $-3x^2 - 9x + 15$

**30.** $-7x^2 - 21x + 14$

**31.** $-12x^2 + 18x + 6$

**32.** $5x(x + 3) + 8y(x + 3)$

**33.** $3x(2x - 1) + 5(2x - 1)$

**34.** $4y(3y - 5) + 7(3y - 5)$

**35.** $7a(a - b) + 2b(a - b)$

**36.** $5m(2m - 3n) - 7n(2m - 3n)$

**37.** $y(y - 2) - 3(y - 2)$

**38.** $3x(2x - 7) - 8(2x - 7)$

**39.** $7y(9y - 2) - 5(9y - 2)$

---

## 10–2 | *Multiplying and Dividing Polynomials*

*Learning Outcomes*

**1** Multiply polynomials.

**2** Use the FOIL method to multiply two binomials.

**3** Multiply polynomials that result in special products.

**4** Divide polynomials using long division.

**1 Multiply Polynomials.**

As we expand our experiences with multiplication, we need to use the appropriate terminology associated with polynomials. Recall:

- A *polynomial* is an expression with constants and variables with whole-number exponents and contains one or more terms with at least one variable term.

    Monomials, binomials, and trinomials are polynomials.

- A *monomial* contains one term, such as $5x^2$.
- A *binomial* contains two terms, such as $3a + 4$.
- A *trinomial* contains three terms, such as $4x^2 + x - 2$.

    We multiply two binomials by using the distributive property. According to the distributive property, each term of the first factor is multiplied by each term of the second factor. This means we are required to use the distributive property more than once.

---

**EXAMPLE** Multiply $(x + 4)(x + 2)$.

$$(x + 4)\,(x + 2) = x\,(x + 2) + 4\,(x + 2)$$     Apply the distributive property.

$$= x^2 + 2x + 4x + 8$$     Combine like terms.

$$= x^2 + 6x + 8$$

---

## To multiply two polynomials:

1. Use the distributive property to multiply each term of the first polynomial times the entire second polynomial.
2. Combine like terms.

Symbolically,

$$(a + b)(c + d) = a(c + d) + b(c + d) = ac + ad + bc + bd$$
$$(a + b + c)(d + e + f) = a(d + e + f) + b(d + e + f) + c(d + e + f)$$
$$= ad + ae + af + bd + be + bf + cd + ce + cf$$

**EXAMPLE**  Multiply $(2x^2 + 3x - 2)(3x^2 - 5x + 6)$.

$$(2x^2 + 3x - 2)(3x^2 - 5x + 6) =$$
$$2x^2(3x^2 - 5x + 6) + 3x(3x^2 - 5x + 6) - 2(3x^2 - 5x + 6) = \quad \text{Distribute.}$$
$$6x^4 - 10x^3 + 12x^2 + 9x^3 - 15x^2 + 18x - 6x^2 + 10x - 12 = \quad \text{Combine like terms.}$$
$$\mathbf{6x^4 - x^3 - 9x^2 + 28x - 12}$$

**TIP!**

### Use Long Multiplication to Multiply Polynomials

The preceding problem can be organized using a procedure similar to the long-multiplication procedure in arithmetic.

1. Multiply each term in the multiplier times the entire multiplicand.
2. Align partial products so that like terms are in columns.
3. Combine like terms.

$$
\begin{array}{r}
3x^2 - 5x + 6 \\
2x^2 + 3x - 2 \\
\hline
-6x^2 + 10x - 12 \\
9x^3 - 15x^2 + 18x \\
6x^4 - 10x^3 + 12x^2 \\
\hline
6x^4 - \phantom{1}x^3 - 9x^2 + 28x - 12
\end{array}
$$

    Multiply by $-2$.
Multiply by $3x$.
Multiply by $2x^2$.
Combine like terms.

Since multiplication is commutative the polynomials can be interchanged.

$$
\begin{array}{r}
2x^2 + 3x - 2 \\
3x^2 - 5x + 6 \\
\hline
+ 12x^2 + 18x - 12 \\
-10x^3 - 15x^2 + 10x \\
6x^4 + 9x^3 - 6x^2 \\
\hline
6x^4 - \phantom{1}x^3 - 9x^2 + 28x - 12
\end{array}
$$

Multiply by 6.
Multiply by $-5x$.
Multiply by $3x^2$.
Combine like terms.

**2**  **Use the FOIL Method to Multiply Two Binomials.**

When multiplying two binomials, we can guide ourselves through the repeated applications of the distributive property by using the acronym FOIL.

Chapter 10 / Products and Factors

1. Write the product of the *first* term of each factor.
2. Write the product of the two *outer* terms of the factors.
3. Write the product of the two *inner* terms of the factors.
4. Write the product of the *last* term of each factor.
5. Combine like terms.

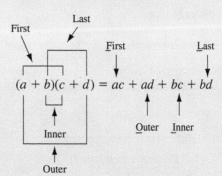

$$(a + b)(c + d) = ac + ad + bc + bd$$

---

**EXAMPLE**  Use the FOIL method to multiply $(2x - 3)(x + 1)$.

$$\begin{array}{c} \text{F} \quad \text{O} \quad \text{I} \quad \text{L} \\ (2x - 3)(x + 1) = 2x^2 + 2x - 3x - 3 \\ = 2x^2 - x - 3 \end{array}$$

---

**3**  **Multiply Polynomials That Result in Special Products.**

When the factors being multiplied have certain characteristics or conditions, we can anticipate the product without going through all the steps. Identifying and using these special characteristics allows us to do many multiplications mentally.

When we multiply the sum of two terms by the difference of the same two terms, we make some special observations about the product.

$$\begin{array}{c} \text{F} \quad \text{O} \quad \text{I} \quad \text{L} \\ (x + 3)(x - 3) = x^2 - 3x + 3x - 9 = x^2 - 9 \\ \underbrace{\qquad\qquad}_{0} \end{array}$$

Examine the final product. Notice that the product has only *two* terms and is a *difference*. The sum of the outer and inner products is zero $(-3x + 3x = 0)$. The terms of the product are *perfect squares*.

Compare the product with its factors. The first term of the product is the *square of the first term* in either of its factors. The second term in the product is the *square of the second term* in either of its factors.

We call pairs of factors like $(a + b)(a - b)$ **the sum and difference of the same two terms,** or **conjugate pairs.** The product, $a^2 - b^2$, is called **the difference of two perfect squares.**

**To mentally multiply the sum and difference of the same two terms (conjugate pairs):**

1. Square the first term of either binomial.
2. Insert a minus sign.
3. Square the second term of either binomial.

Symbolically, $(a + b)(a - b) = a^2 - b^2$

---

**EXAMPLE**   Find the products mentally.

(a) $(x + 2)(x - 2)$        (b) $(2a + y)(2a - y)$

Since each of these is the sum and difference of the same two terms, the product is the difference of two perfect squares.

(a) $(x + 2)(x - 2) = x^2 - 4$        Square $x$. Square 2.

(b) $(2a + y)(2a - y) = 4a^2 - y^2$        Square $2a$. Square $y$.

---

When we multiply the same two binomials, we make some special observations. Because $x \cdot x = x^2$, $(x + 3)(x + 3) = (x + 3)^2$. The quantity $(x + 3)^2$ is called a **binomial square** or the **square of a binomial.**

---

**EXAMPLE**   Find the products of the binomial squares using the FOIL method.

(a) $(x + 2)^2$        (b) $(2x - 5)^2$        (c) $(3a - 2b)^2$

To find the products, write each binomial square as two factors; then use the FOIL method to multiply.

(a)   $(x + 2)^2 = (x + 2)(x + 2)$
$$= x^2 + 4x + 4$$        Outer + Inner: $2x + 2x = 4x$.

(b)   $(2x - 5)^2 = (2x - 5)(2x - 5)$
$$= 4x^2 - 20x + 25$$        Outer + Inner: $-10x - 10x = -20x$.

(c)   $(3a - 2b)^2 = (3a - 2b)(3a - 2b)$
$$= 9a^2 - 12ab + 4b^2$$        Outer + Inner: $-6ab - 6ab = -12ab$.

---

Notice that each product is a trinomial (three terms). These are special trinomials called **perfect-square trinomials.**

**To mentally square a binomial:**

1. Square the first term of the binomial for the first term of the perfect-square trinomial.
2. Double the product of the two terms of the binomial for the middle term of the trinomial.
3. Square the second term of the binomial for the third term of the trinomial.

Symbolically,
$$(a + b)^2 \quad = \quad a^2 + 2ab + b^2$$
$$(a - b)^2 \quad = \quad a^2 - 2ab + b^2$$

binomial square             perfect-square trinomial

---

**EXAMPLE**  Square the binomials mentally.

(a) $(x + 2)^2$     (b) $(2x - 3)^2$     (c) $(5x + 1)^2$

| | **Square First Term** | **Double Product of Terms** | **Square Second Term** |
|---|---|---|---|
| **(a)** $(x + 2)^2 =$ | $(x)^2$ | $2 \cdot (2x)$ | $(2)^2$ |
| | $x^2$ | $+4x$ | $+4$ |
| **(b)** $(2x - 3)^2 =$ | $(2x)^2$ | $2 \cdot (-6x)$ | $(-3)^2$ |
| | $4x^2$ | $-12x$ | $+9$ |
| **(c)** $(5x + 1)^2 =$ | $(5x)^2$ | $2 \cdot (5x)$ | $(1)^2$ |
| | $25x^2$ | $+10x$ | $+1$ |

---

**4**   **Divide Polynomials Using Long Division.**

A polynomial can be divided by a polynomial by using a long division procedure. If the remainder is 0, both the divisor and the quotient are factors of the dividend.

**To divide a polynomial by a polynomial using long division:**

1. Divide the first term of the dividend by the first term of the divisor. The partial quotient is placed above the first term of the dividend.
2. Multiply the partial quotient times the divisor and align the product under like terms of the dividend.
3. Subtract (change subtrahend to opposite and use addition rules).
4. Bring down the next term of the dividend and repeat Steps 1–3.
5. Repeat Steps 1–4 until all terms of the dividend have been brought down. The result of the last subtraction is the remainder.
6. Write the remainder as a fraction with the remainder as the numerator and the divisor as the denominator.

**EXAMPLE**  Is $x - 4$ a factor of $2x^3 - 9x^2 + 7x - 12$?

$$
\begin{array}{r}
2x^2 - \phantom{x}x + 3 \\
x - 4\ \overline{)\ 2x^3 - 9x^2 + 7x - 12} \\
\underline{2x^3 - 8x^2} \phantom{+ 7x - 12} \\
-\ x^2 + 7x \phantom{- 12} \\
\underline{-\ x^2 + 4x} \phantom{- 12} \\
3x - 12 \\
\underline{3x - 12} \\
0
\end{array}
$$

Divide: $\dfrac{2x^3}{x} = 2x^2$.

Multiply: $2x^2(x - 4) = 2x^3 - 8x^2$.

Subtract: $2x^3 - 2x^3 = 0;\ -9x^2 - (-8x^2) = -x^2$.

Divide: $\dfrac{-x^2}{x} = -x$. Multiply: $-x(x - 4)$.

Subtract.

Divide: $\dfrac{3x}{x} = 3$. Multiply: $3(x - 4)$.

Subtract.

Remainder $= 0$

**Yes, $x - 4$ is a factor of $2x^3 - 9x^2 + 7x - 12$, and $(x - 4)(2x^2 - x + 3)$ is the factored form of $2x^3 - 9x^2 + 7x - 12$.**

**EXAMPLE**  Is $x - 1$ a factor of $x^3 + 1$?

$$
\begin{array}{r}
x^2 + \phantom{x}x\ + 1\ + \dfrac{2}{x - 1} \\
x - 1\ \overline{)\ x^3 + 0x^2 + 0x + 1} \\
\underline{x^3 - \phantom{0}x^2} \phantom{+ 0x + 1} \\
x^2 + 0x \phantom{+ 1} \\
\underline{x^2 - \phantom{0}x} \phantom{+ 1} \\
x + 1 \\
\underline{x - 1} \\
2
\end{array}
$$

Represent missing powers of $x$ with terms having a coefficient of 0.

Subtract: $0x^2 - (-x^2) = x^2$.

Subtract: $0x - (-x) = x$.

Subtract: $1 - (-1) = 2$.

**No, $x - 1$ is not a factor of $x^3 + 1$ since the remainder is not zero.**

## SECTION 10–2 SELF-STUDY EXERCISES

**1** Multiply.

1. $(x + 7)(x + 3)$
2. $(x + 8)(x + 5)$
3. $(2x - 1)(x + 2)$
4. $(3x + 7)(x - 5)$
5. $(x + 5)(x^2 + 2x - 1)$
6. $(x + 7)(x^2 - 3x + 2)$
7. $(x - 7)(x^2 - 5x + 2)$
8. $(x - 3)(x^2 - 8x + 1)$
9. $(2x - 5)(x^2 - x + 1)$
10. $(3x - 2)(x^2 + x - 3)$
11. $(2x + 7)(3x^2 - 5x + 2)$
12. $(5x - 3)(4x^2 - 2x - 3)$
13. $(x - 2)(x^2 + 2x + 4)$
14. $(2x - 3)(4x^2 + 6x + 9)$
15. $(x + 3)(x^2 - 3x + 9)$
16. $(3x + 2)(9x^2 - 6x + 4)$
17. $(x + 5)(x^2 - 5x + 25)$
18. $(11x - 3)(121x^2 + 33x + 9)$

**2** Use the FOIL method to find the products. Practice combining the outer and inner products mentally.

19. $(a + 3)(a + 8)$
20. $(x - 4)(x + 5)$
21. $(y - 2)(y - 9)$
22. $(y - 7)(y - 3)$
23. $(2a + 3)(a + 4)$
24. $(3a - 5)(a + 1)$
25. $(3a - 2b)(a - 2b)$
26. $(5x - y)(x - 5y)$
27. $(3x - 4)(2x - 3)$
28. $(a - b)(2a - 5b)$
29. $(7 - m)(3 - 7m)$
30. $(5 - 2x)(8 - x)$
31. $(x + 7)(x + 4)$
32. $(y - 7)(y - 5)$
33. $(m + 3)(m - 7)$
34. $(3b - 2)(x + 6)$
35. $(4r - 5)(3r + 2)$
36. $(5 - x)(7 - 3x)$
37. $(4 - 2m)(1 - 3m)$
38. $(2 + 3x)(3 + 2x)$
39. $(x + 3)(2x - 5)$
40. $(5x - 7y)(4x + 3y)$
41. $(2a + 3b)(7a - b)$
42. $(5a + 2b)(6a - 5b)$
43. $(9x - 2y)(3x + 4y)$
44. $(5x - 8y)(4x - 3y)$
45. $(7m - 2n)(3m + 5n)$

Find the special products using patterns.

**46.** $(a + 3)(a - 3)$      **47.** $(2x + 3)(2x - 3)$      **48.** $(a - y)(a + y)$
**49.** $(4r + 5)(4r - 5)$      **50.** $(5x + 2)(5x - 2)$      **51.** $(7 + m)(7 - m)$
**52.** $(3y - 5)(3y + 5)$      **53.** $(8y + 3)(8y - 3)$      **54.** $(3a - 11b)(3a + 11b)$
**55.** $(5y - 3)(5y + 3)$      **56.** $(x - 7)(x + 7)$      **57.** $(x - 11)(x + 11)$
**58.** $(2 - 3x)^2$      **59.** $(3x + 4)^2$      **60.** $(Q + L)^2$
**61.** $(a^2 + 1)^2$      **62.** $(2d - 5)^2$      **63.** $(3a + 2x)^2$
**64.** $(3x - 7)^2$      **65.** $(6 + Q)^2$      **66.** $(y + 5x)^2$
**67.** $(4 - 3j)^2$      **68.** $(3m - 2p)^2$      **69.** $(m^2 + p^2)^2$

**4** Perform the indicated division and determine if the binomial is a factor of the dividend.

**70.** $x - 3 \overline{)x^2 + 3x - 18}$      **71.** $x + 1 \overline{)x^3 - 2x^2 - x + 2}$

**72.** $x + 2 \overline{)x^3 + 3x^2 - 3x - 10}$      **73.** $x + 3 \overline{)3x^3 + 7x^2 - 3x + 8}$

**74.** $x - 5 \overline{)2x^3 - 3x^2 - 33x + 8}$      **75.** $x + 1 \overline{)x^3 + 2x^2 - 5x - 6}$

**76.** $x + 3 \overline{)x^3 + 2x^2 - 5x - 6}$      **77.** $x - 3 \overline{)x^3 - 27}$

---

## 10–3 | *Factoring Special Products*

*Learning Outcomes*

**1** Recognize and factor the difference of two perfect squares.
**2** Recognize and factor a perfect-square trinomial.

To factor any of the special products of the preceding section, we apply the inverse of the process. That is, we start with the product and "work back" to the factors that produce these special products.

To rewrite a special product in factored form, we must recognize the product as a pattern. Once we identify the special product, then we must know the pattern of the product in factored form.

**1** **Recognize and Factor the Difference of Two Perfect Squares.**

Before we can factor such a special product, we must be able to recognize an expression as a special product.

**To identify a binomial as the difference of two perfect squares:**

1. Verify that the expression is a binomial.
2. Verify that the absolute value of each term is a perfect square.
3. Verify that the second term is subtracted from the first term.

**EXAMPLE**    Identify the special products that are the difference of two perfect squares.

     **(a)** $x^2 - 9$      **(b)** $a^2 + 49$      **(c)** $m^2 - 27$      **(d)** $3y^2 - 25$
     **(e)** $9x^2 - 4$      **(f)** $4x^2 - 4x + 1$

   **(a)** Difference of two perfect squares.
   **(b)** Not the difference of two perfect squares; this is a *sum*, not a difference.
   **(c)** Not the difference of two perfect squares; 27 is not a perfect square.
   **(d)** Not the difference of two perfect squares; in $3y^2$, 3 is not a perfect square.

**(e)** Difference of two perfect squares.

**(f)** Not the difference of two perfect squares; this is a trinomial, not a binomial.

## To factor the difference of two perfect squares:

1. Take the square root of the first term.
2. Take the square root of the second term.
3. Write one factor as the *sum* of the square roots found in Steps 1 and 2, and write the other factor as the *difference* of the square roots from Steps 1 and 2.

Symbolically,

$$a^2 - b^2 = (a + b)(a - b)$$

**EXAMPLE** Factor the special products, which are the differences of two perfect squares.

**(a)** $a^2 - 9$     **(b)** $x^2 - 36$     **(c)** $4x^2 - 1$     **(d)** $-49 + 16m^2$

**(a)** $a^2 - 9 = (a + 3)(a - 3)$     **(b)** $x^2 - 36 = (x + 6)(x - 6)$

**(c)** $4x^2 - 1 = (2x + 1)(2x - 1)$     **(d)** $-49 + 16m^2 = 16m^2 - 49 = (4m + 7)(4m - 7)$

**TIP!**

### Order of Factors

Because multiplication is commutative, the factors given as answers in the preceding example may be expressed in any order, such as $(a + 3)(a - 3)$ or $(a - 3)(a + 3)$, $(x + 6)(x - 6)$ or $(x - 6)(x + 6)$, and so on.

**2** **Recognize and Factor a Perfect-Square Trinomial.**

A trinomial is a **perfect-square trinomial** if the first and last terms are positive perfect squares and the absolute value of the middle term is *twice* the product of the square roots of the first and last terms. Again, we need to be able to distinguish these special products from other expressions before we factor them.

## To identify a perfect-square trinomial:

1. Verify that the expression is a trinomial.
2. Verify that the first and last terms are positive and perfect squares.
3. Mentally take the square root of the first and last terms and multiply the results. Two times this product should equal the absolute value of the middle term of the original trinomial.

**EXAMPLE** Verify that the trinomials are perfect-square trinomials.

(a) $x^2 + 14x + 49$      (b) $4m^2 - 12m + 9$      (c) $9x^2 + 24xy + 16y^2$

(a) The first and last terms, $x^2$ and 49, are positive perfect squares. The middle term, $14x$, has an absolute value that is twice the product of the square roots of $x^2$ and 49. And, $2(7x) = 14x$.

(b) The first and last terms, $4m^2$ and 9, are positive perfect squares. The middle term, $-12m$, has an absolute value that is twice the product of the square roots of $4m^2$ and 9. And, $2(2m \cdot 3) = 12m$.

(c) The first and last terms, $9x^2$ and $16y^2$, are positive perfect squares. The middle term, $24xy$, has an absolute value that is twice the product of the square roots of $9x^2$ and $16y^2$. And, $2(3x \cdot 4y) = 24xy$.

**EXAMPLE** Tell why the trinomials are *not* perfect-square trinomials.

(a) $x^2 + 2x - 1$      (b) $4x^2 + 6x + 9$
(c) $x^2 - 5x + 4$      (d) $-4x^2 - 4x + 1$

(a) The last term, $-1$, is negative. This term must be positive in a perfect-square trinomial.
(b) The middle term, $6x$, is not *twice* the product of $2x$ and 3.
(c) The middle term, $-5x$, is not *twice* the product of $x$ and 2.
(d) The first term, $-4x^2$, is negative. It should be positive.

## To factor a perfect-square trinomial:

1. Write the square root of the first term.
2. Write the sign of the middle term.
3. Write the square root of the last term.
4. Indicate the square of this binomial quantity.

**EXAMPLE** Factor the perfect-square trinomials.

(a) $x^2 + 14x + 49$      (b) $4m^2 - 12m + 9$      (c) $9x^2 + 24xy + 16y^2$

| | Square root of *first* term | Sign of *middle* term | Square root of *last* term | *Square* the quantity |
|---|---|---|---|---|
| (a) $x^2 + 14x + 49$ | $x$ | $+$ | 7 | $(x + 7)^2$ |
| (b) $4m^2 - 12m + 9$ | $2m$ | $-$ | 3 | $(2m - 3)^2$ |
| (c) $9x^2 + 24xy + 16y^2$ | $3x$ | $+$ | $4y$ | $(3x + 4y)^2$ |

**1** Identify the special products that are the difference of two perfect squares.

**1.** $r^2 - s^2$

**2.** $d^2 - 4d + 10$

**3.** $4y^2 - 16$

**4.** $36m^2 - 9n^2$

**5.** $9 + 4a^2$

**6.** $64p^2 - q^2$

Factor the following special products.

**7.** $y^2 - 49$

**8.** $16x^2 - 1$

**9.** $9a^2 - 100$

**10.** $4m^2 - 81n^2$

**11.** $9x^2 - 64y^2$

**12.** $25x^2 - 64$

**13.** $100 - 49x^2$

**14.** $4x^2 - 49y^2$

**15.** $121m^2 - 49n^2$

**16.** $81x^2 - 169$

**17.** $-9 + 4a^2$

**18.** $-16 + 25r^2$

**19.** $36x^2 - 49y^2$

**20.** $49 - 144x^2$

**21.** $16x^2 - 81y^2$

**2** Verify which of the trinomials are perfect-square trinomials.

**22.** $4y^2 + 2y + 16$

**23.** $9m^2 - 24mn + 16n^2$

**24.** $16a^2 + 8a - 1$

**25.** $-9r^2 + 12r + 4$

**26.** $y^2 - 14y + 49$

**27.** $p^2 + 10p + 25$

Factor the special products.

**28.** $x^2 + 6x + 9$

**29.** $x^2 + 14x + 49$

**30.** $x^2 - 12x + 36$

**31.** $x^2 - 16x + 64$

**32.** $4a^2 + 4a + 1$

**33.** $25x^2 - 10x + 1$

**34.** $9m^2 - 48m + 64$

**35.** $4x^2 - 36x + 81$

**36.** $x^2 - 12xy + 36y^2$

**37.** $4a^2 - 20ab + 25b^2$

**38.** $y^2 - 10y + 25$

**39.** $9x^2 + 60xy + 100y^2$

**40.** $-x^2 - 12x - 36$

**41.** $-9x^2 + 6x - 1$

**42.** $-x^2 - 8x - 16$

---

## 10–4 | *Factoring General Trinomials*

*Learning Outcomes*

**1** Factor general trinomials whose squared term has a coefficient of 1.

**2** Remove common factors after grouping an expression.

**3** Factor a general trinomial by grouping.

**4** Factor any binomial or trinomial that is not prime.

Many trinomials do not have a common factor and do not match the pattern of a special product. Some will still factor as the product of two binomials. First, let's examine a trinomial whose squared term has a coefficient of 1.

**1 Factor General Trinomials Whose Squared Term Has a Coefficient of 1.**

Trinomials that are not perfect-square trinomials are **general trinomials.**

**To factor a trinomial with a squared term that has a coefficient of 1:**

1. Ensure the trinomial is arranged in descending powers of one variable.
2. Determine the signs of the second term of each binomial by examining the sign of the third term of the trinomial.

   $+ \Rightarrow$ like signs (both $+$ or both $-$, matching the middle sign)

   $- \Rightarrow$ unlike signs (one $+$ and one $-$, with the sign of the larger absolute value of the factor pair matching the sign of the middle term of the trinomial)

3. Write all factor pairs of the coefficient of the third term.

4. Select the factor pair that adds (algebraically) to the coefficient of the middle term of the trinomial. Include appropriate signs.
5. Write the binomial factors of the trinomial with the square root of the variable as the first term of each factor and the two factors from Step 3 as the second terms of the binomials.

---

**EXAMPLE**  Factor $x^2 + 6x + 5$.

$x^2 + 6x + 5$        Terms are already in descending order.

$( \ + \ )( \ + \ )$        Last term + and middle term + means signs are alike and both plus.

                   Only factor pair of 5 is 1(5). The algebraic sum is $+1 + (+5) = +6$.

$(x + 1)(x + 5)$        First term of each factor is $x$. Second terms are $+1$ and $+5$.

Check by multiplying.

---

**EXAMPLE**  Factor $x^2 + 4x - 12$.

$x^2 + 4x - 12$        Terms are already in descending order. Last term − means signs are unlike.

$( \ + \ )( \ - \ )$        Middle term + means the factor with the larger absolute value in the selected factor pair will be +.

                   Factor pairs of 12:

                   $1 \cdot 12$

                   $\boxed{2 \cdot 6}$ Algebraic sum (difference when signs are unlike) of $-2$ and $+6$ is $+4$.

                   $3 \cdot 4$

$(x - 2)(x + 6)$        First term of each factor is $x$. Second terms are $-2$ and $6$.

---

**EXAMPLE**  Factor $x^2 - 20 - x$.

$x^2 - x - 20$        Arrange in descending order of $x$. Last sign − means signs are unlike.

                   Middle sign − means the factor with the larger absolute value in the selected factor pair will be −.

                   Factor pairs of 20:

                   $1 \cdot 20$

                   $2 \cdot 10$

                   $\boxed{4 \cdot 5}$ Difference (when signs are unlike) of $+4$ and $-5$ is $-1$.

$(x + 4)(x - 5)$        First term of each factor is $x$. Second terms are $+4$ and $-5$.

---

## 2   Remove Common Factors After Grouping an Expression.

Algebraic expressions that have more than three terms and have no factors common to every term in the expression may have common factors for some groups of terms. In these cases, the expression may be written as groupings.

### To remove common factors after grouping an expression:

1. Identify groupings.
2. Factor all common factors from each grouping.
3. Examine each term to see if there are common groupings. If so, factor out the common grouping.

**EXAMPLE** Write the expression $2x^2 - 2xb + ax - ay$ as an alternate expression by grouping pairs of terms. Factor common factors from each grouping.

$2x^2 - 2xb + ax - ay$      Group two terms in each grouping.

$(2x^2 - 2xb) + (ax - ay)$      Factor common factors from each grouping.

$\mathbf{2x(x - b) + a(x - y)}$      There is no common grouping in the terms.

**Groupings Are Not Necessarily Factors**

In the preceding example the groupings $(2x^2 - 2xb)$ and $(ax - ay)$ are not factors. They are groupings that are added. To rewrite the expression as $2x(x - b) + a(x - y)$ is not factored form. The expression has two terms and each term has two or more factors. *An expression in factored form is only one term.*

**EXAMPLE** Write the expressions in factored form by using grouping.

(a) $mx + 2m - 4x - 8$      (b) $y^2 + 2xy + 3y + 6x$

(c) $3x^2 - 9x - 7x + 21$      (d) $2x^2 + 8x + 5y - 15$

**(a)**   $mx + 2m - 4x - 8 =$      Group the four-termed expression into two terms.

$(mx + 2m) + (-4x - 8) =$      Factor any common factor in each of the two terms.

$m(x + 2) + {}^-4(x + 2) =$      Convert double signs to an equivalent single sign.

$m \boxed{(x + 2)} - 4 \boxed{(x + 2)} =$      Factor into one term by factoring out the common binomial factor $(x + 2)$.

$\boxed{(x + 2)} \, \mathbf{(m - 4)}$ or $\mathbf{(m - 4)(x + 2)}$      Check the result by using the FOIL method.

$(x + 2)(m - 4) = xm - 4x + 2m - 8$      Rearrange terms and factors (commutative properties of multiplication and addition).

$= mx + 2m - 4x - 8$      The factoring checks.

**(b)**   $y^2 + 2xy + 3y + 6x =$      Group into two terms.

$(y^2 + 2xy) + (3y + 6x) =$      Factor the common factor in each term.

$y \boxed{(y + 2)} + 3 \boxed{(y + 2)} =$      Factor into one term by factoring the common binomial factor.

$\boxed{(y + 2)} \, \mathbf{(y + 3)}$ or $(y + 3)(y + 2)$

**(c)**   $3x^2 - 9x - 7x + 21 =$      Group into two terms.

$(3x^2 - 9x) + (-7x + 21) =$      Factor each term.

$3x(x - 3) + {}^-7(x - 3) =$      Convert double signs to a single sign.

$3x \boxed{(x - 3)} - 7 \boxed{(x - 3)} =$      Factor into one term by factoring the common binomial factor.

$\boxed{(x - 3)} \, \mathbf{(3x - 7)}$ or $(3x - 7)(x - 3)$

**(d)**   $2x^2 + 8x + 5y - 15 =$      Group into two terms.

$(2x^2 + 8x) + (5y - 15) =$      Factor each term.

$2x(x + 4) + 5(y - 3)$

In this example, the two terms do not have a common factor. Thus, the expression cannot be factored into one term. Even if we rearrange the terms, we will not be able to write the expression as a single term.

$\mathbf{2x^2 + 8x + 5y - 15}$ **is prime.**

**Make Leading Coefficient of Binomial Positive**

When factoring by grouping, manipulate the signs of the common factor so that the leading coefficient of the binomial is positive.

$$-3x + 9$$

Factor as $-3(x - 3)$, *not* $3(-x + 3)$. Parts a and c of the preceding example illustrate this tip.

**3** **Factor a General Trinomial by Grouping.**

We can now use a systematic method to factor general trinomials with leading integral coefficients that are positive and not equal to 1.

**To factor a general trinomial of the form $ax^2 + bx + c$ by grouping:**

1. Verify that $a$ is a positive integer or rewrite with $-1$ as the common factor.
2. Multiply the coefficient of the first term of the trinomial by the coefficient of the last term.
3. Factor the product from Step 2 into a pair of factors:
   **(a)** whose *sum* is the coefficient of the middle term if the sign of the last term is positive, or
   **(b)** whose *difference* is the coefficient of the middle term if the sign of the last term is negative.
4. Rewrite the trinomial as a polynomial with four terms by replacing the middle term with two terms that have the coefficients identified in Step 3.
5. Group the polynomial with four terms from Step 4 into two groups with a common factor in each.
6. Factor the common factors from each of the two groups.
7. Factor the common binomial.

**EXAMPLE** Factor $6x^2 + 19x + 10$ by grouping.

| | |
|---|---|
| $6(10) = 60$ | Multiply the coefficients of the first and third terms. |
| $60 = (1)60$ | List all factor pairs of 60. |
| $(2)30$ | |
| $(3)20$ | |
| $(4)15$ | Identify the pair that *adds* to 19, since the sign of |
| $(5)12$ | the last term is positive. |
| $(6)10$ | |
| $6x^2 + 19x + 10 =$ | Separate $+19x$ into two terms using the coefficients 4 and 15. |
| $6x^2 + 4x + 15x + 10 =$ | Group into two terms. |
| $(6x^2 + 4x) + (15x + 10) =$ | Factor common factors in each term. |
| $2x(3x + 2) + 5(3x + 2) =$ | Factor common binomial. |
| $(3x + 2)(2x + 5)$ or $(2x + 5)(3x + 2)$ | Check using the FOIL method. |

**EXAMPLE** Factor $20x^2 - 23x + 6$ by grouping.

$20(6) = 120$            Find the product of 20 and 6.

List the factor pairs of 120 and select the pair that has a sum of 23.

$120 = (1)120$          List all factor pairs of 120.
$(2)60$
$(3)40$
$(4)30$
$(5)24$
$(6)20$
$(8)15$          Identify the pair that adds to 23, since the sign of the last term is
$(10)12$         positive.

$20x^2 - 23x + 6 =$      Separate $-23x$ into two terms using the coefficients $-8$ and $-15$.

$20x^2 - 8x - 15x + 6 =$      Group into two terms. Be sure the second grouping keeps the
                                   negative sign with $15x$.

$(20x^2 - 8x) + (-15x + 6) =$      Factor common factors in each term.

$4x(5x - 2) - 3(5x - 2) =$      Factor the common binomial.

$(5x - 2)(4x - 3)$      Check using the FOIL method.

If the third term of a trinomial has a negative sign, we use the same procedure but look for two factors whose *difference* is the coefficient of the middle term.

**EXAMPLE** Factor $10x^2 + 19x - 15$ by grouping.

$10(15) = 150$          Find the product of 10 and 15.

List the factor pairs of 150 and select the pair that has a difference of 19.

$150 = (1)150$          List all factor pairs of 150.
$(2)75$
$(3)50$
$(5)30$
$(6)25$          Identify the pair that subtracts to 19, since the last term is negative.
$(10)15$

The factors 25 and 6 have a difference of 19. When we rewrite the trinomial, we write $19x$ as $25x - 6x$.

$10x^2 + 19x - 15 =$      Separate $+19x$ into two terms using the coefficients 25 and $-6$.
                                   The signs will be different and the larger coefficient will be
                                   positive.

$10x^2 + 25x - 6x - 15 =$      Group into two terms.

$(10x^2 + 25x) + (-6x - 15) =$      Factor common factors in each term.

$5x(2x + 5) - 3(2x + 5) =$      Factor the common binomial.

$(2x + 5)(5x - 3)$      Check using the FOIL method.

### Shortening the Process for Factoring by Grouping

The following is a variation of the factor-by-grouping method. The variation is mathematically sound and employs a strategy using the property of 1 that is often overlooked.

**Factor $6x^2 + 7x - 20$.**

| | | | |
|---|---|---|---|
| $6x^2 + 7x - 20$ | | | Multiply the coefficients of the first and third terms: $6(-20) = -120$. |
| $120 =$ | 1 | 120 | A negative product means the factor pair has unlike signs. List all factor pairs of 120. |
| | 2 | 60 | |
| | 3 | 40 | |
| | 4 | 30 | |
| | 5 | 24 | |
| | 6 | 20 | |
| | 8 | 15 | Identify the factor pair with a *difference* of +7. |
| | 10 | 12 | The larger factor will be positive. **$-8 + 15 = 7$** |

**Variation in procedure begins here.**

$$\frac{(6x\quad)(6x\quad)}{6}$$

Use the coefficient of the first term as the coefficient of the first term in *each* binomial. This gives us an extra factor of 6, and we compensate by *dividing the expression by 6*.

$$\frac{(6x - 8)(6x + 15)}{6} =$$

Use the factors of $-120$ that have an algebraic sum of 7 as the second term of each binomial. Signs will be unlike.

$$\frac{2(3x - 4)(3)(2x + 5)}{6} =$$

Factor the common factors from each binomial and simplify the numerical factors.

$$\frac{\cancel{6}(3x - 4)(2x + 5)}{\cancel{6}} =$$

The factors of 6 in the numerator and denominator reduce.

$$(3x - 4)(2x + 5)$$

Check using the FOIL method.

**Now, let's shorten the process again.**

Once we have selected the pair of factors whose sum or difference matches the middle term, *make two fractions using the selected factor pair as the denominators*. The *leading coefficient of the trinomial will be the numerator of each fraction*.

$$6x^2 + 7x - 20 \qquad 6(-20) = -120$$

The factor pair of $-120$ that has a sum of $+7$ is $-8$ and $+15$. Make fractions and reduce each fraction.

$$\frac{6}{-8} = \frac{3}{-4} \qquad \frac{6}{+15} = \frac{2}{+5}$$

The leading coefficient of the trinomial is the numerator of both fractions.

Each reduced fraction gives the coefficients of one of the binomial factors. 3, $-4$ and 2, $+5$.

$$(\mathbf{3x - 4})(\mathbf{2x + 5})$$

**Let's try this process with another example.**

If the trinomial has a common factor, we must factor the common factor before finding the factor pair and writing the fractions. The common factor will be part of the final answer.

**Factor $30x^2 + 8 - 32x$.**

| | |
|---|---|
| $30x^2 - 32x + 8$ | Arrange in descending powers of $x$. |
| $\boxed{2}\,(15x^2 - 16x + 4)$ | Factor any common factors. |
| $15(4) = 60$ | Write the factor pairs of 60. |

| | |
|---|---|
| 1 | 60 |
| 2 | 30 |
| 3 | 20 |
| 4 | 15 |
| 5 | 12 |
| 6 | 10 |

Identify the pair that has a *sum* of $-16$. Signs will be alike. $\boxed{-6 + (-10) = -16}$.

Make fractions with the factor pair $\boxed{-6 \text{ and } -10}$ as the denominators and the leading coefficient of the factored trinomial, 15, as the numerators.

$$\frac{15}{-6} = \frac{5}{-2} \qquad \frac{15}{-10} = \frac{3}{-2} \qquad \text{Reduce each fraction.}$$

Write the factors using the coefficients $5, -2$ and $3, -2$. Don't forget the common factor.

$$\boxed{2}\,(5x - 2)(3x - 2)$$

---

**TIP!**

### Practice Moves You from Systematic Processes to Intuitive Processes

Systematic processes help you understand the concepts and build confidence. They also help you develop your mathematical senses like your number sense and your spatial sense. The more you practice, the more you develop your mathematical senses.

Many students instinctively and automatically move to shortened and mental processes. Many times you can test a few combinations of coefficients and find the correct factors without going through the entire systematic process. With practice, you can write expressions in factored form more readily.

---

**4** **Factor Any Binomial or Trinomial That Is Not Prime.**

Now let's develop a strategy for factoring any binomial or trinomial that can be factored. This strategy allows us to say with confidence that a particular binomial or trinomial will not factor and is prime. When the word **prime** refers to algebraic expressions, it describes algebraic expressions that have only themselves and 1 as factors.

### To factor any binomial or trinomial:

Perform the following steps in order.

1. Factor out the greatest common factor (if any).
2. Check the binomial or trinomial to see if it is a special product.
   (a) If it is the difference of two perfect squares, use the pattern.
   (b) If it is a perfect-square trinomial, use the pattern.
3. If there is no special product, factor by grouping or by any appropriate process.
4. Examine each factor to see if it can be factored further.
5. Check factoring by multiplying.

The steps for factoring any binomial or trinomial are presented visually in the flowchart (Fig. 10–1).

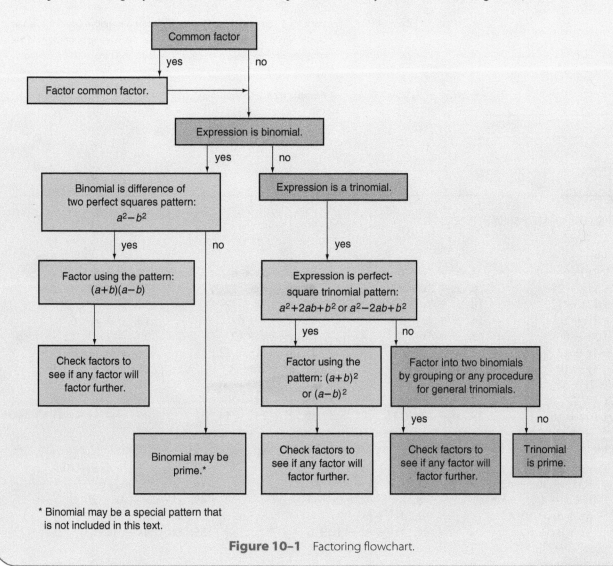

**Figure 10–1** Factoring flowchart.

**EXAMPLE**  Completely factor $4x^3 - 2x^2 - 6x$.

$2x(2x^2 - x - 3)$      2x is the common factor. The trinomial is not a special product.

$\mathbf{2x(2x - 3)(x + 1)}$      Factor the trinomial using any appropriate process but *keep* the 2x factor. Check by multiplying.

**EXAMPLE**  Completely factor $12x^2 - 27$.

$3(4x^2 - 9)$      3 is the common factor. The binomial is the difference of two perfect squares.

$\mathbf{3(2x - 3)(2x + 3)}$      Factor the difference of two perfect squares. Keep the common factor of 3. Check.

**EXAMPLE**  Completely factor $-18x^3 + 24x^2 - 8x$.

$-2x(9x^2 - 12x + 4)$  Look for a common factor: $-2x$. We factor a negative when the leading coefficient is negative. It is preferred that the leading coefficient in the binomial or trinomial be positive. The resulting trinomial is a perfect-square trinomial.

$-2x(3x - 2)^2$  Factor the perfect-square trinomial.

**SECTION 10–4 SELF-STUDY EXERCISES**

**1** Factor.

1. $x^2 + 7x + 6$
2. $x^2 - 7x + 6$
3. $x^2 - 5x + 6$
4. $x^2 + 5x + 6$
5. $x^2 - 11x + 28$
6. $x^2 + 8x + 12$
7. $x^2 - 8x + 12$
8. $x^2 + 13x + 12$
9. $x^2 - 13x + 12$
10. $x^2 + 7x + 12$
11. $x^2 - 7x + 12$
12. $x^2 - 4x + 3$
13. $x^2 + 8x + 7$
14. $x^2 + 7x + 10$
15. $x^2 - x - 6$
16. $x^2 + x - 6$
17. $x^2 - 5x - 6$
18. $x^2 + 5x - 6$
19. $x^2 - x - 12$
20. $x^2 + x - 12$
21. $x^2 + 4x - 12$
22. $x^2 - 4x - 12$
23. $x^2 - 11x - 12$
24. $x^2 + 11x - 12$
25. $y^2 - 3y - 10$
26. $y^2 - y - 6$
27. $b^2 + 2b - 3$
28. $-14 - 5b + b^2$
29. $12 - 7x + x^2$
30. $-30 - x + x^2$
31. $11x + 18 + x^2$
32. $-9x + x^2 + 18$
33. $-7x - 18 + x^2$
34. $-18 + x^2 + 17x$
35. $x^2 + 9x + 20$
36. $x^2 - 12x + 20$
37. $x^2 - 10x + 16$
38. $x^2 - 17x + 16$
39. $x^2 - 13x - 14$
40. $x^2 - 5x - 14$

**2** Factor the polynomials by removing the common factors after grouping.

41. $x^2 + xy + 4x + 4y$
42. $6x^2 + 4x - 3xy - 2y$
43. $3mx + 5m - 6nx - 10n$
44. $30xy - 35y - 36x + 42$
45. $x^2 - 2x + 8x - 16$
46. $6x^2 - 2x - 21x + 7$
47. $x^2 - 4x + x - 4$
48. $8x^2 - 4x + 6x - 3$
49. $x^2 - 5x + 4x - 20$
50. $3x^2 - 6x + 5x - 10$
51. $4x^2 + 8x - 3x - 6$
52. $4x^2 - 8x + 3x - 6$
53. $4x^2 + 8x + 3x + 6$
54. $4x^2 - 8x - 3x + 6$
55. $8x^2 + 4x - 6x - 3$

**3** Factor the trinomials by grouping.

56. $3x^2 + 7x + 2$
57. $3x^2 + 14x + 8$
58. $6x^2 + 13x + 6$
59. $8x^2 + 2x - 3$
60. $6x^2 - 17x + 12$
61. $2x^2 - 9x + 10$
62. $6x^2 - 13x + 5$
63. $8x^2 + 10x + 3$
64. $6x^2 - 11x + 5$
65. $8x^2 + 26x + 15$
66. $15x^2 - 22x - 5$
67. $8x^2 - 10x + 3$
68. $2x^2 - 5x - 7$
69. $12x^2 + 8x - 15$
70. $10x^2 + x - 3$
71. $12x^2 + 5x - 2$
72. $12x^2 - 5x - 2$
73. $12x^2 + 11x + 2$
74. $12x^2 - 11x + 2$
75. $24x^2 + 5x - 1$
76. $24x^2 - 11x + 1$
77. $6x^2 + 7xy - 10y^2$
78. $6a^2 - 17ab - 14b^2$
79. $18x^2 - 3x - 10$

**4** Factor the polynomials completely. Identify common factors first, then identify special cases.

80. $x^2 + x - 6$
81. $2x^2 + x - 3$
82. $x^2 - 9$
83. $4x^2 - 16$
84. $m^2 + 2m - 15$
85. $2a^2 + 6a + 4$
86. $b^2 + 6b + 9$
87. $16m^2 - 8m + 1$
88. $x^2 + 8x + 7$
89. $2m^2 + 5m + 2$
90. $2m^2 - 5m - 3$
91. $2a^2 - 3a - 5$
92. $3x^2 + 10x - 8$
93. $6x^2 + x - 15$
94. $8x^2 + 10x - 3$
95. $-2x^2 + 6x - 4$

| Learning Objectives | What to Remember with Examples |
|---|---|

## Section 10–1

**1** Factor an expression containing a common factor (pp. 410–412).

To factor an expression containing a common factor: **1.** Find the greatest factor common to each term of the expression. **2.** Divide each term by the common factor. **3.** Rewrite the expression as the indicated product of the greatest common factor and the remaining quantity.

Factor $3ab + 9b^2$.

$$3b\left(\frac{3ab}{3b} + \frac{9b^2}{3b}\right) = 3b(a + 3b)$$

## Section 10–2

**1** Multiply polynomials (pp. 413–414).

To multiply two polynomials distribute by multiplying each term of the first factor times each term of the second factor. Combine like terms.

Multiply $(x^2 + 3x + 1)(x^2 - 3x + 2)$.

$$x^2(x^2 - 3x + 2) + 3x(x^2 - 3x + 2) + 1(x^2 - 3x + 2) =$$
$$x^4 - 3x^3 + 2x^2 + 3x^3 - 9x^2 + 6x + x^2 - 3x + 2 =$$
$$x^4 - 6x^2 + 3x + 2$$

**2** Use the FOIL method to multiply two binomials (pp. 414–415).

To multiply two binomials by the FOIL method: Multiply *F*irst terms, *O*uter terms, *I*nner terms, and *L*ast terms. Combine like terms.

Multiply $(3a - 2)(a + 3)$.

$$\begin{array}{cccc} F & O & I & L \\ 3a^2 & + 9a & - 2a & - 6 \end{array} \quad \text{Combine like terms.}$$
$$3a^2 + 7a - 6$$

**3** Multiply polynomials that result in special products (pp. 415–417).

To multiply the sum and difference of the same two terms: **1.** Square the first term. **2.** Insert minus sign. **3.** Square the second term.

Multiply $(3b - 2)(3b + 2)$.

$$9b^2 - 4 \qquad (3b)^2 = 9b^2; (2)^2 = 4$$

To square a binomial: **1.** Square the first term. **2.** Double the product of the two terms. **3.** Square the second term.

Multiply $(2a - 3)^2$.

$$4a^2 - 12a + 9 \qquad (2a)^2 = 4a^2; 2(2a)(-3) = -12a; 3^2 = 9$$

**4** Divide polynomials using long division (pp. 417–418).

To divide a polynomial by a polynomial using long division: **1.** Divide the first term of the dividend by the first term of the divisor. The partial quotient is placed above the first term of the dividend. **2.** Multiply the partial quotient times the divisor and align the product under like terms of the dividend. **3.** Subtract (change subtrahend to opposite and use addition rules). **4.** Bring down the next term of the dividend and repeat Steps 1–3. **5.** Repeat Steps 1–4 until all terms of the dividend have been brought down. The result of the last subtraction is the remainder. **6.** Write the remainder as a fraction with the remainder as the numerator and the divisor as the denominator.

Divide $(3x^2 + 13x - 10) \div (x + 5)$

$$\begin{array}{r} 3x - 2 \\ x + 5 \overline{\smash{\big)}\, 3x^2 + 13x - 10} \\ \underline{3x^2 + 15x} \\ -2x - 10 \\ \underline{-2x - 10} \\ 0 \end{array}$$

## Section 10–3

**1** Recognize and factor the difference of two perfect squares (pp. 419–420).

The difference of two perfect squares is a binomial in the form $a^2 - b^2$.

Identify which expression is the special product, the difference of two squares.

(a) $36 - 27A^2$; no, 27 is not a perfect square.
(b) $9c^2 - 4$; yes, both terms are perfect squares. The terms are subtracted.

To factor the difference of two perfect squares: **1.** Take the square root of the first term. **2.** Take the square root of the second term. **3.** Write one factor as the sum and one factor as the difference of the square roots from Steps 1 and 2. Symbolically, $a^2 - b^2 = (a + b)(a - b)$.

Factor $9c^2 - 4$.

$$(3c + 2)(3c - 2)$$

**2** Recognize and factor a perfect-square trinomial (pp. 420–421).

A perfect-square trinomial is a trinomial in the form $a^2 + 2ab + b^2$. The first and last terms are positive perfect squares and the absolute value of the middle term is twice the product of the square roots of the first and last terms.

Identify which expression is a perfect-square trinomial.

(a) $4x^2 + 18x + 9$; not a perfect-square trinomial. The middle term is not twice the product of the square roots of the first and last terms. $(2)(2x)(3) = 12x$
(b) $9a^2 + 6a + 4$; the first and last terms are perfect squares. The middle term is twice the product of the square roots of the first and last terms, so it is a perfect-square trinomial.

To factor a perfect-square trinomial: **1.** Write the square root of the first term. **2.** Write the sign of the middle term. **3.** Write the square root of the last term. **4.** Indicate the square of the quantity. Symbolically, $a^2 + 2ab + b^2 = (a + b)^2$ and $a^2 - 2ab + b^2 = (a - b)^2$.

Factor $9a^2 + 6a + 4$.

$$(3a + 2)^2$$

## Section 10–4

**1** Factor general trinomials whose squared term has a coefficient of 1 (pp. 422–423).

To factor a general trinomial whose squared term has a coefficient of 1: Factor using the FOIL method in reverse. Use the factors of the third term that will give the desired sum that produces the middle term and sign.

Factor $a^2 - 5a + 6$.

| | |
|---|---|
| $(\quad)(\quad)$ | Factors of 6 that add to 5 are 3 and 2. |
| $(a \quad 3)(a \quad 2)$ | Use terms with like signs that are negative and add to $-5$. |
| $(a - 3)(a - 2)$ | These factors of 6, $-3$ and $-2$, give $-5a$ as the middle term. |

**2** Remove common factors after grouping an expression (pp. 423–425).

Arrange terms of expression into groups of two terms. Factor common factors from each group. Factor the common binomial factor if there is one.

Factor $2x^2 - 4x + xy - 2y$ completely.

$(2x^2 - 4x) + (xy - 2y)$       Arrange into groups of two terms. Factor common factors.
$2x(x - 2) + y(x - 2)$       Factor common binomial.
$(x - 2)(2x + y)$

**3** Factor a general trinomial by grouping (pp. 425–428).

To factor a general trinomial by grouping: **1.** Verify that the first term is positive or rewrite with $-1$ as the common factor. **2.** Multiply the coefficients of the first term and last term. **3.** Factor the product from Step 2 into two factors (a) whose sum is the coefficient of the middle term if the last term is positive, or (b) whose difference is the coefficient of the middle term if the last term is negative. **4.** Rewrite the trinomial as four terms so the coefficients of the middle term use the factors from Step 3. **5.** Group the polynomial from Step 4 in two groups of two terms. **6.** Factor the common factors from each grouping. **7.** Factor the common binomial factor.

Factor $6x^2 - 17x + 12$.

$6 \cdot 12 = 72$       Factor pairs of 72.
$1 \cdot 72$       Find the factor pair that has a sum of $-17$.
$2 \cdot 36$
$3 \cdot 24$
$4 \cdot 18$
$6 \cdot 12$
$\boxed{8 \cdot 9}$       Selected pair: $-8 + (-9) = -17$.

$6x^2 - 8x - 9x + 12$       Rewrite the middle term. $-17x = -8x + (-9x)$.
$(6x^2 - 8x) + (-9x + 12)$       Factor common factor from each binomial.
$2x(3x - 4) - 3(3x - 4)$       Factor the common binomial.
$(3x - 4)(2x - 3)$

**4** Factor any binomial or trinomial that is not prime (pp. 428–430).

To factor any binomial or trinomial: **1.** Factor the greatest common factor (if any). **2.** Check for any special products. **3.** If there is no special product, factor by grouping. **4.** Examine each factor to see if it can be factored further. **5.** Check factoring by multiplying.

Factor $12x^3 + 6x^2 - 18x$ completely.

$$6x(2x^2 + x - 3) =$$
$$6x(2x^2 + 3x - 2y - 3) =$$
$$6x[(2x^2 + 3x) + (-2x - 3)] =$$
$$6x[x(2x + 3) - 1(2x + 3)] =$$
$$6x(2x + 3)(x - 1)$$

(After removing the common factor, procedures for factoring general trinomials can be used to factor the remaining trinomial.)

## CHAPTER REVIEW EXERCISES

### Section 10–1

Factor by removing the greatest common factor.

**1.** $5x + 5y$
**2.** $2x + 5x^2$
**3.** $12m^2 - 8n^2$
**4.** $25x^2y - 10xy^3 + 5xy$
**5.** $2a^3 - 14a^2 - 2a$
**6.** $30a^3 - 18a^2 - 12a$

### Section 10–2

Use the FOIL method to find the products. Combine the outer and inner products mentally.

**7.** $(x + 3)(2x - 5)$
**8.** $(5x - 7y)(4x + 3y)$
**9.** $(2a + 3b)(7a - b)$
**10.** $(5a + 2b)(6a - 5b)$
**11.** $(9x - 2y)(3x + 4y)$
**12.** $(5x - 8y)(4x - 3y)$

Find the special products mentally.

**13.** $(6x - 5)(6x + 5)$      **14.** $(3m + 4)(3m - 4)$      **15.** $(7y + 11)(7y - 11)$

**16.** $(12r - 7s)(12r + 7s)$      **17.** $(x + 9)^2$      **18.** $(x - 7)^2$

**19.** $(x - 3)^2$      **20.** $(2x - 3)^2$      **21.** $(4x - 15)^2$

**22.** $(5 + 3m)^2$      **23.** $(8 + 7m)^2$

Find the quotients.

**24.** $x + 4 \overline{)x^2 - 3x - 28}$      **25.** $x - 9 \overline{)x^2 - 11x + 18}$

**26.** $2x - 3 \overline{)6x^2 - 13x + 21}$      **27.** $5x + 2 \overline{)15x^2 - 4x - 4}$

## Section 10–3

Factor the expressions that are special products. Explain why the other expressions are not special products.

**28.** $x^2 - 81$      **29.** $25y^2 - 4$      **30.** $100a^2 - 8ab^2$

**31.** $a^2b^2 + 49$      **32.** $121 - 9m^2$      **33.** $a^2 + 2a + 1$

**34.** $4x^2 + 12x + 9$      **35.** $n^2 + 169 - 26n$      **36.** $16d^2 - 20d + 25$

**37.** $36a^2 + 84ab + 49b^2$      **38.** $4x^2 - 25y^2$      **39.** $49 - 14x + x^2$

**40.** $9x^2y^2 - 49z^2$      **41.** $64 + 25x^2$      **42.** $4x^2 + 12xy + 9y^2$

**43.** $16x^2 + 24x + 9y^2$      **44.** $36 - x^2$      **45.** $49 - 81y^2$

**46.** $64x^2 - 25y^2$      **47.** $9x^2 - 100y^2$      **48.** $a^2 - 10a + 25$

**49.** $9x^2 - 6xy + y^2$      **50.** $25 - 16a^2b^2$      **51.** $9x^2y^2 - z^2$

## Section 10–4

Factor the trinomials.

**52.** $x^2 - 13x + 40$      **53.** $x^2 - 9x + 8$      **54.** $x^2 - 17x - 18$

**55.** $x^2 - 11x - 26$      **56.** $x^2 + x - 30$      **57.** $x^2 + 5x - 24$

**58.** $6x^2 + 25x + 14$      **59.** $6x^2 + 25x + 4$      **60.** $4x^2 - 23x + 15$

**61.** $5x^2 - 34x + 24$      **62.** $3x^2 - x - 14$      **63.** $6x^2 - x - 35$

Factor the polynomials. Look for common factors and special cases.

**64.** $5mn - 25m$      **65.** $9a^2 - 100$      **66.** $a^2 - b^2$

**67.** $2x^2 - 3x - 2$      **68.** $b^3 - 8b^2 - b$      **69.** $a^2 - 81$

**70.** $x^2 - 14x + 13$      **71.** $y^2 - 14y + 49$      **72.** $m^2 - 3m + 2$

**73.** $b^2 + 8b + 15$      **74.** $x^2 - 13x + 30$      **75.** $169 - m^2$

## TEAM PROBLEM-SOLVING EXERCISES

**1.** A standard-sized rectangular swimming pool is 25 ft long and 15 ft wide. This gives a water-surface area of 375 ft². A customer wants to examine some options for varying the size of the pool.

     **(a)** Write an expression in both factored and expanded form for finding the water-surface area of a pool that is changed by $x$ feet in length and $y$ feet in width.

     **(b)** Write an expression using one variable in both factored and expanded form for the water-surface area of a pool when the length and width increase by the same amount.

**2.** Use the expressions written in Exercise 1 to answer the following.

     **(a)** Will the expressions work for both increasing and decreasing the size of the pool?

     **(b)** Illustrate your answer to part (a) with numerical examples.

Find the products.
1. $(m - 7)(m + 7)$
2. $(3x - 2)(3x + 2)$
3. $(a + 3)^2$
4. $(2x - 7)^2$
5. $(x - 3)(2x - 5)$
6. $(7x - 3)(2x + 1)$

Divide using long division.
7. $x - 3{\overline{\smash{\big)}\,2x^2 + x - 21}}$
8. $x + 3{\overline{\smash{\big)}\,2x^3 + 3x^2 - 7x + 6}}$

Factor by removing the greatest common factor.
9. $7x^2 + 8x$
10. $6ax + 15bx$
11. $7a^2b - 14ab$

Factor into special products.
12. $a^2 - 25$
13. $9x^2 - 25$
14. $x^2 + 4x + 4$
15. $x^2 - 18x + 81$

Factor into two binomials.
16. $x^2 + 5x - 24$
17. $6x^2 - 5x - 6$
18. $y^2 - y - 6$
19. $a^2 + 16ab + 64b^2$
20. $x^2 - 7x + 10$
21. $b^2 - 3b - 10$
22. $3x^2 + x - 4$
23. $3m^2 - 5m + 2$
24. $3x^2 + 11xy + 3y^2$

Factor completely.
25. $3x^2 - 12$
26. $3x^2 - 3x - 18$
27. $5x^2 - 20$
28. $27a^2 - 48$
29. $3x^2 + 12x + 12$
30. $3x^2 - 15x - 18$

## CAREER APPLICATION: INFORMATION SYSTEMS—SECURITY CODING

One of the most important areas of information systems is security for computer programs, files, and databases. Governments, companies, and individuals use sophisticated methods of coding passwords and log-on procedures. In one type of advanced code, a branch of mathematics called *matrix theory* encodes and decodes. A rectangular array of numbers, called a *matrix*, is multiplied by the password matrix after the letters of the password have been assigned numeric values (A = 1, B = 2, for example). The inverse of the encoding matrix is then used to decode the password. It is desirable that the encoding matrix not have a determinant that is a prime number because prime numbers make it much easier for hackers to learn the encoding matrix (and hence the password).

About 250 B.C., a Greek scholar named Eratosthenes devised a method, called a *sieve*, for finding all prime numbers up to any given number $N$. Use his method to find all prime numbers up to 200 (which would be eliminated as choices for encoding matrix determinants). Continue the process shown in the Tip box on page 64 for the numbers 1–200. The best choices for encoding matrix determinants are composite numbers with numerous factors. These numbers would have been crossed out repeatedly.

### Exercises

Use the sieve that you constructed above to answer the following.

1. Count and list the prime numbers less than 200. Remember that 1 is neither prime nor composite because it has exactly one factor; the number 1. (Prime numbers have exactly two factors, and composite numbers have more than two factors.)

2. Which was the first prime number to have no multiples crossed out? Eratosthenes proved that it is never necessary to go higher than the square root of $N$. Did your sieve comply with this rule?

3. Select five good choices for encoding matrix determinants, which would be composite numbers less than or equal to 200 with many factors.

4. Write a computer program to construct the same sieve, and compare results between the computer method and the manual method. If they differ, explain why.

5. Use the program from Exercise 4 to find the primes less than 400. List and count them. Are there double the number of primes less than 200? Why do you think this is so?

6. Alter your program to list the composites with 10 or more factors.

## Answers

1. 46 primes: 2, 3, 5, 7, 11, 13, 17, 19, 23, 29, 31, 37, 41, 43, 47, 53, 59, 61, 67, 71, 73, 79, 83, 89, 97, 101, 103, 107, 109, 113, 127, 131, 137, 139, 149, 151, 157, 163, 167, 173, 179, 181, 191, 193, 197, and 199.

2. 17 was the first prime number to have no multiples crossed out. Note that the square root of 200 is about 14.2, and the next largest prime is 17.

3. Good choices, 48, 60, 72, 90, 96, 120, 144, 168, 180, and 192.

4. The results should be the same, but a common error is to forget to eliminate the number 1 from the list of primes.

5. 78 primes: 2, 3, 5, 7, 11, 13, 17, 19, 23, 29, 31, 37, 41, 43, 47, 53, 59, 61, 67, 71, 73, 79, 83, 89, 97, 101, 103, 107, 109, 113, 127, 131, 137, 139, 149, 151, 157, 163, 167, 173, 179, 181, 191, 193, 197, 199, 211, 223, 227, 229, 233, 239, 241, 251, 257, 263, 269, 271, 277, 281, 283, 293, 307, 311, 313, 317, 331, 337, 347, 349, 353, 359, 367, 373, 379, 383, 389, and 397. There are fewer and fewer primes as the numbers increase because larger numbers are more likely than smaller numbers to have additional factors.

6. Good choices: 240, 252, 264, 270, 288, 300, 336, 360, 384, and 400.

# 11

# Quadratic Equations

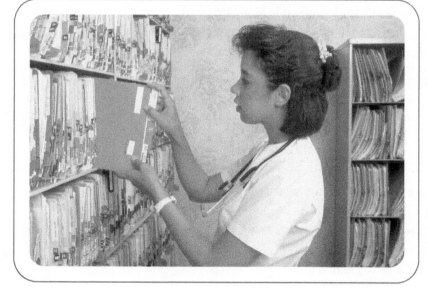

## Focus on Careers

All hospitals and doctor's offices are required to maintain accurate records of observations, medical and surgical procedures and treatments, and outcomes of the procedures and treatments. Medical-records and health-information technicians organize these records and evaluate them for accuracy and completeness.

Medical-records technicians have little or no direct contact with patients. They communicate with physicians and other health-care professionals to clarify diagnoses or obtain additional information to complete the patient's medical record.

These careers require an associate degree from a community or junior college and courses include anatomy, physiology, medical terminology, legal aspects of health information, statistics, database management, and computer science. Job prospects for this career are very good, especially in physicians' offices. Job growth is expected to be much faster than average through 2012.

*(continued)*

Medical-records technicians usually work a 40-hr week and some overtime may be required. Hospitals have medical records technicians on duty 24 hr a day, 7 days a week. Because accuracy is essential, these technicians must be very detail oriented.

Registered health-information technicians must pass a written examination after graduating from a 2-year associate degree program accredited by the Commission on Accreditation of Allied Health Education Programs of the American Medical Association.

In 2002, the highest 10% of medical-records technicians earned in excess of $38,640. The middle 50% of medical-records technicians earned between $19,550 and $30,600 annually. More jobs are found in nursing-care facilities, where the median annual earnings were $25,160.

Source: *Occupational Outlook Handbook* 2004–2005 Edition. U.S. Department of Labor, Bureau of Labor Statistics.

The equations we have solved so far are called linear equations. In such equations, the variable appears only to the first power, such as $x$ or $y$. In this chapter, we examine quadratic equations.

---

## 11–1 | Solving Quadratic Equations by the Square-Root Method

*Learning Outcomes*

**1** Write quadratic equations in standard form.

**2** Identify the coefficients of the quadratic, linear, and constant terms of a quadratic equation.

**3** Solve pure quadratic equations by the square-root method ($ax^2 + c = 0$).

**1** Write Quadratic Equations in Standard Form.

There are several methods for solving quadratic equations. Each method has strengths and weaknesses depending on the characteristics of the equation being solved. First, let's examine the basic characteristics of quadratic equations, starting with quadratic equations having only one variable.

A **quadratic equation** is an equation in which at least one letter term is raised to the second power and no letter term has a power more than 2 or less than 0. The **standard form** for a quadratic equation is $ax^2 + bx + c = 0$, where $a$, $b$, and $c$ are real numbers and $a > 0$.

---

**EXAMPLE** Arrange the quadratic equations in standard form.

(a) $3x^2 - 3 + 5x = 0$      (b) $7x - 2x^2 - 8 = 0$

(c) $12 = 7x^2 - 4x$      (d) $7x^2 = 3$

(a) $3x^2 - 3 + 5x = 0$        Arrange the terms in descending powers of $x$.

$\mathbf{3x^2 + 5x - 3 = 0}$        Standard form.

(b) $7x - 2x^2 - 8 = 0$        Arrange the terms in descending powers of $x$.

$-2x^2 + 7x - 8 = 0$        Multiply both sides by $-1$ to make the coefficient of the $x^2$ term positive.

$-1(-2x^2 + 7x - 8) = -1(0)$

$\mathbf{2x^2 - 7x + 8 = 0}$        Standard form.

(c) $12 = 7x^2 - 4x$        Rearrange the terms so that one side of the equation equals zero.

$0 = 7x^2 - 4x - 12$        Interchange sides of equation.

$\mathbf{7x^2 - 4x - 12 = 0}$

(d) $7x^2 = 3$        Rearrange so that one side equals zero.

$\mathbf{7x^2 - 3 = 0}$

A quadratic equation is a second-degree equation, which means that it can have as many as two real solutions. As we examine the process for solving quadratic equations, we shall see quadratic equations that have two, one, or no solution.

### 2  Identify the Coefficients of the Quadratic, Linear, and Constant Terms of a Quadratic Equation.

Some methods for solving quadratic equations can be used for any type of quadratic equation but are very time consuming. Other methods are quicker but apply only to certain types of quadratic equations. In choosing an appropriate method it is helpful to be able to recognize similar and different characteristics of quadratic equations.

The standard form of quadratic equations, $ax^2 + bx + c = 0$, has three types of terms.

1. $ax^2$ is a **quadratic term;** that is, the degree of the term is 2. In standard form, this term is the **leading term** and $a$ is the **leading coefficient.**

2. $bx$ is a **linear term;** that is, the degree of the term is 1 and the coefficient is $b$.
3. $c$ is a **number term** or **constant term;** that is, the degree of the term is 0. The coefficient is $c$ ($c = cx^0$).

A quadratic equation that has all three types of terms is sometimes referred to as a **complete quadratic equation.**

$$ax^2 + bx + c = 0 \qquad \text{Complete quadratic equation.}$$

A quadratic equation that has a quadratic term ($ax^2$) and a linear term ($bx$) but no constant term ($c$) is sometimes referred to as an **incomplete quadratic equation.**

$$ax^2 + bx = 0 \qquad \text{Incomplete quadratic equation.}$$

A quadratic equation that has a quadratic term ($ax^2$) and a constant ($c$) but no linear term ($bx$) is sometimes referred to as a **pure quadratic equation.**

$$ax^2 + c = 0 \qquad \text{Pure quadratic equation.}$$

It is helpful in planning our strategy for solving a quadratic equation to be able to identify the types of terms in a quadratic equation and to identify the coefficients $a$, $b$, and $c$.

---

**EXAMPLE**  Write each equation in standard form and identify $a$, $b$, and $c$.

(a) $3x^2 + 5x - 2 = 0$  
(b) $5x^2 = 2x - 3$  
(c) $3x = 5x^2$  
(d) $-6x^2 + 2x = 0$  
(e) $5x^2 - 4 = 0$  
(f) $x^2 = 9$  
(g) $x^2 + 3x = 5x$  
(h) $7 - x^2 = 3 + x$  

(a) $3x^2 + 5x - 2 = 0$      In standard form.
$$a = 3, \quad b = 5, \quad c = -2$$

(b) $5x^2 = 2x - 3$      Write in standard form.
$$5x^2 - 2x + 3 = 0$$
$$a = 5, \quad b = -2, \quad c = 3$$

(c) $3x = 5x^2$      Write in standard form.
$$0 = 5x^2 - 3x \qquad \text{Interchange sides of equation.}$$
$$5x^2 - 3x = 0$$
$$a = 5, \quad b = -3, \quad c = 0 \qquad \text{No constant term means } c = 0.$$

(d) $-6x^2 + 2x = 0$      Multiply by $-1$.
$$-1(-6x^2 + 2x = 0)$$
$$6x^2 - 2x = 0 \qquad \text{Standard form.}$$
$$a = 6, \quad b = -2, \quad c = 0 \qquad \text{No constant term.}$$

(e) $5x^2 - 4 = 0$      Standard form.
$$a = 5, \quad b = 0, \quad c = -4 \qquad \text{No linear term means } b = 0.$$

**(f)** $x^2 = 9$                                     Write in standard form.

$x^2 - 9 = 0$

$a = 1, \quad b = 0, \quad c = -9$              No linear term.

**(g)** $x^2 + 3x = 5x$                               Write in standard form.

$x^2 + 3x - 5x = 0$                                 Combine like terms.

$x^2 - 2x = 0$

$a = 1, \quad b = -2, \quad c = 0$              No constant term.

**(h)** $7 - x^2 = 3 + x$                             Rearrange.

$-x^2 - x + 7 - 3 = 0$                              Combine like terms.

$-x^2 - x + 4 = 0$                                   Multiply by $-1$.

$-1(-x^2 - x + 4 = 0)$

$x^2 + x - 4 = 0$                                      Standard form.

$a = 1, \quad b = 1, \quad c = -4$

---

**3**  **Solve Pure Quadratic Equations by the Square-Root Method ($ax^2 + c = 0$).**

To solve pure quadratic equations, solve for the squared letter and then take the square root of both sides of the equation.

**To solve a pure quadratic equation ($ax^2 + c = 0$):**

1.  Rearrange the equation, if necessary, so that quadratic or squared-letter terms are on one side and constants or number terms are on the other side of the equation.
2.  Combine like terms, if appropriate.
3.  Solve for the squared letter so that the coefficient is $+1$.
4.  Apply the square-root principle of equality by taking the square root of both sides.
5.  Check each solution in the original equation.

---

**EXAMPLE**    Solve $3y^2 = 27$.

$3y^2 = 27$        Divide both sides by the coefficient of the quadratic term.

$y^2 = 9$          Take the square root of both sides.

$y = \pm 3$        Solutions are $y = 3$ and $y = -3$.

Check $y = +3$                              Check $y = -3$

$3y^2 = 27$                                     $3y^2 = 27$

$3(\,3\,)^2 = 27$                               $3(\,-3\,)^2 = 27$

$3(9) = 27$                                      $3(9) = 27$

$27 = 27$                                         $27 = 27$        Both solutions check.

---

**EXAMPLE**    Solve $4x^2 - 9 = 0$.

$4x^2 - 9 = 0$          Sort terms.

$4x^2 = 9$              Divide by 4.

$x^2 = \dfrac{9}{4}$    Take the square root of the numerator and denominator.

$x = \pm\dfrac{3}{2}$   Divide if a decimal answer is desired.

$x = \pm 1.5$           Solutions are $x = 1.5$ and $x = -1.5$.

$$\begin{array}{cc}
\text{Check } x = \boxed{+1.5} & \text{Check } x = \boxed{-1.5} \\
4x^2 - 9 = 0 & 4x^2 - 9 = 0 \\
4(\boxed{1.5})^2 - 9 = 0 & 4(\boxed{-1.5})^2 - 9 = 0 \\
4(2.25) - 9 = 0 & 4(2.25) - 9 = 0 \\
9 - 9 = 0 & 9 - 9 = 0 \\
0 = 0 & 0 = 0
\end{array}$$

In some applications, the fractional answer may be more convenient. In other applications, the decimal answer may be more convenient.

**TIP!**

### Roots of Pure Quadratic Equations

In a pure quadratic equation, the value of $b$ is zero: $ax^2 + 0x + c = 0$.

- The equation has real number solutions only if $c$ is negative when the equation is in standard form. Remember, standard form requires that $a$ be positive.
- The two solutions have the same absolute value and are opposites.
- There will be either two solutions or no real solution. There will never be just one solution.

### SECTION 11–1 SELF-STUDY EXERCISES

**1** Arrange the quadratic equations in standard form.

**1.** $5 - 4x + 7x^2 = 0$      **2.** $8x^2 - 3 = 6x$      **3.** $7x^2 = 5$

**4.** $x^2 = 6x - 8$      **5.** $x^2 - 3x = 6x - 8$      **6.** $8 - x^2 + 6x = 2x$

**7.** $3x^2 + 5 - 6x = 0$      **8.** $5 = x^2 - 6x$      **9.** $x^2 - 16 = 0$

**10.** $8x^2 - 7x = 8$      **11.** $8x^2 + 8x - 2 = 8$      **12.** $0.3x^2 - 0.4x = 3$

**2** Write each equation in standard form and identify $a$, $b$, and $c$.

**13.** $5x = x^2$      **14.** $7x - 3x^2 = 5$      **15.** $7x^2 - 4x = 0$

**16.** $8 + 3x^2 = 5x$      **17.** $5x - 6 = x^2$      **18.** $11x^2 = 8x$

**19.** $x = x^2$      **20.** $9x^2 - 7x = 12$      **21.** $5 = x^2$

**22.** $-x^2 - 6x + 3 = 0$      **23.** $0.2x - 5x^2 = 1.4$      **24.** $\dfrac{2}{3}x^2 - \dfrac{5}{6}x = \dfrac{1}{2}$

**25.** $1.3x^2 - 8 = 0$      **26.** $\sqrt{3}x^2 + \sqrt{5}x - 2 = 0$

Write equations in standard form.

**27.** The coefficient of the quadratic term is 8, the coefficient of the linear term is $-2$, the constant term is $-3$.

**28.** The constant is 0, the coefficient of the linear term is 3, and the coefficient of the quadratic term is 1.

**29.** $a = 5$, $b = 2$, $c = -7$

**30.** $a = 2.5$, $c = -0.8$

**3** Solve the equations. Round to thousandths when necessary.

**31.** $x^2 = 9$      **32.** $x^2 - 49 = 0$      **33.** $9x^2 = 64$

**34.** $16x^2 - 49 = 0$      **35.** $0.09y^2 = 0.81$      **36.** $0.04x^2 = 0.81$

**37.** $2x^2 = 10$      **38.** $3x^2 + 4 = 7$      **39.** $5x^2 - 8 = 12$

**40.** $7x^2 - 2 = 19$

41. **AG/H** The label on a new tarpaulin shows it is a square and contains 529 ft². What is the length of each side of the tarpaulin?

42. **AG/H** A farmer has a field designed in the shape of a circle for efficiency in irrigation. The area of the field is known to be 21,000 yd². What length of irrigation line is required to provide water to the entire field? (*Note:* The line moves from the center of the field in a circle around the field. The area of a circle is $A = \pi r^2$.)

43. Describe the process for solving a pure quadratic equation.

44. What is the relationship between the roots of a pure quadratic equation?

---

## 11–2 | Solving Quadratic Equations by Factoring

*Learning Outcomes*    **1** Solve incomplete quadratic equations by factoring ($ax^2 + bx = 0$).

**2** Solve complete quadratic equations by factoring ($ax^2 + bx + c = 0$).

**1** **Solve Incomplete Quadratic Equations by Factoring ($ax^2 + bx = 0$).**

An incomplete quadratic equation always has a variable common factor, so we can factor the expression $x^2 + 2x = 0$ to $x(x + 2) = 0$. We have two factors that have a product of 0. The **zero-product property** shown below can be used to solve the equation.

---

**Zero-product property**

If $ab = 0$, then $a$ or $b$ or both equal zero.

---

**To solve an incomplete quadratic equation by factoring ($ax^2 + bx = 0$):**

1. Sort so that all terms are on one side of the equation in standard form, with zero on the other side.
2. Factor all common factors and set each factor containing a variable equal to zero.
3. Solve for the variable in each equation formed in Step 2.
4. Check each solution in the original equation.

---

**EXAMPLE**    Solve $2x^2 = 5x$ for $x$.

$$2x^2 = 5x$$    Write in standard form.
$$2x^2 - 5x = 0$$    Factor common factor.
$$x\,(2x - 5) = 0$$    Set each factor equal to zero.

$x = 0$      $2x - 5 = 0$    Solve each equation.

$x = \mathbf{0}$      $2x = 5$

$$x = \frac{5}{2}$$

$$\text{Check } x = \boxed{0}$$ $$\text{Check } x = \boxed{\dfrac{5}{2}}$$

$$2x^2 = 5x$$ $$2x^2 = 5x$$

$$2(\,\boxed{0}\,)^2 = 5(\,\boxed{0}\,)$$ $$2\left(\boxed{\dfrac{5}{2}}\right)^2 = 5\left(\boxed{\dfrac{5}{2}}\right)$$

$$2(0) = 5(0)$$ $$\overset{1}{2}\left(\dfrac{25}{\underset{2}{\cancel{4}}}\right) = \dfrac{25}{2}$$

$$0 = 0$$ $$\dfrac{25}{2} = \dfrac{25}{2}$$

---

**EXAMPLE** Solve $4x^2 - 8x = 0$ for $x$.

$$4x^2 - 8x = 0$$ Standard form. Factor common factors.

$$\boxed{4x\ (x - 2)} = 0$$ Set each variable factor equal to zero.

$$\boxed{4x = 0} \quad \boxed{x - 2 = 0}$$ Solve each equation.

$$x = \dfrac{0}{4} \qquad x = \boxed{2}$$

$$x = \boxed{0}$$

$$\text{Check } x = \boxed{0}$$ $$\text{Check } x = \boxed{2}$$
$$4(\,\boxed{0}\,)^2 - 8(\,\boxed{0}\,) = 0$$ $$4(\,\boxed{2}\,)^2 - 8(\,\boxed{2}\,) = 0$$
$$4(0) - 8(0) = 0$$ $$4(4) - 8(2) = 0$$
$$0 - 0 = 0$$ $$16 - 16 = 0$$
$$0 = 0$$ $$0 = 0$$

---

**TIP!**

## Roots of Incomplete Quadratic Equations

In an incomplete quadratic equation, the value of $c = 0$: $ax^2 + bx + 0 = 0$.

$$x(ax + b) = 0$$

$$x = 0 \qquad ax + b = 0$$

$$\dfrac{ax}{a} = \dfrac{-b}{a}$$

$$x = \dfrac{-b}{a}$$

- One root is always zero.
- The other root is $\dfrac{-b}{a}$.

**2** **Solve Complete Quadratic Equations by Factoring ($ax^2 + bx + c = 0$).**

Complete quadratic equations have all three types of terms. If the expression on the left will factor into two binomials, we can apply the zero-product property. If the trinomial will not factor, we must use another method for solving the quadratic equation.

**To solve a complete quadratic equation by factoring:**

1. Arrange the equation in standard form.
2. Factor the trinomial into the product of two binomials, and set each binomial factor equal to zero.
3. Solve for the variable in each equation formed in Step 2.

---

**EXAMPLE** Solve $x^2 + 6x + 5 = 0$ for $x$.

$$x^2 + 6x + 5 = 0 \qquad \text{Standard form. Factor into the product of two binomials.}$$
$$(x + 5)\,(x + 1) = 0 \qquad \text{Set each factor equal to 0.}$$
$$x + 5 = 0 \quad x + 1 = 0 \qquad \text{Solve for } x \text{ in each equation.}$$
$$x = -5 \qquad x = -1$$

$$\text{Check } x = -5 \qquad\qquad\qquad \text{Check } x = -1$$
$$x^2 + 6x + 5 = 0 \qquad\qquad\qquad x^2 + 6x + 5 = 0$$
$$(-5)^2 + 6(-5) + 5 = 0 \qquad\qquad (-1)^2 + 6(-1) + 5 = 0$$
$$25 + (-30) + 5 = 0 \qquad\qquad\qquad 1 + (-6) + 5 = 0$$
$$-5 + 5 = 0 \qquad\qquad\qquad\qquad -5 + 5 = 0$$
$$0 = 0 \qquad\qquad\qquad\qquad\qquad 0 = 0$$

---

**EXAMPLE** Solve $6x^2 + 4 = 11x$ for $x$.

$$6x^2 + 4 = 11x \qquad \text{Write in standard form.}$$
$$6x^2 - 11x + 4 = 0 \qquad \text{Factor trinomial by grouping. Write } -11x \text{ as } -3x - 8x.$$
$$6x^2 - 3x - 8x + 4 = 0 \qquad \text{Group.}$$
$$(6x^2 - 3x) + (-8x + 4) = 0 \qquad \text{Factor common factors.}$$
$$3x(2x - 1) - 4(2x - 1) = 0 \qquad \text{Factor the common grouping.}$$
$$(3x - 4)\,(2x - 1) = 0 \qquad \text{Set each factor equal to 0.}$$
$$3x - 4 = 0 \qquad 2x - 1 = 0 \qquad \text{Solve each equation.}$$
$$3x = 4 \qquad\qquad 2x = 1$$
$$x = \frac{4}{3} \qquad\qquad x = \frac{1}{2}$$

$$\text{Check } x = \frac{4}{3} \qquad\qquad\qquad \text{Check } x = \frac{1}{2}$$
$$6x^2 + 4 = 11x \qquad\qquad\qquad 6x^2 + 4 = 11x$$
$$6\left(\frac{4}{3}\right)^2 + 4 = 11\left(\frac{4}{3}\right) \qquad\qquad 6\left(\frac{1}{2}\right)^2 + 4 = 11\left(\frac{1}{2}\right)$$
$$\overset{2}{\cancel{6}}\left(\frac{16}{\underset{3}{\cancel{9}}}\right) + 4 = \frac{44}{3} \qquad\qquad \overset{3}{\cancel{6}}\left(\frac{1}{\underset{2}{\cancel{4}}}\right) + 4 = \frac{11}{2}$$

$$\frac{32}{3} + \frac{12}{3} = \frac{44}{3} \qquad \left(4 = \frac{12}{3}\right)$$

$$\frac{44}{3} = \frac{44}{3}$$

$$\frac{3}{2} + \frac{8}{2} = \frac{11}{2} \qquad \left(4 = \frac{8}{2}\right)$$

$$\frac{11}{2} = \frac{11}{2}$$

**TIP!**

### Roots of Complete Quadratic Equations

Because all three types of terms are included in complete quadratic equations, the roots do *not* have the same characteristics as the roots of pure or incomplete quadratic equations.

- Zero is not a root.
- The two roots are not opposites.
- There can be just one root (double root).

### SECTION 11–2 SELF-STUDY EXERCISES

**1** Solve by factoring.

1. $x^2 - 3x = 0$
2. $x^2 - 6x = 0$
3. $5x^2 - 10x = 0$
4. $2x^2 + x = 0$
5. $8x^2 - 4x = 0$
6. $5x^2 - 15x = 0$
7. $3x^2 - 7x = 0$
8. $y^2 + 4y = 0$
9. $3x^2 + 2x = 0$
10. $9x^2 = 12x$
11. $6x^2 = 18x$
12. $9x^2 = 6x$

13. The square of a number is 8 times the number. Find the number.

14. A square rug has an area 15 times the length of one of the sides. What is the length of a side?

15. How does an incomplete quadratic equation differ from a pure quadratic equation?

16. Will one of the two roots of an incomplete quadratic equation always be zero? Justify your answer.

**2** Solve the equations by factoring.

17. $x^2 + 5x + 6 = 0$
18. $x^2 - 6x + 9 = 0$
19. $x^2 - 5x - 14 = 0$
20. $x^2 + 3x - 18 = 0$
21. $x^2 + 7x + 12 = 0$
22. $y^2 - 8y = -15$
23. $a^2 - 13a = 14$
24. $b^2 - 9b = -18$
25. $2x^2 - 7x + 3 = 0$
26. $3x^2 + 13x + 4 = 0$
27. $10x^2 - x - 3 = 0$
28. $6x^2 + 11x + 3 = 0$
29. $2x^2 + 13x + 15 = 0$
30. $3x^2 - 10x + 8 = 0$
31. $6x^2 + 17x - 3 = 0$
32. $3x^2 + 14x + 8 = 0$
33. $2x^2 - 13x + 15 = 0$
34. $6x^2 - 7x + 2 = 0$
35. $8x^2 + 3 = 10x$
36. $5x^2 + 3x = 2$
37. $6x^2 = x + 15$
38. $9x^2 + 18x = -5$
39. $6x^2 + 3 = 11x$
40. $5x^2 + 13x = 6$

41. **CON** A rectangular hallway is 6 ft longer than its width. The area is 55 ft². What are the length and width of the hallway?

42. **AUTO** A rectangular metal plate covering a spare tire well that has an opening of 378 in² is broken and must be reconstructed. The width is 3 in. less than the length. What dimensions should the metalsmith use when making the replacement part?

*Learning Outcome*  **1**  Solve quadratic equations using the quadratic formula.

**1**  **Solve Quadratic Equations Using the Quadratic Formula.**

Another method for solving quadratic equations is to use the **quadratic formula.**

### Quadratic formula:

$$x = \frac{-b \pm \sqrt{b^2 - 4ac}}{2a}$$

where $a$, $b$, and $c$ are coefficients of a quadratic equation in the form $ax^2 + bx + c = 0$.

### To evaluate the quadratic formula:

1. Write the equation in standard form.
2. Identify $a$, $b$, and $c$.
3. Substitute numbers for $a$, $b$, and $c$ in the formula.
4. Use the order of operations to simplify the expression under the radical.
5. Simplify the radical.
6. Factor any common factors in the numerator.
7. Simplify by reducing.
8. Write as two distinct solutions.
9. Write as exact solutions or approximate solutions.

**TIP!**

### Common Cause for Error

When you use the quadratic formula to solve problems, begin by writing the formula to help you remember it. When you write the formula, be sure to extend the fraction bar beneath the *entire* numerator. This omission is a common cause for errors.

**EXAMPLE**  Use the quadratic formula to solve $x^2 + 5x + 6 = 0$ for $x$.

$a = 1, \qquad b = 5, \qquad c = 6$    Identify $a$, $b$, and $c$.

$x = \dfrac{-b \pm \sqrt{b^2 - 4ac}}{2a}$    Quadratic formula. Substitute for $a$, $b$, and $c$.

$x = \dfrac{-5 \pm \sqrt{5^2 - 4 \cdot 1 \cdot 6}}{2 \cdot 1}$    Raise to power and multiply in grouping.

$x = \dfrac{-5 \pm \sqrt{25 - 24}}{2}$    Combine terms under radical.

$$x = \frac{-5 \pm \sqrt{1}}{2}$$ Evaluate radical.

$$x = \frac{-5 \pm 1}{2}$$

At this point, we separate the formula into two parts, one using the $+1$ and the other using the $-1$.

$$x = \frac{-5 \boxed{+} 1}{2} \qquad\qquad x = \frac{-5 \boxed{-} 1}{2}$$

$$x = \frac{-4}{2} \qquad\qquad x = \frac{-6}{2}$$

$$x = -2 \qquad\qquad x = -3$$

Check $x = \boxed{-2}$        Check $x = \boxed{-3}$

$$x^2 + 5x + 6 = 0 \qquad\qquad x^2 + 5x + 6 = 0$$

$$(\boxed{-2})^2 + 5(\boxed{-2}) + 6 = 0 \qquad (\boxed{-3})^2 + 5(\boxed{-3}) + 6 = 0$$

$$4 - 10 + 6 = 0 \qquad\qquad 9 - 15 + 6 = 0$$

$$-6 + 6 = 0 \qquad\qquad -6 + 6 = 0$$

$$0 = 0 \qquad\qquad 0 = 0$$

**EXAMPLE**   Use the quadratic formula to solve $3x^2 - 3x - 7 = 0$ for $x$. Round answers to the nearest hundredth.

$a = 3, \quad b = -3, \quad c = -7$      Identify $a$, $b$, and $c$.

$$x = \frac{-b \pm \sqrt{b^2 - 4ac}}{2a}$$      Quadratic formula. Substitute.
$-(b) = -(-3) = +3$.

$$x = \frac{+3 \pm \sqrt{(-3)^2 - 4(3)(-7)}}{2 \cdot 3}$$      Perform calculations under the radical.

$$x = \frac{+3 \pm \sqrt{9 + 84}}{6}$$

$$x = \frac{+3 \pm \sqrt{93}}{6}$$      Evaluate the radical.

$$x = \frac{+3 \pm 9.643650761}{6}$$      Separate two solutions.

$$x = \frac{3 + 9.643650761}{6} \qquad x = \frac{3 - 9.643650761}{6}$$

$$x = \frac{12.643650761}{6} \qquad x = \frac{-6.643650761}{6}$$

$x = \mathbf{2.11}$    Rounded.    $x = \mathbf{-1.11}$    Rounded.

Notice that we round to hundredths *after* making the final calculation.

Check $x = \boxed{2.11}$        Check $x = \boxed{-1.11}$

$$3x^2 - 3x - 7 = 0 \qquad\qquad 3x^2 - 3x - 7 = 0$$

$$3(\boxed{2.11})^2 - 3(\boxed{2.11}) - 7 = 0 \qquad 3(\boxed{-1.11})^2 - 3(\boxed{-1.11}) - 7 = 0$$

$$3(4.4521) - 6.33 - 7 = 0 \qquad 3(1.2321) + 3.33 - 7 = 0$$

$$13.3563 - 6.33 - 7 = 0 \qquad 3.6963 + 3.33 - 7 = 0$$

$$0.0263 \approx 0 \qquad\qquad 0.0263 \approx 0$$

**TIP!**

### Rounding Discrepancies

The symbol ≈ means "is approximately equal to." Another symbol for approximations is ≐. If we had checked *before* rounding in the example on the previous page, the checks would have been "closer" to zero.

**TIP!**

### Check with Calculator

When solving a formula with as many steps as the quadratic formula, it is important to check your solutions.

Use the full calculator value to check. The full calculator value improves the accuracy of the check. Calculate the first root in the previous example.

$$3 \boxed{+} \boxed{\sqrt{\phantom{x}}} \ 93 \boxed{=} \boxed{\div} \ 6 \boxed{=} \Rightarrow 2.107275127$$

Check:

If your calculator allows you to insert the answer of your last calculation, use this function in checking. Otherwise, you can store an answer in memory and recall as appropriate.

$$3 \boxed{\text{ANS}} \boxed{x^2} \boxed{-} 3 \boxed{\text{ANS}} \boxed{-} 7 \boxed{=} \Rightarrow 0$$

Calculate the second root.

$$3 \boxed{-} \boxed{\sqrt{\phantom{x}}} \ 93 \boxed{=} \boxed{\div} \ 6 \boxed{=} \Rightarrow -1.107275127$$

Check:

$$3 \boxed{\text{ANS}} \boxed{x^2} \boxed{-} 3 \boxed{\text{ANS}} \boxed{-} 7 \boxed{=} \Rightarrow 0.$$

Even though quadratic equations have two roots, a root may be disregarded in an applied problem because it is not appropriate within the context of the problem.

**EXAMPLE** Find the length and width of a rectangular table if the length is 8 in. more than the width and the area is 260 in².

| | |
|---|---|
| **Unknown facts** | Length and width of rectangle |
| **Known facts** | Area = 260 in², length = 8 in. more than width |
| **Relationships** | Let $x$ = number of inches in the width |
| | $x + 8$ = number of inches in the length |
| | Area = length times width, or $A = lw$ |
| **Estimation** | The square root of 260 is between 16 and 17; the width should be less than 16 and the length more than 16. |
| **Calculations** | $260 = (x + 8)(x)$ |

Substitute into area formula and distribute.

$260 = x^2 + 8x$
$x^2 + 8x - 260 = 0$
$a = 1, \qquad b = 8, \qquad c = -260$

Write in standard form.

Chapter 11 / Quadratic Equations

$$x = \frac{-b \pm \sqrt{b^2 - 4ac}}{2a}$$

Quadratic formula. Substitute.

$$x = \frac{-8 \pm \sqrt{(8)^2 - 4(1)(-260)}}{2(1)}$$

Simplify the radicand.

$$x = \frac{-8 \pm \sqrt{64 + 1{,}040}}{2}$$

Combine terms in radicand.

$$x = \frac{-8 \pm \sqrt{1{,}104}}{2}$$

$$x = \frac{-8 \pm 33.22649545}{2}$$

Separate into two solutions.

$$x = \frac{-8 + 33.22649545}{2} \qquad x = \frac{-8 - 33.22649545}{2}$$

$$x = \frac{25.22649545}{2} \qquad x = \frac{-41.22649545}{2}$$

$$x = 12.6 \qquad x = -20.6$$

Disregard negative solution.

$$x + 8 = 20.6$$

**Interpretation**   Measurements are positive, so disregard the negative solution.
**The width is 12.6 in. and the length is 20.6 in.**

### SECTION 11–3 SELF-STUDY EXERCISES

**1** Solve the quadratic equations by using the quadratic formula.

**1.** $3x^2 - 7x - 6 = 0$       **2.** $x^2 + x - 12 = 0$       **3.** $5x^2 - 6x = 11$

**4.** $x^2 - 6x + 9 = 0$       **5.** $x^2 - x - 6 = 0$       **6.** $8x^2 - 2x = 3$

Solve the quadratic equations by using the quadratic formula. Round each final answer to the nearest hundredth.

**7.** $x^2 = -9x + 20$       **8.** $3x^2 + 6x + 1 = 0$       **9.** $\dfrac{2}{x} + \dfrac{5}{2} = x$

**10.** $\dfrac{2}{x} + 3x = 8$       **11.** $2x^2 + 3x + 3 = 0$       **12.** $x^2 - x + 2 = 0$

Solve the applied problems.

**13. CAD/ARC**   A rectangular tabletop is 3 cm longer than it is wide. Find the length and width to the nearest hundredth if the area is 47.5 cm². (Area = length × width, or $A = lw$.)

**14. INDTEC**   A rectangular instrument case has an area of 40 in². If the length is 6 in. more than the width, find the dimensions (length and width) of the instrument case.

**15. CON**   A bricklayer plans to build an arch with a span ($s$) of 8 m and a radius ($r$) of 4 m. How high ($h$) is the arch? (Use the formula $h^2 - 2hr + \frac{s^2}{4} = 0$.)

**16. CON**   A rectangular patio slab is 180 ft². If the length is 1.5 times the width, find the width and length of the slab to the nearest whole number.

**17. AG/H**   A farmer normally plants a field 80 ft by 120 ft. This year the government requires the planting area to be decreased by 20%, and the farmer chooses to decrease the width and length of the field by an equal amount. Draw both the original field and the reduced-size field. If $x$ is the amount by which the length and width are decreased, find the length and width of the new field. Round to the nearest whole number.

**18. HLTH/N**   The recommended dosage of a certain type of medicine is determined by the patient's weight. The formula to determine the dosage is given by $D = 0.1w^2 + 5w$, where $D$ is the dosage in milligrams (mg) and $w$ is the patient's body weight in kilograms. A doctor ordered a dosage of 1,800 mg. This dosage is to be administered to what weight patient? Round to the nearest tenth.

**1** Graph quadratic equations using the table-of-solutions method.
**2** Graph quadratic equations by examining properties.
**3** Solve a quadratic equation from a graph.
**4** Graph quadratic equations using a graphing calculator.

### 1 Graph Quadratic Equations Using the Table-of-Solutions Method.

The graphs of linear equations are *straight* lines. We will examine the graphs of some equations that have a degree higher than 1, which are not linear equations.

The graph of an equation of a degree higher than 1 is a *curved* line. The curved line can be a parabola, hyperbola, circle, ellipse, or an irregular curved line. We do not define these terms at this time, but Figure 11–1 illustrates them. The equation for a **parabola** that opens up or down is distinguished from other quadratic equations. The $y$ variable has degree 1. The $x$ variable must have one term with degree 2 and no term with a degree higher than 2. The quadratic equation for a parabola can be written in function notation: $f(x) = ax^2 + bx + c$. However, we write it as $y = ax^2 + bx + c$.

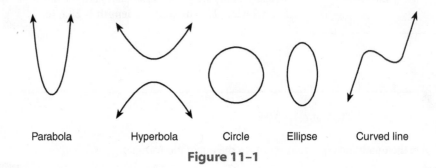

| Parabola | Hyperbola | Circle | Ellipse | Curved line |

**Figure 11–1**

One method of graphing quadratic equations is to form a table of solutions and plot the points. This method requires *more* points than we customarily use in graphing linear equations. Other methods of graphing quadratic equations involve examining characteristics or properties of the equation.

---

**EXAMPLE** Prepare a table of solutions and graph the equation $y = x^2 + x - 6$ using integral values of $x$ between $-3$ and $3$, inclusive. Discuss the characteristics of the domain and range.

$y = x^2 + x - 6$

| $x$ | $y$ |
|----|----|
| $-3$ | $0$ |
| $-2$ | $-4$ |
| $-1$ | $-6$ |
| $0$ | $-6$ |
| $1$ | $-4$ |
| $2$ | $0$ |
| $3$ | $6$ |

At $x = -3$:
$$y = (-3)^2 + (-3) - 6$$
$$y = 9 - 3 - 6$$
$$\mathbf{y = 0}$$

At $x = -2$:
$$y = (-2)^2 + (-2) - 6$$
$$y = 4 - 2 - 6$$
$$\mathbf{y = -4}$$

At $x = -1$:
$$y = (-1)^2 + (-1) - 6$$
$$y = 1 - 1 - 6$$
$$\mathbf{y = -6}$$

At $x = 0$:
$$y = 0^2 + 0 - 6$$
$$y = 0 + 0 - 6$$
$$\mathbf{y = -6}$$

At $x = 1$:
$$y = 1^2 + 1 - 6$$
$$y = 1 + 1 - 6$$
$$\mathbf{y = -4}$$

At $x = 2$:
$$y = 2^2 + 2 - 6$$
$$y = 4 + 2 - 6$$
$$\mathbf{y = 0}$$

At $x = 3$:
$$y = 3^2 + 3 - 6$$
$$y = 9 + 3 - 6$$
$$\mathbf{y = 6}$$

The domain is all real numbers. The range is real numbers greater than or equal to the lowest point of the graph $\left(-6\frac{1}{4}\right)$.

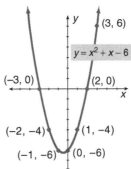

Plot the points indicated in the table of solutions and connect the points with a smooth, continuous curve. The curve is symmetrical (Figure 11–2).

**Figure 11–2**

### 2 Graph Quadratic Equations by Examining Properties.

As with linear equations, graphing quadratic equations that form a parabola by the table-of-solutions method can be time-consuming. The graph can be drawn and other important characteristics determined by examining some key properties of the parabola.

A parabola is **symmetrical;** that is, it can be folded in half and the two halves will match. The fold line is called the **axis of symmetry.** For a parabola in the form $y = ax^2 + bx + c$, the equation of the axis of symmetry is $x = -\dfrac{b}{2a}$.

The point of the graph that crosses the axis of symmetry is the *vertex* of the parabola. Thus, the x-coordinate of the vertex of the parabola is $-\dfrac{b}{2a}$.

**To graph quadratic equations in the form $y = ax^2 + bx + c$:**

1. Find the axis of symmetry: $x = -\dfrac{b}{2a}$.
2. Find the vertex: x-coordinate of vertex $= -\dfrac{b}{2a}$. To find the y-coordinate of the vertex, substitute the value for the x-coordinate into the original equation and solve for y.
3. Find one or two additional points that are to the right of the axis of symmetry.
4. Apply the property of symmetry to find additional points. The x-coordinate will be the same distance from the axis of symmetry on the opposite side. The y-coordinate will be unchanged.
5. Connect the plotted points with a smooth, continuous curved line.

**EXAMPLE** Graph the equation $y = x^2$ by finding the vertex, the axis of symmetry, and some additional points. Discuss the characteristics of the domain and range.

| x | y |
|---|---|
| −3 | 9 |
| −2 | 4 |
| −1 | 1 |
| Vertex 0 | 0 |
| 1 | 1 |
| 2 | 4 |
| 3 | 9 |

**Figure 11–3**

$$y = x^2$$

$$y = x^2 + 0x + 0 \qquad \text{Standard form: } a = 1,\ b = 0,\ c = 0.$$

$$x = -\frac{b}{2a} \qquad \text{Substitute values.}$$

$$x = -\frac{0}{2(1)} \qquad \text{Simplify.}$$

$$x = 0 \qquad \text{Axis of symmetry and x-coordinate of vertex.}$$

$$y = 0^2 \qquad \text{Substitute 0 for x in } y = x^2.$$

$$y = 0 \qquad \text{y-coordinate of vertex.}$$

**Vertex: (0, 0); Axis of symmetry: $x = 0$**

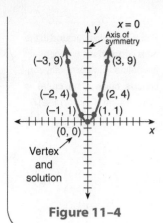

**Figure 11-4**

Find $y$ for $x = 1$: $y = 1^2$
$y = 1$

Find $y$ for $x = 2$: $y = 2^2$
$y = 4$

Find $y$ for $x = 3$: $y = 3^2$
$y = 9$

Apply the principle of symmetry to complete the table (Figure 11–3). Plot the points and connect them with a smooth, continuous curve (Figure 11–4).

**The domain is the set of all real numbers. The range is the set of real numbers greater than or equal to 0.**

**3** **Solve a Quadratic Equation from a Graph.**

As with linear equations, the real-number solutions of a quadratic equation are at the $x$-intercepts of the graph of the equation written as a function.

**To find the solution of a quadratic function from the graph:**

1. Graph the function.
2. Determine the $x$-coordinate of all $x$-intercepts.
3. If there are no $x$-intercepts, there are no real solutions.

If the equation $y = x^2$ is written in function notation as $f(x) = x^2$, *the solutions of the equation are the values (or value) of $x$ that make the function zero.* If $0 = x^2$ or $x^2 = 0$, then $x = 0$ is the only solution of the equation. The solutions of the equation are found where the graph crosses or meets the $x$-axis. When we graph a quadratic equation, we can visually determine the nature of the solutions of the equation. *If the graph does not cross or meet the $x$-axis, the equation does not have any real-number solutions.*

The solutions can be found algebraically and used to graph the equation, or the equation can be graphed to find the solutions.

**EXAMPLE** Find the real-number solutions of the equations by examining the graphs.

**(a)** $y = x^2 - 2x - 3$     **(b)** $y = -3x^2$     **(c)** $y = x^2 + 5$

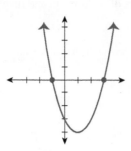

**Figure 11-5**

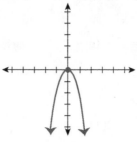

**Figure 11-6**

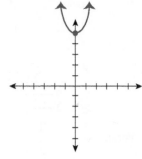

**Figure 11-7**

Solutions: $x = -1$
$x = 3$

Solution: $x = 0$

No real solutions

Chapter 11 / Quadratic Equations

**EXAMPLE**  Graph the equation $y = x^2 + x - 6$ from the example on p. 450 by using the axis of symmetry, the vertex, and the solutions. Discuss the characteristics of the domain and range.

Axis of symmetry:

$$x = -\frac{b}{2a}$$

Substitute values $a = 1$ and $b = 1$.

$$x = -\frac{1}{2(1)}$$

Simplify.

$$x = -\frac{1}{2}$$

Also, $x$-coordinate of vertex.

Vertex: $\left(-\frac{1}{2}, y\right)$

Use the $x$-coordinate, $-\frac{1}{2}$, to find the $y$-coordinate of the vertex.

$$y = x^2 + x - 6 \text{ for } x = -\frac{1}{2}$$

Substitute.

$$y = \left(-\frac{1}{2}\right)^2 + \left(-\frac{1}{2}\right) - 6$$

Raise to power.

$$y = \frac{1}{4} - \frac{1}{2} - 6$$

Write as equivalent fractions with a common denominator.

$$y = \frac{1}{4} - \frac{2}{4} - \frac{24}{4}$$

Combine fractions.

$$y = -\frac{25}{4} \quad \text{or} \quad -6\frac{1}{4}$$

$y$-coordinate of vertex.

$$\left(-\frac{1}{2}, -6\frac{1}{4}\right)$$

Solutions:

$$0 = x^2 + x - 6$$

Substitute $y = 0$ and factor.

$$0 = (x + 3)(x - 2)$$

Set factors equal to 0.

$$x + 3 = 0 \quad x - 2 = 0$$

Solve each equation.

$$x = -3 \quad x = 2$$

$x$-coordinates of $x$-intercepts.

$$(-3, 0); (2, 0)$$

Point notation of solutions.

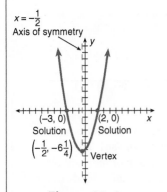

$x = -\frac{1}{2}$
Axis of symmetry

$(-3, 0)$ Solution  $(2, 0)$ Solution

$\left(-\frac{1}{2}, -6\frac{1}{4}\right)$ Vertex

**Figure 11–8**

Using the axis of symmetry, vertex, and two solutions, we can get a general idea of the shape of the graph (Figure 11–8).

**The domain is all real numbers. The range is real numbers greater than or equal to $-6\frac{1}{4}$.**

---

**EXAMPLE**  Graph the equation $y = -x^2 + 3$ by using the axis of symmetry, the vertex, and the solutions. Discuss the characteristics of the domain and range.

Axis of symmetry:  $x = -\frac{b}{2a}$

$y = -x^2 + 0x + 3$. Substitute values. $a = -1$, $b = 0$.

$$x = -\frac{0}{2(-1)}$$

Simplify.

$$x = 0$$

Also, $x$-coordinate of vertex.

Vertex: $(0, y)$

Find $y$-coordinate of vertex.

$$y = -0^2 + 0 + 3$$

Substitute $x = 0$ and solve for $y$.

$$y = 3$$

$y$-coordinate of vertex.

$$(0, 3)$$

| Solutions: | $0 = -x^2 + 3$ | Substitute $y = 0$ and solve for $x$. |
| | $x^2 = 3$ | Use square-root principle. |
| | $x = \pm\sqrt{3}$ | Exact solutions. |
| | $x \approx \pm 1.7$ | Approximate solutions. |
| | $(1.7, 0); (-1.7, 0)$ | Point notation of solutions. |

Plot the points and connect them with a smooth, continuous curve. Two additional points are plotted to give a more complete view of the parabola (Figure 11–9).

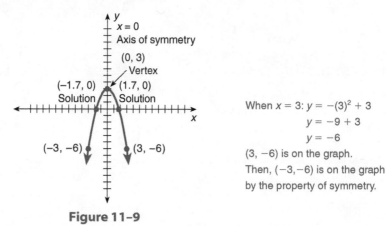

When $x = 3$: $y = -(3)^2 + 3$
$y = -9 + 3$
$y = -6$
$(3, -6)$ is on the graph.
Then, $(-3, -6)$ is on the graph by the property of symmetry.

**Figure 11–9**

**The domain is all real numbers. The range is real numbers less than or equal to 3.**

TIP!

### Tips for Sketching Curves

Positive    Negative

**Figure 11–10**

When the coefficient of the squared letter term is negative as in the preceding example, the graph of the parabola opens downward. When the coefficient of the squared letter term is positive, the graph of the parabola opens upward. Think of the parabola as a cup or glass: positive holds water; negative spills water (Figure 11–10).

**4  Graph Quadratic Equations Using a Graphing Calculator.**

Quadratic equations are graphed on graphing calculators using the same procedure as for linear equations. The equation must first be solved for $y$. Critical values on the graph such as the vertex, $x$-intercepts or solutions, and the $y$-intercepts can be determined using calculator functions.

**To find the vertex of a quadratic equation from the calculator display of the graph:**

1. Graph the equation on the calculator.
2. Adjust the view of the graph by using the Window or Zoom feature so that the vertex is visible. If the graph turns upward, the vertex is a low point or **minimum.** If the graph turns downward, the vertex is a high point or **maximum** (Figure 11–11).

Maximum

Minimum

**Figure 11–11**

3. Select the appropriate minimum or maximum option from the CALC menu.
4. Move the cursor and press $\boxed{\text{ENTER}}$ to mark points to the left of the vertex (left bound?), to the right of the vertex (right bound?), and near the vertex (guess?).
5. Read the coordinates of the vertex at the bottom of the display screen.

**To find the real solutions of a quadratic equation from the calculator display of the graph:**

1. Graph the equation on the calculator.
2. Adjust the view of the graph by using the Window or Zoom feature so that one or two $x$-intercepts are visible or so that it can be determined that there are no $x$-intercepts.
3. If there are no $x$-intercepts, there are no real solutions.
4. If there are one or two intercepts, select the zero option from the CALC menu.
5. Move the cursor and press $\boxed{\text{ENTER}}$ to mark points to the left (left bound?), right (right bound?), and near each $x$-intercept (guess?). Each $x$-intercept is one real solution.
6. Read the coordinates of each $x$-intercept at the bottom of the display screen.

**EXAMPLE** Graph the equation $y = 2x^2 - 5x - 3$ using a graphing calculator. Find the vertex and the solutions of the equation from the graph.

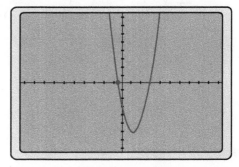

**Figure 11–12**

Find the vertex: The vertex is a minimum.

CALC
$\boxed{\text{2nd}}$ $\boxed{\text{TRACE}}$ 3

**Vertex = (1.25, −6.125)**

Mark left, right, and guess. $x = 1.25$ (rounded), $y = -6.125$.

Find the solutions: There are two real solutions.

CALC
$\boxed{\text{2nd}}$ $\boxed{\text{TRACE}}$ 2

Mark left, right, and guess for leftmost solution.

**Solution 1: $x = -0.5$**

$x = -0.5, y = 0.$

CALC
$\boxed{\text{2nd}}$ $\boxed{\text{TRACE}}$ 2

Mark left, right, and guess for rightmost solution.

**Solution 2: $x = 3$**

$x = 3, y = 0$

SECTION 11–4 SELF-STUDY EXERCISES

**1** Use a table of solutions to graph the quadratic equations. Discuss the characteristics of the domain and range.

**1.** $y = x^2$

**2.** $y = 3x^2$

**3.** $y = \frac{1}{3}x^2$

**4.** $y = -4x^2$

**5.** $y = -\frac{1}{4}x^2$

**6.** $y = x^2 - 4$

**7.** $y = x^2 + 4$

**8.** $y = x^2 - 6x + 9$

**9.** $y = -x^2 + 6x - 9$

**2** Graph the equations using the vertex, solutions, and one other point on the graph. Discuss the characteristics of the domain and range.

**10.** $y = x^2 - 4x + 4$

**11.** $y = x^2 + 4x - 2$

**12.** $y = 2x^2 + 10x + 8$

**13.** $y = -3x^2 - 6x + 9$

**14.** $y = -3x^2 - 6x - 6$

**15.** $y = \frac{1}{2}x^2 - 6x + 3$

**3** Find the real-number solutions of the equations by writing them as functions and graphing the equations.

**16.** $x^2 - 5x + 6 = 0$

**17.** $4x^2 = 0$

**18.** $3x^2 = -5$

**19. BUS** The revenue $R$ for a leather wallet is based on the price $P$ of the item and is given by the formula $R = -4p^2 + 36p$. What prices generate $0 revenue?

**20. TELE** The distance $y$ that an object falls in a vacuum because of gravity in $t$ seconds after it is released is given by the formula $d = 4.9t^2$. How much time is elapsed when $d = 0$?

**4** Graph using a graphing calculator. Adjust the window if necessary to show the vertex of the parabola.

**21.** $y = 3x^2 + 5x - 2$

**22.** $y = (2x - 3)(x - 1)$

**23.** $y = 2x^2 - 9x - 5$

**24.** $y = -2x^2 + 9x + 5$

**25.** $y = 3x^2 + 5x - 2$

**26.** $y = (2x - 3)(x - 1)$

**27.** $y = 2x^2 - 9x - 5$

**28.** $y = -2x^2 + 9x + 5$

---

## 11–5 | Imaginary Numbers

*Learning Outcomes*

**1** Write imaginary numbers using the letter $i$ or $j$.

**2** Raise imaginary numbers to powers.

**3** Determine the nature of the roots of a quadratic equation by examining the discriminant.

### 1 Write Imaginary Numbers Using the Letter *i* or *j*.

Taking the square root of a negative number introduces a new type of number. This new type of number is called an *imaginary number*. An **imaginary number** has a factor of $\sqrt{-1}$, which is represented by $i$ ($i = \sqrt{-1}$).

Imaginary numbers and real numbers combine to form the set of complex numbers (Figure 11–13). A **complex number** is a number that can be written in the form $a + bi$, where $a$ and $b$ are real numbers and $i$ is $\sqrt{-1}$.

The square root of a negative number, say, $\sqrt{-16}$, can be simplified as $\sqrt{-1 \cdot 16}$ or $4\sqrt{-1}$. Because the square root of $-1$ is an *imaginary* factor and 4 is a real factor, we will use the letter $i$ to represent $\sqrt{-1}$ and rewrite $4\sqrt{-1}$ as $4i$. Similarly, $\sqrt{-4} = \sqrt{-1 \cdot 4} = 2\sqrt{-1} = 2i$.

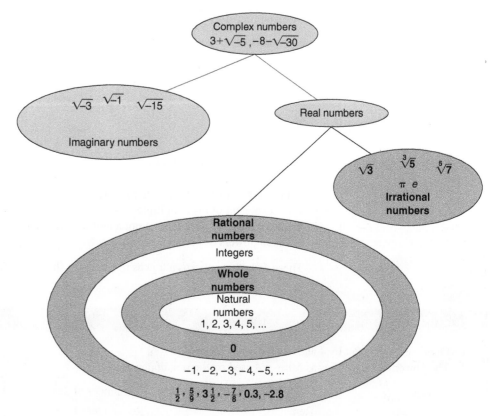

**Figure 11–13** Complex-number system.

TIP!

**Is _i_ Different from _j_?**

Because the letter _i_ is also used to represent current, electronics and other applications of imaginary numbers may use _j_ rather than _i_ to represent $\sqrt{-1}$. Thus, $3 + 4i$ and $3 + 4j$ represent the same amount.

**EXAMPLE**   Rewrite the imaginary numbers using the letter _j_ for $\sqrt{-1}$.

    **(a)** $\sqrt{-9}$       **(b)** $\sqrt{-25}$      **(c)** $\sqrt{-7}$

    **(a)** $\sqrt{-9} = \sqrt{-1 \cdot 9} = 3\sqrt{-1} = \mathbf{3\,j}$

    **(b)** $\sqrt{-25} = \sqrt{-1 \cdot 25} = 5\sqrt{-1} = \mathbf{5\,j}$

    **(c)** $\sqrt{-7} = \sqrt{-1 \cdot 7} = \sqrt{7} \cdot \sqrt{-1} = \mathbf{\sqrt{7}\,j}$ or $\mathbf{j\sqrt{7}}$.

TIP!

**Are Numerical Coefficients Always First?**

$\sqrt{7}j$ and $\sqrt{7j}$ do not represent the same amount. In the first term, $\sqrt{7}j$, _j_ is not under the radical symbol. In the second term, $\sqrt{7j}$, _j_ is under the radical symbol. Because it is easy to confuse the two terms, when the coefficient of an imaginary number is an irrational number, we write the _j_ factor first: $\sqrt{7}j = j\sqrt{7}$.

## 2 Raise Imaginary Numbers to Powers.

Some powers of imaginary numbers are real numbers. Examine the pattern that develops with powers of $j$.

$j = \sqrt{-1} = j$

$j^2 = (\sqrt{-1})^2 = \boxed{-1}$

$j^3 = j^2 \cdot j^1 = -1j = -j$

$j^4 = j^2 \cdot j^2 = -1(-1) = \boxed{1}$

$j^5 = j^4 \cdot j^1 = 1(j) = j$

$j^6 = j^4 \cdot j^2 = 1(-1) = \boxed{-1}$

$j^7 = j^4 \cdot j^3 = 1(-j) = -j$

$j^8 = j^4 \cdot j^4 = 1(1) = \boxed{1}$

$j^9 = j^4 \cdot j^4 \cdot j^1 = 1(1)(j) = j$

$j^{10} = j^4 \cdot j^4 \cdot j^2 = 1(1)(-1) = \boxed{-1}$

$j^{11} = j^4 \cdot j^4 \cdot j^3 = 1(1)(-j) = -j$

$j^{12} = j^4 \cdot j^4 \cdot j^4 = 1(1)(1) = \boxed{1}$

- All powers of $j$ simplify to either $j$, $-1$, $-j$, or 1.
- If the exponent of $j$ is a multiple of 4, the result is 1.
- If the exponent of $j$ is even but not a multiple of 4, the result is $-1$.
- All even powers of $j$ are real numbers.

These and other patterns allow us to develop a shortcut process for simplifying powers of $j$.

**To simplify a power of $j$:**

1. Divide the exponent by 4 and examine the *remainder.*
2. The power of $j$ simplifies as follows, based on the remainder in Step 1.

| Remainder of 0 $\Rightarrow j^0 = 1$ | Remainder of 2 $\Rightarrow j^2 = -1$ |
| Remainder of 1 $\Rightarrow j^1 = j$ | Remainder of 3 $\Rightarrow j^3 = -j$ |

---

**EXAMPLE**  Simplify the powers of $j$.

**(a)** $j^{15}$  **(b)** $j^{20}$  **(c)** $j^{33}$  **(d)** $j^{18}$

**(a)** $j^{15} = -j$  $\quad$ $15 \div 4 = 3$ R3. Remainder of 3 $\Rightarrow j^3 = -j$.

**(b)** $j^{20} = 1$  $\quad$ $20 \div 4 = 5$. Remainder 0 $\Rightarrow j^0 = 1$.

**(c)** $j^{33} = j$  $\quad$ $33 \div 4 = 8$ R1. Remainder 1 $\Rightarrow j^1 = j$.

**(d)** $j^{18} = -1$  $\quad$ $18 \div 4 = 4$ R2. Remainder 2 $\Rightarrow j^2 = -1$.

---

## 3 Determine the Nature of the Roots of a Quadratic Equation by Examining the Discriminant.

In the previous sections, we made various observations about the roots of the different types of quadratic equations. Now, we make additional observations about the roots of quadratic equations.

Which method for solving quadratic equations is best? Since any method that is mathematically sound and produces consistently correct solutions is good, a "best" method may be evaluated differently. Factoring methods are generally quicker, especially when a calculator or computer is unavailable.

In general, follow these suggestions for solving quadratic equations.

1. Write the equation in standard form: $ax^2 + bx + c = 0$.
2. Identify the type of quadratic equation.
3. Solve pure quadratic equations by the square-root method.
4. Solve incomplete quadratic equations by finding common factors.
5. Solve complete quadratic equations by factoring into two binomials, if possible.
6. If factoring is not possible or is difficult, use the quadratic formula to solve complete quadratic equations.

All types of quadratic equations can be solved using the quadratic formula. Only certain types of quadratic equations can be solved by factoring or applying the square-root property. The radicand of the radical portion of the quadratic formula, $b^2 - 4ac$, is called the **discriminant.** The general characteristics of the roots of a quadratic equation can be determined by examining the discriminant.

---

**Properties of the discriminant, $b^2 - 4ac$:**

1. If $b^2 - 4ac \geq 0$, the equation has real-number roots.
   a. If $b^2 - 4ac$ is a perfect square, there are two rational roots.
   b. If $b^2 - 4ac = 0$, there is one rational root (sometimes called a *double root*).
   c. If $b^2 - 4ac$ is not a perfect square, there are two irrational roots.
2. If $b^2 - 4ac < 0$, the equation has no real-number roots. The roots are imaginary or complex.

---

**EXAMPLE**  Examine the discriminant of each equation and determine the nature of the roots. Then solve the equation.

(a) $5x^2 + 3x - 1 = 0$     (b) $3x^2 + 5x = 2$     (c) $4x^2 + 2x + 3 = 0$

(a)  $5x^2 + 3x - 1 = 0$      $a = 5, b = 3, c = -1$.

$\quad b^2 - 4ac = 3^2 - 4(5)(-1)$      Examine the discriminant.

$\quad\quad\quad\quad\quad = 9 + 20$      **There will be two irrational roots.**

$\quad\quad\quad\quad\quad = \boxed{29}$      Use the quadratic formula.

$\quad x = \dfrac{-b \pm \sqrt{b^2 - 4ac}}{2a}$      Substitute 29 for the discriminant.

$\quad x = \dfrac{-3 \pm \sqrt{29}}{2(5)}$      Identify each root.

$\quad x = \dfrac{-3 + \sqrt{29}}{10}$ or $x = \dfrac{-3 - \sqrt{29}}{10}$      Exact irrational roots.

(b)  $\quad 3x^2 + 5x = 2$

$\quad 3x^2 + 5x - 2 = 0$      Standard form: $a = 3, b = 5, c = -2$.

$\quad b^2 - 4ac = 5^2 - 4(3)(-2)$      Examine the discriminant.

$\quad\quad\quad\quad\quad = 25 + 24$

$\quad\quad\quad\quad\quad = 49$      The discriminant is a perfect square. **There are two rational roots** and the trinomial will factor.

$\quad (3x - 1)(x + 2) = 0$      Factor. Set each factor equal to zero.

$\quad 3x - 1 = 0 \quad\quad x + 2 = 0$

$\quad\quad\quad 3x = 1$

$\quad\quad\quad x = \dfrac{1}{3} \quad\quad x = -2$      Exact rational roots.

(c)  $4x^2 + 2x + 3 = 0$      $a = 4, b = 2, c = 3$.

$\quad b^2 - 4ac = 2^2 - 4(4)(3)$      Examine the discriminant.

$\quad\quad\quad\quad\quad = 4 - 48$

$\quad\quad\quad\quad\quad = \boxed{-44}$      The discriminant is negative. There are no real roots. **The roots are imaginary or complex.**

$$x = \frac{-2 \pm \sqrt{-44}}{2 \cdot 4}$$   Substitute $-44$ for the discriminant in the quadratic formula.

$$x = \frac{-2 \pm 2j\sqrt{11}}{8}$$   Simplify the radical and factor.

$$x = \frac{2(-1 \pm j\sqrt{11})}{8}$$   Reduce.

$$x = \frac{-1 \pm j\sqrt{11}}{4}$$   The roots are complex.

## SECTION 11–5 SELF-STUDY EXERCISES

**1** Write the imaginary numbers using the letter $j$. Simplify if possible.

**1.** $\sqrt{-25}$  **2.** $\sqrt{-36}$  **3.** $\sqrt{-64x^2}$  **4.** $\sqrt{-32y^5}$

**2** Simplify the powers of $j$.

**5.** $j^{17}$  **6.** $j^5$  **7.** $j^{20}$  **8.** $j^{10}$
**9.** $j^{24}$  **10.** $j^9$  **11.** $j^{32}$  **12.** $j^{15}$

**3** Examine the discriminant of each equation and determine the nature of the roots. Solve each equation. Round to hundredths if necessary.

**13.** $3x^2 + x - 2 = 0$  **14.** $x^2 - 3x = -1$  **15.** $2x^2 + x = 2$
**16.** $3x^2 - 2x + 1 = 0$  **17.** $x^2 - 3x - 7 = 0$  **18.** $3x^2 + 5x - 6 = 0$

**19.** Describe the discriminant of a quadratic equation that has real roots.

**20.** Describe the discriminant of a quadratic equation that has rational and unequal roots.

## CHAPTER REVIEW OF KEY CONCEPTS

**Learning Outcomes**  **What to Remember with Examples**

### Section 11–1

**1** Write quadratic equations in standard form (pp. 438–439).

Write the equation with all terms on the left side of the equation and zero on the right and arrange with the terms in descending powers. The leading coefficient should be positive.

Write the equation $5x + 3 = 4x^2$ in standard form.

$$5x + 3 = 4x^2$$   Rearrange all terms on the left and arrange in descending order.
$$-4x^2 + 5x + 3 = 0$$   Make leading coefficient positive.
$$-1(-4x^2 + 5x + 3) = -1(0)$$
$$4x^2 - 5x - 3 = 0$$

**2** Identify the coefficients of the quadratic, linear, and constant terms of a quadratic equation (pp. 439–440).

The coefficient of the quadratic term is the coefficient of the squared variable, the coefficient of the linear term is the coefficient of the first-power variable, and the constant term is its own coefficient.

Identify the coefficient of the quadratic and linear terms and identify the constant term in the following:

$$5x = 4x^2 - 3 \qquad \text{Write the equation in standard form.}$$
$$4x^2 - 5x - 3 = 0 \qquad a = 4, b = -5, c = -3.$$

The coefficient of the quadratic term is 4. The coefficient of the linear term is $-5$ and the constant term is $-3$.

**3** Solve pure quadratic equations by the square-root method ($ax^2 + c = 0$) (pp. 440–441).

To solve pure quadratic equations, solve for the squared variable; then take the square root of both sides of the equation. The two roots have the same absolute value.

Solve $2x^2 - 72 = 0$.

$$2x^2 - 72 = 0 \qquad \text{Sort.}$$
$$2x^2 = 72 \qquad \text{Divide by 2.}$$
$$x^2 = 36 \qquad \text{Take the square root of both sides.}$$
$$x = \pm 6$$

## Section 11–2
**1** Solve incomplete quadratic equations by factoring ($ax^2 + bx = 0$) (pp. 442–443).

Arrange the equation in standard form. Factor the common factor. Then set each of the two factors equal to zero and solve for the variable in each equation. Both roots are rational and one root is always zero.

Solve $5x^2 - 15x = 0$.

$$5x^2 - 15x = 0 \qquad \text{Factor common factors.}$$
$$5x(x - 3) = 0 \qquad \text{Set each factor equal to 0.}$$
$$5x = 0 \qquad x - 3 = 0 \qquad \text{Solve each equation.}$$
$$x = 0 \qquad x = 3$$

**2** Solve complete quadratic equations by factoring ($ax^2 + bx + c = 0$) (pp. 444–445).

Arrange the quadratic equation in standard form and factor the trinomial. Then set each factor equal to zero and solve for the variable. If the expression factors, the roots are rational.

Solve $2x^2 - 5x - 3 = 0$.

$$2x^2 - 5x - 3 = 0 \qquad \text{Factor the trinomial.}$$
$$(x - 3)(2x + 1) = 0 \qquad \text{Set each factor equal to 0.}$$
$$x - 3 = 0 \qquad 2x + 1 = 0 \qquad \text{Solve each equation.}$$
$$x = 3 \qquad 2x = -1$$
$$x = -\frac{1}{2}$$

## Section 11–3
**1** Solve quadratic equations using the quadratic formula (pp. 446–449).

In the quadratic formula, $x = \dfrac{-b \pm \sqrt{b^2 - 4ac}}{2a}$, $a$ is the coefficient of the quadratic term, $b$ is the coefficient of the linear term, and $c$ is the constant when the equation is written in standard form. The variable is $x$.

Solve using the quadratic formula:

$$5x^2 + 7x - 6 = 0 \qquad \text{Identify } a, b, \text{ and } c.$$
$$a = 5, \qquad b = 7, \qquad c = -6 \qquad \text{Substitute values.}$$

$$x = \frac{-b \pm \sqrt{b^2 - 4ac}}{2a}$$

$$x = \frac{-7 \pm \sqrt{7^2 - 4(5)(-6)}}{2(5)}$$

Chapter Review of Key Concepts

461

$$x = \frac{-7 \pm \sqrt{49 + 120}}{10}$$   Add in the radicand.

$$x = \frac{-7 \pm \sqrt{169}}{10}$$   Evaluate the square root.

$$x = \frac{-7 \pm 13}{10}$$   Separate into two cases.

$$x = \frac{-7 + 13}{10} \qquad x = \frac{-7 - 13}{10}$$   Solve each equation.

$$x = \frac{6}{10} \qquad x = -\frac{20}{10}$$

$$x = \frac{3}{5} \qquad x = -2$$   Exact roots.

Solving applied problems requires knowledge of other mathematical formulas, for example, $A = lw$ (area of a rectangle).

Find the length and width of a rectangular parking lot if the length is to be 12 m longer than the width and the area is to be 6,205 m$^2$.

Let $x$ = number of meters in width
$x + 12$ = number of meters in length

Estimate: If the parking lot was a square, one side would be $s \doteq \sqrt{A}$ or $s = 78.8$ m. In a rectangle where the length and width are different, the length is more than 78.8 m and the width is less than 78.8 m.

$$lw = A$$
$$x(x + 12) = 6,205$$
$$x^2 + 12x = 6,205$$
$$x^2 + 12x - 6,205 = 0$$

Use a calculator with the formula.

$$x = \frac{-12 \pm \sqrt{12^2 - 4(1)(-6,205)}}{2(1)}$$   Substitute values for $a$, $b$, and $c$.

$$x = \frac{-12 \pm \sqrt{144 + 24,820}}{2}$$   Simplify radicand.

$$x = \frac{-12 \pm \sqrt{24,964}}{2}$$   Evaluate square root.

$$x = \frac{-12 \pm 158}{2}$$   Separate into two cases.

$$x = \frac{-12 + 158}{2} \qquad x = \frac{-12 - 158}{2}$$   Solve each case.

$$x = \frac{146}{2} = 73 \qquad x = \frac{-170}{2} = -85$$   Disregard the negative root for a length measure.

$$x = 73 \text{ width}$$
$$x + 12 = 85 \text{ length}$$

**1.** Prepare a table of solutions of approximately five ordered pairs. **2.** Plot the points on a rectangular coordinate system. **3.** Connect the points with a smooth, continuous curve.

---

Graph $y = x^2 - 6x + 8$ using a table of solutions. Start with three values of $x$, $-2$, 0, and 2.

at $x = -2$:
$y = (-2)^2 - 6(-2) + 8$
$y = 4 + 12 + 8$
$y = 24$

at $x = 0$:
$y = 0^2 - 6(0) + 8$
$y = 0 - 0 + 8$
$y = 8$

at $x = 2$:
$y = 2^2 - 6(2) + 8$
$y = 4 - 12 + 8$
$y = 0$

Try two more, $x = 4$ and $x = 6$.

at $x = 4$
$y = 4^2 - 6(4) + 8$
$y = 16 - 24 + 8$
$y = 0$

at $x = 6$
$y = 6^2 - 6(6) + 8$
$y = 36 - 36 + 8$
$y = 8$

Finally, see what happens between $x = 2$ and $x = 4$.

at $x = 3$
$y = 3^2 - 6(3) + 8$
$y = 9 - 18 + 8$
$y = -1$

| $x$ | $y$ |
|-----|-----|
| $-2$ | 24 |
| 0 | 8 |
| 2 | 0 |
| 3 | $-1$ |
| 4 | 0 |
| 6 | 8 |

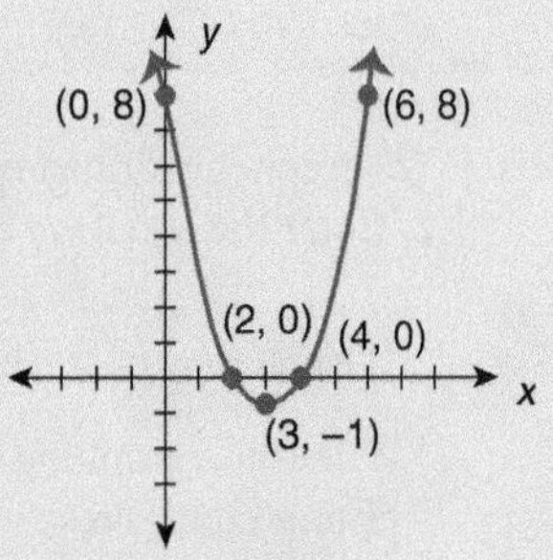

**Figure 11–14**

---

**1.** Determine the $x$-coordinate of the vertex of the parabola by finding the value of $-\frac{b}{2a}$. Use the $x$-coordinate of the vertex to write an equation $x = -\frac{b}{2a}$, which is the axis of symmetry. **2.** Substitute the value as the $x$-coordinate into the equation to find the corresponding $y$-value. **3.** Replace $y$ with zero and solve the resulting equation for $x$. Plot these points if any, which are the solutions for the equation. **4.** Use symmetry to find additional points. **5.** Connect the plotted points and the vertex with a smooth, continuous curved line.

---

Graph $y = x^2 - 6x + 8$ by examining properties.

$x = -\dfrac{b}{2a}$     Substitute $a = 1$ and $b = -6$.

$x = -\dfrac{-6}{2(1)}$     Simplify.

$x = \dfrac{6}{2}$

$x = 3$     Axis of symmetry: $x = 3$.

To find vertex:

$y = (3)^2 - 6(3) + 8$     Substitute $x = 3$.
$y = 9 - 18 + 8$     Simplify.
$y = -1$     Vertex: $(3, -1)$.

To find $x$-intercepts:

$0 = x^2 - 6x + 8$     Substitute $y = 0$.
$0 = (x - 2)(x - 4)$     Factor. Solve for $x$.
$x - 2 = 0 \quad x - 4 = 0$
$\quad x = 2 \quad\quad x = 4$     $x$-intercepts: $(2, 0)$ and $(4, 0)$.

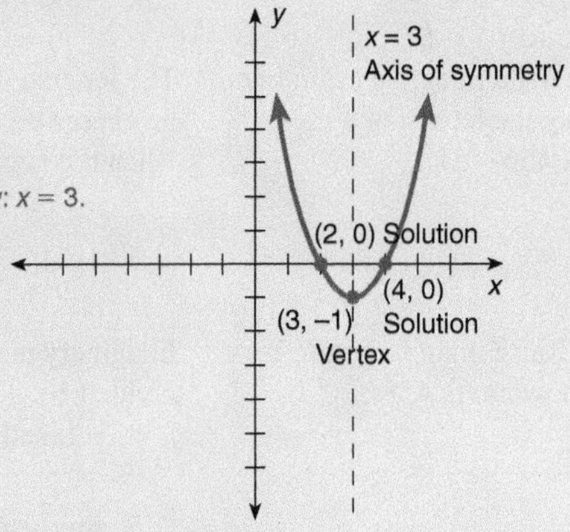

**Figure 11–15**

---

Chapter Review of Key Concepts

| | |
|---|---|
| **3** Solve a quadratic equation from a graph (pp. 452–454). | **1.** Find the *x*-intercepts on the graph. **2.** The solutions of the equation are the *x*-coordinates of the *x*-intercepts. |

Find the solutions for the equation $y = x^2 - 6x + 8$ from the graph (Figure 11–16).

The *x*-intercepts are $(2, 0)$ and $(4, 0)$.
The solutions are $x = 2$ and $x = 4$.

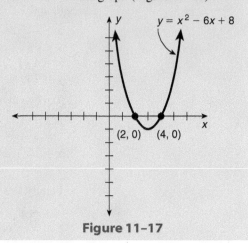

**Figure 11–17**

| | |
|---|---|
| **4** Graph quadratic equations using a graphing calculator (pp. 454–456). | To graph quadratic equations using a graphing calculator: **1.** Clear the graphing screen. **2.** Enter the equation and press the graph function. |

Graph $y = x^2 - 6x + 8$.

Clear the graphing screen.
Enter equation.
Graph.
Change viewing window if appropriate.

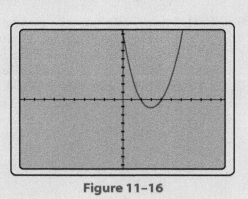

**Figure 11–16**

## Section 11–5

| | |
|---|---|
| **1** Write imaginary numbers using the letter *i* or *j* (pp. 456–457). | The letters *i* and *j* are used to represent $\sqrt{-1}$. Thus, the square root of negative numbers can be expressed as imaginary numbers and simplified using the letter *j*. Be careful to distinguish when the negative is *outside* the radical and when it is *under* the radical. |

Simplify. $\sqrt{-48} = \sqrt{-1 \cdot 16 \cdot 3} = 4j\sqrt{3}$

| | |
|---|---|
| **2** Raise imaginary numbers to powers (p. 458). | Imaginary numbers can be raised to powers by examining the remainder when the exponent is divided by 4. The remainder can be used as a simplified exponent. Then, $j^0 = 1$, $j^1 = j$, $j^2 = -1$, and $j^3 = -j$. |

Simplify. $j^{17} = j^1 = j$    $17 \div 4 = 4R1$

$j^{42} = j^2 = -1$    $42 \div 4 = 10R2$

Chapter 11 / Quadratic Equations

**3** Determine the nature of the roots of a quadratic equation by examining the discriminant (pp. 458–460).

The radicand of the quadratic formula, $b^2 - 4ac$, is the discriminant of the quadratic equation.

1. If $b^2 - 4ac \geq 0$, the equation has real-number solutions.
   a. If $b^2 - 4ac$ is a perfect square, the two solutions are rational.
   b. If $b^2 - 4ac = 0$, there is one rational solution.
   c. If $b^2 - 4ac$ is not a perfect square, the two roots are irrational.
2. If $b^2 - 4ac < 0$, the equation has no real-number solutions or the solutions are complex.

Use the discriminant to determine the characteristics of the roots of the equation $3x^2 - 5x + 7 = 0$.

$a = 3, \quad b = -5, \quad c = 7$    Identify $a$, $b$, and $c$.
$(-5)^2 - 4(3)(7) = 25 - 84$    Substitute values into $b^2 - 4ac$ and simplify.
$\qquad\qquad\qquad = -59$

Since $-59$ is less than zero, the roots are not real or they are complex.    Interpret result.

## CHAPTER REVIEW EXERCISES

### Section 11–1

Identify the quadratic equations as pure, incomplete, or complete.
1. $x^2 = 49$
2. $x^2 - 5x = 0$
3. $5x^2 - 45 = 0$
4. $3x^2 + 2x - 1 = 0$
5. $8x^2 + 6x = 0$
6. $5x^2 + 2x + 1 = 0$
7. $x^2 - 32 = 0$
8. $x^2 + x = 0$
9. $3x^2 + 6x + 1 = 0$

Write the equations in standard form.
10. $x^2 - 7x = 8$
11. $2x^2 - 5 = 8x$
12. $7x = 2x^2 - 3$
13. $5 + x^2 - 7x = 0$
14. $x - 4 = 3x^2$
15. $3x = 1 - 4x^2$

Solve the equations by the square-root method. Round to thousandths when necessary.
16. $x^2 = 121$
17. $x^2 = 100$
18. $x^2 - 64 = 0$
19. $4x^2 = 9$
20. $64x^2 - 49 = 0$
21. $0.36y^2 = 1.09$
22. $0.16x^2 = 0.64$
23. $5x^2 = 40$
24. $2x^2 - 5 = 3$
25. $6x^2 + 4 = 34$
26. $5x^2 - 6 = 19$
27. $3x^2 = 12$
28. $3x^2 - 4 = 8$
29. $2x^2 = 34$
30. $5x^2 - 9 = 30$
31. $3y^2 - 36 = -8$
32. $2x^2 + 3 = 51$
33. $\frac{1}{2}x^2 = 8$
34. $\frac{2}{3}x^2 = 24$
35. $\frac{1}{4}x^2 - 1 = 15$
36. $\frac{2}{5}x^2 + 2 = 8$

37. A circle has an area of 845 cm². What is the radius of the circle? Use the formula $A = \pi r^2$

38. The area of an oil painting that is a square is known to be 9,072 cm². What are the inside dimensions of its picture frame?

### Section 11–2

Solve the equations by factoring.
39. $x^2 - 5x = 0$
40. $4x^2 = 8x$
41. $6x^2 - 12x = 0$
42. $3x^2 + x = 0$
43. $10x^2 + 5x = 0$
44. $3y^2 = 12y$
45. $y^2 - 5y = 0$
46. $x^2 = 16x$
47. $12x^2 + 8x = 0$
48. $8x^2 - 12x = 0$
49. $x^2 + 3x = 0$
50. $4x^2 - 28x = 0$
51. $5x^2 = 45x$
52. $7x^2 = 28x$
53. $y^2 + 8y = 0$
54. $z^2 - 6z = 0$
55. $3m^2 - 5m = 0$
56. $4n^2 - 3n = 0$
57. $2x^2 = x$
58. $5y^2 = y$

**59.** 3 times the square of a number is the same as 12 times the number. What is the number?

**60.** Describe the steps for solving an incomplete quadratic equation. Include a clear description of the nature of the roots.

Solve the equations by factoring.

**61.** $x^2 - 4x + 3 = 0$

**62.** $x^2 + 7x + 12 = 0$

**63.** $x^2 + 3x = 10$

**64.** $x^2 - 7x + 12 = 0$

**65.** $x^2 + 7x = -6$

**66.** $x^2 + 3 = -4x$

**67.** $x^2 - 6x + 8 = 0$

**68.** $6y + 7 = y^2$

**69.** $6y^2 - 5y - 6 = 0$

**70.** $5y^2 + 23y = 10$

**71.** $10y^2 - 21y - 10 = 0$

**72.** $6x^2 - 16x + 8 = 0$

**73.** $4x^2 + 7x + 3 = 0$

**74.** $3x^2 = -7x + 6$

**75.** $12y^2 - 5y - 3 = 0$

**76.** $x^2 - 3x = 18$

**77.** $x^2 + 19x = 42$

**78.** $3x^2 + x - 2 = 0$

**79.** $3y^2 + y - 2 = 0$

**80.** $2x^2 - 4x - 6 = 0$

**81.** $2x^2 - 10x + 12 = 0$

**82.** $y^2 + 18y + 45 = 0$

**83.** $x^2 - 3x - 18 = 0$

**84.** $3x^2 - 9x - 30 = 0$

**85.** $2y^2 + 22y + 60 = 0$

**86.** $x^2 + 13x + 12 = 0$

**87.** $x^2 + 7x - 18 = 0$

**88.** **AG/H** An office building in the shape of a rectangle is known to have 47,500 ft$^2$ of space on the ground floor. The tenant wants to landscape the two longer sides of the building. The tenant also knows that the building is about 60 ft longer than it is wide. How many feet of land along the building need to be landscaped?

**89.** **CAD/ARC** Jerri Amour is an architect who is designing a hospital. She knows that a kidney dialysis machine needs a space that is 7 ft longer than it is wide. The total area needed is 228 ft$^2$ of space. Advise Jerri about the number of feet of space she should include in the length and width of the planned space.

## Section 11–3

Indicate the values for $a$, $b$, and $c$ in the quadratic equations.

**90.** $5x^2 + x + 6 = 0$

**91.** $x^2 - 2x = 8$

**92.** $x^2 - 7x + 12 = 0$

**93.** $x^2 + 3x = 4$

**94.** $3x^2 = 2x + 7$

**95.** $x^2 - 3x = -2$

Solve the quadratic equations by using the quadratic formula.

**96.** $x^2 - 9x + 20 = 0$

**97.** $x^2 - 8x - 9 = 0$

**98.** $x^2 - 5x = -6$

**99.** $x^2 + 2x = 8$

**100.** $x^2 - x - 12 = 0$

**101.** $2x^2 - 3x - 2 = 0$

Solve the quadratic equations by using the quadratic formula. Round each final answer to the nearest hundredth.

**102.** $3x^2 + 6x + 2 = 0$

**103.** $2x^2 - 3x - 1 = 0$

**104.** $5x^2 + 4x - 8 = 0$

**105.** $3x^2 + 5x + 1 = 0$

**106.** **CON** A bricklayer plans to build an arch with a span ($s$) of 10 m and a radius ($r$) of 5 m. How high ($h$) is the arch? (Use the formula $h^2 - 2hr + \frac{s^2}{4} = 0$.)

**107.** **CON** A rectangular kitchen contains 240 ft$^2$. If the length is 2 times the width, find the length and width of the room to the nearest whole number. (Area = length × width, or $A = lw$.)

**108.** **CON** What are the dimensions of a rectangular tool storage room if the area is 45.5 m$^2$ and the room is 0.5 m longer than it is wide? Round to the nearest tenth.

**109.** **INDTEC** Find the length and width of a piece of fiberglass if its length is 3 times the width and the area is 591 in$^2$. Round to the nearest inch.

## Section 11–4

Graph the quadratic equations. Discuss the characteristics of the domain and range.

**110.** $y = x^2 - 1$

**111.** $y = -x^2 - 1$

**112.** $y = x^2 + 3x - 10$

**113.** $y = x^2 - 6x + 8$

**114.** $y = x^2 - 2x + 1$

**115.** $y = -x^2 + 2x - 1$

Find the real-number solutions of the equations by writing them as functions and graphing.

**116.** $x^2 + x - 12 = 0$

**117.** $x^2 - 4x - 12 = 0$

**118.** $x^2 - 5x = 14$

**119.** $x^2 + 8x = -16$

Use a calculator to graph the following. Adjust the window to display the vertex.

**120.** $y = x^2 - 2x + 1$

**121.** $y = -2x^2$

**122.** $y = \frac{1}{2}x^2 - 3$

Write the numbers using the letter $j$. Simplify if possible.

**123.** $\sqrt{-100}$  **124.** $-\sqrt{-16x^2}$  **125.** $\pm\sqrt{-24y^7}$

Simplify the powers of $j$.

**126.** $j^5$  **127.** $j^{14}$  **128.** $j^{98}$  **129.** $j^{77}$

Use the discriminant of the quadratic formula to determine the nature of the roots of the equations.

**130.** $x^2 - 3x + 2 = 0$  **131.** $x^2 + 8x + 16 = 0$  **132.** $2x^2 - 3x - 5 = 0$
**133.** $5x^2 - 100 = 0$  **134.** $3x^2 - 2x + 4 = 0$  **135.** $2x = 5x^2 - 3$

## TEAM PROBLEM-SOLVING EXERCISES

1. Many objects are designed with dimensions according to the *golden ratio*. Objects that have measurements according to this ratio are said to be most pleasing to the eye. The **golden rectangle** has dimensions of length ($l$) and height ($h$) that satisfy the formula

$$\frac{l + h}{l} = \frac{l}{h}$$

When you cross multiply the formula for the golden rectangle, a quadratic equation results.

Determine the largest-size wall hanging that has dimensions of the golden rectangle that can be placed on a wall that is 32 ft long and 20 ft high. The wall hanging must also be at least 2 ft from the ceiling, floor, and each of the side corners.

2. Complex numbers in the form $ax + bj$ can be multiplied using the same procedures as multiplying polynomials. Find the products of the following.
   **(a)** $5j(3 - 4j)$
   **(b)** $12(2 + 2j)$
   **(c)** $(3 + 2j)(3 - 2j)$
   **(d)** $(4 + 5j)(3 - 4j)$
   **(e)** $(6 - 5j)^2$

## PRACTICE TEST

Identify the quadratic equations as pure, incomplete, or complete.
**1.** $3x^2 = 42$  **2.** $7x^2 - 3x + 2 = 0$
**3.** $5x^2 = 7x$  **4.** $4x^2 - 1 = 0$

Solve the quadratic equations using the square-root method. Round to the nearest hundredth.
**5.** $x^2 = 81$  **6.** $81x^2 - 64 = 0$

Solve the equations by factoring.
**7.** $3x^2 - 6x = 0$  **8.** $2x^2 + 3x + 1 = 0$
**9.** $x^2 - 5x + 6 = 0$  **10.** $3x^2 - 2x - 1 = 0$
**11.** $2x^2 + 12 = 11x$  **12.** $x^2 - 3x - 4 = 0$

Solve the equations using the quadratic formula. Round to the nearest hundredth.
**13.** $2x^2 + 3x - 5 = 0$  **14.** $3x^2 - 5x + 4 = 0$
**15.** $x^2 - 3x - 5 = 0$  **16.** $4x^2 - 2x = 3$

Simplify.
**17.** $j^{23}$  **18.** $j^{88}$

Describe the roots of the equations without solving.

**19.** $x^2 - 8x + 12 = 0$

**20.** $3x^2 - 2x + 3 = 0$

Graph the quadratic equations by examining properties. Show the axis of symmetry, vertex, and solutions.

**21.** $y = x^2 + 2x + 1$

**22.** $y = x^2 + 4x + 4$

## CAREER APPLICATION: CIVIL ENGINEERING— FREEWAY SUPPORTS

Safe, dependable, durable freeway-support design is extremely important because highways carry an ever-increasing number of heavy vehicles (such as tandem tractor-trailer trucks) and heavier weight loads. Civil engineers try to design supports that hold up during temperature extremes, earthquakes, and high wind conditions (for example, tornadoes and hurricanes). When designing a rectangular box support, researchers found that support is optimal when the depth is at least one-half of the width and the volume is at least 10 ft$^3$ per ft of support height.

Civil engineers lobby a state government to allot construction money for a freeway passing through an earthquake fault zone at a higher, safer volume rate of 12.5 ft$^3$ per ft of support height. At intersections, rectangular box supports 16 ft high are required.

### Exercises

Use the information for the higher, safer volume rate to answer the questions. Round answers to the nearest tenth.

**1.** Draw a diagram of a rectangular box support with a 16-ft height. Label the sides $W$ for width and $L$ for length.

**2.** What is the required volume of this safer 16-ft-high support?

**3.** Use $L$ to write an expression for the width if it is to be half the length.

**4.** Use the volume formula $V = LWH$ to write an equation to find the width and depth of the support. Solve the equation, and state your answers rounded to the nearest tenth of a foot. What is the volume of this support?

**5.** If a support has a width three-fourths the size of the length, find its dimensions and volume.

**6.** This freeway will be in an earthquake fault zone, so state civil engineers decide to make the depth equal to the width of the support to withstand maximum vibration stress. Find the dimensions and volume of each support.

**7.** Why do the dimensions you found in Exercise 4 have a volume of exactly 200 ft$^3$, while the dimensions you found in Exercises 5 and 6 have volumes over and under the prescribed 200 ft$^3$?

**8.** A recent computer modeling simulation found that supports with the smallest cross-sectional perimeter per volume are best able to endure vibration stress, which means that supports in earthquake zones should be in the shape of a cylinder (a cylinder has a circular cross-sectional perimeter). Use $V = \pi r^2 h$ to find the radius of the support with a height of 16 ft and a volume of 200 ft$^3$. Use the $\pi$ key on your calculator.

**9.** Find the cross-sectional perimeters of the supports designed in Exercises 4, 5, 6, and 8. Does the circular cross section have the smallest perimeter? The perimeter or circumference of a circle is $C = 2\pi r$.

### Answers

**1.** Support diagram:
**2.** At 12.5 ft$^3$ per ft of height, 200 ft$^3$ is allotted.
**3.** $W = (\frac{1}{2})L$ or $W = \frac{L}{2}$ or $W = 0.5L$.
**4.** The length should be 5.0 ft and the width 2.5 ft, with a volume of 200.0 ft$^3$.
**5.** From the equation $16(0.75L)(L) = 200$, the length should be 4.1 ft, the width should be 3.1 ft, and the height should be 16 ft. These rounded dimensions have a volume of 203.4 ft$^3$.

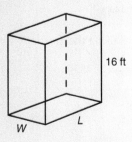

16 ft

W    L

**6.** With $L = W$, both the length and width should be 3.5 ft, and the height should be 16 ft. These rounded dimensions have a volume of 196 ft$^3$.

**8.** The radius should be 2.0 ft, which produces a volume of 201.1 ft$^3$.

**7.** When the pure quadratic equation of Exercise 4 was solved by the square-root method, an exact square root was found. In Exercises 5 and 6, square roots were approximated to the nearest tenth of a foot, which resulted in round-off error propagation in the volume formula.

**9.** The perimeter of the rectangular cross section is 15 ft in Exercise 4 and 14.4 ft in Exercise 5. The square cross section of Exercise 6 is 14 ft, while the perimeter of the circular cross section of Exercise 8 is 12.6 ft. The perimeter of the circular cross section has the smallest perimeter per volume.

## CUMULATIVE PRACTICE TEST FOR CHAPTERS 9–11

Perform the indicated operations.

**1.** $5x^7 \cdot 8x^3$

**2.** $\dfrac{x^3 y^{-3}}{x^4 y^{-5}}$

**3.** $(x^4)^8$

Arrange in descending order.

**4.** $15 - 32x^3 + 6x^4 - 8x^2 + 40x$

Simplify.

**5.** $5x^4 - 3x^2 + 7x^4 + 8x^2$

**6.** $-8x^2(3x^2 - 4x + 7)$

**7.** $\dfrac{10x^5 - 20x^6 + 15x^4}{5x^3}$

Factor by removing the greatest common factor.

**8.** $12x^3 - 18x^2 y$

**9.** $24x^3 - 16x^2 - 8x$

Find the product.

**10.** $(3x - 2)(x + 1)$

**11.** $(4x - 5)(3x - 2)$

Find the quotient.

**12.** $x + 3 \overline{)x^2 - 4x - 21}$

**13.** $x - 5 \overline{)x^2 - 8x + 15}$

Factor.

**14.** $49y^2 - 36$

**15.** $x^2 - 16x + 64$

**16.** $x^2 - 8x + 15$

**17.** $x^2 + x - 42$

**18.** $3x^2 - 17x + 10$

**19.** $6x^2 - 17x + 12$

**20.** $x^3 - 5x^2 - x$

Solve by the square-root method. Round to thousandths if necessary.

**21.** $3x^2 - 7 = 20$

**22.** $4x^2 - 9 = 39$

Solve by factoring.

**23.** $x^2 - 8x = 0$

**24.** $x^2 - x - 12 = 0$

**25.** $2x^2 - 7x + 3 = 0$

**26.** $4x^2 + 5x = 6$

Solve using the quadratic formula. Round to the nearest hundredth.

**27.** $4x^2 - 3x - 2 = 0$

**28.** $x^2 + 6x = 2$

Simplify.

**29.** $j^{17}$

**30.** $j^{30}$

# 12

# Geometry

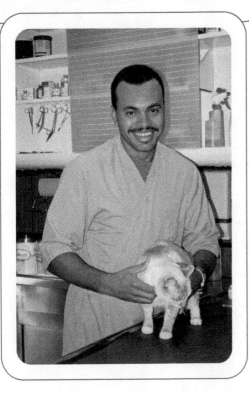

## Focus on Careers

If you love animals, this may be the career for you! Veterinary technicians usually work in a private practice with a supervising veterinarian, in animal hospitals, or in research facilities. They perform laboratory tests, assist with dental prophylaxis, prepare tissue samples, take blood samples, and assist veterinarians with a variety of tests and analyses.

Most entry-level technicians have an associate degree from an accredited community college program in veterinary technology with clinical and laboratory settings using live animals. All states require veterinary technicians and technologists to pass a credentialing exam following their coursework.

The job outlook for veterinary technologists and technicians is expected to grow much faster than average through 2012. Although the top 10% of veterinary technologists and technicians earned more than $33,750 in 2002, the middle 50% earned between $19,210 and $27,890. Economic recession is less likely to cause layoffs than in some other occupations because animals continue to require medical care.

Working conditions in this career are usually indoors but may take place outdoors. The work can be physically and emotionally demanding and sometimes dangerous. Workers risk exposure to small-animal

*(continued)*

bites and scratches and other injuries when working with larger animals. When appropriate safety precautions are taken, injuries can be minimized.

Source: *Occupational Outlook Handbook* 2004–2005 Edition. U.S. Department of Labor, Bureau of Labor Statistics.

Geometry is one of the oldest and most useful of the mathematical sciences. **Geometry** involves the study and measurement of shapes according to their sizes, volumes, and positions. A knowledge of geometry is necessary in many careers. Notice how the meanings of common words with which you are already familiar change when they are precisely defined for geometry.

# 12–1 | Lines and Angles

*Learning Outcomes*

**1** Use various notations to represent points, lines, line segments, rays, planes, and angles.

**2** Classify angles according to size.

**3** Determine the measure of an angle by using relationships among intersecting lines.

**4** Convert angle measures between decimal degrees and degrees, minutes, and seconds.

**1 Use Various Notations to Represent Points, Lines, Line Segments, Rays, Planes, and Angles.**

Geometry is the study of size, shape, position, and other properties of the objects around us. The basic terms used in geometry are *point, line,* and *plane.* Generally, these terms are not defined. Instead, they are only described. Once described, they are used in definitions of other terms and concepts.

A **point** is a location or position that has no size or dimension. A dot is used to represent a point, and a capital letter may be used to label the point (Figure 12–1).

A **line** extends indefinitely in both directions and contains an infinite number of points. It has length but no width. In our discussions, the word *line* always refers to a straight line unless otherwise specified. In Figure 12–1, the line can be identified by naming any two points on the line (such as *A* and *B*).

A **plane** is a flat, smooth surface that extends indefinitely in all directions. A plane contains an infinite number of points and lines (Figure 12–1).

Since a line extends indefinitely in both directions, most geometric applications deal with parts of lines. A part of a line is called a **line segment** or **segment.** A line segment starts and stops at distinct points that we call **endpoints.** A *line segment,* or *segment,* consists of all points on the line between and including the two *endpoints* (Figure 12–2).

The notation for a line that extends through points *A* and *B* is $\overleftrightarrow{AB}$ (read "line *AB*"). The notation for the line segment including points *A* and *B* and all the points between is $\overline{AB}$ (read "line segment *AB*").

Another term used in connection with parts of a line is *ray.* Before we give the definition of a ray, consider the beam of light from a flashlight. The beam is like a ray. It seems to continue indefinitely in only one direction. A **ray** consists of a point on a line and all points of the line on one side of the point (Figure 12–3).

The point from which the ray originates is called the **endpoint,** and all other points on the ray are called **interior points** of the ray. A ray is named by its endpoint and any interior point on the ray. In Figure 12–3, we use the notation $\overrightarrow{RS}$ to denote the ray whose endpoint is *R* and that passes through *S*.

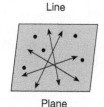

*A*
•

Point

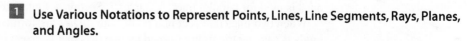

*A*         *B*

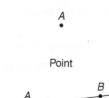

Line

Plane

**Figure 12–1**

*A*     *B*

**Figure 12–2**

*R*       *S*

**Figure 12–3**

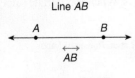

Line *AB*

A        B

$\overleftrightarrow{AB}$

Line segment *AB*

A        B

$\overline{AB}$

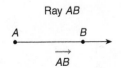

Ray *AB*

A        B

$\overrightarrow{AB}$

**Figure 12–4**

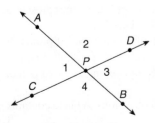

C

E   B

A        D

E   F   G   H

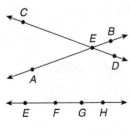

J

I        L

K

**Figure 12–6**

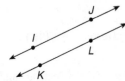

A

2        D

1  P

C  4    3

B

**Figure 12–7**

Figure 12–4 shows the contrast in the notation used for a line, line segment, and ray. To illustrate appropriate notations for lines, segments, and rays, consider a line with several points designated on the line (Figure 12–5).

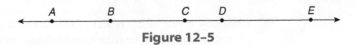

A        B        C   D        E

**Figure 12–5**

Any two points can be used to name the line in Figure 12–5. For example, $\overleftrightarrow{AB}$, $\overleftrightarrow{AC}$, $\overleftrightarrow{CE}$, $\overleftrightarrow{BD}$, and $\overleftrightarrow{BC}$ are some of the possible ways to name the line. However, a segment is named *only* by its endpoints. Thus, in Figure 12–5, $\overline{AB}$ is not the same segment as $\overline{AC}$, but $\overleftrightarrow{AB}$ and $\overleftrightarrow{AC}$ represent the same line.

In Figure 12–5, $\overrightarrow{BC}$ and $\overrightarrow{BD}$ represent the same ray, but $\overline{BC}$ and $\overline{BD}$ do not represent the same segment.

A line can be extended indefinitely in either direction. If two lines are drawn in the same plane, three things can happen:

1. The two lines *intersect* in *one and only one point.* In Figure 12–6, $\overleftrightarrow{AB}$ and $\overleftrightarrow{CD}$ intersect at point *E*.
2. The two lines *coincide;* that is, one line fits exactly on the other. In Figure 12–6, $\overleftrightarrow{EF}$ and $\overleftrightarrow{GH}$ coincide.
3. The two lines never intersect. In Figure 12–6, $\overleftrightarrow{IJ}$ and $\overleftrightarrow{KL}$ are the same distance from each other along their entire lengths and so never touch.

The relationship described in the third situation has a special name, **parallel lines.** The symbol ∥ is used for parallel lines. (Chapter 7, Section 4, Outcome 1).

When two lines intersect in a point, four *angles* are formed, as shown in Figure 12–7. An **angle** is a geometric figure formed by two rays that intersect in a point, and the point of intersection is the endpoint of each ray.

In Figure 12–8, rays $\overrightarrow{AB}$ and $\overrightarrow{AC}$ intersect at point *A*. Point *A* is the endpoint of $\overrightarrow{AB}$ and $\overrightarrow{AC}$. $\overrightarrow{AB}$ and $\overrightarrow{AC}$ are called the **sides** or **legs** of the angle. Point *A* is called the **vertex** of the angle.

Angles can be named in several ways. An angle can be named by using a number or lowercase letter, by using the capital letter that names the vertex point, or by using three capital letters. If three capital letters are used, two of the letters name interior points of each of the two rays, and the middle letter names the vertex point of the angle. Using the symbol ∠ for angle, the angle in Figure 12–9 can be named ∠1, ∠*KLM*, ∠*MLK*, or ∠*L*. One capital letter is used only when it is perfectly clear which angle is designated by the letter. To name the angle in Figure 12–10 with three letters, we write ∠*XZY* or ∠*YZX*. This angle can also be named ∠*a* or ∠*Z*.

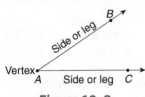

B

Side or leg

Vertex
A   Side or leg   C

**Figure 12–8**

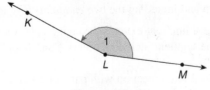

K

1

L        M

**Figure 12–9**

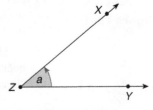

X

Z  a

Y

**Figure 12–10**

Figure 12–11 illustrates how the intersection of two rays actually forms two angles. In this text, we refer to the smaller of the two angles formed by two rays unless the other angle is specifically indicated. Arcs (curved lines) and arrows are often used to clarify which angle we are considering.

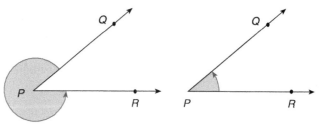

**Figure 12–11**

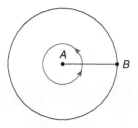

**Figure 12–12**

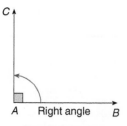

A    Right angle    B

**Figure 12–14**

String

Weight

Vertical line

**Figure 12–15**

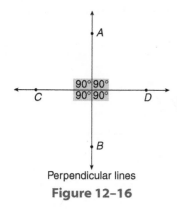

Perpendicular lines

**Figure 12–16**

**2**  **Classify Angles According to Size.**

The measure of an angle is determined by the amount of opening between the two sides of the angle. The length of the sides does not affect the angle measure. Two units are commonly used to measure angles, *degrees* and *radians*. In this section, only degrees are used to measure angles. Radians are discussed in Section 12–3.

Consider the hands of a clock as the sides of an angle. When the two hands both point to the same number, the measure of the angle formed is 0 degrees (0°). An angle of 0 degrees is used in trigonometry but is seldom used in geometric applications. During 1 hour, the minute hand makes one complete revolution. Ignoring the movement of the hour hand, this revolution of the minute hand contains 360 degrees (360°). Figure 12–12 shows a revolution or rotation of 360°. Note that $A$ is kept as a fixed point and $B$ rotates around point $A$. This rotation can be either clockwise or counterclockwise.

The customary rotation around a point progresses *opposite* to the normal rotation of the hands of a clock. The direction is called **counterclockwise** (Figure 12–13).

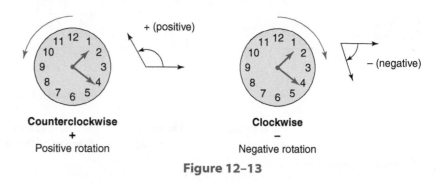

**Figure 12–13**

A **degree** is a unit for measuring angles. It represents $\frac{1}{360}$ of a complete rotation about the vertex. Suppose, in Figure 12–14, that $\overrightarrow{AC}$ rotates from $\overrightarrow{AB}$ through one-fourth of a circle. Then $\overrightarrow{AC}$ and $\overrightarrow{AB}$ form a 90° angle ($\frac{1}{4}$ of 360 = 90). This angle is a **right angle.** The symbol for a right angle is ⌐.

If two lines intersect so that right angles (90° angles) are formed, the lines are **perpendicular** to each other. (See Section 7–5.) The symbol for "perpendicular" is ⊥. However, the right-angle symbol also implies the lines forming the angle are perpendicular.

If a string is suspended at one end and weighted at the other (Figure 12–15), the line it forms is a **vertical line.** A line that is perpendicular to the vertical line is a **horizontal line.** In Figure 12–16, $\overleftrightarrow{AB}$ is a vertical line, and $\overleftrightarrow{AB}$ and $\overleftrightarrow{CD}$ form right angles. Thus, $\overleftrightarrow{AB} \perp \overleftrightarrow{CD}$, and $\overleftrightarrow{CD}$ is a horizontal line.

12–1 Lines and Angles

473

Now look at Figure 12–17. When $\overrightarrow{ML}$ rotates one-half a circle from $\overrightarrow{MN}$, an angle of 180° is formed ($\frac{1}{2}$ of 360 = 180). This angle is a **straight angle.**

An angle that is less than 90° but more than 0° is an **acute angle.** An angle that is more than 90° but less than 180° is an **obtuse angle** (see Figure 12–18).

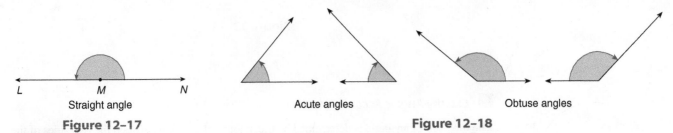

Straight angle

**Figure 12–17**

Acute angles

Obtuse angles

**Figure 12–18**

### To classify angles according to size:

1. Examine the angle to determine the number of degrees it measures.
2. Classify the angle as a straight, right, obtuse, or acute angle based on its degree measure.

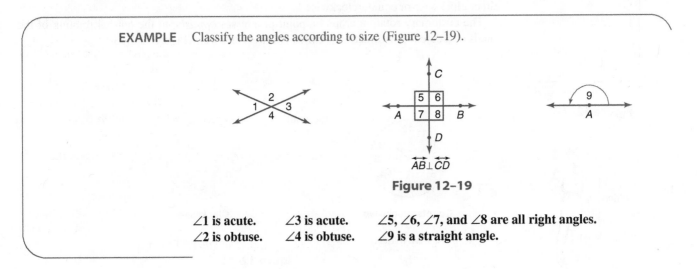

**EXAMPLE**   Classify the angles according to size (Figure 12–19).

$\overleftrightarrow{AB} \perp \overleftrightarrow{CD}$

**Figure 12–19**

∠1 is acute.       ∠3 is acute.       ∠5, ∠6, ∠7, and ∠8 are all right angles.
∠2 is obtuse.      ∠4 is obtuse.            ∠9 is a straight angle.

The angle classifications used so far—right, straight, acute, and obtuse—deal with one angle at a time. If two angles together form a right angle or if their measures total 90°, they are **complementary angles.** If two angles together form a straight angle or if their measures total 180°, they are **supplementary angles** (Figure 12–20).

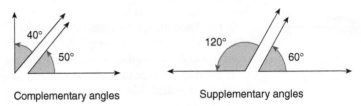

Complementary angles            Supplementary angles

**Figure 12–20**

## To find the complement or supplement of an angle:

1. To find the complement of an angle, subtract its measure from 90°.
2. To find the supplement of an angle, subtract its measure from 180°.

---

**EXAMPLE**  Find the complement and supplement of an angle that measures 57°.

**Complement**

$90° - 57° = 33°$

**The complement of 57° is 33°.**

**Supplement**

$180° - 57° = 123°$

**The supplement of 57° is 123°.**

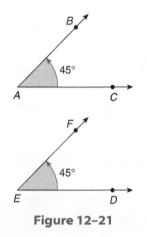

**Figure 12–21**

When the word *equal* is used to describe the relationship between two angles, it implies the measures of the angles are equal. Another word often used in geometry is *congruent*. When geometric figures are **congruent,** one figure can be placed on top of the other, and the two figures will match perfectly. If two angles are congruent, the measures of the angles are equal. Also, if the measures of two angles are equal, the angles are congruent. The symbol for congruence is ≅. For instance, in Figure 12–21, the two angles have equal measures, both 45°. Because they have the same measures, they are congruent; that is, $\angle BAC \cong \angle FED$.

Traditionally, we indicate the measures of angles by placing the letter *m* before the angle symbol. Thus, $m\angle BAC = m\angle FED$ tells us that the measures of angles *BAC* and *FED* are equal. We interpret $\angle BAC = \angle FED$ and $m\angle BAC = m\angle FED$ as giving us the same information.

### 3 Determine the Measure of an Angle by Using Relationships Among Intersecting Lines.

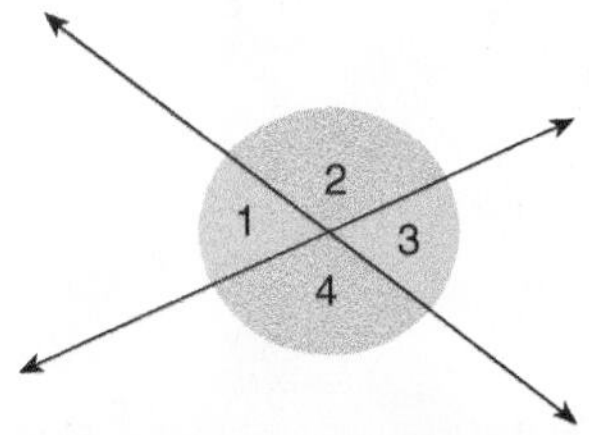

Intersecting lines generate several angle relationships. When two lines intersect, two pairs of **vertical angles** are formed. In Figure 12–22 angles 1 and 3 are vertical angles. Also, angles 2 and 4 are vertical angles. Two angles that share a common side as angles 1 and 2 are **adjacent angles.** Other adjacent angles are 2 and 3, 3 and 4, and 4 and 1.

**Figure 12–22**

---

**Relationships among angles formed by intersecting lines can be used to find missing measures of angles (Figure 12–22).**

| | | |
|---|---|---|
| *Vertical angles are equal.* | $\angle 1 = \angle 3$ | $\angle 2 = \angle 4$ |
| *Adjacent angles are supplementary.* | $\angle 1 + \angle 2 = 180°$ | $\angle 2 + \angle 3 = 180°$ |
| | $\angle 3 + \angle 4 = 180°$ | $\angle 4 + \angle 1 = 180°$ |

---

A **transversal** is a line that intersects two or more lines in different points. When a transversal intersects (cuts) two lines, eight angles are formed (Figure 12–23).

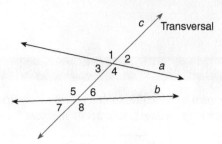

**Figure 12-23**

The four shaded angles *between* lines *a* and *b* (∠3, ∠4, ∠5, and ∠6) are called **interior angles** (Figure 12–24). The four angles *outside* lines *a* and *b* (∠1, ∠2, ∠7, and ∠8) are called **exterior angles** (Figure 12–24).

An exterior angle and an interior angle on the same side of the transversal are **corresponding angles.** Shaded angles ∠1 and ∠5 are corresponding angles (Figure 12–25). Other corresponding angles are ∠2 and ∠6, ∠3 and ∠7, and ∠4 and ∠8. **Alternate angles** are on opposite sides of the transversal. In Figure 12–26, ∠1 and ∠8 are **alternate exterior angles** and ∠3 and ∠6 are **alternate interior angles.** There are other pairs of alternate angles.

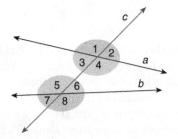 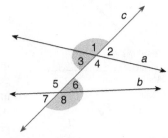

**Figure 12–24** Interior and exterior angles.  **Figure 12–25** Corresponding angles.  **Figure 12–26** Alternate angles.

**Relationships among angles on parallel lines cut by a transversal (Figure 12–27) can be used to find missing measures of angles.**

| | | |
|---|---|---|
| *Corresponding angles are equal.* | $\angle 1 = \angle 5$ <br> $\angle 3 = \angle 7$ | $\angle 2 = \angle 6$ <br> $\angle 4 = \angle 8$ |
| *Adjacent angles are supplementary.* | $\angle 1 + \angle 2 = 180°$ <br> $\angle 3 + \angle 4 = 180°$ | $\angle 2 + \angle 4 = 180°$ <br> $\angle 3 + \angle 1 = 180°$ |
| *Alternate angles are equal.* | $\angle 1 = \angle 8$ <br> $\angle 3 = \angle 6$ | $\angle 2 = \angle 7$ <br> $\angle 4 = \angle 5$ |
| *Interior angles on the same side of the transversal are supplementary.* | $\angle 3 + \angle 5 = 180°$ | $\angle 4 + \angle 6 = 180°$ |
| *Exterior angles on the same side of the transversal are supplementary.* | $\angle 1 + \angle 7 = 180°$ | $\angle 2 + \angle 8 = 180°$ |

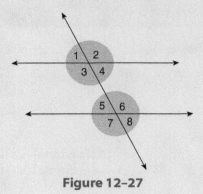

**Figure 12–27**

**EXAMPLE**    In Figure 12–27, ∠2 is 125°. Find the measures of the other angles.

∠1 = 180° − 125°        Adjacent angles are supplementary.
∠1 = 55°

∠3 = ∠2                 Vertical angles are equal.
∠3 = 125°

∠4 = ∠1                 Vertical angles are equal.
∠4 = 55°

∠5 = ∠1                 Corresponding angles are equal.
∠5 = 55°

∠6 = ∠2                 Corresponding angles are equal.
∠6 = 125°

∠7 = ∠3                 Corresponding angles are equal.
∠7 = 125°

∠8 = ∠4                 Corresponding angles are equal.
∠8 = 55°

### 4  Convert Angle Measures Between Decimal Degrees and Degrees, Minutes, and Seconds.

A device used to measure angles is called a **protractor.** The most common protractor is a semicircle with two scales from 0° to 180°. An **index mark** is in the middle of the straight edge of the protractor (Figure 12–28).

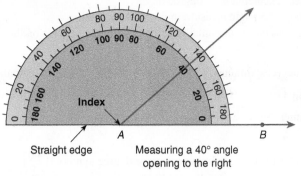

Straight edge          Measuring a 40° angle
                       opening to the right

**Figure 12–28**

To measure an angle that opens to the right, read the degree measure from the lower scale. Notice in Figure 12–28 the lower scale starts with 0 at the right. The index of the protractor is aligned with the vertex of the angle, and the straight edge lies along the lower side of the angle.

To measure an angle that opens to the left, read the degree measure from the upper scale. In Figure 12–29 the upper scale starts with 0 at the left. The index of the protractor is aligned with the vertex of the angle and the straight edge lies along the lower side of the angle.

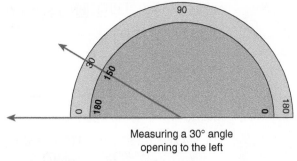

Measuring a 30° angle
opening to the left

**Figure 12–29**

Degrees are divided into 60 equal parts. Each part is called a **minute.**

$$1 \text{ degree } (1°) = 60 \text{ minutes } (60')$$

The symbol ′ is used for minutes. Similarly, 1 minute is divided into 60 equal parts called **seconds.**

$$1 \text{ minute } (1') = 60 \text{ seconds } (60'')$$

The symbol ″ is used for seconds. Thus,

$$1° = 60' = 3{,}600''$$

With the increased popularity of the calculator, it is sometimes desirable to change minutes or seconds to decimal equivalents. Minutes or seconds can be changed to a fractional part of a degree. Then the fraction is changed to its decimal equivalent by dividing the numerator by the denominator. Remember, $1' = \frac{1}{60}$ of a degree, and $1'' = \frac{1}{3{,}600}$ of a degree.

Scientific and graphing calculators have a key or menu choice that automatically converts between degrees, minutes, and seconds and decimal degrees. A key may be labeled $\boxed{° \; ' \; ''}$ or $\boxed{\text{DMS}}$, or a menu choice on the angle menu may be DMS. This function key is also called a **sexagesimal function key** because of its relationship with the number 60. The same key is also used for hours, minutes, and seconds.

TIP!

## To Change the Format of Angle Measures Using a Calculator:

From degrees in decimal notation to degrees, minutes, and seconds (calculator steps will vary with different calculators):

1. Set the calculator to degree mode.
2. Enter the angle measure in decimal notation.
3. Access the sexagesimal function key or menu choice on the calculator (most often labeled $\boxed{° \; ' \; ''}$ or $\boxed{\text{DMS}}$).
4. Press $\boxed{\text{ENTER}}$ or $\boxed{=}$.

From degrees, minutes, and seconds to decimal notation (calculator steps will vary with different calculators):

**Option 1**

1. Enter the series of calculations—degrees + minutes ÷ 60 + seconds ÷ 3,600.

**Option 2**

1. Enter the degrees followed by the degree symbol ($\boxed{2^{\text{nd}}}$, $\boxed{\text{ANGLE}}$, $\boxed{1}$)
2. Enter the minutes followed by the minute symbol ($\boxed{2^{\text{nd}}}$, $\boxed{\text{ANGLE}}$, $\boxed{2}$)
3. Enter the seconds followed by the second symbol ($\boxed{\text{ALPHA}}$,)
4. Press $\boxed{\text{ENTER}}$ or $\boxed{=}$.

**EXAMPLE**  Change 25.375° to degrees, minutes, and seconds.

25.375 $\boxed{2^{\text{nd}}}$ $\boxed{\text{ANGLE}}$ $\boxed{4}$ $\boxed{\text{ENTER}}$  Using a TI-83 or TI-84.

$25.375° = \mathbf{25°22'30''}$

**EXAMPLE**  Change 115°40′15″ to a decimal-degree equivalent. Round to the nearest thousandth.

$115 + 40/60 + 15/3600 = 115.6708333$     Perform the series of calculations.
$115°40'15'' = 115.671°$     Round.

**1** Use Figure 12–30 for Exercises 1–3.

**1.** Name the line in three different ways.
**2.** Name in two different ways the ray with endpoint *P* and with interior points *Q* and *R*.
**3.** Name the segment with endpoints *Q* and *R*.

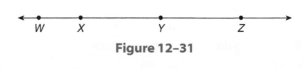

**Figure 12–30**

Use Figure 12–31 for Exercises 4–10.

**4.** Does $\overleftrightarrow{XY}$ represent the same line as $\overleftrightarrow{YZ}$ ?
**5.** Does $\overrightarrow{XY}$ represent the same ray as $\overrightarrow{YZ}$ ?
**6.** Does $\overline{XY}$ represent the same segment as $\overline{YZ}$?
**7.** Is $\overleftrightarrow{WX}$ the same as $\overleftrightarrow{WY}$ ?

**Figure 12–31**

Use Figure 12–32 for Exercises 8–12.

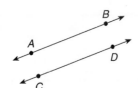

**Figure 12–32**

**8.** $\overleftrightarrow{AB}$ and $\overleftrightarrow{CD}$ are _____ lines.
**10.** $\overleftrightarrow{IJ}$ and $\overleftrightarrow{KL}$ _____.
**9.** $\overleftrightarrow{EF}$ and $\overleftrightarrow{GH}$ _____ at point *O*.
**11.** $\overleftrightarrow{AB}$ and $\overleftrightarrow{CD}$ will never _____.

**12.** Name two lines that intersect in exactly one point.

Use Figure 12–33 for Exercises 13–15.

**13.** Name the angle in two different ways using three capital letters.
**14.** Name the angle using one capital letter.
**15.** Name the angle using a number.

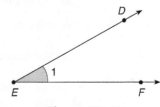

**Figure 12–33**

**2** Fill in the blanks regarding angle rotation.

**16.** One complete rotation is _____°.
**17.** One-half a complete rotation is _____°.
**18.** One-fourth a complete rotation is _____°.
**19.** One-eighth of a complete rotation is _____°.

Classify the angle measures using the terms *right, straight, acute,* or *obtuse*.

**20.** 38°   **21.** 95°   **22.** 90°   **23.** 153°   **24.** 10°   **25.** 180°

Tell whether the angle pairs are complementary, supplementary, or neither.

**26.** 42°, 80°   **27.** 17°, 73°   **28.** 38°, 142°   **29.** 52°, 48°   **30.** 60°, 30°   **31.** 110°, 70°
**32.** 42°, 138°

**3** Use Figure 12–34 to answer Exercises 33–35.

**33.** Name two pairs of equal angles.
**34.** What are angles *a* and *b* called?
**35.** What is the sum of ∠*c* and ∠*d*?

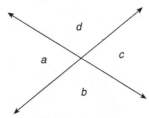

**Figure 12–34**

In Figure 12–35 *l* ∥ *m* and *n* is a transversal.

**36.** Name two pairs of alternate interior angles.

**37.** What is the sum of ∠*a* and ∠*g*?

**38.** What name is given to ∠*d* and ∠*h*?

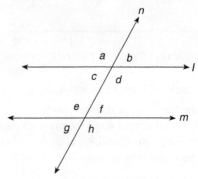

**Figure 12–35**

**4** Change to decimal degree equivalents. Express the decimals to the nearest ten-thousandth.

**39.** 47′    **40.** 36″    **41.** 5′14″    **42.** 10′15″    **43.** 10°18′15″

Change to equivalent minutes and seconds. Round to the nearest second when necessary.

**44.** 0.35°    **45.** 0.20°    **46.** 0.12°    **47.** 0.213°    **48.** 0.3149°

---

## 12–2 | *Perimeter and Area*

*Learning Outcomes*

**1** Find the perimeter of a polygon using the appropriate formula.

**2** Find the area of a polygon using the appropriate formula.

**1** **Find the Perimeter of a Polygon Using the Appropriate Formula.**

A **polygon** is a plane or flat, closed figure described by straight-line segments and angles. Polygons have different numbers of sides and different properties. Some common polygons are the parallelogram, rectangle, square, rhombus, trapezoid, and triangle (Figure 12–36).

The **base** of any polygon is the horizontal side or a side that would be horizontal if the polygon's orientation is modified.

The **adjacent side** of any polygon is the side that has an endpoint in common with the base.

A **parallelogram** is a four-sided polygon with opposite sides that are parallel.

A **rectangle** is a parallelogram with angles that are all right angles.

A **square** is a parallelogram with sides all of equal length and with all right angles. A square can also be described as a rectangle with all sides of equal length.

A **rhombus** is a parallelogram with all sides of equal length.

A **trapezoid** is a four-sided polygon having only two parallel sides.

A **triangle** is a polygon that has three sides.

The **perimeter** is the total length of the sides of a plane figure.

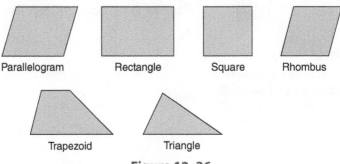

Parallelogram    Rectangle    Square    Rhombus

Trapezoid    Triangle

**Figure 12–36**

When do you need to find the perimeter of a polygon? Some common applications for perimeter are finding the amount of trim molding for a room, determining the amount of fencing for a yard, finding the amount of edging for a flower bed, and so on.

A general procedure for finding the perimeter of any shape is to add the lengths of the sides. However, shortcuts based on the properties of the shape are given as formulas.

## Perimeter

| Parallelogram | $P = 2b + 2s$ or $P = 2(b + s)$ | | $b$ is the base $s$ is the adjacent side |
|---|---|---|---|
| Rectangle | $P = 2l + 2w$ or $P = 2(l + w)$ | | $l$ is length $w$ is width |
| Square | $P = 4s$ | | $s$ is length of each side |
| Rhombus | $P = 4s$ | | $s$ is length of each side |
| Trapezoid | $P = b_1 + b_2 + a + c$ | | $b_1$ and $b_2$ are bases, $a$ and $c$ are the other sides |
| Triangle | $P = a + b + c$ | | $a$, $b$, and $c$ are sides |

### To find the perimeter of a polygon:

1. Select the appropriate formula.
2. Substitute the known values into the formula.
3. Evaluate the formula (perform the indicated operations).

The sides of the parallelogram are called *base* and *adjacent side* (instead of length and width). Notice the locations of the base and adjacent side in Figure 12–37.

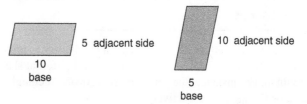

**Figure 12–37**

**EXAMPLE** Find the perimeter of a parallelogram with a base of 16 in. and an adjacent side of 8 in. (Figure 12–38).

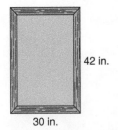

8 in.

16 in.

**Figure 12–38**

Visualize the parallelogram.

$P_{parallelogram} = 2(b + s)$     Select the perimeter of a parallelogram formula.
$P_{parallelogram} = 2(16 + 8)$     Substitute: $b = 16$, $s = 8$.
$P_{parallelogram} = 2(24)$     Evaluate.

$P_{parallelogram} = \textbf{48 in.}$

**EXAMPLE**
**BUS** A shop that makes custom picture frames has an order for a frame whose outside measurements are 42 in. by 30 in. (Figure 12–39). How many inches of picture frame molding are needed for the job?

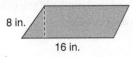

42 in.

30 in.

**Figure 12–39**

$P_{rectangle} = 2(l + w)$     The figure is a rectangle. Select the appropriate formula.
$P_{rectangle} = 2(42 + 30)$     Substitute and evaluate the formula.
$P_{rectangle} = 2(72)$
$P_{rectangle} = 144$

When analyzing the dimensions, inches are added to inches, so the result is written in inches. Then the inches are multiplied by a number and the final result is inches.

**Thus, 144 in. are needed for the project.**

In some applications, we need to decrease the total perimeter to account for doorways and other openings in the perimeter.

**EXAMPLE**
**CON** A chain-link fence is to be installed around a yard measuring 25 m by 30 m (Figure 12–40). A gate 3 m wide will be installed over the driveway. The gate comes preassembled from the manufacturer. How much fencing does the installer need to put up the fence?

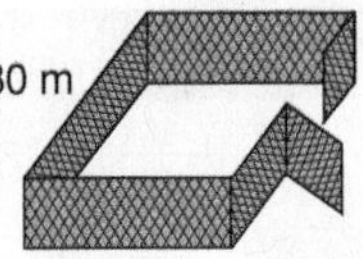

25 m

30 m

**Figure 12–40**

$P_{rectangle} = 2(l + w) - 3$     Select the perimeter of a rectangle formula. Subtract the length of the gate.
$P_{rectangle} = 2(30 + 25) - 3$     Substitute and perform operations.
$P_{rectangle} = 2(55) - 3$     Multiply.
$P_{rectangle} = 110 - 3$     Subtract.
$P_{rectangle} = 107$ m     Perimeter is a linear measure.

**The job requires 107 m of fencing.**

**TIP!**

### Using Subscripts

Subscripted words, letters, or numbers are a handy way to provide additional information. Because we will examine the formulas for the perimeter of several different shapes, we sometimes use subscripts to distinguish among them. For instance, the perimeter of a square, rectangle, or parallelogram can be indicated as $P_{square}$, $P_{rectangle}$, or $P_{parallelogram}$, respectively.

A *trapezoid* is a four-sided polygon having only two parallel sides. Unlike a parallelogram, the trapezoid's four sides may all be of unequal size, as illustrated in Figure 12–41.

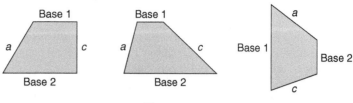

**Figure 12–41**

The *parallel* sides are called **bases** ($b_1$ and $b_2$) with the subscripts 1 and 2 used to distinguish them. The nonparallel sides are designated side $a$ and side $c$.

---

**EXAMPLE**
**AUTO**

The rear window of a pickup truck is a trapezoid whose shorter base is 40 in., whose longer base is 48 in., and whose nonparallel sides are each 16 in. How many inches of rubber gasket are needed to surround the window?

$P = b_1 + b_2 + a + c$        Select appropriate formula.

**Estimation**    By rounding, the perimeter is $40 + 50 + 20 + 20$ or 130 in.

$P = 40 + 48 + 16 + 16$        Substitute in formula.
$P = 120$ in.                          Add.

**Interpretation**    **The rubber gasket to surround the rear window must be 120 in. long.**

---

**Figure 12–42**

A **triangle** is a three-sided polygon. The perimeter of a triangle is the sum of the lengths of the three sides (Figure 12–42). The triangle, like the trapezoid, may have unequal lengths for all three sides.

---

**EXAMPLE**
**CON**

The triangular gables of an apartment unit under construction are outlined in contrasting trim. If each gable is 32 ft wide and 18 ft on each side, how many feet of contrasting wood trim are needed for each gable? Disregard overlap at the corners (Figure 12–43).

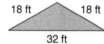

**Figure 12–43**

$P = a + b + c$        Select the appropriate formula and substitute in formula.
$P = 18 + 32 + 18$    Evaluate.
$P = 68$ ft

**Each gable will require 68 linear feet of trim.**

---

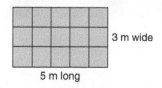

**Figure 12–44**

**2**  **Find the Area of a Polygon Using the Appropriate Formula.**

The area of a polygon is the amount of surface of a plane figure. Area is expressed in square units. For example, if a rectangle is 3 meters wide and 5 meters long, there are 15 square meters in the area (Figure 12–44).

$$\text{area} = 15 \text{ m}^2$$

12–2  Perimeter and Area

**Writing Square Units**

The exponent 2 following a unit of measure indicates *square measure* or area. Thus, 15 m$^2$ = 15 square meters, 23 ft$^2$ = 23 square feet, 120 cm$^2$ = 120 square centimeters, and so on. This is a shortcut way to express a measure that has *already* been "squared."

When do you need the area of a polygon? Some common applications for area are finding the amount of carpeting needed, the amount of paint needed to paint a surface, the amount of fertilizer needed to treat a lawn, and so on.

## Area

| | | | |
|---|---|---|---|
| Rectangle | $A = lw$ | | $l$ is length<br>$w$ is width |
| Square | $A = s^2$ | | $s$ is length of a side |
| Parallelogram | $A = bh$ | | $b$ is base<br>$h$ is height |
| Rhombus | $A = bh$ | | $b$ is base<br>$h$ is height |
| Trapezoid | $A = \frac{1}{2}h(b_1 + b_2)$ | | $b_1$ and $b_2$ are bases<br>$h$ is height |
| Triangle | $A = \frac{1}{2}bh$ | | $b$ is base<br>$h$ is height |

$$A = \sqrt{s(s - a)(s - b)(s - c)}$$
(Heron's formula)

$$s = \frac{1}{2}(a + b + c)$$
$a$, $b$, and $c$ are sides

## To find the area of a polygon:

1. Visualize the polygon.
2. Select the appropriate formula.
3. Substitute the known values into the formula.
4. Evaluate the formula.

EXAMPLE
CON

A carpet installer is carpeting a room measuring 16 ft by 20 ft. Projecting out from one wall is a fireplace whose hearth measures 3 ft by 6 ft (Figure 12–45). How many square yards of carpet does the installer require for the job? How much is wasted?

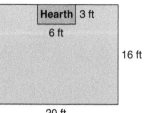

Figure 12–45

First, compute the area of the room without considering the area of the hearth. This is the amount of carpet needed. Second, compute the area of the fireplace hearth. This is the amount wasted. Finally, analyze the dimensions to see if the result is expressed in the desired unit. If not, convert to the desired unit.

Let $A_{room}$ = the area of the room and $A_{hearth}$ = the area of the hearth.

| | |
|---|---|
| $A_{room} = lw$ | Select the appropriate formula and substitute. |
| $A_{room} = 20 \times 16$ | Multiply. |
| $A_{room} = 320 \text{ ft}^2$ | Area is a square measure. |
| $A_{hearth} = lw$ | Select the appropriate formula and substitute. |
| $A_{hearth} = 3 \times 6$ | Multiply. |
| $A_{hearth} = 18 \text{ ft}^2$ | Area is a square measure. |

Because carpet is sold in square yards, we convert the square footage to square yards. Since 9 ft$^2$ = 1 yd$^2$, we can use a *unity ratio*.

A unity ratio is a fraction with one unit of measure in the numerator and a different, but equivalent, unit of measure in the denominator. A unity ratio contains the original unit and the new unit. Use the unity ratio $\frac{1 \text{ yd}^2}{9 \text{ ft}^2}$.

$$\frac{320 \text{ ft}^2}{1} \times \frac{1 \text{ yd}^2}{9 \text{ ft}^2} = \frac{320 \text{ yd}^2}{9} = 35\frac{5}{9} \text{ yd}^2 \text{ or } 36 \text{ yd}^2 \text{ of carpet needed}$$

$$\frac{\overset{2}{\cancel{18}} \text{ ft}^2}{1} \times \frac{1 \text{ yd}^2}{\underset{1}{\cancel{9}} \text{ ft}^2} = 2 \text{ yd}^2 \text{ carpet wasted on the hearth}$$

**This job requires $35\frac{5}{9}$ yd$^2$ or 36 yd$^2$ of carpet. Of this, 2 yd$^2$ is waste from the hearth.**

---

EXAMPLE

Find the area of a parallelogram with a base of 16 in., an adjacent side of 8 in., and a height of 7 in. (Figure 12–46).

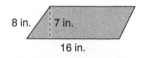

Figure 12–46

Visualize the parallelogram.

| | |
|---|---|
| $A_{parallelogram} = bh$ | Substitute. |
| $A_{parallelogram} = 16 \times 7$ | Multiply. |
| $A_{parallelogram} = \textbf{112 in.}^2$ | |

---

EXAMPLE

Find the area of the trapezoid in Figure 12–47.

Figure 12–47

| | |
|---|---|
| $A = \frac{1}{2}h(b_1 + b_2)$ | Select the appropriate formula and substitute the known values. |
| $A = \frac{1}{2}(4)(9 + 13)$ | Add inside grouping. |
| $A = \frac{1}{2}(4)(22)$ | Multiply. |
| $A = \textbf{44 cm}^2$ | cm × cm = cm$^2$. |

**EXAMPLE** Find the area of the triangle in Figure 12–48.

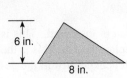

**Figure 12–48**

$$A = \frac{1}{2}bh$$      Select the appropriate formula and substitute the known values.

$$A = \frac{1}{2}(8)(6)$$      Multiply.

$$A = 24 \text{ in}^2$$      in. × in. = in².

In some triangles, the height is measured along an imaginary line outside the triangle from the base to the highest point of the triangle (the *apex*) (Figure 12–49).

**EXAMPLE** The base of $\Delta DEF$ in Figure 12–49 is 15 cm and the height is 21 cm. Find the area.

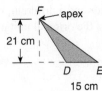

**Figure 12–49**

$$A = \frac{1}{2}bh$$      Substitute into the formula.

$$A = \frac{1}{2}(15)(21)$$      Multiply.

$$A = \frac{1}{2}(315)$$

$$A = 157.5 \text{ cm}^2$$

**The area of $\Delta DEF$ is 157.5 cm².**

When the three sides of a triangle are known and the height is not known, the area of a triangle can be found using *Heron's formula* (page 484).

**EXAMPLE** Find the area of the triangle in Figure 12–50 by using Heron's formula.

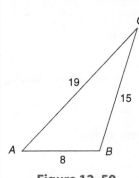

**Figure 12–50**

$$s = \frac{1}{2}(a + b + c)$$      Find *s*. Substitute length of sides.

$$s = \frac{1}{2}(19 + 15 + 8)$$      Add, then multiply.

$$s = 21$$

$$\text{Area} = \sqrt{s(s - a)(s - b)(s - c)}$$      Heron's formula.

$$\text{Area} = \sqrt{21(21 - 19)(21 - 15)(21 - 8)}$$      Substitute for *s*, *a*, *b*, and *c*, and perform each subtraction.

$$\text{Area} = \sqrt{21(2)(6)(13)}$$      Multiply.

$$\text{Area} = \sqrt{3,276}$$      Take the square root using a calculator.

$$\text{Area} = 57.23635209$$      Principal square root.

**Area = 57.24 square units**      Round to four significant digits.

**1**

1. Find the perimeter of the parallelogram in Figure 12–51.

2. Find the perimeter of the parallelogram in Figure 12–52.

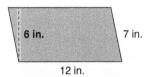

**Figure 12–51**

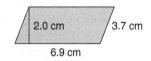

**Figure 12–52**

Solve the problems involving perimeter.

3. An illuminated sign in the main entrance of a hospital is a parallelogram with a base of 48 in. and an adjacent side of 30 in. How many feet of aluminum molding are needed to frame the sign?

4. A customized van has a window cut in each side in the shape of a parallelogram with a base of 20 in. and an adjacent side of 11 in. How many inches of trim are needed to surround the two windows?

5. **CON** A contemporary building has a window in the shape of a parallelogram with a base of 50 in. and an adjacent side of 30 in. How many inches of trim are needed to surround the window?

6. **CON** A table for a reading lab has a top in the shape of a parallelogram with a base of 36 in. and an adjacent side of 18 in. How many inches of edge trim are needed to surround the tabletop?

7. Find the perimeter of the rectangle in Figure 12–53.

8. Find the perimeter of the rectangle in Figure 12–54.

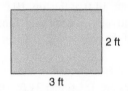

**Figure 12–53**

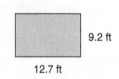

**Figure 12–54**

9. **CON** A rectangular parking lot is 340 ft by 125 ft. Find the perimeter of the parking lot.

10. **CON** A room is 15 ft by 12 ft. How many feet of chair rail are needed for the room? Disregard openings.

11. **CON** How many feet of quarter-round molding are needed to finish around the baseboard after sheet vinyl flooring is installed if the room is 16 ft by 18 ft and there are three 3-ft-wide doorways?

12. **CON** The swimming pool in Figure 12–55 measures 32 ft by 18 ft. How much fencing is needed, including material for a gate, if the fence is to be built 7 ft from each side of the pool?

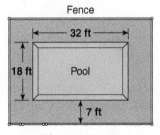

**Figure 12–55**

13. **CON** A Formica tabletop measures 40 in. by 62 in. How many feet of edge trim are needed (12 in. = 1 ft)?

14. **CON** A countertop requires rolled edging to be installed on all four sides. How much rolled edging material is needed if the countertop measures 25 in. by 40 in.?

**15.** Find the perimeter of Figure 12–56.

3 cm

3 cm

**Figure 12–56**

**16.** Find the perimeter of Figure 12–57.

8.9 cm

8.9 cm

**Figure 12–57**

**17. CON** The square parking lot of a doctor's office is to have curbs built on all four sides. If the lot is 150 ft on each side, how many feet of curb are needed? Allow 10 ft for a driveway into the parking lot.

**18. CON** A border of 4-in. × 4-in. wall tiles surrounds the floor of a shower stall that is 48 in. × 48 in. How many tiles are needed for this border? Disregard spaces for grout (connecting material between the tiles).

Find the perimeter of Figures 12–58 and 12–59. Round to hundredths.

**19.**

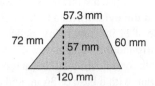

57.3 mm

72 mm    57 mm    60 mm

120 mm

**Figure 12–58**

**20.**

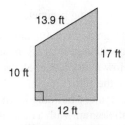

13.9 ft

17 ft

10 ft

12 ft

**Figure 12–59**

**21. CON** The six glass panes in a kitchen light fixture each measure $4\frac{1}{2}$ in. along the top and 10 in. along the bottom. The top and bottom are parallel. The two nonparallel sides of each pane are 10 in. What is the combined perimeter of the six trapezoidal panes?

**22. CON** A section of a hip roof is a trapezoid measuring 38 ft at the bottom, 14 ft at the top, 10 ft high, and 11 and 12 ft, respectively, on each side. Find the perimeter of this section of the roof.

**23.** A lot in an urban area has two nonparallel sides that are each 64 ft. The parallel sides are 120 ft and 154 ft. Find the perimeter of the trapezoidal property.

**24.** A swimming pool is fashioned in a trapezoidal design. The parallel sides are $18\frac{1}{2}$ ft and 31 ft. The other sides are $24\frac{1}{2}$ ft and 24 ft. What is the perimeter of the pool?

Find the perimeter of the triangles in Figures 12–60 and 12–61. Round to hundredths.

**25.**

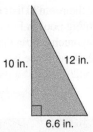

10 in.    12 in.

6.6 in.

**Figure 12–60**

**26.**

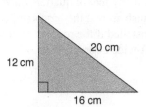

20 cm

12 cm

16 cm

**Figure 12–61**

**27. CON** Selena Henson is planning a patio that will adjoin the sides of her L-shaped home. One side of the patio is 24 ft and the other is 18 ft. The shape of the patio is triangular. Draw a representation of the patio and find the perimeter if the length of the third side is 30 ft.

**28.** Find the perimeter of a triangle that has sides measuring 2 ft 6 in., 1 yd 8 in., and 4 ft 6 in.

**29.** Find the area of the shape in Figure 12–62.

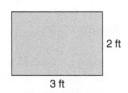

**Figure 12–62**

**30.** Find the area of the parallelogram in Figure 12–63.

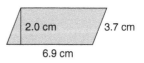

**Figure 12–63**

**31.** Find the area of the shape in Figure 12–64.

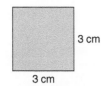

**Figure 12–64**

**32.** Find the area of the shape in Figure 12–65.

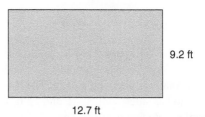

**Figure 12–65**

**33.** A rectangular parking lot is 340 ft by 125 ft. Find the number of square feet in the parking lot.

**35.** Find the area of Figure 12–66.

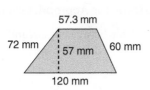

**Figure 12–66**

**34.** A room is 15 ft by 12 ft. How many square feet of flooring are needed for the room?

**36.** Find the area of the square in Figure 12–67.

**Figure 12–67**

**37.** Madison Duke is wallpapering a laundry room 8 ft by 8 ft by 8 ft high. How many square feet of paper will she need if there are 63 ft$^2$ of openings in the room?

**38.** Making no allowances for bases, the pitcher's mound, or the home plate area, how many square yards of artificial turf are needed to resurface an infield at an indoor baseball stadium? The infield is 90 ft on each side (9 ft$^2$ = 1 yd$^2$).

**39.** Ted Davis is a farmer who wants to apply fertilizer to a 40-acre field with dimensions $\frac{1}{4}$ mi × $\frac{1}{4}$ mi. Find the area in square miles.

**40.** **CON**  A 36-in. × 36-in. ceramic tile shower stall is being installed. How many 4-in. × 4-in. tiles are needed to cover the floor? Disregard the drain opening and grout spaces.

**41.** **CON**  Tiles that are 6 in. × 6 in. cover the floor of a shower. How many tiles are needed for the floor if the shower measures 4.5 ft by 6 ft?

**42.** **INDTR**  A 20-in. × 20-in. central heating and air conditioning return air vent is being installed in a wall. Find the area of the wall opening.

**43.** Find the area of Figure 12–68. Round to hundredths.

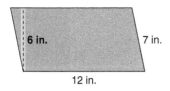

**Figure 12–68**

**44.** Find the area of Figure 12–69. Round to hundredths.

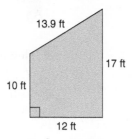

**Figure 12–69**

**45.** The six glass panes in a kitchen light fixture each measure $4\frac{1}{2}$ in. along the top and 10 in. along the bottom. The top and bottom are parallel. The height of each pane is 8 in. What is the combined area of the six trapezoidal panes?

**46. CON** A section of a hip roof is a trapezoid measuring 38 ft at the bottom, 14 ft at the top, and 10 ft high. Find the area of this section of the roof in square feet.

**47.** Find the area of the triangle in Figure 12–70.

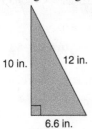

10 in.    12 in.

6.6 in.

**Figure 12–70**

**48.** Find the area of the triangle in Figure 12–71.

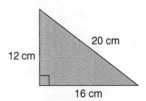

20 cm

12 cm

16 cm

**Figure 12–71**

**49.** Carlee McAnally is planning a patio. The triangular-shaped patio has sides 24 ft, 18 ft, and 22 ft. Find the number of square feet of surface area to be covered with concrete. Round to the nearest square foot.

**50. CON** If aluminum siding costs $6.75 a square yard installed, how much does it cost to put the siding on the two triangular gable ends of a roof under construction? Each gable has a span (base) of 30 ft 6 in. and a rise (height) of 7 ft 6 in. Any portion of a square yard is rounded to the next highest square yard for each gable.

---

## 12-3 | Circles and Radians

*Learning Outcomes*

**1** Find the circumference or area of a circle using the appropriate formula.

**2** Convert angle measures between degrees and radians.

**3** Find the arc length of a sector.

**4** Find the area of a sector or segment.

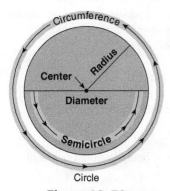

**Figure 12–72**

A **circle** is a closed curved line with points that lie in a plane and are the same distance from the *center* of the figure (Figure 12–72).

The **center** of a circle is the point that is the same distance from every point on the circle.

The **radius** (plural: *radii*, pronounced "ray · dē · ī") is a straight line segment from the center of a circle to a point on the circle. It is half the diameter.

The **diameter** of a circle is a straight line segment from a point on the circle through the center to another point on the circle.

The **circumference** of a circle is the perimeter or length of the closed curved line that forms the circle.

A **semicircle** is half a circle and is created by drawing a diameter.

**1** **Find the Circumference or Area of a Circle Using the Appropriate Formula.**

The circle is a geometric form with a special relationship between its circumference and its diameter. If we divide the circumference of any circle by its diameter, the quotient is always the same number.

$$\pi = \frac{\text{circumference}}{\text{diameter}} = \frac{C}{d}$$

This number is a nonrepeating, nonterminating decimal approximately equal to 3.1415927 to seven decimal places. The Greek letter $\pi$ (pronounced "pie") represents this value. Convenient approximations often used in calculations involving $\pi$ are $3\frac{1}{7}$ and 3.14. Many calculators have a $\pi$ key.

The formulas for the circumference and area of a circle are:

| Circumference (C) | Area (A) |
| --- | --- |

$C = \pi d$
$C = 2\pi r$
$A = \pi r^2$

$d$ is diameter ($d = 2r$)
$r$ is radius ($r = \frac{1}{2}d$)

### To find the circumference or area of a circle:

1. Select the appropriate formula.
2. Substitute values for $r$ or $d$ as appropriate.
3. Evaluate the formula. Use the calculator value for $\pi$. 3.14 can be used as a less-accurate approximation.

**EXAMPLE** Find the circumference (to the nearest tenth of a meter) of a circle that has a diameter of 1.3 m (Figure 12–73).

1.3 m

$C = \pi d$      Select circumference formula with diameter $d$.
$C = \pi(1.3)$      Use $\pi$ key on calculator.
$C = 4.08407045$ m      Evaluate.

**Figure 12–73**      **The circumference is 4.1 m (rounded).**      Circumference is a linear measure.

TIP!

#### Calculator Values of $\pi$

Calculations involving $\pi$ are always approximations. Scientific or graphing calculators include a $\pi$ function or menu option where $\pi$ to seven or more decimal places is computed by pressing a single key. Other calculators require the use of more than one key to activate the $\boxed{\pi}$ key. However, for computations by hand or a calculator without the $\boxed{\pi}$ function, 3.14 is sometimes adequate. **We use the calculator value 3.141592654 for $\pi$ in all examples and exercises unless stated otherwise.**

TIP!

#### Fractions versus Decimals

When you use your calculator, you will find it easier if you convert mixed U.S. customary linear measurements to their decimal equivalents. For instance, if a diameter is 7 ft 6 in., convert it to 7.5 ft (from $7\frac{6}{12}$ ft, in which $\frac{6}{12} = 0.5$).

**EXAMPLE**  Find to the nearest hundredth the circumference of a circle with a radius of 1 ft 9 in. (Figure 12–74).

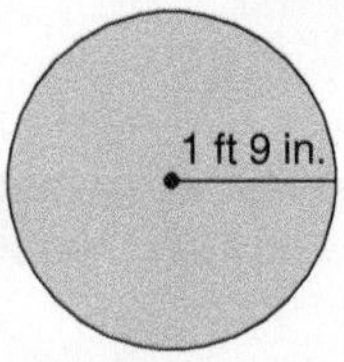

$$C = 2\pi r$$

Substitute for $\pi$ and $r$: $1\frac{9}{12}$ ft = 1.75 ft

$$C = 2\pi(1.75)$$

$$C = 10.99557429$$

1 ft 9 in.

**Figure 12–74**

$$C = \textbf{11.00 ft (rounded)}$$

Circumference is a linear measure.

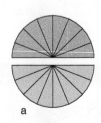

a

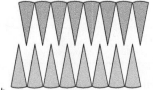

b

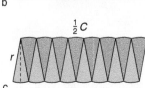

c

**Figure 12–75**

The area of a circle, like the circumference, is obtained from relationships within the circle. If we divide a circle into two semicircles and then subdivide each semicircle into pie-shaped pieces, we get something like Figure 12–75a. If we then spread the upper and lower pie-shaped pieces, we get Figure 12–75b. Now if we push the upper and lower pieces together, the result approximates the rectangle in Figure 12–75c, whose length is $\frac{1}{2}$ the circumference and whose width is the radius. Thus, the area of the circle is approximately the area of a rectangle, that is, length times width. Since the length of the rectangle is one-half the circumference and the width is the radius, the area of a circle equals one-half the circumference times the radius.

$$A = \frac{1}{2} C \times r$$

The formula for circumference is $C = 2\pi r$, so we substitute $2\pi r$ for $C$.

$$A = \frac{1}{2}(2\pi r)(r)$$

Multiply. Reduce where possible.

$$A = \frac{1}{\overset{1}{2}}(\overset{1}{2}\pi r)(r)$$

Area is a square measure: $r \cdot r = r^2$

$$A = \pi r^2$$

---

**EXAMPLE**  Find the area of a circle whose radius is 8.5 m (Figure 12–76). Round to tenths.

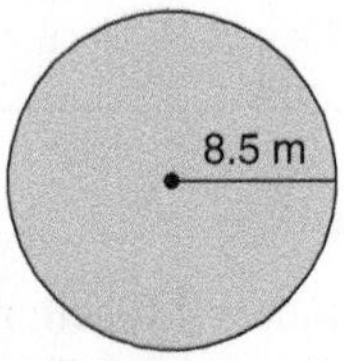

8.5 m

**Figure 12–76**

$$A = \pi r^2$$  Select appropriate formula and substitute for $r$.
$$A = \pi(8.5)^2$$  Square the radius.
$$A = \pi(72.25)$$  Multiply by $\pi$. Use the $\pi$ key on your calculator.
$$A = 226.9800692 \text{ m}^2$$  Area is a square measure.

**The area of the circle is 227.0 m².**

---

**EXAMPLE**  Find the area to the nearest hundredth of the top of a circular tank with a diameter of 12 ft 8 in.

$$A = \pi r^2$$

$$A = (\pi)\left(\frac{12 + \frac{8}{12}}{2}\right)^2$$

Follow the order of operations. 8 in. = $\frac{8}{12}$ ft.
The diameter, $12 + \frac{8}{12}$, divided by 2 is the radius.

$$A = 126.012772 \text{ ft}^2$$  Calculator result

$$A = 126.01 \text{ ft}^2 \text{ (rounded)}$$  Area is a square measure.

**The area of the top of the tank is 126.01 ft² .**

Chapter 12 / Geometry

Shapes called **composite** figures are figures made up of two or more geometric figures.

1. Determine how the missing dimension is related to known dimensions.
2. Make a calculation of known dimensions according to the relationship found in Step 1.

**EXAMPLE** Find the missing dimensions $x$ and $y$ on the slab foundation (Figure 12–77).

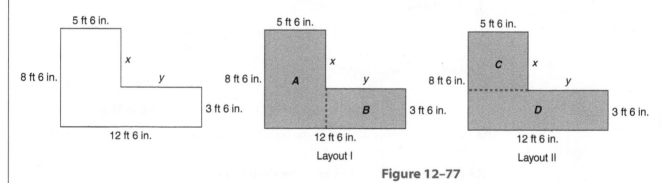

**Figure 12–77**

Separate the figure into parts.

In layout I (Figure 12–77) the side of $B$ opposite its 3′6″ side is also 3′6″ because opposite sides of a rectangle are equal. The side of $A$ opposite its 8′6″ side is, for the same reason, 8′6″. Dimension $x$ must therefore be the difference between 8′6″ and 3′6″.

$$x = 8'6'' - 3'6''$$

$$x = 5'$$

If we think of layout II as two horizontal rectangles, we can find dimension $y$. The side opposite the 5′6″ side of $C$ must be 5′6″. The side of $D$ opposite the 12′6″ side must also be 12′6″. Dimension $y$ must therefore be the difference between 12′6″ and 5′6″.

$$y = 12'6'' - 5'6''$$

$$y = 7'$$

**The missing dimensions are $x = 5'$ and $y = 7'$.**

*Perimeter* is the sum of the lengths of the sides of a figure. The number and the length of the sides vary from one composite figure to the next, so no specific formula covers the entire variety of composite shapes that exist. However, we can use a general formula.

$$P = a + b + c + \cdots$$

where $a, b, c, \ldots$ are lengths of all the sides.

12–3 Circles and Radians

493

EXAMPLE CON Find the number of feet of 4-in. stock needed for the base plates of a room that has the layout shown in Figure 12–78. Make no allowances for openings when calculating the linear footage of the base plates.

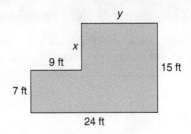

**Figure 12–78**

1. Find the missing dimensions.

$$x = 15 \text{ ft} - 7 \text{ ft} \qquad y = 24 \text{ ft} - 9 \text{ ft}$$
$$x = 8 \text{ ft} \qquad\qquad y = 15 \text{ ft}$$

2. Apply the general formula for the perimeter of a polygon.

$$P = a + b + c + \cdots$$
$$P = 7 \text{ ft} + 24 \text{ ft} + 15 \text{ ft} + 15 \text{ ft} + 8 \text{ ft} + 9 \text{ ft} = 78 \text{ ft}$$

3. Count the number of sides on the layout to make sure that each is substituted into the formula.

**The room needs 78 ft of 4-in. stock for the base plates.**

As with the perimeter, the *area* of a composite figure is found by finding the sum of the areas of the parts of the figure.

**Area of a composite figure:**

$$A = A_1 + A_2 + A_3 + \cdots$$

where $A_1, A_2, A_3, \ldots$ are the areas of all the parts of the composite figure.

EXAMPLE Find the number of square yards of carpeting required for the room in the preceding example ($9 \text{ ft}^2 = 1 \text{ yd}^2$).

1. Divide the composite shape into two polygons with areas we can compute (Figure 12–79). In this case, $A$ is a rectangle and $B$ is a square.

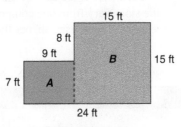

**Figure 12–79**

**2.** Find the areas of the smaller polygons and add them.

| Rectangle *A* | Square *B* |
|---|---|
| $A_1 = lw$ | $A_2 = s^2$ |
| $A_1 = 9 \times 7$ | $A_2 = 15^2$ |
| $A_1 = 63 \text{ ft}^2$ | $A_2 = 225 \text{ ft}^2$ |

$$A_1 + A_2 = \text{total area}$$
$$63 + 225 = 288 \text{ ft}^2$$

**3.** Convert square feet to square yards using a unity ratio.

$$\frac{\overset{32}{\cancel{288}} \text{ ft}^2}{1} \times \frac{1 \text{ yd}^2}{\underset{1}{\cancel{9}} \text{ ft}^2} = 32 \text{ yd}^2$$

**The room requires 32 yd² of carpeting.**

---

**EXAMPLE** A 15-in.-diameter wheel has a 3-in. hole in the center. Find the area of a side of the wheel to the nearest tenth (Figure 12–80).

We are asked to find the area of the colored portion of the wheel in Figure 12–80. To do so, we find $A_{\text{outside}}$, the area of the larger circle (diameter 15 in.) and *subtract* the area of $A_{\text{inside}}$, the smaller circle (diameter 3 in.). The colored portion is called a *ring*.

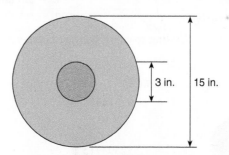

**Figure 12–80**

$$A_{\text{outside}} = \pi r^2 \quad \left( r = \frac{15}{2} = 7.5 \right) \qquad A_{\text{inside}} = \pi r^2 \quad \left( r = \frac{3}{2} = 1.5 \right)$$

$A_{\text{outside}} = \pi(7.5)^2$          $A_{\text{inside}} = \pi(1.5)^2$

$A_{\text{outside}} = \pi(56.25)$         $A_{\text{inside}} = \pi(2.25)$

$A_{\text{outside}} = 176.7145868 \text{ in}^2$     $A_{\text{inside}} = 7.068583471 \text{ in}^2$

area of wheel (ring) $= A_{\text{outside}} - A_{\text{inside}}$

$A_{\text{wheel}} = 176.7145868 - 7.068583471 = 169.6460033$     or     $169.6 \text{ in}^2$

**The area of the wheel (ring) is 169.6 in².**

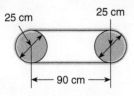

25 cm       25 cm

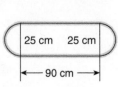

25 cm     25 cm

← 90 cm →

**Figure 12–81**

**EXAMPLE**
**INDTR**

A bandsaw has two 25-cm wheels spaced 90 cm between centers (Figure 12–81). Find the length of the saw blade.

This layout is a composite figure which consists of a semicircle at each end and a rectangle in the middle. It is called a *semicircular-sided* figure. The two semicircles equal one whole circle, so we need to find the circumference of one circle (wheel) and add it to the lengths of the two sides of the rectangle.

| | |
|---|---|
| $C = \pi d$ | total length of blade $= C + 2l$ |
| $C = \pi(25)$ | total length of blade $= 78.53981634 + 2(90)$ |
| $C = 78.53981634$ cm | total length of blade $= 258.5$ cm (rounded) |

**The bandsaw blade is 258.5 cm in length.**

Specific applications for a particular industry or career often use the formulas for area, perimeter, or circumference.

**EXAMPLE**
**INDTEC**

Find the cutting speed of a lathe if a piece of work that has a 7-in. diameter turns on a lathe at 75 revolutions per minute (rpm) (Figure 12–82).

The *cutting speed* is the speed of a tool that passes over the work, such as the speed of a lathe or a sander as it sands (passes over) a piece of wood. If the cutting speed is too fast or too slow, safety and quality are impaired. The formula for cutting speed is CS = C (in feet) × rpm.

CS = cutting speed
C = circumference or one revolution (in feet)
rpm = revolutions per minute

Cutting speed is measured in *feet per minute* (ft/min)

$C = \pi d$

$C = \pi(7)$

$C = 21.99114858$ in.          Convert 21.99114858 in. into feet using a unity ratio.

$C = 1.832595715$ ft      $\dfrac{21.99114858 \text{ in.}}{1} \times \dfrac{1 \text{ ft}}{12 \text{ in.}} = 1.832595715 \text{ ft}$

$\text{CS} = C \times \text{rpm} = 1.832595715 \times 75 = 137$ ft/min    Round.

**Figure 12–82**

**The cutting speed of the lathe is approximately 137 ft/min.**

### 2 Convert Angle Measures Between Degrees and Radians.

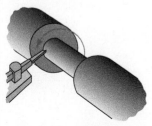

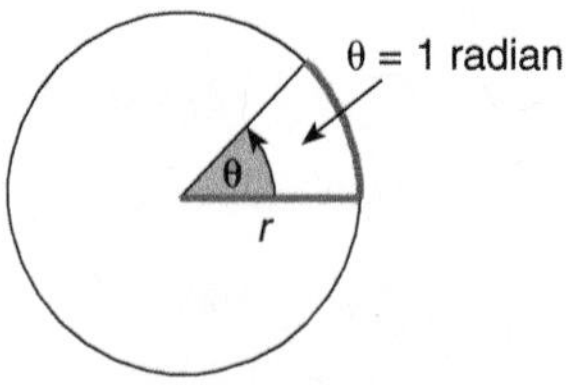

θ = 1 radian

θ
r

**Figure 12–83**

In Section 12–1, we learned that angles can be measured in units called *degrees*. Angles can also be measured in *radians*. A **radian** is the measure of a central angle of a circle whose intercepted arc is equal in length to the radius of the circle (Figure 12–83). The abbreviation for radian is **rad.**

The circumference of a circle is related to the radius by the formula $C = 2\pi r$. Thus, the ratio of the circumference to the radius of any circle is $\frac{C}{r} = 2\pi$; that is, the radius could be measured off $2\pi$ times (about 6.28 times) along the circumference. A complete rotation is $2\pi$ radians (see Figure 12–84).

| 1 radian | 3 radians | 3.14 or π radians | 5 radians | 6.28 or 2π radians |

**Figure 12–84**

How are radians and degrees related? A central angle measuring 1 radian makes an arc length equal to the radius.

**TIP!**

### Degree and Radian Notation

Angles measured in degrees always require the word *degree* or the degree symbol to be written. No comparable symbol exists for the radian. The abbreviation *rad* or no unit at all indicates that radian is the unit. Some calculators use a raised *r* to indicate a radian unit.

A complete rotation is 360°, or 2π rad. To convert from one unit of angle measure to another, we multiply by a *unity ratio* that relates degrees and radians. Because 360° = 2π rad, we can simplify the relationship to $180° = \pi$ rad $\left(\dfrac{360°}{2} = \dfrac{2\pi}{2} \text{ rad}\right)$.

### To convert degrees to radians:

1. Multiply degrees by the unity ratio $\dfrac{\pi \text{ rad}}{180°}$.
2. Write the product in simplest form.

Using a calculator (steps may vary):

1. Set angle MODE to radian.
2. Enter the degree measure followed by the degree symbol. Use $\boxed{° \ ' \ ''}$, $\boxed{\text{DMS}}$, or the angle menu, option 1.
3. Enter the minute measure followed by the minute symbol. Use $\boxed{° \ ' \ ''}$, $\boxed{\text{DMS}}$, or the angle menu, option 1.
4. Enter the second measure followed by the second symbol. Use $\boxed{° \ ' \ ''}$, $\boxed{\text{DMS}}$, or $\boxed{\text{ALPHA}}$ $\boxed{''}$ .
5. Press $\boxed{\text{ENTER}}$.

**EXAMPLE** Convert the angle measures to radians. Use the calculator value for π to change to a decimal equivalent rounded to the nearest hundredth.

**(a)** 20°  **(b)** 175°

**(a)** $20° \times \dfrac{\pi \text{ rad}}{180°} = \dfrac{20(\pi)}{180} = \textbf{0.35 rad}$    Multiply by $\dfrac{\pi \text{ rad}}{180°}$.

**(b)** $175° = 3.054326191 = \textbf{3.05 rad}$    Set calculator in radian mode.
175 $\boxed{2^{nd}}$ $\boxed{\text{ANGLE}}$ $\boxed{1}$ $\boxed{\text{ENTER}}$

When converting from radians to degrees, we multiply by a unity ratio so that the radians cancel and are replaced by degrees.

## To convert radians to degrees:

1. Multiply radians by the unity ratio $\dfrac{180°}{\pi \text{ rad}}$.

2. Write the product in simplest form.

Using a calculator (steps may vary):

1. Set angle MODE to degree.
2. Enter radian amount and radian symbol (angle menu, option 3).
3. If degrees, minutes, and seconds are desired, use DMS function (angle menu, option 4).

---

**EXAMPLE**  Convert to degrees. Round to the nearest ten-thousandth of a degree.

(a) 2 rad    (b) $\dfrac{\pi}{2}$ rad

(a) $2 \text{ rad} \times \dfrac{180°}{\pi \text{ rad}} = \dfrac{360°}{\pi} = \mathbf{114.5916°}$   Multiply by $\dfrac{180°}{\pi \text{ rad}}$. Use calculator value

of $\pi$ and round.

(b) $\dfrac{\pi}{2} \text{ rad} = 90°$

In degree mode:

$\boxed{(}\ \boxed{\pi}\ \boxed{\div}\ \boxed{2}\ \boxed{)}$ [angle menu, option 3] $\boxed{\text{ENTER}}$.

---

**EXAMPLE**  Convert to degrees, minutes, and seconds. Round to the nearest second.

(a) 1 rad    (b) 3.2 rad

(a) $1 \text{ rad} = 57°17'\,44.806''$    In degree mode:
Press 1 [angle menu, option 3] [angle menu, option 4]
$\boxed{\text{ENTER}}$.

$1 \text{ rad} = \mathbf{57°17'45''}$    Round.

(b) $3.2 \text{ rad} = 183°20'47.38''$    In degree mode:
Press 3.2 [angle menu, option 3] [angle menu, option 4]
$\boxed{\text{ENTER}}$.

$3.2 \text{ rad} = \mathbf{183°20'47''}$    Round.

**Thus, 3.2 rad $= \mathbf{183°20'47''}$**

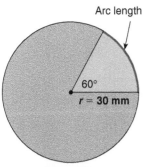

Arc length

60°

$r = 30$ mm

**Figure 12–85**

### 3 Find the Arc Length of a Sector.

We often work with figures that are less than a whole circle. For example, earlier we worked with the semicircle in composite figures. A **sector** of a circle is the portion of the area of a circle cut off by two radii. These two radii form a **central angle**. The sector is formed by a central angle of the circle and the arc connecting the sides of the angle (radii). See Figure 12–85.

The **arc length** of a sector is the portion of the circumference intercepted by the sides of the sector (see Figure 12–85).

---

#### To find the arc length of a sector:

Using degrees:

1. Substitute known values into the formula.

$$s = \frac{\theta}{360}(2\pi r) \quad \text{or} \quad s = \frac{\theta}{360}(\pi d)$$

where $\theta$ is a central angle measured in degrees and $r$ is the radius of the circle.
2. Evaluate.

Using radians:

1. Substitute the given radian measure of the central angle and the radius in the formula

$$s = \theta r$$

where $\theta$ is the central angle measured in radians and $r$ is the radius of the circle.
2. Simplify the expression.

---

**EXAMPLE** Find the arc length of the sector formed by a 60° central angle if the radius of the circle is 30 mm (Figure 12–85).

$s = \dfrac{\theta}{360}(2\pi r)$      Select the formula for degrees. Substitute values.

$s = \dfrac{60}{360}(2)(\pi)(30)$      Evaluate.

$s = 31.41592654$

$s = \mathbf{31.42}$ **mm**      Round.

---

12-3 Circles and Radians

**EXAMPLE** Find the arc length intercepted on the circumference of the circle in Figure 12–86 by a central angle of $\frac{\pi}{3}$ radians (rad) if the radius of the circle is 10 cm.

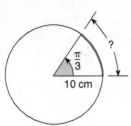

$s = \theta r$ — Select the formula for radians. Substitute $\frac{\pi}{3}$ for $\theta$ and 10 cm for $r$.

$s = \dfrac{\pi}{3}(10\ \text{cm})$ — Evaluate.

$s = \dfrac{\pi(10\ \text{cm})}{3}$ — Use calculator value for $\pi$.

$s = 10.47\ \text{cm}$ — To the nearest hundredth.

**Figure 12–86**

**Thus, the arc length of the intercepted arc is 10.47 cm.**

**TIP!**

### Analyzing Arc Length Dimensions

In the preceding example, $\left(\dfrac{\pi}{3}\ \text{rad}\right)(10\ \text{cm}) = 10.47\ \text{cm}$, what happened to the radians? In the definition of a radian, we relate the measure of an arc connecting the endpoints of a central angle to the measure of the radius of the circle. Therefore, the arc length will have the same measuring unit as the radius.

### To find the central angle or radius of a sector given the arc length:

1. Substitute given values in the formula $s = \theta r$, where $s$ is the arc length, $\theta$ is the central angle measured in radians, and $r$ is the radius of the circle.
2. Solve for the missing value.

**EXAMPLE** Find the radian measure of an angle at the center of a circle of radius 5 m. The angle intercepts an arc length of 12.5 m (see Figure 12–87).

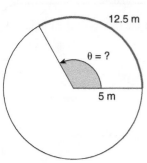

$s = \theta r$ — Formula for arc length. Solve for $\theta$.

$\theta = \dfrac{s}{r}$ — Substitute 12.5 m for $s$ and 5 m for $r$. Arc length and radius measuring units are compatible.

$\theta = \dfrac{12.5\ \text{m}}{5\ \text{m}}$

$\theta = 2.5\ \text{rad}$

**Figure 12–87**   **The angle is 2.5 rad.**

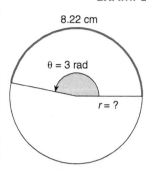

8.22 cm

θ = 3 rad

r = ?

**Figure 12–88**

**EXAMPLE** Find the radius of an arc if the length of the arc is 8.22 cm and the intercepted central angle is 3 rad (see Figure 12–88).

$s = \theta r$      Formula for arc length. Solve for *r*.

$r = \dfrac{s}{\theta}$      Substitute 8.22 cm for *s* and 3 rad for θ.

$r = \dfrac{8.22 \text{ cm}}{3}$      Simplify.

$r = 2.74 \text{ cm}$      Same measuring unit as arc length.

**The radius of the arc is 2.74 cm.**

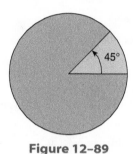

45°

**Figure 12–89**

### 4 Find the Area of a Sector or Segment.

We use the Greek letter *theta* (θ) to represent the measure of the angle of the sector.

**To find the area of a sector:**

Using degrees:

1. Calculate the portion of the circle included in the sector. θ is a central angle measured in degrees.

$$\frac{\theta}{360} = \text{fractional part of circle}$$

2. Find the fractional part of the area of the circle.

$$A = \frac{\theta}{360} \pi r^2 \text{ where } \pi r^2 = \text{area of circle}$$

Using radians:

1. Substitute known values into the formula $A = \dfrac{1}{2}\theta r^2$, where θ is the central angle measured in radians and *r* is the radius of the circle.

2. Solve for the missing value.

**EXAMPLE** Find the area of a sector with a central angle of 45° in a circle with a radius of 10 in. Round to hundredths.

$$A = \frac{\theta}{360}\pi r^2$$  Substitute for θ and r.

$$A = \frac{45}{360}(\pi)(10)^2$$  Perform the indicated operations.

$$A = 0.125(\pi)(100)$$

$$A = 39.26990817 \text{ in}^2$$  Round.

**The area of the sector is 39.27 in².**

---

**EXAMPLE**
**INDTR** A cone is made from sheet metal. To form a cone, a sector with a central angle of 40°20′ is cut from a metal circle whose diameter is 20 in. Find the area of the stretchout (portion of the circle) that is used to form the cone to the nearest hundredth (Figure 12–90).

area used for cone = area of circle − area of sector

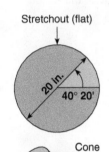

Stretchout (flat)

20 in.
40° 20′

Cone

**Figure 12–90**

| **Circle** | **Sector** | |
|---|---|---|
| $A_1 = \pi r^2$ | $A_2 = \frac{\theta}{360}\pi r^2$ | Substitute known values. Convert 40°20′ to 40.333333°. |
| $A_1 = \pi(10)^2$ | $A_2 = \frac{40.3\overline{3}}{360}(314.1592654)$ | Substitute $A_1$ for $\pi r^2$. |
| $A_1 = \pi(100)$ | $A_2 = 0.112037037(314.1592654)$ | |
| $A_1 = 314.1592654 \text{ in}^2$ | $A_2 = 35.19747324 \text{ in}^2$ | |

area used for cone $= A_1 - A_2$

$$A_3 = 314.1592654 - 35.19747324$$

$$A_3 = 278.96 \text{ in}^2$$  Round.

**The area of the metal sector used to form the cone is 278.96 in².**

---

**EXAMPLE** Find the area of a sector that has a central angle of 5 rad and has a radius of 7.2 in.

$$A = \frac{1}{2}\theta r^2$$  Substitute θ = 5 and r = 7.2.

$$A = \frac{1}{2}(5)(7.2 \text{ in.})^2$$  Simplify.

$$A = 129.6 \text{ in}^2$$  Area of sector.

**The area of the sector is 129.6 in².**

If a line segment (called a *chord*) joins the endpoints of the radii that form a sector, the sector is divided into two figures, a triangle and a *segment* (Figure 12–91). A **chord** is a line segment joining two points on the circumference of a circle. The portion of the circumference cut

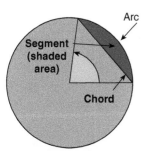

**Figure 12–91**

off by a chord is an **arc.** A **segment** is the portion of the area of a circle bounded by a chord and an arc.

Because the chord divides the sector into a triangle and a segment, we can calculate the area of the segment by subtracting the area of the triangle from the area of the sector.

## To find the area of a segment:

1. Substitute known values into the formula.

Using degrees:

$$A = \frac{\theta}{360}\pi r^2 - \frac{1}{2}bh$$

Using radians:

$$A = \frac{1}{2}\theta r^2 - \frac{1}{2}bh$$

where $\frac{\theta}{360}\pi r^2$ or $\frac{1}{2}\theta r^2$ is the area of the sector and $\frac{1}{2}bh$ is the area of the triangle.

2. Evaluate.

---

**EXAMPLE**   Find the area to the nearest hundredth of the segment in Figure 12–92.

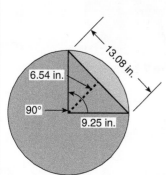

**Figure 12–92**

$$A = \frac{\theta}{360}\pi r^2 - \frac{1}{2}bh \qquad \text{Substitute in formula.}$$

$$A = \frac{90}{360}(\pi)(9.25)^2 - \frac{1}{2}(13.08)(6.54) \qquad \text{Evaluate.}$$

$$A = 67.20063036 - 42.7716$$

$$A = 24.43 \text{ in}^2 \qquad \text{Round.}$$

**The area of the segment is 24.43 in².**

---

**EXAMPLE**
**INDTR**
A segment in Figure 12–93 is removed so that a template for a cam is made from the rest of the circle. What is the area of the template? Give the answer to the nearest hundredth.

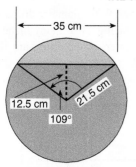

**Figure 12–93**

area of template = area of circle − area of segment

**Circle**

$$A_1 = \pi r^2$$

$$A_1 = \pi(21.5)^2$$

$$A_1 = 1,452.201204 \text{ cm}^2$$

**Segment**

$$A_2 = \frac{\theta}{360}\pi r^2 - \frac{1}{2}bh$$

$$A_2 = \frac{109}{360}(1,452.201204) - \frac{1}{2}(35)(12.5)$$

$$A_2 = 220.9442535 \text{ cm}^2$$

12-3 Circles and Radians

$$\text{area}_3 \text{ (template)} = A_1 - A_2$$

$$A_3 = 1{,}452.201204 - 220.9442535$$

$$A_3 = 1{,}231.26 \text{ cm}^2 \qquad \text{Round.}$$

**The area of the template is 1,231.26 cm$^2$.**

SECTION 12–3 SELF-STUDY EXERCISES

**1** Find the circumference of circles with the following dimensions. Round to tenths.

**1.** Diameter = 8 cm

**2.** Diameter = 15 m

**3.** Radius = 3 in.

**4.** Radius = 1.5 ft

**5.** Radius = $8\frac{1}{2}$ ft

**6.** Diameter = 5.5 m

Find the area of circles with the following dimensions. Round to tenths.

**7.** Diameter = 8 cm

**8.** Diameter = 15 m

**9.** Radius = 3 in.

**10.** Radius = 16 yd

**11.** Radius = 1.5 ft

**12.** Radius = $8\frac{1}{2}$ ft

**13.** Diameter = 5.5 m

**14.** Diameter = $5\frac{1}{4}$ in.

Find the color area of Figures 12–94 through 12–97 to the nearest tenth.

**15.**

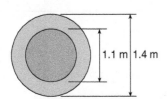

1.1 m  1.4 m

**Figure 12–94**

**16.**

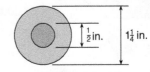

2 ft

$1\frac{1}{2}$ ft

**Figure 12–95**

**17.**

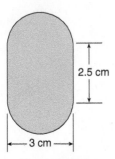

2.5 cm

3 cm

**Figure 12–96**

**18.**

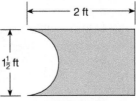

$\frac{1}{2}$ in.   $1\frac{1}{4}$ in.

**Figure 12–97**

**19.** A swimming pool is in the form of a semicircular-sided figure. Its width is 20 ft and the parallel portions of the sides are each 20 ft (Figure 12–98). What is the area of a 5-ft-wide walk surrounding the pool? Round to tenths.

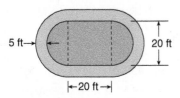

5 ft→    20 ft

|←20 ft→|

**Figure 12–98**

**20.** INDTR   A belt connecting two 9-in.-diameter drums on a conveyor system needs replacing. How many inches must the new belt be if the centers of the drums are 10 ft apart (Figure 12–99)? Round to tenths.

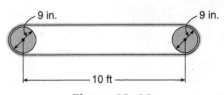

9 in.                9 in.

10 ft

**Figure 12–99**

21. **INDTR** A 2-in.-inside-diameter pipe and a 4-in.-inside-diameter pipe empty into a third pipe whose inside diameter is 5 in. (Figure 12–100). Is the third pipe large enough for the combined flow? (Justify your answer.)

22. **INDTR** A large pipe whose interior cross-sectional area is 20 in² empties into two smaller pipes that each have an interior diameter of 4 in. (Figure 12–101). Are the smaller pipes together large enough to carry off the flow from the larger pipe? (Justify your answer.)

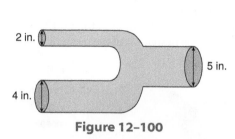

**Figure 12–100**

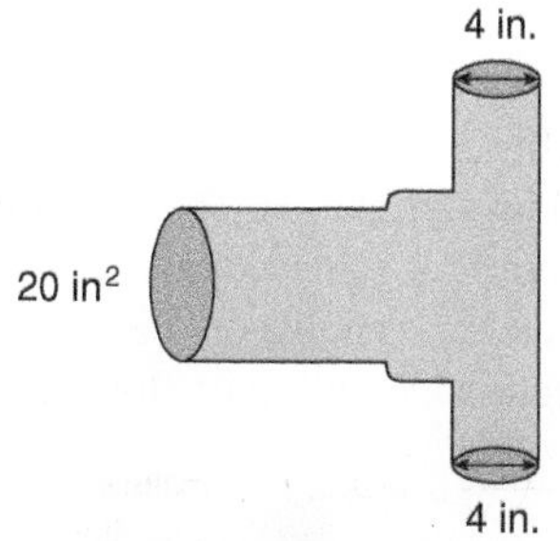

4 in.

20 in²

4 in.

**Figure 12–101**

23. **INDTR** Cutting speed, when applied to a grinding wheel, is called *surface speed.* What is the surface speed in ft/min of a 9-in.-diameter grinding wheel revolving at 1,200 rpm? (Surface speed = circumference in feet × rpm.)

24. **INDTR** A 12-in.-diameter polishing wheel revolves at 500 rpm. What is the surface speed? (See Exercise 23 for formula.)

**2** Convert the measures to radians rounded to the nearest hundredth.

25. 45°

26. 56°

27. 140°

Convert the measures to degrees rounded to the nearest ten-thousandth. Then convert to radians to the nearest hundredth.

28. 21°45′

29. 177°33′

30. 44°54′12″

Convert the measures to degrees. Round to the nearest ten-thousandth of a degree.

31. $\frac{\pi}{4}$ rad

32. $\frac{\pi}{6}$ rad

33. 2.5 rad

Convert the measures to degrees, minutes, and seconds. Round to the nearest second.

34. 0.5 rad

35. $\frac{\pi}{8}$ rad

36. 0.75 rad

**3** Find the arc length of the sectors of a circle. Round to hundredths.

37. ∠ = 54°
    r = 30 mm

38. ∠ = 150°
    r = 5 in.

39. ∠ = 120°30′
    r = 1.52 ft

40. ∠ = 25°16′
    r = 16 cm

Solve the following problems. Round answers to hundredths if necessary.

41. Find the arc length intercepted on the circumference of a circle by a central angle of 2.15 rad if the radius of the circle is 3 in.

42. Find the arc length intercepted on the circumference of a circle by a central angle of 4 rad if the radius of the circle is 3.5 cm.

43. Use radians to find an angle at the center of a circle of radius 2 in. if the angle intercepts an arc length of 8.5 in.

44. Use radians to find an angle at the center of a circle of radius 4.3 cm if the angle intercepts an arc length of 15 cm.

45. Find the radius of an arc if the length of the arc is 14.7 cm and the intercepted central angle is 2.1 rad.

46. Find the radius of an arc if the length of the arc is 12.375 in. and the intercepting central angle is 2.75 rad.

12-3 Circles and Radians

**4** Find the area of the sector of a circle using Figure 12–102. Round to hundredths.

**47.** $\angle = 54°$
$r = 16$ cm

**48.** $\angle = 25°16'$
$r = 30$ mm

**49.** $\angle = 120°30'$
$r = 1.52$ ft

**50.** $\angle = 65°$
$r = 5$ in.

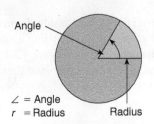

$\angle =$ Angle
$r =$ Radius

**Figure 12–102**

**51.** **CON** The library of a contemporary elementary school is circular (Figure 12–103). The floor plan includes sectors reserved for science materials, literary materials, reference materials, and so on. Find the area of the reference section excluding its storage area.

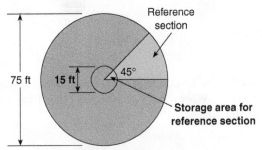
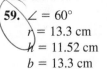

**Figure 12–103**

**52.** **CON** A mason lays a tile mosaic featuring a four-sector design (color portion of Figure 12–104). What is the area of the design to the nearest hundredth?

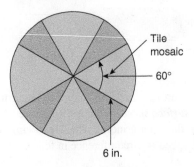

**Figure 12–104**

**53.** Find the area of a sector whose central angle is 2.14 rad and whose radius is 4 in.

**54.** Find the area of a sector that has a central angle of 6 rad and a radius of 1.2 cm.

**55.** Find the radius of a sector if the area of the sector is 7.5 cm² and the central angle is 3 rad.

**56.** How many radians does the central angle of a sector measure if its area is 1.7 in² and its radius is 2 in.?

**57.** Find the area of a sector whose central angle is 1.83 rad and whose radius is 7.2 cm. Round to hundredths.

**58.** Find the number of radians of a central angle of a sector whose area is 5.6 cm² and whose radius is 4 cm.

Find the area of the segments of a circle using Figure 12–105. Round to hundredths.

**59.** $\angle = 60°$
$r = 13.3$ cm
$h = 11.52$ cm
$b = 13.3$ cm

**60.** $\angle = 110°$
$r = 10$ in.
$h = 5.74$ in.
$b = 16.38$ in.

**61.** $\angle = 105°$
$r = 11.25$ in.
$h = 6.85$ in.
$b = 17.85$ in.

**62.** $\angle = 60°$
$r = 24$ cm
$h = 20.8$ cm
$b = 24$ cm

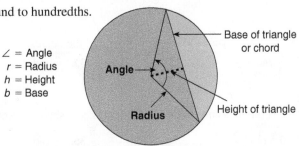

$\angle =$ Angle
$r =$ Radius
$h =$ Height
$b =$ Base

**Figure 12–105**

**63.** **INDTEC** A motor shaft has milled on it a flat for a setscrew to rest so that it can hold a pulley on the shaft (Figure 12–106). What is the cross-sectional area of the shaft after being milled? Round to hundredths.

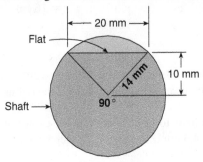

**Figure 12–106**

**64.** **CON** A contractor pours a concrete patio in the shape of a circle except where the patio touches the exterior wall of the house (Figure 12–107). What is the area of the patio in square feet? Round to hundredths.

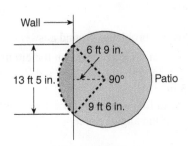

**Figure 12–107**

*Learning Outcome*

**1** Find the volume of three-dimensional objects.

**2** Find the surface area of three-dimensional objects.

Common household items like cardboard storage boxes and toy building blocks are examples of three-dimensional geometric figures classified generally as *prisms*. Cans and pipes are examples of *cylinders*.

**1 Find the Volume of Three-Dimensional Objects.**

A **prism** is a three-dimensional figure with polygonal bases (ends) that are parallel and faces (sides) that are parallelograms, rectangles, or squares (Figure 12–108). In a **right prism,** the faces are perpendicular to the bases.

A **right circular cylinder** is a three-dimensional figure with a curved surface and two circular bases such that the height is perpendicular to the bases.

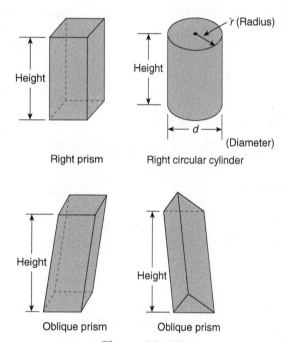

**Figure 12–108**

In an **oblique prism** or an **oblique cylinder** the faces are *not* perpendicular to the bases.

The **height** of a three-dimensional figure with two bases is the shortest distance between the two bases.

In right circular cylinders and in right prisms, the height is the same as the length of a side or face. However, in oblique prisms and cylinders the height is the perpendicular distance between the bases and is different from the length of a side or face.

The *volume* of an object, such as a container, is used to estimate how many containers can be loaded into a given-size storage area or shipped in a container of certain dimensions.

The **volume** of a three-dimensional geometric figure is the amount of space it occupies, measured in terms of three dimensions (length, width, and height).

If we have a rectangular box measuring 1 ft long, 1 ft wide, and 1 ft high (Figure 12–109), it will be a cube representing 1 cubic foot ($ft^3$). Its volume is calculated by multiplying length × width × height, or 1 ft × 1 ft × 1 ft = 1 $ft^3$. We indicate a cubic measure with an exponent 3 after the unit of measure, meaning that the measure is "*cubed.*"

From this concept of 1 $ft^3$ or 1 cubic foot comes the formula for the volume of a rectangular box as length × width × height or $V = lwh$. Note that $l \times w$ is the formula for the area of the

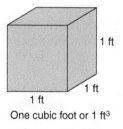

One cubic foot or 1 $ft^3$

**Figure 12–109**

rectangle (or square) that forms the base of the rectangular box. If the base of the prism is a triangle, trapezoid, or other polygon, we use the appropriate formula for the area of the base. A general formula can be used for the volume of *any* right prism or right cylinder.

**Formula for the volume of right prism or right cylinder:**

$$V = Bh$$

where $B$ is the area of the base and $h$ is the height of the prism or cylinder.

**EXAMPLE** Find the volume of the triangular prism in Figure 12–110 if the height is 15 cm and the bases are triangles 3 cm on a side and 2.6 cm in height.

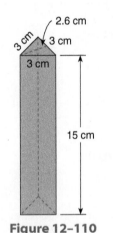

Figure 12–110

$V = Bh$      Substitute the formula for the area of the triangular base for *B*.

$V = \left(\dfrac{1}{2} bh_1\right)h_2$      $h_1 = 2.6$ cm (height of prism base), $h_2 = 15$ cm (height of prism).

$V = \left[\dfrac{1}{2}(3)(2.6)\right]15$      Substitute values and evaluate.

$V = 58.5 \text{ cm}^3$

**The volume of the prism is 58.5 cm³.**

**EXAMPLE** What is the cubic-inch displacement (space occupied) of a cylinder (Figure 12–111) whose diameter is 5 in. and whose height is 4 in.?

Figure 12–111

$V = Bh$      Substitute the formula for the area of the circular base for *B*.
$V = \pi r^2 h$      Substitute values; $r = \frac{1}{2}$ diameter, or 2.5.
$V = \pi(2.5)^2(4)$      Evaluate.
$V = 78.5 \text{ in}^3$      Round.

**The cylinder displacement is 78.5 in³.**

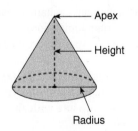

**Figure 12–112** Circular cone.

A right **cone** is a three-dimensional figure whose base is a circle and whose side surface tapers to a point, called the **vertex** or **apex,** and whose height is a perpendicular line between the base and apex (Figure 12–112). One example of a *cone* is the funnel or the circular rain cap placed on top of stove vent pipes extending through the roof of some homes.

A **pyramid** is a three-dimensional geometric shape that has a polygon for a **base** and **lateral faces** that are triangles with a common vertex (apex) (Figure 12–113). The *height or altitude* of the pyramid is the perpendicular distance from the vertex to the base. If the base of the pyramid is a polygon that has all sides equal, such as a square, the height meets the base at the center of the base and is a **right** or **regular pyramid.** Figure 12–113 shows right pyramids with 3-, 4-, and 5-sided bases.

Chapter 12 / Geometry

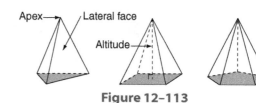

**Figure 12–113**

---

**Volume of any cone or pyramid:**

$$V = \frac{1}{3} Bh$$

where $V$ is the volume, $B$ is the area of the base, and $h$ is the height of the cone or pyramid.

---

**EXAMPLE** Find the volume of a pyramid with a square base that is 15 cm on a side and a height of 28 cm (Figure 12–114).

**Figure 12–114**

$V = \dfrac{1}{3} Bh$      Find the area of the base and substitute values for $B$ and $h$.
$B = s^2 = 15^2 = 225$ cm²

$V = \dfrac{1}{3} (225)(28)$      Multiply.

$V = \mathbf{2{,}100}$ **cm³**      Volume is cubic units.

A **frustum of a cone** is a part of a cone between the base and a plane passing through the cone parallel to the base (Figure 12–115).

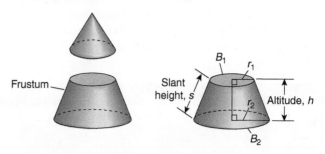

**Figure 12–115**

Similarly, a **frustum of a pyramid** is a part of a pyramid between the base and a plane passing through the pyramid that is parallel to the base (Figure 12–116).

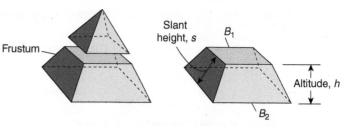

**Figure 12–116**

12–4 Volume and Surface Area

$$V = \frac{1}{3} h \left( B_1 + B_2 + \sqrt{B_1 B_2} \right)$$

where $V$ is the volume, $B_1$ is the area of the base, $B_2$ is the area of the top, and $h$ is the height of the frustum.

**EXAMPLE**  Find the volume of the frustum of a pyramid with a square base that is 4 in. on each side, a square top that is 2 in. on each side, and a height of 3 in. (Figure 12–117).

$B_2$

$h$

$B_1$

**Figure 12–117**

$V = \frac{1}{3} h \left( B_1 + B_2 + \sqrt{B_1 B_2} \right)$   Substitute the values for $B_1$, $B_2$, and $h$.
$B_1 = 4^2 = 16$ in$^2$; $B_2 = 2^2 = 4$ in$^2$

$V = \frac{1}{3} (3)(16 + 4 + \sqrt{16 \cdot 4})$   Simplify the radicand.

$= \frac{1}{3} (3)(16 + 4 + \sqrt{64})$   Take the square root.

$= \frac{1}{3} (3)(16 + 4 + 8)$   Simplify the grouping.

$= \frac{1}{3} (3)(28)$   Multiply.

$V = \textbf{28 in}^3$   Volume is cubic units.

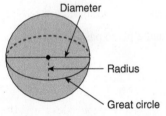

Diameter

Radius

Great circle

**Figure 12–118**  Sphere.

Soccer balls, golf balls, tennis balls, baseballs, and ball bearings are *spheres*. Spheres are also used as tanks to store gas and water because spheres hold the greatest volume for a specified amount of surface area. A **sphere** is a three-dimensional figure formed by a curved surface with points that are all equidistant from a point inside called the center (Figure 12–118). A **great circle** divides the sphere in half and is formed by a plane passing through the center of the sphere.

A sphere does not have bases like prisms and cylinders. Because of the relationship of the sphere to the circle, the formula for the volume of a sphere includes elements of the formula for the area of a circle.

**Volume of a sphere:**

$$V = \frac{4 \pi r^3}{3}$$

where $r =$ radius.

Note that the radius is *cubed*, or raised to the power of 3, indicating volume.

**EXAMPLE**  Find the volume of a sphere that has a diameter of 90 cm.

$V = \frac{4 \pi r^3}{3}$   Substitute values.

$V = \frac{4(\pi)(45)^3}{3}$   Cube 45, multiply, and divide.

$V = \textbf{381,704 cm}^3$   Round.

Chapter 12 / Geometry

**EXAMPLE**
**INDTR**

Find the weight of the cast-iron object shown in Figure 12–119 if cast iron weighs 0.26 lb per cubic inch. Round to the nearest whole pound.

The solution requires finding the volume of the cone that forms the top of the object, the volume of the cylinder that forms the middle portion of the object, and the volume of the **hemisphere** (half sphere) that forms the bottom of the object.

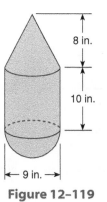

**Figure 12–119**

$$V_{cone} = \frac{\pi r^2 h}{3} \qquad V_{cylinder} = \pi r^2 h \qquad V_{hemisphere} = \frac{1}{2}\left[\frac{4\pi r^3}{3}\right]$$

$$V_{cone} = \frac{(\pi)(4.5)^2(8)}{3} \qquad V_{cylinder} = (\pi)(4.5)^2(10) \qquad V_{hemisphere} = \frac{1}{2}\left[\frac{4(\pi)(4.5)^3}{3}\right]$$

$$V_{cone} = 169.6460033 \text{ in}^3 \qquad V_{cylinder} = 636.1725124 \text{ in}^3 \qquad V_{hemisphere} = 190.8517537 \text{ in}^3$$

$$\text{total volume} = V_{cone} + V_{cylinder} + V_{hemisphere}$$

$$\text{total volume} = 996.6702694 \text{ in}^3$$

Convert to pounds:

$$\frac{996.6702694 \text{ in}^3}{1} \times \frac{0.26 \text{ lb}}{1 \text{ in}^3} = 259 \text{ lb} \qquad \text{Round.}$$

**To the nearest whole pound, the cast-iron object weighs 259 lb.**

**2** **Find the Surface Area of Three-Dimensional Objects.**

The surface area of a three-dimensional figure can refer to just the area of the *sides* of the figure. Or surface area can refer to the overall area, including the bases along with the sides.

The **lateral surface area (LSA)** of a three-dimensional figure is the area of its sides only.

The **total surface area (TSA)** of a three-dimensional figure is the area of the sides plus the area of its base or bases.

To find the lateral surface area, we find the sum of the area of each of the sides using the formula for the area of the appropriate plane surface. We can also find the lateral surface area (LSA) of a right prism or right circular cylinder (Figure 12–120) by multiplying the perimeter of the base times the height of the figure.

**Lateral surface area of a right prism or a right circular cylinder:**

$$LSA = ph$$

where $p$ is the perimeter of the base and $h$ is the height of the three-dimensional figure.

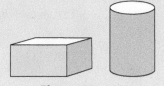

**Figure 12–120**

To get the total surface area (TSA) (Figure 12–121), add the areas of the two bases to the lateral surface area.

Total surface area of a right prism or a right circular cylinder:

$$\text{TSA} = ph + 2B$$

where $p$ is the perimeter of the base, $h$ is the height of the
three-dimensional figure, and $B$ is the area of the base.

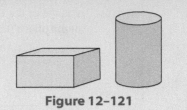

**Figure 12–121**

---

**EXAMPLE**
**INDTR**

Find the lateral surface area of a rectangular shipping carton measuring 24 in. in length,
12 in. in width, and 20 in. in height (Figure 12–122).

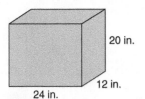

20 in.

12 in.

24 in.

**Figure 12–122**

| | |
|---|---|
| $\text{LSA} = ph$ | Substitute the formula for the perimeter of a rectangle, $p = 2l + 2w$. |
| $\text{LSA} = (2l + 2w)h$ | Substitute numerical values. |
| $\text{LSA} = [2(24) + 2(12)]20$ | Perform calculations inside grouping. |
| $\text{LSA} = [72]20$ | Multiply. |
| $\text{LSA} = 1{,}440 \text{ in}^2$ | Area requires square units. |

**The lateral surface area of the carton is 1,440 in².**

---

**EXAMPLE**
**INDTR**

How many square centimeters of sheet metal are required to manufacture a can that has a
radius of 4.5 cm and height of 9 cm? Assume no waste or overlap (Figure 12–123).

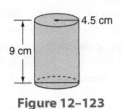

4.5 cm

9 cm

**Figure 12–123**

| | |
|---|---|
| $\text{TSA} = ph + 2B$ | Total surface area is needed. Substitute formulas $p = 2\pi r$, $B = \pi r^2$. |
| $\text{TSA} = 2\pi r h + 2\pi r^2$ | Substitute values. |
| $\text{TSA} = 2(\pi)(4.5)(9) + 2(\pi)(4.5)^2$ | Square 4.5 and perform multiplications. |
| $\text{TSA} = 254.4690049 + 127.2345025$ | Add. |
| $\text{TSA} = 381.70 \text{ cm}^2$ | Round. Area requires square units. |

**The can requires 381.70 cm² of sheet metal.**

---

**EXAMPLE**

Find the total surface area of the triangular prism shown in Figure 12–124.

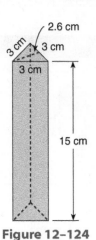

2.6 cm

3 cm   3 cm

3 cm

15 cm

**Figure 12–124**

| | |
|---|---|
| $\text{TSA} = ph + 2B$ | Perimeter of triangular base is $a + b + c$. Area of triangular base $B$ is $\frac{1}{2}bh_1$, where $h_1$ is the height of the triangular base. |
| $\text{TSA} = (a + b + c)h + 2\left(\dfrac{1}{2}bh_1\right)$ | Substitute values. |
| $\text{TSA} = (3 + 3 + 3)(15) + 2\left(\dfrac{1}{2}\right)(3)(2.6)$ | Add inside grouping. |
| $\text{TSA} = 9(15) + (2)\left(\dfrac{1}{2}\right)(3)(2.6)$ | Perform multiplications. |
| $\text{TSA} = 135 + 7.8$ | Add. |
| $\text{TSA} = 142.8 \text{ cm}^2$ | Area requires square units. |

**The total surface area of the triangular prism is 142.8 cm².**

A sphere does not have bases like prisms and cylinders. The surface area of a sphere includes *all* the surface so there is only one formula. Because of the relationship of the sphere to the circle, the formula includes elements of the formula for the area of a circle. The total surface area of the sphere is 4 times the area of a circle with the same radius.

**Total surface area of a sphere:**

$$\text{TSA} = 4\pi r^2$$

where $r$ = radius.

---

**EXAMPLE** Find the surface area of a sphere that has a diameter of 90 cm.

$\text{TSA} = 4\pi r^2$          Substitute values. $\frac{1}{2}d = r$, so $r = 45$ cm.

$\text{TSA} = 4(\pi)(45)^2$       Square 45 and multiply.

$\textbf{TSA = 25,447 cm}^2$     Round.

---

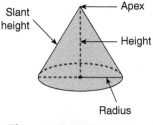

**Figure 12–125** Circular cone.

The surface area of a sphere includes *all* the surface, so there is only one formula. The total surface area of the sphere is 4 times the area of a circle with the same radius.

The **slant height** of a cone is the distance along the side from the base to the apex. Figure 12–125 shows the *perpendicular height* and the *slant height* of a cone.

The lateral surface area of a cone equals the circumference of the base times $\frac{1}{2}$ the slant height, or $\text{LSA} = 2\pi r \left( \dfrac{1}{2}s \right)$, which simplifies as $\pi rs$.

**Lateral surface area of a cone:**

$$\text{LSA} = \pi rs$$

where $r$ is the radius and $s$ is the slant height.

---

The total surface area, then, is the lateral surface area plus the area of the base.

**Total surface area of a cone:**

$$\text{TSA} = \pi rs + \pi r^2$$

where $r$ is the radius of the circular base, $s$ is the slant height, and $\pi r^2$ is the area of the base.

---

**EXAMPLE** Find the lateral surface area and total surface area of a cone that has a diameter of 8 cm, height of 6 cm, and slant height of 7 cm. Round to hundredths.

$\text{LSA} = \pi rs$                Substitute values: $r = \frac{1}{2}d$, or 4.

$\text{LSA} = (\pi)(4)(7)$         Multiply.

$\textbf{LSA = 87.96 cm}^2$      Rounded from 87.9645943.

$\text{TSA} = \pi rs + \pi r^2$        Substitute values. Use full calculator value for $\pi rs$.

$\text{TSA} = 87.9645943 + (\pi)(4)^2$    Perform operations using the proper order of operations.

$\textbf{TSA = 138.23 cm}^2$     Round.

---

**1** Find the volume in Figures 12–126 and 12–127. Round to the nearest hundredth if necessary.

**1.**

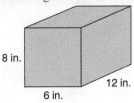

8 in.

12 in.

6 in.

**Figure 12–126**

**2.**

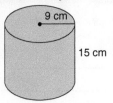

9 cm

15 cm

**Figure 12–127**

**3.** A right pentagonal (five-sided) prism is 10 cm high. If the area of each pentagonal base is 32 cm², what is the volume of the prism?

**5.** How many cubic inches are in an aluminum can with a $2\frac{1}{2}$ in. diameter and $4\frac{3}{4}$ in. height? Round to tenths.

**4.** What is the volume of a triangular prism that has a height of 8 in., a triangular base that measures 4 in. on each side, and a height of 3.46 in.? Round to hundredths.

**6.** What is the volume of a cylindrical oil storage tank that has a 40 ft diameter and 15 ft height? Round to the nearest whole number.

Solve. Round to tenths.

**7.** Find the volume of the cone in Figure 12–128.

**8.** **CON** How many cubic feet are in a conical pile of sand that is 30 ft in diameter and is 20 ft high?

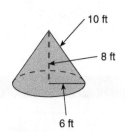

10 ft

8 ft

6 ft

**Figure 12–128**

**9.** **BUS** A cone-shaped storage container holds a photographic chemical. If the container is 80 cm wide and 30 cm high, how many liters of the chemical does it hold if 1 L = 1,000 cm³?

**11.** Find the volume of a pyramid that has a square base of 48 m on a side and a height of 100 m.

**13.** A frustum of a pyramid has a square base that is 18 in. on each side, a square top that is 10 in. on each side, and a height of 13 in. Find the volume of the frustum.

**10.** Find the volume of a pyramid that has a square base of 30 cm on a side and a height of 42 cm.

**12.** Find the volume of a pyramid that has an equilateral triangular base with altitude 10.39 m and a side of 12 m. The height is 20 m.

**14.** A frustum of a pyramid has a triangular base that has an area of 32 cm² and a triangular top that has a surface area of 28 cm². The height of the frustum is 81 cm. Find the volume of the frustum.

15. **CON** Cap blocks for a fence are molded in the shape of a frustum of a pyramid that has a square base and top. The base of the frustum is 30 in. on each side and the top is 24 in. on each side. The cap block is 5 in. thick (height of frustum). What is the volume of the frustum?

16. Find the volume of a sphere with a radius of 5 cm.

17. Find the volume of a sphere with a radius of 6 in.

18. If 1 ft$^3$ = 7.48 gal, how many gallons can a spherical water tank hold if its diameter is 45 ft?

19. **INDTR** If a spherical propane tank is filled to 90% of its capacity, how many gallons of propane does the tank hold if its diameter is 4 ft? (1 ft$^3$ = 7.48 gal.)

**2** Solve. Round to tenths.

20. **INDTEC** How many square inches are in the lateral surface area and total surface area of an aluminum can with a $2\frac{1}{2}$ in. diameter and $4\frac{3}{4}$ in. height?

21. What are the lateral surface area and total surface area of a cylindrical oil storage tank that has a 40 ft diameter and 15 ft height?

22. A right pentagonal prism is 10 cm high. If the area of each pentagonal base is 32 cm$^2$ and the perimeter is 20 cm, what are the lateral and total surface areas of the prism?

23. What are the lateral and total surface areas of a triangular prism that has a height of 8 in. and a triangular base that measures 4 in. on each side with an altitude of 3.46 in.?

24. **CON** A cylindrical water well is 1,200 ft deep and 6 in. across. Find the lateral surface area of the well. Round to the nearest square foot. (*Hint:* Convert measures to a common unit.)

25. Find the lateral surface area and total surface area of the cone that has a radius of 6 ft, slant height of 10 ft and height of 8 ft.

26. **INDTR** How many square centimeters of sheet metal are needed to form a conical rain cap 25 cm in diameter if the slant height is 15 cm?

27. Find the total surface area of a conical tank that has a radius of 15 ft and a slant height of 20 ft.

28. Find the surface area of a sphere with a radius of 5 cm.

29. **INDTR** How many square feet of steel are needed to manufacture a spherical water tank with a diameter of 45 ft?

30. **INDTR** A spherical propane tank has a diameter of 4 ft. How many square feet of surface area need to be painted?

31. **INDTR** A cylindrical water tower with a conical top and hemispheric bottom (see Figure 12–129) needs to be painted. If the cost is $2.19 per square foot, how much does it cost (to the nearest dollar) to paint the tank?

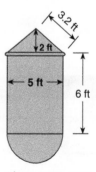

**Figure 12–129**

*Learning Outcomes*

**1** Classify triangles by the relationship of the sides or angles.

**2** Use the Pythagorean theorem to find the missing side of a right triangle.

We have examined various relationships among lines and angles in the previous sections. In this section, we study one of the most useful mathematical figures for any technician who uses geometry—the *triangle*. The relationships among the sides and angles in the triangle enable us to obtain much information that is implied but not always expressed in certain applications.

Triangles can be classified according to the relationship of their sides. The symbol $\Delta$ is used to indicate a triangle.

**1 Classify Triangles by the Relationship of the Sides or Angles.**

Three relationships are possible among the three sides of a triangle and result in special names for each type of triangle.

An **equilateral triangle** is a triangle with three equal sides. The three angles of an equilateral triangle are also equal. Each angle measures 60° (Figure 12–130, left).

An **isosceles triangle** is a triangle with *exactly* two equal sides. The angles opposite these equal sides are also equal (Figure 12–130, middle).

A **scalene triangle** is a triangle with *all* three sides unequal (Figure 12–130, right).

An equilateral (three sides equal) triangle also has three equal angles. The two equal sides of an isosceles triangle have opposite angles that are equal (Figure 12–130).

In a scalene triangle, where no sides are equal, we can state an important relationship between the sides and their opposite angles (Figure 12–130).

Equilateral $\Delta ABC$
$AB = BC = AC$
$\angle A = \angle B = \angle C$

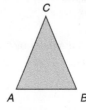

Isosceles $\Delta ABC$
$AC = BC$
$\angle A = \angle B$

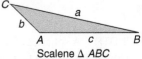

Scalene $\Delta ABC$
Longest side *BC* or *a*
Middle side *AB* or *c*
Shortest side *AC* or *b*
Largest angle $\angle A$
Middle angle $\angle C$
Smallest angle $\angle B$

**Figure 12–130**

**To determine the longest and shortest sides of a triangle:**

If the three sides of a triangle are unequal, the *largest* angle is opposite the *longest* side and the *smallest* angle is opposite the *shortest* side.

**EXAMPLE** Identify the longest and shortest sides of the triangle in Figure 12–131.

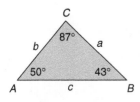

**Figure 12–131**

**The longest side is *AB* or *c*** (87° is the largest angle). **The shortest side is *AC* or *b*** (43° is the smallest angle).

EXAMPLE   Identify the largest and smallest angles in the triangle in Figure 12–132.

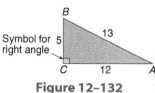

**Figure 12–132**

**The largest angle is** $\angle C$ (13 is the longest side). **The smallest angle is** $\angle A$ (5 is the shortest side).

### 2  Use the Pythagorean Theorem to Find the Missing Side of a Right Triangle.

One of the most famous and useful theorems in mathematics is the Pythagorean theorem. It is named for the Greek mathematician, Pythagoras. The theorem applies to a right triangle.

A **right triangle** has two sides that form a right angle (square corner). These two sides are called the **legs** of the triangle. The side opposite the right angle, or the third side of the triangle, is called the **hypotenuse:** The **Pythagorean theorem** states that the square of the hypotenuse of a right triangle is equal to the sum of the squares of the two legs of the triangle.

**Formula for Pythagorean theorem:**

$$c^2 = a^2 + b^2$$

Where $c$ is the hypotenuse of a right triangle and; $a$ and $b$ are the legs (Figure 12–133).

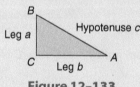

**Figure 12–133**

### To find a leg or the hypotenuse of a right triangle:

1. Identify the two known sides of the triangle and the missing side, and state the theorem symbolically.
2. Substitute known values in the Pythagorean theorem. $c^2 = a^2 + b^2$
3. Solve for the missing value.
4. The solution is the principal square root.

EXAMPLE   Find $b$ when $a = 8$ mm and $c = 17$ mm (Figure 12–134).

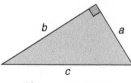

**Figure 12–134**

| | |
|---|---|
| $c^2 = a^2 + b^2$ | State the theorem symbolically and substitute known values. |
| $17^2 = 8^2 + b^2$ | |
| $289 = 64 + b^2$ | Sort terms to isolate $b^2$. |
| $289 - 64 = b^2$ | Combine like terms. |
| $225 = b^2$ | Take the square root of both sides. |
| $\sqrt{225} = b$ | Solution is the principal square root. |
| $\mathbf{15\ mm = b}$ | |

Many times we are given problems that contain "hidden" triangles. In these cases, we need to visualize the triangle or triangles in the problems. Drawing one or more of the sides of the "hidden" triangle helps solve the problem.

**EXAMPLE**
**INDTR**

Find the center-to-center distance between pulleys $A$ and $C$ (Figure 12–135).

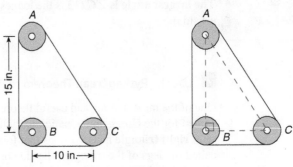

**Figure 12–135**

Connect the center points of the three pulleys to form a right triangle. The hypotenuse is the distance between the centers of pulleys $A$ and $C$. Use the Pythagorean theorem and substitute the given values for the two known sides.

| | |
|---|---|
| $(AC)^2 = (AB)^2 + (BC)^2$ | State theorem symbolically and substitute values. |
| $(AC)^2 = 15^2 + 10^2$ | Square both constants. |
| $(AC)^2 = 225 + 100$ | Combine like terms. |
| $(AC)^2 = 325$ | Take the square root of both sides. |
| $AC = \sqrt{325}$ | Find the approximate principal square root. |
| $AC = 18.0277564$ | |

**The distance from pulley $A$ to pulley $C$ is 18.0 in. (to the nearest tenth).**

**EXAMPLE**
**INDTR**

The head of a bolt is a square 0.5 in. on a side (distance across flats). What is the distance from corner to corner (distance across corners)? (See Figure 12–136.)

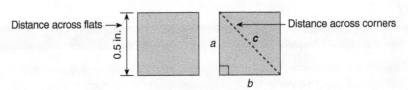

**Figure 12–136**

The sides of the bolt head form the legs of a triangle if a diagonal line is drawn from one corner to the opposite corner. This diagonal forms the hypotenuse of the right triangle. Because the legs of the triangle are known to be 0.5 in. each, we can substitute in the Pythagorean theorem to find the diagonal line, the hypotenuse, which is the distance across corners.

| | |
|---|---|
| $c^2 = a^2 + b^2$ | State theorem symbolically. |
| $c^2 = 0.5^2 + 0.5^2$ | Substitute values. |
| $c^2 = 0.25 + 0.25$ | |
| $c^2 = 0.5$ | Take the square root of both sides. |
| $c = \sqrt{0.5}$ | Evaluate. |
| $c = 0.707106781$ | |

**The distance across corners is 0.71 in. (to the nearest hundredth).**

Sometimes right triangles are used to represent certain relationships, such as forces acting on an object at right angles or electrical and electronic phenomena related in the way the three sides of a right triangle are related. Let's look at an example.

EXAMPLE  Forces $A$ and $B$ come together at a right angle to produce force $C$ (Figure 12–137). If force $A$ is 74.8 lb and the resulting force $C$ is 91.5 lb, what is force $B$?

Since the forces are related in the way the sides of a right triangle are related, we may use the Pythagorean theorem to find the missing force $B$, a leg of the triangle.

Force B

**Figure 12-137**

| | |
|---|---|
| $A^2 + B^2 = C^2$ | State theorem symbolically and substitute values. |
| $74.8^2 + B^2 = 91.5^2$ | Square both constants. |
| $5,595.04 + B^2 = 8,372.25$ | Sort terms to isolate $B^2$. |
| $B^2 = 8,372.25 - 5,595.04$ | Combine constants. |
| $B^2 = 2,777.21$ | Take the square root of both sides. |
| $B = \sqrt{2,777.21}$ | Find the approximate principal square root. |
| $B = 52.7$ | Round to the nearest tenth. |

**Force $B$ is 52.7 lb.**

SECTION 12–5 SELF-STUDY EXERCISES

**1** Fill in the blanks.

**1.** A triangle with no equal sides is called a(n) _____ triangle.

**2.** A triangle with three equal sides is called a(n) _____ triangle.

**3.** A triangle with only two equal sides is called a(n) _____ triangle.

Identify the longest and shortest sides in Figures 12–138 and 12–139.

**4.**

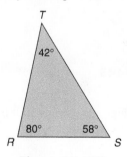

**Figure 12-138**

**5.**

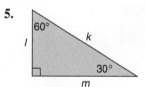

**Figure 12-139**

List the angles in order of size from largest to smallest in Figures 12–140 and 12–141.

**6.**

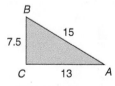

**Figure 12-140**

**7.**

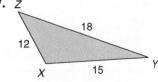

**Figure 12-141**

12–5  Special Triangle Relationships

519

**2** Find the missing side of the right triangle. Round the final answers to the nearest thousandth. Use Figure 12–142 for Exercises 8–9.

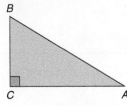

**Figure 12–142**

**8.** $AC = 24$ mm
   $BC = ?$
   $AB = 26$ mm

**9.** $AC = 15$ yd
   $BC = 8$ yd
   $AB = ?$

Use Figure 12–143 for Exercises 10–12. Find the missing dimensions. Round to thousandths where appropriate.

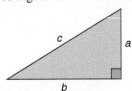

**Figure 12–143**

**10.** $a = 5$ cm
   $b = 4$ cm
   $c = ?$

**11.** $a = ?$
   $b = 9$ m
   $c = 11$ m

**12.** $a = 9$ ft
   $b = ?$
   $c = 15$ ft

Solve the problems. Round final answers to the nearest thousandth if necessary.

**13. CON** A light pole will be braced with a wire that is to be tied to a stake in the ground 18 ft from the base of the pole, which extends 26 ft above the ground. If the wire is attached to the pole 2 ft from the top, how much wire must be used to brace the pole? (See Figure 12–144.)

**14. INDTR** Find the center-to-center distance between holes $A$ and $C$ in a sheet-metal plate if the distance between the centers of $A$ and $B$ is 16.5 cm and the distance between the centers of $B$ and $C$ is 36.2 cm (Figure 12–145).

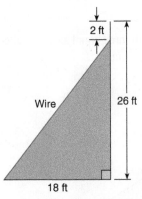

**Figure 12–144**

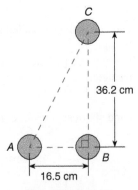

**Figure 12–145**

**15. CON** Find the length of a rafter that has a 10-in. overhang if the rise of the roof is 10 ft and the joists are 48 ft long (Figure 12–146).

**16. CON** A stair stringer is 8 ft high and extends 10 ft from the wall (Figure 12–147). How long will the stair stringer be?

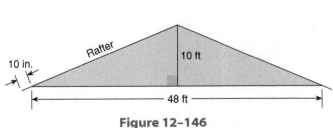

**Figure 12–146**

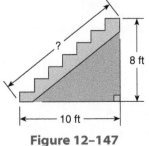

**Figure 12–147**

Chapter 12 / Geometry

17. **INDTR** A machinist wishes to strengthen an L bracket that is 5 cm by 12 cm by welding a brace (hypotenuse of the triangle) to each end of the bracket. How much metal rod is needed for the brace?

18. **CON** To make a rectangular table more stable, a diagonal brace is attached to the underside of the table surface. If the table is 27 dm by 36 dm, how long is the brace?

19. Find the length of the side of the largest square nut that can be milled from a piece of round stock whose diameter is 15 mm (Figure 12–148).

20. **ELEC** A rigid length of electrical conduit must be shaped as shown in Figure 12–149 to clear an obstruction. What total length of the conduit is needed? (*Hint:* Don't forget to include *AB* and *CD* in the total length.)

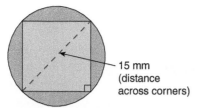

15 mm
(distance
across corners)

**Figure 12–148**

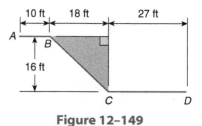

**Figure 12–149**

---

## CHAPTER REVIEW OF KEY CONCEPTS

**Learning Outcomes**

**What to Remember with Examples**

### Section 12–1

**1** Use various notations to represent points, lines, line segments, rays, and planes and angles (pp. 471–473).

A dot represents a point and a capital letter names the point. A line extends in both directions and is named by any two points on the line (with double arrow above the letters, such as $\overleftrightarrow{AB}$). A line segment has a beginning point and an ending point, which are named by letters (with a bar above the letters, such as $\overline{CD}$). Line segments are sometimes named by letters only, such as *CD*. A ray extends from a point on a line and includes all points on one side of the point and is named by the endpoint and any other point on the line forming the ray (with an arrow above the letters, such as $\overrightarrow{AC}$). A plane contains an infinite number of points and lines on a flat surface.

> Use proper notation for the following:
>
> (a) Line *GH*   (b) Line segment *OP*   (c) Ray *ST*
>
> (a) $\overleftrightarrow{GH}$   (b) $\overline{OP}$   (c) $\overrightarrow{ST}$

Lines that intersect meet at only one point. Lines that coincide fit exactly on top of one another. Lines that are parallel are the same distance from one another and never intersect.

> Classify the pairs of lines in Figure 12–150. 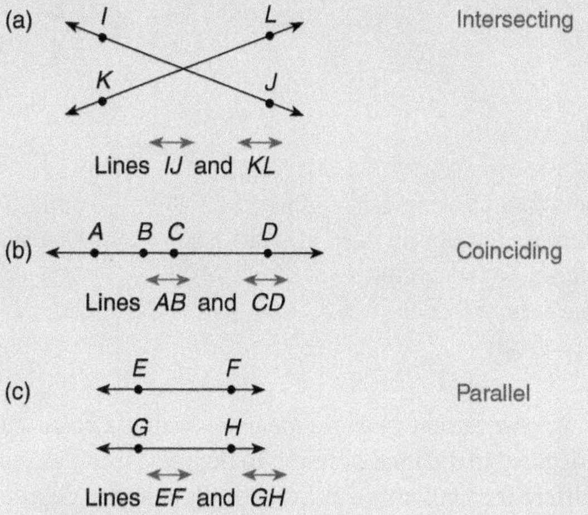
>
> (a) Lines $\overleftrightarrow{IJ}$ and $\overleftrightarrow{KL}$     Intersecting
>
> (b) Lines $\overleftrightarrow{AB}$ and $\overleftrightarrow{CD}$     Coinciding
>
> (c) Lines $\overleftrightarrow{EF}$ and $\overleftrightarrow{GH}$     Parallel
>
> **Figure 12–150**

When two rays intersect in a point (endpoint of rays), an angle is formed. Angles may be named with three capital letters (endpoint in middle and one point on each ray), such as ∠ABC. They may be named by only the middle letter (vertex of the angle), such as ∠B. They may be assigned a number or lowercase letter placed within the vertex, such as ∠2 or ∠d.

Name the angle in Figure 12–151 three ways.

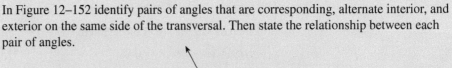

**Figure 12–151**

∠DEF or ∠FED, ∠E, ∠a

**2** Classify angles according to size (pp. 473–475).

A right angle contains 90°, or one-fourth a rotation. A straight angle contains 180°, or one-half a rotation. An acute angle contains less than 90° but more than 0°. An obtuse angle contains more than 90° but less than 180°. The sum of the measures of complementary angles is 90°. The sum of the measures of supplementary angles is 180°.

Identify the following angles:

(a) 30°        (b) 100°        (c) 90°        (d) 180°

(a) acute      (b) obtuse      (c) right       (d) straight

**3** Determine the measure of an angle by using relationships among intersecting lines (pp. 475–477).

Relationships of angles formed by intersecting lines and by two lines cut by are transversal are: Vertical angles are equal. Adjacent angles are supplementary. When two parallel lines are cut by a transversal: Corresponding angles are equal. Alternate interior angles are equal. Alternate exterior angles are equal. Exterior angles on the same side of the transversal are supplementary.

In Figure 12–152 identify pairs of angles that are corresponding, alternate interior, and exterior on the same side of the transversal. Then state the relationship between each pair of angles.

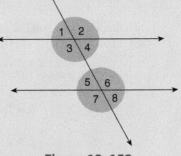

**Figure 12–152**

| | | |
|---|---|---|
| Corresponding angles: | ∠1 and ∠5 | Corresponding angles are equal. |
| Alternate interior angles: | ∠3 and ∠6 | Alternate interior angles are equal. |
| Exterior angles on the same side of the transversal: | ∠1 and ∠7 | Exterior angles on the same side of the transversal are supplementary. |

**4** Convert angle measures between decimal degrees and degrees, minutes, and seconds (pp. 477–478).

To change the format of angle measures using a calculator:
From degrees in decimal notation to degrees, minutes, and seconds (calculator steps will vary with different calculators): **1.** Set the calculator to degree mode. **2.** Enter the angle measure in decimal notation. **3.** Access the sexagesimal function key or menu choice on the calculator (most often labeled $\boxed{° \, ' \, ''}$ or $\boxed{\text{DMS}}$). **4.** Press $\boxed{\text{ENTER}}$ or $\boxed{=}$.

From degrees, minutes, and seconds to decimal notation (calculator steps will vary with different calculators):

**Option 1**

**1.** Enter the series of calculations—degrees + minutes ÷ 60 + seconds ÷ 3,600.

**Option 2**

**1.** Enter the degrees followed by the degree symbol ( $\boxed{2^{nd}}$ , $\boxed{\text{ANGLE}}$ , 1) **2.** Enter the minutes followed by the minute symbol ( $\boxed{2^{nd}}$ , $\boxed{\text{ANGLE}}$ , 2 **3.** Enter the seconds followed by the second symbol ( $\boxed{\text{ALPHA}}$ , $\boxed{"}$ **4.** Press $\boxed{\text{ENTER}}$ .

---

Change 110.625° to degrees, minutes, and seconds.

110.625 $\boxed{2^{nd}}$ $\boxed{\text{ANGLE}}$ $\boxed{4}$ $\boxed{\text{ENTER}}$
110.625° = 110°37′30″

Change 45°15′45″ to a decimal degree equivalent. Round to the nearest hundredth.

45 + 15/60 + 45/3600 = 45.2625
45°15′45″ = 45.2625°

---

**Section 12–2**

**1** Find the perimeter of a polygon using the appropriate formula (pp. 480–483).

**1.** Select the appropriate formula. **2.** Substitute the known values into the formula. **3.** Evaluate the formula (perform the indicated operations). Formulas for perimeter:

$P_{\text{parallelogram}} = 2(b + s)$: $b$ is the base, or the length of a side that is or can be rotated to a horizontal position; and $s$ is the length of a side that joins with the base.
$P_{\text{rectangle}} = 2(l + w)$: $l$ is length of the long side; $w$ is width of the short side.
$P_{\text{square}} = 4s$: $s$ is the length of a side.
$P_{\text{rhombus}} = 4s$, where $s$ is the length of a side.
$P_{\text{trapezoid}} = b_1 + b_2 + a + c$, where $b_1$ and $b_2$ are the lengths of the bases or parallel sides and $a$ and $c$ are the lengths of the sides adjacent to the bases.
$P_{\text{triangle}} = a + b + c$, where $a$, $b$, and $c$ are the lengths of the three sides.

---

What is the perimeter ($P$) of a triangular roof vent that has sides $a$, $b$, and $c$ of 6 ft, 4 ft, and 4 ft, respectively?

$P_{\text{triangle}} = a + b + c$      Select the appropriate formula and substitute values.
$P_{\text{triangle}} = 6 + 4 + 4$      Add.
$P_{\text{triangle}} = 14$ ft

---

**2** Find the area of a polygon using the appropriate formula (pp. 483–486).

**1.** Select the appropriate formula. **2.** Substitute the known values into the formula. **3.** Evaluate the formula (perform the indicated operations). Formulas for area:

$A_{\text{rectangle}} = lw$, where $l$ is the length and $w$ the width.
$A_{\text{square}} = s^2$, where $s$ is the length of a side.
$A_{\text{parallelogram}} = bh$: $b$ is the base; $h$ is the height, or length of a perpendicular distance between the bases.
$A_{\text{rhombus}} = bh$, where $b$ is the length of a side and $h$ is the height.
$A_{\text{trapezoid}} = \frac{1}{2}h(b_1 + b_2)$, where $h$ is the height between the parallel bases and $b_1$ and $b_2$ are the lengths of the bases or two parallel sides.
$A_{\text{triangle}} = \frac{1}{2}(bh)$, where $b$ is the base and $h$ is the height.
Heron's formula for the area of a triangle: $A = \sqrt{s(s - a)(s - b)(s - c)}$, where $s = \frac{1}{2}(a + b + c)$ and $a$, $b$, and $c$ are sides.

---

Find the area of a trapezoid that has bases of 11 in. and 18 in. and height of 12 in.

$A_{\text{trapezoid}} = \frac{1}{2}h(b_1 + b_2)$      Select the appropriate formula and substitute values.

$A_{\text{trapezoid}} = \frac{1}{2}(12)(11 + 18)$      Perform operation in grouping.

---

$$A_{\text{trapezoid}} = \frac{1}{2}(12)(29) \qquad \text{Multiply (in.} \times \text{in.} = \text{in}^2\text{).}$$

$$A_{\text{trapezoid}} = 174 \text{ in}^2$$

## Section 12–3

**1** Find the circumference or area of a circle using the appropriate formula (pp. 490–496).

**1.** Select the appropriate formula. **2.** Substitute the known values into the formula. **3.** Evaluate the formula (perform the indicated operations). Formula for circumference: $C = \pi d$, or $C = 2\pi r$; $d$ is the diameter or distance across the center of a circle; $r$ is the radius, or half the diameter; $\pi$ is approximated on a calculator as 3.141592654. Formula for area: $A_{\text{circle}} = \pi r^2$; $r$ is the radius.

Find the area of a circle whose diameter is 3 m.

First, find the radius: $r = \dfrac{d}{2}$; 3 m $\div$ 2 = 1.5 m.

$$A_{\text{circle}} = \pi r^2$$
$$A_{\text{circle}} = \pi(1.5 \text{ m})^2$$
$$A_{\text{circle}} = 7.07 \text{ m}^2 \text{ (rounded)}$$

**2** Convert angle measures between degrees and radians (pp. 496–498).

Degrees to radians: Using a calculator (steps may vary): **1.** Set angle MODE to radian. **2.** Enter degree measure followed by the degree symbol. Use $\boxed{\circ \, ' \, ''}$, $\boxed{\text{DMS}}$, or the angle menu, option 1. **3.** Enter minute measure followed by the minute symbol. Use $\boxed{\circ \, ' \, ''}$, $\boxed{\text{DMS}}$, or the angle menu, option 1. **4.** Enter second measure followed by the second symbol. Use $\boxed{\circ \, ' \, ''}$, $\boxed{\text{DMS}}$, or $\boxed{\text{ALPHA}}$ $\boxed{''}$. **5.** $\boxed{2^{\text{nd}}}$ $\boxed{\text{ALPHA}}$ $\boxed{1}$ **6.** Press $\boxed{\text{ENTER}}$.

Change 125°30′30″ to radians to the nearest hundredth.

125 [ $\boxed{2^{\text{nd}}}$ $\boxed{\text{ANGLE}}$ menu, option 1] 30 [ $\boxed{2^{\text{nd}}}$ $\boxed{\text{ANGLE}}$ menu, option 2] 30 $\boxed{\text{ALPHA}}$ $\boxed{''}$ $\boxed{2^{\text{nd}}}$ $\boxed{\text{ALPHA}}$, menu option 1 $\boxed{\text{ENTER}}$ = 2.190533655 = 2.19 (rounded)

Radians to degrees: Using a calculator (steps may vary): **1.** Set angle MODE to degree. **2.** Enter radian amount and radian symbol (angle menu, option 3). **3.** If degrees, minutes, and seconds are desired, use DMS function (angle menu, option 4).

Change 1.245 rad to a decimal degree to the nearest thousandth.

1.245 [ $\boxed{2^{\text{nd}}}$ $\boxed{\text{ANGLE}}$ menu, option 3] $\boxed{\text{ENTER}}$ = 71.33324549 = 71.333° (rounded)

**3** Find the arc length of a sector (pp. 499–501).

Arc length is the portion of the circumference formed by the sides of a sector.

Using degrees: arc length $(s) = \dfrac{\theta}{360}(2\pi r)$ or $s = \dfrac{\theta}{360}(\pi d)$.

Using radians: $s = \theta r$.

Find the arc length of a sector formed by a 50° central angle if the radius is 20 cm (Figure 12–153).

$$s = \frac{\theta}{360}(2\pi r) \qquad \text{Substitute } \theta = 50° \text{ and } r = 20 \text{ cm.}$$

$$s = \frac{50}{360}(2)(\pi)(20) \qquad \text{Evaluate.}$$

$$s = 17.45329252 \qquad \text{or} \qquad 17.45 \text{ cm (rounded)}$$

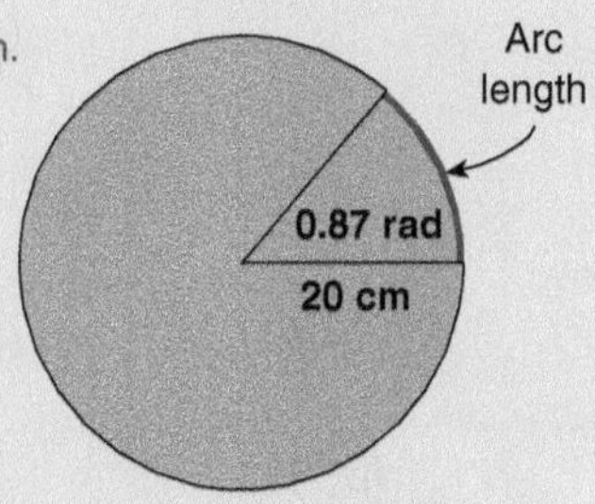

**Figure 12–153**

**4** Find the area of a sector or segment (pp. 501–504).

A sector is a portion of a circle cut off by two radii (*r*).

Using degrees: Area of a sector $= \dfrac{\theta}{360}\pi r^2$, where $\theta$ is the degree measure of the central angle formed by the radii.

Using radians: $A = \dfrac{1}{2}\theta r^2$, where $\theta$ is the radian measure of the central angle formed by the radii.

Find the area of a sector that has a central angle of 0.87 rad and a radius of 20 cm (Figure 12–154).

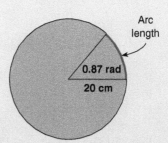

Arc length

0.87 rad

20 cm

**Figure 12–154**

$\text{Area} = \dfrac{1}{2}\theta r^2$      Formula for area of a sector of a circle. Substitute $\theta = 0.87$ rad and $r = 20$ cm.

$\text{Area} = \dfrac{1}{2}(0.87)(20)^2$      Evaluate. (cm × cm = cm²)

$\text{Area} = 174 \text{ cm}^2$      Round.

A chord is a line segment joining two points on a circle. An arc is the portion of the circumference cut off by a chord. A segment is the portion of a circle bounded by a chord and an arc.

The area of a segment is the area of the sector minus the area of the triangle formed by the chord and the central angle of the sector.

Using degrees: $A = \dfrac{\theta}{360}\pi r^2 - \dfrac{1}{2}bh$. Using radians: $A = \dfrac{1}{2}\theta r^2 - \dfrac{1}{2}bh$.

Find the area of the segment formed by a 12-cm chord if the height of the triangle part of the sector is 6 cm and the radius is 8.5 cm. The central angle is 90° (Figure 12–155).

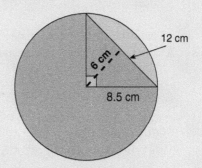

12 cm

6 cm

8.5 cm

**Figure 12–155**

$A_{\text{segment}} = \dfrac{\theta}{360}\pi r^2 - \dfrac{1}{2}bh$      Substitute $\theta = 90°$, $r = 8.5$, $b = 12$, $h = 6$.

$A_{\text{segment}} = \dfrac{90}{360}(\pi)(8.5)^2 - \dfrac{1}{2}(12)(6)$      Evaluate.

$A_{\text{segment}} = 0.25(\pi)(72.25) - 0.5(72)$

$A_{\text{segment}} = 20.74501731 \quad \text{or} \quad 20.75 \text{ cm}^2$      Round.

Chapter Review of Key Concepts

## Section 12–4

**1** Find the volume of three-dimensional objects (pp. 507–511).

**Formulas:**

Volume of a prism or cylinder: $V = Bh$, where $B$ is the area of the base and $h$ is the height.

Volume of a cone or pyramid: $V = \dfrac{1}{3}Bh$, where $B$ is the area of the base and $h$ is the height of the cone or pyramid.

Volume of a frustum of a cone or pyramid: $V = \frac{1}{3}h(B_1 + B_2 + \sqrt{B_1 B_2})$, where $h$ is the height of the frustum, $B_1$ is the area of the base, and $B_2$ is the area of the top.

Volume of a sphere: $V = \dfrac{4\pi r^3}{3}$, where $r$ is the radius.

Find the volume of a cylinder that has a diameter of 20 mm and a height of 80 mm.

$V = Bh$
$V = \pi r^2 h$
$V = \pi(10)^2(80)$
$V = \pi(100)(80)$
$V = 25,132.74123 \text{ mm}^3 = 25,132 \text{ mm}^3$     Round.

$B = \pi r^2$
$r = \frac{1}{2}d = 10 \text{ mm}$

**2** Find the surface area of three-dimensional objects (pp. 511–514).

**Formulas:**

Lateral surface area (area of sides) of a right prism or cylinder: $LSA = ph$, where $p$ is the perimeter of the base and $h$ is the height.

Total surface area (sides plus bases) of a right prism or cylinder: $TSA = ph + 2B$, where $p$ is the perimeter of the base, $h$ is the height, and $B$ is the area of a base.

Total surface area of a sphere: $TSA = 4\pi r^2$, where $r$ is the radius.

Lateral surface area of a cone: $LSA = \pi rs$, where $r$ is the radius and $s$ is the slant height.

Total surface area of a cone: $TSA = \pi rs + \pi r^2$, where $r$ is the radius and $s$ is the slant height. $\pi r^2$ is the area of the base.

A conical pile of gravel has a diameter of 30 ft and a slant height of 40 ft. What is the lateral surface area?

$LSA = \pi rs$
$LSA = \pi(15)(40)$
$LSA = 1,884.96 \text{ ft}^2$     Rounded from 1,884.955592.

$r = \frac{1}{2}d = \frac{1}{2}(30) = 15.$

## Section 12–5

**1** Classify triangles by sides or angles (pp. 516–517).

An equilateral triangle has three equal sides. An isosceles triangle has exactly two equal sides. A scalene triangle has three unequal sides.

Identify the following triangles by the measures of their sides:

(a) 6 cm, 6 cm, 6 cm       Equilateral
(b) 12 in., 10 in., 12 in.       Isosceles
(c) 10 cm, 12 cm, 15 cm       Scalene

An equilateral triangle has three equal angles. The equal sides of an isosceles triangle have opposite angles that are equal. If the three sides of a triangle are unequal, the largest angle is opposite the longest side, and the smallest angle is opposite the shortest side. In a right triangle, the hypotenuse is always the longest side.

Identify the largest and the smallest angles in a triangle with the following sides: 12 in., 10 in., 9 in.
The largest angle is opposite the 12-in. side; the smallest angle is opposite the 9-in. side.

**2** Find the missing side of a right triangle using the Pythagorean theorem (pp. 517–519).

The square of the hypotenuse of a right triangle equals the sum of the squares of the other two sides. $c^2 = a^2 + b^2$

The hypotenuse ($c$) of a right triangle is 10 in. If side $b$ is 6 in., find side $a$ (Figure 12–156).

| | |
|---|---|
| $c^2 = a^2 + b^2$ | Substitute values. |
| $10^2 = a^2 + 6^2$ | Raise to powers. |
| $100 = a^2 + 36$ | Sort. |
| $100 - 36 = a^2$ | Combine. |
| $64 = a^2$ | Take square root of both sides. |
| $\sqrt{64} = a$ | Find the principal square root. |
| $a = 8$ in. | |

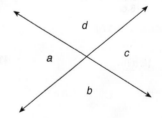

**Figure 12–156**

# CHAPTER REVIEW EXERCISES

## Section 12–1

Use Figure 12–157 for Exercises 1–3.
1. Which lines are parallel?
2. Which lines coincide?
3. Which lines intersect?

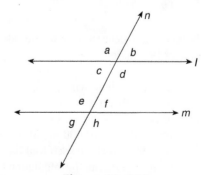

**Figure 12–157**

Use Figure 12–158 for Exercises 4–5.
4. Name $\angle a$ using three capital letters.
5. Name $\angle a$ using one capital letter.

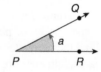

**Figure 12–158**

Classify the angle measures using the terms *right, straight, acute,* or *obtuse.*
6. $50°$     7. $90°$     8. $120°$     9. $180°$

State whether the angle pairs are complementary, supplementary, or neither.
10. $98°, 62°$     11. $135°, 45°$     12. $45°, 35°$     13. $21°, 79°$     14. $90°, 90°$

Use Figure 12–159 to answer Exercises 15–16.
15. Name two pairs of supplementary angles.
16. What are angles $a$ and $c$ called?

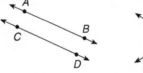

**Figure 12–159**

In Figure 12–160, $l \parallel m$ and $n$ is a transversal.
17. Name two pairs of alternate exterior angles.
18. If $\angle a = 147°$ what is the measure of $\angle e$?
19. If $\angle d = 135°$ what is the measure of $\angle f$?

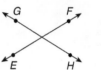

**Figure 12–160**

Change to decimal degree equivalents. Express the decimals to the nearest ten-thousandth.
**20.** 32°14′15″  **21.** 29′  **22.** 47″  **23.** 7′34″

Change to equivalent minutes and seconds. Round to the nearest second when necessary.
**24.** 20.6°  **25.** 0.75°  **26.** 0.46°  **27.** 0.2176°

## Section 12–2

Find the perimeter of Figures 12–161 through 12–165.

**28.**

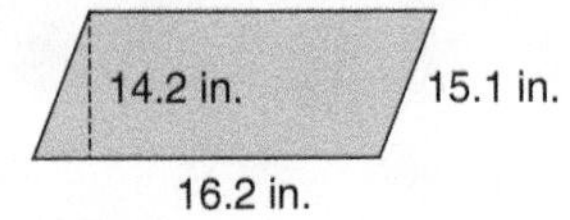

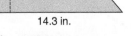

14.2 in.  15.1 in.

16.2 in.

**Figure 12–161**

**29.**

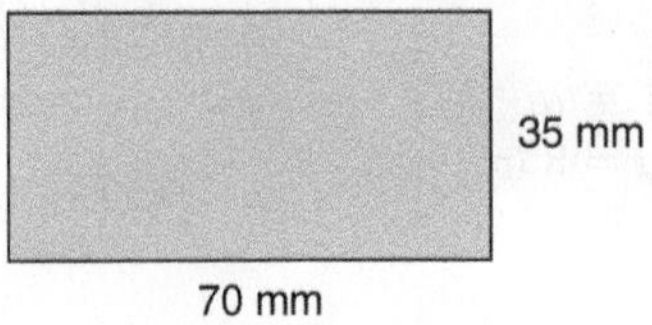

35 mm

70 mm

**Figure 12–162**

**30.**

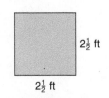

2½ ft

2½ ft

**Figure 12–163**

**31.**

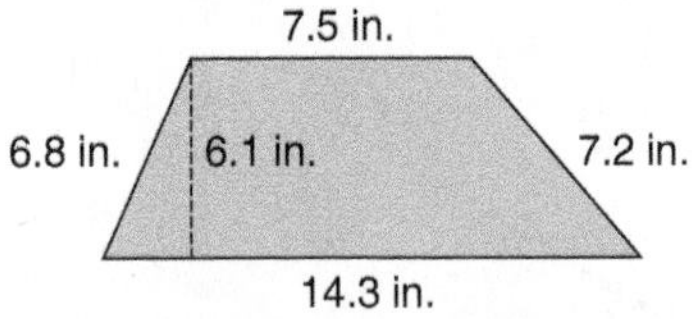

7.5 in.

6.8 in.  6.1 in.  7.2 in.

14.3 in.

**Figure 12–164**

**32.**

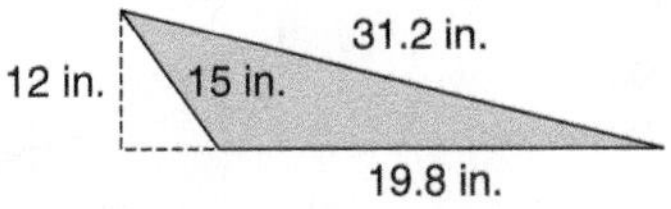

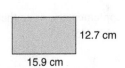

12 in.  15 in.  31.2 in.

19.8 in.

**Figure 12–165**

**33.** The Tennessee Highway Department has signs in the form of a parallelogram. One set of parallel sides each measures 15 ft and one set of parallel sides each measures 18 feet. Find the perimeter of the sign.

**34.** Antique tiles were often made in the form of a square that is 6 in. on each side. What is the perimeter of a tile?

**35.** A rectangular tablecloth measures 84 in. by 60 in. What length of lace is required to trim the edges of the cloth?

**36.** A classroom table in the form of a trapezoid has parallel sides that measure 26 in. and 48 in. The two nonparallel sides both measure 21.1 in. Find the length of trim needed to encase the edges of the table.

Find the area of Figures 12–166 through 12–170.

**37.**

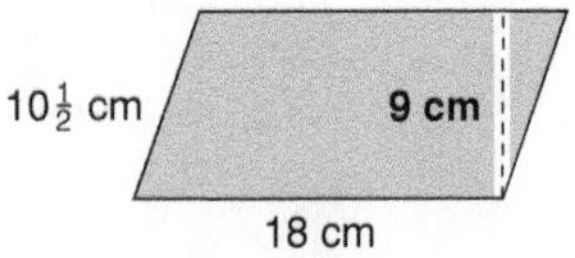

10½ cm  **9 cm**

18 cm

**Figure 12–166**

**38.**

12.7 cm

15.9 cm

**Figure 12–167**

**39.**

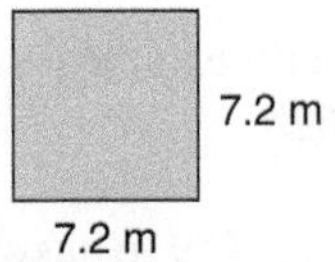

7.2 m

7.2 m

**Figure 12–168**

**40.**

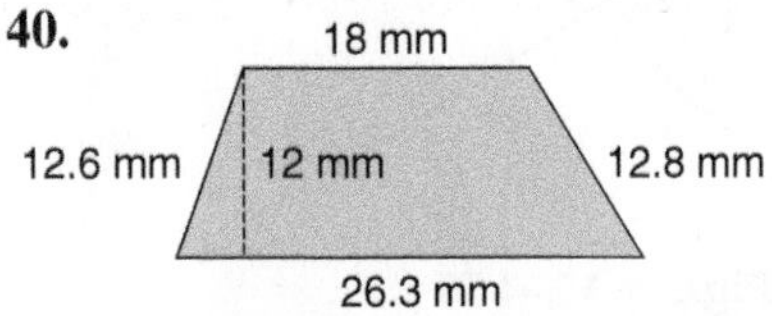

18 mm

12.6 mm  12 mm  12.8 mm

26.3 mm

**Figure 12–169**

**41.**

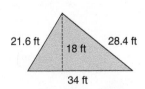

21.6 ft  18 ft  28.4 ft

34 ft

**Figure 12–170**

**42.** Find the area of a scalene triangle whose sides are 15 in., 31.2 in., and 19.8 in. long.

**43.** CON  If a parking lot for a new hospital in the shape of a parallelogram measures 275 ft by 150 ft and has a height of 120 ft, how many square feet need to be paved?

**44.** CON  A hall wall with no windows or doors measures 25 ft long by 8 ft high. Find the number of square feet to be covered if paneling is installed on the two walls.

**45.** CON  A den 18 ft by 16½ ft is to be carpeted. How many square yards of carpeting are needed?

**46.** A square area of land is 10,000 m². If a baseball field must be at least 99.1 m along each foul line to the park fence, is the square adequate for regulation baseball?

**47.** CON  Debbie Murphy is building a contemporary home with four front windows, each in the form of a parallelogram. If each window has a 5-ft base and a height of 2 ft, how many square feet of the 25 ft × 11 ft wall will require stain?

**48. CON** If the stain used on the wall in Exercise 47 is applied at a cost of $2.75 per square yard, find the cost to the nearest dollar of staining the front wall.

**Section 12–3**

Find the circumference (perimeter) and area of Figures 12–171 through 12–174. Round to hundredths.

**49.**

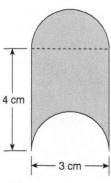

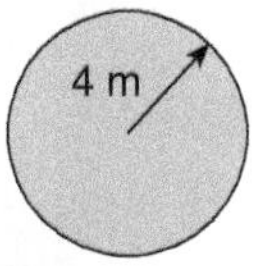

4 m

**Figure 12–171**

**50.**

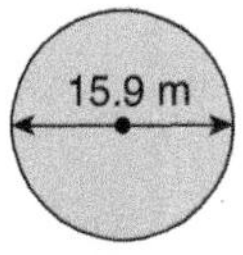

15.9 m

**Figure 12–172**

**51.**

4 cm

3 cm

**Figure 12–173**

**52.**

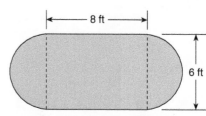

8 ft

6 ft

**Figure 12–174**

**53. INDTR** A $\frac{1}{4}$-in. electric drill with variable speed control turns as slowly as 25 rpm. If an abrasive disk with a 5-in. diameter is attached to the drill driveshaft, what is the disk's slowest cutting speed in ft/min? Round to the nearest whole number. (Cutting speed = circumference measured in feet × rpm.)

**54. CON** What is the cross-sectional area of the opening in a round flue tile whose inside diameter is 8 in.? Round any part of an inch to the next tenth of an inch.

Convert the degree measures to radians rounded to the nearest hundredth.
**55.** 60°
**56.** 212°

Convert the degree, minute, and second measures to degrees rounded to the nearest ten-thousandth. Then convert to radians to the nearest hundredth.
**57.** 99°45′
**58.** 120°20′40″

Convert the radian measures to degrees. Round to the nearest ten-thousandth of a degree.
**59.** $\frac{5\pi}{6}$ rad
**60.** 2.4 rad

Find the area of the sectors of a circle using Figure 12–175. Round to hundredths.
**61.** ∠ = 45°9′
   $r$ = 2.58 cm
**62.** ∠ = 2.88 rad
   $r$ = 15 in.
**63.** ∠ = 40°
   $r$ = 2$\frac{1}{2}$ ft

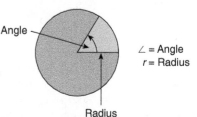

Angle

∠ = Angle
$r$ = Radius

Radius

Find the arc length of the sectors of a circle using Figure 12–175. Round to hundredths.
**64.** ∠ = 0.7 rad
   $r$ = 1.45 ft
**65.** ∠ = 180°
   $r$ = 10 in.

**Figure 12–175**

Find the area of the segments of a circle using Figure 12–176. Round to hundredths.

**66.** $\angle = 55°$
    $r = 10''$
    $h = 8.9''$
    $b = 9.2''$

**67.** $\angle = 30°$
    $r = 24$ cm
    $h = 23.2$ cm
    $b = 12.4$ cm

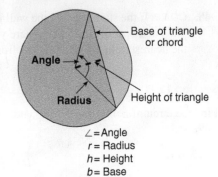

$\angle$ = Angle
$r$ = Radius
$h$ = Height
$b$ = Base

**Figure 12–176**

Solve.

**68. INDTR** A machine cuts a 12-in.-diameter frozen pizza into slices with sides that form 72° angles at the center of the pizza. What is the surface area of each slice to the nearest square inch?

**69. CON** A drain pipe with a 20-in diameter has 4 in. of water in it (Figure 12–177). What is the cross-sectional area of the water in the pipe to the nearest hundredth?

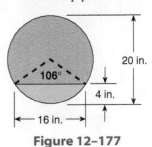

**Figure 12–177**

### Section 12–4

**70.** Find the volume of the prism in Figure 12–178.

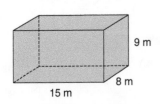

**Figure 12–178**

**71.** Find the volume of the cylinder in Figure 12–179.

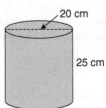

**Figure 12–179**

**72. CON** If concrete weighs 160 lb per cubic foot, what is the weight of a concrete circular slab 4 in. thick and 15 ft across? Round to the nearest pound.

**73. CON** How many cubic yards of topsoil are needed to cover an 85-ft by 65-ft area for landscaping if the topsoil is 6 in. deep? Round to the nearest whole number. (27 ft$^3$ = 1 yd$^3$)

**74. CON** An interstate highway is repaired in one section 48 ft across, 25 ft long, and 8 in. deep. If concrete costs $25.50 per cubic yard, what is the cost of the concrete needed to repair the highway rounded to the nearest dollar? (27 ft$^3$ = 1 yd$^3$)

**75.** Find the total surface area of the triangular prism in Figure 12–180.

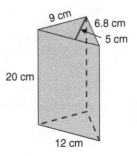

**Figure 12–180**

**76.** Find the lateral surface area of the cylinder in Figure 12–181. Round to the nearest cm². 

**77.** Find the total surface area of the cylinder in Figure 12–181. Round to the nearest cm².

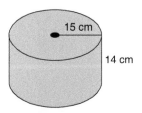

**Figure 12–181**

**78.** Find the total surface area of a cylinder that has a radius of 10 cm and a height of 30 cm. Round to the nearest cm².

**79. CON** A pipeline to carry oil between two towns 5 mi apart has an inside diameter of 18 in. If 1 mi = 5,280 ft, find the lateral surface area to the nearest ft².

**80.** How many barrels of oil will a conical tank hold if its height is $65\frac{1}{2}$ ft and its radius is 20 ft? Round to the nearest whole barrel. (31.5 gal = 1 barrel and 1 ft³ = 7.48 gal.)

**81.** Find the volume of a pyramid that has a square base of 12 m on a side and a height of 36 m.

**82.** Find the volume of a frustum of a pyramid that has a square base of 15 yd on a side and a top of 12 yd on a side. The height is 12 yd.

Solve. Round the final answer to the nearest tenth unless otherwise specified.

**83.** Find the total surface area of a sphere with a radius of 9 m.

**84.** Find the volume of a sphere that has a diameter of 30 cm.

**85.** Find the lateral surface area of a cone with a radius of 6 cm and a slant height of 9 cm.

**86.** Find the total surface area of a cone that has a radius of 4 m and a slant height of 8 m.

**87. CON** The entire exterior surface of a conical tank with a slant height of 12 ft and a diameter of 18 ft is being painted. If the paint covers at a rate of 350 ft² per gallon, how many gallons of paint are needed for the job? Round any fraction of a gallon to the next whole gallon.

**88. AG/H** A hopper deposits grain in a cone-shaped pile with a diameter of 9′6″ and a height of 8′3″. To the nearest cubic foot, how much grain is deposited?

**Section 12–5**

Identify the longest and shortest sides in Figure 12–182.

**89.**

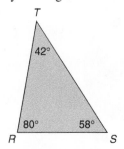

**Figure 12–182**

List the angles in order of size from largest to smallest in Figure 12–183.

**90.**

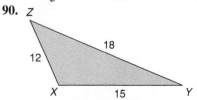

**Figure 12–183**

Use Figure 12–184 to solve the following exercises. Round the final answers to the nearest thousandth if necessary.

**91.** $a = 9$ in.
$b = 12$ in.
$c = ?$

**92.** $a = 8$ mm
$b = ?$
$c = 17$ mm

**93.** $a = ?$
$b = 15$ yd
$c = 17$ yd

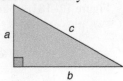

**Figure 12–184**

Solve the following problems. Round the final answers to the nearest thousandth if necessary.

**94. CON** If the base of a ladder is placed on the ground 4 ft from a house, how tall must the ladder be (to the nearest foot) to reach the chimney top that extends $18\frac{1}{2}$ ft above the ground?

**95. AUTO** In an automobile, three pulleys are connected by one belt. The center-to-center distance between the pulleys farthest apart cannot be measured conveniently. The other center-to-center distances are 12 in. and 18 in. (Figure 12–185). Find the distance between the pulleys that are farthest apart.

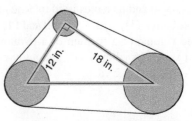

**Figure 12–185**

## TEAM PROBLEM-SOLVING EXERCISES

**1.** A *polygon is inscribed in a circle* if each vertex lies on the circumference of the circle. If a radius from the center of the circle is drawn to each vertex of the inscribed polygon, the central angles formed are equal.

The difference between the area of the circle and the area of the inscribed polygon can be found by calculating the areas of the circle and the polygon and subtracting. Another approach is to find the sum of the areas of the segments of the circle. There will be the same number of segments as sides of the polygon.

For a circle with a radius of 5 cm:
**(a)** Find the difference between the areas of the circle and an inscribed equilateral triangle and find the area of the triangle.
**(b)** Find the difference between the areas of the circle and an inscribed square and find the area of the square.
**(c)** Find the difference between the areas of the circle and the inscribed hexagon (six equal sides) in Figure 12–186 and find the area of the hexagon.

**2.** Using the information found in Exercise 1:
**(a)** Estimate the area of the regular pentagon (five equal sides) in Figure 12–187 by giving two values the area is between. Present a convincing argument for your estimate.
**(b)** Estimate the area of an inscribed regular octagon (eight equal sides) that has a radius of 5 cm and give an argument for your estimate.

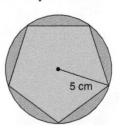

**Figure 12–187**

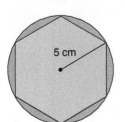

**Figure 12–186**

**1.** Change 0.3125° to minutes and seconds.

**2.** Change 15′32″ to a decimal degree to the nearest ten-thousandth.

Convert the degree measures to radians rounded to the nearest hundredth.

**3.** 35°

**4.** 122°

Convert the radian measures to degrees. Round to the nearest ten-thousandth of a degree.

**5.** $\dfrac{5\pi}{8}$ rad

**6.** 3.1 rad

Find the perimeter and area of Figures 12–188 through 12–191. Round to the nearest tenth.

**7.**

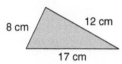

24 ft 6 in.

21 ft 0 in.    21 ft 0 in.

24 ft 6 in.

**Figure 12–188**

**8.**

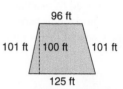

96 ft

101 ft    100 ft    101 ft

125 ft

**Figure 12–189**

**9.**

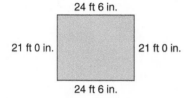

8 cm    12 cm

17 cm

**Figure 12–190**

**10.**

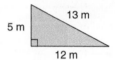

13 m

5 m

12 m

**Figure 12–191**

Find the circumference and area (Figures 12–192 and 12–193).

**11.**

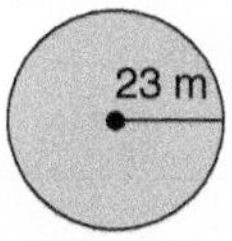

23 m

**Figure 12–192**

**12.**

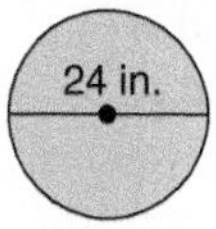

24 in.

**Figure 12–193**

Use the relationships of arc length *s*, area *A*, central angle (in radians) θ, and radius *r* to solve the problems relating to sectors. Round to hundredths if necessary.

**13.** Find *s* if θ = 0.5 and *r* = 2 in.

**14.** Find θ if *s* = 5.3 m and *r* = 7 m.

**15.** Find *r* if θ = 1.7 and *s* = 2.9 m.

**16.** Find *A* if θ = 0.6 and *r* = 7.3 cm.

**17.** How many square feet of floor space are there in the plan shown in Figure 12–194?

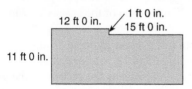

12 ft 0 in.    1 ft 0 in.
             15 ft 0 in.

11 ft 0 in.

**Figure 12–194**

Solve and when necessary, round to hundredths.

**18.** Find the area of the colored portion of the tiled walk that surrounds a rectangular swimming pool (Figure 12–195).

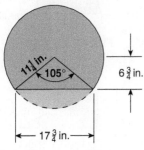

**Figure 12–195**

**19.** Find the area of the composite figure in Figure 12–196.

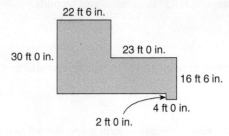

**Figure 12–196**

**20.** A section of a hip roof is a trapezoid measuring 35 ft at the bottom, 15 ft at the top, and 10 ft high. Find the area of this section of the roof in square feet.

**22.** A segment is removed from a flat metal circle so that the piece rests on a horizontal base (Figure 12–197). Find the area of the segment that is removed.

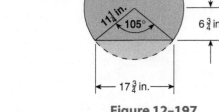

**Figure 12–197**

**21.** Find the area of a sector of a circle that has a radius of 14 cm if the sector has an angle of 42°. Round to hundredths.

Use Figure 12–198 and the Pythagorean theorem to find the missing value. Round to the nearest whole number.

**23.** $a = 20$, $b = 48$, find $c$.

**24.** $b = 10$, $c = 18$, find $a$.

**Figure 12–198**

**25.** Three pulleys are designed so their centers form a right triangle when connecting lines are drawn. The distances between the centers forming the sides of the triangle are 8 in. and 15 in., respectively. What is the distance between the centers of the pulleys that form the hypotenuse?

**27.** A pentagonal prism (five sides) measures 1 in. on each side of its base and has a height of 10 in. What is its lateral surface area?

**29.** A spherical tank 12 ft in diameter can hold how many gallons of fluid if 1 ft$^3$ = 7.48 gal? Answer to the nearest whole gallon.

**26.** Find the total surface area of a sphere with a radius of 8.2 m. Round to the nearest hundredth.

**28.** Find the total surface area of the pentagonal prism in Exercise 27 if the area of the base is 2.17 in$^2$.

**30.** The base of a brass pyramid is an equilateral triangle with sides of 3 in. and altitude of 2.6 in. If the pyramid's height is 8 in., what is the volume of the pyramid?

Suppose you work for an interior design company. After the decorator helps the customer choose paint, wallpaper, and border, your job is to go to the site, measure the room(s), and order the correct amounts. Your company sometimes orders an extra gallon of paint, one extra single roll of wallpaper, and 20 extra feet of wallpaper border.

   Your first job is a child's rectangular room measuring 12 ft by 16 ft with 8-ft ceilings. Three walls will be wallpapered and the fourth short wall will be painted. A wallpaper border 6 in. wide will circle the room. The painted wall and one long wall both have a centered window 6 ft wide by 5 ft tall. The other long wall has a 30-in. × 80-in. entry door and a 60-in. × 80-in. closet door. The doors will not be papered or painted.

### Exercises

Use the given measurements to find the answers. Show computations. Round all answers to the next whole unit.

1. Using a ruler, draw a two-dimensional diagram of this room looking down from the ceiling. Label the length of each wall, and place the doors and windows correctly.

2. Using a ruler, draw a two-dimensional diagram of the two short walls looking at the walls while standing inside the room. Label the length of each side, and place the window correctly.

3. Using a ruler, draw a two-dimensional diagram of the two long walls looking at the walls while standing inside the room. Label the length of each side, and place the doors and the window correctly.

4. Using the diagram from Exercise 2, calculate the amount of paint needed to give the short wall without the window two coats. Each gallon of paint covers 300 ft².

5. Using the diagram from Exercise 1, calculate the length of wallpaper border needed to circle the entire room. Keep in mind that the doors and windows do not reach the ceiling. If the wallpaper border comes only in 20-ft rolls, how many rolls will you need? If the customer wants border seams only at the corners of the room, how many rolls will it take?

6. Use the diagrams from Exercises 2 and 3 to calculate the area to be covered with wallpaper on the three walls. The wallpaper is put up first, then the border covers the wallpaper at the ceiling. Don't forget to subtract the area of the doors and windows. (*Note:* 1 ft² = 144 in².)

7. The customer has chosen a uniform wallpaper pattern that does not require matching. How many rolls of wallpaper will it take if each single roll of wallpaper covers 28 ft²? If this pattern is sold only in double rolls, how many double rolls will you need?

8. Most wallpaper comes in 27-in.-wide rolls. Paperers cut panels equal to the height of the room and cut out doors and windows from each panel. To find the number of single rolls needed, professional wallpaper hangers typically divide the total area to be covered by 22 ft². This technique ensures the minimum number of seams after allowing for door and window waste and pattern matching. How many single and double rolls will be needed using this method of calculation?

9. Place the order for the paint, rolls of wallpaper border, and double rolls of wallpaper using the professional wallpaper hanger's method for calculating the amount of paper needed.

### Answers

1. Ceiling-view diagram:

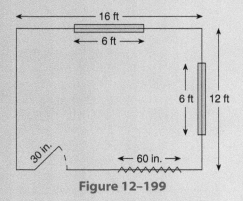

**Figure 12–199**

**2.** Short-wall diagrams:

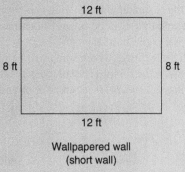

Wallpapered wall
(short wall)

**Figure 12–200**

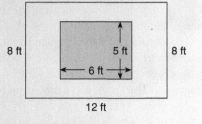

Painted wall with window
(short wall)

**Figure 12–201**

**3.** Long-wall diagrams:

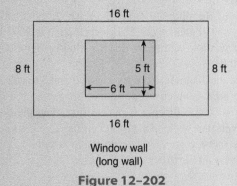

Window wall
(long wall)

**Figure 12–202**

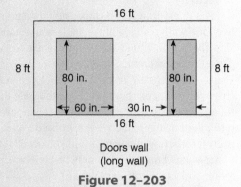

Doors wall
(long wall)

**Figure 12–203**

**4.** The painted wall's area is 66 ft². Double that for two coats equals 132 ft², so 1 gal is plenty.

**6.** Short wall with window: 96 ft².
Long wall with window: 98 ft².
Long wall with doors: 48 ft².
Total wallpaper area: 242 ft².

**8.** 11 single rolls; 6 double rolls

**5. a.** 56 ft of wallpaper border are needed.
**b.** 3 rolls are needed for seams anywhere.
**c.** 4 rolls are needed for corner seams only.

**7.** 9 single rolls; 5 double rolls

**9.** Order 1 gal of paint, 4 rolls of wallpaper border, and 6 double rolls of wallpaper.

Thanks to Bettye Yates, owner of Interior Creations, Collierville, TN.

# 13

# Right-Triangle Trigonometry

## Focus on Careers

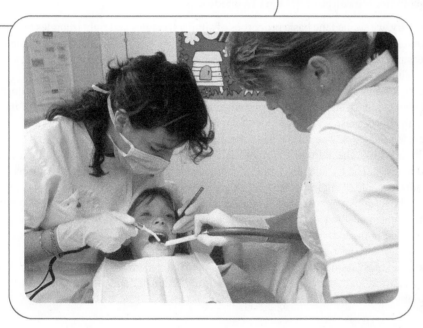

Want a flexible schedule? This may be the job for you! Dental hygienists clean patients' teeth and teach patients how to practice good oral hygiene. They take and develop dental X-rays and apply cavity-preventive agents. In some states, hygienists may administer anesthetics, remove sutures, and prepare clinical and laboratory diagnostic tests for dentists.

Flexible scheduling is an attractive feature of this career. Dental hygienists can find full-time or part-time work. Evening and weekend schedules are available as well as daytime weekday schedules. Dentists may hire hygienists to work only 2 or 3 days a week, so hygienists who desire full-time work may work in more than one dentist's office.

The number of jobs exceeds the number of hygienists because many hygienists hold more than one job. More than half of all dental hygienists worked less than 35 hours a week in 2002.

*(continued)*

Dental hygienists must be licensed by the state in which they practice. For licensure, they must graduate from an accredited dental hygiene school and pass both a written and a clinical examination. Jobs are expected to grow much faster than average through 2012 because of increasing demand for dental care and because hygienists increasingly perform services previously performed by dentists.

Although earnings vary by geographic location, employment requirements, and years of experience, the highest 10% of dental hygienists earned more than $39.24 per hour in 2002. The middle 50% earned between $21.96 and $32.48 per hour in 2002.

Source: *Occupational Outlook Handbook* 2004–2005 Edition. U.S. Department of Labor, Bureau of Labor Statistics.

## 13–1 | *Trigonometric Functions*

*Learning Outcomes*

**1** Find the sine, cosine, and tangent of angles of right triangles, given the measures of at least two sides.

**2** Find trigonometric values for the sine, cosine, and tangent using a calculator.

**3** Find the angle measure, given a trigonometric value.

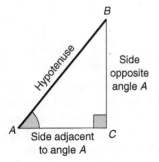

**Figure 13–1**

### **1** Find the Sine, Cosine, and Tangent of Angles of Right Triangles, Given the Measures of at Least Two Sides.

In geometry, we studied the basic properties of similar triangles and right triangles that allowed us to find missing measures of the sides of the triangle. Using trigonometry, we can determine the measure of either acute angle of a right triangle if we know the measure of at least two sides of the right triangle.

Figure 13–1 shows a right triangle, *ABC,* with the sides of the triangle labeled according to their relationship to angle *A.* The **hypotenuse** is the side opposite the right angle of the triangle and the hypotenuse forms one side of angle *A.* The other side that forms angle *A* is the adjacent side of angle *A.* The third side of the triangle is the opposite side of angle *A.*

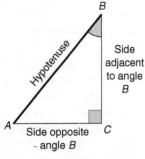

**Figure 13–2**

In Figure 13–2, the sides of the right triangle *ABC* are labeled according to their relationship to angle *B.* The hypotenuse forms one side of angle *B,* the other side that forms angle *B* is the adjacent side of angle *B,* and the third side is the opposite side of angle *B.* In Chapter 12, we used the term hypotenuse, but the terms adjacent side and opposite side are also very important in understanding trigonometric functions. The **adjacent side** of an acute angle of a right triangle is the side that forms the angle with the hypotenuse. The **opposite side** of an acute angle of a right triangle is the side that does not form the given angle.

The three most commonly used trigonometric functions are the **sine, cosine,** and **tangent.** The sine, cosine, and tangent of angle *A* in Figure 13–3 are defined as ratios of the sides of the right triangle.

**Trigonometric functions of angle *A* in a standard right triangle:**

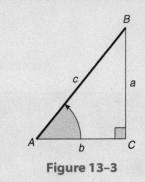

$$\sin A = \frac{\text{side opposite} \angle A}{\text{hypotenuse}} = \frac{a}{c}$$

$$\cos A = \frac{\text{side adjacent to} \angle A}{\text{hypotenuse}} = \frac{b}{c}$$

$$\tan A = \frac{\text{side opposite} \angle A}{\text{side adjacent to} \angle A} = \frac{a}{b}$$

**Figure 13–3**

$$\text{sine of angle } A = \frac{\text{side opposite angle } A}{\text{hypotenuse}}$$

$$\text{cosine of angle } A = \frac{\text{side adjacent to angle } A}{\text{hypotenuse}}$$

$$\text{tangent of angle } A = \frac{\text{side opposite angle } A}{\text{side adjacent to angle } A}$$

For convenience, the sine, cosine, and tangent functions are abbreviated as **sin, cos,** and **tan,** respectively. We will use letters to designate the sides of the **standard right triangle,** as shown in Figure 13–3. In the standard right triangle, side $a$ is opposite angle $A$, side $b$ is opposite angle $B$, and side $c$ (hypotenuse) is opposite angle $C$ (right angle).

Similarly, we can identify the sine, cosine, and tangent relationships for the other acute angle, angle $B$ (Figure 13–4).

---

**Trigonometric functions of angle $B$ in a standard right triangle:**

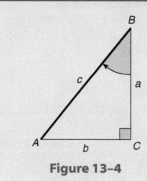

$$\sin B = \frac{\text{side opposite } \angle B}{\text{hypotenuse}} = \frac{b}{c}$$

$$\cos B = \frac{\text{side adjacent to } \angle B}{\text{hypotenuse}} = \frac{a}{c}$$

$$\tan B = \frac{\text{side opposite } \angle B}{\text{side adjacent to } \angle B} = \frac{b}{a}$$

**Figure 13–4**

---

**TIP!**

**Words Are Easier to Remember Than Letters**

Use these common tips for remembering the trigonometric relationships of sin, cos, and tan.

• "**O**scar **h**ad **a** **h**eap **o**f **a**pples."

$$\sin = \frac{\text{o}\,\text{pposite}}{\text{h}\,\text{ypotenuse}} = \frac{\text{O}\,\text{scar}}{\text{h}\,\text{ad}}$$

$$\cos = \frac{\text{a}\,\text{djacent}}{\text{h}\,\text{ypotenuse}} = \frac{\text{a}}{\text{h}\,\text{eap}}$$

$$\tan = \frac{\text{o}\,\text{pposite}}{\text{a}\,\text{djacent}} = \frac{\text{o}\,\text{f}}{\text{a}\,\text{pples}}$$

• "Chief **Soh** - **Cah** - **Toa**" or "**S**ome **o**ld **h**ogs **c**ome **a**round **h**ere **t**asting **o**ur **a**pples."

$$\text{s}\,\text{in} = \frac{\text{o}\,\text{pposite}}{\text{h}\,\text{ypotenuse}} \qquad \text{c}\,\text{os} = \frac{\text{a}\,\text{djacent}}{\text{h}\,\text{ypotenuse}} \qquad \text{t}\,\text{an} = \frac{\text{o}\,\text{pposite}}{\text{a}\,\text{djacent}}$$

## To write the trigonometric ratios for a right triangle given the measures of two sides:

1. Identify the acute angle that is being used.
2. Identify the hypotenuse, opposite side, and adjacent side in relation to the acute angle selected in Step 1.
3. Write the appropriate trigonometric ratio.
4. Simplify the ratio or convert to a decimal equivalent.

**EXAMPLE** Find the sine, cosine, and tangent of angles $A$ and $B$ in Figure 13–5. Express answers as decimals. Round to the nearest ten-thousandth.

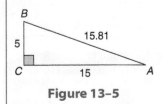

**Figure 13–5**

$$\sin A = \frac{\text{opposite}}{\text{hypotenuse}} = \frac{5}{15.81} = \textbf{0.3163} \qquad \sin B = \frac{\text{opposite}}{\text{hypotenuse}} = \frac{15}{15.81} = \textbf{0.9488}$$

$$\cos A = \frac{\text{adjacent}}{\text{hypotenuse}} = \frac{15}{15.81} = \textbf{0.9488} \qquad \cos B = \frac{\text{adjacent}}{\text{hypotenuse}} = \frac{5}{15.81} = \textbf{0.3163}$$

$$\tan A = \frac{\text{opposite}}{\text{adjacent}} = \frac{5}{15} = \textbf{0.3333} \qquad \tan B = \frac{\text{opposite}}{\text{adjacent}} = \frac{15}{5} = \textbf{3}$$

When both terms of a ratio are expressed in the same unit of measure, that is, *like* units, then the ratio itself is unitless because the common units cancel, for example, $\frac{3 \text{ in.}}{5 \text{ in.}} = \frac{3}{5}$. Thus, *trigonometric ratios are numerical values with no unit of measure.*

**EXAMPLE** Find the sine, cosine, and tangent of angles $A$ and $B$ in Figure 13–6. Leave the answers as fractions in lowest terms.

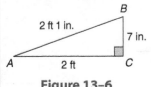

**Figure 13–6**

$$\sin A = \frac{\text{opposite}}{\text{hypotenuse}} = \frac{7 \text{ in.}}{2 \text{ ft } 1 \text{ in.}} = \frac{7 \text{ in.}}{25 \text{ in.}} = \frac{\textbf{7}}{\textbf{25}}$$

$$\cos A = \frac{\text{adjacent}}{\text{hypotenuse}} = \frac{2 \text{ ft}}{2 \text{ ft } 1 \text{ in.}} = \frac{24 \text{ in.}}{25 \text{ in.}} = \frac{\textbf{24}}{\textbf{25}}$$

$$\tan A = \frac{\text{opposite}}{\text{adjacent}} = \frac{7 \text{ in.}}{2 \text{ ft}} = \frac{7 \text{ in.}}{24 \text{ in.}} = \frac{\textbf{7}}{\textbf{24}}$$

$$\sin B = \frac{\text{opposite}}{\text{hypotenuse}} = \frac{2 \text{ ft}}{2 \text{ ft } 1 \text{ in.}} = \frac{24 \text{ in.}}{25 \text{ in.}} = \frac{\textbf{24}}{\textbf{25}}$$

$$\cos B = \frac{\text{adjacent}}{\text{hypotenuse}} = \frac{7 \text{ in.}}{2 \text{ ft } 1 \text{ in.}} = \frac{7 \text{ in.}}{25 \text{ in.}} = \frac{\textbf{7}}{\textbf{25}}$$

$$\tan B = \frac{\text{opposite}}{\text{adjacent}} = \frac{2 \text{ ft}}{7 \text{ in.}} = \frac{24 \text{ in.}}{7 \text{ in.}} = \frac{\textbf{24}}{\textbf{7}}$$

**2** **Find Trigonometric Values for the Sine, Cosine, and Tangent Using a Calculator.**

In studying similar triangles, we found that corresponding angles of similar triangles are equal and that corresponding sides are proportional. These properties of similar triangles lead us to a very important property of trigonometric functions.

Chapter 13 / Right-Triangle Trigonometry

Every angle of a specified measure has a specific set of values for its trigonometric functions.

We recommend the use of a calculator to find trigonometric values. To find trigonometric values on your calculator, the keys SIN, COS, and TAN are used. The angle measure can be entered in degrees or radians, depending on the selected mode.

### To find trigonometric values using the calculator:

1. Set your calculator to the desired mode of angle measure. Calculators generally have at least two angle modes: degrees and radians.
2. Press the appropriate function key (sin, cos, tan) and then enter the angle measure.
3. Display the result by pressing the $=$ or ENTER key.

**EXAMPLE** Using a calculator, find the trigonometric values.

(a) $\sin 27°$        (b) $\sin 20°30'$

(c) $\cos 1.34$        (d) $\tan \dfrac{\pi}{4}$

(a) Be sure your calculator is in degree mode.
   SIN 27 $=$ $\Rightarrow$ **0.4539904997**     Keystrokes may vary.

(b) Be sure your calculator is in degree mode.
   SIN 20.5 $=$ $\Rightarrow$ **0.3502073813**     $20°30' = 20.5°$.

   With some calculators, angle measures can be entered using degrees, minutes, and seconds *or* using the decimal equivalent.

(c) Reset your calculator to radian mode.
   COS 1.34 $=$ $\Rightarrow$ **0.2287528078**

(d) Be sure your calculator is in radian mode.
   TAN ( ( $\pi$ $\div$ 4 ) ) $=$ $\Rightarrow$ **1**

### Automatic Parentheses in Calculator Functions

Some calculators automatically insert an open parenthesis after certain functions. If a close parenthesis is not entered before the calculation is executed, the close parenthesis is automatically interpreted at the end of the entry. Test your calculator with and without the close parenthesis to verify the interpretation made by your calculator.

Entry:     SIN 27 ENTER              Entry:     SIN 27 ) ENTER
Display:   SIN (27                     Display:   SIN (27)
                .4539904997                                 .4539904997

**3** **Find the Angle Measure, Given a Trigonometric Value.**

Finding one of the acute angle measures of a right triangle when given the trigonometric value is an **inverse operation** to finding the trigonometric value when given the angle measure. The notation most commonly used is $\sin^{-1}$, $\cos^{-1}$, or $\tan^{-1}$.

**To find angle measures using the calculator:**

1. Set your calculator to the desired mode of angle measure (degrees or radians).
2. Select the appropriate inverse trigonometric function key or menu option ($\boxed{\text{SIN}^{-1}}$, $\boxed{\text{COS}^{-1}}$, or $\boxed{\text{TAN}^{-1}}$). Enter the trigonometric value and $\boxed{=}$ or $\boxed{\text{ENTER}}$.

Some calculators use an inverse key $\boxed{\text{INV}}$, shift key $\boxed{\text{SHIFT}}$ or second $\boxed{\text{2}^{\text{nd}}}$ key to access the inverse keys. Test your calculator with a known value. For example, in the previous example we found $\sin 27° = 0.4539904997$. $\text{Sin}^{-1}$ 0.4539904997 should be 27°.

---

EXAMPLE  Find the angle in degrees given the trigonometric values in parts (a) and (b). $\theta$ represents the unknown angle measure. Round to the nearest tenth of a degree. For part (c), find the radians to the nearest thousandth.

  **(a)** $\sin \theta = 0.6561$   **(b)** $\cos \theta = 0.4226$   **(c)** $\tan \theta = 2.825$

  **(a)** Be sure your calculator is in degree mode.
    $\boxed{\text{SIN}^{-1}}$ .6561 $\boxed{=}$ $\Rightarrow 41.0031105 \approx \mathbf{41.0°}$    Round.

  **(b)** Be sure your calculator is in degree mode.
    $\boxed{\text{COS}^{-1}}$ .4226 $\boxed{=}$ $\Rightarrow 65.00115448 \approx \mathbf{65.0°}$    Round.

  **(c)** Be sure your calculator is in radian mode.
    $\boxed{\text{TAN}^{-1}}$ 2.825 $\boxed{=}$ $\Rightarrow 1.230578215 \approx \mathbf{1.231\ rad}$    Round.

---

SECTION 13–1 SELF-STUDY EXERCISES

**1** Use Figure 13–7 to find the indicated trigonometric ratios.

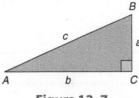

**Figure 13–7**

Use $a = 5$, $b = 12$, and $c = 13$ in Exercises 1–6 and express the ratios as fractions in lowest terms.

  **1.** $\sin A$   **2.** $\cos A$   **3.** $\tan A$   **4.** $\sin B$   **5.** $\cos B$   **6.** $\tan B$

Use $a = 9$, $b = 12$, and $c = 15$ in Exercises 7–12 and express the ratios as fractions in lowest terms.

  **7.** $\sin A$   **8.** $\cos A$   **9.** $\tan A$   **10.** $\sin B$   **11.** $\cos B$   **12.** $\tan B$

Use $a = 16$, $b = 30$, and $c = 34$ in Exercises 13–18 and express the ratios as fractions in lowest terms.

**13.** $\sin A$      **14.** $\cos A$      **15.** $\tan A$      **16.** $\sin B$      **17.** $\cos B$      **18.** $\tan B$

Use $a = 9$, $b = 14$, and $c = 16.64$ in Exercises 19–24 and express the ratios as decimals to the nearest ten-thousandth.

**19.** $\sin A$      **20.** $\cos A$      **21.** $\tan A$      **22.** $\sin B$      **23.** $\cos B$      **24.** $\tan B$

**2** Use your calculator to find the trigonometric values. Express your answers in ten-thousandths.

| | | | |
|---|---|---|---|
| **25.** $\sin 21°$ | **26.** $\cos 3.5°$ | **27.** $\tan 47°$ | **28.** $\cos 52.5°$ |
| **29.** $\sin 0.5498$ | **30.** $\cos 21°30'$ | **31.** $\cos 1.1519$ | **32.** $\cos 0.3665$ |
| **33.** $\sin 53°30'$ | **34.** $\tan 42.5°$ | **35.** $\tan 47.7°$ | **36.** $\sin 62°10'$ |
| **37.** $\cos 12°40'$ | **38.** $\cos 1.0530$ | **39.** $\tan 73°14'$ | **40.** $\sin 1.2363$ |
| **41.** $\cos 46.8°$ | **42.** $\cos 0.3549$ | **43.** $\cos 1.1636$ | **44.** $\tan 12.4°$ |

**3** Find the angles of the trigonometric values in degrees. $\theta$ represents the unknown angle measure. Express each answer to the nearest tenth of a degree.

| | | |
|---|---|---|
| **45.** $\sin \theta = 0.3420$ | **46.** $\cos \theta = 0.9239$ | **47.** $\tan \theta = 2.356$ |
| **48.** $\cos \theta = 0.4617$ | **49.** $\cos \theta = 0.540$ | **50.** $\sin \theta = 0.5712$ |
| **51.** $\tan \theta = 1.265$ | **52.** $\cos \theta = 0.137$ | **53.** $\sin \theta = 0.6298$ |

Find the angles of the trigonometric values in radians. Express each answer to the nearest ten-thousandth.

| | | |
|---|---|---|
| **54.** $\tan \theta = 0.8098$ | **55.** $\cos \theta = 0.6947$ | **56.** $\cos \theta = 0.3907$ |
| **57.** $\cos \theta = 0.968$ | **58.** $\sin \theta = 0.9959$ | **59.** $\tan \theta = 0.3160$ |
| **60.** $\tan \theta = 2.430$ | **61.** $\cos \theta = 0.9610$ | **62.** $\cos \theta = 0.4210$ |

---

## 13–2 | Solving Right Triangles Using the Sine, Cosine, and Tangent Functions

*Learning Outcomes*

**1** Find the missing parts of a right triangle using the sine function.

**2** Find the missing parts of a right triangle using the cosine function.

**3** Find the missing parts of a right triangle using the tangent function.

**4** Select the most direct method for solving right triangles.

**5** Solve applied problems using right-triangle trigonometry.

We often use the sine, cosine, and tangent functions to find unknown parts of a right triangle. To do this, we manipulate formulas and use other algebraic principles depending on what information is given and what information needs to be found.

**1** Find the Missing Parts of a Right Triangle Using the Sine Function.

Using the relationship, $\sin \theta = \dfrac{\text{opp}}{\text{hyp}}$, we can find parts of right triangles when we know any two parts that involve the sine function: one acute angle, the side opposite the known acute angle, and the hypotenuse.

**Use the sine function to find unknown parts of a right triangle:**

**1.** Two of these three parts must be known:
    **(a)** One acute angle
    **(b)** The side opposite the known acute angle
    **(c)** Hypotenuse

**2.** Substitute the two known values in $\sin \theta = \dfrac{\text{opp}}{\text{hyp}}$.

**3.** Solve for the missing part.

**Does It Matter Which Acute Angle Is Known?**

No. The acute angles of a right triangle are complementary. Therefore, if we know either acute angle (*a*), we can find the other one ($90° - a$) or ($\frac{\pi}{2} - a$). Thus, we can use the sine function if we know *either* acute angle and any side.

**EXAMPLE**  Find angle *A* to the nearest tenth of a degree if $a = 7$ and $c = 21$ (see Figure 13–8).

The two known values are the side opposite angle *A* and the hypotenuse, so we use the sine function for the acute angle *A*.

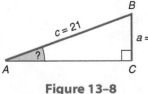

**Figure 13-8**

$$\sin A = \frac{\text{opp}}{\text{hyp}}$$  Substitute the known values. opp = 7, hyp = 21.

$$\sin A = \frac{7}{21}$$  Convert ratio to a decimal equivalent.

$$\sin A = 0.3333333333$$  Find $\boxed{\sin^{-1}}$ of 0.3333333333.

$$A = 19.47122063°  \quad \textbf{or} \quad  19.5°$$  Round to the nearest tenth of a degree.

**How Do I Round My Answers?**

Rounding practices are generally dictated by the context of the problem or industry standards; however, **for consistency, we round all trigonometric ratios to four significant digits and all angle values to the nearest 0.1° or the nearest thousandth radian throughout Chapters 13 and 14 unless otherwise indicated.**

**EXAMPLE**  Find side *a* in triangle *ABC* (see Figure 13–9).

We are given $\angle A$ and the hypotenuse and are asked to find the measure of the side opposite $\angle A$, so we use the sine function.

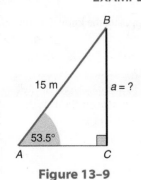

**Figure 13-9**

$$\sin A = \frac{\text{opp}}{\text{hyp}}$$  Substitute known values. $\angle A = 53.5°$, hyp = 15 m.

$$\sin 53.5° = \frac{a}{15}$$  sin 53.5° = 0.8038568606.

$$0.8038568606 = \frac{a}{15}$$  Solve for *a*.

$$15(0.8038568606) = a$$

$$a = 12.05785291 \quad \textbf{or} \quad 12.06 \text{ m}$$  Round to four significant digits.

### Rearrange the Formula Before You Calculate

In the preceding example and those to follow, we can visualize a continuous sequence of calculator steps if we rearrange the formula for the missing part *before* we make any calculations.

$$\sin 53.5° = \frac{a}{15}$$   Rearrange for *a*.

$$15(\sin 53.5°) = a$$   Evaluate.

Then a continuous series of calculations is made. In degree mode,

$$15 \boxed{\times} \boxed{\text{SIN}} 53.5 \boxed{=} \Rightarrow 12.05785291$$

---

**EXAMPLE**   Find the hypotenuse in triangle *RST* (see Figure 13–10).

We are given an acute angle and the side opposite the angle and are asked to find the hypotenuse, so we use the sine function.

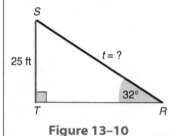

**Figure 13–10**

$$\sin R = \frac{\text{opp}}{\text{hyp}}$$   Substitute known values. $\angle R = 32°$, opp = 25 ft.

$$\sin 32° = \frac{25}{t}$$   Rearrange for *t*.

$$t(\sin 32°) = 25$$

$$t = \frac{25}{\sin 32°}$$   Evaluate.

$$t = 47.17699787$$

$$\boldsymbol{t = 47.18 \text{ ft}}$$   Round to four significant digits.

---

### Choose Given Values over Calculated Values Whenever Possible

It is best to use given values rather than calculated values when finding missing parts of a triangle. Because the rounded value for one missing part is sometimes used to find other missing parts, final answers may vary slightly due to rounding discrepancies. For instance, the sum of the angles of a triangle may be as little as 179° or as much as 181°. The length of a side may be slightly different in the last significant digit. If the full calculator value of a side or angle is used to find other missing parts, the rounding discrepancy is reduced.

To **solve** a triangle means to find the measures of all sides and all angles. A right triangle can be solved if we know one side and any other part besides the right angle.

**EXAMPLE**  Solve triangle *DEF* (see Figure 13–11). One side and one other part besides the right angle are known.

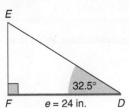

**Figure 13–11**

Because we have one acute angle and a side that is not opposite the known angle, we must find the other angle of the triangle.

$$\angle E = 90° - 32.5° = \mathbf{57.5°}$$    Find the complement of ∠*D*.

To find the hypotenuse *f*:

$$\sin E = \frac{\text{opp}}{\text{hyp}}$$    Use the sine function and substitute *E* = 57.5° and opp = 24.

$$\sin 57.5° = \frac{24}{f}$$    Solve for *f*.

$$f(\sin 57.5°) = 24$$

$$f = \frac{24}{\sin 57.5°}$$    Evaluate.

$$f = 28.45653714$$

$$\mathbf{f = 28.46 \text{ in.}}$$    Round to four significant digits.

To find side *d*:

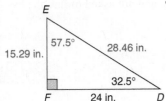

**Figure 13–12**

$$\sin D = \frac{\text{opp}}{\text{hyp}}$$    Substitute *D* = 32.5° and hyp = 28.45653714. Use full calculator value for *f* to get the most accurate result.

$$\sin 32.5° = \frac{d}{28.45653714}$$    Solve for *d*.

$$28.45653714(\sin 32.5°) = d$$    Evaluate.

$$15.28968626 = d$$

$$\mathbf{15.29 \text{ in.} = d}$$    Round to four significant digits.

**The solved triangle is shown in Figure 13–12.**

**TIP!**

### Check Computations Using the Pythagorean Theorem

In this and other problems involving right triangles, you can check your computations using the Pythagorean theorem, $(\text{hyp})^2 = (\text{leg})^2 + (\text{leg})^2$. Let's check the solution to the preceding example.

| **Using Values to the Nearest Hundredth** | **Using Full Calculator Values** |
|---|---|
| $(28.46)^2 = (15.29)^2 + (24)^2$ | $(28.45653714)^2 = (15.28968626)^2 + (24)^2$ |
| $809.9716 = 233.7841 + 576$ | $809.774506 = 233.7745059 + 576$ |
| $809.9716 = 809.7841$ | $809.774506 = 809.7745059$ |

Rounding discrepancies are minimized when more significant digits are used.

### See the BIG Picture; Then Focus on the Little Parts

In solving a right triangle, we are generally given the values of three parts and are asked to find the values of the three missing parts. Look at the big picture first.

- Identify the three given parts, one of which is the right angle.
- Identify the missing parts.
- Plan a strategy to find each missing part.
- Focus on one part at a time.
- Check by using the Pythagorean theorem and the property that the three angles of a triangle add to 180°.

**2** **Find the Missing Parts of a Right Triangle Using the Cosine Function.**

In some of the previous examples, when we found a side not opposite the given angle, we had to find the other angle first by subtracting the given acute angle from 90°. If we use the cosine function, however, we can find the desired side by using the given angle, rather than its complement.

### Use the cosine function to find unknown parts of a right triangle:

1. Two of these three parts of a right triangle must be known:
   (a) One acute angle
   (b) The side adjacent to the known acute angle
   (c) Hypotenuse
2. Substitute two known values in the ratio $\cos \theta = \dfrac{\text{adj}}{\text{hyp}}$.
3. Solve for the missing part.

**EXAMPLE** Find angle $A$ of Figure 13–13.

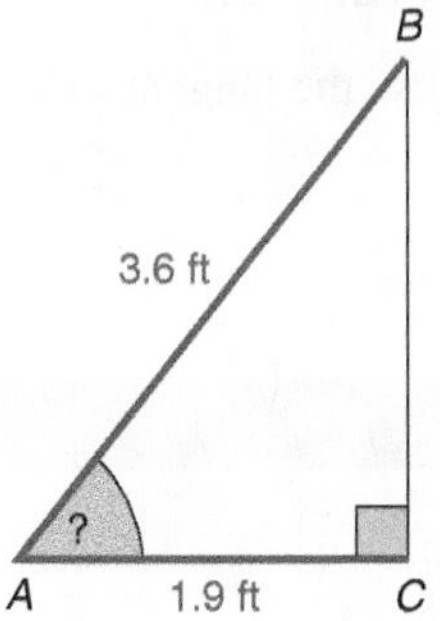

We are given the hypotenuse and the side adjacent to the angle, so we use the cosine function to find the desired angle.

$$\cos A = \frac{\text{adj}}{\text{hyp}}$$ 
Substitute known values.

$$\cos A = \frac{1.9}{3.6}$$ 
Use the inverse cosine function. Evaluate.

$$\cos^{-1}\left(\frac{1.9}{3.6}\right) = A$$

$$A = 58.14456918$$

$$A = \mathbf{58.1°}$$ 
Round to nearest 0.1°.

**Figure 13–13**

B

3.6 ft

?

A    1.9 ft    C

**EXAMPLE**  Find side $b$ of Figure 13–14.

We can use either the sine or the cosine function because we are given the hypotenuse and an angle. However, we do not have to find the complement of the given angle if we use the cosine function.

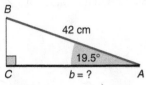

**Figure 13–14**

$$\cos A = \frac{adj}{hyp}$$  Substitute known values.

$$\cos 19.5° = \frac{b}{42}$$  Solve for $b$.

$$42(\cos 19.5°) = b$$  Evaluate.

$$39.59094263 = b$$

$$\mathbf{39.59\ cm = b}$$  Four significant digits.

---

**EXAMPLE**  Find side $t$ of Figure 13–15.

The cosine function is the most efficient function to use because we are finding the hypotenuse and given an acute angle and its adjacent side.

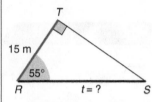

**Figure 13–15**

$$\cos R = \frac{adj}{hyp}$$  Substitute known values.

$$\cos 55° = \frac{15}{t}$$  Solve for $t$.

$$t(\cos 55°) = 15$$

$$t = \frac{15}{\cos 55°}$$  Evaluate.

$$t = 26.15170193$$

$$t = \mathbf{26.15\ m}$$  Four significant digits.

---

**3**  **Find the Missing Parts of a Right Triangle Using the Tangent Function.**

If we know only the length of the two legs of a triangle, we should use the tangent function.

**Use the tangent function to find unknown parts of a right triangle:**

1. Two of the following three parts of a right triangle must be known:
   (a) An acute angle
   (b) The side opposite the acute angle
   (c) The side adjacent to the acute angle

2. Substitute two known values in the formula $\tan \theta = \dfrac{opp}{adj}$.

3. Solve for the missing part.

**EXAMPLE**  Find angle $A$ of Figure 13–16.

We are looking for an angle and given the side opposite and the side adjacent to the angle, so we use the tangent function.

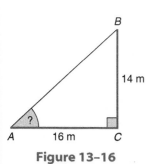

**Figure 13–16**

$$\tan A = \frac{\text{opp}}{\text{adj}}$$  Substitute known values.

$$\tan A = \frac{14}{16}$$  Use the inverse tangent function. Evaluate.

$$\tan^{-1}\left(\frac{14}{16}\right) = A$$

$$A = 41.18592517°$$

$$A = \mathbf{41.2°}$$  Round to nearest 0.1°.

---

**EXAMPLE**  Find side $a$ of Figure 13–17.

We are given an acute angle and the side adjacent to the acute angle. We are looking for the opposite side so use the tangent function.

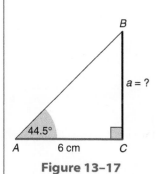

**Figure 13–17**

$$\tan A = \frac{\text{opp}}{\text{adj}}$$  Substitute known values.

$$\tan 44.5° = \frac{a}{6}$$  Solve for $a$.

$$6(\tan 44.5°) = a$$  Evaluate.

$$5.896183579 = a$$

$$\mathbf{5.896 \text{ cm}} = a$$  Four significant digits.

---

**EXAMPLE**  Find side $b$ of Figure 13–18.

We know an acute angle and its opposite side. We are looking for the side adjacent to the acute angle.

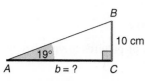

**Figure 13–18**

$$\tan A = \frac{\text{opp}}{\text{adj}}$$  Substitute known values.

$$\tan 19° = \frac{10}{b}$$  Solve for $b$.

$$b(\tan 19°) = 10$$

$$b = \frac{10}{\tan 19°}$$  Evaluate.

$$b = 29.04210878$$

$$b = \mathbf{29.04 \text{ cm}}$$  Four significant digits.

**4** **Select the Most Direct Method for Solving Right Triangles.**

Our first task in solving any problem involving right triangles is to *select the most convenient and efficient function.*

> **To select the most direct method for solving a right triangle:**

1. Where possible, choose the function that uses given parts rather than unknown parts that must be calculated.
2. Where possible, choose the function that gives the desired part directly, that is, without having to find other parts first.

---

**EXAMPLE** In $\triangle ABC$ of Figure 13–19, find $c$.

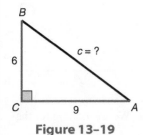

**Figure 13–19**

The most direct way of finding side $c$ is to use the *Pythagorean theorem.* The two legs of a right triangle are given.

| | |
|---|---|
| $c^2 = a^2 + b^2$ | Substitute known values. |
| $c^2 = 6^2 + 9^2$ | Solve for $c$. |
| $c^2 = 36 + 81$ | |
| $c^2 = 117$ | Take square root of both sides. |
| $c = 10.81665383$ | |
| $\boldsymbol{c = 10.82}$ | Four significant digits. |

---

**EXAMPLE** In $\triangle ABC$ of Figure 13–20, find angle $B$.

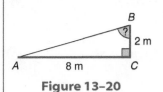

**Figure 13–20**

We are given the side adjacent to and the side opposite angle $B$, so we use the tangent function for a direct solution.

| | |
|---|---|
| $\tan \theta = \dfrac{\text{opp}}{\text{adj}}$ | Substitute known values. |
| $\tan B = \dfrac{8}{2}$ | Reduce. |
| $\tan B = 4$ | Use the inverse tangent function. |
| $\tan^{-1} 4 = B$ | Evaluate. |
| $B = 75.96375653$ | |
| $\boldsymbol{B = 76.0°}$ | Round to nearest 0.1°. |

---

**5** **Solve Applied Problems Using Right-Triangle Trigonometry.**

In solving technical problems, it's a good idea to draw diagrams or pictures to visualize the various relationships.

---

**EXAMPLE** A jet takes off at a 30° angle (see Figure 13–21). If the runway (from takeoff) is 875 ft long,
**AVIA** find the altitude of the airplane as it flies over the end of the runway.

**Known facts** One acute angle = 30°; adjacent side = 875 ft

**Unknown fact** Plane's altitude at end of runway or opposite side

---

Chapter 13 / Right-Triangle Trigonometry

| Relationship | $\tan\theta = \dfrac{\text{opp}}{\text{adj}}$ |
|---|---|

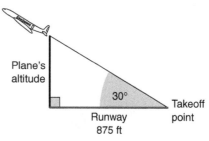

**Figure 13–21**

| | |
|---|---|
| **Estimation** | When the legs of a right triangle are equal, the acute angles are equal to 45° each. Because 30° is less than 45°, the side opposite the 30° angle should be less than 875 ft. |
| **Calculations** | $\tan\theta = \dfrac{\text{opp}}{\text{adj}}$ |

$\tan\theta = \dfrac{\text{opp}}{\text{adj}}$      Substitute known values.

$\tan 30° = \dfrac{a}{875}$      Solve for *a*.

$875(\tan 30°) = a$      Evaluate.

$505.1814855 = a$

| **Interpretation** | **505.2 ft = plane's altitude**      Four significant digits. |
|---|---|

Many right-triangle applications use the terminology **angle of elevation** and **angle of depression.** See Figure 13–22. The angle of elevation is generally used when we are looking *up* at an object. We use the angle of depression to describe the location of an objective *below* our eye level. *Both* angles are formed by a line of sight and a horizontal line from the point of sight.

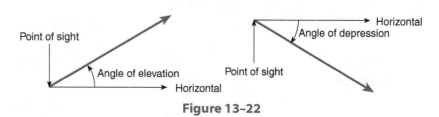

**Figure 13–22**

**EXAMPLE CON** A stretch of roadway drops 30 ft for every 300 ft of road (see Fig. 13–23). Find the *angle of declination* of the road.

The **angle of declination** is the angle of depression. The opposite side and the hypotenuse are given.

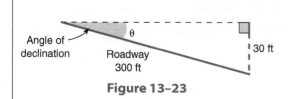

**Figure 13–23**

$\sin\theta = \dfrac{\text{opp}}{\text{hyp}}$      Substitute known values.

$\sin\theta = \dfrac{30}{300}$      Reduce.

$\sin\theta = 0.1$      Use the inverse sine function.

$\sin^{-1} 0.1 = \theta$

$\theta = 5.739170477°$

**The angle of declination of the road is 5.7° rounded to the nearest 0.1°.**

A surveyor locates two points on a steel column so that it can be set plumb (perpendicular to the horizon). If the angle of elevation is 15° and the surveyor's transit is 175 ft from the column (see Figure 13–24), find the distance from the transit to the upper point (point B) on the column. (A **transit** is a surveying instrument used for measuring angles.)

An acute angle and the adjacent side are given. To find the distance from the transit to point *B* on the column (hypotenuse), we use the *cosine function.*

$$\cos \theta = \frac{\text{adj}}{\text{hyp}}$$  Substitute known values.

$$\cos 15° = \frac{175}{\text{hyp}}$$  Solve for hypotenuse.

$$\text{hyp}(\cos 15°) = 175$$

$$\text{hyp} = \frac{175}{\cos 15°}$$  Evaluate.

$$\text{hyp} = 181.1733316$$

$$\text{hyp} = 181.2 \text{ ft}$$  Four significant digits.

**Point *B* is 181.2 ft from the transit.**

**Figure 13–24**

---

Find the angle a rafter makes with a joist of a house if the rise is 12 ft and the span is 30 ft (see Figure 13–25). Also, find the length of the rafter.

The span is twice the distance from the outside end to the center point of the joist. Therefore, to solve the right triangle for the desired angle, we draw the triangle shown in Figure 13–26. Because the legs of a right triangle are given, the tangent function is used.

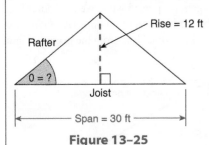

**Figure 13–25**

Find θ.

$$\tan \theta = \frac{\text{opp}}{\text{adj}}$$  Substitute known values.

$$\tan \theta = \frac{12}{15}$$  Use the inverse tangent function.

$$\tan^{-1} \frac{12}{15} = \theta$$

$$\theta = 38.65980825$$  Evaluate.

$$\theta = 38.7°$$  Round to nearest 0.1°

Find the length of the rafter (hypotenuse).

We use the *Pythagorean theorem* to find the length of the rafter directly.

$$(\text{hyp})^2 = (\text{leg})^2 + (\text{leg})^2$$  Substitute known values.

$$(\text{hyp})^2 = 12^2 + 15^2$$  Evaluate.

$$(\text{hyp})^2 = 144 + 225$$

$$(\text{hyp})^2 = 369$$  Take square root of both sides.

$$\text{hyp} = 19.20937271$$

$$\text{hyp} = 19.21 \text{ ft}$$  Four significant digits.

**The angle the rafter makes with the joist is 38.7° and the rafter is 19.21 ft.**

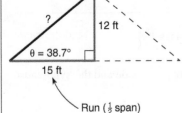

**Figure 13–26**

Chapter 13 / Right-Triangle Trigonometry

Find the angle formed by the rod in the mechanical assembly shown in Figure 13–27.

Given the hypotenuse and the side opposite the desired angle, we use the sine function.

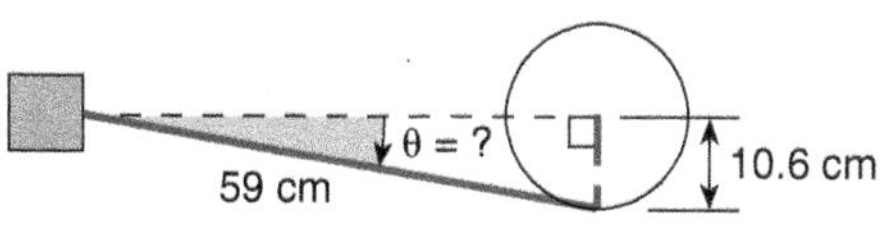

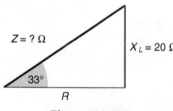

**Figure 13–27**

$$\sin \theta = \frac{\text{opp}}{\text{hyp}}$$   Substitute known values.

$$\sin \theta = \frac{10.6}{59}$$   Use the inverse sine function. Evaluate.

$$\sin^{-1} \frac{10.6}{59} = \theta$$

$$\theta = 10.35001563$$   Round.

$$\theta = 10.4°$$

**The angle formed by the rod is 10.4°.**

EXAMPLE
ELEC

Find the impedance $Z$ of a circuit with 20 Ω of reactance $X_L$ represented by the vector diagram in Figure 13–28.

Because we are looking for the hypotenuse $Z$ and are given an acute angle and the opposite side, we use the sine function.

**Figure 13–28**

$$\sin \theta = \frac{\text{opp}}{\text{hyp}}$$   Substitute known values.

$$\sin 33° = \frac{20}{Z}$$   Solve for $Z$.

$$Z (\sin 33°) = 20$$

$$Z = \frac{20}{\sin 33°}$$   Evaluate.

$$Z = 36.72156918$$

$$Z = 36.72 \ \Omega$$   Four significant digits.

**The impedance, $Z$, is 36.72 Ω.**

SECTION 13–2 SELF-STUDY EXERCISES

**1** Use the sine function to find the indicated parts of the triangle $LMN$ in Fig. 13–29. Round lengths of sides to four significant digits and angles to the nearest 0.1°.

**1.** Find $M$ if $n = 15$ m and $m = 7$ m.
**2.** Find $l$ if $n = 13$ in. and $L = 32°$.
**3.** Find $m$ if $l = 15$ m and $L = 28°$.
**4.** Find $n$ if $l = 12$ ft and $M = 42°$.
**5.** Find $M$ if $m = 13$ cm and $n = 19$ cm.
**6.** Find $M$ if $n = 3.7$ in. and $l = 2.4$ in.

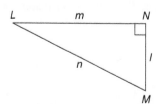

**Figure 13–29**

Solve triangle *STU* in Figure 13–30 for the given values using the sine function. Check the measures of the sides by using the Pythagorean theorem. Round as above.

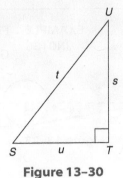

**Figure 13–30**

7. Solve if $t = 18$ yd and $s = 14$ yd.
8. Solve if $U = 45°$ and $u = 4.7$ m.
9. Solve if $S = 34.5°$ and $t = 8.5$ mm.
10. Solve if $S = 16°$ and $s = 14$ m.

**2** Use the cosine function to find the indicated parts of triangle *KLM* in Figure 13–31. Round sides to four significant digits and angles to the nearest $0.1°$.

11. Find $M$ if $k = 13$ m and $l = 16$ m.
12. Find $k$ if $l = 11$ cm and $M = 24°$.
13. Find $l$ if $M = 31°$ and $k = 27$ ft.
14. Find $l$ if $m = 15$ dm and $M = 25°$.
15. Find $k$ if $K = 72°$ and $l = 16.7$ mm.
16. Find $l$ if $K = 67°$ and $k = 13$ yd.

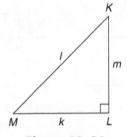

**Figure 13–31**

Use the sine *or* cosine function to solve triangle *QRS* in Figure 13–32. Round as above.

17. Solve if $s = 23$ ft and $q = 16$ ft.
18. Solve if $s = 17$ cm and $R = 46°$.
19. Solve if $q = 14$ dkm and $Q = 73.5°$.
20. Solve if $R = 59.5°$ and $q = 8$ m.

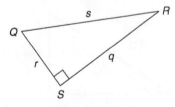

**Figure 13–32**

**3** Use the tangent function to find the indicated parts of triangle *ABC* in Figure 13–33. Round sides to four significant digits and angles to the nearest $0.1°$.

21. Find $A$ if $b = 11$ cm and $a = 6$ cm.
22. Find $b$ if $a = 1.9$ m and $A = 25°$.
23. Find $a$ if $A = 40.5°$ and $b = 7$ ft.
24. Find $A$ if $b = 10.8$ m and $a = 4.7$ m.
25. Find $a$ if $A = 43°$ and $b = 0.05$ cm.
26. Find $a$ if $B = 68°$ and $b = 0.03$ m.

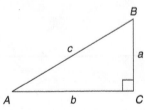

**Figure 13–33**

Solve triangle *DEF* in Figure 13–34. Round sides to four significant digits and angles to the nearest $0.1°$.

27. Solve if $e = 4.6$ m and $d = 3.2$ m.
28. Solve if $D = 42°$ and $e = 7$ ft.
29. Solve if $E = 73.5°$ and $e = 20.13$ in.
30. Solve if $d = 11$ ft and $e = 8$ ft.

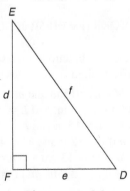

**Figure 13–34**

**4** Find the indicated part of the right triangles in Figures 13–35 through 13–39, by the most direct method.

**31.**

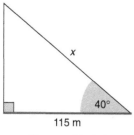

Figure 13–35

**32.**

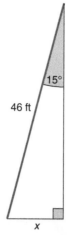

Figure 13–36

**33.**

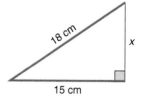

Figure 13–37

**34.**

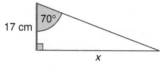

Figure 13–38

**35.**

Figure 13–39

**5** Use the trigonometric functions to solve the problems. Round side lengths to four significant digits and angles to 0.1°.

**36. CON** A sign is attached to a building by a triangular brace. If the horizontal length of the brace is 48 in. and the angle at the sign is 25° (see Figure 13–40), what is the length of the wall support piece?

**37. CON** To measure a property line that crosses a pond, a surveyor sights to a point across the pond, makes a right angle, measures 50 ft, and sights the point across the pond with a 47° angle (see Figure 13–41). Find the distance across the pond from the initial point.

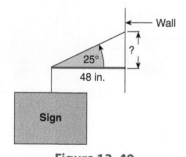

Figure 13–40

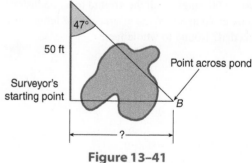

Figure 13–41

**38. AVIA** At what angle must a jet descend if it is 900 ft above the end of the runway and must touch down 1,500 ft from the runway's end?

**39. CON** A 50-ft wire is used to brace a utility pole. If the wire is attached 4 ft from the top of the 35-ft pole, how far from the base of the pole will the wire be attached to the ground?

**40. CON** A roadway rises 4 ft for every 15 ft along the road. What is the angle of inclination of the roadway?

**41. AG/H** A shadow cast by a tree is 32 ft long when the angle of inclination of the sun is 36°. How tall is the tree?

42. **ELEC** The vector diagram of the ac circuit in Figure 13–42 shows the relationship between the impedance Z, the resistance R, the reactance X and the phase angle θ. Find the reactance X. All units are in ohms. The impedance is 20 Ω and the phase angle is 30°.

43. **ELEC** Using Figure 13–42, find resistance R, when the impedance is 20 Ω and the phase angle is 30°.

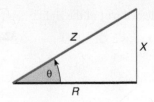

**Figure 13–42**

44. **ELEC** An ac circuit has a resistance of 30 Ω and an impedance of 38 Ω. What is the phase angle (Figure 13–42)?

45. **ELEC** Find the impedance and reactance of an ac circuit that has a resistance of 26 Ω and a phase angle of 25° (Figure 13–42). Round to the nearest tenth.

46. **INDTR** A piston assembly at the midpoint of its stroke forms a right triangle (see Figure 13–43). Find the length of R.

47. **CON** Find the angle a rafter makes with a joist of a house if the rise is 18 ft and the span is 50 ft. Refer to Figure 13–25 on page 552.

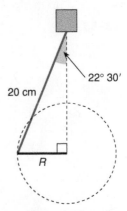

**Figure 13–43**

48. **AVIA** A plane takes off from Reagan National Airport at an angle of elevation of 20°. If the plane travels 4,000 ft/min, how far above the ground will the plane be after 2 min (Figure 13–44)?

49. **AVIA** How far above the ground will a plane be if its angle of elevation is 22° and it travels 10,000 ft (in the air) from takeoff?

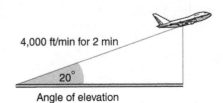

Angle of elevation

**Figure 13–44**

50. **CON** A 20-ft ladder is set against a building at a 70° angle to the ground in Figure 13–45. How high up the wall does the ladder touch? Round to whole feet.

51. **CON** A cell tower is 100 m high. A guy wire that makes a 60° angle with the ground must be fastened on the tower 90 m above the ground. What length guy wire is needed? Round to whole feet.

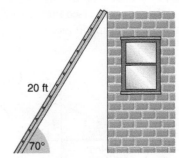

**Figure 13–45**

**52. INDTR** An L brace is installed as shown in Figure 13–46. Find the length of $x$ and angle A in degrees. Round the length to the nearest tenth inch and the angle to the nearest tenth degree.

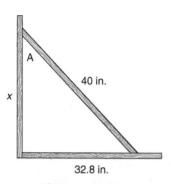

**Figure 13–46**

**53. INDTEC** Three holes must be drilled in a plastic plate, as shown in Figure 13–47. Find distance $a$.

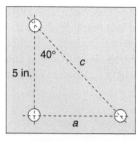

**Figure 13–47**

Use Figure 13–48 for Exercises 54–56.

**54. CON** A sewage drain pipe has a 60° offset angle and a run of 24 ft. Find the length of the offset and the diagonal. Round to the nearest foot.

**55. ELEC** An electrical conduit has an offset of 46 in. and an offset angle of 56°. Find the length of the run and of the diagonal. Round to the nearest inch.

**56. CON** A 2-in. diameter pipe has a 15-in. offset and a 40° bend (angle of offset). Find the run and diagonal. Round to the nearest inch.

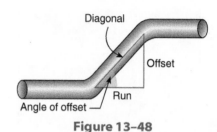

**Figure 13–48**

**57. INDTEC** A V-slot gauge is constructed to enable machinists to check the diameter of bearings. Find the radius of the bearing ($AC$) and the distance ($BC$) from the vertex ($B$) to the point where the bearing touches the gauge ($C$) in Figure 13–49 if $AB = 2.8$ in. and angle $ABC$ is 45°. Round to tenths.

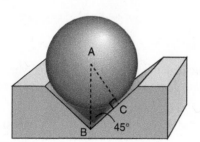

**Figure 13–49**

**58. INDTEC** Find the angle of taper for the tapered steel rod shown in Figure 13–50. Round to the nearest tenth degree.

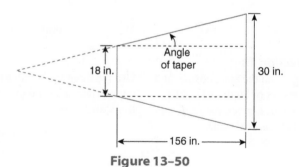

**Figure 13–50**

**59. INDTEC** Find the angle of taper for a flag pole that is 5 m tall and measures 90 cm on one end and 10 cm on the other end. Round to the nearest tenth degree.

**60. INDTEC** Bolt circles are used on a wide variety of structures such as hubs, wheels, cranks, and metal and plastic covers. A *bolt circle* is the visualized circle that runs through the centers of bolt holes in a piece of metal or plastic. The *bolt distance* is the distance between the centers of two bolt holes. Find the bolt distance $E$ if the diameter of the bolt circle is 12 inches and the angle formed by the bolt circle diameter and bolt distance is 67.5° (Figure 13–51). Round to tenths.

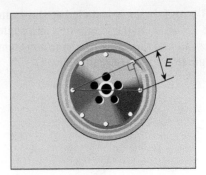

**Figure 13–51**

**61. INDTEC** Find the diameter of the bolt circle in Figure 13–52 if the bolt distance is 8 cm and the angle formed by the diameter and the bolt distance is 72°. Round to tenths.

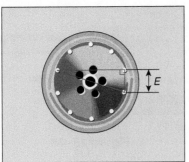

**Figure 13–52**

**62. CON** What is the depth of a dovetail in Figure 13–53 if the larger opening is 5.2 cm, the smaller opening is 3.6 cm, and the cut angle is 42°? Round to tenths.

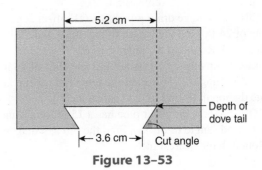

**Figure 13–53**

---

## CHAPTER REVIEW OF KEY CONCEPTS

**Learning Objectives**

**What to Remember with Examples**

### Section 13–1

**1** Find the sine, cosine, and tangent of angles of right triangles, given the measures of at least two sides (pp. 538–540).

Use the three trigonometric functions to calculate the value of the function when the appropriate sides are given. The Pythagorean theorem may be needed to find the length of a third side.

$$\text{sine } A = \frac{\text{opposite side}}{\text{hypotenuse}} \qquad \text{cosine } A = \frac{\text{adjacent side}}{\text{hypotenuse}} \qquad \text{tangent } A = \frac{\text{opposite side}}{\text{adjacent side}}$$

Find the sine, cosine, and tangent of angle $A$ in Figure 13–54 and round to four significant digits.

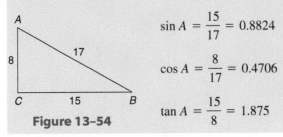

**Figure 13–54**

$$\sin A = \frac{15}{17} = 0.8824$$

$$\cos A = \frac{8}{17} = 0.4706$$

$$\tan A = \frac{15}{8} = 1.875$$

**2** Find trigonometric values for the sine, cosine, and tangent using a calculator (pp. 540–541).

Use the $\boxed{\text{SIN}}$, $\boxed{\text{COS}}$, and $\boxed{\text{TAN}}$ keys to find the trigonometric value of a specified angle. Be sure your calculator is set to the appropriate mode: degree or radian.

Use your calculator to find the following: sin 35°, cos 18°, tan 55°, sin 1.3, cos 0.87, tan 1.1.

*Most calculators:*
Be sure your calculator is in degree mode:

$\boxed{\text{SIN}}$   35 $\boxed{=}$ $\Rightarrow$ 0.5735764364
$\boxed{\text{COS}}$   18 $\boxed{=}$ $\Rightarrow$ 0.9510565163
$\boxed{\text{TAN}}$   55 $\boxed{=}$ $\Rightarrow$ 1.428148007

Change to radian mode:

$\boxed{\text{SIN}}$   1.3 $\boxed{=}$ $\Rightarrow$ 0.9635581854
$\boxed{\text{COS}}$   0.87 $\boxed{=}$ $\Rightarrow$ 0.6448265472
$\boxed{\text{TAN}}$   1.1 $\boxed{=}$ $\Rightarrow$ 1.964759657

**3** Find the angle measure, given a trigonometric value (p. 542).

To find the angle measure when the trigonometric value is given, use the inverse trigonometric function key, which may be accessed by pressing the $\boxed{\text{SHIFT}}$, $\boxed{\text{INV}}$ or $\boxed{2^{\text{nd}}}$ key.

Find the value of $x$ in degrees: $\cos x = 0.906307787$ or $x = \cos^{-1}(0.906307787)$.

$\boxed{\text{COS}^{-1}}$ 0.906307787 $\boxed{=}$     Be sure calculator is in degree mode.

$$x = 25°$$

## Section 13–2

**1** Find the missing parts of a right triangle using the sine function (pp. 543–547).

$$\sin \theta = \frac{\text{opposite side}}{\text{hypotenuse}}$$

Use the sine function to find side $x$ (Figure 13–55).

$\sin \theta = \dfrac{\text{opp}}{\text{hyp}}$     Substitute known values.

$\sin 25° = \dfrac{x}{12}$     Solve for $x$.

$x = 12 \sin 25°$     Evaluate.

$x = 5.071419141$

$x = 5.071$ in.     Four significant digits.

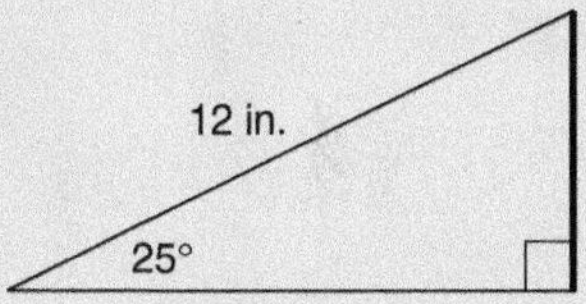

**Figure 13–55**

**2** Find the missing parts of a right triangle using the cosine function (pp. 547–548).

$$\cos \theta = \frac{\text{adjacent side}}{\text{hypotenuse}}$$

Use the cosine function to find side $x$ (Figure 13–56).

$\cos \theta = \dfrac{\text{adj}}{\text{hyp}}$     Substitute known values.

$\cos 37° = \dfrac{26}{x}$     Solve for $x$.

$x = \dfrac{26}{\cos 37°}$     Evaluate.

$x = 32.55552711$

$x = 32.56$ m

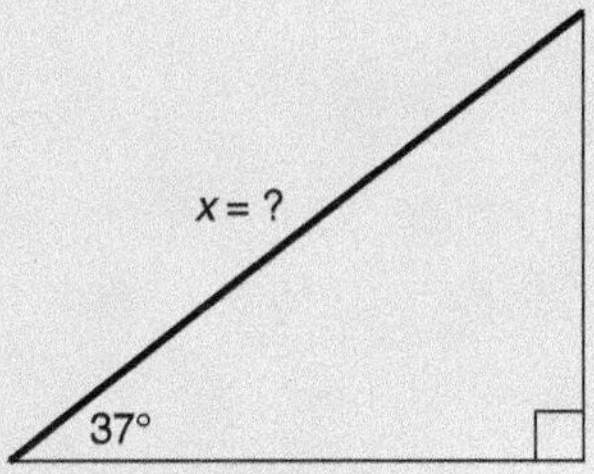

**Figure 13–56**

Chapter Review of Key Concepts

**3** Find the missing parts of a right triangle using the tangent function (pp. 548–549).

$$\tan\theta = \frac{\text{opposite side}}{\text{adjacent side}}$$

Use the tangent function to find θ in degrees (Figure 13–57).

| | |
|---|---|
| $\tan\theta = \dfrac{\text{opp}}{\text{adj}}$ | Substitute known values. |
| $\tan\theta = \dfrac{8}{11}$ | Use inverse tangent function. |
| $\theta = \tan^{-1}\left(\dfrac{8}{11}\right)$ | Evaluate. |
| $\theta = 36.02737339$ | |
| $\theta = 36.0°$ | Round. |

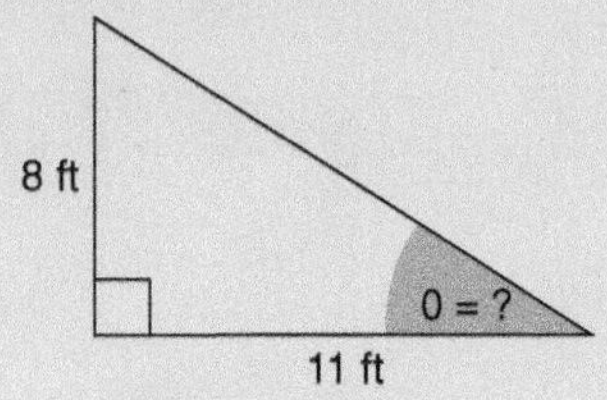

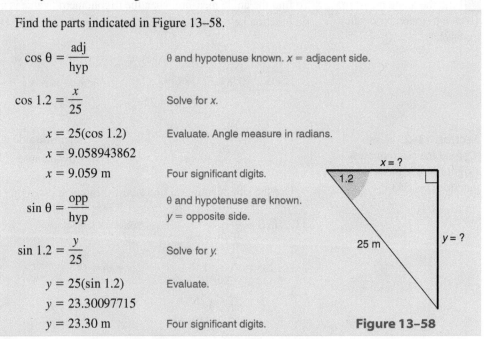

Figure 13–57

**4** Select the most direct method for solving right triangles (p. 550).

Whenever possible, choose the function that uses parts given in the problem rather than parts that are not given and must be calculated.

Whenever possible, choose the function that gives the desired part directly; that is, find the desired part without having to find other parts first.

Find the parts indicated in Figure 13–58.

| | |
|---|---|
| $\cos\theta = \dfrac{\text{adj}}{\text{hyp}}$ | θ and hypotenuse known. $x$ = adjacent side. |
| $\cos 1.2 = \dfrac{x}{25}$ | Solve for $x$. |
| $x = 25(\cos 1.2)$ | Evaluate. Angle measure in radians. |
| $x = 9.058943862$ | |
| $x = 9.059$ m | Four significant digits. |
| $\sin\theta = \dfrac{\text{opp}}{\text{hyp}}$ | θ and hypotenuse are known. $y$ = opposite side. |
| $\sin 1.2 = \dfrac{y}{25}$ | Solve for $y$. |
| $y = 25(\sin 1.2)$ | Evaluate. |
| $y = 23.30097715$ | |
| $y = 23.30$ m | Four significant digits. |

Figure 13–58

**5** Solve applied problems using right-triangle trigonometry (pp. 550–553).

To solve a problem using the trigonometric functions, first determine the given parts and the missing part or parts; then identify the trigonometric function that relates the given and missing parts. Use this function and the given information to find the missing information.

A jet takes off at a 25° angle. Find the distance traveled by the plane from the takeoff point to the end of the runway if the runway is 950 ft long (Figure 13–59).

| | |
|---|---|
| $\cos\theta = \dfrac{\text{adj}}{\text{hyp}}$ | θ and adjacent side are known. Find the hypotenuse. |
| $\cos 25° = \dfrac{950}{x}$ | Solve for $x$. |
| $x = \dfrac{950}{\cos 25°}$ | Evaluate. |
| $x = 1{,}048.209023$ | |
| $x = 1{,}048$ ft | Four significant digits. |

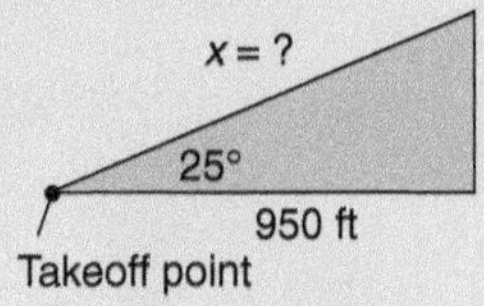

Figure 13–59

## Section 13–1

Find the trigonometric ratios using Figure 13–60.

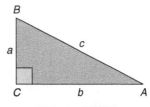

**Figure 13–60**

Use $a = 15$, $b = 20$, $c = 25$ in Exercises 1–6 and express the ratios as fractions in lowest terms.

**1.** $\sin A$      **2.** $\cos A$      **3.** $\tan A$      **4.** $\tan B$      **5.** $\sin B$      **6.** $\cos B$

Use $a = 2$ ft, $b = 10$ in., $c = 2$ ft 2 in. in Exercises 7–10 and express the ratios as fractions in lowest terms.

**7.** $\sin A$      **8.** $\tan A$      **9.** $\tan B$      **10.** $\cos A$

Use $a = 7$, $b = 10.5$, $c = 12.62$ in Exercises 11–16 and express the ratios in decimals to the nearest ten-thousandth.

**11.** $\cos B$      **12.** $\cos A$      **13.** $\tan A$      **14.** $\sin B$      **15.** $\tan B$      **16.** $\sin A$

## Section 13–2

Use your calculator to find the trigonometric values. Round to the nearest ten-thousandth.

**17.** $\cos 32°50'$      **18.** $\cos 42.5°$      **19.** $\sin 0.4712$      **20.** $\tan 1.0210$
**21.** $\tan 47°$      **22.** $\tan 15.6°$      **23.** $\sin 0.8610$      **24.** $\tan 25°40'$

Find an angle (in degrees) having the trigonometric values. $\theta$ represents the unknown angle measure. Express answers to the nearest tenth of a degree.

**25.** $\sin \theta = 0.5446$      **26.** $\tan \theta = 0.8720$      **27.** $\cos \theta = 0.6088$      **28.** $\cos \theta = 0.8897$

Find an angle (in radians) having the trigonometric values. Round to the nearest ten-thousandth.

**29.** $\tan \theta = 2.723$      **30.** $\sin \theta = 0.9205$      **31.** $\tan \theta = 0.3440$      **32.** $\cos \theta = 0.9450$

## Section 13–3

Find the indicated parts of the triangles in Figures 13–61 through 13–66. Round sides to four significant digits and round angles to the nearest 0.1°.

**33.** Find $A$.            **34.** Find $r$.            **35.** Find $h$.

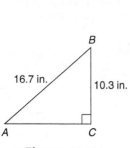

**Figure 13–61**

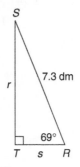

**Figure 13–62**

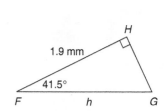

**Figure 13–63**

**36.** Find $c$.

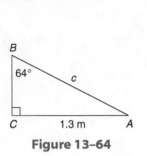

**Figure 13–64**

**37.** Find $y$.

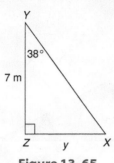

**Figure 13–65**

**38.** Find $B$.

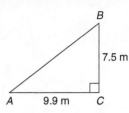

**Figure 13–66**

Solve the triangles in Figures 13–67 through 13–68. Round sides to four significant digits and round angles to the nearest $0.1°$.

**39.**

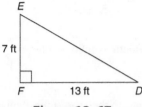

**Figure 13–67**

**40.**

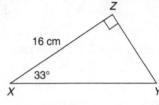

**Figure 13–68**

Find the indicated parts of the triangles in Figures 13–69 through 13–71. Round sides to four significant digits and angle measures to the nearest $0.1°$.

**41.** Find $A$.

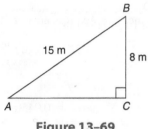

**Figure 13–69**

**42.** Find $A$.

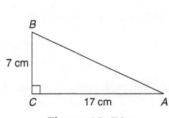

**Figure 13–70**

**43.** Find $a$.

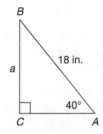

**Figure 13–71**

**44. CON** A railway inclines $14°$. How many feet of track must be laid if the hill is 15 ft high?

**46. CON** From a point 5 ft above the ground and 20 ft from the base of a building, a surveyor uses a transit to sight an angle of $38°$ to the top of the building. Find the height of the building.

**45. CON** A corner shelf is cut so that the sides placed on the wall are 37 in. and 42 in. What are the measures of the acute angles? Round to tenths.

**47. CON** A surveyor makes the measures indicated in Figure 13–72. Solve the triangle. All parts of the triangle should be included in a report. Right sides to four significant digits.

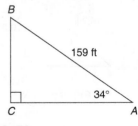

**Figure 13–72**

Chapter 13 / Right-Triangle Trigonometry

**48. CON** What length of rafter is needed for a roof if the rafters form an angle of 35.5° with a joist and the rise is 8 ft? Round to hundredths.

**49. INDTEC** Find the angle formed by the rod and the horizontal in the mechanical assembly shown in Figure 13–73. Round to tenths.

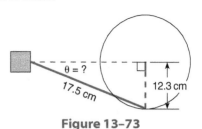

**Figure 13–73**

**50. CON** A utility pole is 40 ft above ground level. A wire must be attached to the pole 3 ft from the top to give it support. If the wire forms a 20° angle with the ground, how long is the wire? Disregard the length needed for attaching the wire to the pole or ground. Round to hundredths.

**51. ELEC** Solve for reactance $X_L$ and resistance $R$ in Figure 13–74. All units are in ohms. Round to hundredths.

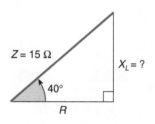

**Figure 13–74**

---

## TEAM PROBLEM-SOLVING EXERCISES

**1.** Draw triangle $ABC$ so that angle $C$ is 90° and angle $A$ is 70°. Make a point $E$ on side $AB$ that is 10 cm from $A$. Make a point $D$ on $BC$ so that $DE$ is parallel to $CA$. If $CA$ is 24 cm, find the length of $BD$.

**2.** You are designing a stairway for a home and need to determine how many steps the stairway will have. If the stairway is 8 ft high and the floor space allotted is 14 ft, determine how many steps are needed and determine the optimum measure (rise and run) for each step.

**3.** Pairs of trigonometric functions have special relationships.
   **(a)** Is there an angle between 0° and 360° for which the sine and cosine of an angle are equal? Justify your answer by making a table of values for the sine and cosine of angles between 0° and 360° with 30° increments and by graphing sin $x$ and cos $x$ for these values.
   **(b)** Investigate the relationship for the sine and tangent functions and justify your answer with a table of values and graph for values between 0° and 360° with 30° increments.

---

## PRACTICE TEST

Write the trigonometric ratios as fractions in lowest terms for the following trigonometric functions using triangle $ABC$ in Figure 13–75. $a = 10$, $b = 24$, $c = 26$.

**1.** sin $A$
**2.** tan $B$

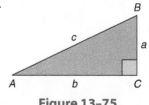

**Figure 13–75**

Express the trigonometric values in decimals to the nearest ten-thousandth. Refer to Figure 13–75 and use $a = 5$, $b = 11.5$, $c = 12.54$.

**3.** $\cos A$

**4.** $\sin B$

Use your calculator to find the trigonometric values. Round to the nearest ten-thousandth.

**5.** $\sin 53°$

**6.** $\sin 61°10'$

Find an angle (in degrees) having the indicated trigonometric values; $\theta$ represents the unknown angle measure. Express your answers to the nearest tenth of a degree.

**7.** $\sin \theta = 0.2756$

**8.** $\tan \theta = 1.280$

Find an angle (in radians) having the indicated trigonometric values. Round to the nearest ten-thousandth.

**9.** $\sin \theta = 0.7660$

**10.** $\cos \theta = 0.8387$

**11.** Solve triangle $ABC$ in Figure 13–76.

**12.** Solve triangle $DEF$ in Figure 13–77. Round as above.

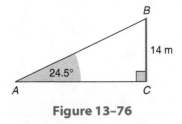

**Figure 13–76**

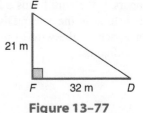

**Figure 13–77**

**13.** A stair has a rise of 4 in. for every 5 in. of run. What is the angle of inclination of the stair? Round to tenths.

**14.** A surveyor uses indirect measurement to find the width of a river at a certain point. The surveyor marks off 50 ft along the river bank at right angles with the river and then sights an angle of 43° to point $A$ across the river (Fig. 13–78). Find the width of the river. Round to hundredths.

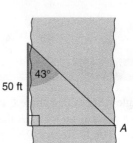

**Figure 13–78**

**15.** Steel girders are reinforced by placing steel supports between two runners so that right triangles are formed (see Figure 13–79). If the runners are 24 in. apart and a 30° angle is desired between the support and a runner, find the length of the support to be placed at a 30° angle.

**16.** In the circuit represented by the diagram of Figure 13–80, find the total current $I_t$. All units are in amps.

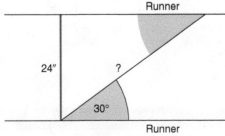

**Figure 13–79**

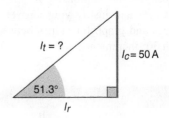

**Figure 13–80**

17. Find the current in the resistance branch $I_r$ of the circuit in Figure 13–80 of Exercise 16. Round to four significant digits.

18. The minimum clearances for the installation of a metal chimney pipe are shown in Figure 13–81. How far from the ridge should the hole be cut for the pipe to pass through the roof? Round to hundredths.

19. Refer to Figure 13–81. What angle is formed by the chimney pipe and the roof where the pipe passes through the roof?

20. Elbows are used to form bends in rigid pipe and are measured in degrees. What is the angle of bend of the elbow in the installation shown in Figure 13–82 to the nearest whole degree?

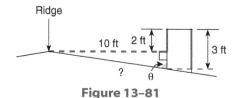

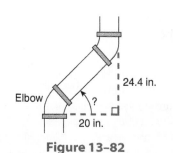

**Figure 13–81**

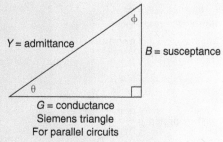

**Figure 13–82**

## CAREER APPLICATION: ELECTRONICS—SERIES CIRCUITS AND PARALLEL CIRCUITS

Electronic technicians use many right triangles, but the two most common ones are the ohms triangle and the siemens triangle. **Ohms** and **siemens** are reciprocals of each other. **Series circuits** use the ohms triangle (Fig. 13–83) and **parallel circuits** use the siemens triangle (Fig. 13–84).

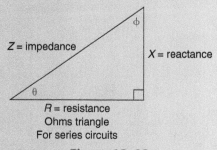

$Z$ = impedance
$X$ = reactance
$R$ = resistance
Ohms triangle
For series circuits
**Figure 13–83**

$Y$ = admittance
$B$ = susceptance
$G$ = conductance
Siemens triangle
For parallel circuits
**Figure 13–84**

For each triangle, the units are the same on all three legs. The hypotenuse and the angle of rotation combine to give the **polar notation**, and the horizontal side and the vertical side combine to give the coordinates of a point in rectangular notation. A polar number shows the location of a point differently from the rectangular coordinate system. The coordinates of a polar number are the length of the hypotenuse ($r$) and the angle of rotation ($\theta$).

Most analyses are done with rectangular numbers. All measurements are done with polar numbers.

Some basic facts about circuits will help us in interpreting a problem.

- All three sides in the siemens triangle are measured in siemens, the reciprocal of ohms.
- If a component has a high resistance, it has a low conductance, and vice versa.
- $R$ and $G$ are used for both dc (direct current) and ac (alternating current) analyses.
- $X$, $Z$, $B$, and $Y$ are used only for ac analyses.

The Pythagorean theorem applies to any right triangle. The formula can be rearranged to solve for any missing side if the other two sides are known. The following equations give a variation of the Pythagorean theorem for each side in both the ohms and siemens triangles.

**Ohms Triangle for Series Circuits**

$$Z = \sqrt{R^2 + X^2}$$
$$R = \sqrt{Z^2 - X^2}$$
$$X = \sqrt{Z^2 - R^2}$$

**Siemens Triangle for Parallel Circuits**

$$Y = \sqrt{G^2 + B^2}$$
$$G = \sqrt{Y^2 - B^2}$$
$$B = \sqrt{Y^2 - G^2}$$

The following are some additional basic facts in electronics:

- The sides for $R$ and $G$ are always positive and rest on the horizontal axis.
- $X$ and $B$ are always perpendicular to the horizontal axis, and they may be positive or negative depending on whether the circuit is inductive or capacitive.
- The hypotenuse, $Z$ or $Y$, is considered positive because it is a measure of magnitude only rather than magnitude *and* direction found in the vertical and horizontal measures.
- The angle $\theta$ is also called the **phase angle** and is either positive or negative, depending on the slope of the hypotenuse.
- The angle $\phi$ is the angle between the hypotenuse and the vertical side.
- $\theta$ and $\phi$ are complementary angles. $\phi = 90° - \theta$ and $\theta = 90° - \phi$.

The angles $\theta$ and $\phi$ can be calculated in several different ways, depending on which sides or angles you were given originally. Another standard notation for $\sin^{-1}$, $\cos^{-1}$, and $\tan^{-1}$ is **arcsin, arccos,** and **arctan,** respectively.

### Ohms

$$\theta = \arcsin\left(\frac{\text{opp}}{\text{hyp}}\right) \qquad \theta = \arcsin\left(\frac{X}{Z}\right) \qquad \theta = \sin^{-1}\left(\frac{X}{Z}\right)$$

$$\theta = \arccos\left(\frac{\text{adj}}{\text{hyp}}\right) \qquad \theta = \arccos\left(\frac{R}{Z}\right) \qquad \theta = \cos^{-1}\left(\frac{R}{Z}\right)$$

$$\theta = \arctan\left(\frac{\text{opp}}{\text{adj}}\right) \qquad \theta = \arctan\left(\frac{X}{R}\right) \qquad \theta = \tan^{-1}\left(\frac{X}{R}\right)$$

### Siemens

$$\theta = \arcsin\left(\frac{\text{opp}}{\text{hyp}}\right) \qquad \theta = \arcsin\left(\frac{B}{Y}\right) \qquad \theta = \sin^{-1}\left(\frac{B}{Y}\right)$$

$$\theta = \arccos\left(\frac{\text{adj}}{\text{hyp}}\right) \qquad \theta = \arccos\left(\frac{G}{Y}\right) \qquad \theta = \cos^{-1}\left(\frac{G}{Y}\right)$$

$$\theta = \arctan\left(\frac{\text{opp}}{\text{adj}}\right) \qquad \theta = \arctan\left(\frac{B}{G}\right) \qquad \theta = \tan^{-1}\left(\frac{B}{G}\right)$$

Finally,

The calculator sequence for finding an arcfunction is the same as the inverse function.
Find all indicated missing sides and angles in Figs. 13–85 and 13–86.

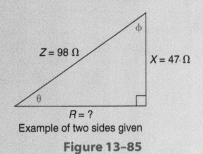

Example of two sides given

**Figure 13–85**

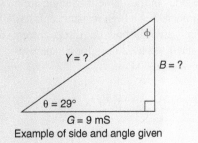

Example of side and angle given

**Figure 13–86**

Chapter 13 / Right-Triangle Trigonometry

$$Z^2 = R^2 + X^2 \qquad \text{Solve for } R. \qquad \cos \theta = \frac{G}{Y} \qquad\qquad \tan \theta = \frac{B}{G}$$

$$R^2 = Z^2 - X^2 \qquad\qquad\qquad Y = \frac{G}{\cos \theta} \qquad\qquad B = G \tan \theta$$

$$R = \sqrt{Z^2 - X^2} \qquad\qquad\qquad Y = \frac{9}{\cos 29°} \qquad\qquad B = 9 \tan 29°$$

$$R = \sqrt{98^2 - 47^2} \qquad\qquad\qquad Y = 10.29018661 \qquad\qquad B = 4.988781463$$

$$R = 85.99418585 \qquad\qquad\qquad \mathbf{Y = 10.29 \text{ mS}} \qquad\qquad \mathbf{B = 4.99 \text{ mS}}$$

$$\mathbf{R = 85.99 \ \Omega} \qquad\qquad\qquad \mathbf{\phi = 90° - 29° = 61°}$$

$$\sin \theta = \frac{\text{opp}}{\text{hyp}} = \frac{X}{Z}$$

$$\theta = \arcsin\left(\frac{X}{Z}\right) = \sin^{-1}\left(\frac{X}{Z}\right)$$

$$\theta = \arcsin\left(\frac{47}{98}\right) = \sin^{-1}\left(\frac{47}{98}\right)$$

$$\theta = 28.65874762$$

$$\mathbf{\theta = 28.7°}$$

$$\mathbf{\phi = 90° - 28.7° = 61.3°}$$

Fill in all answers on the triangles in Figures 13–87 and 13–88 with the correct units.

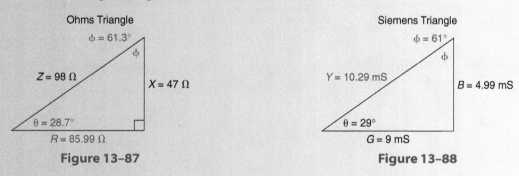

Ohms Triangle

$\phi = 61.3°$  
$Z = 98 \ \Omega$  
$X = 47 \ \Omega$  
$\theta = 28.7°$  
$R = 85.99 \ \Omega$

**Figure 13–87**

Siemens Triangle

$\phi = 61°$  
$Y = 10.29 \text{ mS}$  
$B = 4.99 \text{ mS}$  
$\theta = 29°$  
$G = 9 \text{ mS}$

**Figure 13–88**

*Proof:* $Y^2 = G^2 + B^2$

$10.29^2 \overset{?}{=} 9^2 + 4.99^2$

$105.8841 \overset{?}{=} 81 + 24.9001$

$105.8841 \overset{?}{=} 105.9001$ \qquad True, except for rounding discrepancy.

$\phi + \theta = 90°$

$61 + 29 \overset{?}{=} 90°$ \qquad True.

*Proof:* $\tan 28.7° \overset{?}{=} \dfrac{47}{85.99}$

$0.547484008 \approx 0.546575183$ \qquad True, except for rounding discrepancy.

## Exercises

Redraw the triangles in Figures 13–89 to 13–96. Fill in all missing sides and angles. Include correct measuring units. Round angle measures to the nearest 0.1° and side measures to hundreths.

**1.**

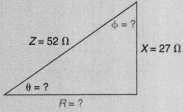

**Figure 13–89**

**2.**

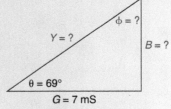

**Figure 13–90**

**3.**

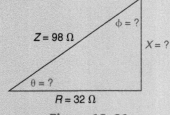

**Figure 13–91**

**4.**

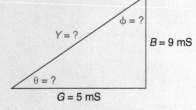

**Figure 13–92**

**5.**

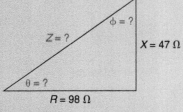

**Figure 13–93**

**6.**

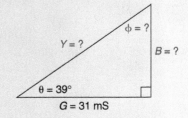

**Figure 13–94**

**7.**

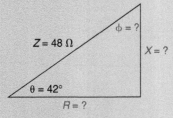

**Figure 13–95**

**8.**

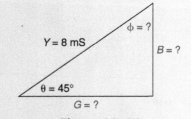

**Figure 13–96**

## Answers for Exercises

**1.**

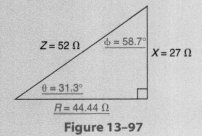

**Figure 13–97**

**2.**

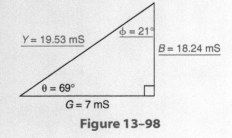

**Figure 13–98**

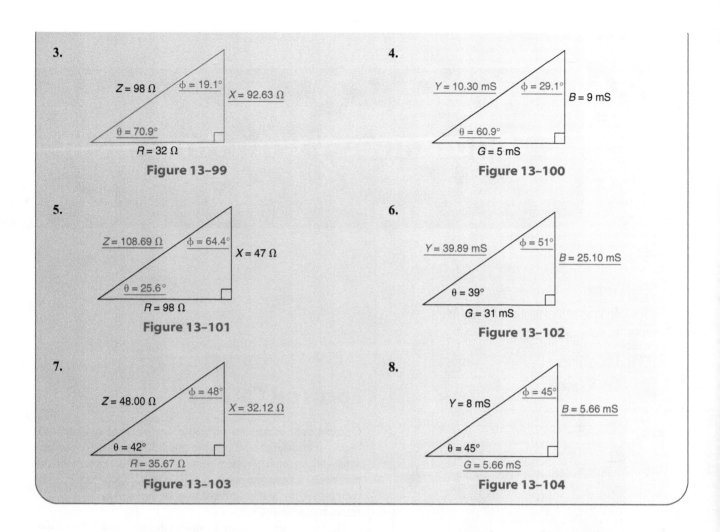

**3.**

$Z = 98\ \Omega$   $\phi = 19.1°$   $X = 92.63\ \Omega$
$\theta = 70.9°$
$R = 32\ \Omega$

**Figure 13–99**

**4.**

$Y = 10.30\ mS$   $\phi = 29.1°$   $B = 9\ mS$
$\theta = 60.9°$
$G = 5\ mS$

**Figure 13–100**

**5.**

$Z = 108.69\ \Omega$   $\phi = 64.4°$   $X = 47\ \Omega$
$\theta = 25.6°$
$R = 98\ \Omega$

**Figure 13–101**

**6.**

$Y = 39.89\ mS$   $\phi = 51°$   $B = 25.10\ mS$
$\theta = 39°$
$G = 31\ mS$

**Figure 13–102**

**7.**

$Z = 48.00\ \Omega$   $\phi = 48°$   $X = 32.12\ \Omega$
$\theta = 42°$
$R = 35.67\ \Omega$

**Figure 13–103**

**8.**

$Y = 8\ mS$   $\phi = 45°$   $B = 5.66\ mS$
$\theta = 45°$
$G = 5.66\ mS$

**Figure 13–104**

# 14

# Trigonometry with Any Angle

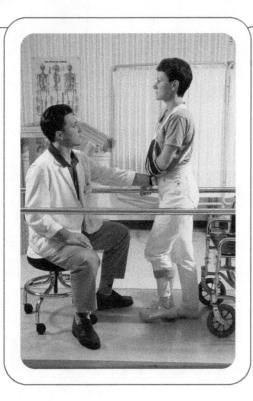

## Focus on Careers

Occupational therapists supervise the work of occupational therapist assistants who provide rehabilitative services to persons with mental, physical, emotional, or developmental impairments. These assistants also help injured workers become reemployed by teaching them how to perform tasks with different motor skills.

Because more and more hands-on therapy work is being delegated to occupational therapist assistants and because the number of people with disabilities is increasing, persons choosing this career may expect excellent opportunities for a rewarding career of service to others.

An associate degree or certificate from an accredited community college or technical school is usually required for occupational therapist assistants. Students take courses in areas of basic medical terminology, anatomy, physiology, mental health, adult physical disabilities, gerontology, and pediatrics. Most states regulate occupational therapist assistants and require them to pass a national certification examination after graduation.

Employment for this career is expected to grow much faster than most other occupations through 2012. Job growth will result from the increasing number of individuals with disabilities and from an aging population who will need these services.

*(continued)*

The highest 10% of occupational therapist assistants earned more than $48,480 in 2002 and the middle 50% earned between $31,090 and $43,030 in the same year.

Source: *Occupational Outlook Handbook* 2004–2005 Edition. U.S. Department of Labor, Bureau of Labor Statistics.

Although right triangles are perhaps the most frequently used triangles, we often work with other kinds of triangles. In this chapter, we apply the laws of sines and cosines to oblique triangles and use *vectors* for applications with angles.

## 14–1 | Vectors

*Learning Outcomes*

1. Find the magnitude of a vector in standard position, given the coordinates of the endpoint.
2. Find the direction of a vector in standard position, given the coordinates of the endpoint.
3. Find the sum of vectors.

### 1 Find the Magnitude of a Vector in Standard Position, Given the Coordinates of the Endpoint.

Quantities that we have discussed so far in this text have been described by specifying their size or magnitude. Quantities such as area, volume, length, and temperature, which are characterized by magnitude only, are called **scalars.** Other quantities such as electrical current, force, velocity, and acceleration are called *vectors.* A quantity described by magnitude (or length) and direction is a **vector.** A vector represents a shift or movement from one point to another.

When we consider the speed of a plane to be $500 \frac{\text{mi}}{\text{h}}$, we are considering a scalar quantity. However, when we consider the speed of a plane *traveling northeast from a given location,* we are concerned with both the distance traveled and the direction. This is a vector quantity. We have seen such quantities in the vector diagrams used to solve electronic problems by means of right triangles in Chapter 13 and elsewhere.

Vectors are represented by straight arrows. The length of the arrow represents the **magnitude** of the vector (see Figure 14–1). The curved arrow shows the counterclockwise **direction** of the vector (see Figure 14–2), with the end of the arrow being the beginning point and the arrowhead being the end point.

In relating trigonometric functions to vector quantities, we will place the vectors in standard position. **Standard position** places the beginning point of the vector at the origin of a rectangular coordinate system. The direction of the vector is the counterclockwise angle measured from the positive x-axis (horizontal axis).

In Figure 14–2, vector *P* is a first-quadrant vector that has a direction of $\theta_1$ and an end point at the point *P*. The magnitude of a vector can be determined if the vector is in standard position and the x- and y-coordinates of the endpoint are known. Vector *R* is a third-quadrant vector that has a direction of $\theta_2$ and an endpoint at point *R*. Its magnitude may be determined similarly.

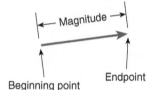

**Figure 14–1**

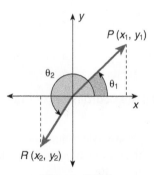

**Figure 14–2**

### Magnitude of a vector in standard position:

The *magnitude* of a vector in standard position is the length of the hypotenuse of the right triangle formed by the vector, the x-axis, and the vertical line from the endpoint of the vector to the x-axis.

**To find the magnitude of a vector in standard position:**

1. Substitute into the Pythagorean theorem the coordinates of the endpoints of the vector $(x_1, y_1)$ as the legs of a right triangle.
2. Solve for the hypotenuse.

---

**EXAMPLE**  Find the magnitude of vector $p$ in Figure 14–3 if the coordinates of $P$ are (4, 3).

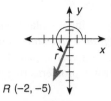

**Figure 14–3**

$p^2 = x^2 + y^2$      Substitute $x = 4$ and $y = 3$.
$p^2 = 4^2 + 3^2$      Solve for $p$.
$p^2 = 16 + 9$
$p^2 = 25$
$p = \pm\sqrt{25}$
$p = \pm 5$      We use only +5 because length or magnitude is positive.

**The magnitude of vector $p$ is 5 units.**

---

**EXAMPLE**  Find the magnitude of vector $r$ in Figure 14–4 if the coordinates of point $R$ are $(-2, -5)$.

The magnitude of a vector is always positive; however, the $x$- and $y$-coordinates can be negative.

$R\ (-2, -5)$

**Figure 14–4**

$r^2 = x^2 + y^2$      Substitute $x = -2$ and $y = -5$.
$r^2 = (-2)^2 + (-5)^2$      Solve for $r$.
$r^2 = 4 + 25$
$r^2 = 29$
$r = \sqrt{29}$      Principal square root.
$r = 5.385$      Round.

**The magnitude of vector $r$ is 5.385.**

---

**2**  **Find the Direction of a Vector in Standard Position, Given the Coordinates of the Endpoint.**

If the vector in standard position falls in quadrant I, we can use right-triangle trigonometry to find the direction of the vector. For a vector in standard position with an endpoint in quadrant I, the tangent function can be used to find the direction or angle of the vector.

**To find the direction of a quadrant I vector in standard position:**

1. Identify the coordinates of the endpoint of the vector.
2. Substitute the coordinates of the endpoint into the tangent function: $\tan\theta = \dfrac{\text{opp}}{\text{adj}}$ or $\tan\theta = \dfrac{y}{x}$.
3. Solve for $\theta$.

**EXAMPLE** Find the direction of vector $p$ in Figure 14–5 if the coordinates of $P$ are (4, 3).

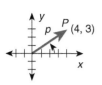

**Figure 14–5**

$$\tan \theta = \frac{y}{x}$$  Substitute values for endpoint of vector $P$: $x = 4$ and $y = 3$.

$$\tan \theta = \frac{3}{4}$$  Use the inverse tangent function and the decimal equivalent of $\frac{3}{4}$.

$$\tan^{-1} 0.75 = \theta$$  Evaluate.

$$\theta = 36.9°$$  Nearest 0.1°.

**The direction of vector $p$ is 36.9°.**

---

**3** **Find the Sum of Vectors.**

Two vectors with the same direction can be added by aligning the beginning point of one vector with the end point of the other. The vector represented by the sum is called the **resultant vector,** or **resultant.** The resultant has the same direction as the vectors being added and a magnitude that is the sum of the two magnitudes.

## To add vectors of the same direction:

1. Add the magnitudes of the vectors to get the magnitude of the resultant vector.
2. The resultant vector will have the same direction as the original vectors.

---

**EXAMPLE** Add two vectors with a direction of 35° if the magnitudes of the vectors are 4 and 5, respectively (see Figure 14–6).

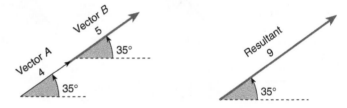

**Figure 14–6**

**The resultant vector has a magnitude of 9 and a direction of 35°.**

---

Two vectors have opposite directions if the directions of the vectors differ by 180°. Two vectors with opposite directions can be added by aligning the beginning point of the vector with the smaller magnitude, with the endpoint of the vector with the larger magnitude. The resultant has the same direction as the vector with the larger magnitude. The magnitude of the resultant is the difference of the two magnitudes.

## To add vectors of opposite direction:

1. Subtract the magnitude of the smaller vector from the magnitude of the larger vector to get the magnitude of the resultant vector.
2. The resultant vector has the direction of the vector with the larger magnitude.

EXAMPLE  Add a vector with a direction of 60° and a magnitude of 6 to a vector with a direction of 240° and a magnitude of 4 (see Figure 14–7).

Subtract the magnitudes: $6 - 4 = 2$.

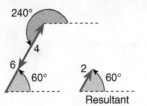

**Figure 14–7**

**The resultant has a direction of 60° and a magnitude of 2.**

Adding any two vectors that have different directions is accomplished by performing the shifts of each vector in succession.

EXAMPLE  Show a graphical representation of the sum of a 45° vector with a magnitude of 5 (vector $A$) and a 60° vector with a magnitude of 6 (vector $B$) (see Figure 14–8).

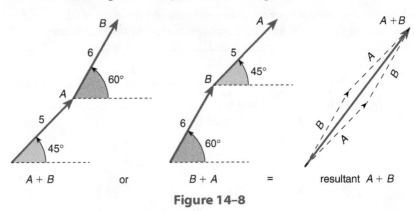

**Figure 14–8**

**To add vectors with different directions:**

1. Determine the $x$- and $y$-coordinates of the endpoints of each vector in standard position.
2. Add the $x$-coordinates of the endpoints for the $x$-coordinate of the endpoint of the resultant vector.
3. Add the $y$-coordinates of the endpoints for the $y$-coordinate of the endpoint of the resultant vector.
4. Find the magnitude using the Pythagorean theorem and the endpoint of the resultant vector.
5. Find the direction using the tangent function and the endpoint of the resultant vector.

Complex numbers in the form $a + bj$ are sometimes used to represent vectors. The $x$-coordinate of a vector in standard position is represented as the real component $a$. The $y$-coordinate of a vector in standard position is represented as the imaginary component $bj$.

## To write a vector in complex notation:

1. Find the *x*-coordinate of the endpoint of the vector (Figure 14–9). Use the relationship

$$\cos \theta = \frac{\text{adj}}{\text{hyp}} = \frac{x}{r} \text{ and solve for } x. \qquad r = \text{magnitude}$$

2. Find the *y*-coordinate of the endpoint of the vector. Use the relationship

$$\sin \theta = \frac{\text{opp}}{\text{hyp}} = \frac{y}{r} \text{ and solve for } y. \qquad r = \text{magnitude}$$

3. Write the vector in complex notation as *x* + *yj*.

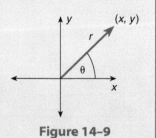

**Figure 14–9**

---

**EXAMPLE**   Write the vectors *A*, *B*, and *A* + *B* from Figure 14–8 in complex notation.

| **Vector *A*:** | ***x*-coordinate** | ***y*-coordinate** | |
|---|---|---|---|
| magnitude = 5 | $\cos 45° = \dfrac{x}{5}$ | $\sin 45° = \dfrac{y}{5}$ | Solve for *x* or *y*. |
| direction = 45° | $5(\cos 45°) = x$ | $5(\sin 45°) = y$ | Evaluate. |
| | $3.535533906 = x$ | $3.535533906 = y$ | |

**Vector *A* in complex form: 3.535533906 + 3.535533906*j***

| **Vector *B*:** | ***x*-coordinate** | ***y*-coordinate** | |
|---|---|---|---|
| magnitude = 6 | $\cos 60° = \dfrac{x}{6}$ | $\sin 60° = \dfrac{y}{6}$ | Solve for *x* or *y*. |
| direction = 60° | $6(\cos 60°) = x$ | $6(\sin 60°) = y$ | Evaluate. |
| | $3 = x$ | $5.196152423 = y$ | |

**Vector *B* in complex form: 3 + 5.196152423*j***

Resultant *A* + *B* = ( 3.535533906 + 3.535533906 *j*)          Add *x* values.

          + ( 3 + 5.196152423 *j*)          Add *y* values.

**Resultant *A* + *B* = 6.535533906 + 8.731686329*j***     or     **6.536 + 8.732*j***

---

**EXAMPLE**   Find the magnitude and direction of the resultant *A* + *B* in the preceding example.

The resultant in complex form is 6.535533906 + 8.731686329*j;* thus, the *x*-coordinate of the end point is 6.535533906 and the *y*-coordinate is 8.731686329.

$$\text{magnitude} = \sqrt{x^2 + y^2} \qquad\qquad \text{Substitute for } x \text{ and } y.$$

$$\text{magnitude} = \sqrt{(6.535533906)^2 + (8.731686329)^2} \qquad\qquad \text{Evaluate.}$$

$$\text{magnitude} = \sqrt{118.9555496}$$

**magnitude = 10.91**     Round to 4 significant digits.

14–1  Vectors

$$\tan \theta = \frac{y}{x}$$

Substitute for *x* and *y*. Evaluate.

$$\tan \theta = \frac{8.731686329}{6.535533906}$$

$$\tan \theta = 1.336032596$$

Use inverse tangent function.

$$\tan^{-1} 1.336032596 = \theta$$

Evaluate.

$$\theta = 53.18570658$$

**Direction: $\theta = 53.2°$**

Nearest 0.1°.

Vectors written in complex form are often used in electronics. The imaginary part is referred to as the *j*-factor.

Resultant vector $A + B$ from the first example on the preceding page would be written as $6.536 + j8.732$. The customary notation is for *j* to be followed by the coefficient.

## SECTION 14–1 SELF-STUDY EXERCISES

**1** Find the magnitude of the vectors in standard position with endpoints at the indicated points. Round to the nearest thousandth if necessary.

**1.** $(5, 12)$  **2.** $(-12, 9)$  **3.** $(2, -7)$  **4.** $(-8, -3)$  **5.** $(1.5, 2.3)$

**2** Find the direction of the vectors in standard position with endpoints at the indicated points. Round to the nearest hundredth.

**6.** $(5, 12)$  **7.** $(6, 8)$  **8.** $(8, 3)$  **9.** $(2, 5)$  **10.** $(1, 4)$

**3**

**11.** Find the resultant vector of two vectors that have a direction of 42° and magnitudes of 7 and 12, respectively.

**12.** Two vectors have a direction of 72°. Find the sum of the vectors if their magnitudes are 1 and 7, respectively.

**13.** Find the sum of two vectors if one has a direction of 45° and a magnitude of 7 and the other has a direction of 225° and a magnitude of 8.

**14.** Find the sum of two vectors if one has a direction of 75° and a magnitude of 15 and the other has a direction of 255° and a magnitude of 9.

Find the magnitude and direction of the resultant of the sum of the two given vectors in complex notation.

**15.** $5 + 3j$ and $7 + 2j$

**16.** $1 + 2j$ and $5 + 2j$

## 14–2 | Trigonometric Functions for Any Angle

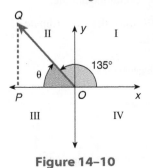

**Figure 14–10**

**Learning Outcomes**

**1** Find related acute angles for angles or vectors in quadrants II, III, and IV.

**2** Determine the signs of trigonometric values of angles of more than 90°.

**3** Find the trigonometric values of angles of more than 90° using a calculator.

**1** **Find Related Acute Angles for Angles or Vectors in Quadrants II, III, and IV.**

The direction of any vector in standard position can be determined, if the coordinates of the endpoint are known, by applying our knowledge of trigonometric functions. A right triangle that we will refer to as a **reference triangle** can be formed by drawing a vertical line from the endpoint of the vector to the *x*-axis. See the example of a reference triangle in Figure 14–10. The angle $\theta$ is called the *related angle*. The **related angle** is the acute angle formed by the *x*-axis and the vector.

Chapter 14 / Trigonometry with Any Angle

In quadrant I, the related angle is the same as the direction of the vector. Therefore, the direction of the vector in quadrant I is always less than 90°. For vectors in quadrants II, III, and IV, see Figures 14–10, 14–11, and 14–12.

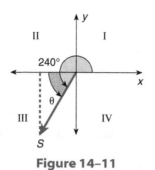

**Figure 14–11**

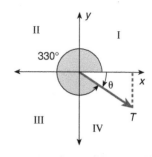

**Figure 14–12**

In Figure 14–10, $\overline{PQ}$, the $x$-axis, and vector $Q$ form a right triangle. The direction of the vector is 135°; therefore, the related angle is 45° (180° − 135°). 135° is a second-quadrant angle. Second-quadrant angles are more than 90° and less than 180°, or more than $\frac{\pi}{2}$ rad (1.57) and less than $\pi$ rad (3.14). The related angle for any second-quadrant vector can be found by subtracting the direction of the vector from 180° (or $\pi$ radians).

Third-quadrant angles are more than 180° and less than 270°, or more than $\pi$ radians (3.14) and less than $\frac{3\pi}{2}$ radians (4.71). In Figure 14–11, vector $S$ is a third-quadrant vector. The related angle $\theta$ is 60° (240° − 180°). The related angle for any third-quadrant vector is found by subtracting 180° (or $\pi$ radians) from the direction of the vector.

Vector $T$ in Figure 14–12 is a fourth-quadrant vector. Fourth-quadrant vectors or angles are more than 270° and less than 360°, or more than $\frac{3\pi}{2}$ rad (4.71) and less than $2\pi$ rad (6.28). The related angle $\theta$ is 30° (360° − 330°). The related angle for any fourth-quadrant vector is found by subtracting the direction of the vector from 360° (or $2\pi$ rad).

**To find related angles for vectors that are more than 90° ($\frac{\pi}{2}$ rad):**

The related angle for quadrant I angles and vectors is equal to the direction (angle) of the vector. The related angle for angles and vectors more than 90° can be found as follows (Figure 14–13):

Quadrant II angle (90° < $\theta_2$ < 180°)   or   $\frac{\pi}{2}$ rad < $\theta_2$ < $\pi$ rad):   180° − $\theta_2$   or   $\pi - \theta_2$

Quadrant III angle (180° < $\theta_3$ < 270°   or   $\pi$ rad < $\theta_3$ < $\frac{3\pi}{2}$ rad):   $\theta_3$ − 180°   or   $\theta_3 - \pi$

Quadrant IV angle (270° < $\theta_4$ < 360°   or   $\frac{3\pi}{2}$ rad < $\theta_4$ < $2\pi$ rad):   360° − $\theta_4$   or   $2\pi - \theta_4$

Related angles are always angles less than 90° or $\frac{\pi}{2}$ rad (1.57 rad).

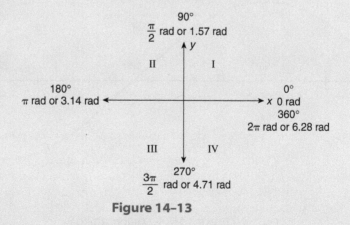

**Figure 14–13**

**EXAMPLE** Find the related angle for the following angles.

(a) 210°  (b) 1.93 rad

(a) 210° is between 180° and 270° and is a quadrant III angle.

$$\theta_3 - 180° =$$
$$210° - 180° = 30°$$

**The related angle is 30°.**

(b) 1.93 rad is between $\frac{\pi}{2}$ (1.57) and $\pi$ (3.14) rad and is a quadrant II angle.

$$\pi - 1.93 = 1.211592654 \text{ rad.} \qquad \text{Use calculator value for } \pi.$$

**The related angle is 1.21 rad to the nearest hundredth.**

**2 Determine the Sign of Trigonometric Values of Angles of More Than 90°.**

**Signs of Trigonometric Functions**

Knowing the signs of the trigonometric functions in the various quadrants is helpful in checking calculations and making estimations. As you examine the graphs of the sine, cosine, and tangent functions, pay particular attention when the graph is above the *x*-axis (positive) and below the *x*-axis (negative).

The graphical representation of trigonometric functions can help you visualize the sign patterns of the various functions and draws attention to other properties of trigonometric functions. To graph these functions, we make a table of values from 0° to 360°, or from 0 rad to 2π rad. See Figures 14–14 through 14–16.

| | | Quadrant I | | | | Quadrant II | | | | Quadrant III | | | | Quadrant IV | | | |
|---|---|---|---|---|---|---|---|---|---|---|---|---|---|---|---|---|---|
| sin θ | 0 | 0.5 | 0.71 | 0.87 | 1 | 0.87 | 0.71 | 0.5 | 0 | −0.5 | −0.71 | −0.87 | −1 | −0.87 | −0.71 | −0.5 | 0 |
| θ | 0 | $\frac{\pi}{6}$ | $\frac{\pi}{4}$ | $\frac{\pi}{3}$ | $\frac{\pi}{2}$ | $\frac{2\pi}{3}$ | $\frac{3\pi}{4}$ | $\frac{5\pi}{6}$ | $\pi$ | $\frac{7\pi}{6}$ | $\frac{5\pi}{4}$ | $\frac{4\pi}{3}$ | $\frac{3\pi}{2}$ | $\frac{5\pi}{3}$ | $\frac{7\pi}{4}$ | $\frac{11\pi}{6}$ | $2\pi$ |
| θ° | 0 | 30 | 45 | 60 | 90 | 120 | 135 | 150 | 180 | 210 | 225 | 240 | 270 | 300 | 315 | 330 | 360 |

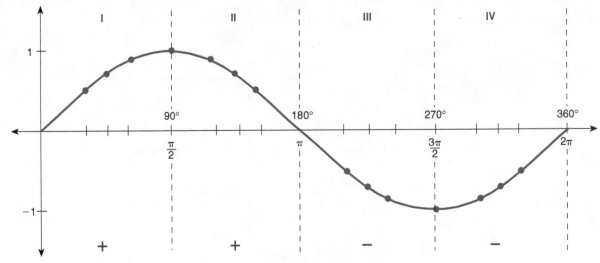

**Figure 14–14** Graph of sine θ.

| cos θ | 1 | Quadrant I | | | 0 | Quadrant II | | | −1 | Quadrant III | | | 0 | Quadrant IV | | | 1 |
|---|---|---|---|---|---|---|---|---|---|---|---|---|---|---|---|---|---|
| cos θ | 1 | 0.87 | 0.71 | 0.5 | 0 | −0.5 | −0.71 | −0.87 | −1 | −0.87 | −0.71 | −0.5 | 0 | 0.5 | 0.71 | 0.87 | 1 |
| θ | 0 | $\dfrac{\pi}{6}$ | $\dfrac{\pi}{4}$ | $\dfrac{\pi}{3}$ | $\dfrac{\pi}{2}$ | $\dfrac{2\pi}{3}$ | $\dfrac{3\pi}{4}$ | $\dfrac{5\pi}{6}$ | π | $\dfrac{7\pi}{6}$ | $\dfrac{5\pi}{4}$ | $\dfrac{4\pi}{3}$ | $\dfrac{3\pi}{2}$ | $\dfrac{5\pi}{3}$ | $\dfrac{7\pi}{4}$ | $\dfrac{11\pi}{6}$ | 2π |
| θ° | 0 | 30 | 45 | 60 | 90 | 120 | 135 | 150 | 180 | 210 | 225 | 240 | 270 | 300 | 315 | 330 | 360 |

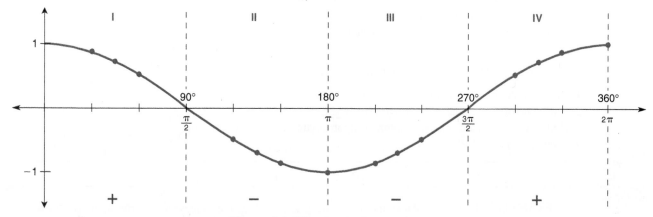

**Figure 14–15**   Graph of cosine θ.

| tan θ | 0 | Quadrant I | | | ∞ | Quadrant II | | | 0 | Quadrant III | | | ∞ | Quadrant IV | | | 0 |
|---|---|---|---|---|---|---|---|---|---|---|---|---|---|---|---|---|---|
| tan θ | 0 | 0.58 | 1 | 1.7 | ∞ | −1.7 | −1 | −0.58 | 0 | 0.58 | 1 | 1.7 | ∞ | −1.7 | −1 | −0.58 | 0 |
| θ | 0 | $\dfrac{\pi}{6}$ | $\dfrac{\pi}{4}$ | $\dfrac{\pi}{3}$ | $\dfrac{\pi}{2}$ | $\dfrac{2\pi}{3}$ | $\dfrac{3\pi}{4}$ | $\dfrac{5\pi}{6}$ | π | $\dfrac{7\pi}{6}$ | $\dfrac{5\pi}{4}$ | $\dfrac{4\pi}{3}$ | $\dfrac{3\pi}{2}$ | $\dfrac{5\pi}{3}$ | $\dfrac{7\pi}{4}$ | $\dfrac{11\pi}{6}$ | 2π |
| θ° | 0 | 30 | 45 | 60 | 90 | 120 | 135 | 150 | 180 | 210 | 225 | 240 | 270 | 300 | 315 | 330 | 360 |

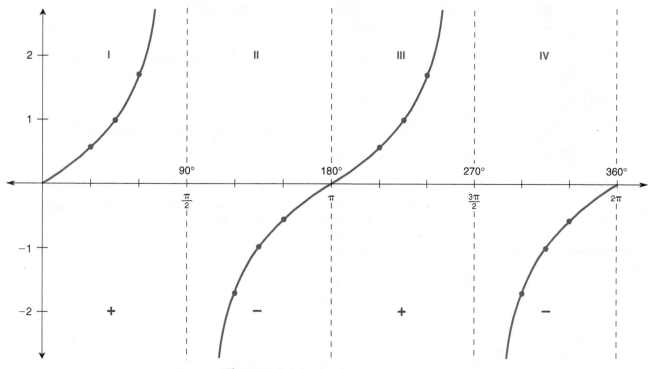

**Figure 14–16**   Graph of tangent θ.

14–2  Trigonometric Functions for Any Angle

**3** **Find the Trigonometric Values of Angles of More Than 90° Using a Calculator.**

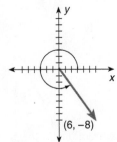

### General Tips for Using the Calculator

To find the value of the trigonometric function of an angle that is more than 90° using a calculator:

1. Set the calculator to degree or radian mode as desired.
2. Enter the trigonometric function.
3. Enter the angle measure in degrees or radians.
4. Press $=$ or ENTER.

To find the related angle, given the endpoint of a vector:

1. Set the calculator to degree or radian mode as desired.
2. Press the inverse tangent TAN⁻¹ or ARCTAN function.
3. Enter the coordinates of the endpoint as appropriate in the tangent ratio $\left(\dfrac{y}{x}\right)$.
4. Press $=$ or ENTER.

---

**EXAMPLE** Use a calculator, including the π key, to find the values of the trigonometric functions. Round the final answer to four significant digits.

(a) sin 155°    (b) cos 3    (c) tan 208°    (d) sin 4.2

(e) cos 304.5°    (f) $\tan \dfrac{5\pi}{3}$

| | | | |
|---|---|---|---|
| (a) sin 155° = **0.4226** | degree mode | (b) cos 3 = **−0.9900** | radian mode |
| (c) tan 208° = **0.5317** | degree mode | (d) sin 4.2 = **−0.8716** | radian mode |
| (e) cos 304.5° = **0.5664** | degree mode | (f) $\tan\dfrac{5\pi}{3}$ = **−1.732** | radian mode |

---

**EXAMPLE** Find the direction in degrees of a vector in standard position if the coordinates of its endpoint are (6, −8) (Figure 14–17).

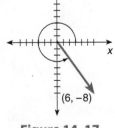

**Figure 14–17**

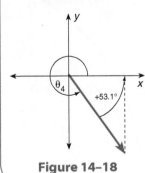

**Figure 14–18**

$$\tan \theta = \frac{y}{x}$$    Substitute values of the endpoints.

$$\tan \theta = \frac{-8}{6}$$    Quadrant IV.

$$\tan \theta = -1.333333333$$    Use inverse tangent function.

$$\tan^{-1}(-1.333333333) = \theta$$    Evaluate.

$$\theta = -53.1°$$    To the nearest tenth degree.

related angle = 53.1°    Figure 14–18. Positive direction from vector to *x* axis.
$$360° - \theta_4 = 53.1°$$    Solve for $\theta_4$, direction of vector.
$$360° - 53.1° = \theta_4$$
$$306.9° = \theta_4$$

**Thus, the direction of the vector is 306.9°.**

**SECTION 14–2 SELF-STUDY EXERCISES**

**1** Find the related angle for the angles. Use the calculator value for $\pi$ for radian measures and round to hundredths.

| | | | | |
|---|---|---|---|---|
| **1.** 120° | **2.** 195° | **3.** 290° | **4.** 345° | **5.** 148° |
| **6.** 250° | **7.** 212° | **8.** 118° | **9.** 2.18 rad | **10.** 5.84 rad |

**2** Give the signs of the sine, cosine, and tangent functions of vectors with the indicated endpoints.

| | | | | |
|---|---|---|---|---|
| **11.** $(3, 5)$ | **12.** $(-2, 6)$ | **13.** $(-4, -2)$ | **14.** $(5, -3)$ | **15.** $(5, 0)$ |

**3** Using a calculator and the $\pi$ key, evaluate the trigonometric functions. Round to four significant digits.

| | | | | |
|---|---|---|---|---|
| **16.** $\sin 210°$ | **17.** $\tan 140°$ | **18.** $\cos 2.5$ | **19.** $\cos 4$ | **20.** $\sin 300°$ |
| **21.** $\tan 6$ | **22.** $\cos 100°$ | **23.** $\sin \dfrac{5\pi}{6}$ | | |

**24.** Find the direction in degrees of a vector in standard position if the coordinates of its endpoint are $(-3, 2)$. Round to the nearest 0.1°.

**25.** Find the direction in radians of a vector in standard position if the coordinates of its endpoint are $(-2, -1)$. Round to the nearest hundredth.

---

## 14–3 | *Period and Phase Shift*

*Learning Objectives*
**1** Graph a sine or cosine function using a calculator.
**2** Find the period of a sine or cosine function.
**3** Find the phase shift of a sine or cosine function.

### **1** Graph a Sine or Cosine Function Using a Calculator.

The graphs of sine and cosine functions continue to repeat in both directions as the graphs are extended to values less than 0° and more than 360°. One complete cycle of the graph is called the **period** of the function. The period of the sine and cosine functions is 360° or $2\pi$ rad.

**To graph two or more periods or repetitions of a sine or cosine function using a calculator:**

1. Set MODE to degrees or radians as desired.
2. From the ZOOM function select the option Ztrig or manually set the window as desired.
3. Enter the function using the $Y =$ function.
4. View the graph using GRAPH function.

**EXAMPLE** Use a calculator to graph the functions $y = \sin x$ and $y = \cos x$. See Figures 14–19 and 14–20.

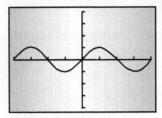

**Figure 14–19** $y = \sin x$.

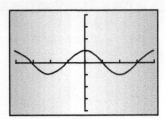

**Figure 14–20** $y = \cos x$.

We refer to the repetitions of the graph as **waves.** The **wavelength** is the *distance* along the x-axis of one repetition of a wave. Wavelength is represented by the Greek letter **lambda,** $\lambda$. If the x-axis is a *time* variable, the time it takes for one complete wave to pass through a given point is the **period,** $T$. The number of waves that pass through a given point in one second is called **frequency,** $f$. Frequency and period are reciprocals; that is,

$$f = \frac{1}{T} \quad \text{or} \quad T = \frac{1}{f}$$

Frequency is measured in hertz (Hz). That is 1 Hz = 1 wave or cycle per second. Common measuring units for frequency are kilohertz (1 kHz = $10^3$ Hz), megahertz (1 MHz = $10^6$ Hz), and gigahertz (1 GHz = $10^9$ Hz).

**EXAMPLE** Find the frequency of a wave that passes through a given point once each 0.00008 sec.

$$f = \frac{1}{T}$$

$$f = \frac{1}{0.00008} \qquad \text{The period is 0.00008 sec per cycle.}$$

$f = 12,500$ Hz or 12.5 kHz

**12,500 waves pass through the point each second.**

**Wave velocity,** $v$, is the rate at which a position on the wave changes or moves. Wave velocity is related to frequency or period and wavelength by the formulas

$$v = \lambda f \quad \text{and} \quad v = \frac{\lambda}{T}$$

where $v$ is the wave velocity, $\lambda$ is the wavelength, $f$ is the frequency, and $T$ is the period.

**EXAMPLE**
**ELEC** A National Public Radio (NPR) station has an FM band (frequency) of 91.1 MHz. The speed of a radio wave is the same as the speed of light, which is 300,000 km/s or 300,000,000 m/s. Find the station's wavelength in meters.

$$v = \lambda f \qquad\qquad\qquad \text{Solve for } \lambda. \text{ Divide both sides by } f.$$

$$\lambda = \frac{v}{f} \qquad\qquad\qquad \text{Substitute 300,000,000 m/s and 91.1 MHz.}$$

Chapter 14 / Trigonometry with Any Angle

$$\lambda = \frac{300,000,000 \text{ m/s}}{91.1 \text{ MHz}}$$

1 MHz = $10^6$ Hz, or $10^6$ waves per second.

$$\lambda = \frac{300,000,000 \text{ m/s}}{91,100,000 \text{ waves/s}}$$

$$\lambda = 3.293084523 \text{ m/wave}$$

Round to hundredths.

**The length of each wave is 3.29 m.**

One of the most common applications of waves is in alternating current. For example, in a generator, the current $i$ changes according to the equation

$$i = I \sin x$$

where $I$ is the maximum current and $x$ is the angle through which the coil rotates in a magnetic field to generate an electric current. The voltage $v$ changes according to the equation

$$v = V \sin x$$

where the maximum voltage is $V$ and the angle through which the coil rotates is $x$. Both these equations are examples of an equation in the form

$$y = A \sin x \quad \text{for } A > 0$$

In this equation, $A$ represents the **amplitude,** or maximum value that the graph will reach on the y-axis. The minimum value is represented by $-A$.

**EXAMPLE**
**ELEC**

The maximum voltage $V$ in a simple generator is 30 V. As the coil rotates, the voltage changes according to the equation $v = 30 \sin x$. Use your calculator to graph the voltage changes for one complete revolution. Set the minimum Y-value to be less than the amplitude and the maximum Y-value to be greater than the amplitude to show the graph in the calculator window. Determine the voltage at 210°.

Set MODE to degree measure for angles.

Set Window:  Xmin = 0                    Settings for one period.
             Xmax = 360
             Xscl = 30
             Ymin = $-40$                 Set Ymin to be less than amplitude of $-30$.
             Ymax = 40                    Set Ymax to be greater than amplitude of 30.
             Yscl = 1
             Xres = 1

Enter function: Y = 30 sin $x$           Use Y = function.
View graph.                              Use GRAPH function.
Find voltage at 210° (Figure 14–21).     Use CALC function and VALUE option.

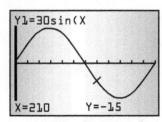

**Figure 14–21**

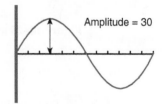

Amplitude = 30

**Figure 14–22**

**The voltage at 210° is $-15$. Figure 14–22 shows the amplitude.**

14–3  Period and Phase Shift

## 2  Find the Period of a Sine or Cosine Function.

In the equations $y = \sin x$ and $y = \cos x$, the coefficient of the variable $x$ is 1, which indicates that the period of the function is 360°, or $2\pi$ rad. A change in the coefficient of $x$ will result in a change of the period. In the equations $y = A \sin Bx$ and $y = A \cos Bx$, the period $P$ is found by the formula $P = \dfrac{360°}{B}$.

**To find the period of a function in the form $y = A \sin Bx$ or $y = A \cos Bx$:**

1. Divide 360° or $2\pi$ radians by the coefficient of $x$.

$$P = \frac{360°}{B} \text{ for degrees} \quad \text{or} \quad P = \frac{2\pi}{B} \text{ for radians}$$

**EXAMPLE**  Find the amplitude and period in degrees of the function $y = 3 \cos 2x$. Graph the equation to show two periods.

Amplitude = 3                          $A = 3$ (coefficient of $\cos 2x$).

Period $= \dfrac{360°}{2} = 180°$           $B = 2$ (coefficient of $x$).

**The amplitude is 3 and the period is 180°.**

Set calculation to degree mode.

Set Window:   Xmin = 0              Set X minimum and maximum to show two periods;
                      Xmax = 360           multiply degrees in period by 2. 180° × 2 = 360°
                      Xscl = 30             Set X scl (tick marks) to 30° each.
                      Ymin = −4            Set Ymin to be < −3.
                      Ymax = 4             Set Ymax to be > 3.
                      Yscl = 1              Set Y scl (tick marks) each 1 unit.

Enter equation: $y = 3 \cos 2x$
Graph (Figure 14–23).

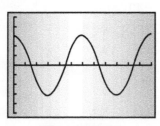

**Figure 14–23**

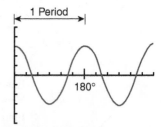

**Figure 14–24**

**Figure 14–24 shows the period.**

## 3  Find the Phase Shift of a Sine or Cosine Function.

All graphs in the form $y = A \sin Bx$ pass through the origin and all graphs in the form $y = A \cos Bx$ pass through the point $(0, A)$. If the respective graphs do not pass through the appropriate point, the function is *out of phase*. The **phase shift** is the distance between two corresponding points on the appropriate reference graph of $y = A \sin Bx$ or $y = A \cos Bx$ and the out-of-phase function. The equations for out-of-phase functions are

$$y = A \sin (Bx + C) \text{ or } y = A \cos (Bx + C)$$

where $C \neq 0$.

**To find the phase shift for the function $y = A \sin (Bx + C)$ or $y = A \cos (Bx + C)$:**

1. Find the value $\dfrac{C}{B}$.

2. If $\dfrac{C}{B}$ is positive, shift the graph $\dfrac{C}{B}$ units to the *left*.

3. If $\dfrac{C}{B}$ is negative, shift the graph $\dfrac{C}{B}$ units to the *right*.

**EXAMPLE**   Graph $y = 5 \sin (2x + 90°)$.

Amplitude $= 5$                    Coefficient of $\sin (2x + 90°)$.

Period $= \dfrac{360°}{2} = 180°$          $\dfrac{360°}{B}$.

Phase shift $= \dfrac{90°}{2} = 45°$        $\dfrac{C}{B}$ is positive.

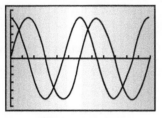

**Figure 14–25**

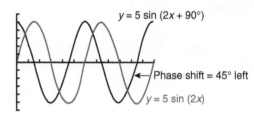

**Figure 14–26**

**Graph of $y = 5 \sin (2x + 90°)$ shifts 45° to the left from the graph $y = 5 \sin 2x$ (Figures 14–25 and 14–26).**

SECTION 14–3 SELF-STUDY EXERCISES

**1**  Use the Ztrig function to graph the functions.

**1.** $y = \sin 2x$         **2.** $y = \sin 3x$         **3.** $y = \sin 4x$         **4.** $y = \sin 8x$
**5.** $y = \cos 2x$         **6.** $y = \cos 4x$         **7.** $y = \cos 5x$         **8.** $y = \cos 6x$

**9.** Compare the graphs for Exercises 1–4.          **10.** Compare the graphs for Exercises 5–8.

Graph using a graphing calculator with $x$-values between $-90°$ and $360°$ in multiples of $30°$ ($x$-scale) and $y$-values between $-8$ and $8$ and a $y$-scale of 1 unit.

**11.** $y = 2 \sin x$                    **12.** $y = 4 \sin x$                    **13.** $y = 6 \sin x$
**14.** $y = 3 \cos x$                    **15.** $y = 4 \cos x$                    **16.** $y = 5 \cos x$

17. Compare the graphs for Exercises 11–13.
18. Compare the graphs for Exercises 14–16.
19. Use the CALC function and VALUE option to find the value of $y$ in the function $y = 10 \sin x$ when $x = 100°$.
20. Find the frequency of a wave that passes through a given point once each 0.045 s. Round to thousandths.
21. Find the frequency of a wave that has a period of 0.006 s.
22. What is the period of a wave if $25 \times 10^4$ waves pass through a given point each second?
23. What is the period of a wave that has a frequency of 99.8 MHz? Round to thousandths.
24. **ELEC** WHQB AM broadcasts at 1390 KHz. What is the wavelength of its radio waves? Radio waves travel at the speed of light, 300,000 km/sec.
25. **ELEC** Cellular phone waves travel at the same speed of light, 300,000 km/s. What is the frequency of a cellular phone wave if the phone operates at a wavelength of 300 mm?
26. **ELEC** What is the velocity of a wavelength that has a period of 0.00004 s and a wavelength of 0.82 m?
27. **ELEC** What is the velocity of a wave that has a wavelength of 1.2 m and a frequency of 88.9 MHz?

Use a graphing calculator to show the changing voltage $v$ for a simple generator that has the given maximum voltage $V$; $v = V \sin x$.
28. **ELEC** $V = 42$ V
29. **ELEC** $V = 32$ V
30. **ELEC** Use the CALC function and VALUE option to estimate to the nearest tenth the changing voltage $v$ when $x = 45°$ and $x = 60°$ if the maximum voltage of the generator is $V = 42$ V.
31. **ELEC** Use the CALC function and VALUE option to estimate to the nearest tenth the changing voltage $v$ when $x = 120°$ and $x = 300°$ if the maximum voltage of the generator is $V = 32$ V.

Use the equation $i = I \sin x$ to graph the changing current in a simple generator that has the given maximum current.
32. **ELEC** $I = 3.5$ A
33. **ELEC** $I = 8.4$ A
34. **ELEC** Use the CALC function and VALUE option to estimate to the nearest tenth the changing current $i$ when $x = 225°$ and $x = 315°$ if the maximum voltage of the generator is $I = 3.5$ A.
35. **ELEC** Use the CALC function and VALUE option to estimate to the nearest thenth the changing current $i$ when $x = 115°$ and $x = 270°$ if the maximum voltage of the generator is $I = 8.4$ A.

**2** Find the amplitude and period in degrees for the functions. Graph each function to show two periods.
36. $y = 4 \sin 6x$
37. $y = 30 \sin 10x$
38. $y = 3 \cos 5x$
39. $y = 4 \cos 8x$
40. $y = 2.5 \cos \dfrac{3}{4} x$
41. $y = 1.9 \sin \dfrac{8}{5} x$

**3** Find the amplitude, period, and phase shift for the functions. Graph each function to show two periods.
42. $y = \sin (x + 45°)$
43. $y = \cos (x - 180°)$
44. $y = 2 \sin (x + 120°)$
45. $y = 6 \sin (3x - 240°)$
46. $y = 10 \cos (4x + 270°)$
47. $y = 8 \cos (6x - 45°)$

---

## 14-4 | Law of Sines

*Learning Outcomes*  **1** Find the missing parts of an oblique triangle, given two angles and a side.
**2** Find the missing parts of an oblique triangle, given two sides and an angle opposite one of them.

An **oblique triangle** is a triangle that does not contain a right angle. Because these triangles do not have right angles, we cannot use the Pythagorean theorem or trigonometric functions directly as we did previously to find sides and angles of right triangles. However, two formulas based on the trigonometric functions of right triangles can be used to solve oblique triangles. In this section, we examine one of these formulas, the **law of sines.**

**Law of sines:**

The **law of sines** states that the ratios of the sides of a triangle to the sines of the angles opposite these respective sides are equal (see Figure 14–27).

$$\frac{a}{\sin A} = \frac{b}{\sin B} = \frac{c}{\sin C}$$

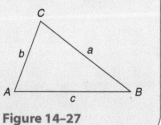

**Figure 14–27**

TIP!

**Conditions for Using the Law of Sines**

The law of sines is used to solve triangles when either of these conditions exists:

1. Two angles and a side are known.
2. Two sides and the angle opposite one of the sides are known.

When using the law of sines you must have at least one complete ratio—an angle and side opposite. However, if we know two angles of a triangle, we can always find the third angle. Then, we can form a complete ratio.

**To find the missing parts of an oblique triangle, given two angles and a side:**

1. Find the third angle of the triangle by adding the two known angles and subtracting the sum from 180° or 2π radians.
2. Set up a ratio with the given side and sine of the opposite angle.
3. Set up a second ratio with a missing side and the sine of the opposite angle.
4. Form a proportion using the ratios in Steps 2 and 3 and solve the proportion for the missing side.
5. Repeat Steps 2–4 if the value of the third side is needed.

In the examples that follow, all digits of the calculated value that show in the calculator display will be given. Rounding should be done after the last calculation has been made.

**EXAMPLE**   Solve triangle $ABC$ if $A = 50°$, $B = 75°$, and $b = 12$ ft.

Sketch the triangle and label the parts, as shown in Figure 14–28.

Find $\angle C$.

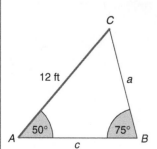

**Figure 14–28**

| | |
|---|---|
| $A + B + C = 180°$ | Substitute $A = 50°$ and $B = 75°$. |
| $50° + 75° + C = 180°$ | Solve for C. $50° + 75° = 125°$ |
| $C = 180° - 125°$ | |
| $\mathbf{C = 55°}$ | |

Find *a*.

$$\frac{a}{\sin A} = \frac{b}{\sin B}$$

Side *b* and ∠*B* form the ratio with both values known.
Substitute *b* = 12, *A* = 50°, *B* = 75°.

$$\frac{a}{\sin 50°} = \frac{12}{\sin 75°}$$

Solve for *a*.

$$a \sin 75° = 12 \sin 50°$$

$$a = \frac{12 \sin 50°}{\sin 75°}$$

Perform calculations.

$$a = 9.516810781$$

$$\mathbf{a = 9.517 \ ft}$$

Four significant digits.

Find *c*.

$$\frac{b}{\sin B} = \frac{c}{\sin C}$$

Substitute *b* = 12, *B* = 75°, and *C* = 55°.

$$\frac{12}{\sin 75°} = \frac{c}{\sin 55°}$$

Solve for *c*.

$$12 \sin 55° = c \sin 75°$$

$$\frac{12 \sin 55°}{\sin 75°} = c$$

Perform calculations.

$$c = 10.1765832$$

$$\mathbf{c = 10.18 \ ft}$$

Four significant digits.

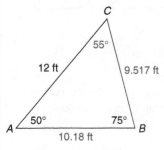

**Figure 14–29**

As a quick check for our work, the longest side should be opposite the largest angle, and the shortest side should be opposite the smallest angle (see Figure 14–29).

---

**EXAMPLE**  Solve triangle *ABC* if *A* = 35°, *a* = 7 cm, and *B* = 40°.

Sketch the triangle and label its parts (see Figure 14–30).

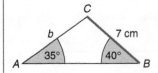

**Figure 14–30**

Find ∠*C*.

$$A + B + C = 180°$$

Substitute *A* = 35° and *B* = 40°.

$$35° + 40° + C = 180°$$

Solve for *C*.

$$C = 180° - 35° - 40°$$

$$\mathbf{C = 105°}$$

Find *b*.

$$\frac{a}{\sin A} = \frac{b}{\sin B}$$

Side *a* and ∠*A* form the ratio with both values known.
Substitute *a* = 7, *A* = 35°, and *B* = 40°.

$$\frac{7}{\sin 35°} = \frac{b}{\sin 40°}$$

Solve for *b*.

$$7 \sin 40° = b \sin 35°$$

$$\frac{7 \sin 40°}{\sin 35°} = b$$

Evaluate.

$$b = 7.844661989$$

$$\mathbf{b = 7.845 \ cm}$$

Four significant digits.

Chapter 14 / Trigonometry with Any Angle

Find $c$.

$$\frac{a}{\sin A} = \frac{c}{\sin C}$$

Substitute $a = 7$, $A = 35°$, and $C = 105°$.

$$\frac{7}{\sin 35°} = \frac{c}{\sin 105°}$$

Solve for $c$.

$$7 \sin 105° = c \sin 35°$$

$$\frac{7 \sin 105°}{\sin 35°} = c$$

Evaluate.

$$c = 11.78828201$$

$$c = 11.79 \text{ cm}$$

Four significant digits.

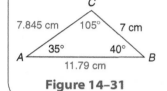

**Figure 14–31**

As a quick check for our work, the longest side should be opposite the largest angle, and the shortest side should be opposite the smallest angle (see Figure 14–31).

## 2 Find the Missing Parts of an Oblique Triangle, Given Two Sides and an Angle Opposite One of Them.

When two sides of a triangle and an angle opposite one of the sides are known, we do not always have a single triangle. If the given sides are $a$ and $b$ and the given angle is $B$, three possibilities may exist (see Figure 14–32).

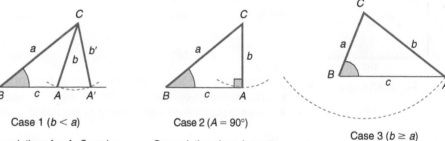

Case 1 ($b < a$)

Two solutions for $A$, $C$, and $c$ since $b$ can meet side $c$ in either of two points, $A$ and $A'$.

Case 2 ($A = 90°$)

One solution since $b$ meets side $c$ in exactly one point.

Case 3 ($b \geq a$)

One solution since $b$ meets side $c$ in only one point.

**Figure 14–32**

- If $b < a$ and $\angle A \neq 90°$, we have two possible solutions (case 1).
- If $b < a$ and $\angle A = 90°$, we have one solution (case 2).
- If $b \geq a$ and $\angle A \neq 90°$, we have one solution (case 3).

Note that side $b$ above could be any side of the triangle that is opposite the given angle. Side $a$ is then the other given side. Because of the lack of clarity when two sides and an angle opposite one of them are given, the situation is called the **ambiguous case.** (*Ambiguous* means that the given information can be interpreted in more than one way.)

**To find the missing parts of an oblique triangle, given two sides and an angle opposite one of them:**

1. Use the law of sines to find the measure of the angle opposite the other given side.
2. Determine if there are two angle values that are possible for the solution.
3. Find the third angle by subtracting from 180° the sum of the two known angles.
4. Find the third side using the law of sines.
5. If two angles were found in Step 2, find the second possible solution by repeating Steps 3 and 4.

**EXAMPLE** Solve triangle $ABC$ if $A = 35°$, $a = 8$, and $b = 11.426$ (see Figure 14–33).

Find $\angle B$.

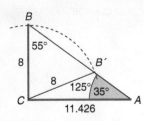

**Figure 14–33**

$$\frac{a}{\sin A} = \frac{b}{\sin B}$$

Substitute $A = 35°$, $a = 8$, and $b = 11.426$. $b > a$. There will be two solutions.

$$\frac{8}{\sin 35°} = \frac{11.426}{\sin B}$$

Solve for $B$.

$$8 \sin B = 11.426 \sin 35°$$

$$\sin B = \frac{11.426 \sin 35°}{8}$$

Evaluate.

$$\sin B = 0.8192105452$$

$$B = \sin^{-1} 0.8192105452$$

Use inverse sine function.

$$B = 55.00584421°$$   or   $B' = 180° - 55.00584421°$
$$B' = 124.99415579°$$
$$B' = 125.0°$$

$$\boldsymbol{B = 55.0°}$$

Nearest 0.1°.

**Possible solution 1:**

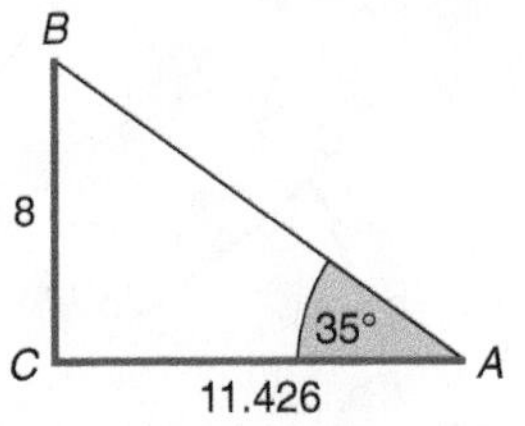

Find $\angle C$.

$$A + B + C = 180°$$

Substitute $A = 35°$ and $B = 55°$.

$$35° + 55° + C = 180°$$

Solve for $C$.

$$C = 180° - (55° + 35°)$$

$$\boldsymbol{C = 90°}$$

$ABC$ is approximately a right triangle (Figure 14–34).

**Figure 14–34**

Even if $<C$ had been exactly $90°$, we still have two possible solutions. The angle opposite the second given side is not the $90°$ angle. To find the third side of this right triangle, we can use the Pythagorean theorem or the law of sines.

Find $c$.

$$\frac{a}{\sin A} = \frac{c}{\sin C}$$

Substitute $a = 8$, $A = 35°$, and $C = 90°$.

$$\frac{8}{\sin 35°} = \frac{c}{\sin 90°}$$

Solve for $c$.

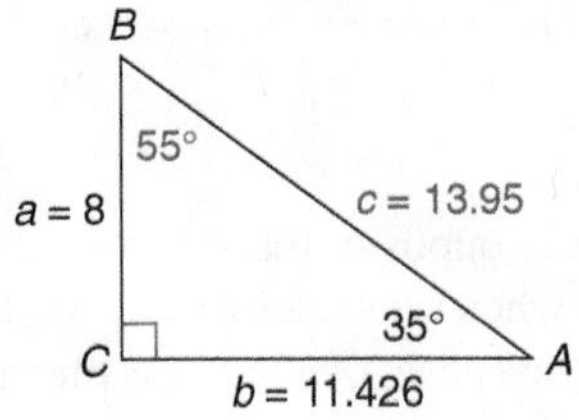

$$8 \sin 90° = c \sin 35°$$

$$\frac{8 \sin 90°}{\sin 35°} = c$$

Evaluate.

$$c = 13.94757436$$

$$\boldsymbol{c = 13.95}$$

Four significant digits.

The solved triangle for this possibility is shown in Figure 14–35.

**Figure 14–35**

**Possible solution 2:**

Find $\angle C'$.

$$C' = 180° - (35° + 124.99415579°) = 20.00584421°$$

$$\boldsymbol{C' = 20.0°}$$

See Figure 24–33.

Find $c'$.

$$\frac{a}{\sin A} = \frac{c'}{\sin C'}$$

Substitute $a = 8$, $A = 35°$, and $C' = 20.00584421°$.

$$\frac{8}{\sin 35°} = \frac{c'}{\sin 20.00584421°}$$

Solve for $c'$.

$$c' \sin 35° = 8 \sin 20.00584421°$$

$$c' = \frac{8 \sin 20.00584421°}{\sin 35°}$$

$$\mathbf{c' = 4.771688224 \quad or \quad c' = 4.772}$$

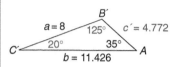

**Figure 14–36**

The solved triangle for this possibility is shown in Figure 14–36.

---

**EXAMPLE**   Solve triangle $ABC$ if $b = 10$, $c = 12$, and $C = 40°$ (Figure 14–37).

Find $B$.

Because $c$, the side opposite the given angle $C$, is *longer* than the other given side, $b$, we expect *one* solution.

$$\frac{c}{\sin C} = \frac{b}{\sin B}$$

Substitute $c = 12$, $C = 40°$, and $b = 10$.

$$\frac{12}{\sin 40°} = \frac{10}{\sin B}$$

Solve for $B$.

$$12 \sin B = 10 \sin 40°$$

$$\sin B = \frac{10 \sin 40°}{12}$$

Evaluate.

$$\sin B = 0.5356563414$$

Use inverse sine function.

$$B = \sin^{-1} 0.5356563414$$

$$B = 32.38843382°$$

$$\mathbf{B = 32.4°}$$

Nearest 0.1°.

Both $32.38843382°$ and $147.6115662°$ have a sine of approximately $0.5356563414$; however, we cannot have an angle of $147.6115662°$ in this triangle because the triangle already has a $40°$ angle. $147.6115662° + 40° = 187.6115662°$. The *three* angles of a triangle total only $180°$. Therefore, we have only *one* solution.

Find $A$.

$$A = 180° - 40° - 32.38843382°$$

Use full calculator value for $B$.

$$A = 107.6115662°$$

$$\mathbf{A = 107.6°}$$

Find $a$.

$$\frac{a}{\sin A} = \frac{c}{\sin C}$$

Substitute full calculator value for $A$.

$$\frac{a}{\sin 107.6115662°} = \frac{12}{\sin 40°}$$

Solve for $a$.

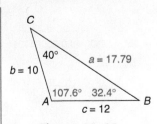

**Figure 14–38**

$$a \sin 40° = 12 \sin 107.6115662°$$

$$a = \frac{12 \sin 107.6115662°}{\sin 40°}$$    Evaluate.

$$a = 17.79367732$$

$$\boldsymbol{a = 17.79}$$    Four significant digits.

The solved triangle is shown in Figure 14–38.

---

**EXAMPLE**
**CON**

A technician checking a surveyor's report is given the information shown in Figure 14–39. Calculate the missing information.

Find B.

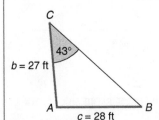

**Figure 14–39**

$$\frac{c}{\sin C} = \frac{b}{\sin B}$$    Substitute given values.

$$\frac{28}{\sin 43°} = \frac{27}{\sin B}$$    Solve for B.

$$28 \sin B = 27 \sin 43°$$

$$\sin B = \frac{27 \sin 43°}{28}$$

$$\sin B = 0.6576412758$$    Use the inverse sine function.

$$B = \sin^{-1} 0.6576412758$$

$$B = 41.12023021°$$

$$\boldsymbol{B = 41.1°}$$    Nearest 0.1°.

There is only one case here because we are given the triangle. Also, the other angle whose sine is 0.6576412758 is 138.8797698°, which we exclude because 138.8797698° + 43° = 181.8797698°.

Find angle A.

$$A = 180° - (43° + 41.12023021°)$$
$$A = 95.87976979°$$
$$\boldsymbol{A = 95.9°}$$    Nearest 0.1°

Find a.

$$\frac{a}{\sin A} = \frac{c}{\sin C}$$    Substitute.

$$\frac{a}{\sin 95.87976979°} = \frac{28}{\sin 43°}$$

$$a \sin 43° = 28 \sin 95.87976979°$$

$$a = \frac{28 \sin 95.87976979°}{\sin 43°}$$    Evaluate.

$$a = 40.83982458$$

$$\boldsymbol{a = 40.84 \text{ ft}}$$    Four significant digits.

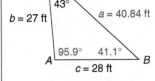

**Figure 14–40**

The completed survey should have the measures shown in Figure 14–40.

Chapter 14 / Trigonometry with Any Angle

**1** Solve each of the oblique triangles using the law of sines. Round the final answer for sides to four significant digits and angles to the nearest 0.1°.

1. $a = 46$ m, $A = 65°$, $B = 52°$
2. $b = 7.2$ mi, $B = 58°$, $C = 72°$
3. $a = 65$ cm, $B = 60°$, $A = 87°$
4. $c = 3.2$ ft, $A = 120°$, $C = 30°$
5. $b = 12$ km, $A = 95°$, $B = 35°$
6. $b = 142$ dkm, $A = 135°$, $B = 25°$
7. Find $b$ if $a = 14$ in., $B = 72°$, and $C = 60°$.
8. Find $a$ if $c = 148$ m, $A = 100°$, and $B = 30°$.
9. Find $C$ if $c = 2.8$ m, $A = 42°$, and $B = 62°$.
10. Find $c$ if $a = 5\frac{3}{8}$ ft, $B = 61.7°$, and $C = 72.5°$.
11. Find $b$ if $a = 7,896$ ft, $B = 42.8°$, and $C = 51.9°$.
12. Find $a$ if $c = 14.32$ km, $A = 38.2°$, and $B = 42.6°$.
13. **AVIA** A satellite sends a signal to cellular towers A and B in two cities and makes an angle of elevation of 88° with tower A and an angle of 89° with tower B. The satellite is 18,000 mi from tower A. What is the approximate distance between the two cities?

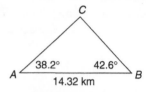

**Figure 14-41**

**2** Use the law of sines to solve the triangles. If a triangle has two possibilities, find both solutions. Round sides to four significant digits and angles to the nearest 0.1°.

14. $a = 42$ in., $b = 24$ in., $A = 40°$
15. $b = 15$ yd, $c = 3$ yd, $B = 70°$
16. $a = 18$ ft, $c = 9$ ft, $C = 20°$
17. $a = 8$ m, $b = 4$ m, $A = 30°$
18. $a = 65.2$ cm, $b = 42.5$ cm, $A = 30°$
19. $b = 22.9$ m, $c = 45.6$ m, $C = 120°$
20. **AG/H** Find the missing angle and sides of the plot of land described by Figure 14-42.
21. **CON** Find the distance from $A$ to $B$ on the surveyed plot shown in Figure 14-43.

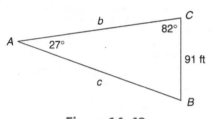

**Figure 14-42**

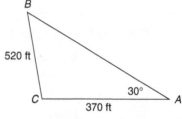

**Figure 14-43**

# 14-5 | Law of Cosines

*Learning Outcomes*

**1** Find the missing parts of an oblique triangle, given three sides of the triangle.

**2** Find the missing parts of an oblique triangle, given two sides and the included angle of the triangle.

**1** **Find the Missing Parts of an Oblique Triangle, Given Three Sides of the Triangle.**

In some cases, our given information does not allow us to use the law of sines. For example, if we know all three sides of a triangle, we cannot use the law of sines to find the angles. In a case such as this, however, we can use the **law of cosines.** This law is based on the trigonometric functions just as the law of sines is.

## Law of cosines:

The square of any side of a triangle equals the sum of the squares of the other sides minus twice the product of the other two sides and the cosine of the angle opposite the first side. For triangle *ABC*:

$$a^2 = b^2 + c^2 - 2\,bc\cos A$$
$$b^2 = a^2 + c^2 - 2\,ac\cos B$$
$$c^2 = a^2 + b^2 - 2\,ab\cos C$$

To use the law of cosines efficiently, we are given

1. Three sides of a triangle

   or

2. Two sides and the included angle of a triangle.

We can sometimes use *either* the law of sines or the law of cosines to solve certain triangles. However, whenever possible, the law of sines is generally preferred because it involves fewer calculations.

### To find the missing parts of an oblique triangle, given three sides of the triangle:

1. Substitute the given values of the three sides into any version of the law of cosines.
2. Solve for one of the unknown angles.
3. Use the angle measure found in Step 2 and find a second angle using a different version of the law of cosines.
4. Find the third angle of the triangle by subtracting from 180° the sum of the two angles found in Steps 2 and 3.

**EXAMPLE** Find the angles in triangle *ABC* (see Figure 14–44).

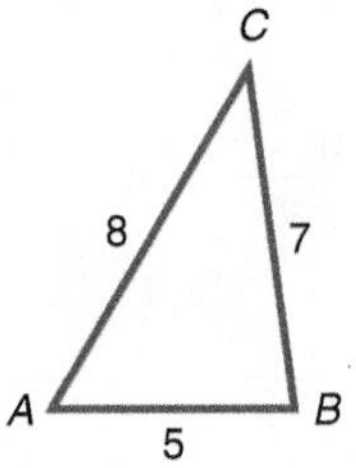

We may find any one of the angles first. If we choose to find angle *A* first, we must use the formula that contains cos *A*: $a^2 = b^2 + c^2 - 2bc\cos A$. Rearrange the formula to solve for cos *A*.

Find *A*.

$$a^2 - b^2 - c^2 = -2bc\cos A$$   Multiply each term on both sides by −1 to reduce the number of negative signs.

$$-a^2 + b^2 + c^2 = 2bc\cos A$$   Solve for cos *A*.

$$\frac{-a^2 + b^2 + c^2}{2bc} = \cos A$$   Substitute *a* = 7, *b* = 8, and *c* = 5.

$$\frac{-(7)^2 + 8^2 + 5^2}{2(8)(5)} = \cos A$$   Solve for *A*.

$$\frac{-49 + 64 + 25}{80} = \cos A$$

$$\frac{40}{80} = \cos A$$

$$0.5 = \cos A$$

$$A = \cos^{-1} 0.5$$   Use inverse cosine function.

$$\mathbf{60° = A}$$

**Figure 14–44**

Find *B*.

To find angle *B*, we use the law of cosines and only given values. The law of sines could be used but would involve using a calculated value.

$$b^2 = a^2 + c^2 - 2ac \cos B \qquad \text{Solve for cos } B.$$

$$2ac \cos B = a^2 + c^2 - b^2$$

$$\cos B = \frac{a^2 + c^2 - b^2}{2ac} \qquad \text{Substitute } a = 7, b = 8, \text{ and } c = 5.$$

$$\cos B = \frac{7^2 + 5^2 - (8)^2}{2(7)(5)} \qquad \text{Solve for } B.$$

$$\cos B = \frac{49 + 25 - 64}{70}$$

$$\cos B = \frac{10}{70}$$

$$\cos B = 0.1428571429 \qquad \text{Use inverse cosine function.}$$

$$B = \cos^{-1} 0.1428571429$$

$$B = 81.7867893°$$

$$\boldsymbol{B = 81.8°} \qquad \text{Nearest 0.1°.}$$

Find *C*.

The law of cosines or the law of sines can be used to find the third angle. However, the quickest way to find this angle is to subtract the sum of *A* and *B* from 180°.

$$C = 180° - (A + B)$$

$$C = 180° - 60° - 81.7867893°$$

$$C = 180° - 141.7867893°$$

$$C = 38.2132107°$$

$$\boldsymbol{C = 38.2°} \qquad \text{Nearest 0.1°.}$$

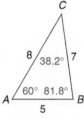

**Figure 14–45**

The solved triangle is shown in Figure 14–45.

**EXAMPLE** Find the angles in triangle *ABC* (see Figure 14–46).

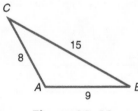

**Figure 14–46**

Find *A*.

$$a^2 = b^2 + c^2 - 2bc \cos A \qquad \text{Solve for cos } A.$$

$$\cos A = \frac{b^2 + c^2 - a^2}{2bc} \qquad \text{Substitute.}$$

$$\cos A = \frac{8^2 + 9^2 - 15^2}{2(8)(9)} \qquad \text{Evaluate.}$$

$$\cos A = \frac{64 + 81 - 225}{144}$$

$$\cos A = \frac{-80}{144}$$

$$\cos A = -0.5555555556 \qquad \text{Use inverse cosine function.}$$

$$A = \cos^{-1} -0.5555555556$$

$$A = 123.7489886°$$

$$\boldsymbol{A = 123.7°} \qquad \text{Nearest 0.1°.}$$

14–5 Law of Cosines

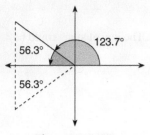

**Figure 14-47**

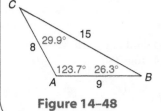

**Figure 14-48**

Recall from Section 14–2 that the cosine is *negative* in the second and third quadrants. Because $A$ is either acute ($<90°$) or obtuse ($>90°$ but $<180°$) and because its cosine is negative (see Figure 14–47), it must be in the second quadrant. If a calculator is used and $-0.5555555556$ is entered, $123.7489886°$ will appear in the display. We can use the law of sines to find the second angle.

Find $B$.

$$\frac{a}{\sin A} = \frac{b}{\sin B} \qquad \text{Substitute.}$$

$$\frac{15}{\sin 123.7489886°} = \frac{8}{\sin B} \qquad \text{Solve for } B.$$

$$15 \sin B = 8 \sin 123.7489886°$$

$$\sin B = \frac{8 \sin 123.7489886°}{15}$$

$$\sin B = 0.4434556903 \qquad \text{Use the inverse sine function.}$$

$$B = \sin^{-1} 0.4434556903$$

$$B = 26.32457654°$$

$$\boldsymbol{B = 26.3°} \qquad \text{Rounded to 0.1°.}$$

Find $C$.

$$C = 180° - A - B \qquad \text{Substitute.}$$

$$C = 180° - 123.7489886° - 26.32457654° \qquad \text{Evaluate.}$$

$$C = 29.92643486°$$

$$\boldsymbol{C = 29.9°}$$

The solved triangle is given in Figure 14–48.

---

**2** **Find the Missing Parts of an Oblique Triangle, Given Two Sides and the Included Angle of the Triangle.**

The law of cosines is needed to solve oblique triangles if two sides and the included angle are given.

**To find the missing parts of an oblique triangle, given two sides and the included angle of the triangle:**

1. Find the side opposite the given angle using the law of cosines.
2. Use the law of sines to find one of the two unknown angles.
3. Find the third angle by subtracting from 180° the sum of the two known angles.

---

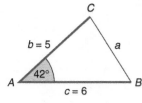

**Figure 14-49**

**EXAMPLE**  Solve triangle $ABC$ in Figure 14–49.

Find $a$.

$$a^2 = b^2 + c^2 - 2bc \cos A \qquad \text{Substitute into the appropriate version of the law of cosines.}$$

$$a^2 = 5^2 + 6^2 - 2(5)(6) \cos 42° \qquad \text{Evaluate.}$$

$$a^2 = 25 + 36 - 60 \cos 42° \qquad \cos 42° = 0.7431448255$$

$$a^2 = 25 + 36 - 44.58868953 \qquad \text{Simplify.}$$

$$a^2 = 16.41131047 \qquad \text{Solve for } a.$$

$$a = 4.051087566$$

$$\boldsymbol{a = 4.051} \qquad \text{Four significant digits.}$$

Find *B*.

$$\frac{a}{\sin A} = \frac{b}{\sin B}$$   Substitute $a = 4.051087566$, $b = 5$, and $A = 42°$.

$$\frac{4.051087566}{\sin 42°} = \frac{5}{\sin B}$$   Solve for *B*.

$$4.051087566 \sin B = 5 \sin 42°$$

$$\sin B = \frac{5 \sin 42°}{4.051087566}$$   Evaluate.

$$\sin B = 0.8258653947$$   Use the inverse sine function.

$$B = 55.67632739°$$

$$\boldsymbol{B = 55.7°}$$   Nearest 0.1°.

Find *C*.

$$\frac{a}{\sin A} = \frac{c}{\sin C}$$   Substitute $a = 4.051087566$, $c = 6$, and $A = 42°$.

$$\frac{4.051087566}{\sin 42°} = \frac{6}{\sin C}$$   Solve for *C*.

$$4.051087566 \sin C = 6 \sin 42°$$

$$\sin C = \frac{6 \sin 42°}{4.051087566}$$   Evaluate.

$$\sin C = 0.9910384737$$   Use inverse sine function.

$$C = \sin^{-1} 0.9910384737$$

$$C = 82.32367267°$$

$$\boldsymbol{C = 82.3°}$$   Nearest 0.1°.

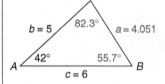

**Figure 14–50**

The solved triangle is given in Figure 14–50.

To check, verify that the sum of the angles adds to 180°, the longest side is opposite the largest angle, and the shortest side is opposite the smallest angle.

---

**EXAMPLE**

**CON**

A vertical 45-ft pole is placed on a hill that is inclined 17° to the horizontal (see Figure 14–51). How long a wire is needed if the wire is placed 5 ft from the top of the pole and attached to the ground at a point 32 ft uphill from the base of the pole?

Point *A* is where the pole enters the ground. Point *B* is where the wire is attached to the ground. Point *C* is where the wire is attached to the pole (5 ft from the top of the pole). Side *b* is the length of the pole from the wire to the ground ($45 - 5 = 40$ ft). The length of the wire is side *a* in the triangle. Because *AC* makes a 90° angle with the horizontal, we know that angle *A* is $90° - 17°$, or 73°. Using the law of cosines, we have

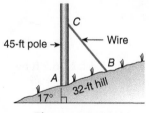

**Figure 14–51**

$$a^2 = b^2 + c^2 - 2bc \cos A$$   Substitute $b = 40$, $c = 32$, and $A = 73°$.

$$a^2 = 40^2 + 32^2 - 2(40)(32) \cos 73°$$   Evaluate.

$$a^2 = 1{,}600 + 1{,}024 - 2{,}560(0.2923717047)$$

$$a^2 = 1{,}600 + 1{,}024 - 748.4715641$$

$$a^2 = 1{,}875.528436$$

$$a = 43.30737161$$

$$a = 43.31 \text{ ft}$$   Four significant digits.

**The length of the wire is 43.31 ft.**

**1** Solve the triangles in Figures 14–52 and 14–53. Round angles to the nearest 0.1°.

**1.**

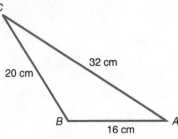

**Figure 14–52**

**2.**

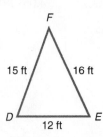

**Figure 14–53**

Find the three angles of each triangle. Round angle measures to the nearest 0.1°.

**3.** $a = 26$ cm, $b = 28$ cm, $c = 32$ cm

**4.** $a = 12.5$ m, $b = 16.1$ m, $c = 14.6$ m

**5.** $a = 8.6$ ft, $b = 9.8$ ft, $c = 7.2$ ft

**6.** $a = 128.5$ in., $b = 120.4$ in., $c = 114.3$ in.

**2** Solve the triangles in Figures 14–54 and 14–55. Round sides to four significant digits. Round angle to the nearest 0.1°.

**7.**

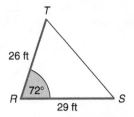

**Figure 14–54**

**8.**

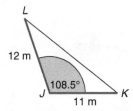

**Figure 14–55**

**9.**

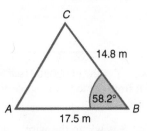

**Figure 14–56**

**10.**

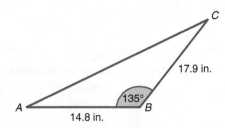

**Figure 14–57**

**11. CON** A hill is inclined 20° to the horizontal. A pole stands vertically on the side of the hill with 35 ft above the ground. How much wire will it take to reach from a point 2 ft from the top of the pole to a point on the ground 27 ft downhill from the base of the pole?

**12. CON** A triangular tabletop is to be 8.4 ft by 6.7 ft by 9.3 ft. What angles must be cut?

| Learning Outcomes | What to Remember with Examples |
|---|---|

**Section 14–1**

**1** Find the magnitude of a vector in standard position, given the coordinates of the endpoint (pp. 571–572).

Use the Pythagorean theorem to find the magnitude of a vector in standard position: $r = \sqrt{x^2 + y^2}$, where $r$ = magnitude, $x$ = $x$-coordinate of the endpoint, and $y$ = $y$-coordinate of the endpoint.

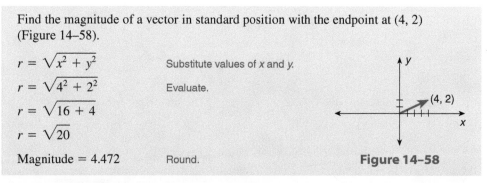

Find the magnitude of a vector in standard position with the endpoint at (4, 2) (Figure 14–58).

$r = \sqrt{x^2 + y^2}$    Substitute values of $x$ and $y$.

$r = \sqrt{4^2 + 2^2}$    Evaluate.

$r = \sqrt{16 + 4}$

$r = \sqrt{20}$

Magnitude = 4.472    Round.

**Figure 14–58**

**2** Find the direction of a vector in standard position, given the coordinates of the endpoint (pp. 572–573).

Use the tangent function to find the direction of a vector in standard position: $\tan \theta = \frac{y}{x}$, where $\theta$ is the direction of the vector, $x$ is the $x$-coordinate of the endpoint and $y$ is the $y$-coordinate of the endpoint.

Find the direction of a vector in standard position with the endpoint at (2, 4).

$\tan \theta = \frac{y}{x}$    Substitute values of $x$ and $y$.

$\tan \theta = \frac{4}{2}$    Simplify.

$\tan \theta = 2$    Use inverse tangent function.

$\tan^{-1} 2 = 63.43494882°$

$\theta = 63.4°$    Round.

Direction = 63.4°

**3** Find the sum of vectors (pp. 573–576).

**1.** Represent the vectors in the form of complex numbers, where the $x$-coordinate of the endpoint of a vector in standard position is the real component and the $y$-coordinate of the endpoint of a vector is the imaginary component. **2.** Add the like components.

Add vectors $A$ and $B$. See Figure 14–59.

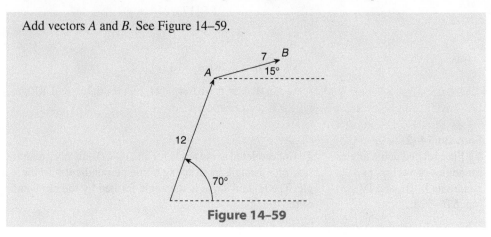

**Figure 14–59**

Vector *A:*

*x*-coordinate:     $\cos \theta = \dfrac{x}{r}$          $r$ = magnitude. Substitute $\theta = 70°$ and $r = 12$.

$\cos 70° = \dfrac{x}{12}$          Solve for *x*.

$12(\cos 70°) = x$

$4.10424172 = x$

$x = 4.104$          Round.

*y*-coordinate:     $\sin \theta = \dfrac{y}{r}$          $r$ = magnitude.

$\sin 70° = \dfrac{y}{12}$          Solve for *y*.

$12(\sin 70°) = y$

$11.27631145 = y$

$y = 11.28$          Round.

vector $A = 4.104 + 11.28i$

vector *B:*

*x*-coordinate:     $\cos \theta = \dfrac{x}{r}$          Substitute $\theta = 15°$ and $r = 7$.

$\cos 15° = \dfrac{x}{7}$          Solve for *x*.

$7(\cos 15°) = x$

$6.761480784 = x$

$x = 6.761$          Round.

*y*-coordinate:     $\sin \theta = \dfrac{y}{r}$          Substitute $\theta = 15°$ and $r = 7$.

$\sin 15° = \dfrac{y}{7}$          Solve for *y*.

$7(\sin 15°) = y$

$1.811733316 = y$

$y = 1.812$          Round.

vector $B = 6.761 + 1.812j$

vector $A + B = (4.104 + 11.28j) + (6.761 + 1.812j)$          Add like vector components.
$\qquad\qquad = 10.87 + 13.09j$

## Section 14–2

**1** Find related acute angles for angles or vectors in quadrants II, III, and IV (pp. 576–578).

To find a related acute angle for an angle ($\theta$) in any quadrant, draw the angle and form the third side of a triangle by drawing a line perpendicular to the *x*-axis from the line forming the angle. The related angle is the angle formed by the ray forming the angle and the *x*-axis portion of the triangle.

Quadrant II angle (Figure 14–60):
$\theta = 180° - \theta_2$   or   $\theta = \pi - \theta_2$

Quadrant III angle (Figure 14–61):
$\theta = \theta_3 - 180°$   or   $\theta = \theta_3 - \pi$

Quadrant IV angle (Figure 14–62):
$\theta = 360° - \theta_4$   or   $\theta = 2\pi - \theta_4$

**Figure 14–60    Figure 14–61    Figure 14–62**

Find the related angle for an angle of 156°.

The angle is in quadrant II, thus, $180° - 156° = 24°$.

**2** Determine the signs of trigonometric values of angles of more than 90° (pp. 578–579).

The memory device ALL-SIN-TAN-COS can be used to remember the signs of the trigonometric functions in the various quadrants. The signs of *all* trigonometric functions are positive in the first quadrant. The sign of the SIN function is positive in the second quadrant (COS and TAN are negative). The sign of the TAN function is positive in the third quadrant. The sign of the COS function is positive in the fourth quadrant.

What is the sign of cos 145°?

145° is a quadrant II angle and the cosine function is negative in quadrant II. Thus, the sign of cos 145° is negative.

**3** Find the trigonometric values of angles of more than 90° using a calculator (p. 580).

Most calculators give the trigonometric value of an angle regardless of the quadrant in which it is found.

Use a calculator to find the value of the following trigonometric functions. Round to four significant digits.

$$\sin 125° = 0.8192$$
$$\cos 215° = -0.8192$$
$$\tan 335° = -0.4663$$

**Section 14–3**

**1** Graph a sine or cosine function using a calculator (pp. 581–583).

**1.** Set MODE to degrees or radians as desired. **2.** From the ZOOM function select the option Ztrig or manually set the window as desired. **3.** Enter the function using the Y = function. **4.** View the graph using GRAPH function.

Graph $y = 2 \cos x$.
Set ZOOM to option Ztrig.
Enter $y = 2 \cos x$.
Press GRAPH (Figure 14–63).

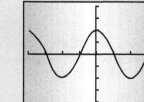

**Figure 14–63**

**2** Find the period of a sine or cosine function (p. 584).

To find the period of a function in the form $y = A \sin Bx$ or $y = A \cos Bx$: **1.** Divide by 360° or $2\pi$ radians by the coefficient of $x$.

$$P = \frac{360°}{B} \quad \text{or} \quad P = \frac{2\pi}{B}$$

Chapter Review of Key Concepts

Find the amplitude and period in degrees of the function $y = 2 \cos 3x$. Graph the equation from $-90°$ to $270°$.

Amplitude = 2 $\qquad\qquad$ $A = 2$ (coefficient of $\cos 3x$).

$$\text{Period} = \frac{360°}{3} = 120° \qquad B = 3 \text{ (coefficient of } x).$$

Set calculation to Degree mode.
Set Window; Xmin = $-90$
$\qquad\qquad$ Xmax = 270
$\qquad\qquad$ Xscl = 30
$\qquad\qquad$ Ymin = $-3$
$\qquad\qquad$ Ymax = 3
$\qquad\qquad$ Yscl = 1

Enter equation: $\quad y = 2 \cos 3x$

Graph (Figure 14–64).

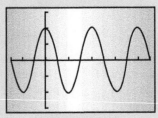

**Figure 14–64**

**3** Find the phase shift of a sine or cosine function (pp. 584–585).

**1.** Find the value $\dfrac{C}{B}$. **2.** If $\dfrac{C}{B}$ is positive, shift the curve $\dfrac{C}{B}$ units to the *left*. **3.** If $\dfrac{C}{B}$ is negative, shift the curve $\dfrac{C}{B}$ units to the *right*.

Graph $y = 3 \sin (3x - 180°)$

Amplitude = 3 $\qquad\qquad\qquad$ coefficient of sine

$$\text{Period} = \frac{360°}{3} = 120° \qquad \frac{360°}{B}$$

$$\text{Phase shift} = \frac{-180°}{3} = -60° \qquad \frac{C}{B} \text{ is negative.}$$

Graph of $y = 3 \sin (3x - 180°)$ shifts 60° to the right from the graph $y = 3 \sin 3x$ (Figure 14–65).

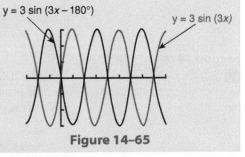

$y = 3 \sin (3x - 180°)$ $\qquad\qquad$ $y = 3 \sin (3x)$

**Figure 14–65**

## Section 14–4

**1** Find the missing parts of an oblique triangle, given two angles and a side (pp. 587–589).

Use the law of sines to find the missing part of an oblique triangle when two angles and a side opposite one of them are given.

$$\frac{a}{\sin A} = \frac{b}{\sin B} = \frac{c}{\sin C}$$

Find side $AB$ in the triangle of Figure 14–66.

$$\frac{a}{\sin A} = \frac{c}{\sin C} \qquad\qquad \text{Substitute } a = 13, A = 22°, \text{ and } C = 97°.$$

$$\frac{13}{\sin 22°} = \frac{c}{\sin 97°} \qquad\qquad \text{Solve for } c.$$

$$c \sin 22° = 13 \sin 97°$$

$$c = \frac{13 \sin 97°}{\sin 22°} \qquad\qquad \text{Evaluate.}$$

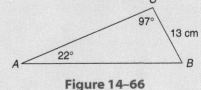

**Figure 14–66**

$$c = 34.44440167 \text{ cm}$$

$$AB = c = 34.44 \text{ cm} \qquad \text{Round.}$$

**2** Find the missing parts of an oblique triangle, given two sides and an angle opposite one of them (pp. 589–592).

Use the law of sines to find the missing part of an oblique triangle when two sides and an angle opposite one of them are given.

Find the measure of angle $A$ in the triangle of Figure 14–67.

$$\frac{a}{\sin A} = \frac{b}{\sin B} \qquad \text{Substitute } a = 18, b = 16, \text{ and } B = 32°.$$

$$\frac{18}{\sin A} = \frac{16}{\sin 32°} \qquad \text{Solve for } A.$$

$$16 \sin A = 18 \sin 32°$$

$$\sin A = \frac{18 \sin 32°}{16} \qquad \text{Evaluate.}$$

$$\sin A = 0.5961591723 \qquad \text{Use the inverse sine function.}$$

$$\sin^{-1} 0.5961591723 = 36.59531105°$$

$$A = 36.6° \qquad \text{Round.}$$

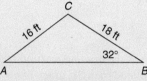

**Figure 14–67**

Given $\triangle ABC$ and $b$ is a known side and $a$ is a known side opposite a known angle, if $b < a$ and $\angle A \neq 90°$, we have two possible solutions. If $b < a$ and $\angle A = 90°$, we have one solution. If $b \geq a$, we have one solution.

Determine how many solutions exist for the triangle in Figure 14–68 based on the given facts. Explain your response.

$B = 87°$
$b = 15$ cm (known side opposite known angle)
$a = 8$ cm (known side opposite unknown angle)

Because $b > a$, there is one possible solution.

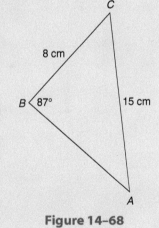

**Figure 14–68**

## Section 14–5

**1** Find the missing parts of an oblique triangle, given three sides of the triangle (pp. 593–596).

To find an angle when three sides are given, we need to use the law of cosines. One variation of the law of cosines is $a^2 = b^2 + c^2 - 2bc \cos A$. Rearrange the formula to solve for $\cos A$. The rearranged formula is

$$\cos A = \frac{-a^2 + b^2 + c^2}{2bc}$$

Find the measure of angle $A$ in the triangle of Figure 14–69.

$a = 20$ cm
$b = 9$ cm
$c = 15$ cm

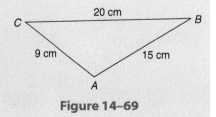

**Figure 14–69**

$$\cos A = \frac{-a^2 + b^2 + c^2}{2bc} \qquad \text{Substitute known values.}$$

$$\cos A = \frac{-20^2 + 9^2 + 15^2}{2(9)(15)} \qquad \text{Evaluate.}$$

$$\cos A = \frac{-400 + 81 + 225}{270}$$

$$\cos A = -0.3481481481 \qquad \text{Use inverse cosine function.}$$

$$\cos^{-1}(-0.3481481481) = 110.3740893°$$

$$A = 110.4° \qquad \text{Round.}$$

**2** Find the missing parts of an oblique triangle, given two sides and the included angle of the triangle (pp. 596–597).

Use the law of cosines to find the missing parts of an oblique triangle when given two sides of the triangle and the included angle. Solve a variation of the law of cosines for the side opposite the given angle.

$$a = \sqrt{b^2 + c^2 - 2bc \cos A}$$

Find the side opposite the given angle of the triangle in Figure 14–70.

$$a = \sqrt{b^2 + c^2 - 2bc \cos A} \qquad \text{Substitute } A = 70°, b = 17, \text{ and } c = 23.$$

$$a = \sqrt{17^2 + 23^2 - 2(17)(23) \cos 70°}$$

$$a = \sqrt{289 + 529 - 267.4597521}$$

$$a = \sqrt{550.5402479}$$

$$a = 23.4635941$$

$$a = 23.46 \text{ cm} \qquad \text{Round.}$$

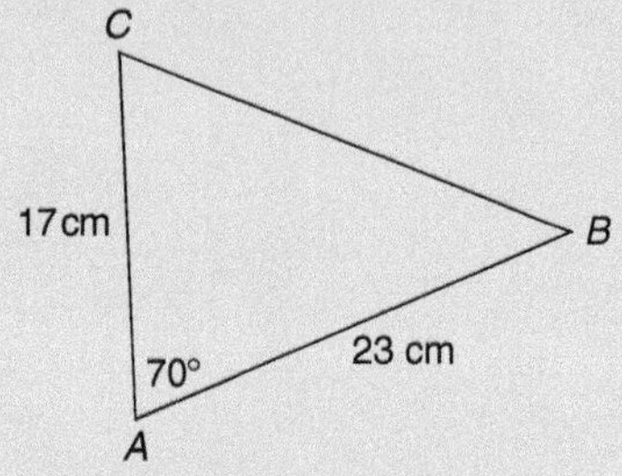

**Figure 14–70**

Find as many parts as possible without using information that you have calculated.

Find the measures of all the angles and sides of the triangle in Figure 14–71.

$$\frac{a}{\sin A} = \frac{c}{\sin C} \qquad \text{Substitute } a = 6.8, A = 48°, \text{ and } c = 7.5.$$

$$\frac{6.8}{\sin 48°} = \frac{7.5}{\sin C} \qquad \text{Solve for } C.$$

$$\sin C = \frac{7.5 \sin 48°}{6.8}$$

$$\sin C = 0.8196450281 \qquad \text{Use inverse sine function.}$$

$$\sin^{-1} 0.8196450281 = 55.04927548°$$

$$C = 55.0° \qquad \text{Round.}$$

$$\angle B = 180° - \angle A - \angle C$$

$$\angle B = 180° - 48° - 55.04927548°$$

$$\angle B = 76.95072452°$$

$$\angle B = 77.0° \qquad \text{Round.}$$

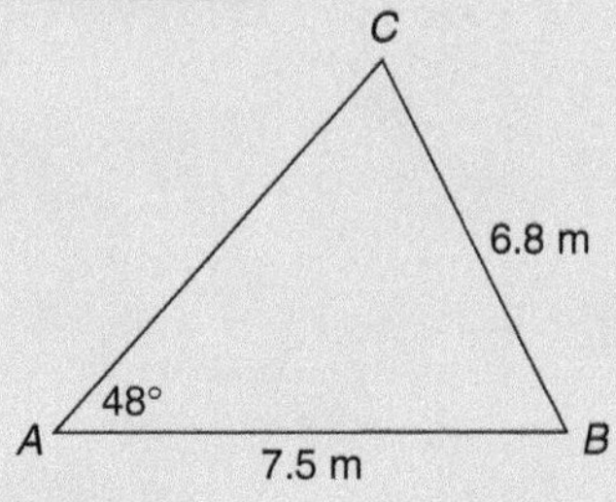

**Figure 14–71**

Chapter 14 / Trigonometry with Any Angle

Use the law of sines or law of cosines to find $b$ or $AC$.

$$\frac{a}{\sin A} = \frac{b}{\sin B}$$    Using the law of sines, solve for $b$.

$$b = \frac{a \sin B}{\sin A}$$    Substitute $a = 6.8$, $B = 76.95072452°$, and $A = 48°$.

$$b = \frac{6.8 \sin 76.95072452°}{\sin 48°}$$

$$b = 8.914007364$$

$$b = 8.914 \text{ m}$$

## CHAPTER REVIEW EXERCISES

### Section 14–1

Find the direction and magnitude of vectors in standard position with the indicated endpoints. Round lengths to four significant digits and angles to the nearest tenth of a degree.

**1.** $(4, 6)$      **2.** $(2, 8)$      **3.** $(6, 1)$

Find the magnitude and direction of the resultant of the sum of the two given vectors in complex notation. Round as above.

**4.** $3 + 5j$ and $2 + 3j$      **5.** $1 + 4j$ and $6 + 8j$

### Section 14–2

Find the related angle for the angles. Express radian measures to the nearest ten-thousandth.

**6.** $115°$    **7.** $3.04$ rad    **8.** $4.75$ rad    **9.** $221°$
**10.** $305°$    **11.** $5.4$ rad    **12.** $138.5°$    **13.** $212°15'10''$

Using the calculator, evaluate the trigonometric functions. Round to four significant digits.

**14.** $\cos 250°$    **15.** $\sin 2.1$    **16.** $\tan 175°$    **17.** $\sin 340°$

**18.** $\tan 4.5$    **19.** $\cos 290°$    **20.** $\sin \dfrac{3\pi}{4}$    **21.** $\tan \dfrac{5\pi}{4}$

**22. ELEC** Find the magnitude and direction of a vector of an electrical current in standard position if the coordinates of the end point of the vector are $(2, -3)$. Round the magnitude (in amps) to the nearest hundredth. Express the direction in degrees to the nearest $0.1°$.

**23.** Find the direction and magnitude of a vector in standard position if the coordinates of the end point of the vector are $(-2, 2)$. Express its direction in radians and round its direction and its magnitude to the nearest hundredth.

### Section 14–3

**24. ELEC** What is the frequency of a wave that passes through a given point once each $0.025$ sec?

**25. ELEC** What is the period of a wave that has a frequency of $15,000$ Hz?

**26. ELEC** WMC AM has a frequency of $790$ KHz. Find the station's wavelength in meters.

**27.** Find the amplitude and period of the function $y = 5 \cos 4x$. Graph the function to show two periods.

**28.** Find the amplitude, period, phase shift and graph the function $y = \frac{1}{2} \sin (8x - 180°)$.

**29. ELEC** The maximum voltage $V$ in a generator is $50$ V. The voltage charges according to the equation $v = 50 \sin x$. Use a calculator to find the voltage at $120°$.

## Section 14–4

Solve the triangles. If a triangle has two possible solutions, find both solutions. Round sides to tenths and angles to the nearest 0.1°.

**30.** $A = 60°$, $B = 40°$, $b = 20$ cm

**32.** $A = 60°$, $B = 60°$, $a = 10$ ft

**34.** $b = 10$ m, $c = 8$ m, $B = 52°$

**36.** A surveyor needs the measure of $JK$ in Figure 14–72. Find $JK$ to the nearest foot.

**31.** $B = 120°$, $C = 20°$, $a = 8$ m

**33.** $a = 5$ dkm, $c = 7$ dkm, $C = 45°$

**35.** $a = 9.2$ cm, $b = 6.8$ cm, $B = 28°$

**37.** In Figure 14–73, find $RS$.

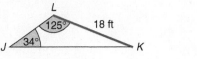

**Figure 14–72**

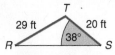

**Figure 14–73**

## Section 14–5

Solve the triangles in Figures 14–74 to 14–79. Round sides to four significant digits and angles to the nearest 0.1°.

**38.**

**Figure 14–74**

**39.**

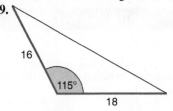

**Figure 14–75**

**40.**

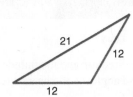

**Figure 14–76**

**41.**

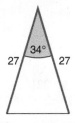

**Figure 14–77**

**42.**

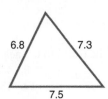

**Figure 14–78**

**43.**

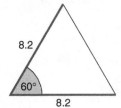

**Figure 14–79**

Solve the exercises. Round sides to four significant digits and angles to the nearest 0.1°.

**44. CON** A triangular lot has sides 180 ft long, 160 ft long, and 123.5 ft long. Find the angles of the lot.

**46. AVIA** A ship sails from a harbor 35 nautical miles east, and then 42 nautical miles in a direction 32° south of east (Figure 14–80). How far is the ship from the harbor?

**45. CON** A vertical 50-ft pole stands on top of a hill inclined 18° to the horizontal. What length of wire is needed to reach from a point 6 ft from the pole's top to a point 75 ft downhill from the base of the pole?

**47. CON** A hill with a 35° grade (inclined to the horizontal) is cut down for a roadbed to a 10° grade. If the distance from the base to the top of the original hill is 800 ft (Figure 14–81), how many vertical feet will be removed from the top of the hill, and what is the distance from the bottom to the top of the hill for the roadbed?

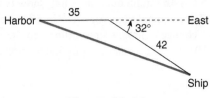

**Figure 14–80**

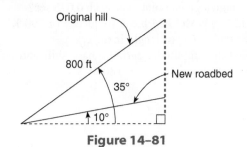

**Figure 14–81**

Chapter 14 / Trigonometry with Any Angle

1. Use a calculator to graph the following in radian angle mode:
   (a) $y = \sin x$
   (b) $y = \sin (x + 1)$
   (c) $y = \sin (x + 2)$
   (d) $y = \sin (x - 1)$
   (e) $y = \sin (x - 2)$
   (f) Discuss the similarities and differences of the graphs.

2. Use a calculator to graph the following in radian angle mode:
   (a) $y = \sin x$
   (b) $y = \sin (x) + 1$
   (c) $y = \sin (x) + 2$
   (d) $y = \sin (x) - 1$
   (e) $y = \sin (x) - 2$
   (f) Discuss the similarities and differences of the graphs.
   (g) Discuss how these graphs are different from the graphs in Exercise 1.

## PRACTICE TEST

Find the values. Round to four significant digits.
1. $\sin 125°$
2. $\tan 140°$
3. $\cos 160°$

4. Find the length of the vector that has endpoint coordinates of (8, 15).

5. Find the angle of the vector whose endpoint coordinates are $(-8, 8)$.

Use the law of sines or the law of cosines to find the side or angle indicated in Figures 14–82 to 14–89. Round sides to four significant digits and angles to the nearest 0.1°.

6.

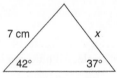

Figure 14–82

7.

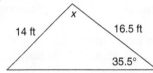

Figure 14–83

8.

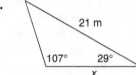

Figure 14–84

9.

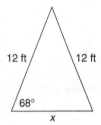

Figure 14–85

10.

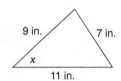

Figure 14–86

11.

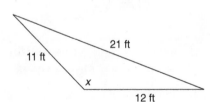

Figure 14–87

12.

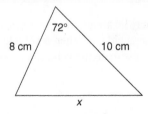

Figure 14–88

13.

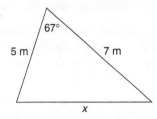

Figure 14–89

**14.** Find the *AC* of triangle *ABC* in Figure 14–90. Round to four significant digits.

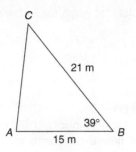

**Figure 14–90**

**15.** Find ∠*R* of triangle *RST* in Figure 14–91. Round to the nearest 0.1 degree.

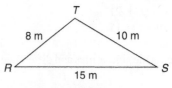

**Figure 14–91**

**16.** Find the direction in degrees (to the nearest 0.1°) of the vector $I_t$ (total current) in Figure 14–92.

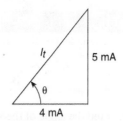

**Figure 14–92**

**17.** A surveyor measures two sides of a triangular lot and the angle formed by these two sides. What is the length of the third side if the two measured sides are 76 ft and 110 ft and the angle formed by these two sides is 107°?

**18.** The sides of a triangle cast measure 31 mm, 42 mm, and 27 mm. Find the measure of the angle opposite the 31 mm side.

**19.** Find the third side of a triangular flower bed if two sides measure 12 ft and 15 ft and the included angle measures 48°.

**20.** A plane is flown from an airport due east for 32 mi, and then turns 15° north of east and travels 72 mi. How far is the plane from the airport?

**21.** Find the magnitude to the nearest tenth of the vector $I_t$ in Exercise 16.

**22.** A rod 11 cm long joins a crank 20 cm long to form a triangle with a third, imaginary line (Figure 14–93). If the crank and the rod form an angle of 150°, find the angle formed by the rod and the imaginary line.

**23.** Find the magnitude of vector *Z* in Figure 14–94 if vectors *R* and $X_L$ form a right angle. All units are in ohms.

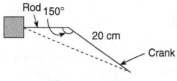

**Figure 14–93**

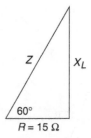

**Figure 14–94**

**24.** Find the amplitude, period, and phase shift, and graph two periods of the function $y = 4 \sin (x - 90°)$.

---

### CAREER APPLICATION: LAW ENFORCEMENT—VEHICLE VAULTS

Today's traffic accident investigators often act as traffic reconstructionists because drivers may lie about their driving speed, be seriously injured (and so be unable to think clearly and answer questions), or die in the accident. When a vehicle becomes airborne with an uphill takeoff angle of at least 6°, investigators use trigonometry to estimate the vehicle's speed at takeoff with the following vault formula based on an oblique triangle:

$$S = \frac{2.73 \cdot D}{\sqrt{D \cdot \cos \theta \cdot \sin \theta \pm [H \cdot (\cos \theta)^2]}}$$

where $S$ = speed in miles per hour, $D$ = horizontal distance from takeoff point to first contact, $H$ = vertical distance from takeoff point to first contact, and $\theta$ = takeoff angle in degrees. Use + in the formula when the vehicle lands at a point lower than takeoff, and − when it lands at a point higher than takeoff.

The takeoff angle is usually the same as the road grade. The grade of a road is the slope of the road written as a percent, or

$$\text{grade} = \frac{\text{vertical rise}}{\text{horizontal run}} \cdot 100\%$$

The takeoff angle in degrees is $\tan^{-1}$ of the grade (in decimal form). In the case of a motorcycle driver vaulting off his vehicle, the takeoff angle is assumed to be 45°, and the driver's vertical drop on level ground is assumed to be 3 ft.

## Exercises

Use the preceding information to estimate the speed of the vehicle in each of the following vault accidents. Round answers to the nearest unit.

1. The driver of an empty tractor-trailer truck falls asleep at the wheel, "straightens the curve" on an uphill 7% grade road, and vaults off a cliff. The truck's first contact point is 40 vertical feet and 55 horizontal feet below its takeoff point. The driver of the car behind the truck said the truck was going between 20 and 25 $\frac{\text{mi}}{\text{h}}$ when it left the road. Is this correct?

2. On a level street, a driver runs a red light at a T intersection and is hit by a motorcycle whose driver vaults over the top of the car. The motorcycle driver hits the street 57 horizontal feet from the takeoff point. How fast was the motorcycle traveling when it hit the car?

3. A car traveling on a divided highway blows a tire crosses the median ditch (which dips 5 vertical feet and then rises sharply), and vaults into the oncoming traffic lanes. It makes first contact at a point 12 ft higher than the takeoff point and 162 ft horizontally from the takeoff point. Combining the 6% uphill road grade with the ditch angle gives a takeoff angle of 32°. Find the car's speed.

4. A drunk driver in a high-rise sport utility vehicle with big wheels drifts into the median strip of a divided highway at an overpass. He hits a large mound of dirt piled in the median, and the vehicle is launched upward at a 50° angle. The vehicle vaults over the oncoming traffic lane, and falls 30 ft to the ground below. Find the takeoff speed if the horizontal distance is 153 ft.

5. Do you think the takeoff speed of the vehicle from Exercise 4 is the same speed the vehicle was going just before it hit the median strip? Explain.

## Answers

1. The driver of the car estimated the truck's speed correctly at about 22 $\frac{\text{mi}}{\text{h}}$.

2. The motorcycle was traveling at least 25 $\frac{\text{mi}}{\text{h}}$.

3. The car's speed was at least 55 $\frac{\text{mi}}{\text{h}}$.

4. When the truck left the top of the dirt mound, its speed was at least 44 $\frac{\text{mi}}{\text{h}}$.

5. Climbing the dirt mound would have slowed the vehicle considerably, so the truck's speed when it left the road was greater.

---

### CUMULATIVE PRACTICE TEST FOR CHAPTERS 12–14

Classify the angle measures using the terms *right, straight, acute,* and *obtuse.*

1. 28°
2. 150°
3. 180°

Find the measure of the complementary angle.

4. 46°
5. 28°

Find the measure of the supplementary angle.
  **6.** 52°                                              **7.** 143°

Change to decimal degree equivalent rounded to the nearest ten-thousandth.
  **8.** 42′15″                                           **9.** 30°12′20″

Change to minutes and seconds. Round to the nearest second when necessary.
  **10.** 0.86°                                           **11.** 0.352°

Find the perimeter and area to the nearest whole unit.
  **12.** A rectangle with length of 30 cm and width of 18 cm.   **13.** A triangle with sides that measure 42 m, 36 m, and 30 m.
  **14.** Find the circumference and area of a circle that has a radius of 40 cm.   **15.** Find the circumference and area of a circle that has a diameter of 150 cm.

Convert to radians. Round to the nearest ten-thousandth radian.
  **16.** 45°                                             **17.** 15°22′

Convert to degrees. Round to the nearest ten-thousandth of a degree.
  **18.** $\dfrac{3\pi}{4}$ rad                           **19.** 1.8 rad

  **20.** Find the area of a sector of a circle that has a radius of 12 cm if the angle of the sector if 60°.   **21.** Find the volume to the nearest centimeter of a cylinder that has a circular base with a radius of 12 cm if the cylinder has a height of 40 cm.

  **22.** Find the lateral surface area of a cone with a radius of 72 cm and a slant height of 100 cm.   **23.** Use the Pythagorean theorem to find the hypotenuse to the nearest inch of a right triangle that has legs measuring 12 in. and 15 in.

Use the sine, cosine, or tangent to find the sides and angles. Round sides to four significant digits and round angles to the nearest 0.1°. In $\triangle ABC$, $\angle C$ is a right angle.
  **24.** $BC = 4.6$ cm; $\angle A = 40°$. Find $AB$.   **25.** $AC = 23.5$ cm; $\angle A = 60°$. Find $AB$.
  **26.** $AC = 32.1$ cm; $BC = 28.9$ cm. Find $\angle A$.   **27.** $AB = 15.8$ cm; $AC = 12.5$ cm. Find $\angle B$.

Find the magnitude and direction of the vector in standard position with the indicated endpoint. Round the magnitude to the nearest thousandth and the direction to the nearest tenth of a degree.
  **28.** $(3, 4)$                                        **29.** $(5, 2)$
  **30.** Find the resultant vector of the vectors that have a direction of 30° and magnitudes of 8 and 10, respectively.

Find the amplitude, period, and phase shift for the functions.
  **31.** $y = \cos(x + 90°)$                             **32.** $y = 5 \sin(x - 180°)$
  **33.** $y = 3 \sin(2x - 240°)$                         **34.** A triangle has sides that measure 8.3 cm, 8.9 cm, and 9.2 cm. Find the measure of each of the three angles. Round to the nearest 0.1°.

  **35.** In $\triangle ABC$, $\angle C$ measures 38°. $AC = 34$ cm and $BC = 36$ cm. Find $AB$ to the nearest centimeter.

# Statistics

## Focus on Careers

Want a high-paying job? Get an education! An associate or bachelor's degree is the most significant source of education for 10 of the 20 fastest growing occupations.

High school dropouts have the lowest expected lifetime earnings. As the successive level of schooling increases, so do the lifetime earnings. Persons with an associate degree are expected to earn 1.6 times as much as a person with some high school but no diploma. And a person with a bachelor's degree can expect to earn 2.1 times as many dollars as a person with some high school but no diploma.

For a number of years, employment in the United States has been shifting to service-providing jobs and that is expected to continue through 2012. Jobs in education and health services are expected to grow faster and add more jobs than any other employment sector. Approximately 25% (over 5 million) of new jobs created in the U.S. economy are expected to be in the health-care, social assistance, or private educational services sector.

Professional and business services are expected to add nearly 5 million new jobs by 2012. In this sector,

*(continued)*

the fastest-growing industry is expected to be employment services. Management, scientific, and technical consulting services are expected to grow very rapidly because of increased use of new technology and computer software.

Software publishers, Internet publishing, Internet service providers, and data processing services are some careers in the information sector. Telecommunications, broadcasting, and newspaper, periodical, book, and directory publishers will also fuel job growth in the information sector.

Sources: *Occupational Outlook Handbook* 2004–2005 Edition. U.S. Department of Labor, Bureau of Labor Statistics. "The Big Payoff: Educational Attainment and Synthetic Estimates of Work-Life Earnings," *Current Population Reports,* pp. 23–210, by Jennifer Cheeseman Day and Eric C. Newburger.

A **graph** shows information visually. Graphs may show how our tax dollars are divided among various government services, trace the fluctuations in a patient's temperature, or illustrate regional planting seasons. Other graphs may show equations, inequalities, and their solutions. Tables, on the other hand, usually list data. For example, income tax tables list taxes due on different incomes.

# 15–1 | Reading Circle, Bar, and Line Graphs

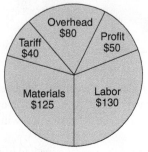

**Figure 15–1** Distribution of wholesale price for a $425 color television.

*Learning Outcomes*

**1** Read circle graphs.
**2** Read bar graphs.
**3** Read line graphs.

Graphs give us useful information at a glance if we interpret them properly. Three common graphs used to represent data are the circle graph, the bar graph, and the line graph.

## **1** Read Circle Graphs.

A **circle graph** uses a divided circle to show pictorially how a whole quantity is divided into parts.

The complete circle represents one whole quantity. The circle is divided into parts so that the sum of all the parts equals the whole quantity. These parts can be expressed as fractions, decimals, or percents. Figure 15–1 is a circle graph.

When we "read" a graph, we examine the information on the graph.

### To read a circle, bar, or line graph:

1. Examine the title of the graph to find out what information is shown.
2. Examine the parts to see how they relate to one another and to the whole.
3. Examine the labels for each part of the graph and any explanatory remarks that may be given.
4. Use the given parts to calculate additional amounts.

---

**EXAMPLE**  Use Figure 15–1 to answer these questions.

**(a)** What percent of the wholesale price is the cost of labor?
**(b)** What percent of the wholesale price is the cost of materials?
**(c)** What would the wholesale price be if no tariff (tax) was paid on imported parts?

**(a)** $R = \dfrac{P}{B}$   Use the precentage formula to find *R*.

$R = \dfrac{130}{425}$   *R* is the percent of the wholesale price ($425) that is attributed to labor cost. The labor cost is $130.

$R = 0.3058823529$    Change to a percent.

$R = 30.6\%$ (labor)

**(b)** $R = \dfrac{P}{B}$    Use the percentage formula.

$R = \dfrac{125}{425}$    $R$ is the percent of the wholesale price ($425) that is attributed to materials cost. The materials cost is $125.

$R = 0.2941176471$    Change to a percent.

$R = 29.4\%$ (materials)

**(c)** Price − tariff = ⬚425⬚ − 40 = $385 (cost without tariff)

### 2 Read Bar Graphs.

Different types of graphs allow us to access different types of information. A **bar graph** uses two or more bars to compare two or more amounts.

The bar lengths represent the amounts being compared. Bars can be drawn either horizontally or vertically.

The **axis,** or **reference line,** that runs along the length of the bars is a scale of the amounts being compared; in Figure 15–2 this line is horizontal. The other reference line (vertical in this case) labels the bars. Figure 15–2 is a horizontal bar graph.

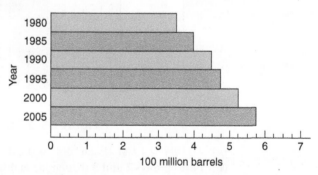

**Figure 15–2**   Company oil production.

**EXAMPLE**   Use Figure 15–2 to answer these questions.

**(a)** How many more 100 million barrels of oil are indicated for the company in 2005 than in 1985?

**(b)** Judging from the graph, should company oil production in 2010 be more or less than in 2005?

**(c)** How many 100 million barrels of oil did the company produce in 1985?

**(a) 2005 production − 1985 production = 5.75 − 4.00 = 1.75 hundred million barrels.**
**(b) More, because the trend has been toward greater production.**
**(c) Four hundred million barrels in 1985.**

### 3  Read Line Graphs.

Line graphs are encountered in industrial reports, handbooks, and the like. A **line graph** uses one or more lines to show changes in data.

The horizontal axis on a line graph usually represents periods of time or specific times. The vertical axis represents numerical amounts. Line graphs show trends in data and high and low values at a glance. Figure 15–3 is a line graph.

**EXAMPLE**   Use Figure 15–3 to answer these questions regarding a patient's temperature.

**(a)** On what date and time of day did the patient's temperature first drop to within 0.2 degrees of normal (98.6° F)?

**(b)** On which post-op (post-operative) days did the patient's temperature remain within 0.2 degrees of normal?

**(c)** What was the highest temperature recorded for the patient?

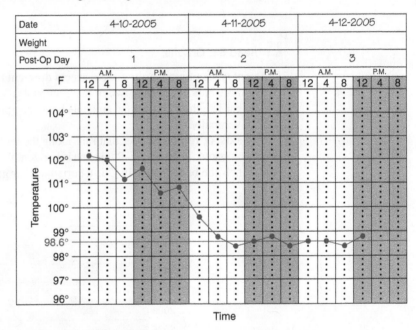

**Figure 15–3**   Graphic temperature chart.

**(a)** **4-11-04 at 4 A.M.** (Each "dot" is 0.2 degrees, so the temperature was 98.8° F.)

**(b)** **Post-op days 2 and 3** (beginning at 4 A.M. on day 2)

**(c)** **102.2 degrees** (recorded at 12 A.M. on 4-10-05)

### SECTION 15–1 SELF-STUDY EXERCISES

**1**  Use Figure 15–4 to answer Exercises 1–3.

**1.** What percent of the gross salary goes into retirement?

**2.** What percent of the take-home pay is federal income tax? Round to tenths.

**3.** What percent of the gross pay is the take-home pay? Round to tenths.

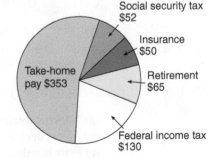

**Figure 15–4**   Distribution of weekly salary of $650.

Use Figure 15–5 to answer Exercises 4–6. Round to tenths.

4. What percent of the day is spent working?
5. What percent of the day is spent sleeping?
6. The amount of time spent studying is what percent of the time spent in class?

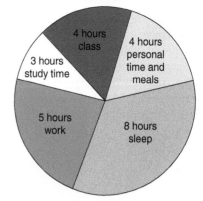

**Figure 15–5** Distribution of a student's typical day.

---

**2** Use Figure 15–6 to answer Exercises 7–9.

7. **BUS** What expenditure is expected to be the same next year as this year?
8. **BUS** What two expenditures are expected to increase next year?
9. **BUS** What two expenditures are expected to decrease next year?

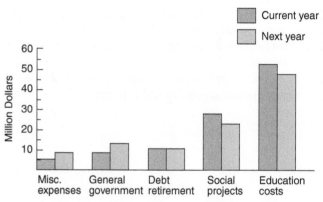

**Figure 15–6** Distribution of local tax dollars.

---

Use Figure 15–7 to answer Exercises 10–13.

10. What year had the largest daily number of barrels of oil produced on shore?
11. What was the total number of barrels of oil produced daily in 1990? In 2000?
12. By what percent did daily offshore oil production increase from 1960 to 2000? Round to tenths.
13. In what year was the total daily oil production at its lowest level for the 5 decades reported?

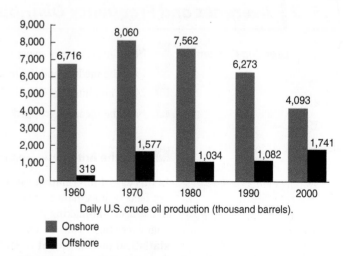

**Figure 15–7** Daily U.S. crude oil production (thousand barrels).
*Source:* Energy Information Administration, U.S. Government.

Use Figure 15–8 to answer Exercises 14–17.

14. **ELEC** How many amperes of current are produced by 50 V when the resistance is 10 Ω?
15. **ELEC** How many volts are needed to produce 2 A of current when the resistance is 25 Ω?
16. **ELEC** Approximately how many volts are required to produce a current of 3.5 A when the resistance is 10 ohms?
17. **ELEC** Find the resistance when 100 V is needed to produce a current of 4 A.

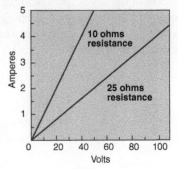

**Figure 15–8** Amperage produced by voltage across two resistances.

Use Figure 15–9 to answer Exercises 18–22.

18. What year was the motor gasoline supply 6,580,000 barrels of oil?
19. Which petroleum product consistently had the greatest supply?
20. For which 10-year time period was the increase in supply of motor gasoline the greatest?
21. What time period saw the greatest percentage of increase in jet fuel?
22. In 2000 the supply of jet fuel was what percent of the motor gasoline supply? Round to tenths.

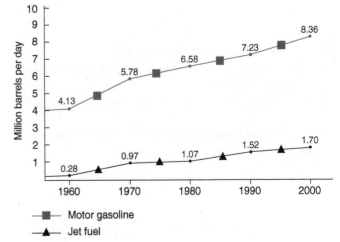

**Figure 15–9** Petroleum products supplied by type in the United States.
*Source:* Energy Information Administration, U.S. Government.

## 15–2 | *Averages and Frequency Distributions*

*Learning Outcomes*

1. Find the arithmetic mean or arithmetic average.
2. Find the median and the mode.
3. Make and interpret frequency distributions.
4. Find the mean of grouped data.

### 1 Find the Arithmetic Mean or Arithmetic Average.

In this age of information explosion we have massive amounts of data available to us. **Data** are facts or information from which conclusions may be drawn. To use these data effectively in the decision-making process, we need to examine the data and summarize key trends and characteristics. This summary is generally in the form of statistical measurements. A **statistical measurement** or **statistic** is a standardized, meaningful measure of a set of data that reveals a certain feature or characteristic of the data. Such statistics are called **descriptive statistics**.

An **average** is an approximate number that is a central value of a set of data. The most common average is the arithmetic mean or arithmetic average. The **arithmetic mean** or **statistical mean** is the sum of the quantities in the data set divided by the number of quantities.

Symbols can be used to write the procedures for statistical measures. The formula in symbols for the mean for a set of data is

$$\bar{x} = \frac{\Sigma x_i}{n}$$

The formula is read as "the mean $\bar{x}$ (read "x-bar") equals the sum $\Sigma$ of each value of $x$ (read "x sub $i$") divided by the number of values $n$. The Greek capital letter **sigma, $\Sigma$**, is a summation symbol and indicates the addition of a set of values. The notation $x_i$ identifies each data value with a subscript: $x_1$ is the first value, $x_2$ is the second value, and $x_n$ is the $n$th value.

## To find the arithmetic mean or arithmetic average:

Find the mean of 22, 31, and 37.

$$x_1 = 22, x_2 = 31, x_3 = 37$$

**1.** Add the quantities.

$\Sigma x_i$

$$\Sigma x_i = 22 + 31 + 37$$
$$\Sigma x_i = 90$$

**2.** Divide the sum by the number of quantities.

$$\bar{x} = \frac{\Sigma x_i}{n} = \frac{90}{3}$$
$$\bar{x} = 30$$

Symbolically, $\bar{x} = \frac{\Sigma x_i}{n}$, where $\bar{x}$ = arithmetic mean; $\Sigma$ means sum;

$x_i$ represents each $x$ value; $n$ = number of values.

---

**EXAMPLE**   Find the average or mean of each set of quantities.

**HLTH/N**   **(a)** Pulse rates: 68, 84, 76, 72, 80

There are five pulse rates, so we find their sum and divide by 5:

$$\bar{x} = \frac{\Sigma x_i}{n} = \frac{68 + 84 + 76 + 72 + 80}{5} = \frac{380}{5} = 76$$

**The mean pulse rate is 76.**

**(b)** Pounds: 21, 33, 12.5, 35.2 (to the nearest whole number)

There are four weights, so we find their sum and divide by 4.

$$\bar{x} = \frac{\Sigma x_i}{n} = \frac{21 + 33 + 12.5 + 35.2}{4} = \frac{101.7}{4} = 25.425 = 25$$

**The mean weight is 25 lb (to the nearest pound).**

---

**EXAMPLE**   An automobile used 41 gal of regular gasoline on a trip of 876 mi. What was the average miles per gallon ($\frac{mi}{gal}$) to the nearest gallon?

**AUTO**   The number of miles traveled on the 41 gal of gasoline is 876 and represents the total miles. Divide the total miles by 41.

$$\frac{876 \text{ mi}}{41 \text{ gal}} = 21.36585366 = 21 \frac{mi}{gal} \qquad \text{To nearest gallon.}$$

**The car averaged 21 $\frac{mi}{gal}$ on the trip.**

**EXAMPLE**

Table 15–1 lists the final grades of a horticulture student. Find the student's QPA (quality point average) to the nearest hundredth based on a 4-point system.

**AG/H**

To find the QPA for a term, the quality points for the letter grade of each course are multiplied by the credit hours of each course to obtain the total quality points for each course. This total is divided by the total credit hours earned. The quality points awarded are A = 4, B = 3, and C = 2.

| Table 15–1 | Quality Points for Final Grades | | | | | | |
|---|---|---|---|---|---|---|---|
| **Subject** | **Grade** | **Credit Hours** | | **Quality Points per Hour** | | | **Total Points** |
| Algebra 101 | B | 3 | × | 3 | = | | 9 |
| Spray chemicals 102 | A | 3 | × | 4 | = | | 12 |
| Landscape 301 | A | 3 | × | 4 | = | | 12 |
| English 101 | C | 4 | × | 2 | = | | 8 |
| | | 13 hr | | | | | 41 points |

$$\frac{41}{13} = 3.153846154 \text{ or } 3.15 \qquad \text{To nearest hundredth.}$$

**Thus, the student's QPA for the term is 3.15.**

---

**EXAMPLE**

A community college student has the following grades in Physics 101: 73, 84, 80, 62, and 70. What grade is needed on the last test for the student to get a C on the final grade or a 75 average?

Let $x$ = the score on the last test.

$$\frac{73 + 84 + 80 + 62 + 70 + x}{6} = 75 \qquad \text{Average = 75. Add like terms in the numerator.}$$

$$\frac{369 + x}{6} = 75 \qquad \text{Clear the fraction.}$$

$$6\left(\frac{369 + x}{6}\right) = 75(6)$$

$$369 + x = 450 \qquad \text{Solve for } x.$$

$$x = 450 - 369$$

$$x = 81 \qquad \text{Score needed on last test.}$$

**The student must earn a score of 81 or higher on the last test to have an average of 75 or higher.**

$$\text{Check:} \qquad \frac{73 + 84 + 80 + 62 + 70 + 81}{6} = \frac{450}{6} = 75$$

---

**2   Find the Median and the Mode.**

Besides the mean or arithmetic average, we also use the *median* and the *mode* to describe groups of data. These are measures of *central tendency*. **Measures of central tendency** are descriptive statistics that determine the center of a data set from a different perspective.

The **median** is the middle value when the data values are arranged in order of size. The median is frequently used for data relating to annual incomes, housing price ranges, etc.

## To find the median:

1. Arrange the values in order of size, either smaller to larger or larger to smaller.
2. If the number of values is odd, the median is the middle value.
3. If the number of values is even, the median is the average of the two middle values.

---

**EXAMPLE**

**HLTN/N**

A TPR chart shows a patient's temperature, pulse rate, and respiration rate. The following pulse rates were recorded on a TPR chart: 68, 88, 76, 64, 72. What is the patient's median pulse rate?

In descending order of size:   88
                               76
                               72   ← median or middle value in an odd number of values
                               68
                               64

**The median pulse rate is 72.**

---

**EXAMPLE**

The following temperatures were recorded: 56°, 48°, 66°, and 62°. What is the median temperature?

The number of temperatures is even, so we find the average of the two middle values.

In ascending order of size   48
                             56 $\Big\}$ $\dfrac{56 + 62}{2} = \dfrac{118}{2} = 59$
                             62
                             66

**The median temperature is 59°.**

---

The **mode** is the most frequently occurring value in the data set.

## To find the mode:

1. Identify the value or values that occur with the greatest frequency as the mode.
2. If no value occurs more than another value, there is *no mode* for the data set.
3. If more than one value occurs with the greatest frequency, the modes of the data set are the values that have the greatest frequency.

---

**EXAMPLE**

**BUS**

The hourly pay rates at a local fast-food restaurant are as follows: cooks, $8.50; servers, $7.15; bussers, $7.15; dishwashers, $7.25; managers, $10.50. Find the mode.

Identify the value or values that appear most often.

**The hourly pay rate of $7.15 occurs more than any other rate. It is the mode.**

**EXAMPLE** The daily work shifts at a mall clothing store are 4, 6, and 8 hours. Find the mode.

**No shift occurs more frequently than another, so there is no mode.**

### 3 Make and Interpret a Frequency Distribution.

Suppose for a class of 25 students the instructor records the following grades:

76  91  71  83  97  87  77  88  93  77  93  81  63
79  74  77  76  97  87  89  68  90  84  88  91

It is difficult to make sense of all these numbers as they appear here. But the instructor can arrange the scores into several smaller groups, called **class intervals.** The word *class* means a special category.

These scores can be grouped into class intervals of 5, such as 60–64, 65–69, 70–74, 75–79, 80–84, 85–89, 90–94, and 95–99. Each class interval has an odd number of scores. The *middle score* of each interval is a **class midpoint.**

The instructor can now *tally* the number of scores that fall into each class interval to get a **class frequency,** the number of scores in each class interval.

A compilation of class intervals, midpoints, tallies, and class frequencies is called a **grouped frequency distribution.**

**EXAMPLE** Examine the grouped frequency distribution in Table 15–2, and answer the following questions.

#### Table 15–2   Frequency Distribution of 25 Scores

| Class Interval | Midpoint | Tally | Class Frequency |
|:---:|:---:|:---:|:---:|
| 60–64 | 62 | / | 1 |
| 65–69 | 67 | / | 1 |
| 70–74 | 72 | // | 2 |
| 75–79 | 77 | ⧼⧽ / | 6 |
| 80–84 | 82 | /// | 3 |
| 85–89 | 87 | ⧼⧽ | 5 |
| 90–94 | 92 | ⧼⧽ | 5 |
| 95–99 | 97 | // | 2 |

**(a)** How many students scored 70 or above?

$2 + 6 + 3 + 5 + 5 + 2 = 23$   Add the frequencies for class intervals with scores 70 or higher.

**23 students scored 70 or above.**

**(b)** How many students made As (90 or higher)?

$5 + 2 = 7$   Add the frequencies for class intervals 90–94 and 95–99.

**7 students made As (90 or higher).**

**(c)** What percent of the total grades were As (90s)?

$\dfrac{7 \text{ As}}{25 \text{ total}} = \dfrac{7}{25} = 0.28 = \mathbf{28\% \text{ As}}$   The portion or part is 7 and the base or total is 25.

**(d)** Were the students prepared for the test or was the test too difficult?

The relatively high number of 90s (7) compared to the relatively low number of 60s (2) suggests that **in general, most students were prepared for the test.**

**(e)** What is the ratio of As (90s) to Fs (60s)?

$$\frac{7 \text{ As}}{2 \text{ Fs}} = \frac{7}{2}$$

**The ratio is** $\dfrac{7}{2}$**.**

---

**EXAMPLE**  Students in a history class reported their credit-hour loads as shown. Make a grouped frequency distribution of their credit hours. Credit hours carried: 3, 12, 15, 3, 6, 6, 12, 9, 12, 9, 6, 3, 12, 18, 6, 9.

To establish a class interval with an easy-to-find midpoint, use an odd number of points in the interval. Here, an interval of 5 is used; that is, 0–4 contains five possibilities: 0, 1, 2, 3, and 4. The middle number is the midpoint, 2. Make a tally mark for each time the credit hours of a student falls in the interval. Then count the tally marks to get the class frequency (Table 15–3).

**Table 15–3  Frequency Distribution of Credit-Hour Loads**

| Class Interval | Midpoint | Tally | Class Frequency |
|:---:|:---:|:---:|:---:|
| 0–4 | 2 | /// | 3 |
| 5–9 | 7 | #### // | 7 |
| 10–14 | 12 | //// | 4 |
| 15–19 | 17 | // | 2 |

**4  Find the Mean of Grouped Data.**

When data are grouped, it may be desirable to find the mean of the grouped data. To do this we extend our frequency distribution.

**To find the mean of grouped data:**

1. Make a frequency distribution.
2. Find the product ($xf$) of the midpoint ($x$) of the interval and the frequency ($f$) for the interval.
3. Find the sum of the frequencies ($\Sigma f$).
4. Find the sum of the products ($\Sigma xf$).
5. Divide the sum of the products by the sum of the frequencies.

Symbolically, mean$_{\text{grouped}} = \dfrac{\Sigma xf}{\Sigma f}$.

**EXAMPLE** Find the mean to the nearest whole number of the grouped data in the frequency distribution in Table 15–4.

**Table 15–4   Frequency Distribution of Credit-Hour Loads**

| Class Interval | Midpoint $x$ | Frequency $f$ | Product $xf$ |
|:---:|:---:|:---:|:---:|
| 0–4 | 2 | 3 | 6 |
| 5–9 | 7 | 7 | 49 |
| 10–14 | 12 | 4 | 48 |
| 15–19 | 17 | 2 | 34 |
| Total | | 16 | 137 |

$$\Sigma f = 16 \qquad \text{Add the frequencies.}$$

$$\Sigma xf = 137 \qquad \text{Add the products.}$$

$$\text{mean}_{\text{grouped}} = \frac{\Sigma xf}{\Sigma f} \qquad \text{Substitute.}$$

$$\text{mean}_{\text{grouped}} = \frac{137}{16} \qquad \text{Divide.}$$

$$\text{mean}_{\text{grouped}} = 8.5625$$

$$\text{mean}_{\text{grouped}} = 8.6 \qquad \text{Round.}$$

**TIP!**

## Is the Mean of Grouped Data Exact?

No. The mean of grouped data is based on the assumption that all the data in an interval have a mean that is exactly equal to the midpoint of the interval. Because this is usually not the case, the mean of grouped data is a reasonable approximation.

SECTION 15–2 SELF-STUDY EXERCISES

**1** Find the mean of the given values. Round to hundredths if necessary.

**1.** 12, 14, 16, 18, 20
**2.** 13, 15, 17, 19, 21
**3.** 68, 54, 73, 69
**4.** 85, 68, 77, 65
**5.** 37.6, 29.8
**6.** 65.3, 67.9
**7.** 32°F, 41°F, 54°F
**8.** 10°C, 13°C, 15°C
**9.** $27, $32, $65, $29, $21
**10.** $32, $43, $22, $63, $36
**11.** 11 in., 17 in., 16 in.
**12.** 9 in., 7 in., 8 in.
**13.** Respiration rates: 16, 24, 20
**14.** Pulse rates: 68, 84, 76

Solve the problems.

**15.** A baseball player batted 276 home runs over a 16-year period. What was the average number of home runs per year to the nearest tenth?

**16.** Noel Womack scored 87, 96, 86, 92, and 93 in English 101. What must he score on the last test to earn an average score of 92?

17. **AUTO** An automobile used 32 gal of regular gasoline on a 786-mi trip. What was the average miles per gallon to the nearest tenth?

18. **AUTO** A pickup truck used 25 gal of regular gasoline on a 256-mi trip. What was the average miles per gallon to the nearest tenth?

**2** Find the median for each data set.

19. 32, 56, 21, 44, 87

20. 78, 23, 56, 43, 38

21. 12, 21, 14, 18, 15, 16

22. 21, 33, 18, 32, 19, 44

23. $22, $35, $45, $30, $29

24. $66, $54, $76, $55, $69

25. **BUS** The following hourly pay rates are used at fast-food restaurants: cooks, $8.15; servers, $8.25; bussers, $8.15; dishwashers, $8.25; managers, $11.25. Find the median pay rate.

26. **BUS** The following hourly pay rates are used at a locally owned store: clerks, $8.45; bookkeepers, $9.25; operators, $8.15; assistant managers, $10.95. Find the median pay rate.

Find the mode for each data set.

27. 2, 4, 6, 2, 8, 2

28. 5, 12, 5, 5, 20

29. 21, 32, 67, 34, 23, 22

30. 32, 45, 41, 23, 56, 77

31. $56, $67, $32, $78, $67, $20, $67, $56

32. $32, $87, $67, $32, $32, $87, $77, $22

33. **BUS** These weekend work shifts are in effect at a mall clothing store: 4 hours in A.M., 6 hours in P.M., 4 hours in P.M. Find the mode for the number of hours.

34. **BUS** These special prices are in effect at a fast-food restaurant: $1.75, hamburgers; $1.97, hot ham sandwiches; $2.38, chicken fillet sandwiches; $1.97, roast beef sandwiches. Find the mode.

**3** Use Table 15–5 to answer these questions. The frequency distribution shows the ages of 25 college students in a landscaping class.

35. How many students are 22 or younger?

36. How many students are older than 34?

37. What is the ratio of the number of students 38–40 to the number of students 17–19?

38. What is the ratio of the smallest class frequency to the largest class frequency?

39. What percent of the total class are students age 17–19?

40. What percent of the total class are students age 20–22?

**Table 15–5 Frequency Distribution of 25 Ages**

| Class Interval | Midpoint | Tally | Class Frequency |
|---|---|---|---|
| 38–40 | 39 | / | 1 |
| 35–37 | 36 | / | 1 |
| 32–34 | 33 | // | 2 |
| 29–31 | 30 | /// | 3 |
| 26–28 | 27 | // | 2 |
| 23–25 | 24 | ⵝ / | 6 |
| 20–22 | 21 | ⵝ // | 7 |
| 17–19 | 18 | /// | 3 |

41. What two age groups make up the smallest number of students in the class?

42. What two age groups make up the largest number of students in the class?

43. How many students are over age 28?

44. How many students are under age 26?

Use the given hourly pay rates (rounded to the nearest whole dollar) for 33 support employees in a private college to complete a frequency distribution using the format shown in Table 15–6.

**Table 15–6 Pay Rates of 33 Support Employees**

| | Class Interval | Midpoint | Tally | Class Frequency |
|---|---|---|---|---|
| 45. | $14–16 | _____ | ____ | _____ |
| 46. | $11–13 | _____ | ____ | _____ |
| 47. | $8–10 | _____ | ____ | _____ |
| 48. | $5–7 | _____ | ____ | _____ |

| | | | | | | |
|---|---|---|---|---|---|---|
| $6 | $6 | $10 | $7 | $6 | $6 | $6 |
| $6 | $7 | $7 | $8 | $8 | $6 | $6 |
| $11 | $10 | $7 | $11 | $8 | $16 | $6 |
| $6 | $9 | $6 | $7 | $9 | $6 | $6 |
| $12 | $13 | $7 | $15 | $5 | | |

Use the given 40 test scores of two physics classes to complete a frequency distribution using the format in Table 15–7.

Table 15–7   Test Scores of 40 Physics Students

| | Class Interval | Midpoint | Tally | Class Frequency |
|---|---|---|---|---|
| 49. | 91–95 | _____ | ____ | _____ |
| 50. | 86–90 | _____ | ____ | _____ |
| 51. | 81–85 | _____ | ____ | _____ |
| 52. | 76–80 | _____ | ____ | _____ |
| 53. | 71–75 | _____ | ____ | _____ |
| 54. | 66–70 | _____ | ____ | _____ |
| 55. | 61–65 | _____ | ____ | _____ |
| 56. | 56–60 | _____ | ____ | _____ |

| 57 | 91 | 76 | 89 | 82 | 59 | 72 | 88 |
| 76 | 84 | 67 | 59 | 77 | 66 | 56 | 76 |
| 77 | 84 | 85 | 79 | 69 | 88 | 75 | 58 |
| 85 | 65 | 67 | 66 | 93 | 83 | 69 | 81 |
| 80 | 64 | 78 | 76 | 72 | 90 | 79 | 90 |

57. Students recorded the number of hours they studied each week as: 3, 15, 18, 0, 2, 9, 12, 16, 7, 8, 5, 10, 14, 9, 7, 3, 4, 14, 17, 16, 6, 4, 8, 11, 14, 13, 10, 15, 16, 5.

Create a grouped frequency distribution with four classes beginning with the class 0–4. Show the class interval, midpoint, tally, and class frequency for each class interval.

58. The number of animals at the city animal shelter varies daily. Use the data to make a frequency distribution with five classes beginning with the class 11–20. Show the class interval, midpoint, tally, and class frequency for each class interval: 22, 31, 32, 27, 29, 16, 12, 18, 21, 30, 46, 52, 43, 51, 42, 26, 42, 17, 19, 25.

**4**  Round grouped mean to tenths.

59. Find the grouped mean for the data in Table 15–5.
60. Find the grouped mean for the data in Table 15–6.
61. Find the grouped mean for the data in Table 15–7.
62. Find the grouped mean for the data in Exercise 57.

# 15-3 | Range and Standard Deviation

*Learning Outcomes*   **1** Find the range.
**2** Find the standard deviation.

**1** Find the Range.

The mean, the median, and the mode are *measures of central tendency*. Other statistical measures are **measures of variation** or **dispersion**. The variation or dispersion of a set of data may also be referred to as the **spread**. One of these measures of dispersion is the **range**. The range is the difference between the highest value and the lowest value in a set of data.

**To find the range:**

1. Find the highest and lowest values.
2. Find the difference between the highest and lowest values.

EXAMPLE   Find the range for the data described in the example for fast-food restaurant hourly pay rates
BUS   on page 619.

The high value is $10.50. The low value is $7.15.

$$\text{range} = \$10.50 - \$7.15 = \mathbf{\$3.35}$$

**Use More Than One Statistical Measure**

A common mistake when making conclusions or inferences from statistical measures is to examine only one statistic, such as the range. To obtain a complete picture of the data requires looking at more than one statistic.

**2** **Find the Standard Deviation.**

Although the range gives us some information about dispersion, it does not tell us whether the highest or lowest values are typical values or extreme **outliers.** We can get a clearer picture of the data set by examining how much each data point *differs* or *deviates* from the mean.

The **deviation from the mean** of a data value is the difference between the value and the mean.

**To find the deviations from the mean:**

Data set: 38, 43, 45, 44.

1. Find the mean of a set of data.

$$\bar{x} = \frac{\text{sum of data values}}{\text{number of values}} = \frac{\Sigma x_i}{n}$$

$$\frac{38 + 43 + 45 + 44}{4} = \frac{170}{4} = 42.5$$

2. Find the amount that each data value deviates or is different from the mean.

deviation from the mean =
data value − mean = $x_i - \bar{x}$

$38 - 42.5 = -4.5$ (below the mean)
$43 - 42.5 = 0.5$ (above the mean)
$45 - 42.5 = 2.5$ (above the mean)
$44 - 42.5 = 1.5$ (above the mean)

When the value is smaller than the mean, the difference is represented by a *negative* number indicating the value is *below* or less than the mean. When the value is larger than the mean, the difference is represented by a positive number indicating the value is *above* or greater than the mean. *The absolute value of the sum of the deviations below the mean should equal the sum of the deviations above the mean.* In the example in the box, only one value is below the mean, and its deviation is −4.5. Three values are above the mean, and the sum of these deviations is $0.5 + 2.5 + 1.5 = 4.5$. We say that *the sum of all deviations from the mean is zero.* This is true for all sets of data.

We have not gained any statistical insight or new information by analyzing the sum of the deviations from the mean or even by analyzing the average of the deviations.

$$\text{average deviation} = \frac{\text{sum of deviations}}{\text{number of values}} = \frac{0}{n} = 0$$

**EXAMPLE** Find the deviations from the mean for the set of data 45, 63, 87, and 91.

$$\bar{x} = \frac{\Sigma x_i}{n} = \frac{45 + 63 + 87 + 91}{4} = \frac{286}{4} = 71.5 \quad \text{Mean.}$$

To find the deviation from the mean, subtract the mean, $\bar{x}$, from each value of $x$. We arrange these values in a table.

| Values $x_i$ | Deviations $x_i - \bar{x}$ |
|---|---|
| 45 | $45 - 71.5 = -26.5$ |
| 63 | $63 - 71.5 = -8.5$ |
| 87 | $87 - 71.5 = 15.5$ |
| 91 | $91 - 71.5 = 19.5$ |
| Sum        286 | 0 |

As we might expect, the sum of the deviations in the example equals zero because the sum of the negative deviations $(-26.5 + -8.5 = -35)$ equals the sum of the positive deviations $(15.5 + 19.5 = 35)$. To compensate for this situation, mathematicians employ a statistical measure called the **standard deviation,** which uses the square of each deviation from the mean. The square of a negative value is always positive. The squared deviations are averaged (mean), and the result is called the **variance:**

$$\text{variance} = v = \frac{\Sigma(x_i - \bar{x})^2}{n - 1}$$

The square root is taken of the variance so that the result can be interpreted within the context of the problem. Various formulas exist for finding the standard deviation of a set of values, but we examine only one formula. This formula averages the values by dividing by one less than the number of values $(n - 1)$. Several calculations are necessary and are best organized in a table.

$$\text{standard deviation} = s = \sqrt{\frac{\Sigma(x_i - \bar{x})^2}{n - 1}}$$

**To find the standard deviation of a set of data:**

1. Find the mean, $\bar{x}$.
2. Find the deviation of each value from the mean: $(x_i - \bar{x})$
3. Square each deviation: $(x_i - \bar{x})^2$
4. Find the sum of the squared deviations: $\Sigma(x_i - \bar{x})^2$
5. Divide the sum of the squared deviations by *one less than* the number of values in the data set. This amount is called the *variance:* $v = \dfrac{\Sigma(x_i - \bar{x})^2}{n - 1}$.
6. Find the standard deviation by taking the square root of the variance: $s = \sqrt{\dfrac{\Sigma(x_i - \bar{x})^2}{n - 1}}$.

The formula we are using deals with a sample of an entire population or a small set of data. In this formula we divide by $n - 1$ instead of $n$. In very large sets of data $n$ is used. There is very little difference in the calculations using $n$ versus $n - 1$. The logical basis for using $n - 1$ assumes that one data value is *exactly* the mean (which may or may not be true) and there are $n - 1$ data values that deviate from the mean.

**EXAMPLE**  Find the standard deviation for the values 45, 63, 87, and 91.

From the previous example the mean is 71.5 and the number of values is 4.

$$\bar{x} = 71.5, n = 4$$

Chapter 15 / Statistics

| Values $x_i$ | Deviations from the mean $x_i - \bar{x}$ | Squares of the deviations from the mean $(x_i - \bar{x})^2$ |
|---|---|---|
| 45 | $45 - 71.5 = -26.5$ | $(-26.5)^2 = 702.25$ |
| 63 | $63 - 71.5 = -8.5$ | $(-8.5)^2 = 72.25$ |
| 87 | $87 - 71.5 = 15.5$ | $(15.5)^2 = 240.25$ |
| 91 | $91 - 71.5 = 19.5$ | $(19.5)^2 = 380.25$ |
| Sum of values 286 | Sum of deviations 0 | Sum of squared deviations 1,395 |

$$v = \frac{\Sigma(x_i - \bar{x})^2}{n-1} \qquad \frac{\text{Sum of squared deviations}}{n-1}.$$

$$v = \frac{1,395}{3} = 465$$

$$s = \sqrt{v} \qquad \text{Square root of variance.}$$

$$s = \sqrt{\frac{1,395}{3}} = \sqrt{465} = 21.56385865 \ or \ \mathbf{21.56 \ standard \ deviation}$$

A small standard deviation indicates that the mean is a typical value in the data set. A large standard deviation indicates that the mean is not typical, and other statistical measures should be examined to better understand the characteristics of the data set.

Let's examine the various statistics for the data set on a number line (Figure 15–10).

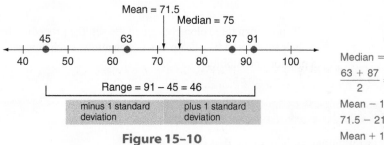

**Figure 15–10**

Median $=$
$$\frac{63 + 87}{2} = \frac{150}{2} = 75.$$
Mean $-$ 1 standard deviation $=$
$71.5 - 21.6 = 49.9.$
Mean $+$ 1 standard deviation $=$
$71.5 + 21.6 = 93.1.$

We can confirm visually that the dispersion of the data is broad and the mean is not a typical value in the data set.

SECTION 15–3 SELF-STUDY EXERCISES

**1** Find the range for each data set.

**1.** 22, 36, 41, 41, 17
**2.** 28, 33, 36, 13, 28
**3.** 10, 23, 12, 17, 13, 16
**4.** 23, 23, 18, 32, 29, 14
**5.** $25, $15, $25, $40, $19
**6.** $36, $44, $26, $52, $19
**7.** 23°F, 37°F, 29°F, 54°F, 46°F, 71°F, 67°F

**2** Find the standard deviation for each data set. Round to the nearest hundredth.

**8.** 12, 14, 16, 18, 20
**9.** 68, 54, 73, 69
**10.** 32°F, 41°F, 54°F
**11.** $27, $32, $65, $29, $21
**12.** Respiration rates: 16, 24, 20
**13.** Pulse rates: 68, 84, 76

---

*Learning Outcomes*

**1** Count the number of ways objects in a set can be arranged.

**2** Determine the probability of an event occurring if an activity is repeated over and over.

**1** **Count the Number of Ways Objects in a Set Can Be Arranged.**

A **set** is a well-defined group of objects or **elements.** The numbers 2, 4, 6, 8, and 10 can be a set of even numbers between 1 and 12. Women, men, and children can be a set of people. A, B, and C can be a set of the first three capital letters in the alphabet.

**Counting,** in this section, means determining all the possible ways the elements in a set can be arranged. One way to count is to *list* all possible arrangements and then count the number of arrangements.

**EXAMPLE** List and count the ways the elements in the set *A*, *B*, and *C* can be arranged.

| *A* is first, 2 choices | *B* is first, 2 choices | *C* is first, 2 choices |
|---|---|---|
| *A BC* | *B AC* | *C AB* |
| *A CB* | *B CA* | *C BA* |

**Therefore, *A*, *B*, and *C* can be arranged in six ways.** Each of these arrangements can also be called a set.

If more than three elements are in the set, the procedure becomes more challenging. It may be helpful to use a **tree diagram,** which allows each new set of possibilities to branch out from a previous possibility.

**To make a tree diagram of possible arrangements of items in a set:**

1. List the choices for putting an item in the first slot.
2. From each choice, list the remaining choices for the second slot (first branch).
3. From the end of each branch, make a branch to the remaining choices for the third slot.
4. Continue the process until there are no remaining choices.
5. To list an arrangement, start with the first slot and follow a branch to its end and record the choices.
6. To count the total number of possible arrangements, count the number of branch ends in the last slot.

**EXAMPLE**   Make a tree diagram and count the number of ways the elements in the set containing letters *W, X, Y,* and *Z* can be arranged.

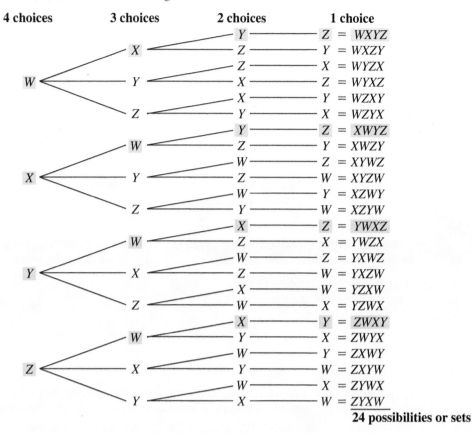

As evident in the previous example, the greater the number of elements in a set, the greater the complexity and time required to list all possible arrangements. We can use logic and common sense to obtain a count of the possible arrangements of a set of elements.

There are four possibilities for the first letter, *W, X, Y,* or *Z.* For each of the four possible first letters, we have three choices for second letters. Now, we have $4 \times 3$ or 12 possibilities. For each of the 12 possibilities two choices remain for the third letter: $12 \times 2 = 24$. Then, for each of the 24 three-letter combinations, only one choice is left: $24 \times 1 = 24$. So, there are a total of 24 possibilities.

Another way to visualize this is to think of drawing letters from a bag or bowl (Figure 15–11).

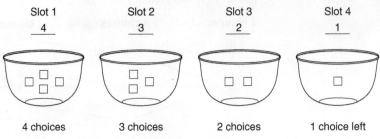

Figure 15–11

By multiplying the number of possible choices for each position, we can determine the total number of possibilities without listing them: $4 \cdot 3 \cdot 2 \cdot 1 = 24$.

## To determine the number of choices for arranging a specified number of items:

1. Determine the number of slots to be filled.
2. Determine the number of choices for each slot.
3. Multiply the numbers from Step 2.

**EXAMPLE** A coin is tossed three times. With each toss, the coin falls heads up or tails up. How many possible outcomes of heads and tails are there with three tosses of the coin?

There are three tosses. Each toss has only two possibilities, heads or tails; that is,

| 1st toss | 2nd toss | 3rd toss |
|---|---|---|
| 2 possibilities | 2 possibilities | 2 possibilities |

By multiplying the number of possible outcomes for each toss, we get $2 \cdot 2 \cdot 2 = 8$.

**So there are 8 possible outcomes.**

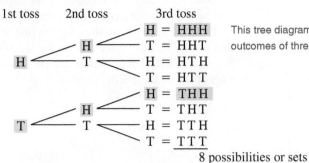

This tree diagram lists the possible outcomes of three tosses.

8 possibilities or sets

TIP!

### To Repeat or Not Repeat

In some situations, as in arranging W, X, Y, and Z, once a choice or selection is made, that choice cannot be repeated. In these situations the number of choices decreases with each selection.

Number of possible arrangements of W, X, Y, Z = $4 \cdot 3 \cdot 2 \cdot 1 = 24$

Chapter 15 / Statistics

In other situations, as in tossing a coin, every coin toss can result in a head or a tail. The result of any coin toss can repeat the result of a previous toss.

$$\text{number of possible results of three coin tosses} = 2 \cdot 2 \cdot 2 = 2^3 = 8$$

When counting the number of possible outcomes, *first determine if repeats are allowed or not.*

**EXAMPLE**  Henry has three ties: a red tie, a blue tie, and a green tie. He also has three shirts: a white shirt, a pink shirt, and a yellow shirt. How many sets of shirts and ties are possible?

If we start with the shirts, there are three possibilities (white, pink, and yellow). For each shirt, there are three possible ties (red, blue, and green). **So we have 3 · 3 = 9 possible outcomes.**

**EXAMPLE**  Given the digits 1, 2, 3, 4, 5, and 6, how many three-digit numbers can be made without repeating a digit?

The numbers are to contain three digits, so there are three positions to fill. We have six digits to work with. The first digit can be one of six. For each of these six digits, there are five possible second digits. For each of the five second digits, there are four possible third digits.

| Positions: | 1st | 2nd | 3rd |
|------------|-----|-----|-----|
| Possibilities: | 6 | 5 | 4 |

By multiplying the possibilities for each position, we get 6 · 5 · 4 = **120 possible outcomes or ways to make a three-digit number.**

## 2  Determine the Probability of an Event Occurring If an Activity Is Repeated Over and Over.

**Probability** means the chance of an event occurring if an activity is repeated over and over. The probability of an event occurring is expressed as a ratio or a percent.

Weather forecasters use percents when they forecast a 60% chance of rain or a 20% chance of snow. This text will use ratios like $\frac{3}{5}$ for 3 chances out of 5 or $\frac{2}{3}$ for 2 chances out of 3. The decimal equivalent of a ratio can also be used to express a probability: $\frac{3}{5} = 0.6$.

When a coin is tossed, two outcomes are possible, heads or tails. But only one side will be up. The probability of tossing heads is 1 out of 2, $\frac{1}{2}$, 0.5, or 50%.

### To express the probability of an event occurring successfully:

1. Determine the total number of elements in the set (total possible outcomes).
2. Determine the number of elements that are defined as successful.
3. Make a ratio with the number of choices that are defined as successful divided by the total number of elements in the set.

$$\frac{\text{number of successful possibilities}}{\text{number of total possibilities}}$$

4. Express the ratio in lowest terms. The ratio can be expressed as a decimal or percent if desired.

The probability of an event occuring ranges from 0 for an impossible event to 1 for a certain event.

**EXAMPLE**

When a die is rolled, what is the probability that a 3 will appear on the top face (Figure 15–12)?

**Figure 15–12**

A die has six sides where dots represent 1–6. Each side is an element in the set of six sides, so there is a total of six elements in the set.
Only one element is a 3.

**The probability of rolling a 3 is $\frac{1}{6}$.**

---

**EXAMPLE**

**BUS**

A holiday gift shopper wrapped eight men's ties in separate boxes. There were two solid-color ties and six striped ties. If the gift boxes were given at random to eight men, what is the probability of a man receiving a solid-color tie?

eight ties         Total possibilities
two solid-color ties      Successful possibilities

$$\text{probability} = \frac{\text{successful possibilities}}{\text{total possibilities}} = \frac{2}{8} = \frac{1}{4}$$

**The probability of receiving a solid-color tie is $\frac{1}{4}$ or 0.25 or 25%.**

---

**EXAMPLE**

**HLTH/N**

A practical nurse has a box of 144 syringes in individual sterile packets. Three of the syringes have torn packets. What is the probability that the first syringe picked will have a torn packet? If the packet for the first syringe selected is torn, what is the probability of picking a second syringe with a torn packet?

**On the first pick, the probability of picking one syringe with a torn packet is $\frac{3}{144}$, which reduces to $\frac{1}{48}$.** We assume that the first pick was a syringe with a torn packet, so this leaves a total of 143 syringes and now only 2 have torn packets. **On the second pick, the probability of picking a syringe with a torn packet is $\frac{2}{143}$.**

---

**SECTION 15–4 SELF-STUDY EXERCISES**

**1** Complete the exercises using counting techniques.

**1.** List and count all the possible arrangements for Keaton, Brienne, and Renee to be seated in three adjacent seats at a basketball game.

**2.** List and count all the possible outcomes for arranging books A, B, C, and D on a shelf.

**3.** Count all the possible outcomes for Jim Riddle to arrange a T-shirt, a sport shirt, a dress shirt, and a sweater on a shelf for display.

**4.** **HLTH/N** How many outcomes are possible for arranging containers of cotton balls, gauze pads, swabs, tongue depressors, and adhesive tape in a row on a shelf in a doctor's examining room?

**5.** **AG/H** A landscaping process involves five steps. The steps can be arranged in any order. The landscaping company efficiency officer wants to determine the most efficient order for the five steps. How many arrangements of steps are possible?

**2** Use simple probability to complete the exercises.

6. A drawing will be held to award door prizes. If the names of 24 people (no repeated names) are in the pool for the drawing, what is the probability that David's name will be pulled at random for the first prize? If his name is pulled and not replaced, what is the probability that Gaynell's name will be pulled next?

7. A box of greeting cards contains 20 friendship cards, 10 get-well cards, and 10 congratulations cards. What is the probability of picking a get-well card at random?

8. Mimi tosses 21 pennies, 16 nickels, and 11 dimes into a container. What is the probability of reaching in and picking a dime?

9. A jar holds 10 lock washers and 15 flat washers. What is the probability of drawing a lock washer at random?

10. A TV quiz program puts all questions in a box. If the box contains five hard questions, five average questions, and five easy questions, what is the probability of being asked an easy question?

11. When a single die is rolled, what is the probability that a 1 will appear?

---

## CHAPTER REVIEW OF KEY CONCEPTS

**Learning Outcomes**
**Section 15–1**

**1** Read circle graphs (pp. 612–613).

**What to Remember with Examples**

A circle graph compares parts to a whole.

If the total monthly revenue at a used car dealership is $75,000, what is the revenue from trucks? (See Figure 15–13.)

35% of 75,000 =
  0.35 × 75,000 = $26,250

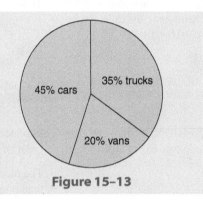

**Figure 15–13**

**2** Read bar graphs (p. 613).

Bar graphs are used to compare values to each other.

What is the ratio of men's salaries to women's salaries in the pants department? (See Figure 15–14.)

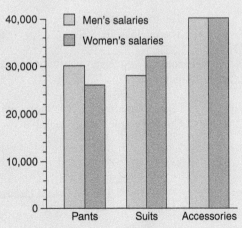

**Figure 15–14**

Men's salary in the pants department = $30,000
Women's salary in the pants department = $26,000
Ratio of men's salaries to women's salaries = $\frac{30}{26} = \frac{15}{13}$

**3** Read line graphs (p. 614).

A line graph shows how an item changes with time.

Use the line graph of the average prices for textbooks to find the year when textbooks averaged $67 per book. (See Figure 15–15.)

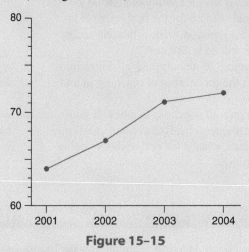

**Figure 15–15**

In 2002, the average price of textbooks was $67.

## Section 15–2

**1** Find the arithmetic mean or arithmetic average (pp. 616–618).

To find the mean, add the values; divide by the number of values.

Find the arithmetic mean of the test scores: 76, 86, 93, 87, 68, 76, 88

$$76 + 86 + 93 + 87 + 68 + 76 + 88 = 574$$

$$574 \div 7 = 82$$

The mean is 82.

**2** Find the median and the mode (pp. 618–620).

The median of an odd number of values is the middle value when the values are arranged in order of size. For an even number of values, average the two middle values.

The mode is the value that occurs most frequently. A set of values may have no mode or more than one mode.

Find the median and mode of the set of test scores:

68, 76, 76, 86 , 87, 88, 93          Already arranged in order.
      ↑   ↑
The median is 86.          Middle value.
The mode is 76.          Most frequent value.

**3** Make and interpret a frequency distribution (pp. 620–621).

To make a frequency distribution, determine the appropriate interval for classifying the data. Tally the data.

Make a frequency distribution with the following data, indicating leave days for State College employees. (See Table 15–8.)

| 2 | 2 | 4 | 4 | 4 | 5 | 5 | 6 | 6 | 8 |
|---|---|---|---|---|---|---|---|---|---|
| 8 | 8 | 9 | 12 | 12 | 12 | 14 | 15 | 20 | 20 |

**Table 15–8    Annual Leave Days of 20 State College Employees**

| Class Interval | Midpoint | Tally | Class Frequency |
|---|---|---|---|
| 16–20 | 18 | // | 2 |
| 11–15 | 13 | ⟶HH | 5 |
| 6–10 | 8 | ⟶HH / | 6 |
| 1–5 | 3 | ⟶HH // | 7 |

**4** Find the mean of grouped data (pp. 621–622).

To find the mean of grouped data: **1.** Make a frequency distribution. **2.** Find the product $(xf)$ of the midpoint $(x)$ of the interval and the frequency $(f)$ for the interval. **3.** Find the sum of the frequencies $(\Sigma f)$. **4.** Find the sum of the products $(\Sigma xf)$. **5.** Divide the sum of the products by the sum of the frequencies.

Symbolically, $\text{mean}_{\text{grouped}} = \dfrac{\Sigma xf}{\Sigma f}$.

Find the mean of the grouped data in the frequency distribution in Table 15–9.

**Table 15–9  Annual Leave Days of 20 State College Employees**

| Class Interval | Midpoint $x$ | Frequency $f$ | Product $xf$ |
|---|---|---|---|
| 16–20 | 18 | 2 | 36 |
| 11–15 | 13 | 5 | 65 |
| 6–10 | 8 | 6 | 48 |
| 1–5 | 3 | 7 | 21 |
| Total | | 20 | 170 |

$$\Sigma f = 20 \qquad \text{Add the frequencies.}$$

$$\Sigma xf = 170 \qquad \text{Add the products.}$$

$$\text{mean}_{\text{grouped}} = \frac{\Sigma xf}{\Sigma f} \qquad \text{Substitute.}$$

$$\text{mean}_{\text{grouped}} = \frac{170}{20} \qquad \text{Divide.}$$

$$\text{mean}_{\text{grouped}} = 8.5$$

## Section 15–3

**1** Find the range (pp. 624–625).

The range for a set of data is the difference between the largest value and the smallest value.

Find the range of the set of test scores:

$$68, 76, 76, 86, 87, 88, 93$$

$$\text{range} = 93 - 68 = 25$$

**2** Find the standard deviation (pp. 625–627).

To find the standard deviation from the mean: **1.** Find the mean, $\bar{x}$. **2.** Find the deviation of each value from the mean: $(x_i - \bar{x})$. **3.** Square each deviation: $(x_i - \bar{x})^2$. **4.** Find the sum of the squared deviations: $\Sigma(x_i - \bar{x})^2$. **5.** Divide the sum of the squared deviations by one less than the number of values in the data set. This is called the variance, $v = \dfrac{\Sigma(x_i - \bar{x})^2}{n - 1}$. **6.** Find the standard deviation by taking the square root of the variance.

$$s = \sqrt{\frac{\Sigma(x_i - \bar{x})^2}{n - 1}}$$

Find the standard deviation of the test scores: 68, 76, 76, 86, 87, 88, 93

$$\bar{x} = \frac{68 + 76 + 76 + 86 + 87 + 88 + 93}{7} = \frac{574}{7} = 82$$

| $x_i - \bar{x}$ | $(x_i - \bar{x})^2$ |
|---|---|
| $68 - 82 = -14$ | 196 |
| $76 - 82 = -6$ | 36 |
| $76 - 82 = -6$ | 36 |
| $86 - 82 = 4$ | 16 |
| $87 - 82 = 5$ | 25 |
| $88 - 82 = 6$ | 36 |
| $93 - 82 = 11$ | 121 |

$$\Sigma(x_i - \bar{x})^2 = 466$$

$$v = \frac{\Sigma(x_i - \bar{x})^2}{n-1} = \frac{466}{6} = 77.66666667$$

$$s = \sqrt{\Sigma\frac{(x_i - \bar{x})^2}{n-1}} = \sqrt{77.66666667} = 8.812869378 = 8.8 \quad \text{Round.}$$

### Section 15–4

**1** Count the number of ways objects in a set can be arranged (pp. 628–631).

Multiply the number of choices for each position in the set.

Renee Smith's closet has two new blazers (navy and red) and four new skirts (gray, black, tan, and brown). How many outfits can she make from the new clothes?

$$1 \text{ blazer } + 1 \text{ skirt } = 1 \text{ outfit}$$

$$\begin{pmatrix} \text{blazer} \\ \text{choices} \end{pmatrix} \cdot \begin{pmatrix} \text{skirt} \\ \text{choices} \end{pmatrix} = \begin{pmatrix} \text{possible} \\ \text{outfits} \end{pmatrix}$$

$$2 \qquad \cdot \qquad 4 \qquad = \qquad 8$$

**2** Determine the probability of an event occurring if an activity is repeated over and over (pp. 631–632).

The probability of an event occurring is the ratio of the number of possible successful outcomes to the number of possible outcomes.

A box in a doctor's office contains thirty $\frac{1}{2}$-in. adhesive strips and seventy $\frac{3}{4}$-in. adhesive strips. What is the probability of picking a $\frac{3}{4}$-in. adhesive strip at random?

$$\frac{70}{100} = \frac{7}{10} \qquad \text{Total items from which to select} = 30 + 70 = 100$$

The probability of picking a $\frac{3}{4}$-in. adhesive strip is $\frac{7}{10}$ or 0.7 or 70%.

---

## CHAPTER REVIEW EXERCISES

### Section 15–1

Use Figure 15–16 to answer Exercises 1–4.

1. In what year(s) did women use more sick days than men?

2. In what year(s) did men use about five sick days?

3. In what year(s) did men use more sick days than women?

4. What was the greatest number of sick days for men?

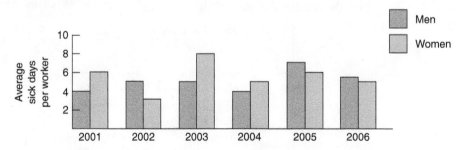

**Figure 15–16**  Comparison of sick days for men and women.

Use Figure 15–17 to answer Exercises 5–7.

5. What percent of the total cost is the cost of the lot? Round to the nearest tenth.

6. What percent of the total cost is the cost of the house? Round to the nearest tenth.

7. The cost of the lot and landscaping represents what percent of the total cost? Round to the nearest tenth.

Use Figure 15–18 to answer Exercises 8–9.

8. **HLTH/N** What was the patient's highest pulse rate? The highest respiration rate?

9. **HLTH/N** On what date and at what time were the patient's pulse rate and respiration rate at their highest level?

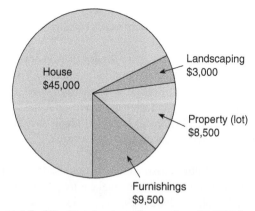

**Figure 15–17** Distribution of costs for a $66,000 home.

### Section 15-2

Solve the problems.

10. Jim Smith made 176 baskets over a 16-game period. What was the average number of baskets per game to the nearest tenth?

11. Lee Vance sold 63 new cars over a 5-month period. What was the average number of cars sold per month to the nearest tenth?

12. An automobile used 22 gal of regular gasoline on a trip of 358 mi. What was the average miles per gallon to the nearest tenth?

13. A delivery truck used 21 gal of regular gasoline on a trip of 289 mi. What was the average miles per gallon to the nearest tenth?

14. These hourly pay rates are used at fast-food restaurants: cooks, $8.25; servers, $8.95; bussers, $7.20; dishwashers, $7.20; managers, $10.25. Find the median pay rate.

15. These hourly pay rates are used at a locally owned store: office assistants, $9.85; bookkeepers, $10.20; cashiers, $9.45; assistant managers, $11.90. Find the median pay rate.

16. These weekend work shifts are in effect in a mall clothing store: 3 hours in A.M., 6 hours in P.M., 3 hours in P.M. Find the mode.

17. These special prices are in effect at a fast-food restaurant: $1.85, hamburgers; $1.98, hot ham sandwiches; $2.28, chicken sandwiches; $1.85, roast beef sandwiches. Find the mode.

18. Cody Collier, a student at a technical college, earned the final grades listed in Table 15–11 for the past term. Find Cody's QPA (quality point average) to the nearest hundredth.

**Table 15–11    Grade Distribution**

| Subject | Grade | Hours | Quality Points per Hour |
|---------|-------|-------|-------------------------|
| Electronics 101 | A | 4 | 4 |
| Circuits 201 | A | 4 | 4 |
| Algebra 101 | B | 4 | 3 |

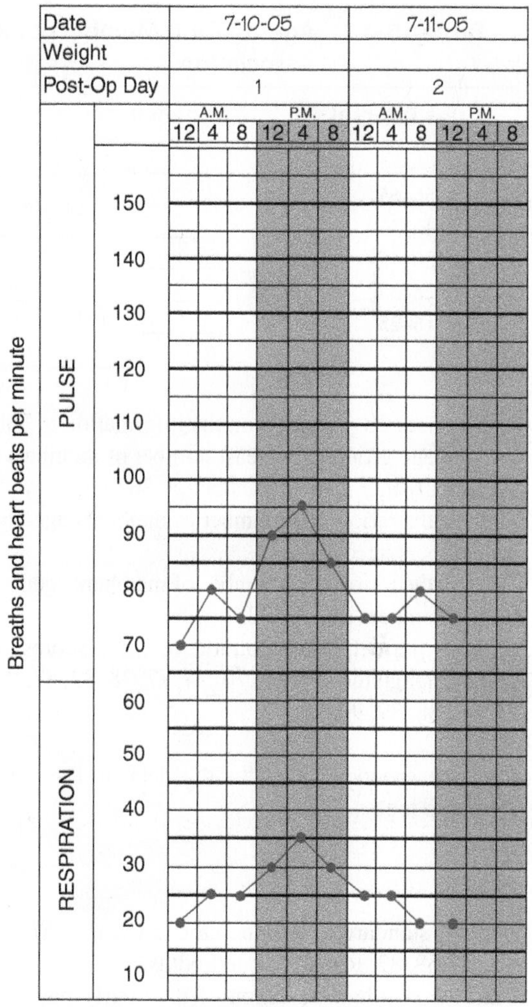

**Figure 15–18**   Graphic respiration/pulse chart.

19. Tami Murphy earned the final grades listed in Table 15–12 for the past term. Find Tami's QPA (quality point average) to the nearest hundredth.
20. Janice Van Dyke has these grades in Algebra 102: 98, 82, 87, 72, and 82. What minimum grade does she need on the last test to have a B or 85 average?
21. Sarah Smith has the following grades in American History: 99, 93, 91, 88, and 86. What grade does she need on the last test to get an A or 90 average?

**Table 15–12   Grade Distribution**

| Subject | Grade | Hours | Quality Points per Hour |
|---------|-------|-------|-------------------------|
| English 201 | A | 3 | 4 |
| History 202 | B | 3 | 3 |
| Philosophy 101 | A | 3 | 4 |

A regional horticultural association is composed of 54 members of several clubs. Use the format shown in Table 15–13 and make a frequency distribution of the members' ages: 17, 18, 20, 21, 21, 24, 24, 29, 29, 29, 31, 31, 33, 33, 34, 35, 35, 38, 38, 38, 39, 41, 42, 43, 43, 43, 43, 45, 45, 47, 47, 48, 48, 48, 49, 50, 51, 51, 52, 56, 56, 58, 58, 60, 60, 62, 64, 64, 65, 66, 68, 70, 71, 71.

**Table 15–13   Ages of Club Members of Regional Horticultural Association**

|  | Class Interval | Midpoint | Tally | Class Frequency |
|------|----------------|----------|-------|-----------------|
| 22. | 66–75 | _____ | ____ | _____ |
| 23. | 56–65 | _____ | ____ | _____ |
| 24. | 46–55 | _____ | ____ | _____ |
| 25. | 36–45 | _____ | ____ | _____ |
| 26. | 26–35 | _____ | ____ | _____ |
| 27. | 16–25 | _____ | ____ | _____ |

Answer Exercises 28–35 based on the information in Table 15–13.
28. In what age group is the least number of members?
29. In what age group is the greatest number of members?
30. How many members are under age 36?
31. How many members are over age 55?
32. What is the ratio of the number of members aged 66–75 to members aged 16–25?
33. What is the ratio of the number of members aged 66–75 to the number of members aged 46–55?
34. What is the ratio of the number of members aged 46–55 to the members aged 16–25?
35. What percent (to the nearest tenth of a percent) of the total number of members are the members aged 46–55?
36. Make a frequency distribution of these scores on an English language test: 68, 70, 74, 77, 78, 82, 82, 84, 86, 86, 86, 88, 89, 90, 90, 93.
37. Make a frequency distribution of these miles per gallon reported in one week by customers to an automotive rental company for six-cylinder cars: 22, 22, 23, 23, 23, 24, 24, 25, 26, 26, 27, 29, 29, 30, 30, 31, 31.
38. Find the grouped mean for the data in Table 15–13. Round to tenths.
39. Find the grouped mean for the data in Exercise 37. Round to tenths.

## Section 15–3

40. Find the standard deviation of the test scores: 67, 87, 76, 89, 70, 69, 82. Round to hundredths.
41. Find the range for Marcus Johnson's test scores of 92, 83, 39, 98, 88, 90.
42. Find the standard deviation of the hourly pay rates in Exercise 14. Round to hundredths.
43. Find the standard deviation of the test scores in Exercise 41. Round to hundredths.

## Section 15–4

44. For her first professional job, Martha Deskin, a recent college graduate, purchased four pairs of shoes, three business suits, and five blouses. How many outfits are possible?
45. There are three magazines on horses and ten magazines on fashion in Dr. Nelson Company's waiting room. If Shirley Riddle sends her toddler to get two magazines, what is the probability that he will randomly pick a fashion magazine on the first draw? If he is successful, what is the probability of his picking a fashion magazine on the second draw?

**46.** Two coins are tossed three times. Count the number of possible outcomes of heads and tails with three tosses of the two coins.

**47.** If four coins are tossed, what is the number of possible outcomes of heads and tails? *Hint:* HHHT and HTHH are different outcomes.

**48.** Cy Pipkin is puzzled over a true-false question on a test and does not know the answer. What is the probability that he will pick the right answer by chance?

## TEAM PROBLEM-SOLVING EXERCISES

**1.** Statistical reports are an important part of a statistical study. Conduct a study on a topic of interest to the team and prepare a statistical report that includes the following components:
   **(a)** Question or questions that you are attempting to answer
   **(b)** Description of the process for collecting data
   **(c)** Summary of the data with appropriate statistics
   **(d)** Conclusion

**2.** Most statistical reports include graphical representations of the data collected. Computer software such as Excel™ allows you to prepare graphs or charts from the data entered into a spreadsheet.
   **(a)** Prepare spreadsheets with appropriate formulas to make the necessary calculations for the data in Exercise 1.
   **(b)** Prepare charts to illustrate the data collected in Exercise 1.

## PRACTICE TEST

**1.** A _____ graph shows how different data values relate to each other.

**3.** A _____ graph shows how an item or items change over time.

**2.** A _____ graph shows how a whole quantity is related to its parts.

Answer Questions 4 and 5 from the line graphs in Figure 15–19.

**4.** How many degrees warmer was it indoors at midnight than outdoors?

**5.** What was the change in outdoor temperature between 11:00 A.M. and noon?

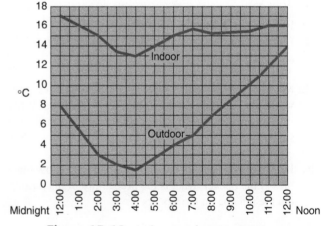

**Figure 15–19** Indoor-outdoor temperature.

Answer the questions about the manufacturing costs for the electronic game as shown in the circle graph in Figure 15–20.

**6.** What percent of the total cost is materials?

**7.** What would the profit be if there were no tariff on imported parts?

**8.** How much are overhead and materials together?

**9.** What percent of the total cost is the profit?

**10.** What is the ratio of labor to total cost?

**11.** What is the ratio of the tariff to total cost?

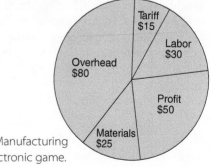

**Figure 15–20** Manufacturing costs of a $200 electronic game.

Use the bar graph in Figure 15–21 to answer the Questions 12–15 about the academic-year starting salaries of women and men college professors in various academic departments of a college.

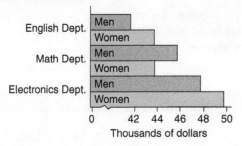

**Figure 15–21** Salaries of women and men college professors.

12. In what departments do men make more than women?
13. In what departments do women make more than men?
14. What percent of women's salaries are men's salaries in the English Department (to the nearest tenth of a percent)?
15. What percent of men's salaries are women's salaries in the Electronics Department (to the nearest tenth of a percent)?

Use the frequency distribution shown in Table 15–14 to answer Questions 19–22. The distribution shows the number of correct answers on a 30-question test in a science class.

16. How many students scored more than 25 correct?
17. What is the ratio of students scoring 6–10 correct to those scoring 21–25 correct?
18. Find the grouped mean for the data in Table 15–14.
19. Find the range, mean, median, and mode for these test scores: 81, 78, 69, 75, 81, 93, 68. Round to tenths.
20. Find the standard deviation of the data: 77, 87, 77, 89, 70, 69, 82.
21. Find the standard deviation of the data in Exercise 19. Round to hundredths.

| Table 15–14 | Frequency Distribution of Correct Answers | | |
|---|---|---|---|
| **Class Interval** | **Midpoint** | **Tally** | **Class Frequency** |
| 26–30 | 28 | /// | 3 |
| 21–25 | 23 | //// | 4 |
| 16–20 | 18 | ₩₩ // | 7 |
| 11–15 | 13 | /// | 3 |
| 6–10 | 8 | // | 2 |
| 1–5 | 3 | / | 1 |

22. List and count the ways the elements in the set *L, M, N,* and *O* can be arranged.
23. Rayford has two ties: a red tie and a blue tie. He also has three shirts: a white shirt, a green shirt, and a yellow shirt. How many combinations of shirts and ties are possible?
24. Millie has in her purse three green eye shadows, four white eye shadows, and two black eye shadows. What is the probability of her pulling out a black eye shadow?
25. An envelope contains the names of two men and three women to be interviewed for a promotion at Washington's Landscape Service. The interviewer draws names to determine the order of the interviews. What is the probability of drawing a woman's name first?
26. If a small boy has one red marble and three yellow marbles in his pocket, what is the probability of his pulling out the red one on the first try?

## Exercises

For each class, find the

1. mean score
3. median
5. standard deviation

2. range
4. mode
6. Explain why you agree or disagree with Ms. Duke's speculation.

## Answers

1. Mean of morning class = 73.9; mean of afternoon class = 73.9.
3. Median of morning class = 80; median of afternoon class = 66.5.
5. Standard deviation of morning class = 15.4; standard deviation of afternoon class = 18.3.

2. Range of morning class = 50; range of afternoon class = 50.
4. Mode for morning class = 80; mode for afternoon class = 99.
6. The high range for each class indicates that the scores have a wide dispersion. In examining the median for each set of scores, you see that the median for the morning class is significantly higher than that for the afternoon class. The mode for the afternoon class is not at all descriptive of class scores as a whole. The standard deviation for the afternoon class is larger than that of the morning class, which indicates that the scores of the afternoon class are not as closely clustered about the mean as those of the morning class. In looking at the total picture, you see that the median and the standard deviation give you the most useful information in agreeing with Ms. Duke that the morning class as a whole has a better understanding of probability than the afternoon class.

# Selected Answers to Student Exercise Material

Answers to all Section Self-Study Exercises and answers to odd-numbered exercises in the Chapter Review Exercises and Practice Tests are included here. Solutions for the Chapter Review Exercises and Practice Tests are available separately in the Student Solutions Manual (odd-numbered) or the Instructor's Resource Manual (even-numbered).

## Section 1–1 Self-Study Exercises

**1** **1.** $6 < 8$; $8 > 6$ **2.** $32 < 42$; $42 > 32$ **3.** $148 < 196$; $196 > 148$ **4.** $2,517 < 2,802$; $2,802 > 2,517$
**5.** $7,809 < 8,902$; $8,902 > 7,809$ **6.** $42,999 < 44,000$; $44,000 > 42,999$ **7.** $\$183,500 < \$198,900$; $\$198,900 > \$183,500$
**8.** 786 flats > 583 flats; 583 flats < 786 flats **9.** 758 rooms < 893 rooms; 893 rooms > 758 rooms
**10.** 5,982 e-mails > 2,807 e-mails; 2,807 e-mails < 5,982 e-mails

**2** **11.** 0.5 **12.** 0.23 **13.** 0.07 **14.** 6.83 **15.** 0.079 **16.** 0.468

**3** **17.** 3.72 **18.** 7.08 **19.** 0.3 **20.** 1.87, 1.9, 1.92 **21.** 72.07, 72.1, 73 **22.** 0.837 in. **23.** the micrometer
reading **24.** yes **25.** 0.04 in. **26.** No. 10 wire **27.** 72.3 kg **28.** $4.2 > 3.8$; $3.8 < 4.2$ **29.** $1.68 > 1.6$;
$1.6 < 1.68$ **30.** 0.026 kW > 0.003 kW; 800-W toaster **31.** $0.394 < 0.621$; 0.621

**4** **32.** 500 **33.** 430,000 **34.** 83,000,000,000 **35.** 300,000,000 **36.** 0.784 in. **37.** 2.8 A **38.** \$3
**39.** Drivers A, B, and D

**5** **40.** 97,614 **41.** 7,007 **42.** 30,133 **43.** 15.7 **44.** 34.18 **45.** 129.97 **46.** 3.077 in. **47.** 15.503 A
**48.** \$181.25 **49.** 391.4 ft **50.** 100.9° **51.** \$28,334.01 **52.** 0; 56.365 **53.** 84,200; 84,213 **54.** 200; 183.405
**55.** 402,300; 402,199 **56.** 0; 81.401 **57.** \$3,440; \$3,443.60 **58.** \$3,400; \$3,365 **59.** 159 lb **60.** Yes, total
capacity available is 115 gal. **61.** Yes, 479 pages are needed. **62.** 350 ft **63.** 1,020 **64.** 115 **65.** 53,036
**66.** 22 bags **67.** 341 boxes **68.** 291.82 **69.** 7.5 **70.** 310.8 **71.** 4.4 **72.** 5°F **73.** 15.1 lb
**74.** 12.08 in., 12.10 in. **75.** 59.83 cm **76.** 4.189 in., 4.201 in. **77.** 186 bricks **78.** 213.8 in. **79.** 125.5 in.
**80.** 45 m

**6** **81.** 378 **82.** 630,000,000 **83.** 102,612 **84.** 4,096 washers **85.** 84 automobiles **86.** 700 tickets
**87.** \$288 **88.** 56.55 **89.** 3.2445 **90.** 0.05805 **91.** 0.08672 **92.** 170.12283 **93.** 0.38381871
**94.** 4.9386274 **95.** 30.66 **96.** 5.25 in. **97.** \$27.48 **98.** 151.2 in. **99.** \$64.20 **100.** \$10,728
**101.** $P = 40$ ft **102.** $P = 61$ m **103.** 123.54 ft$^2$ **104.** 1,164 m$^2$ **105.** 43 **106.** 47 **107.** 32 ft
**108.** \$1,245 **109.** 6 in. **110.** 10.9 **111.** 0.19 **112.** 25 **113.** 6 lb **114.** 23 rolls **115.** 600 revolutions
**116.** 0.7 ft **117.** 0.16 **118.** 9 **119.** \$575 **120.** \$1.98 **121.** 1,070 bricks **122.** \$10,500 **123.** \$104.75
**124.** \$1,950 **125.** 85 **126.** 454 lb **127.** \$765.32 **128.** 1.69 in. **129.** 3.5 A

## Section 1–2 Self-Study Exercises

**1** **1.** 4; 3 **2.** 9; 4 **3.** 2.7; 9 **4.** 15; 2 **5.** 1,000 **6.** 16 **7.** 11.56 **8.** 15 **9.** 8 **10.** 81 **11.** $8^1$
**12.** $14.5^1$ **13.** $12^1$ **14.** $23^1$ **15.** Leaves base unchanged **16.** Use the base as a factor 3 times.

**2** **17.** 64 **18.** 324 **19.** 1.96 **20.** 169 **21.** 10,000 **22.** 14,641 **23.** 12,996 **24.** 14.44 **25.** 64 **26.** 144 **27.** 324 **28.** 10,201 **29.** 5 **30.** 7 **31.** 9 **32.** 6 **33.** 14 **34.** 15 **35.** 11 **36.** 12 **37.** Use the number as a factor 2 times. **38.** Find a number that is used as a factor 2 times to give the desired number.

**3** **39.** 1,000 **40.** 1,200,000 **41.** 20,000 **42.** 10,200 **43.** 25 **44.** 21 **45.** 3 **46.** 250 **47.** Shift the decimal to the right in the number being multiplied by the power of 10 as indicated by the exponent. **48.** Shift the decimal to the left in the number being divided by the power of 10 as indicated by the exponent. **49.** 225 **50.** 343 **51.** 78,125 **52.** 20,736 **53.** 18 **54.** 28 **55.** 33 **56.** 14

## Section 1–3 Self-Study Exercises

**1** **1.** 26 **2.** 18 **3.** 9 **4.** 12.5 **5.** 24 **6.** 32 **7.** 18 **8.** 9 **9.** 10 **10.** 9 **11.** 30 **12.** 15 **13.** 156 **14.** 109 **15.** 23 **16.** 145 **17.** 7 **18.** 15 **19.** 160 **20.** 1,458 **21.** 80 **22.** 70 **23.** 13 **24.** 11 **25.** 29 **26.** 5 **27.** 39 **28.** 20 **29.** 30 **30.** 20.52

**2** **31.** 38 in.; 72 in$^2$ **32.** 21.2 cm; 13.8 cm$^2$ **33.** 156 in. **34.** 124 in. **35.** 160 in. **36.** 108 in. **37.** 10 ft; 6 ft$^2$ **38.** 43.8 ft; 116.84 ft$^2$ **39.** 930 ft **40.** 54 ft **41.** 59 ft **42.** 156 ft **43.** 12 cm; 9 cm$^2$ **44.** 35.6 cm; 79.21 cm$^2$ **45.** 590 ft **46.** 48 tiles

**3** **47.** 29 boxes **48.** $26,178.44 **49.** 5,660 people **50.** 1,080 boxes of cards

## Chapter Review Exercises, Chapter 1

**1.** (a) 0.3 (b) 0.15 (c) 0.04 **3.** (a) 500 (b) 50,000 (c) 41.4 (d) 6.90 (e) 23.4610 **5.** 4.79 **7.** 0.02; 0.021; 0.0216 **9.** $\frac{7}{8}$ **11.** (a) 28 (b) 25 **13.** 34.9 kW **15.** (a) 4.61 (b) 3.127 (c) 204.899 (d) 12,140 **17.** 8.291 in.; 8.301 in. **19.** 0.43 in. **21.** 8.930 in.; 8.940 in. **23.** 84 **25.** 13,725 **27.** 394,254,080 **29.** $1,407 **31.** $43,920 **33.** $14,800, $13,140 **35.** 1,140,000 ft$^2$, 1,204,010.28 ft$^2$ **37.** 0.12096 in. **39.** 13 **41.** 2,008.4 **43.** 23 envelopes, 11 left over **45.** 48.79 ft **47.** (a) 7, 3, 343 (b) 2.3, 4, 27.9841 (c) 8, 4, 4,096 **49.** (a) 1 (b) 15,625 (c) 31.36 (d) 441 **51.** (a) 10$^1$ (b) 10$^3$ (c) 10$^4$ (d) 10$^5$ **53.** (a) 7 (b) 0.04056 (c) 0.605 (d) 2.3079 **55.** 75 **57.** 54 **59.** 15 **61.** 15 **63.** 13 **65.** 13.6 **67.** 113.608 **69.** 49 boxes **71.** $12,880 **73.** $52.60 **75.** 57 cm **77.** 210 mm **79.** 28.8 m **81.** 66 ft **83.** 288 in. **85.** 162 cm$^2$ **87.** 2,450 mm$^2$ **89.** 51.84 m$^2$ **91.** 33,000 ft$^2$ **93.** 33 yd$^2$ **95.** $14.25

## Practice Test, Chapter 1

**1.** 5.09 **3.** 48.3 **5.** 1,007 **7.** $9,271,314 **9.** 134 **11.** 106 **13.** $310, $310 **15.** $10, $14 **17.** 42,730 **19.** 11.6 **21.** 83 **23.** 0.6 **25.** $17,500 **27.** $P = 91$ ft; $A = 514.5$ ft$^2$ **29.** $P = 12.2$ in.; $A = 6.08$ in$^2$

## Section 2–1 Self-Study Exercises

**1** **1.** $5 = 5 \times 1, 10 = 5 \times 2, 15 = 5 \times 3, 20 = 5 \times 4, 25 = 5 \times 5, 30 = 5 \times 6$ **2.** $9 = 9 \times 1, 18 = 9 \times 2, 27 = 9 \times 3, 36 = 9 \times 4, 45 = 9 \times 5, 54 = 9 \times 6$ **3.** $10 = 10 \times 1, 20 = 10 \times 2, 30 = 10 \times 3, 40 = 10 \times 4, 50 = 10 \times 5, 60 = 10 \times 6$ **4.** $12 \times 1 = 12, 12 \times 2 = 24, 12 \times 3 = 36, 12 \times 4 = 48, 12 \times 5 = 60$; answers may vary. **5.** $13 \times 1 = 13, 13 \times 2 = 26, 13 \times 3 = 39, 13 \times 4 = 52, 13 \times 5 = 65, 13 \times 6 = 78$; answers may vary. **6.** $3 \times 1 = 3, 3 \times 2 = 6, 3 \times 3 = 9, 3 \times 4 = 12, 3 \times 5 = 15$; answers may vary. **7.** no; not divisible by 3 **8.** yes; ends in zero **9.** no; last two digits not divisible by 4 **10.** $1 \cdot 4, 2 \cdot 2$ **11.** $1 \cdot 15, 3 \cdot 5$ **12.** $1 \cdot 24, 2 \cdot 12, 3 \cdot 8, 4 \cdot 6$ **13.** $1 \cdot 30, 2 \cdot 15, 3 \cdot 10, 5 \cdot 6$ **14.** $1 \cdot 36, 2 \cdot 18, 3 \cdot 12, 4 \cdot 9, 6 \cdot 6$ **15.** 1, 2, 23, 46 **16.** 1, 2, 4, 13, 26, 52 **17.** 1, 3, 9, 27, 81 **18.** 1, 5, 17, 85 **19.** 1, 2, 7, 14, 49, 98 **20.** prime **21.** composite **22.** composite **23.** composite **24.** composite **25.** prime **26.** $2 \cdot 3 \cdot 3$ or $2 \cdot 3^2$ **27.** $2 \cdot 2 \cdot 5$ or $2^2 \cdot 5$ **28.** $3 \cdot 3 \cdot 3$ or $3^3$ **29.** 29 **30.** $2 \cdot 2 \cdot 2 \cdot 5$ or $2^3 \cdot 5$ **31.** 6 **32.** 60 **33.** 24 **34.** 700 **35.** 27 **36.** 16 **37.** 9 **38.** 5 **39.** 2 **40.** 6 **41.** 5 **42.** 3

**43.** $\frac{8}{10}, \frac{12}{15}, \frac{16}{20}, \frac{20}{25}, \frac{24}{30}$; answers may vary.  **44.** $\frac{14}{20}, \frac{21}{30}, \frac{28}{40}, \frac{35}{50}, \frac{42}{60}$; answers may vary.  **45.** $\frac{12}{21}$  **46.** $\frac{36}{44}$  **47.** $\frac{5}{15}$

**48.** $\frac{3}{5}$  **49.** $\frac{3}{4}$  **50.** $\frac{5}{16}$  **51.** $\frac{1}{2}$  **52.** $\frac{7}{8}$  **53.** $4\frac{9}{16}$ in.  **54.** $4\frac{1}{16}$ in.  **55.** $2\frac{1}{4}$ in.  **56.** 2 in.  **57.** $\frac{3}{4}$ in.

**58.** $\frac{3}{8}$ in.

**2**  **59.** $2\frac{2}{5}$  **60.** $1\frac{3}{7}$  **61.** 1  **62.** $4\frac{4}{7}$  **63.** 4

**3**  **64.** $\frac{15}{8}$  **65.** $\frac{77}{12}$  **66.** $\frac{77}{8}$  **67.** $\frac{31}{8}$  **68.** $\frac{89}{12}$

**4**  **69.** $\frac{1}{10}$  **70.** $\frac{1}{4}$  **71.** $\frac{1}{40}$  **72.** $3\frac{9}{10}$  **73.** $\frac{7}{8}$  **74.** $\frac{3}{8}$  **75.** $\frac{5}{8}$  **76.** $\frac{43}{100}$  **77.** $\frac{83}{100}$ lb  **78.** $\frac{5}{16}$ in.  **79.** 0.3

**80.** 0.21  **81.** 3.875  **82.** 1.4375  **83.** 4.5625  **84.** 0.44  **85.** 1.43  **86.** 3.45  **87.** 0.045  **88.** 0.125 in.

**5**  **89.** 72  **90.** 30  **91.** 50  **92.** 48  **93.** 24  **94.** $\frac{2}{3}$  **95.** $\frac{7}{16}$  **96.** $\frac{15}{32}$  **97.** $\frac{7}{12}$  **98.** $\frac{4}{5}$  **99.** $\frac{4}{5}$

**100.** $\frac{5}{12}$  **101.** $\frac{1}{2}$  **102.** 0.34  **103.** Yes, $\frac{7}{16}$ is greater.  **104.** yes  **105.** no  **106.** yes

## Section 2–2 Self-Study Exercises

**1**  **1.** $\frac{3}{8}$  **2.** $1\frac{3}{8}$  **3.** $\frac{5}{8}$  **4.** $\frac{25}{32}$  **5.** $\frac{9}{16}$  **6.** $1\frac{7}{16}$  **7.** $\frac{11}{64}$  **8** $1\frac{19}{40}$  **9.** $1\frac{23}{36}$  **10.** $1\frac{1}{12}$  **11.** $1\frac{2}{5}$  **12.** $\frac{19}{21}$

**13.** $\frac{15}{16}$ in.  **14.** $1\frac{27}{32}$ in.  **15.** $1\frac{15}{16}$ in.  **16.** $1\frac{7}{8}$ in.  **17.** $1\frac{1}{4}$ in.  **18.** $6\frac{4}{5}$  **19.** $4\frac{1}{8}$  **20.** $16\frac{13}{16}$  **21.** $1\frac{11}{18}$

**22.** $4\frac{13}{16}$  **23.** $5\frac{17}{32}$  **24.** $4\frac{11}{16}$  **25.** $13\frac{1}{2}$  **26.** $8\frac{13}{16}$ in.  **27.** $3\frac{15}{16}$ in.  **28.** $11\frac{5}{8}$ gal  **29.** $24\frac{5}{32}$ in.

**30.** $22\frac{7}{8}$ in.  **31.** $5\frac{7}{8}$ c

**2**  **32.** $\frac{3}{16}$  **33.** $\frac{1}{16}$  **34.** $\frac{1}{8}$  **35.** $\frac{9}{64}$  **36.** $17\frac{3}{4}$  **37.** $4\frac{15}{16}$  **38.** $5\frac{21}{32}$  **39.** $1\frac{5}{7}$  **40.** $10\frac{1}{8}$ in.  **41.** $3\frac{1}{10}$ lb

**42.** $\frac{3}{16}$ in.  **43.** $\frac{29}{32}$

## Section 2–3 Self-Study Exercises

**1**  **1.** $\frac{3}{32}$  **2.** $\frac{7}{32}$  **3.** $\frac{7}{16}$  **4.** $\frac{7}{12}$  **5.** $\frac{1}{3}$  **6.** $\frac{5}{32}$  **7.** $21\frac{7}{8}$  **8.** 75  **9.** $4\frac{1}{8}$  **10.** $36\frac{1}{10}$  **11.** $1\frac{21}{40}$

**12.** $2\frac{1}{6}$  **13.** $18\frac{3}{4}$ L  **14.** 90 in.  **15.** 264 kg copper, 84 kg tin, 36 kg zinc

**2**  **16.** $\frac{1}{16}$  **17.** $\frac{27}{125}$  **18.** $\frac{1}{27}$  **19.** $\frac{49}{81}$

**3**  **20.** $\frac{8}{5}$  **21.** $\frac{5}{11}$  **22.** $\frac{1}{8}$  **23.** $\frac{10}{9}$  **24.** $\frac{5}{9}$  **25.** $\frac{6}{7}$  **26.** $\frac{9}{10}$  **27.** $\frac{11}{12}$  **28.** 2  **29.** $4\frac{2}{3}$  **30.** $13\frac{1}{3}$

**31.** $12\frac{1}{2}$    **32.** $\frac{5}{8}$    **33.** 4    **34.** $2\frac{1}{2}$    **35.** 12 shovels    **36.** $23\frac{11}{12}$ in.    **37.** 20 ft × 15 ft    **38.** 4 whole pieces

**39.** 7 strips    **40.** 23 straws, $3\frac{3}{4}$ in. left

## Section 2–4 Self-Study Exercises

**1** **1.** $\frac{2\,pt}{1\,qt}; \frac{1\,qt}{2\,pt}$    **2.** $\frac{5{,}280\,ft}{1\,mi}; \frac{1\,mi}{5{,}280\,ft}$    **3.** $\frac{12\,in.}{1\,ft}; \frac{1\,ft}{12\,in.}$    **4.** $\frac{3\,ft}{1\,yd}; \frac{1\,yd}{3\,ft}$    **5.** 48 in.    **6.** 21 ft    **7.** 4,400 yd

**8.** $9\frac{1}{3}$ yd    **9.** $2\frac{3}{10}$, or 2.3, lb    **10.** 732.8 oz    **11.** 20 qt    **12.** 13 pt

**2** **13.** feet to inches = 12; inches to feet = $\frac{1}{12}$, or 0.083    **14.** quarts to gallons = $\frac{1}{4}$ or 0.25; gallons to quarts = 4

**15.** 28.75 pt    **16.** 92 pt    **17.** 36.25 lb    **18.** 153 in.    **19.** 7,920 ft    **20.** 28 qt    **21.** 3 ft 8 in.    **22.** 2 mi 1,095 ft

**23.** $3\,lb\,3\frac{1}{2}$ oz    **24.** 2 gal 1 qt    **25.** 2 ft 10 in.    **26.** 6 lb 9 oz    **27.** 4 gal 2 qt 16 oz    **28.** 6 qt 20 oz

**3** **29.** 2 lb 5 oz or 37 oz    **30.** 4 ft 7 in. or 55 in.    **31.** 15 lb 11 oz    **32.** 18 ft 3 in.    **33.** $8\,qt\,\frac{1}{2}$ pt    **34.** 14 gal 1 qt

**35.** 9 yd 1 ft    **36.** 7 yd 1 ft 5 in.    **37.** 2 ft 7 in. or 31 in.    **38.** 11 ft    **39.** 5 lb 15 oz    **40.** 12 lb    **41.** 6 in.    **42.** 1 pt

**43.** 2 ft    **44.** 4 lb 14 oz    **45.** 1 lb 6 oz    **46.** 6 lb 11 oz    **47.** 2 in.    **48.** $3\,gal\,3\,qt\,1\frac{1}{2}$ pt or 3 gal 3 qt 1 pt 1 c

**49.** 4 ft 2 in.    **50.** 2 ft 3 in.    **51.** 45 s    **52.** 39 s    **53.** 69 lb 7 oz    **54.** 16 lb 14 oz

**4** **55.** 60 mi    **56.** 108 gal    **57.** 57 lb 8 oz    **58.** 58 ft    **59.** 36 lb    **60.** 7 qt 1 pt or 1 gal 3 qt 1 pt    **61.** 35 in$^2$
**62.** 108 ft$^2$    **63.** 180 yd$^2$    **64.** 108 mi$^2$    **65.** 378 tiles    **66.** 46 gal 18 oz    **67.** 7 qt 1 pt or 1 gal 3 qt 1 pt
**68.** 6 lb 4 oz    **69.** 1 day 15 h    **70.** 10 yd 1 ft 3 in.    **71.** 1 yd 1 ft 7 in.    **72.** 7 ft 6 in.    **73.** 3 gal 2 qt
**74.** 9 pieces    **75.** 5 ft    **76.** 3 gal 1 qt 5 oz    **77.** 24 lb 3 oz    **78.** 3    **79.** 18    **80.** 3    **81.** 8 pieces    **82.** 9 boxes

**83.** 24 cans    **84.** 9 tickets    **85.** $\frac{3}{4}\frac{lb}{min}$    **86.** $15{,}840\frac{ft}{h}$    **87.** $2{,}304\frac{oz}{min}$    **88.** $2\frac{qt}{s}$    **89.** $3\frac{qt}{min}$    **90.** $53\frac{1}{3}\frac{lb}{min}$

## Section 2–5 Self-Study Exercises

**1** **1.** 40%    **2.** 70%    **3.** 62.5%    **4.** 0.3%    **5.** 20%    **6.** 14%    **7.** 500%    **8.** 800%    **9.** 133.3%    **10.** 350%
**11.** 305%    **12.** 720%

**2** **13.** $\frac{9}{25}$; 0.36    **14.** $\frac{3}{4}$; 0.75    **15.** $\frac{1}{16}$; 0.0625    **16.** $\frac{5}{8}$; 0.625    **17.** $\frac{2}{3}$; 0.6667    **18.** $\frac{3}{500}$; 0.006    **19.** $\frac{1}{500}$; 0.002

**20.** $\frac{1}{2{,}000}$; 0.0005    **21.** $\frac{1}{12}$; 0.0833    **22.** $\frac{3}{16}$; 0.1875    **23.** 8    **24.** 4    **25.** $2\frac{1}{2}$; 2.5    **26.** $4\frac{1}{4}$; 4.25    **27.** $1\frac{19}{25}$; 1.76

**28.** $3\frac{4}{5}$; 3.8    **29.** $1\frac{3}{8}$; 1.375    **30.** $3\frac{7}{8}$; 3.875    **31.** $1\frac{2}{3}$    **32.** $3\frac{1}{6}$    **33.** 1.153    **34.** 2.125

## Section 2–6 Self-Study Exercises

**1** **1.** $R$ missing; $B = 10$; $P = 2$    **2.** $P = 2$; $R = 20\%$; $B$ missing    **3.** $P = 3$; $R$ missing; $B = 4$    **4.** $R$ missing; $B = 25$; $P = 5$    **5.** $P = 6$; $B = 15$; $R$ missing    **6.** $P$ missing; $R = 20\%$; $B = 15$

**2** **7.** 5.4    **8.** 252    **9.** 1.8    **10.** 26    **11.** 50%    **12.** $333\frac{1}{3}\%$    **13.** 87.5%    **14.** 150%    **15.** 64

**16.** 70,000    **17.** 1.0625 lb    **18.** 3%    **19.** 200 hp    **20.** 0.021 lb    **21.** 5%    **22.** $575    **23.** 1,200 lb    **24.** 11%

Selected Answers to Student Exercise Material    645

**25.** 9,625 swabs   **26.** 100,880 welds   **27.** $4.50   **28.** $6.00   **29.** $3,928.57   **30.** $1,364.16   **31.** $4.55, $80.38
**32.** $829.06   **33.** $23.30   **34.** 25%   **35.** 23%   **36.** $41.93   **37.** $11,832   **38.** $17.52   **39.** $3,866.00

**40.** $944.50   **41.** $434.84   **42.** $3,205.75   **43.** 6.8%   **44.** 30.1%   **45.** 111 lb   **46.** $8\frac{3}{4}$ lb   **47.** $32.91

**48.** 21,600 lb   **49.** 170 lb   **50.** $47.94

## Section 2–7 Self-Study Exercises

**1**   **1.** 108   **2.** 109.2   **3.** 10.2   **4.** 33.75   **5.** 672,830; 2,878,830   **6.** 630,504; 2,576,504

**2**   **7.** 60%   **8.** 82%   **9.** 13.7%   **10.** $66\frac{2}{3}$%   **11.** 2,393 board feet   **12.** 25,908 bricks   **13.** $53,350.20

**14.** $2,041.46   **15.** $3,961.82   **16.** $58.80

**3**   **17.** 7%   **18.** 45%   **19.** 20%   **20.** 5%   **21.** 350 hp   **22.** 10.5 yd$^3$   **23.** 3,019 board feet   **24.** 28.5 lb
**25.** 40 in.   **26.** 20%

## Chapter Review Exercises, Chapter 2

**1.** 20, 30, 40, 50, 60; answers may vary.   **3.** yes; sum of digits divisible by 3   **5.** $1 \times 48, 2 \times 24, 3 \times 16, 4 \times 12, 6 \times 8$;

1, 2, 3, 4, 6, 8, 12, 16, 24, 48   **7.** composite   **9.** $2 \cdot 7 \cdot 7$ or $2 \cdot 7^2$   **11.** 360   **13.** 6   **15.** $\frac{15}{24}$   **17.** $\frac{25}{60}$   **19.** $\frac{1}{2}$

**21.** $\frac{3}{4}$   **23.** $3\frac{9}{16}$ in.   **25.** $3\frac{3}{5}$   **27.** $\frac{57}{8}$   **29.** $\frac{23}{5}$   **31.** $\frac{20}{10}$   **33.** $\frac{7}{200}$   **35.** $\frac{17}{20}$   **37.** 0.625 or 0.63   **39.** 0.64

**41.** 60   **43.** no   **45.** smaller   **47.** $\frac{3}{4}$   **49.** $\frac{21}{64}$   **51.** $10\frac{9}{32}$   **53.** $18\frac{1}{16}$ mi   **55.** $12\frac{19}{32}$ in.   **57.** $\frac{7}{8}$ in.

**59.** $\frac{1}{3}$   **61.** $34\frac{3}{4}$   **63.** $\frac{7}{16}$ in.   **65.** $1\frac{29}{64}$ in.   **67.** $1\frac{1}{5}$   **69.** 2   **71.** $2\frac{3}{4}$ c   **73.** 150 cm   **75.** $\frac{9}{16}$   **77.** $\frac{1}{8}$

**79.** $\frac{1}{4}$   **81.** $1\frac{1}{6}$   **83.** 2   **85.** $2\frac{55}{64}$ in.   **87.** 36 lengths   **89.** $\frac{2,000 \text{ lb}}{1 \text{ T}}; \frac{1 \text{ T}}{2,000 \text{ lb}}$   **91.** 80 oz   **93.** $42\frac{1}{2}$ lb

**95.** 6,600 ft   **97.** 13 lb $1\frac{1}{2}$ oz   **99.** 8 gal 2 qt   **101.** 14 oz   **103.** 17 in. or 1 ft 5 in.   **105.** 63 in$^2$

**107.** 10 yd 1 ft 3 in.   **109.** 4 lb 8 oz   **111.** 3.5   **113.** 300 mi/h   **115.** $1\frac{1}{2}$ qt   **117.** 72%   **119.** 70%

**121.** 320%   **123.** 0.125 or $\frac{1}{8}$   **125.** 2.75 or $2\frac{3}{4}$   **127.** 2.272   **129.** 0.83   **131.** $R = 5\%$; $B = 180$; $P$ missing

**133.** $R = 45\%$; $B$ missing; $P = \$36$   **135.** 24   **137.** 60%   **139.** 15.3%   **141.** 1.14%   **143.** $365.66
**145.** 1.75%   **147.** $355   **149.** 87.1%   **151.** 14%   **153.** 142.1 kg   **155.** 12.5%   **157.** 4.8%   **159.** 86%

## Practice Test, Chapter 2

**1.** $\frac{17}{25}$   **3.** $1\frac{1}{4}$ in.   **5.** $10\frac{1}{2}$   **7.** $\frac{7}{10}$   **9.** $7\frac{17}{30}$   **11.** $3\frac{8}{9}$   **13.** $3\frac{1}{2}$   **15.** $13\frac{1}{2}$   **17.** 4 technicians   **19.** 72

**21.** $\frac{2}{7}$   **23.** $3\frac{5}{6}$ c   **25.** 8 yd   **27.** 12%   **29.** $525   **31.** $2,500   **33.** 21.14%   **35.** 13.17%   **37.** 5%

**39.** $104.87

# Section 3-1 Self-Study Exercises

**1.** (a) 1,000 m  (b) 10 L  (c) $\frac{1}{10}$ g  (d) $\frac{1}{1,000}$ m  (e) 100 g  (f) $\frac{1}{100}$ L   **2.** b   **3.** b   **4.** c   **5.** b   **6.** a
**7.** c   **8.** b   **9.** c   **10.** a   **11.** b

**12.** 70   **13.** 580   **14.** 80   **15.** 2.5   **16.** 210   **17.** 85   **18.** 142   **19.** 400   **20.** 8,000   **21.** 250
**22.** 800   **23.** 102,500   **24.** 8,330   **25.** 2,000,000   **26.** 70   **27.** 236 L   **28.** 467 cm   **29.** 38,000 dg
**30.** 13,000 cm   **31.** 2.8   **32.** 23.8   **33.** 10.1   **34.** 6   **35.** 2.9   **36.** 19.25   **37.** 1.7   **38.** 0.225 dL
**39.** 2.743

**40.** 12 hL   **41.** 6 cg   **42.** 2.4 dm or 24 cm   **43.** 5.9 cL or 59 mL   **44.** cannot add   **45.** 10.1 kL or 101 hL
**46.** 0.55 g or 55 cg   **47.** cannot subtract   **48.** 7.002 km or 7,002 m   **49.** 1,000 mL   **50.** 1.47 kL or 147 dkL
**51.** 516 m   **52.** 40.8 m   **53.** 150.96 dm   **54.** 969.5 m   **55.** 4,680 mm   **56.** 6 g   **57.** 13 m   **58.** 9 cL
**59.** 163 g   **60.** 0.4 m or 4 dm   **61.** 5   **62.** 30   **63.** 16 prescriptions   **64.** 80 containers   **65.** 7.5 dm
**66.** 50 mL   **67.** 21,250 containers   **68.** 19 vials

# Section 3-2 Self-Study Exercises

**1.** 1.5 da   **2.** 9.67 min   **3.** 150 min   **4.** 318 s   **5.** 210 min   **6.** 3.03 min   **7.** 432 min   **8.** 1.47 min
1.467 min

**9.** 4,320 lb/h   **10.** 2,400 gal/h   **11.** 10,950 lb/yr   **12.** 1,680 vehicles/h   **13.** 0.0167 mi/s   **14.** 2.0833 gal/min
**15.** 60 lb/min   **16.** 1.6 gal/s

**17.** 35°C   **18.** 0°C   **19.** 45°C   **20.** 5°C   **21.** 15°C   **22.** 10°C   **23.** 65°C   **24.** 50°C   **25.** 80°C
**26.** 120°C

**27.** 158°F   **28.** 59°F   **29.** 113°F   **30.** 122°F   **31.** 68°F   **32.** 419°F   **33.** 590°F   **34.** 770°F
**35.** 365°F   **36.** 32°F

**37.** 0.0005 Hz   **38.** 420,000 H   **39.** 1,400,000 W   **40.** 0.045 s

# Section 3-3 Self-Study Exercises

**1.** 354.33 in.   **2.** 131.232 yd   **3.** 26.0988 mi   **4.** 6.3402 liq qt   **5.** 9.463 L   **6.** 59.5242 lb   **7.** 22.68 kg
**8.** 17.78 cm   **9.** 5.4864 m   **10.** 310.7 mi   **11.** 711.2 mm   **12.** 91.44 cm   **13.** 1.2192 m   **14.** 1,609.344 m
**15.** 91.44 m   **16.** 321.86 km   **17.** 52.835 qt   **18.** 94.63 L   **19.** 30.48 m   **20.** 27.216 kg   **21.** 241.395 km
**22.** 32.808 yd   **23.** 236.132 mi   **24.** 7.4568 mi   **25.** 11.3556 L   **26.** 328.08 ft   **27.** 22   **28.** 18

# Section 3-4 Self-Study Exercises

**1.** 3   **2.** 4   **3.** 2   **4.** 4   **5.** 4   **6.** 2   **7.** 3   **8.** 5   **9.** 5   **10.** 3

**11.** $\frac{1}{32}$ in.   **12.** 0.05 mm   **13.** $\frac{1}{8}$ in.   **14.** $\frac{1}{2}$ or 0.5 oz   **15.** 0.5 L   **16.** $\frac{1}{16}$ in.   **17.** $\frac{1}{64}$ in.   **18.** $\frac{1}{16}$ oz
**19.** 0.05 mi   **20.** 0.05 cm   **21.** 0.005 dg   **22.** 0.05 cg   **23.** 0.05 cL   **24.** 0.05 km   **25.** 0.05 km   **26.** 0.05 km

**27.** absolute error = 0.02 cm   **28.** absolute error = 0.2 mm   **29.** absolute error = 1.5 cm
relative error = 0.00038   relative error = 0.0038   relative error = 0.0308
percent error = 0.038%   percent error = 0.38%   percent error = 3.08%

**30.** absolute error = 0.29 in.    **31.** absolute error = 0.2 L    **32.** absolute error = 0.4 in.
relative error = 0.0204           relative error = 0.0040           relative error = 0.0267
percent error = 2.04%             percent error = 0.40%             percent error = 2.67%

**4**   **33.** 215.4 m   **34.** 64.4 g   **35.** 43.4 kg   **36.** 18.11 kg   **37.** 900,000 m³   **38.** 17,000,000 m³   **39.** 43 g
**40.** 70 km

## Section 3–5 Self-Study Exercises

**1**   **1.** 115 mm or 11.5 cm   **2.** 102 mm or 10.2 cm   **3.** 96 mm or 9.6 cm   **4.** 85 mm or 8.5 cm
**5.** 57 mm or 5.7 cm   **6.** 50 mm or 5 cm   **7.** 44 mm or 4.4 cm   **8.** 30.5 mm or 3.05 cm   **9.** 19 mm or 1.9 cm
**10.** 10 mm or 1 cm

**2**   **11.** 115.90 mm; 4.560 in.   **12.** 8.00 mm; 0.312 in.   **13.** 27.15 mm; 1.069 in.   **14.** 45.10 mm; 1.777 in.

**3**   **15.** 0.175 in.   **16.** 4.225 in.   **17.** 2.275 in.   **18.** 4.359 in.   **19.** 1.242 in.   **20.** 0.2510 in.   **21.** 2.2508 in.
**22.** 1.2503 in.   **23.** 4.2673 in.   **24.** 2.2558 in.

**4**   **25.** 419,850 ft³   **26.** 787,516 ft³   **27.** 145,556 ft³   **28.** 328,153 ft³   **29.** 138,444 ft³   **30.** 287,747 ft³
**31.** 7,555 kWh   **32.** 3,421 kWh   **33.** 6.16 mm   **34.** 6.36 mm   **35.** 3.17 mm   **36.** −8.61 mm   **37.** 135 V
**38.** 210 V   **39.** 0.8 Ω   **40.** 5.5 Ω   **41.** 0.403002 in.

**42.** 1.0 in.
0.03 in.
0.004 in.
0.0005 in.
0.00006 in.

## Chapter Review Exercises, Chapter 3

**1.** kilo-   **3.** milli-   **5.** 10 times   **7.** $\frac{1}{1,000}$ of   **9.** 1,000 times   **11.** a   **13.** c   **15.** 6.71 dkm   **17.** 2,300 mm

**19.** 12,300 mm   **21.** 413.27 km   **23.** 30.00974 kg   **25.** cannot add unlike measures   **27.** 61.47 cg   **29.** 8.5 hL
**31.** 18.9 m   **33.** 245 mL   **35.** 6 m   **37.** 40 shirts   **39.** 3 days   **41.** 2.63 h   **43.** 4 days   **45.** 3.25 yr
**47.** 203°F   **49.** 104°F   **51.** 185°C   **53.** 100.2°F   **55.** 235.124 yd   **57.** 15.8505 liq qt   **59.** 70.5472 lb

**61.** 22.86 cm   **63.** 60.96 m   **65.** 21   **67.** 6   **69.** 4   **71.** $\frac{1}{4}$ in.   **73.** 0.05 cm   **75.** 1%   **77.** 4 significant digits

**79.** 0.05 cg   **81.** 46.5 m or 4,650 cm   **83.** 117 mm or 118 mm   **85.** 60 mm   **87.** 34.70 mm or 1.367 in.
**89.** 2.550 in.   **91.** 4.3801 in.   **93.** absolute error = 0.52; relative error = 0.0419; percent error = 4.19%
**95.** 441,822 ft³   **97.** 5.14 mm   **99.** 300 Ω

## Practice Test, Chapter 3

**1.** 0.298 km   **3.** 9.48 L or 94.8 dL   **5.** 120.6975 km   **7.** 8.4536 pt   **9.** 9°C   **11.** 2,520 min   **13.** 235 h   **15.** 2
**17.** relative error = 0.0012; percent error = 0.12%   **19.** 99 mm or 9.9 cm   **21.** 2.912 in.   **23.** 9.27 mm

## Section 4–1 Self-Study Exercises

**1**   **1.** <   **2.** <   **3.** >   **4.** <   **5.** <   **6.** >   **7.** <   **8.** >   **9.** 7   **10.** 17   **11.** 8   **12.** 7   **13.** −42
**14.** 17   **15.** 78   **16.** −57

**2**   **17.** 23   **18.** −16   **19.** −13   **20.** −29   **21.** −55   **22.** −124   **23.** 29   **24.** −20   **25.** 99   **26.** −59
**27.** −48   **28.** −161   **29.** −53   **30.** 232   **31.** −1,005

**32.** 2  **33.** 2  **34.** 4  **35.** −4  **36.** 6  **37.** −6  **38.** −2  **39.** 4  **40.** −2  **41.** 2  **42.** −10
**43.** −40  **44.** 7  **45.** 15  **46.** −8  **47.** −33  **48.** 9  **49.** −11  **50.** 13  **51.** 9  **52.** 3  **53.** 0  **54.** 0
**55.** 0  **56.** 0  **57.** −3  **58.** −7  **59.** 0  **60.** −3  **61.** −12  **62.** −14  **63.** 13  **64.** −12  **65.** 24
**66.** −98  **67.** 2  **68.** 10  **69.** $66  **70.** $36  **71.** −295°C  **72.** −23°

## Section 4–2 Self-Study Exercises

**1**  **1.** −12  **2.** 6  **3.** −6  **4.** 4  **5.** −25  **6.** −3  **7.** 8  **8.** −3  **9.** −9  **10.** 13  **11.** −1  **12.** −8
**13.** 62  **14.** −11  **15.** −58  **16.** −57  **17.** −15  **18.** −105  **19.** 18  **20.** −17  **21.** −167  **22.** 83
**23.** 1,413  **24.** −103  **25.** 15  **26.** −8  **27.** −12  **28.** 8  **29.** 7  **30.** 10  **31.** 56  **32.** −92  **33.** 14
**34.** −36  **35.** 0  **36.** 0

**2**  **37.** 4  **38.** 7  **39.** 6  **40.** −4  **41.** 1  **42.** 14  **43.** −3  **44.** −9  **45.** −5  **46.** 13  **47.** 2
**48.** −1  **49.** 1  **50.** −20  **51.** 26  **52.** −99  **53.** −45  **54.** −32  **55.** −60  **56.** −80  **57.** −35
**58.** 50  **59.** −21  **60.** 186  **61.** 107°F  **62.** 107°F  **63.** $115,054  **64.** Subtracting zero from a number results
in the same number with the same sign. Subtracting a number from zero results in the opposite of the number.

## Section 4–3 Self-Study Exercises

**1**  **1.** 40  **2.** 12  **3.** 35  **4.** 21  **5.** 24  **6.** 6  **7.** −15  **8.** −10  **9.** −32  **10.** −12  **11.** −56
**12.** −24  **13.** 4(−$28) = −$112  **14.** (7)(−$40) = −$280  **15.** 336  **16.** −60  **17.** 36  **18.** 0  **19.** 90
**20.** 0  **21.** −42  **22.** 54  **23.** −210  **24.** 2,268  **25.** −20,160  **26.** 840  **27.** −240  **28.** 0  **29.** 0
**30.** 0  **31.** 0  **32.** 0  **33.** 0  **34.** 0  **35.** 0  **36.** 0  **37.** 0  **38.** −30  **39.** 168  **40.** 42  **41.** −8
**42.** 0  **43.** −16  **44.** The multiplicative inverse of a number is the number that, when multiplied by the original number,
results in 1, the multiplicative identity.  **45.** Answers will vary, 5(−3) = −3(5) = −15.

**2**  **46.** 9  **47.** −8  **48.** 25  **49.** 0  **50.** −8  **51.** −512  **52.** 625  **53.** 81  **54.** 2,401  **55.** 1,764
**56.** 5  **57.** −28  **58.** 12  **59.** −8  **60.** −10  **61.** −8  **62.** −27  **63.** −121  **64.** 121  **65.** −125

**3**  **66.** −3  **67.** −4  **68.** −4  **69.** −4  **70.** 8  **71.** 5  **72.** −6  **73.** 8  **74.** 5

**75.** −$1,800; $\dfrac{-1,800}{6}$ = −$300  **76.** −6°C per hour  **77.** undefined  **78.** undefined  **79.** 0  **80.** 0  **81.** 0

**82.** undefined  **83.** undefined  **84.** undefined  **85.** 0  **86.** undefined  **87.** multiplication and division
**88.** −10°F  **89.** $3,472.50  **90.** The numerator must be zero and the denominator must be any number except zero.

## Section 4–4 Self-Study Exercises

**1**  **1.** $-\dfrac{-5}{8}, -\dfrac{5}{-8}, \dfrac{-5}{-8}$  **2.** $-\dfrac{-3}{-4}, \dfrac{-3}{4}, \dfrac{3}{-4}$  **3.** $-\dfrac{2}{-5}, -\dfrac{-2}{5}, \dfrac{2}{5}$  **4.** $-\dfrac{7}{8}, \dfrac{-7}{8}, \dfrac{7}{-8}$  **5.** $-\dfrac{-7}{8}, -\dfrac{7}{-8}, \dfrac{-7}{-8}$

**2**  **6.** $-\dfrac{1}{4}$  **7.** $-1\dfrac{1}{10}$  **8.** $-\dfrac{1}{16}$  **9.** $\dfrac{3}{16}$  **10.** $-\dfrac{2}{9}$  **11.** $-10\dfrac{5}{8}$  **12.** $-2\dfrac{13}{24}$  **13.** $-3\dfrac{7}{8}$  **14.** $1\dfrac{1}{10}$

**15.** $-1\dfrac{3}{8}$  **16.** $-\dfrac{1}{16}$  **17.** $2\dfrac{17}{24}$  **18.** $-\dfrac{1}{2}$  **19.** $\dfrac{6}{11}$  **20.** $\dfrac{10}{27}$  **21.** 30  **22.** $-\dfrac{1}{32}$  **23.** −2  **24.** $-\dfrac{1}{10}$

**25.** $-\dfrac{1}{7}$  **26.** $-\dfrac{25}{32}$  **27.** −26.297  **28.** −1.11  **29.** −91.44  **30.** −110.72  **31.** −59.04  **32.** 340.71

**33.** −27.73  **34.** 0.41  **35.** 4.6  **36.** 0.03  **37.** −2.3  **38.** −5.5

**3**  **39.** $-\dfrac{1}{5}$  **40.** $-5\dfrac{7}{8}$  **41.** −7  **42.** $-3\dfrac{3}{5}$  **43.** $\dfrac{37}{144}$  **44.** $-\dfrac{127}{245}$  **45.** −10.38  **46.** −3.992  **47.** −11.88

## Section 4–5 Self-Study Exercises

**1**   **1.** 45,300   **2.** 27   **3.** 5,820   **4.** 0.897   **5.** 5.23   **6.** 0.00806   **7.** 0.573   **8.** 0.00293   **9.** 0.0457
**10.** 857.9   **11.** 4,370   **12.** 8,370   **13.** 37   **14.** 1,820   **15.** 0.56   **16.** 1.42   **17.** 780,000   **18.** 62
**19.** 0.00046   **20.** 0.61   **21.** 0.72   **22.** 42   **23.** $10^{12}$   **24.** 10   **25.** $10^2$   **26.** $10^4$   **27.** 1

**28.** $10^{-2}$ or $\dfrac{1}{10^2}$   **29.** $10^{-12}$ or $\dfrac{1}{10^{12}}$   **30.** $10^{-6}$ or $\dfrac{1}{10^6}$   **31.** $10^{10}$   **32.** $10^{-7}$ or $\dfrac{1}{10^7}$   **33.** $10^{-3}$ or $\dfrac{1}{10^3}$   **34.** 10

**35.** $10^3$   **36.** $10^2$   **37.** $10^{-3}$ or $\dfrac{1}{10^3}$   **38.** 1   **39.** $10^{-5}$ or $\dfrac{1}{10^5}$   **40.** $10^{-1}$ or $\dfrac{1}{10}$

**2**   **41.** $10^6$   **42.** $10^7$   **43.** $10^{-4}$ or $\dfrac{1}{10^4}$   **44.** $10^{10}$   **45.** $10^{-8}$ or $\dfrac{1}{10^8}$

## Section 4–6 Self-Study Exercises

**1**   **1.** 430   **2.** 0.0065   **3.** 2.2   **4.** 73   **5.** 0.093   **6.** 83,000   **7.** 0.0058   **8.** 80,000   **9.** 6.732   **10.** 0.00589
**11.** 78.3   **12.** 1,590   **13.** 397,000   **14.** 0.0004723   **15.** 0.00000991   **16.** 1,030,000

**2**   **17.** $3.92 \times 10^2$   **18.** $2 \times 10^{-2}$   **19.** $7.03 \times 10^0$   **20.** $4.2 \times 10^4$   **21.** $8.1 \times 10^{-2}$   **22.** $2.1 \times 10^{-3}$   **23.** $2.392 \times 10^1$
**24.** $1.01 \times 10^{-1}$   **25.** $1.002 \times 10^0$   **26.** $7.21 \times 10^2$   **27.** $4.2 \times 10^5$   **28.** $3.26 \times 10^4$   **29.** $2.13 \times 10^1$   **30.** $6.2 \times 10^{-6}$
**31.** $5.6 \times 10^1$   **32.** $1.97 \times 10^{-6}$   **33.** $7.45 \times 10^1$   **34.** $1.8 \times 10^4$   **35.** $7.01 \times 10^1$   **36.** $7.25 \times 10^{-1}$

**3**   **37.** $2.144 \times 10^7$   **38.** $5.6 \times 10^3$   **39.** $2.36 \times 10^{-2}$   **40.** $4.73 \times 10^{-3}$   **41.** $7 \times 10^2$   **42.** $6.5 \times 10^{-5}$   **43.** $7 \times 10^2$
**44.** $8 \times 10^{-3}$   **45.** $2.45 \times 10^1$   **46.** $3.2285 \times 10^{13}$ mi   **47.** $4.2 \times 10^2$ Å

**4**   **48.** $5.7 \times 10^8$   **49.** $2 \times 10^{-5}$   **50.** $3.72 \times 10^{-8}$   **51.** $5.0 \times 10^8$   **52.** $3.6 \times 10^{-19}$   **53.** $4 \times 10^8$   **54.** $5 \times 10^{10}$
**55.** $7 \times 10^{-1}$   **56.** $2.7 \times 10^9$   **57.** $5.9 \times 10^7$   **58.** $2.9 \times 10^8$   **59.** $1.9 \times 10^{17}$   **60.** $2 \times 10^0$   **61.** $2 \times 10^4$   **62.** $1 \times 10^9$
**63.** $2.9 \times 10^8$   **64.** $2.3 \times 10^{-9}$

## Chapter Review Exercises, Chapter 4

**1.** 5   **3.** 7   **5.** 12   **7.** 2   **9.** $-87$   **11.** $-7$   **13.** $-6$   **15.** gain of 8 yd; $+8$ yd   **17.** $569   **19.** $-13$
**21.** 14   **23.** $-15$   **25.** $70°$   **27.** $-5.4°F$   **29.** $36°C$   **31.** $-14$   **33.** 0   **35.** $-168$   **37.** 343   **39.** $-16$
**41.** 25   **43.** $-10°$   **45.** higher, 20, $-4 \times 5 = -20$ points   **47.** 4   **49.** 4   **51.** 17   **53.** undefined   **55.** $-3$

**57.** 56   **59.** 1   **61.** 73   **63.** $1\dfrac{5}{7}$   **65.** $-62$   **67.** $-1\dfrac{7}{24}$   **69.** $-11\dfrac{13}{16}$   **71.** 23.76   **73.** 1,000,000,000,000

**75.** $\dfrac{1}{1,000}$ or 0.001   **77.** 8,730   **79.** 375,000   **81.** 0.0000387   **83.** $5.2 \times 10^4$   **85.** $1.7 \times 10^{-4}$   **87.** $8 \times 10^{-9}$

**89.** $3.4 \times 10^{10}$ (rounded to tenths)   **91.** $5.7 \times 10^{-4}$   **93.** $2.5 \times 10^8$

## Practice Test, Chapter 4

**1.** $<$   **3.** $>$   **5.** $-8$   **7.** 4   **9.** $-48$   **11.** 2.1   **13.** 0   **15.** undefined   **17.** $-2$   **19.** 4   **21.** $10^6$
**23.** 42,000   **25.** 0.059   **27.** $2.1 \times 10^{-2}$   **29.** $7.83 \times 10^{-3}$   **31.** $3.5 \times 10^2$   **33.** $1.5 \times 10^6$ ohms   **35.** $175°F$

## Cumulative Practice Test for Chapters 1–4

**1.** 14   **3.** $P = 128$ cm   **5.** 42.82   **7.** Factored form $= (2)(2)(3)(5)(7)$; Exponential notation $= 2^2(3)(5)(7)$

**9.** $19\dfrac{1}{8}$   **11.** 2   **13.** $1\dfrac{4}{17}$   **15.** $R = 83\%$ (rounded)   **17.** Percent of increase $= 7.14\%$

**19.** 1.43 m   **21.** $30°C$   **23.** $\dfrac{1}{8}$ cm   **25.** 30.2 cm   **27.** $-14$   **29.** $-6$   **31.** $-1\dfrac{1}{4}$
**33.** 0.053   **35.** $2 \times 10^6$

## Section 5–1 Self-Study Exercises

**1**   **1.** $-15 = -15$   **2.** $11 = 11$   **3.** $2 = 2$   **4.** $-7 = -7$   **5.** $n = 11$   **6.** $m = 6$   **7.** $y = -4$   **8.** $x = 6$
**9.** $p = 9$   **10.** $b = -9$   **11.** $\boxed{7} + \boxed{c}$   **12.** $\boxed{4a} - \boxed{7}$   **13.** $\boxed{3x} - \boxed{2(x+3)}$   **14.** $\boxed{\dfrac{a}{3}}$

**15.** $\boxed{7xy} + \boxed{3x} - \boxed{4} + \boxed{2(x+y)}$   **16.** $\boxed{14x} + \boxed{3}$   **17.** $\boxed{\dfrac{7}{(a+5)}}$   **18.** $\boxed{\dfrac{4x}{7}} + \boxed{5}$

**19.** $\boxed{11x} - \boxed{5y} + \boxed{15xy}$ (Answers may vary.)   **20.** 5   **21.** $-4$   **22.** $\dfrac{1}{5}$   **23.** $\dfrac{2}{7}$   **24.** 6   **25.** $-\dfrac{4}{5}$

**26.** $-15c$ (Answers may vary.)

**2**   **27.** Four more than a number is 7. (Answers may vary.)   **28.** Five less than a number is 2. (Answers may vary.)
**29.** Three times a number is 15. (Answers may vary.)   **30.** One more than 3 times a number is 7. (Answers may vary.)

**3**   **31.** $x + 5 = 12$   **32.** $\dfrac{x}{6} = 9$   **33.** $4(x - 3) = 12$   **34.** $4x - 3 = 12$   **35.** $12 + 7 + x = 17$   **36.** $2x + 7 = 21$
**37.** $x + 15° = 48°$   **38.** $x + 15 \text{ mL} = 45 \text{ mL}$   **39.** $x \cdot 5 = 45$   **40.** $\dfrac{18 - 3}{5} = x$

**4**   **41.** $-3a$   **42.** $-8x - 2y$   **43.** $7x - 2y$   **44.** $-3a + 6$   **45.** 10   **46.** $3a + 6b + 8c + 1$   **47.** $10x - 20$
**48.** $6x + 13$   **49.** $-4x + 5$   **50.** $-12a + 3$

## Section 5–2 Self-Study Exercises

**1**   **1.** $x = 8$   **2.** $x = 15$   **3.** $x = -3$   **4.** $x = -8$   **5.** $x = 4$   **6.** $x = -11$   **7.** $x = -26$   **8.** $x = 3$   **9.** $x = 5$
**10.** $x = -7$   **11.** $x = 16$   **12.** $x = 9$   **13.** $x = -2$   **14.** $x = 4$

**2**   **15.** $x = 8$   **16.** $x = 9$   **17.** $x = 5$   **18.** $n = 30$   **19.** $a = -9$   **20.** $b = -2$   **21.** $c = 7$   **22.** $x = -9$
**23.** $x = \dfrac{32}{5}$   **24.** $y = \dfrac{9}{2}$   **25.** $n = -\dfrac{5}{2}$   **26.** $b = \dfrac{28}{3}$   **27.** $x = -7$   **28.** $y = 2$   **29.** $x = 15$   **30.** $y = 8$
**31.** $x = \dfrac{10}{3}$   **32.** $y = -\dfrac{20}{3}$   **33.** $x = 12$   **34.** $n = 56$

**3**   **35.** $x = 6$   **36.** $m = 7$   **37.** $a = -3$   **38.** $m = -\dfrac{1}{2}$   **39.** $y = 8$   **40.** $x = 0$   **41.** $y = -7$   **42.** $y = -4$

**4**   **43.** $x = -6$   **44.** $b = -1$   **45.** $x = 8$   **46.** $t = 6$   **47.** $x = 3$   **48.** $y = 5$   **49.** $x = -\dfrac{7}{2}$   **50.** $a = -2$

**51.** all real numbers   **52.** no solution   **53.** $x = -8$   **54.** $t = 7$   **55.** $x = -1$   **56.** $y = 11$   **57.** $y = -2$
**58.** $x = 27$   **59.** $y = 13$   **60.** $R = -68$   **61.** $P = \dfrac{5}{6}$   **62.** $x = 14$   **63.** $x = -\dfrac{4}{15}$   **64.** $m = \dfrac{49}{18}$   **65.** $s = \dfrac{8}{5}$
**66.** $m = \dfrac{8}{3}$   **67.** $T = \dfrac{68}{9}$   **68.** $x = 2$   **69.** $R = 0.8$   **70.** $x = 6.03$   **71.** $x = 0.08$   **72.** $R = 0.34$

## Section 5–3 Self-Study Exercises

**1**   **1.** $x = 7$   **2.** $y = 2$   **3.** $x = -5$   **4.** $x = -2$   **5.** $x = 5$   **6.** $x = -6$   **7.** $x = 1$   **8.** $x = 8$   **9.** $x = 1$
**10.** $x = 0$   **11.** no solution   **12.** all real numbers   **13.** no solution   **14.** no solution   **15.** $x = 2$   **16.** $x = 5$
**17.** $x = \dfrac{1}{2}$   **18.** $x = -15$   **19.** $x = -9$   **20.** $x = 1$   **21.** $x = -2$   **22.** $x = -1$   **23.** $x = \dfrac{16}{5}$   **24.** $x = 5$

**25.** $x + 4 = 12; x = 8$   **26.** $2x - 4 = 6; 5$   **27.** $x + x + (x - 3) = 27$; two parts weigh 10 lb; third part weighs 7 lb
**28.** $24 + x = 60; 36$ gal   **29.** $4.03 - 3.97 = x$; 0.06 kg   **30.** $x + 3x = 400$; 100 ft of solid pipe; 300 ft of perforated
pipe   **31.** $x + 2x + \$70 = \$235$; Spanish text: \$55; calculator: \$110   **32.** $C = \$14(4)(150) = \$8,400$   **33.** $x + 2x = 325$;

tank 1: $216\dfrac{2}{3}$ gal   tank 2: $108\dfrac{1}{3}$ gal   **34.** $w = 6$ ft

## Section 5–4 Self-Study Exercises

**1**   **1.** 28   **2.** $-11$   **3.** $\dfrac{18}{7}$   **4.** 0   **5.** $-\dfrac{27}{5}$   **6.** $-\dfrac{5}{27}$   **7.** $\dfrac{8}{5}$   **8.** $\dfrac{11}{15}$   **9.** $-350$   **10.** 48   **11.** 27

**12.** $\dfrac{1}{3}; Q \neq 0$   **13.** 13   **14.** $\dfrac{25}{8}; p \neq 4$   **15.** 0   **16.** $\dfrac{1}{9}$   **17.** 576   **18.** 4   **19.** 28   **20.** $-68$   **21.** $\dfrac{27}{35}$

**22.** $\dfrac{108}{5}$   **23.** $\dfrac{217}{24}$   **24.** 8   **25.** $-5; R \neq 0$   **26.** $\dfrac{3}{10}$   **27.** 70   **28.** 0   **29.** $\dfrac{32}{63}$   **30.** $\dfrac{32}{5}$   **31.** $-\dfrac{36}{11}$

**32.** $-\dfrac{1}{108}; x \neq 0$   **33.** 21   **34.** $\dfrac{4}{25}$   **35.** $\dfrac{2}{21}$   **36.** $\dfrac{9}{7}$

**2**   **37.** 80 vases   **38.** $1\dfrac{1}{3}$ h   **39.** $3\dfrac{1}{13}$ h   **40.** $3\dfrac{3}{7}$ days   **41.** $\dfrac{2}{5}$ h or 24 min   **42.** 3 min   **43.** 15 min

**44.** $2\dfrac{1}{3}$ min   **45.** 8.57 Ω   **46.** 5 Ω   **47.** 7.38 Ω   **48.** 3.6 Ω

**3**   **49.** 2   **50.** 0.8   **51.** 6.03   **52.** 16   **53.** 4.7   **54.** 0.08   **55.** 0.035   **56.** 0.34   **57.** $-3.3$   **58.** 38.8
**59.** 3.3   **60.** $-0.2$   **61.** 0.6   **62.** 1.128   **63.** 16   **64.** 87.5 lb   **65.** 4.71 in.   **66.** 12 h   **67.** 6.5 h   **68.** 8.3 Ω
**69.** $11.30   **70.** 156.25 V   **71.** $264   **72.** $136   **73.** $1,925   **74.** 12.25%   **75.** 1.2%

## Section 5–5 Self-Study Exercises

**1**   **1.** 280 mi   **2.** $120   **3.** 2,826 in$^2$   **4.** 117.6 m$^2$   **5.** 138 cm   **6.** $310   **7.** 14.25%   **8.** 3 years
**9.** $2,600   **10.** 90 lb   **11.** 8%   **12.** $3,378.38   **13.** 13.6   **14.** 17%   **15.** 66 in.   **16.** 9,326.6 cm$^2$
**17.** $10.50   **18.** 4 mi   **19.** 153.9 in.$^2$   **20.** 6.25 km$^2$   **21.** $48.75   **22.** 4.8 Ω   **23.** 61.6%   **24.** 35 mi/h
**25.** 1.5 A   **26.** 6 ft$^3$   **27.** 3 A   **28.** 2 cylinders   **29.** 1,600 rpm   **30.** 71.71 in.   **31.** 17 Ω

**2**   **32.** $S = I - I_n$   **33.** $r = \dfrac{S}{2\pi h}$   **34.** $b = y - mx$   **35.** $h = \dfrac{V}{\pi r^2}$   **36.** $C = S - M$   **37.** $R = \dfrac{100P}{B}$

**38.** $b = \dfrac{P - 2s}{2}$   **39.** $r = \dfrac{C}{2\pi}$   **40.** $l = \dfrac{A}{w}$   **41.** $C = \dfrac{R - B}{A}$   **42.** $R = \dfrac{E}{I}$   **43.** $R = \dfrac{D}{T}$   **44.** $D = P - S$

**45.** $d = \dfrac{C}{\pi}$   **46.** $r = \dfrac{C}{2\pi}$   **47.** $I = A - P$   **48.** $T = \dfrac{I}{PR}$

## Chapter Review Exercises, Chapter 5

**1.** $12 = 12$   **3.** $8 = 8$   **5.** $x = 14$   **7.** $y = -8$   **9.** $\boxed{15x} - \boxed{\dfrac{3a}{7}} + \boxed{\dfrac{(x-7)}{5}}$
**11.** A number increased by 5 equals 2.

Answers will vary.   **13.** The quotient of a number and 8 equals 7. Answers will vary.   **15.** $2x + 7 = 11$
**17.** $2(x + 8) = 40$   **19.** $5a$   **21.** $7y - 12$   **23.** $-3a - 11$   **25.** $x = 13$   **27.** $x = -2$   **29.** $x = 19$   **31.** $x = 7$

**33.** $b = -\dfrac{15}{2}$   **35.** $x = 15$   **37.** $x = 5$   **39.** $x = -4$   **41.** $x = 2$   **43.** no solution   **45.** $y = -3$   **47.** $x = \dfrac{20}{13}$

**49.** $x = -9$   **51.** $x = 3$   **53.** $x = 2$   **55.** $x = -4$   **57.** $x = 0$   **59.** $x = 3$   **61.** $x = 3$   **63.** $x = -1$

**65.** $x = -\dfrac{1}{8}$   **67.** $5(x + 6) = 42 + x; x = 3$   **69.** $x + (x - 3) = 51; 27$ h, 24 h   **71.** $720 = 2(w + 2w); w = 120$ ft;

$l = 2w = 240$ ft   **73.** $\dfrac{1}{2}$   **75.** $\dfrac{5}{6}$   **77.** $-\dfrac{4}{15}$   **79.** $\dfrac{49}{18}$   **81.** $\dfrac{8}{3}$   **83.** 6   **85.** $2\dfrac{1}{10}$ h   **87.** 25 fixtures   **89.** 1.33 Ω

**91.** 2   **93.** $-11.8$   **95.** 17 Ω   **97.** 2 years   **99.** 2.75 A   **101.** 7 A   **103.** $w = \dfrac{V}{lh}$   **105.** $r = s + d$

**107.** $t = \dfrac{v_0 - v}{32}$   **109.** $P = S + D$

## Practice Test, Chapter 5

**1.** 14     **3.** 10     **5.** 3     **7.** 2     **9.** $-\dfrac{22}{7}$     **11.** $\dfrac{7}{25}$     **13.** $\dfrac{10}{21}$     **15.** 6.17     **17.** 254.24     **19.** $-0.15$     **21.** $I = 2.2\,\text{A}$

**23.** $L = \dfrac{AR}{P}$     **25.** $s = \dfrac{d}{\pi r^2 n}$     **27.** 12,000 cal

## Section 6–1 Self-Study Exercises

**1 1.** 3     **2.** 2     **3.** 29     **4.** $\dfrac{5}{8}$ or 0.628     **5.** $\dfrac{34}{5}$ or 6.8     **6.** $\dfrac{17}{14}$ or 1.214     **7.** 3     **8.** 21     **9.** 4     **10.** $-\dfrac{4}{5}$ or $-0.8$

**11.** $x = 1$     **12.** $x = \dfrac{9}{5}$ or 1.8     **13.** $x = 30$     **14.** $x = 2.698$     **15.** $x = 836.9$     **16.** $x = -2$     **17.** $x = 2.835$

**18.** $x = 1.174$     **19.** $x = 2.769$     **20.** $x = 3.25$     **21.** $x = 214.7$     **22.** $x = 1.438$     **23.** $x = 29.18$     **24.** $x = 10^2$ or 100     **25.** $x = 10^{12}$     **26.** $x = 1.875$     **27.** $x = 1.25$     **28.** $x = 2.556$     **29.** $x = 10$ in.     **30.** $x = 22.5$ ft **31.** $x = 9\,\text{k}\Omega$     **32.** $x = 257.1\,\text{W}$     **33.** $x = 0.26\,\text{V}$     **34.** $x = 0.7317\,\text{mA}$     **35.** $x = 1.346$     **36.** $x = 1$

**37.** $x = \dfrac{5}{6}$ or 0.8333     **38.** $x = \dfrac{5}{24}$ or 0.2083     **39.** $x = \dfrac{10}{7}$ or $1\dfrac{3}{7}$ or 1.429     **40.** $x = \dfrac{28}{5}$ or 5.6     **41.** $x = 8\,\text{h}$

**42.** $x = 0.76\,\text{h}$     **43.** $x = \$98$     **44.** $x = \$95.19$     **45.** $x = \$0.21$     **46.** $x = 0.375\,\text{gal}$     **47.** $x = 44.87\,\text{mg}$ **48.** $x = 5.833\,\text{mg}$     **49.** $x = 0.25\,\text{gal}$     **50.** $x = 132\,\text{gal}$     **51.** $x = 555.6\,\text{mi}$     **52.** $x = 200\,\text{mg}$     **53.** $x = 8.333\,\text{mg}$ **54.** $x = 8.036\,\text{mg}$

## Section 6–2 Self-Study Exercises

**2 1.** $\$5.90$     **2.** $\$74.13$     **3.** 48 engines     **4.** 15 headpieces     **5.** 1,425 mi     **6.** 3,360 mi     **7.** 550 lb     **8.** 1.7 gal **9.** $\$0.0019$ per gram     **10.** $\$0.0022$ per gram     **11.** larger can has lower cost per gram     **12.** 320 bottles per hour **13.** 12 teeth     **14.** 32 teeth     **15.** 4 cm     **16.** 384 cm     **17.** 7 ft     **18.** 6 ft     **19.** 10 ft     **20.** 14 ft     **21.** 355 mg

**22.** 277.5 mg     **23.** 10 mg     **24.** 20 mL     **25.** 15 ft     **26.** $11\dfrac{2}{3}$ ft     **27.** 30 in.     **28.** 40 in.     **29.** $0.02325\,\Omega$

**30.** $0.576\,\Omega$     **31.** 48 mi     **32.** 35 mi     **33.** 23 mi     **34.** 30 in.     **35.** 12.5 in.     **36.** 7.5 in.

**37.**

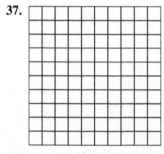

10 × 10

**38.**

4 × 3

**39.**

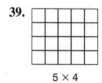

5 × 4

**40.**

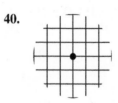

Diameter = 6 squares

**41.**

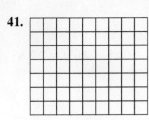

$9 \times 7$

**42.** 1 mL    **43.** 250 mL    **44.** 0.4 mL    **45.** 15 oz    **46.** 12.5 V

**2**    **47.** $\angle A = \angle F, \angle B = \angle E, \angle C = \angle D$    **48.** $PQ = ST, \angle P = \angle S, \angle Q = \angle T$    **49.** $JL = MP, LK = NP, \angle L = \angle P$

**50.** $a = 20, d = 9$    **51.** $DC = 6, DE = 5\frac{1}{3}$    **52.** 65 ft

## Section 6–3 Self-Study Exercises

**1**    **1.** 360    **2.** 960    **3.** 900 rpm    **4.** 200 in³    **5.** 20 in.    **6.** 18 painters    **7.** 10 h    **8.** 50 rpm    **9.** 5 in.
**10.** 4 h    **11.** 5 machines    **12.** 120 rpm    **13.** 4 helpers (5 workers total)    **14.** 9 in.    **15.** 240 rpm    **16.** 225 rpm
**17.** 700 cm    **18.** 4 cm    **19.** 36 cm    **20.** 600 rpm    **21.** 400 rpm    **22.** 140 rpm    **23.** 300 teeth    **24.** 35 teeth
**25.** 37.5 rpm    **26.** 21 teeth    **27.** 10 rpm    **28.** 38 rpm    **29.** 180 teeth    **30.** 108 teeth

## Chapter Review Exercises, Chapter 6

**1.** $\frac{7}{6}$    **3.** $\frac{1}{2}$    **5.** $\frac{49}{12}$    **7.** $-\frac{21}{2}$    **9.** $\frac{3}{14}$    **11.** $-\frac{21}{23}$    **13.** 4,500 women    **15.** $4\frac{1}{5}$ ft    **17.** 129.7 gal    **19.** 5.0 h

**21.** 1,351 ft    **23.** $\frac{AB}{RT} = \frac{BC}{ST} = \frac{AC}{RS}$    **25.** 10    **27.** 23 machines    **29.** $2\frac{1}{2}$ h    **31.** 1,500 rpm    **33.** 3 days

**35.** 60 rpm

## Practice Test, Chapter 6

**1.** $\frac{15}{2}$    **3.** 9,600    **5.** $\frac{32}{5}$    **7.** 200 rpm    **9.** 168.75 rpm    **11.** $x = \$88.74$    **13.** 54.7 L    **15.** $x = 8.8$ A

**17.** 107 lb    **19.** 5

## Section 7–1 Self-Study Exercises

**1**    **1.** *R:* horizontal, 4; vertical, 2    **2.** *S:* horizontal, −4; vertical, 3    **3.** *T:* horizontal, 5; vertical, −3
**4.** *U:* horizontal, −2; vertical, −5    **5.** *V:* horizontal, 7; vertical, 0    **6.** *W:* horizontal, 0; vertical, 5
**7.** *X:* horizontal, 0; vertical, 0    **8.** *Y:* horizontal, −3; vertical, 0
**9.** $A = (4, 2)$   $B = (−3, 2)$   $C = (−2, −1)$   $D = (3, −2)$
**16.** *y*-value = 0    **17.** *x*-value = 0    **18.** $(−x, +y)$    **19.** $(+x, −y)$

**10.–15.**

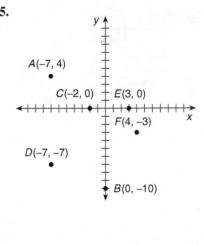

**20.**

| $x$ | $y$ |
|---|---|
| $-2$ | $-13$ |
| $-1$ | $-10$ |
| $0$ | $-7$ |
| $1$ | $-4$ |
| $2$ | $-1$ |

**21.**

| $x$ | $y$ |
|---|---|
| $-1$ | $5$ |
| $0$ | $3$ |
| $1$ | $1$ |
| $2$ | $-1$ |

**22.**

| $x$ | $y$ |
|---|---|
| $-2$ | $13$ |
| $-1$ | $9$ |
| $0$ | $5$ |
| $1$ | $1$ |
| $2$ | $-3$ |
| $3$ | $-7$ |
| $4$ | $-11$ |

**23.**

| $x$ | $y$ |
|---|---|
| $-6$ | $0$ |
| $-2$ | $2$ |
| $0$ | $3$ |
| $2$ | $4$ |
| $6$ | $6$ |

**24.**

| $x$ | $y$ |
|---|---|
| $-6$ | $3$ |
| $-3$ | $1$ |
| $0$ | $-1$ |
| $3$ | $-3$ |
| $6$ | $-5$ |

**25.**

| $x$ | $y$ |
|---|---|
| $-1$ | $-3$ |
| $0$ | $2$ |
| $1$ | $7$ |

**26.**

| $x$ | $y$ |
|---|---|
| $-2$ | $3$ |
| $0$ | $2$ |
| $2$ | $1$ |
| $4$ | $0$ |

**27.**

| $x$ | $y$ |
|---|---|
| $-6$ | $0$ |
| $-3$ | $1$ |
| $0$ | $2$ |
| $3$ | $3$ |
| $6$ | $4$ |

**3**

**28.**
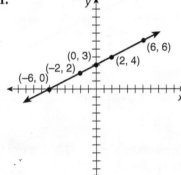

**29.**
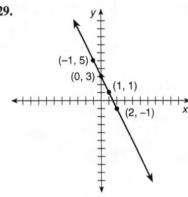

**30.**
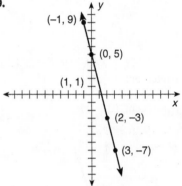

**31.**
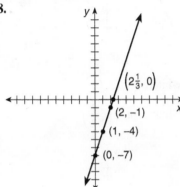

**32.**
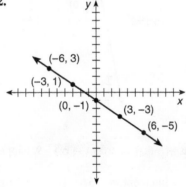

**33.**
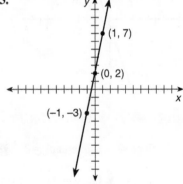

**34.**
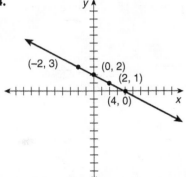

**35.**
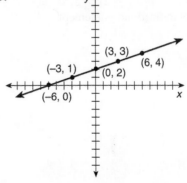

**36.** no  **37.** yes  **38.** yes  **39.** no  **40.** yes  **41.** yes  **42.** yes  **43.** no  **44.** yes  **45.** no  **46.** yes
**47.** yes  **48.** (8, 5)  **49.** (2, 5)  **50.** (−6, 1)  **51.** (21, 3)  **52.** (5, 15)  **53.** (2, 0)  **54.** (1, 3); (1, −2); (1, 0); (1, 1)  **55.** (−1, 4); (3, 4); (0, 4); (2, 4)  **56.** $12  **57.** $2  **58.** $34  **59.** $13,800

## Section 7–2 Self-Study Exercises

**1.** $x$-intercept, (5, 0)
$y$-intercept, (0, 5)

**2.** $x$-intercept, (5, 0)
$y$-intercept, $\left(0, \frac{5}{3}\right)$

**3.** $x$-intercept, (−4, 0)
$y$-intercept, (0, 8)

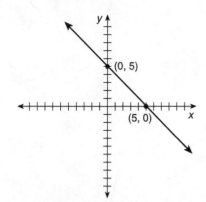

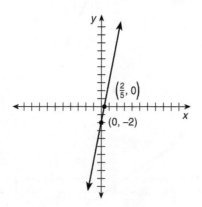

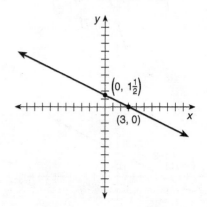

**4.** $x$-intercept, $\left(\frac{1}{3}, 0\right)$
$y$-intercept, (0, −1)

**5.** $x$-intercept, $\left(\frac{2}{5}, 0\right)$
$y$-intercept, (0, −2)

**6.** $x$-intercept, (3, 0)
$y$-intercept, $\left(0, \frac{3}{2}\right)$

**7.** slope = 4; $y$-intercept = 3  **8.** slope = −5; $y$-intercept = 6  **9.** slope = $-\frac{7}{8}$; $y$-intercept = −3  **10.** slope = 0;
$y$-intercept = 3  **11.** $y = 2x - 5$; slope = 2; $y$-intercept = −5  **12.** $y = \frac{5}{2}$; slope = 0; $y$-intercept = $\frac{5}{2}$

**13.** $y = -2x + 6$; slope = −2; $y$-intercept = 6  **14.** $y = 2x + 5$; slope = 2; $y$-intercept = 5  **15.** $x = 4$; slope not defined; no $y$-intercept  **16.** $x = 9$; slope not defined; no $y$-intercept

**17.** slope $= \dfrac{2}{1}$; y-intercept, $(0, -3)$ **18.** slope $= -\dfrac{1}{2}$; y-intercept, $(0, -2)$ **19.** slope $= -\dfrac{3}{5}$; y-intercept, $(0, 0)$

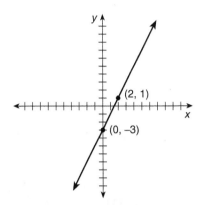 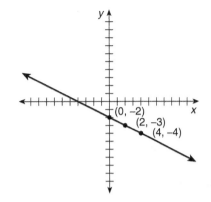 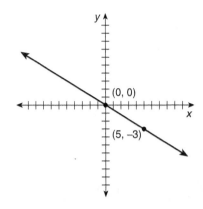

**20.** slope $= \dfrac{1}{2}$; y-intercept, $\left(0, -1\dfrac{1}{2}\right)$ **21.** slope $= -\dfrac{2}{1}$; y-intercept, $(0, 1)$ **22.** slope $= \dfrac{3}{4}$; y-intercept, $(0, 2)$

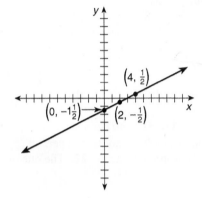 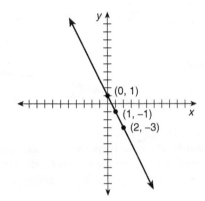 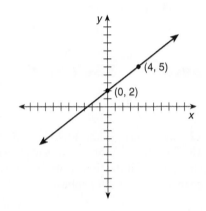

**3**

**23.** **24.** 0.38 **25.**

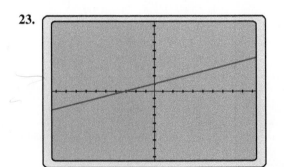 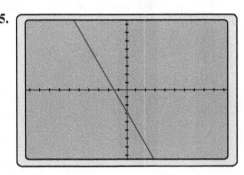

**26.** $-5.08$ **27.** **28.** 1,817

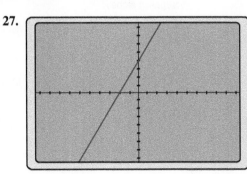

**29.**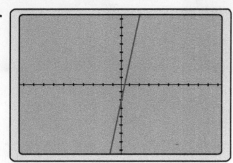

**30.** 136

**31.** \$35,000; \$95,000    **32.** \$332; \$17,264    **33.** 195; 155    **34.** 1,096 puzzles

4  **35.** $y = 3x - 5$

$x = \dfrac{5}{3}$ or 1.66667

**36.** $y = 5x - 6$

$x = 1.2$

**37.** $y = 6x - 5$

$x = 0.83333$

**38.** $y = x - 1$

$x = 1$

**39.** $y = x + 9$

$x = -9$

**40.** $y = 2x + 16$

$x = -8$

## Section 7–3 Self-Study Exercises

1  **1.** $\dfrac{1}{2}$  **2.** $-\dfrac{5}{3}$  **3.** 3  **4.** 1  **5.** $\dfrac{3}{4}$  **6.** 0  **7.** $-\dfrac{5}{4}$  **8.** $-1$  **9.** undefined  **10.** $-3$  **11.** 0

**12.** not defined    **13.** \$1,400    **14.** \$10    **15.** 2,000 ft/min    **16.** \$113.20 per year    **17.** \$123.40 per year
**18.** \$316.00 per year    **19.** \$32.50 per year    **20.** $-$\$2.00 per year    **21.** The data in this table do not form a perfect straight line. If the slope (rate of change) remains the same for such data, the graph will be a straight line.    **22.** The rate of change is greater for public 4-year colleges than it is for public 2-year colleges.

2

**23.**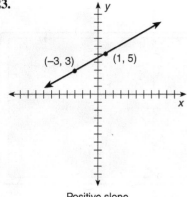

Positive slope

**24.**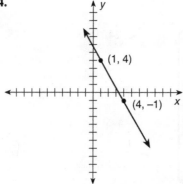

Negative slope

**25.**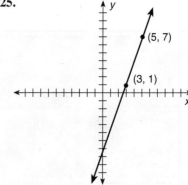

Positive slope

**26.**

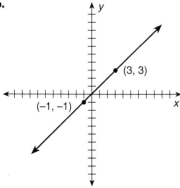

(3, 3)

(−1, −1)

Positive slope

**27.**

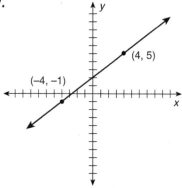

(4, 5)

(−4, −1)

Positive slope

**28.**

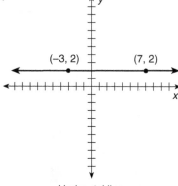

(−3, 2)    (7, 2)

Horizontal line,
zero slope

**29.**

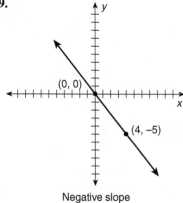

(0, 0)

(4, −5)

Negative slope

**30.**

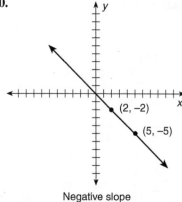

(2, −2)

(5, −5)

Negative slope

**31.**

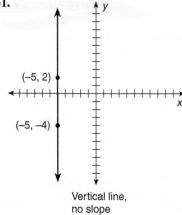

(−5, 2)

(−5, −4)

Vertical line,
no slope

**32.**

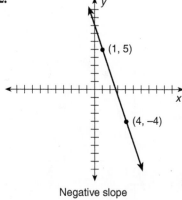

(1, 5)

(4, −4)

Negative slope

**33.**

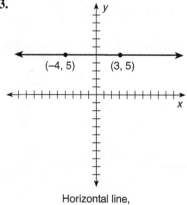

(−4, 5)    (3, 5)

Horizontal line,
zero slope

**34.**

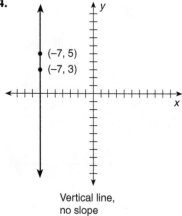

(−7, 5)
(−7, 3)

Vertical line,
no slope

## Section 7–4 Self-Study Exercises

1  **1.** $y = \frac{2}{3}x + \frac{25}{3}$   **2.** $y = -\frac{1}{2}x + 3$   **3.** $y = 2x + 1$   **4.** $y = x - 1$

2  **5.** $y = -\frac{5}{3}x + \frac{38}{3}$   **6.** $x = -1$   **7.** $y = -3$   **8.** $x = -4$   **9.** $y = 2x$   **10.** $y = 0$   **11.** $y = 5x + 1{,}935$

**12.** $y = 22x + 50{,}000$

**3**    **13.** $y = \frac{1}{4}x + 7$    **14.** $y = -8x - 4$    **15.** $y = -2x + 3$    **16.** $y = \frac{3}{5}x - 2$    **17.** $y = x$    **18.** $y = 5x - \frac{1}{5}$

**19.** $y = 2x - 2$    **20.** $y = -\frac{3}{4}x$    **21.** $y = 0.2x + 3$; $12    **22.** $y = 8x + 12{,}000$; $76,000    **23.** $y = -3x + 1$

**24.** $y = \frac{2}{3}x - 2$    **25.** $y = 4$    **26.** $y = 2x + 3$    **27.** $23

**4**    **28.** $x + y = -1$    **29.** $2x + y = 9$    **30.** $x - 3y = 4$    **31.** $3x - y = -7$    **32.** $x + 3y = 7$    **33.** $3x - y = -3$
**34.** $2x + 3y = 5$    **35.** $3x + 2y = 6$    **36.** $2x - 5y = 11$    **37.** $6x - 8y = 3$    **38.** $y = 12x + 20{,}000$
**39.** $y = 0.5x + 45$

**5**    **40.** $x - y = -1$    **41.** $x - 2y = -13$    **42.** $3x + y = 12$    **43.** $x + 3y = -9$    **44.** $3x - y = 11$
**45.** $2x - y = -7$    **46.** $3x - 2y = 11$    **47.** $x - 4y = 0$    **48.** $2x - 2y = -3$    **49.** $x + 5y = 4$

## Chapter Review Exercises, Chapter 7

**1.–6.** See figure to the right.

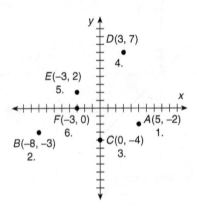

**7.** $(0, 0)$    **9.** $A(3, 0)$; $B(2, 2)$; $C(2, -5)$; $D(-4, -1)$; $E(-3, 1)$

Since plotted solutions will vary, check graphs by comparing $x$- and $y$-intercepts.

**11.**

| $x$ | $y$ |
|---|---|
| $-1$ | $-5$ |
| $1$ | $-1$ |
| $3$ | $3$ |

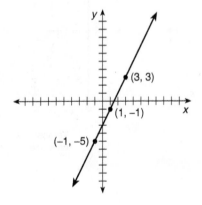

**13.**

| $x$ | $y$ |
|---|---|
| $-1$ | $-3$ |
| $0$ | $0$ |
| $1$ | $3$ |

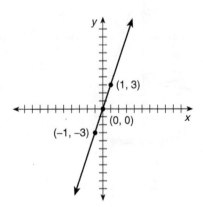

**15.**

| $x$ | $y$ |
|---|---|
| $-1$ | $3$ |
| $0$ | $0$ |
| $1$ | $-3$ |

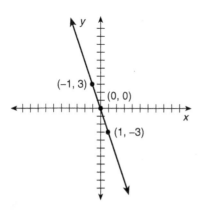

**17.**

| $x$ | $y$ |
|---|---|
| $-1$ | $-1$ |
| $0$ | $1$ |
| $1$ | $3$ |

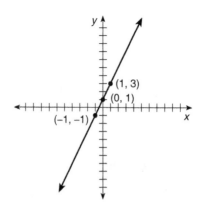

**19.**

| $x$ | $y$ |
|---|---|
| $-2$ | $-10$ |
| $-1$ | $-6$ |
| $0$ | $-2$ |
| $1$ | $2$ |
| $3$ | $10$ |

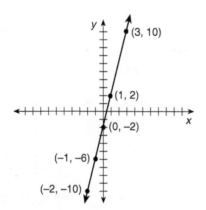

**21.**

| $x$ | $y$ |
|---|---|
| $-2$ | $-3$ |
| $0$ | $-2$ |
| $2$ | $-1$ |
| $4$ | $0$ |

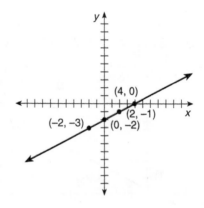

**23.** no    **25.** yes    **27.** yes    **29.** $y = -7$    **31.** $y = 1$

**33.** $x$-intercept, $(-1, 0)$;

$y$-intercept, $\left(0, -\dfrac{1}{4}\right)$

**35.** $3x - y = 1$; $x$-intercept,

$\left(\dfrac{1}{3}, 0\right)$; $y$-intercept, $(0, -1)$

**37.** $\dfrac{1}{2}x + \dfrac{1}{3}y = 1$

$x$-intercept $(2, 0)$; $y$-intercept $(0, 3)$

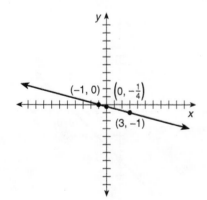

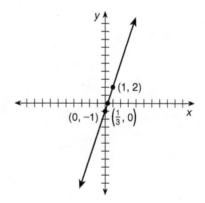

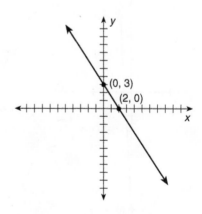

**39.** $y = 5x - 2; m = \dfrac{5}{1}; b = -2$

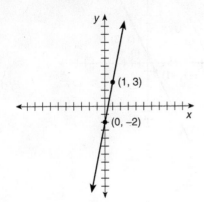

**41.** $y = -3x - 1; m = \dfrac{-3}{1}; b = -1$

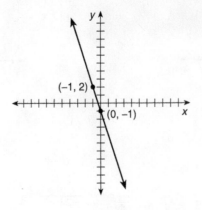

**43.** $y = x - 4; m = \dfrac{1}{1}; b = -4$

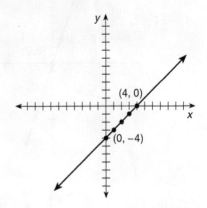

**45.** $y = \dfrac{1}{2}x + \dfrac{1}{2}; m = \dfrac{1}{2}; b = \dfrac{1}{2}$

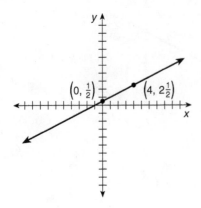

**47.** $y = 2x - 2; m = 2; b = -2$

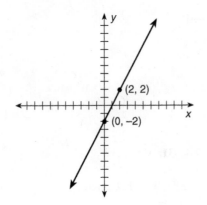

**49.** $y = 0.5x - 3; m = 0.5 = \dfrac{1}{2}; b = -3$

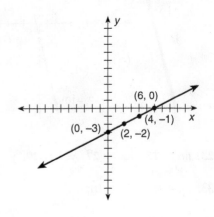

**51.**

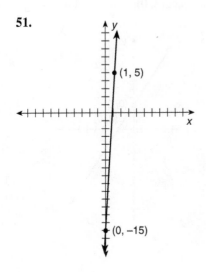

**53.** $58,000

**55.** $y = x - 6$

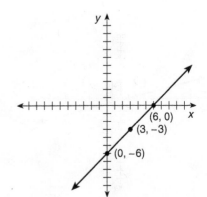

**57.** $y = 3x + 6$

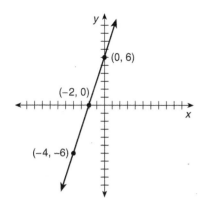

**59.** $y = 2x - 2$

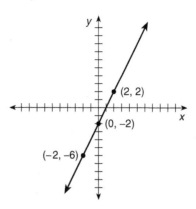

**61.** $m = 3; b = \dfrac{1}{4}$   **63.** $m = -5; b = 4$   **65.** undefined slope; no $y$-intercept   **67.** $m = \dfrac{1}{8}; b = -5$

**69.** $y = -2x + 8; m = -2; b = 8$   **71.** $y = \dfrac{3}{2}x - 3; m = \dfrac{3}{2}; b = -3$   **73.** $y = \dfrac{3}{5}x - 4; m = \dfrac{3}{5}; b = -4$

**75.** $y = \dfrac{5}{3}; m = 0; b = \dfrac{5}{3}$   **77.** $\dfrac{1}{3}$   **79.** 2   **81.** $\dfrac{5}{8}$   **83.** $-\dfrac{4}{3}$   **85.** undefined   **87.** $-\dfrac{4}{7}$   **89.** $\dfrac{7}{2}$   **91.** 0

**93.** undefined   **95.** $(-1, 2)$ $(3, 2)$; answers will vary.   **97.** $-\$43$   **99.** \$124   **101.** The table of values is not a

perfect linear function.   **103.** $y = \dfrac{1}{3}x + 4$   **105.** $y = \dfrac{3}{4}x - 3$   **107.** $y = 4x - 5$   **109.** $y = -\dfrac{1}{11}x + \dfrac{17}{11}$

**111.** $y = \dfrac{7}{4}x - \dfrac{5}{4}$   **113.** $y = \dfrac{9}{5}x + \dfrac{3}{5}$   **115.** $y = x - 3$   **117.** $y = x - 1$   **119.** $y = -2$   **121.** $S = 10x + 3{,}000$

**123.** $y = 3x - 2$   **125.** $y = 2x - 2$   **127.** $x + y = 7$   **129.** $x - 2y = 8$   **131.** $x - 3y = 20$   **133.** $x + 3y = -10$
**135.** $3x - 4y = -7$   **137.** $x - y = -4$   **139.** $2x - y = -4$   **141.** $x - 5y = -11$   **143.** $2x + 10y = 31$
**145.** $x + 4y = 2$

## Practice Test, Chapter 7

Since plotted solutions will vary, check graphs by comparing $x$- and $y$-intercepts.

**1.**

| $x$ | $y$ |
|----|----|
| $-2$ | $-1$ |
| $0$ | $0$ |
| $2$ | $1$ |

 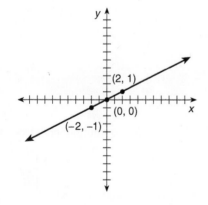

**3.**

| $x$ | $y$ |
|----|----|
| $-1$ | $-6$ |
| $0$ | $-4$ |
| $1$ | $-2$ |
| $2$ | $0$ |

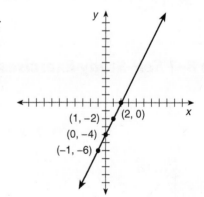

**5.** $y = 3x - 3; x = 1$

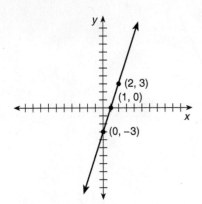

**7.** (2, 5)

**9.** $23,200

**11.** $x$-intercept $(-5, 0)$
$y$-intercept $(0, -5)$

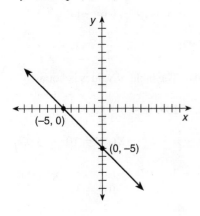

**13.** $x$-intercept $(8, 0)$
$y$-intercept $(0, 4)$

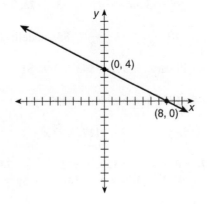

**15.** $y$-intercept $= (0, -3)$

slope $= \dfrac{-2}{1}$

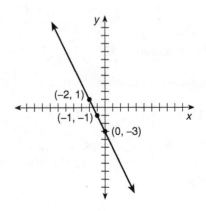

**17.** $-\dfrac{2}{3}$    **19.** $y = 2x + 34; m = 2; b = 34$    **21.** $y = \dfrac{1}{4}x; m = \dfrac{1}{4}; b = 0$    **23.** $y = \dfrac{2}{3}x - 7$    **25.** $y = \dfrac{2}{3}x + \dfrac{7}{3}$

**27.** $y = 2$    **29.** $y = \dfrac{3}{2}x + 3$    **31.** $2x + y = 5$    **33.** $x - 2y = 10$

## Section 8–1 Self-Study Exercises

**1.** $x = 7, y = 5$

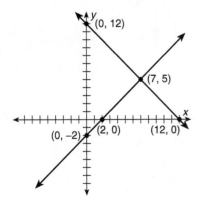

**2.** $x = 3, y = 3$

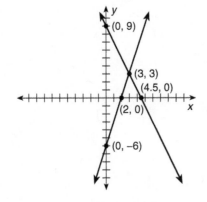

**3.** no solution; no intersection

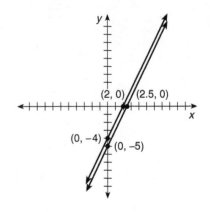

**4.** many solutions; lines coincide

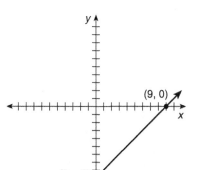

**5.** $x = 4, y = -2$

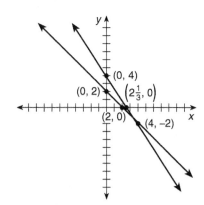

**6.** $x = 3, y = 1$

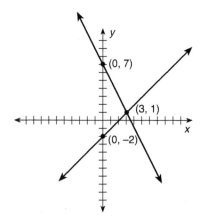

**7.** $x = 4, y = -3$

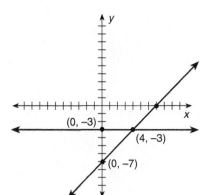

**8.** $x = -1, y = -5$

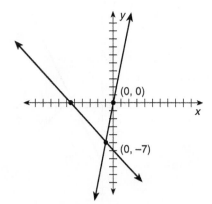

**9.** $x = 1, y = -2$

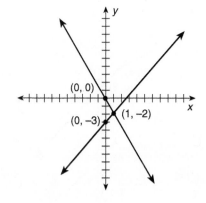

**10.** $x = 4, y = 1$

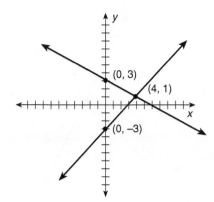

**11.** $x = 1, y = -7$

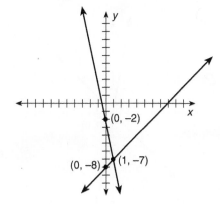

**12.** $x = 0, y = 3$

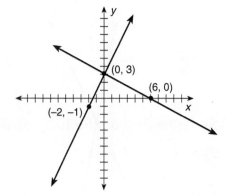

**13.** $x = 1, y = -5$

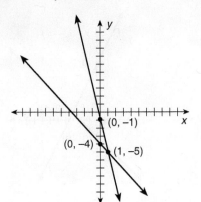

**14.** $x = 2, y = 1$

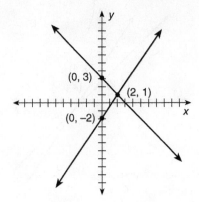

**15.** $x = 7, y = 9$

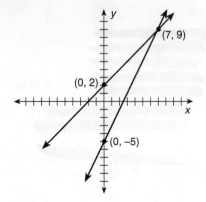

**16.** $x = -2, y = 4$

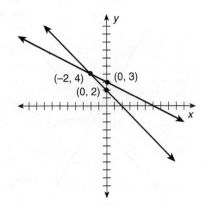

**17.** $x = 4, y = 1$

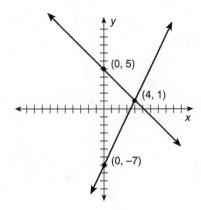

**18.** $x = 3, y = -4$

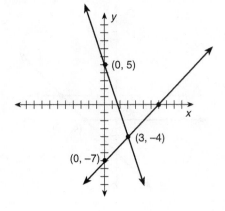

**19.** $x = 2, y = -2$

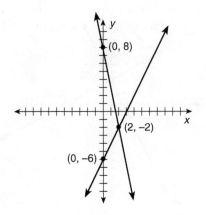

**20.** $x = 0, y = 0$

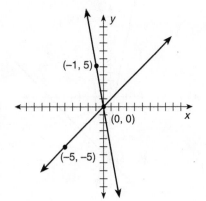

**21.** $x = 3, y = 1$

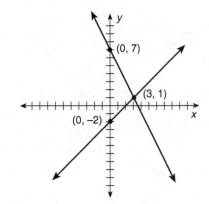

Selected Answers to Student Exercise Material

## Section 8–2 Self-Study Exercises

**1.** $a = 5, b = -1$    **2.** $m = 4, n = -1$    **3.** $x = 1, y = -1$    **4.** $a = 3, b = -3$    **5.** $x = 2, y = \dfrac{3}{2}$

**6.** $x = 4, y = -\dfrac{3}{5}$    **7.** $x = -\dfrac{13}{5}, y = 5$    **8.** $x = -\dfrac{11}{4}, y = 5$    **9.** $x = -4, y = 3$

**10.** $x = 3, y = 0$    **11.** $x = 1, y = 5$    **12.** $a = 2, b = \dfrac{8}{3}$    **13.** $x = -1, y = 2$    **14.** $a = 6, y = 1$

**15.** $x = 1, y = 3$    **16.** $x = 2, y = -2$    **17.** $x = -4, y = 1$    **18.** $x = 5, y = -2$    **19.** $x = 0, y = 3$
**20.** $x = -1, y = 3$    **21.** $x = 13, y = 3$

**22.** inconsistent; no solution    **23.** inconsistent; no solution    **24.** dependent; many solutions
**25.** dependent; many solutions    **26.** inconsistent; no solution    **27.** dependent; many solutions
**28.** inconsistent; no solution    **29.** inconsistent; no solution    **30.** inconsistent; no solution

## Section 8–3 Self-Study Exercises

**1.** $a = 20, b = 10$    **2.** $r = 5, c = 7$    **3.** $x = 13, y = 11$    **4.** $x = 7, y = 5$    **5.** $p = \dfrac{1}{2}, k = \dfrac{1}{3}$    **6.** $x = 3, y = 1$

**7.** $x = 13, y = 5$    **8.** $x = -2, a = -7$    **9.** $x = 1, y = 1$    **10.** $x = 1, y = 6$    **11.** $x = 8, y = 1$    **12.** $x = 4, y = 10$

**13.** $x = 3, y = 2$    **14.** $x = 3, y = 3$    **15.** $x = 1, y = \dfrac{3}{2}$    **16.** $x = \dfrac{1}{3}, y = \dfrac{1}{2}$    **17.** $x = \dfrac{1}{4}, y = \dfrac{2}{3}$    **18.** $x = -\dfrac{1}{2}, y = \dfrac{1}{2}$

**19.** $x = -2, y = 2$    **20.** $x = 2, y = -1$    **21.** $x = -8, y = 3$

## Section 8–4 Self-Study Exercises

**1.** short 15.5 in.
    long 32.5 in.

**2.** $18,000 at 4%
    $17,000 at 5%

**3.** $21 per shirt
    $16 per hat

**4.** seven 8-cylinder jobs
    three 4-cylinder jobs

**5.** resistor $0.25
    capacitor $0.30

**6.** rate of plane = 130 mi/h
    rate of wind = 10 mi/h

**7.** $3,000 at 7%
    $5,000 at 9%

**8.** rate of current = 1.67 mi/h
    rate of motorboat = 11.67 mi/h

**9.** dark roast, $4.10
    with chicory, $3.75

**10.** 75% mixture, 123 lb
    10% mixture, 77 lb

**11.** scientific $8.00
    graphing $72.00

**12.** $4,000 at 10%
    $6,000 at 15%

**13.** reserved $20.00
    general $15.00

**14.** 4 pt at 75%
    4 pt at 25%

**15.** $R_1 = 17\ \Omega, R_2 = 4\ \Omega$

**16.** potassium = 120 lb, nitrogen = 360 lb

**17.** holly = 40 shrubs, nandina = 20 shrubs

**18.** water = 2 ft$^3$, sand = 10 ft$^3$

## Chapter Review Exercises, Chapter 8

**1.**

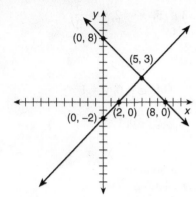

**3.**

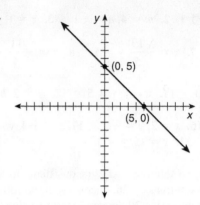

dependent; many solutions, lines coincide

**5.** $(3, 0)$  **7.** $(-2, 4)$  **9.** $(2, -2)$  **11.** $(6, 3)$  **13.** dependent, many solutions  **15.** $(5, 1)$  **17.** $\left(2, \dfrac{8}{3}\right)$

**19.** $(2, 0)$  **21.** $(3, 2)$  **23.** $(4, 4)$  **25.** $(7, 5)$  **27.** $(4, -6)$  **29.** $(3, -4)$  **31.** $\left(\dfrac{4}{5}, \dfrac{2}{5}\right)$  **33.** $(28, 44)$

**35.** $\left(-3, \dfrac{5}{2}\right)$  **37.** $(4, -6)$  **39.** $(12, 2)$

**41.** electrician = $75
apprentice = $35

**43.** shellac = $2.50
thinner = $3.50

**45.** $119°, 56°$

**47.** $2,000 at 5%
$3,000 at 6%

**49.** Colombian = $5
blended = $4

**51.** Ohio = $1.95
Alaska = $2.10

**53.** name brand = $320
generic = $175

**55.** telephone = $15,000
showroom = $25,000

## Practice Test, Chapter 8

**1.** $(5, 0)$

**3.** $(14, -9)$

**5.** $(-6, -6)$

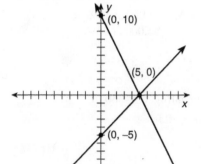

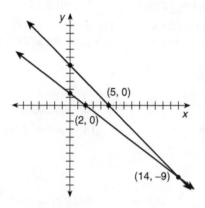

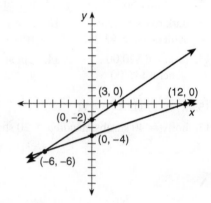

**7.** $(0, 0)$  **9.** $(1, 2)$  **11.** $(2, 2)$  **13.** $(-3, 15)$  **15.** $(-9, -11)$  **17.** 20 A, 15 A

**19.** $L = 51$ in.
$W = 34$ in.

**21.** $20,000 at 3.5%
$5,000 at 4%

**23.** 0.000215 F
0.000055 F

# Cumulative Practice Test for Chapters 5–8

**1.** $-6x + 19$    **3.** $x = 1$    **5.** $x = \dfrac{31}{39}$    **7.** $Z = \sqrt{65}\ \Omega$ or $8.1\ \Omega$    **9.** $x = \dfrac{21}{10}$    **11.** 160 teeth

**13.**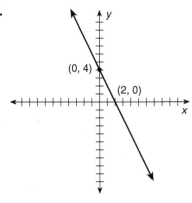

**15.**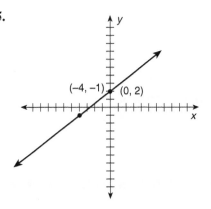

**17.** slope $= \dfrac{3}{2}$; $y$-intercept $= -5$    **19.** $y = -\dfrac{2}{3}x - \dfrac{2}{3}$    **21.** $y = 2x - 4$    **23.** $y = 4x + 3$    **25.** $x - y = 1$
**27.** $x = 1$; $y = 3$    **29.** larger resistance $= 20\ \Omega$; smaller resistance $= 16\ \Omega$

## Section 9–1 Self-Study Exercises

**1** **1.** $x^7$    **2.** $m^4$    **3.** $a^2$    **4.** $x^{10}$    **5.** $y^6$    **6.** $a^2b$    **7.** $12a^5b^7$    **8.** $10x^{11}$    **9.** $-10x^3y^3z^4$    **10.** $24x^3y^4$

**11.** $\dfrac{2}{7}a^3b^5$    **12.** $\dfrac{3}{8}a^4b^3$    **13.** $\dfrac{4.2n^{10}}{m^2}$    **14.** $-72a^6$    **15.** $24x^7$    **16.** $24ab^4$

**2** **17.** $y^5$    **18.** $x^4$    **19.** $\dfrac{1}{a}$    **20.** $b$    **21.** 1    **22.** $\dfrac{1}{x^2}$    **23.** $y^4$    **24.** $n^7$    **25.** $\dfrac{1}{x^{10}}$    **26.** $\dfrac{2}{n}$    **27.** $\dfrac{3x^7}{2}$    **28.** $\dfrac{1}{x^5}$

**29.** $\dfrac{x}{y^6}$    **30.** $\dfrac{a^3}{b^2}$    **31.** $xy^2$    **32.** $\dfrac{a}{b}$    **33.** $\dfrac{3a}{b^5}$    **34.** $\dfrac{3rs^{12}}{2}$    **35.** $\dfrac{2y^2}{3x^4}$    **36.** $\dfrac{3m^6n^3}{2}$

**3** **37.** 4,096    **38.** $x^8$    **39.** 1    **40.** $x^{40}$    **41.** $a^{21}$    **42.** $-\dfrac{1}{8}$    **43.** $\dfrac{4}{49}$    **44.** $\dfrac{a^4}{b^4}$    **45.** $8m^6n^3$    **46.** $\dfrac{x^6}{y^3}$

**47.** $x^6y^{12}$    **48.** $4a^2$    **49.** $x^6y^3$    **50.** $-27a^3b^6$    **51.** $-16,807x^{10}$    **52.** $16x^4y^{16}$    **53.** $\dfrac{-7}{x^{17}}$    **54.** $-x^4$    **55.** $x^4$

**56.** $-6x^8$    **57.** $-x^{13}y$    **58.** $-3x^3$    **59.** $\dfrac{-x^2}{y^7}$    **60.** $-27a^6b^3c^3$

## Section 9–2 Self-Study Exercises

**1** **1.** binomial    **2.** binomial    **3.** monomial    **4.** monomial    **5.** monomial    **6.** monomial    **7.** trinomial
**8.** trinomial    **9.** monomial    **10.** binomial    **11.** binomial    **12.** binomial

**2** **13.** 1    **14.** 2    **15.** 2, 1, 0    **16.** 3, 1, 0    **17.** 1, 0    **18.** 2, 0    **19.** 0    **20.** 0    **21.** 1, 0    **22.** 2, 0
**23.** 5, 2    **24.** 5, 8    **25.** 2    **26.** 3    **27.** 6    **28.** 5    **29.** 2    **30.** 1    **31.** 6    **32.** 6

**3** **33.** $-3x^2 + 5x$; 2; $-3x^2$; $-3$    **34.** $-x^3 + 7$; 3; $-x^3$; $-1$    **35.** $9x^2 + 4x - 8$; 2; $9x^2$; 9    **36.** $5x^2 - 3x + 8$; 2; $5x^2$; 5
**37.** $7x^3 + 8x^2 - x - 12$; 3; $7x^3$; 7    **38.** $-15x^4 + 12x + 7$; 4; $-15x^4$; $-15$    **39.** $8x^6 - 7x^3 - 7x$; 6; $8x^6$; 8
**40.** $-14x^8 + x + 15$; 8; $-14x^8$; $-14$    **41.** $8x^4 + 15x^3 + 12x$; 4; $8x^4$; 8    **42.** $-7x^4 + 5x^3 + 3x - 8$; 4; $-7x^4$; $-7$
**43.** $-5x^4 + 2x^2 + 3x - 5$; 4; $-5x^4$; $-5$    **44.** $-8x^4 - x^3 + 7x - 3$; 4; $-8x^4$; $-8$

## Section 9–3 Self-Study Exercises

**1** **1.** $7a^2$ **2.** $3x^3$ **3.** $-2a^2 + 3b^2$ **4.** $2a - 2b$ **5.** $-5x^2 - 2x$ **6.** $5a^2$ **7.** $-x^2 - 2y$ **8.** $2m^2 + n^2$
**9.** $9a + 2b + 10c$ **10.** $-2x + 3y + 2z$ **11.** $-4$ **12.** $3x + 1$

**2** **13.** $14x^3$ **14.** $2m^3$ **15.** $-21m^2$ **16.** $-2y^6$ **17.** $8x^3 - 28x^2$ **18.** $-6a^3b^2 + 15a^2b^3$ **19.** $35x^4 - 21x^3 - 49x$
**20.** $6x^3y^2 - 10x^2y^3$ **21.** $3x^2 - 18x$ **22.** $12x^3 - 28x^2 + 32x$ **23.** $-8x^2 + 12x$ **24.** $10x^2 + 4x^3$

**3** **25.** $2x^2$ **26.** $-\dfrac{a}{2}$ **27.** $\dfrac{1}{2x^2}$ **28.** $-\dfrac{3x^3}{4}$ **29.** $3x - 2$ **30.** $4x^3 - 2x - 1$ **31.** $7x - \dfrac{1}{x}$ **32.** $4x^2 + 3x$
**33.** $\dfrac{x^2}{3} + \dfrac{4}{9}$ **34.** $5ab^2 - 3b - \dfrac{7}{ab}$ **35.** $\dfrac{5ab^2}{2} - \dfrac{b}{2}$ **36.** $\dfrac{3x}{y} - \dfrac{2y}{x} - 1$ **37.** $5x - 14x^4$ **38.** $24x^9y^6 + 4$
**39.** $-4 - 4x^2$ **40.** $\dfrac{x}{5y} - \dfrac{y}{3x} - 2x^2$

## Chapter Review Exercises, Chapter 9

**1.** $x^{10}$ **3.** $21x^9$ **5.** $x^3$ **7.** $7x^3$ **9.** $\dfrac{x}{y^3}$ **11.** $x^{12}$ **13.** $x^{15}$ **15.** $-27x^6$ **17.** $\dfrac{1}{y^2}$ **19.** $40x^2$ **21.** $4$ **23.** $5$

**25.** $3x^3 + x^2 + 5x - 8; 3; 3x^3, 3$ **27.** $2x^4 + 5x^2 - 12x - 32; 4; 2x^4; 2$ **29.** $-2x^4 - 2x^3 + 3x$ **31.** $-3x^2 + 5y^2$

**33.** $21x^6$ **35.** $2x^3 + 6x^2 - 10x$ **37.** $-2x^4 + 14x^3 - 30x$ **39.** $-\dfrac{2x^3}{3}$ **41.** $-\dfrac{14}{5y^2}$ **43.** $2x^2 - 4x + 7$

**45.** $\dfrac{x^3}{2} - 6x^6$ **47.** $-4x - 3x^2 + 5x^3$

## Practice Test, Chapter 9

**1.** $x^5$ **3.** $\dfrac{16}{49}$ **5.** $\dfrac{1}{x^{10}}$ **7.** $\dfrac{x^4}{y^2}$ **9.** $12a^3 - 8a^2 + 20a$ **11.** $x^2 - 5x$ **13.** $56x^7$ **15.** $4xy - 3x + 6$ **17.** $1$

**19.** $-3x^4 - 2x^3 + 6$

## Section 10–1 Self-Study Exercises

**1** **1.** $7(a + b)$ **2.** $12(x + y)$ **3.** $m(m + 2)$ **4.** $y^2(5y + 8)$ **5.** $3x(2x + 1)$ **6.** $6y^3(2 + 3y)$ **7.** $6x^4(2x - 1)$
**8.** $5x(1 - 3y)$ **9.** prime **10.** prime **11.** $5(ab + 2a + 4b)$ **12.** $2ax(2x + 3a + 5ax)$ **13.** prime
**14.** $3(4a^2 - 5a + 2)$ **15.** $3x(x^2 - 3x - 2)$ **16.** $2ab(4a + 7b^2 + 14a^2b^2)$ **17.** $3m^2(1 - 2m + 4m^2)$
**18.** $6xy(2x - 3y^2 + 4xy)$ **19.** $3a^2bc(5 + 6abc^2 - 7a^2c^4)$ **20.** $5x^2y^2z(4y - 7x - 8)$ **21.** $4x^2y^2(2x^2 - 3y^2 - 1)$
**22.** $-1(x + 7)$ **23.** $-1(3x + 8)$ **24.** $-1(5x - 2)$ **25.** $-1(12x - 7)$ **26.** $-1(x^2 - 3x + 8)$ **27.** $-1(2x^2 + 7x + 11)$
**28.** $-2(x^2 - 3x + 4)$ **29.** $-3(x^2 + 3x - 5)$ **30.** $-7(x^2 + 3x - 2)$ **31.** $-6(2x^2 - 3x - 1)$ **32.** $(x + 3)(5x + 8y)$
**33.** $(2x - 1)(3x + 5)$ **34.** $(3y - 5)(4y + 7)$ **35.** $(7a + 2b)(a - b)$ **36.** $(2m - 3n)(5m - 7n)$ **37.** $(y - 2)(y - 3)$
**38.** $(2x - 7)(3x - 8)$ **39.** $(9y - 2)(7y - 5)$

## Section 10–2 Self-Study Exercises

**1** **1.** $x^2 + 10x + 21$ **2.** $x^2 + 13x + 40$ **3.** $2x^2 + 3x - 2$ **4.** $3x^2 - 8x - 35$ **5.** $x^3 + 7x^2 + 9x - 5$
**6.** $x^3 + 4x^2 - 19x + 14$ **7.** $x^3 - 12x^2 + 37x - 14$ **8.** $x^3 - 11x^2 + 25x - 3$ **9.** $2x^3 - 7x^2 + 7x - 5$
**10.** $3x^3 + x^2 - 11x + 6$ **11.** $6x^3 + 11x^2 - 31x + 14$ **12.** $20x^3 - 22x^2 - 9x + 9$ **13.** $x^3 - 8$ **14.** $8x^3 - 27$
**15.** $x^3 + 27$ **16.** $27x^3 + 8$ **17.** $x^3 + 125$ **18.** $1{,}331x^3 - 27$

**2** **19.** $a^2 + 11a + 24$ **20.** $x^2 + x - 20$ **21.** $y^2 - 11y + 18$ **22.** $y^2 - 10y + 21$ **23.** $2a^2 + 11a + 12$
**24.** $3a^2 - 2a - 5$ **25.** $3a^2 - 8ab + 4b^2$ **26.** $5x^2 - 26xy + 5y^2$ **27.** $6x^2 - 17x + 12$ **28.** $2a^2 - 7ab + 5b^2$
**29.** $21 - 52m + 7m^2$ **30.** $40 - 21x + 2x^2$ **31.** $x^2 + 11x + 28$ **32.** $y^2 - 12y + 35$ **33.** $m^2 - 4m - 21$
**34.** $3bx + 18b - 2x - 12$ **35.** $12r^2 - 7r - 10$ **36.** $35 - 22x + 3x^2$ **37.** $4 - 14m + 6m^2$ **38.** $6 + 13x + 6x^2$
**39.** $2x^2 + x - 15$ **40.** $20x^2 - 13xy - 21y^2$ **41.** $14a^2 + 19ab - 3b^2$ **42.** $30a^2 - 13ab - 10b^2$
**43.** $27x^2 + 30xy - 8y^2$ **44.** $20x^2 - 47xy + 24y^2$ **45.** $21m^2 + 29mn - 10n^2$

**3**    **46.** $a^2 - 9$    **47.** $4x^2 - 9$    **48.** $a^2 - y^2$    **49.** $16r^2 - 25$    **50.** $25x^2 - 4$    **51.** $49 - m^2$    **52.** $9y^2 - 25$
**53.** $64y^2 - 9$    **54.** $9a^2 - 121b^2$    **55.** $25y^2 - 9$    **56.** $x^2 - 49$    **57.** $x^2 - 121$    **58.** $4 - 12x + 9x^2$
**59.** $9x^2 + 24x + 16$    **60.** $Q^2 + 2QL + L^2$    **61.** $a^4 + 2a^2 + 1$    **62.** $4d^2 - 20d + 25$    **63.** $9a^2 + 12ax + 4x^2$
**64.** $9x^2 - 42x + 49$    **65.** $36 + 12Q + Q^2$    **66.** $y^2 + 10xy + 25x^2$    **67.** $16 - 24j + 9j^2$    **68.** $9m^2 - 12mp + 4p^2$
**69.** $m^4 + 2m^2p^2 + p^4$

*707-1300*

**4**    **70.** $x + 6$, yes    **71.** $x^2 - 3x + 2$, yes    **72.** $x^2 + x - 5$, yes    **73.** $3x^2 - 2x + 3 - \dfrac{1}{x + 3}$, no

**74.** $2x^2 + 7x + 2 + \dfrac{18}{x - 5}$, no    **75.** $x^2 + x - 6$, yes    **76.** $x^2 - x - 2$, yes    **77.** $x^2 + 3x + 9$, yes

## Section 10–3 Self-Study Exercises

**1**    **1.** difference    **2.** not difference    **3.** difference    **4.** difference    **5.** not difference    **6.** difference
**7.** $(y + 7)(y - 7)$    **8.** $(4x + 1)(4x - 1)$    **9.** $(3a + 10)(3a - 10)$    **10.** $(2m + 9n)(2m - 9n)$    **11.** $(3x + 8y)(3x - 8y)$
**12.** $(5x + 8)(5x - 8)$    **13.** $(10 + 7x)(10 - 7x)$    **14.** $(2x + 7y)(2x - 7y)$    **15.** $(11m + 7n)(11m - 7n)$
**16.** $(9x + 13)(9x - 13)$    **17.** $(2a + 3)(2a - 3)$    **18.** $(5r + 4)(5r - 4)$    **19.** $(6x + 7y)(6x - 7y)$
**20.** $(7 - 12x)(7 + 12x)$    **21.** $(4x + 9y)(4x - 9y)$

**2**    **22.** not perfect square    **23.** perfect square    **24.** not perfect square    **25.** not perfect square    **26.** perfect square
**27.** perfect square    **28.** $(x + 3)^2$    **29.** $(x + 7)^2$    **30.** $(x - 6)^2$    **31.** $(x - 8)^2$    **32.** $(2a + 1)^2$    **33.** $(5x - 1)^2$
**34.** $(3m - 8)^2$    **35.** $(2x - 9)^2$    **36.** $(x - 6y)^2$    **37.** $(2a - 5b)^2$    **38.** $(y - 5)^2$    **39.** $(3x + 10y)^2$    **40.** $-1(x + 6)^2$
**41.** $-1(3x - 1)^2$    **42.** $-(x + 4)^2$

## Section 10–4 Self-Study Exercises

**1**    **1.** $(x + 6)(x + 1)$    **2.** $(x - 6)(x - 1)$    **3.** $(x - 2)(x - 3)$    **4.** $(x + 3)(x + 2)$    **5.** $(x - 7)(x - 4)$
**6.** $(x + 6)(x + 2)$    **7.** $(x - 6)(x - 2)$    **8.** $(x + 12)(x + 1)$    **9.** $(x - 12)(x - 1)$    **10.** $(x + 4)(x + 3)$
**11.** $(x - 4)(x - 3)$    **12.** $(x - 3)(x - 1)$    **13.** $(x + 7)(x + 1)$    **14.** $(x + 5)(x + 2)$    **15.** $(x - 3)(x + 2)$
**16.** $(x + 3)(x - 2)$    **17.** $(x - 6)(x + 1)$    **18.** $(x + 6)(x - 1)$    **19.** $(x - 4)(x + 3)$    **20.** $(x + 4)(x - 3)$
**21.** $(x + 6)(x - 2)$    **22.** $(x - 6)(x + 2)$    **23.** $(x - 12)(x + 1)$    **24.** $(x + 12)(x - 1)$    **25.** $(y - 5)(y + 2)$
**26.** $(y - 3)(y + 2)$    **27.** $(b + 3)(b - 1)$    **28.** $(b - 7)(b + 2)$    **29.** $(x - 4)(x - 3)$    **30.** $(x - 6)(x + 5)$
**31.** $(x + 9)(x + 2)$    **32.** $(x - 6)(x - 3)$    **33.** $(x - 9)(x + 2)$    **34.** $(x + 18)(x - 1)$    **35.** $(x + 5)(x + 4)$
**36.** $(x - 10)(x - 2)$    **37.** $(x - 8)(x - 2)$    **38.** $(x - 16)(x - 1)$    **39.** $(x - 14)(x + 1)$    **40.** $(x - 7)(x + 2)$

**2**    **41.** $(x + y)(x + 4)$    **42.** $(3x + 2)(2x - y)$    **43.** $(3x + 5)(m - 2n)$    **44.** $(6x - 7)(5y - 6)$    **45.** $(x - 2)(x + 8)$
**46.** $(3x - 1)(2x - 7)$    **47.** $(x - 4)(x + 1)$    **48.** $(2x - 1)(4x + 3)$    **49.** $(x - 5)(x + 4)$    **50.** $(x - 2)(3x + 5)$
**51.** $(x + 2)(4x - 3)$    **52.** $(x - 2)(4x + 3)$    **53.** $(x + 2)(4x + 3)$    **54.** $(x - 2)(4x - 3)$    **55.** $(2x + 1)(4x - 3)$

**3**    **56.** $(3x + 1)(x + 2)$    **57.** $(3x + 2)(x + 4)$    **58.** $(3x + 2)(2x + 3)$    **59.** $(4x + 3)(2x - 1)$    **60.** $(3x - 4)(2x - 3)$
**61.** $(2x - 5)(x - 2)$    **62.** $(3x - 5)(2x - 1)$    **63.** $(4x + 3)(2x + 1)$    **64.** $(6x - 5)(x - 1)$    **65.** $(4x + 3)(2x + 5)$
**66.** $(3x - 5)(5x + 1)$    **67.** $(2x - 1)(4x - 3)$    **68.** $(2x - 7)(x + 1)$    **69.** $(6x - 5)(2x + 3)$    **70.** $(5x + 3)(2x - 1)$
**71.** $(3x + 2)(4x - 1)$    **72.** $(3x - 2)(4x + 1)$    **73.** $(4x + 1)(3x + 2)$    **74.** $(4x - 1)(3x - 2)$    **75.** $(8x - 1)(3x + 1)$
**76.** $(3x - 1)(8x - 1)$    **77.** $(6x - 5y)(x + 2y)$    **78.** $(3a + 2b)(2a - 7b)$    **79.** $(6x - 5)(3x + 2)$

**4**    **80.** $(x + 3)(x - 2)$    **81.** $(2x + 3)(x - 1)$    **82.** $(x + 3)(x - 3)$    **83.** $4(x + 2)(x - 2)$    **84.** $(m + 5)(m - 3)$
**85.** $2(a + 2)(a + 1)$    **86.** $(b + 3)^2$    **87.** $(4m - 1)^2$    **88.** $(x + 7)(x + 1)$    **89.** $(2m + 1)(m + 2)$
**90.** $(2m + 1)(m - 3)$    **91.** $(2a - 5)(a + 1)$    **92.** $(3x - 2)(x + 4)$    **93.** $(3x + 5)(2x - 3)$    **94.** $(4x - 1)(2x + 3)$
**95.** $-2(x - 2)(x - 1)$

## Chapter Review Exercises, Chapter 10

**1.** $5(x + y)$    **3.** $4(3m^2 - 2n^2)$    **5.** $2a(a^2 - 7a - 1)$    **7.** $2x^2 + x - 15$    **9.** $14a^2 + 19ab - 3b^2$
**11.** $27x^2 + 30xy - 8y^2$    **13.** $36x^2 - 25$    **15.** $49y^2 - 121$    **17.** $x^2 + 18x + 81$    **19.** $x^2 - 6x + 9$
**21.** $16x^2 - 120x + 225$    **23.** $64 + 112m + 49m^2$    **25.** $x - 2$    **27.** $3x - 2$    **29.** $(5y - 2)(5y + 2)$

**31.** *NSP*, this is a sum of two squares, not a difference. **33.** $(a + 1)^2$ **35.** $(n - 13)^2$ **37.** $(6a + 7b)^2$ **39.** $(7 - x)^2$
**41.** *NSP*, this is a sum of two squares, not a difference. **43.** *NSP*, the middle term needs a $y$ factor **45.** $(7 + 9y)(7 - 9y)$
**47.** $(3x + 10y)(3x - 10y)$ **49.** $(3x - y)^2$ **51.** $(3xy + z)(3xy - z)$ **53.** $(x - 8)(x - 1)$ **55.** $(x - 13)(x + 2)$
**57.** $(x + 8)(x - 3)$ **59.** $(6x + 1)(x + 4)$ **61.** $(5x - 4)(x - 6)$ **63.** $(3x + 7)(2x - 5)$ **65.** $(3a + 10)(3a - 10)$
**67.** $(2x + 1)(x - 2)$ **69.** $(a + 9)(a - 9)$ **71.** $(y - 7)^2$ **73.** $(b + 5)(b + 3)$ **75.** $(13 + m)(13 - m)$

## Practice Test, Chapter 10

**1.** $m^2 - 49$ **3.** $a^2 + 6a + 9$ **5.** $2x^2 - 11x + 15$ **7.** $2x + 7$ **9.** $x(7x + 8)$ **11.** $7ab(a - 2)$
**13.** $(3x - 5)(3x + 5)$ **15.** $(x - 9)^2$ **17.** $(3x + 2)(2x - 3)$ **19.** $(a + 8b)^2$ **21.** $(b - 5)(b + 2)$
**23.** $(3m - 2)(m - 1)$ **25.** $3(x - 2)(x^2 + 2x + 4)$ **27.** $5(x + 2)(x - 2)$ **29.** $3(x + 2)^2$

## Section 11–1 Self-Study Exercises

**1** **1.** $7x^2 - 4x + 5 = 0$ **2.** $8x^2 - 6x - 3 = 0$ **3.** $7x^2 - 5 = 0$ **4.** $x^2 - 6x + 8 = 0$ **5.** $x^2 - 9x + 8 = 0$
**6.** $x^2 - 4x - 8 = 0$ **7.** $3x^2 - 6x + 5 = 0$ **8.** $x^2 - 6x - 5 = 0$ **9.** $x^2 - 16 = 0$ **10.** $8x^2 - 7x - 8 = 0$
**11.** $8x^2 + 8x - 10 = 0$ **12.** $0.3x^2 - 0.4x - 3 = 0$

**2** **13.** $x^2 - 5x = 0$  **14.** $3x^2 - 7x + 5 = 0$  **15.** $7x^2 - 4x = 0$
$a = 1, b = -5, c = 0$      $a = 3, b = -7, c = 5$      $a = 7, b = -4, c = 0$

**16.** $3x^2 - 5x + 8 = 0$  **17.** $x^2 - 5x + 6 = 0$  **18.** $11x^2 - 8x = 0$
$a = 3, b = -5, c = 8$      $a = 1, b = -5, c = 6$      $a = 11, b = -8, c = 0$

**19.** $x^2 - x = 0$  **20.** $9x^2 - 7x - 12 = 0$  **21.** $x^2 - 5 = 0$
$a = 1, b = -1, c = 0$      $a = 9, b = -7, c = -12$      $a = 1, b = 0, c = -5$

**22.** $x^2 + 6x - 3 = 0$  **23.** $5x^2 - 0.2x + 1.4 = 0$
$a = 1, b = 6, c = -3$      $a = 5, b = -0.2, c = 1.4$      **24.** $\frac{2}{3}x^2 - \frac{5}{6}x - \frac{1}{2} = 0$

$a = \frac{2}{3}, b = -\frac{5}{6}, c = -\frac{1}{2}$

**25.** $1.3x^2 - 8 = 0$  **26.** $\sqrt{3}x^2 + \sqrt{5}x - 2 = 0$  **27.** $8x^2 - 2x - 3 = 0$
$a = 1.3, b = 0, c = -8$      $a = \sqrt{3}, b = \sqrt{5}, c = -2$

**28.** $x^2 + 3x = 0$  **29.** $5x^2 + 2x - 7 = 0$  **30.** $2.5x^2 - 0.8 = 0$

**3** **31.** $x = \pm 3$ **32.** $x = \pm 7$ **33.** $x = \pm\frac{8}{3}$ **34.** $x = \pm\frac{7}{4}$ **35.** $y = \pm 3$ **36.** $x = \pm\frac{9}{2}$ **37.** $x = \pm\sqrt{5}$ or $\pm 2.236$

**38.** $x = \pm 1$ **39.** $x = \pm 2$ **40.** $x = \pm 1.732$ **41.** 23 ft **42.** 81.759 yd **43.** Isolate the squared letter, then take the
square root of both sides of the equation. **44.** opposites

## Section 11–2 Self-Study Exercises

**1** **1.** $x = 3, 0$ **2.** $x = 0, 6$ **3.** $x = 0, 2$ **4.** $x = 0, -\frac{1}{2}$ **5.** $x = 0, \frac{1}{2}$ **6.** $x = 0, 3$ **7.** $x = 0, \frac{7}{3}$

**8.** $y = 0, -4$ **9.** $x = 0, -\frac{2}{3}$ **10.** $x = 0, \frac{4}{3}$ **11.** $x = 0, 3$ **12.** $x = 0, \frac{2}{3}$ **13.** 8 or 0 **14.** 15 units

**15.** An incomplete quadratic equation is missing the constant or number term while a pure quadratic equation is missing the
linear term. **16.** Yes, the common factor of $x$ will be set equal to zero.

**17.** $-3, -2$ **18.** 3, 3 (double root) **19.** 7, $-2$ **20.** 3, $-6$ **21.** $-4, -3$ **22.** 5, 3 **23.** 14, $-1$ **24.** 3, 6

**25.** $\frac{1}{2}, 3$ **26.** $-\frac{1}{3}, -4$ **27.** $\frac{3}{5}, -\frac{1}{2}$ **28.** $-\frac{1}{3}, -\frac{3}{2}$ **29.** $-\frac{3}{2}, -5$ **30.** $\frac{4}{3}, 2$ **31.** $\frac{1}{6}, -3$ **32.** $-4, -\frac{2}{3}$ **33.** $5, \frac{3}{2}$

**34.** $\frac{1}{2}, \frac{2}{3}$ **35.** $\frac{3}{4}, \frac{1}{2}$ **36.** $\frac{2}{5}, -1$ **37.** $\frac{5}{3}, -\frac{3}{2}$ **38.** $-\frac{5}{3}, -\frac{1}{3}$ **39.** $\frac{1}{3}, \frac{3}{2}$ **40.** $-3, \frac{2}{5}$ **41.** $w = 5$
$\qquad\qquad\qquad\qquad\qquad\qquad\qquad\qquad\qquad\qquad\qquad\qquad\qquad\qquad\qquad\qquad\qquad\qquad x + 6 = 5 + 6 = 11$ ft

**42.** $w = 18$ in.
$\quad\;\; x = 21$ in.

## Section 11–3 Self-Study Exercises

**1** **1.** $3, -\frac{2}{3}$ **2.** $3, -4$ **3.** $\frac{11}{5}, -1$ **4.** $3, 3$ **5.** $3, -2$ **6.** $\frac{3}{4}, -\frac{1}{2}$ **7.** $1.84, -10.84$ **8.** $-0.18, -1.82$

**9.** $3.14, -0.64$ **10.** $2.39, 0.28$ **11.** $-0.75 \pm 0.97j$ **12.** $0.5 \pm 1.32j$ **13.** width $= 5.55$ cm, length $= 8.55$ cm
**14.** width $= 4$ in., length $= 10$ in. **15.** 4 m **16.** width $= 11$ ft, length $= 16$ ft **17.** length $= 110$ ft, width $= 70$ ft
(nearest ft); see figure below. **18.** $111.5$ kg

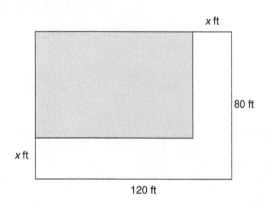

## Section 11–4 Self-Study Exercises

**1** **1.** $y = x^2$. The domain is all real numbers. The range is all real numbers greater than or equal to zero.

**2.** $y = 3x^2$. The domain is all real numbers. The range is all real numbers greater than or equal to zero.

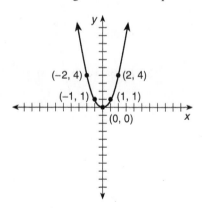

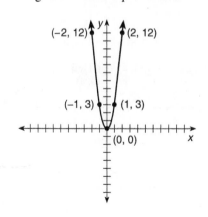

**3.** $y = \frac{1}{3}x^2$. The domain is all real numbers. The range is all real numbers greater than or equal to zero.

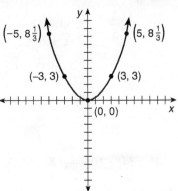

**4.** $y = -4x^2$. The domain is all real numbers. The range is all real numbers less than or equal to zero. ·

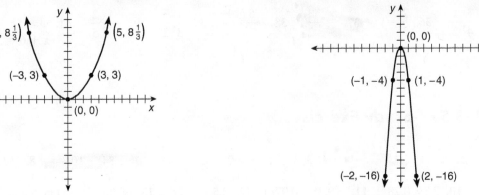

**5.** $y = -\frac{1}{4}x^2$. The domain is all real numbers. The range is all real numbers less than or equal to zero.

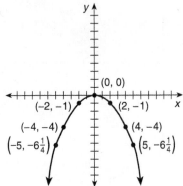

**6.** $y = x^2 - 4$. The domain is all real numbers. The range is all real numbers greater than or equal to $-4$.

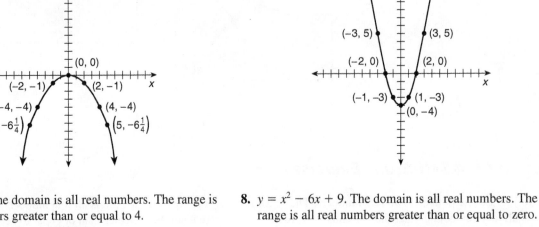

**7.** $y = x^2 + 4$. The domain is all real numbers. The range is all real numbers greater than or equal to 4.

**8.** $y = x^2 - 6x + 9$. The domain is all real numbers. The range is all real numbers greater than or equal to zero.

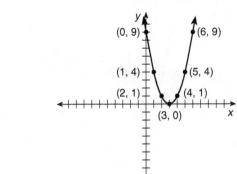

**9.** $y = -x^2 + 6x - 9$. The domain is all real numbers. The range is all real numbers less than or equal to 0.

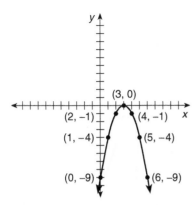

**2** **10.** The domain is all real numbers. The range is all real numbers greater than or equal to 0.

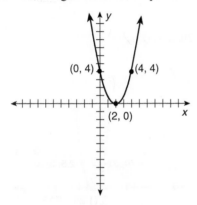

vertex: $(2, 0)$
solution: $(2, 0)$ or $x = 2$

**11.** The domain is all real numbers. The range is all real numbers greater than or equal to $-6$.

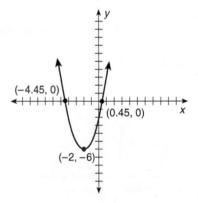

vertex: $(-2, -6)$
solutions: $(-4.45, 0)$; $(0.45, 0)$

**12.** The domain is all real numbers. The range is all real numbers greater than or equal to $-4.5$.

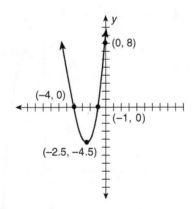

vertex: $(-2.5, -4.5)$
solutions: $(-4, 0)$; $(-1, 0)$

**13.** The domain is all real numbers. The range is all real numbers less than or equal to 12.

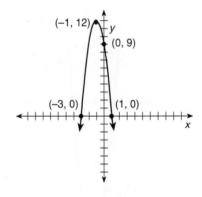

vertex: $(-1, 12)$
solutions: $(-3, 0)$; $(1, 0)$

**14.** The domain is all real numbers. The range is all real numbers less than or equal to −3.

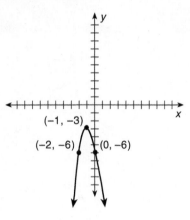

vertex: (−1, −3)
no solution

**15.** The domain is all real numbers. The range is all real numbers greater than or equal to −15.

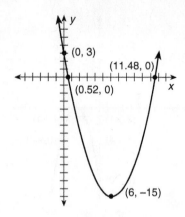

vertex: (6, −15)
solutions: (0.52, 0); (11.48, 0)

**3**   **16.** $x = 2, x = 3$    **17.** $x = 0$    **18.** no real solution    **19.** $0, $9    **20.** $t = 0$ sec

**4**   **21.** $y = 3x^2 + 5x - 2$

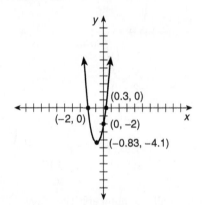

**22.** $y = (2x - 3)(x - 1)$

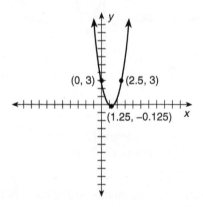

**23.** $y = 2x^2 - 9x - 5$

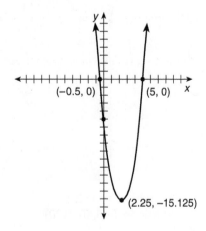

**24.** $y = -2x^2 + 9x + 5$

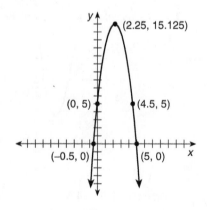

**25.**

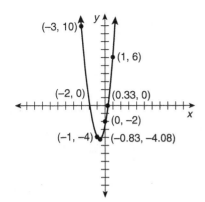

**26.**

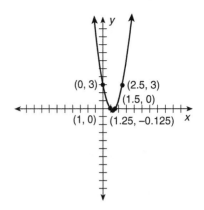

**27.**

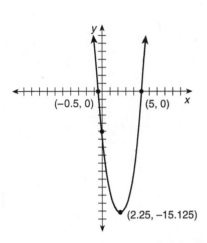

**28.**

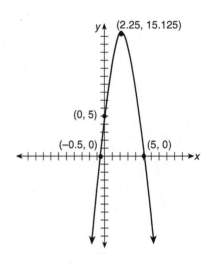

## Section 11–5 Self-Study Exercises

**1**    **1.** $5j$    **2.** $6j$    **3.** $8xj$    **4.** $4y^2j\sqrt{2y}$

**2**    **5.** $j$    **6.** $j$    **7.** $1$    **8.** $-1$    **9.** $1$    **10.** $j$    **11.** $1$    **12.** $-j$

**3**    **13.** real, rational, $x = \dfrac{2}{3}, -1$    **14.** real, irrational, $x = \dfrac{3 \pm \sqrt{5}}{2}$ or $2.62, 0.38$    **15.** real, irrational, $x = \dfrac{-1 \pm \sqrt{17}}{4}$

or $0.78, -1.28$    **16.** no real solutions or $x = \dfrac{1 \pm j\sqrt{2}}{3}$ or $0.33 \pm 0.47j$    **17.** real, irrational, $x = \dfrac{3 \pm \sqrt{37}}{2}$ or $4.54, -1.54$

**18.** real, irrational, $x = \dfrac{-5 \pm \sqrt{97}}{6}$ or $0.81$ or $-2.47$    **19.** The discriminant must be greater than or equal to zero.

**20.** The discriminant must be greater than zero and a perfect square.

## Chapter Review Exercises, Chapter 11

**1.** pure    **3.** pure    **5.** incomplete    **7.** pure    **9.** complete    **11.** $2x^2 - 8x - 5 = 0$    **13.** $x^2 - 7x + 5 = 0$

**15.** $4x^2 + 3x - 1 = 0$    **17.** $x = \pm 10$    **19.** $x = \pm\dfrac{3}{2}$    **21.** $y = \pm 1.740$    **23.** $x = \pm 2.828$    **25.** $x = \pm 2.236$

**27.** $x = \pm 2$    **29.** $x = \pm 4.123$    **31.** $y = \pm 3.055$    **33.** $x = \pm 4$    **35.** $x = \pm 8$    **37.** 16 cm    **39.** $0, 5$    **41.** $0, 2$

**43.** $0, -\dfrac{1}{2}$    **45.** $0, 5$    **47.** $0, -\dfrac{2}{3}$    **49.** $0, -3$    **51.** $0, 9$    **53.** $0, -8$    **55.** $0, \dfrac{5}{3}$    **57.** $0, \dfrac{1}{2}$    **59.** $0, 4$

**61.** $3, 1$    **63.** $-5, 2$    **65.** $-6, -1$    **67.** $2, 4$    **69.** $-\dfrac{2}{3}, \dfrac{3}{2}$    **71.** $-\dfrac{2}{5}, \dfrac{5}{2}$    **73.** $-\dfrac{3}{4}, -1$    **75.** $\dfrac{3}{4}, -\dfrac{1}{3}$    **77.** $-21, 2$

**79.** $\frac{2}{3}, -1$   **81.** $3, 2$   **83.** $6, -3$   **85.** $-6, -5$   **87.** $-9, 2$   **89.** width = 12 ft
length = 19 ft

**91.** $a = 1$
$b = -2$
$c = -8$

**93.** $a = 1$
$b = 3$
$c = -4$

**95.** $a = 1$
$b = -3$
$c = 2$

**97.** $9, -1$   **99.** $2, -4$   **101.** $2, -\dfrac{1}{2}$   **103.** $1.78, -0.28$   **105.** $-0.23, -1.43$   **107.** $w = 11$ ft, $l = 22$ ft

**109.** $w = 14$ in., $l = 42$ in.

**111.** $y = -x^2 - 1$

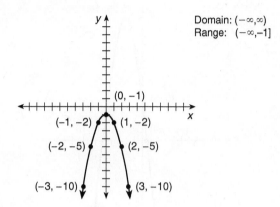

Domain: $(-\infty, \infty)$
Range: $(-\infty, -1]$

**113.** $y = x^2 - 6x + 8$

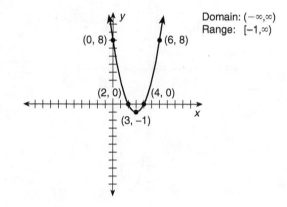

Domain: $(-\infty, \infty)$
Range: $[-1, \infty)$

**115.** $y = -x^2 + 2x - 1$

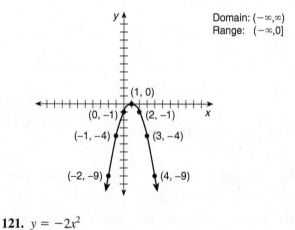

Domain: $(-\infty, \infty)$
Range: $(-\infty, 0]$

**117.** $x = -2, 6$   **119.** $x = -4$ (double root)

**121.** $y = -2x^2$

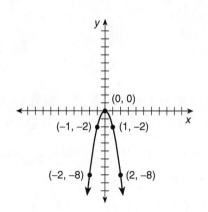

**123.** $10j$     **125.** $\pm 2y^3 j\sqrt{6y}$     **127.** $-1$     **129.** $j$     **131.** real, rational, equal     **133.** real, irrational, unequal

**135.** real, rational, unequal

## Practice Test, Chapter 11

**1.** pure     **3.** incomplete     **5.** $\pm 9$     **7.** $0, 2$     **9.** $3, 2$     **11.** $\dfrac{3}{2}, 4$     **13.** $1, -\dfrac{5}{2}$     **15.** $\dfrac{3 + \sqrt{29}}{2}$ or 4.19, $\dfrac{3 - \sqrt{29}}{2}$ or $-1.19$

**17.** $-j$     **19.** real, rational, unequal     **21.**

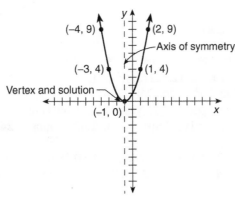

**21.** $y = x^2 + 2x + 1$

## Cumulative Practice Test for Chapters 9–11

**1.** $40x^{10}$     **3.** $x^{32}$     **5.** $12x^4 + 5x^2$     **7.** $2x^2 - 4x^3 + 3x$     **9.** $8x(3x^2 - 2x - 1)$     **11.** $12x^2 - 23x + 10$     **13.** $x - 3$

**15.** $(x - 8)^2$     **17.** $(x + 7)(x - 6)$     **19.** $(2x - 3)(3x - 4)$     **21.** $x = \pm 3$     **23.** $x = 0; x = 8$     **25.** $x = \frac{1}{2}; x = 3$

**27.** $x = 1.18; x = -0.43$     **29.** $j$

## Section 12–1 Self-Study Exercises

**1**    **1.** $\overleftrightarrow{PQ}, \overleftrightarrow{QR}, \& \overleftrightarrow{PR}$ or $\overleftrightarrow{QP}, \overleftrightarrow{RQ},$ and $\overleftrightarrow{RP}$     **2.** $\overrightarrow{PQ}, \overrightarrow{PR}$     **3.** $\overline{QR}$     **4.** Yes     **5.** No, endpoints are different.     **6.** no

**7.** yes     **8.** parallel     **9.** intersect     **10.** coincide     **11.** intersect     **12.** $\overleftrightarrow{EF}$ and $\overleftrightarrow{GH}$     **13.** $\angle DEF, \angle FED$     **14.** $\angle E$
**15.** $\angle 1$

**2**    **16.** $360°$     **17.** $180°$     **18.** $90°$     **19.** $45°$     **20.** acute     **21.** obtuse     **22.** right     **23.** obtuse     **24.** acute
**25.** straight     **26.** neither     **27.** complementary     **28.** supplementary     **29.** neither     **30.** complementary
**31.** supplementary     **32.** supplementary

**3**    **33.** $\angle a$ and $\angle c; \angle b$ and $\angle d$     **34.** supplementary angles     **35.** $180°$     **36.** $\angle c$ and $\angle f; \angle d$ and $\angle e$     **37.** $180°$
**38.** corresponding angles

**4**    **39.** $0.7833°$     **40.** $0.01°$     **41.** $0.0872°$     **42.** $0.1708°$     **43.** $10.3042°$     **44.** $21'$     **45.** $12'$     **46.** $7'12''$
**47.** $12'47''$     **48.** $18'54''$

## Section 12–2 Self-Study Exercises

**1**   **1.** 38 in.   **2.** 21.2 cm   **3.** 156 in.   **4.** 124 in.   **5.** 160 in.   **6.** 108 in.   **7.** 10 ft   **8.** 43.8 ft   **9.** 930 ft   **10.** 54 ft   **11.** 59 ft   **12.** 156 ft   **13.** 17 ft   **14.** 130 in.   **15.** 12 cm   **16.** 35.6 cm   **17.** 590 ft   **18.** 48 tiles   **19.** 309.3 mm   **20.** 52.9 ft   **21.** 207 in.   **22.** 75 ft   **23.** 402 ft   **24.** 98 ft   **25.** 28.6 in.   **26.** 48 cm   **27.** 72 ft   **28.** 4 ft 8 in.

**2**   **29.** 72 $in^2$   **30.** 13.8 $cm^2$   **31.** 6 $ft^2$   **32.** 116.84 $ft^2$   **33.** 42,500 $ft^2$   **34.** 180 $ft^2$   **35.** 9 $cm^2$   **36.** 79.21 $cm^2$   **37.** 193 $ft^2$   **38.** 900 $yd^2$   **39.** $\frac{1}{16}$ $mi^2$   **40.** 81 tiles   **41.** 108 tiles   **42.** 400 $in^2$   **43.** 5,053.05 $mm^2$   **44.** 162 $ft^2$   **45.** 348 $in^2$   **46.** 260 $ft^2$   **47.** 33 $in^2$   **48.** 96 $cm^2$   **49.** 189 $ft^2$   **50.** $351

## Section 12–3 Self-Study Exercises

**1**   **1.** 25.1 cm   **2.** 47.1 m   **3.** 18.8 in.   **4.** 9.4 ft   **5.** 53.4 ft   **6.** 17.3 m   **7.** 50.3 $cm^2$   **8.** 176.7 $m^2$   **9.** 28.3 $in^2$   **10.** 804.2 $yd^2$   **11.** 7.1 $ft^2$   **12.** 227.0 $ft^2$   **13.** 23.8 $m^2$   **14.** 21.6 $in^2$   **15.** 0.6 $m^2$   **16.** 2.1 $ft^2$   **17.** 14.6 $cm^2$   **18.** 1.0 $in^2$   **19.** 592.7 $ft^2$   **20.** 268.27 in.   **21.** Yes, the cross-sectional area of the third pipe is larger than the combined area of the other two pipes.   **22.** Yes, the combined cross-sectional area of the two pipes is 25.1 $in^2$, which is greater than 20 $in^2$—the area of the large pipe.   **23.** 2,827 ft/min   **24.** 1,571 ft/min

**2**   **25.** 0.79 rad   **26.** 0.98 rad   **27.** 2.44 rad   **28.** 21.75°; 0.38 rad   **29.** 177.55°; 3.10 rad   **30.** 44.9033°; 0.78 rad   **31.** 45°   **32.** 30°   **33.** 143.2394°   **34.** 28°38′52″   **35.** 22°30′   **36.** 42°58′19″

**3**   **37.** 28.27 mm   **38.** 13.09 in.   **39.** 3.20 ft   **40.** 7.06 cm   **41.** 6.45 in.   **42.** 14 cm   **43.** 4.25 rad   **44.** 3.49 rad   **45.** 7 cm   **46.** 4.5 in.

**4**   **47.** 120.64 $cm^2$   **48.** 198.44 $mm^2$   **49.** 2.43 $ft^2$   **50.** 14.18 $in^2$   **51.** 530.14 $ft^2$   **52.** 75.40 $in^2$   **53.** 17.12 $in^2$   **54.** 4.32 $cm^2$   **55.** 2.24 cm   **56.** 0.85 rad   **57.** 47.43 $cm^2$   **58.** 0.7 rad   **59.** 16.01 $cm^2$   **60.** 48.98 $in^2$   **61.** 54.83 $in^2$   **62.** 51.99 $cm^2$   **63.** 561.81 $mm^2$   **64.** 257.93 $ft^2$

## Section 12–4 Self-Study Exercises

**1**   **1.** 576 $in^3$   **2.** 3,817.04 $cm^3$   **3.** 320 $cm^3$   **4.** 55.36 $in^2$   **5.** 23.3 $in^3$   **6.** 18,850 $ft^3$   **7.** $V = 301.6$ $ft^3$   **8.** 4,712.4 $ft^3$   **9.** 50.3 L   **10.** 12,600 $cm^3$   **11.** 76,800 $m^3$   **12.** 415.6 $m^3$   **13.** 2,617.3 $in^3$   **14.** 2,428.2 $cm^3$   **15.** 3,660 $in^3$   **16.** 523.6 $cm^3$   **17.** 904.8 $in^3$   **18.** 356,892.8 gal   **19.** 225.6 gal   **20.** $LSA = 37.3$ $in^2$; $TSA = 47.1$ $in^2$   **21.** $LSA = 1,885.0$ $ft^2$; $TSA = 4,398.2$ $ft^2$   **22.** $LSA = 200$ $cm^2$; $TSA = 264$ $cm^2$   **23.** $LSA = 96$ $in^2$; $TSA = 109.8$ $in^2$   **24.** 1,885.0 $ft^2$   **25.** $LSA = 188.5$ $ft^2$; $TSA = 301.6$ $ft^2$   **26.** 589.0 $cm^2$   **27.** 1,649.3 $ft^2$   **28.** 314.2 $cm^2$   **29.** 6,361.7 $ft^2$   **30.** 50.3 $ft^2$   **31.** $347

## Section 12–5 Self-Study Exercises

**1**   **1.** scalene   **2.** equilateral   **3.** isosceles   **4.** longest $\overline{TS}$, shortest $\overline{RS}$   **5.** longest $k$, shortest $l$   **6.** $\angle C$, $\angle B$, $\angle A$   **7.** $\angle X$, $\angle Z$, $\angle Y$

**2**   **8.** $BC = 10$ mm   **9.** $AB = 17$ yd   **10.** $c = 6.403$ cm   **11.** $a = 6.325$ m   **12.** $b = 12$ ft   **13.** 30 ft   **14.** $AC = 39.783$ cm   **15.** 26 ft 10 in.   **16.** 12.806 ft   **17.** 13 cm   **18.** 45 dm   **19.** 10.607 mm   **20.** 61.083 ft

## Chapter Review Exercises, Chapter 12

**1.** $\overleftrightarrow{AB}$ & $\overleftrightarrow{CD}$   **3.** $\overleftrightarrow{EF}$ & $\overleftrightarrow{GH}$   **5.** $\angle P$   **7.** right   **9.** straight   **11.** supplementary   **13.** neither   **15.** $\angle a$ and $\angle b$; $\angle b$ and $\angle c$; $\angle c$ and $\angle d$; $\angle d$ and $\angle a$   **17.** $\angle a$ and $\angle h$; $\angle b$ and $\angle g$   **19.** 45°   **21.** 0.4833°   **23.** 0.1261°   **25.** 45′   **27.** 13′3″   **29.** 210 mm   **31.** 35.8 in.   **33.** 66 ft   **35.** 288 in.   **37.** 162 $cm^2$   **39.** 51.84 $m^2$   **41.** 306 $ft^2$   **43.** 33,000 $ft^2$   **45.** 33 $yd^2$   **47.** 235 $ft^2$   **49.** $A = 50.27$ $m^2$   **51.** $A = 12$ $cm^2$   **53.** 33 ft/min   **55.** 1.05 rad

**57.** 1.74 rad    **59.** 150°    **61.** 2.62 cm$^2$    **63.** 2.18 ft$^2$    **65.** 31.42 in.    **67.** 6.96 cm$^2$    **69.** 44.50 in$^2$    **71.** 7,854 cm$^3$
**73.** 102 yd$^3$    **75.** 616 cm$^2$    **77.** 2,733 cm$^2$    **79.** 124,407 ft$^2$    **81.** 1,728 m$^3$    **83.** 1,017.9 m$^2$    **85.** 169.6 cm$^2$
**87.** 2 gal    **89.** longest $TS$, shortest $RS$    **91.** 15 in.    **93.** 8 yd    **95.** 21.633 in.

## Practice Test, Chapter 12

**1.** 18′45″    **3.** 0.61 rad    **5.** 112.5°    **7.** $P = 91$ ft; $A = 514.5$ ft$^2$    **9.** $P = 37$ cm; $A = 43.5$ cm$^2$    **11.** $C = 144.5$ m;
$A = 1,661.9$ m$^2$    **13.** 1 in.    **15.** 1.71 m    **17.** 282 ft$^2$    **19.** 1,016.5 ft$^2$    **21.** $A = 71.84$ cm$^2$    **23.** $c = 52$
**25.** 17 in.    **27.** 50 in$^2$    **29.** 9,768 gal

## Section 13–1 Self-Study Exercises

**1**
**1.** $\dfrac{5}{13}$    **2.** $\dfrac{12}{13}$    **3.** $\dfrac{5}{12}$    **4.** $\dfrac{12}{13}$    **5.** $\dfrac{5}{13}$    **6.** $\dfrac{12}{5}$    **7.** $\dfrac{9}{15} = \dfrac{3}{5}$    **8.** $\dfrac{12}{15} = \dfrac{4}{5}$    **9.** $\dfrac{9}{12} = \dfrac{3}{4}$    **10.** $\dfrac{12}{15} = \dfrac{4}{5}$

**11.** $\dfrac{9}{15} = \dfrac{3}{5}$    **12.** $\dfrac{12}{9} = \dfrac{4}{3}$    **13.** $\dfrac{16}{34} = \dfrac{8}{17}$    **14.** $\dfrac{30}{34} = \dfrac{15}{17}$    **15.** $\dfrac{16}{30} = \dfrac{8}{15}$    **16.** $\dfrac{30}{34} = \dfrac{15}{17}$    **17.** $\dfrac{16}{34} = \dfrac{8}{17}$

**18.** $\dfrac{30}{16} = \dfrac{15}{8}$    **19.** $\dfrac{9}{16.64} = 0.5409$    **20.** $\dfrac{14}{16.64} = 0.8413$    **21.** $\dfrac{9}{14} = 0.6429$    **22.** $\dfrac{14}{16.64} = 0.8413$

**23.** $\dfrac{9}{16.64} = 0.5409$    **24.** $\dfrac{14}{9} = 1.5556$

**2**    **25.** 0.3584    **26.** 0.9981    **27.** 1.0724    **28.** 0.6088    **29.** 0.5225    **30.** 0.9304    **31.** 0.4068    **32.** 0.9336
**33.** 0.8039    **34.** 0.9163    **35.** 1.0990    **36.** 0.8843    **37.** 0.9757    **38.** 0.4950    **39.** 3.3191    **40.** 0.9446
**41.** 0.6845    **42.** 0.9377    **43.** 0.3960    **44.** 0.2199

**3**    **45.** 20.0°    **46.** 22.5°    **47.** 67.0°    **48.** 62.5°    **49.** 57.3°    **50.** 34.8°    **51.** 51.7°    **52.** 82.1°    **53.** 39.0°
**54.** 0.6807 rad    **55.** 0.8028 rad    **56.** 1.1694 rad    **57.** 0.2537 rad    **58.** 1.4802 rad    **59.** 0.3061 rad
**60.** 1.1804 rad    **61.** 0.2802 rad    **62.** 1.1362 rad

## Section 13–2 Self-Study Exercises

**1**    **1.** 27.8°    **2.** 6.889 in.    **3.** 28.21 m    **4.** 16.15 ft    **5.** 43.2°    **6.** 49.6°

**7.**  $S = 51.1°$
$\quad U = 38.9°$
$\quad u = 11.31$ yd
$\quad 18^2 \approx 14^2 + (11.31)^2$
$\quad 324 \approx 323.9161$

**8.**  $\qquad S = 45°$
$\qquad t = 6.647$ m
$\qquad s = 4.700$ m
$\qquad (6.647)^2 = (4.7)^2 + (4.7)^2$
$\qquad 44.18 = 44.18$

**9.**  $\quad U = 55.5°$
$\qquad s = 4.814$ mm
$\qquad u = 7.005$ mm
$\quad (8.5)^2 \approx (4.814)^2 + (7.005)^2$
$\quad 72.25 \approx 23.17 + 49.07$
$\quad 72.25 \approx 72.24$

**10.**  $\qquad U = 74°$
$\qquad t = 50.79$ m
$\qquad u = 48.82$ m
$\qquad (50.79)^2 \approx (48.82)^2 + (14)^2$
$\qquad 2{,}579.62 \approx 2{,}383.39 + 196$
$\qquad 2.579.62 \approx 2{,}579.39$

**2**    **11.** 35.7°    **12.** 10.05 cm    **13.** 31.50 ft    **14.** 35.49 dm    **15.** 15.88 mm    **16.** 14.12 yd

**17.** $R = 45.9°$
$\quad Q = 44.1°$
$\quad r = 16.52$ ft

**18.** $Q = 44°$
$\quad r = 12.23$ cm
$\quad q = 11.81$ cm

**19.** $R = 16.5°$
$\quad r = 4.147$ dkm
$\quad s = 14.60$ dkm

**20.** $Q = 30.5°$
$\quad r = 13.58$ m
$\quad s = 15.76$ m

**3**   **21.** 28.6°   **22.** 4.075 m   **23.** 5.979 ft   **24.** 23.5°   **25.** 0.04663 cm   **26.** 0.01212 m

**27.** $D = 34.8°$      **28.** $E = 48°$      **29.** $D = 16.5°$      **30.** $D = 54.0°$
$E = 55.2°$           $d = 6.303$ ft        $d = 5.963$ in.       $E = 36.0°$
$f = 5.604$ m       $f = 9.419$ ft        $f = 20.99$ in.      $f = 13.60$ ft

**4**   **31.** 150.1 m   **32.** 11.91 ft   **33.** 9.950 cm   **34.** 46.71 cm   **35.** 21.09 in.

**5**   **36.** 22.38 in.   **37.** 53.62 ft   **38.** 59.0°   **39.** 39.23 ft   **40.** 15.5°   **41.** 23.25 ft   **42.** 10 Ω   **43.** 17.32 Ω
**44.** 37.9°   **45.** impedance = 28.7 Ω; reactance = 12.1 Ω   **46.** 7.654 cm   **47.** 35.8°   **48.** 2,736 ft   **49.** 3,746 ft
**50.** 19 ft   **51.** 104 m   **52.** $x = 22.9$ in; 55.1°   **53.** 4.2 in.   **54.** offset = 42 ft; diagonal = 48 ft   **55.** run = 31 in;
diagonal = 55 in   **56.** run = 18 in.; diagonal = 23 in.   **57.** $AC = 2.0$ in.; $BC = 2.0$ in.   **58.** 4.4°   **59.** 9.1°
**60.** 4.6 in.   **61.** 25.9 in.   **62.** 0.7 cm

## Chapter Review Exercises, Chapter 13

**1.** $\dfrac{15}{25} = \dfrac{3}{5}$   **3.** $\dfrac{15}{20} = \dfrac{3}{4}$   **5.** $\dfrac{20}{25} = \dfrac{4}{5}$   **7.** $\dfrac{2\text{ ft}}{2\text{ ft }2\text{ in.}} = \dfrac{24\text{ in.}}{26\text{ in.}} = \dfrac{12}{13}$   **9.** $\dfrac{10\text{ in.}}{2\text{ ft}} = \dfrac{10\text{ in.}}{24\text{ in.}} = \dfrac{5}{12}$   **11.** $\dfrac{7}{12.62} = 0.5547$

**13.** $\dfrac{7}{10.5} = 0.6667$   **15.** $\dfrac{10.5}{7} = 1.5$   **17.** 0.8403   **19.** 0.4540   **21.** 1.0724   **23.** 0.7585   **25.** 33.0°

**27.** 52.5°   **29.** 1.2188 rad   **31.** 0.3313 rad   **33.** $A = 38.1°$   **35.** $h = 2.537$ mm   **37.** $y = 5.469$ m
**39.** $D = 28.3°$, $E = 61.7°$, $f = 14.76$ ft   **41.** 32.2°   **43.** 11.57 in.   **45.** 48.6° and 41.4°   **47.** $B = 56°$, $a = 88.91$ ft,
$b = 131.8$ ft   **49.** 44.7°   **51.** $X_L = 9.64$ Ω; $R = 11.49$ Ω

## Practice Test, Chapter 13

**1.** $\dfrac{10}{26} = \dfrac{5}{13}$   **3.** $\dfrac{11.5}{12.54} = 0.9171$   **5.** 0.7986   **7.** 16.0°   **9.** 0.8726 rad   **11.** $B = 65.5°$, $b = 30.72$ m, $c = 33.76$ m

**13.** 38.7°   **15.** 48 in.   **17.** 40.06 A   **19.** 84.3°

## Section 14–1 Self-Study Exercises

**1**   **1.** 13   **2.** 15   **3.** 7.280   **4.** 8.544   **5.** 2.746

**2**   **6.** 67.38°   **7.** 53.13°   **8.** 20.56°   **9.** 68.20°   **10.** 75.96°

**3**   **11.** 19; 42°   **12.** 8; 72°   **13.** 1; 225°   **14.** 6; 75°   **15.** $12 + 5j$; 13; 22.62°   **16.** $6 + 4j$; 7.211; 33.69°

## Section 14–2 Self-Study Exercises

**1**   **1.** 60°   **2.** 15°   **3.** 70°   **4.** 15°   **5.** 32°   **6.** 70°   **7.** 32°   **8.** 62°   **9.** 0.96 rad   **10.** 0.44 rad

**2**   **11.** all positive   **12.** sin positive; others negative   **13.** tan positive; others negative   **14.** cos positive; others
negative   **15.** cos positive; tan, sin zero

**3**   **16.** −0.5000   **17.** −0.8391   **18.** −0.8011   **19.** −0.6536   **20.** −0.8660   **21.** −0.2910   **22.** −0.1736
**23.** 0.5000   **24.** 146.3°   **25.** 3.61 rad

## Section 14–3 Self-Study Exercises

**1.** $y = \sin 2x$

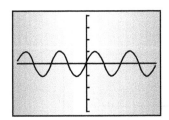

**2.** $y = \sin 3x$

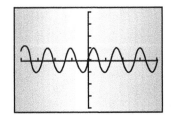

**3.** $y = \sin 4x$

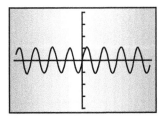

**4.** $y = \sin 8x$

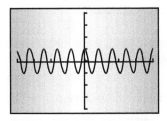

**5.** $y = \cos 2x$

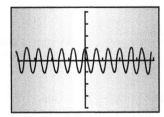

**6.** $y = \cos 4x$

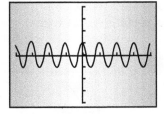

**7.** $y = \cos 5x$

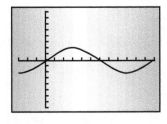

**8.** $y = \cos 6x$

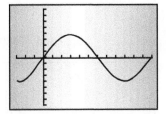

**9.** Answers will vary. The amplitude remains the same, but the period decreases as the coefficient of $x$ increases. All graphs cross the $y$-axis at the orgin.   **10.** Answers will vary. Each graph has the same maximum and minimum, but the number of repeats of the graph increases as the coefficient of $x$ increases. All graphs cross the $y$-axis at 1.

**11.** $y = 2 \sin x$

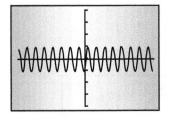

**12.** $y = 4 \sin x$

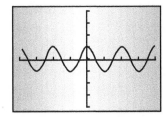

**13.** $y = 6 \sin x$

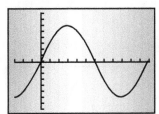

**14.** $y = 3 \cos x$

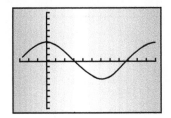

**15.** $y = 4 \cos x$

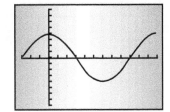

**16.** $y = 5 \cos x$

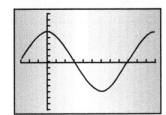

**17.** Answers will vary. As the coefficient of sin *x* increases, the amplitude increases but the period remains unchanged.
**18.** Answers will vary. As the coefficient of cos *x* increases, the amplitude increases but the period remains unchanged.
**19.** 9.8480775   **20.** 22.222 Hz   **21.** 166.667 Hz   **22.** $4 \times 10^{-6}$ s   **23.** $1.002 \times 10^{-8}$ s   **24.** 215.8 m
**25.** $1 \times 10^9$ Hz   **26.** $v = 20{,}500$ m/s   **27.** $v = 106{,}680{,}000$ m/s or $1.0668 \times 10^8$ m/s

**28.** $v = 42 \sin x$; $y = \pm 50$

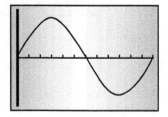

**29.** $v = 32 \sin x$; $y = \pm 50$

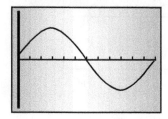

**30.** $x = 45°$; $v = 29.7$ volts
$x = 60°$; $v = 36.4$ volts

**31.** $x = 120°$; $v = 27.7$ volts
$x = 300°$; $v = -27.7$ volts

**32.** $i = 3.5 \sin x$; $y = \pm 4$

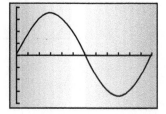

**33.** $i = 8.4 \sin x$; $y = \pm 10$

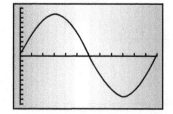

**34.** $x = 225°$; $i = -2.5$ amps
$x = 315°$; $i = -2.5$ amps

**35.** $x = 115°$; $i = 7.6$ amps
$x = 270°$; $i = -8.4$ amps

**36.** amplitude = 4; period = 60°

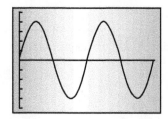

**37.** amplitude = 30; period = 36°

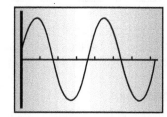

**38.** amplitude = 3; period = 72°

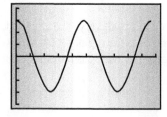

**39.** amplitude = 4; period = 45°

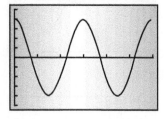

**40.** amplitude = 2.5; period = 480°

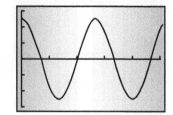

**41.** amplitude = 1.9; period = 225°

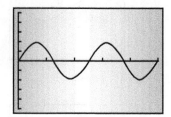

**42.** amplitude = 1; period = 360°; phase shift = 45° left

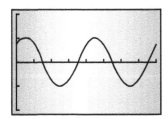

**43.** amplitude = 1; period = 360°; phase shift = 180° right

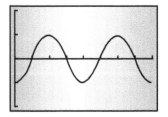

**44.** amplitude = 2; period = 360°; phase shift = 120° left

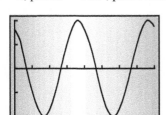

**45.** amplitude = 6; period = 120°; phase shift = 80° right

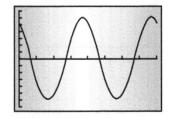

**46.** amplitude = 10; period = 90°; phase shift = 67.5° left

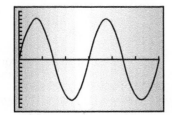

**47.** amplitude = 8; period = 60°; phase shift = 7.5° right

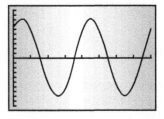

## Section 14–4 Self-Study Exercises

Rounding discrepancies will be minimized if full calculator values are used in all calculations.

**1** **1.** $C = 63°$; $b = 40.00$ m; $c = 45.22$ m    **2.** $A = 50°$; $a = 6.504$ mi; $c = 8.075$ mi    **3.** $C = 33°$; $b = 56.37$ cm; $c = 35.45$ cm    **4.** $B = 30°$; $a = 5.543$ ft; $b = 3.200$ ft    **5.** $C = 50°$; $a = 20.84$ km; $c = 16.03$ km    **6.** $C = 20°$; $c = 114.9$ dkm; $a = 237.6$ dkm    **7.** $b = 17.92$ in.    **8.** $a = 190.3$ m    **9.** $C = 76°$    **10.** $c = 7.150$ ft    **11.** $b = 5,383$ ft    **12.** 8.971 km    **13.** 942 mi

**2** **14.** $B = 21.5°$; $C = 118.5°$; $c = 57.42$ in.    **15.** $C = 10.8°$; $A = 99.2°$; $a = 15.76$ yd    **16.** $A = 43.2°$; $B = 116.8°$; $b = 23.48$ ft or $A = 136.8°$; $B = 23.2°$; $b = 10.3$ ft    **17.** $B = 14.5°$; $C = 135.5°$; $c = 11.21$ m    **18.** $B = 19.0°$; $C = 131.0°$; $c = 98.45$ cm    **19.** $B = 25.8°$; $A = 34.2°$; $a = 29.60$ m    **20.** $B = 71°$; $b = 189.5$ ft; $c = 198.5$ ft    **21.** 805.9 ft

## Section 14–5 Self-Study Exercises

**1** **1.** $A = 30.8°$                                   **2.** $D = 71.7°$
     $B = 125.1°$                                              $E = 62.9°$
     $C = 24.1°$                                              $F = 45.4°$

**3.** $A = 50.8°$; $B = 56.6°$; $C = 72.6°$    **4.** $A = 47.7°$; $B = 72.4°$; $C = 59.8°$    **5.** $A = 58.4°$; $B = 76.1°$; $C = 45.5°$
**6.** $A = 66.3°$; $B = 59.1°$; $C = 54.6°$

**2** **7.** $r = 32.42$ ft                                  **8.** $j = 18.68$ m
     $S = 49.7°$                                          $K = 37.5°$
     $T = 58.3°$                                          $L = 34.0°$

**9.** $b = 15.88$ m; $A = 52.4°$; $C = 69.4°$   **10.** $A = 24.7°$; $C = 20.3°$; $b = 30.23$ m   **11.** 49.27 ft
**12.** $60.8°$; $44.1°$; $75.1°$

## Chapter Review Exercises, Chapter 14

**1.** 7.211; 56.3°   **3.** 6.083; 9.5°   **5.** 13.89; 59.7°   **7.** 0.1016 rad   **9.** 41°   **11.** 0.8832 rad   **13.** 32°15′10″
**15.** 0.8632   **17.** −0.3420   **19.** 0.3420   **21.** 1.000   **23.** 2.83; 2.36 rad   **25.** 0.000067 s

**27.** $y = 5 \cos 4x$; amplitude = 5; period = 90°

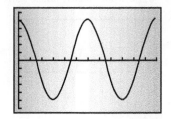

**29.** 43.3 V   **31.** $A = 40°$; $b = 10.8$; $c = 4.3$   **33.** $A = 30.3°$; $B = 104.7°$; $b = 9.6$ dkm   **35.** 1st solution: $A = 39.4°$;
$C = 112.6°$; $c = 13.4$ cm; 2nd solution: $A = 140.6°$; $C = 11.4°$; $c = 2.9$ cm   **37.** 42.0 ft

**39.** 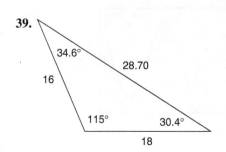   **41.**   **43.**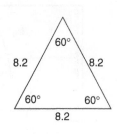

**45.** 97.98 ft   **47.** 343.3 ft; 665.4 ft

## Practice Test, Chapter 14

**1.** 0.8192   **3.** −0.9397   **5.** 135°   **7.** 101.3°   **9.** 8.991 ft   **11.** 131.8°   **13.** 6.830 m   **15.** 38.0°
**17.** 150.9 ft   **19.** 11.32 ft   **21.** 6.4 mA   **23.** 30 Ω

## Cumulative Practice Test for Chapters 12–14

**1.** acute   **3.** straight   **5.** 62°   **7.** 37°   **9.** 30.2056°   **11.** 21′7″   **13.** $P = 108$ m; $A = 529$ m$^2$   **15.** $C = 471$ cm;
$A = 17,671$ cm$^2$   **17.** 0.2682 rad   **19.** 103.1324°   **21.** 18,096 cm$^3$   **23.** 19 in.   **25.** $AB = 47$ cm
**27.** $\angle B = 52.3°$   **29.** magnitude = 5.385; direction = 21.8°   **31.** amplitude = 1; period = 360°; phase shift = 90° left
**33.** amplitude = 3; period = 180°; phase shift = 120° left   **35.** $AB = 23$ cm

## Section 15–1 Self-Study Exercises

**1**   **1.** 10%   **2.** 36.8%   **3.** 54.3%   **4.** 20.8%   **5.** 33.3%   **6.** 75%

**2**   **7.** debt retirement   **8.** misc. expenses and general government   **9.** social projects and education costs   **10.** 1970
**11.** 7,355,000 barrels; 5,834,000 barrels   **12.** 445.8%   **13.** 2000

**14.** 5 A    **15.** 50 V    **16.** 35 V    **17.** 25 Ω    **18.** 1980    **19.** motor gasoline    **20.** 1960–1970
**21.** 1960–1970    **22.** 20.3%

## Section 15–2 Self-Study Exercises

**1**  **1.** 16    **2.** 17    **3.** 66    **4.** 73.75    **5.** 33.7    **6.** 66.6    **7.** 42.33°F    **8.** 12.67°C    **9.** $34.80    **10.** $39.20

**11.** 14.67 in.    **12.** 8 in.    **13.** 20    **14.** 76    **15.** 17.3 runs    **16.** 98    **17.** $24.6\,\frac{\text{mi}}{\text{gal}}$    **18.** $10.2\,\frac{\text{mi}}{\text{gal}}$

**2**  **19.** 44    **20.** 43    **21.** 15.5    **22.** 26.5    **23.** $30    **24.** $66    **25.** $8.25    **26.** $8.85    **27.** 2    **28.** 5
**29.** no mode    **30.** no mode    **31.** $67    **32.** $32    **33.** 4 h    **34.** $1.97

**3**  **35.** 10    **36.** 2    **37.** $\frac{1}{3}$    **38.** $\frac{1}{7}$    **39.** 12%    **40.** 28%    **41.** 35–37 and 38–40    **42.** 20–22 and 23–25
**43.** 7    **44.** 16

| | Midpoint | Tally | Class Frequency |
|---|---|---|---|
| **45.** | 15 | || | 2 |
| **46.** | 12 | |||| | 4 |
| **47.** | 9 | ⅧⅠⅠ | 7 |
| **48.** | 6 | Ⅷ Ⅷ Ⅷ Ⅷ | 20 |

| | Midpoint | Tally | Class Frequency |
|---|---|---|---|
| **49.** | 93 | || | 2 |
| **50.** | 88 | Ⅷ | 5 |
| **51.** | 83 | Ⅷ || | 7 |
| **52.** | 78 | Ⅷ Ⅷ | 10 |
| **53.** | 73 | ||| | 3 |
| **54.** | 68 | Ⅷ | | 6 |
| **55.** | 63 | || | 2 |
| **56.** | 58 | Ⅷ | 5 |

| | Class Interval | Midpoint | Tally | Class Frequency |
|---|---|---|---|---|
| **57.** | 15–19 | 17 | Ⅷ || | 7 |
| | 10–14 | 12 | Ⅷ ||| | 8 |
| | 5–9 | 7 | Ⅷ |||| | 9 |
| | 0–4 | 2 | Ⅷ | | 6 |

| | Class Interval | Midpoint | Tally | Class Frequency |
|---|---|---|---|---|
| **58.** | 51–60 | 55.5 | || | 2 |
| | 41–50 | 45.5 | |||| | 4 |
| | 31–40 | 35.5 | || | 2 |
| | 21–30 | 25.5 | Ⅷ || | 7 |
| | 11–20 | 15.5 | Ⅷ | 5 |

**59.** 25.2    **60.** 7.9    **61.** 75.8    **62.** 9.7

## Section 15–3 Self-Study Exercises

**1**  **1.** 24    **2.** 23    **3.** 13    **4.** 18    **5.** $25    **6.** $33    **7.** 48°F

**2**  **8.** 3.16    **9.** 8.29    **10.** 11.06    **11.** 17.36    **12.** 4    **13.** 8

## Section 15–4 Self-Study Exercises

**1**   **1.** Keaton Brienne Renee
     Keaton Renee Brienne
     Brienne Keaton Renee
     Brienne Renee Keaton
     Renee Keaton Brienne
     Renee Brienne Keaton
     $3 \cdot 2 \cdot 1 = 6$ ways

**2.**   ABCD    BACD
     ABDC    BADC
     ACBD    BCAD
     ACDB    BCDA
     ADBC    BDAC
     ADCB    BDCA
     CABD    DABC
     CADB    DACB
     CBAD    DBAC
     CBDA    DBCA
     CDAB    DCAB
     CDBA    DCBA
     $4 \cdot 3 \cdot 2 \cdot 1 = 24$ ways

**3.** $4 \cdot 3 \cdot 2 \cdot 1 = 24$ ways    **4.** $5 \cdot 4 \cdot 3 \cdot 2 \cdot 1 = 120$ ways    **5.** $5 \cdot 4 \cdot 3 \cdot 2 \cdot 1 = 120$ ways

**2**   **6.** $\dfrac{1}{24}, \dfrac{1}{23}$   **7.** $\dfrac{1}{4}$   **8.** $\dfrac{11}{48}$   **9.** $\dfrac{2}{5}$   **10.** $\dfrac{1}{3}$   **11.** $\dfrac{1}{6}$

## Chapter Review Exercises, Chapter 15

**1.** 2001, 2003, 2004    **3.** 2002, 2005, 2006    **5.** 12.9%    **7.** 17.4%    **9.** 7-10-04 @ 4:00 P.M.    **11.** 12.6 cars

**13.** $13.8 \dfrac{\text{mi}}{\text{gal}}$    **15.** $10.03    **17.** $1.85    **19.** 3.67    **21.** 83

| | Midpoint | Tally | Class Frequency |
|---|---|---|---|
| **23.** | 60.5 | ⦀⦀ ⦀⦀ | 10 |
| **25.** | 40.5 | ⦀⦀ ⦀⦀ ‖ | 12 |
| **27.** | 20.5 | ⦀⦀ ‖ | 7 |

**29.** 36–45    **31.** 15    **33.** $\dfrac{1}{2}$    **35.** $\dfrac{10}{54} = 18.5\%$

| **37.** | Miles per Gallon | Midpoint | Tally | Frequency | (Answers may vary.) |
|---|---|---|---|---|---|
| | 20–24 | 22 | ⦀⦀ ‖ | 7 | |
| | 25–29 | 27 | ⦀⦀ ‖ | 6 | |
| | 30–34 | 32 | ‖‖ | 4 | |

**39.** 26.4    **41.** range: 59    **43.** 21.47    **45.** $\dfrac{10}{13}; \dfrac{3}{4}$    **47.** 16 combinations

## Practice Test, Chapter 15

**1.** bar    **3.** line    **5.** 2°C    **7.** $65    **9.** 25%    **11.** $\dfrac{15}{200} = \dfrac{3}{40}$    **13.** English Dept., electronics Dept.    **15.** 104.2%

**17.** $\dfrac{1}{2}$    **19.** range: 25    **21.** 8.49    **23.** 6    **25.** $\dfrac{3}{5}$
            mean: 77.9
            median: 78
            mode: 81

# *Glossary and Index*

Absolute error, 164

**Absolute value** (| |): the number of units of distance the number is from 0, 200

Accuracy, 162

**Acute angle:** an angle that is less than 90° but more than 0°, 474

**Addends:** in addition, the numbers being added, 8

Addition, 8

associative property, 8

axiom or property, 251

check, 10

commutative property, 8

decimals, 9

estimation, 10

fractions and mixed numbers, 79, 80

integers, 202

key words, 248

metric system, 151

polynomials, 399

signed numbers, 202

U.S. customary system, 99

vectors, 573

whole numbers, 9

zero property, 9, 204

**Addition axiom:** the same quantity can be added to both sides of an equation without changing the equality of the two sides. Casually, we describe applying the addition axiom as sorting or transposing terms, 251

Addition method, 369

**Additive identity:** Zero is the additive identity because $a + 0 = a$ for all values of $a$, 9, 204

**Additive inverse:** The opposite of a number is the additive inverse of the number because $a + (-a) = 0$ for all values of $a$, 201, 204

**Adjacent side:** the adjacent side of a given acute angle of a right triangle is the side that forms the angle with the hypotenuse, 538

**Adjacent side of a polygon:** the adjacent side of a polygon is the side that has an endpoint in common with the base, 38, 480, 481

Altitude, 42

Ambiguous case, 589

Amperes or amps, A, 56

Amplitude, 583

**Angle:** a geometric figure formed by two rays that intersect in a point, and the point of

intersection is the end point of each ray, 472

acute, 474

adjacent, 475

alternate, 475

alternate exterior, 476

alternate interior, 476

complementary, 474

corresponding, 476

degrees to radians, 473, 497

exterior, 476

interior, 476

notation, 472

obtuse, 474

radians to degrees, 496, 498

related, 576

right, 473

sides or legs, 472

straight, 474

supplementary, 474

vertex, 472

vertical, 475

Angle of declination, 551

Angle of depression, 551

Angle of elevation, 551

Apex, 508

**Approximate amount:** another name for the rounded amount, 6, 8

Approximate equivalents, 74

Approximate measures, 167

**Arc:** the portion of the circumference cut off by a chord or a central angle, 499, 503

Arc length, 499

Arccosine, 566

Arcsine, 566

Arctangent, 566

**Area:** the amount of surface of a plane figure, 19, 102, 484

circle, 491

composite figure, 494

lateral area, prism or cylinder, 511

parallelogram, 41, 484

polygon, 484

rectangle, 20, 41, 484

sector, 501

segment, 503

square, 41, 484

surface area, cone, 513

surface area, prism or cylinder, 511

surface area, sphere, 513

trapezoid, 484

triangle, 484

**Arithmetic mean or arithmetic average:** the sum of the quantities in the data set divided by the number of quantities, 26, 616

**Ascending order:** arranging the terms of the polynomial beginning with the term with the lowest degree, 397

**Associative property of addition:** The associative property of addition means that values being added may be grouped in any manner, 8

**Associative property of multiplication:** The associative property of multiplication means that values being multiplied may be grouped in any manner, 15

**Average, numerical:** a measure of central tendency that represents a quantity considered typical or representative of several related quantities, 26, 617

Axis, 370, 613

**Axis of symmetry:** the fold line of a parabola, when the parabola is folded in half and the two halves match, 451

**Bar graph:** a graph that uses two or more bars to show pictorially how two or more amounts compare to each other rather than to a total, 613

Base, 508

**Base in exponential notation:** the value used in repeated multiplication when raising to a power, 31

**Base in a problem involving percent:** the number ($B$) that represents the original or total amount, 113

**Base of any polygon:** the horizontal side of any polygon or a side that would be horizontal if the polygon's orientation were modified, 38, 480, 481

Base unit, 143, 158

**Basic principle of equality:** To preserve equality if we perform an operation on one side of an equation, we must perform the same operation on the other side, 250

Billion, 2

**Binary operation:** an operation that involves working with two numbers at a time, 8, 202

**Binomial:** a polynomial containing two terms. Examples include $2x + 4$ and $7xy - 21a$, 396, 413

Binomial square, 416

of the first fraction times the denominator of the second, and the product of the denominator of the first fraction times the numerator of the second. In the proportion $a/b = c/d$, the cross products are $a \cdot d$ and $c \cdot d$, 295

**Cube:** a number raised to the third power, 31

**Cubed:** another term for "raised to the third power," 31

**Cubic polynomial:** a polynomial that has degree 3, 397

**Cubic term:** a term that has degree 3, 397

Cup, 94

Current, $I$ or $i$, 56

**Cylinder:** a three-dimensional figure with a curved surface and two circular bases such that the height is perpendicular to the bases, 507
- oblique, 507
- right, 507
- surface area, 511
- volume, 508

Data, 616

Day, 104

DC circuit, 56

**Deci-:** the prefix used for a unit that is 1/10 of the standard unit, 143

**Decimal:** a fractional notation based on place values for fractions with a denominator of 10 or a power of 10, 14
- clearing decimals in equations, 271
- comparing decimal numbers, 5
- division, 24
- fractional equivalent, 72
- notation, 4
- repeating, 74

**Decimal fraction:** a fraction whose denominator is always ten or some power of 10; a fractional notation that uses the decimal point and the place values to its right to represent a fraction whose denominator is 10 or some power of 10, such as 100, 1000, and so on. A decimal fraction is also referred to as a decimal, a decimal number, or number using decimal notation, 4

**Decimal number system:** the system of numbers that uses ten individual figures called digits, 0, 1, 2, 3, 4, 5, 6, 7, 8, 9 and place values of powers of 10, 2

**Decimal number:** an alternate name for a decimal, 4

Decimal numbers, *see* Whole numbers and decimals

**Decimal point:** the symbol (period) placed between the ones and the tenths place to identify the place value of each digit, 4, 11

Decrease, 121, 124

**Degree:** a unit for measuring angles. A degree represents 1/360 of a complete rotation around the vertex of the angle, 473
- decimal to minutes and seconds, 478
- degrees to radians, 497
- minutes and seconds to decimal, 478
- radians to degrees, 498

**Degree of a polynomial:** the degree of a polynomial that has only one variable and only positive integral exponents is the degree of the term with the largest exponent, 397

**Degree of a term:** the degree of a term that has only one variable and only positive integral exponents is the same as the exponent of the variable, 396

**Deka-:** the prefix used for a unit that is 10 times larger than the standard unit, 143

Delta, $\Delta$, 338

**Denominator:** The denominator of a fraction is the number of parts one unit has been divided into. It is the bottom number of a fraction or the divisor of the indicated division, 4, 60
- common, 74

**Dependent system:** when solving a system of equations, if both variables are illuminated and the resulting statement is true, then the equations are dependent and have many solutions, 368, 373

**Dependent variable:** the variable that represents the results of calculations. In function notation, the dependent variable is $y$, 297

**Descending order:** arranging the terms of the polynomial beginning with a term with the highest degree, 397

Deviation, 625

Descriptive statistics, 616

**Diameter:** the straight line segment from a point on the circle through the center to another point on the circle, 490
- inside, 17
- outside, 17

**Difference:** the answer to a subtraction problem, 11

Difference of two perfect squares, 415, 419

Digits, 2
- nonzero, 32

Dimension, 96

Dimension analysis, 96

Direct current (dc) circuit, 56

Direct measurement, 94, 144
- metric rule, 167
- metric system (SI), 143
- U.S. customary rule, 94
- U.S. customary system, 94
- U.S.-metric conversions, 160
- *see also* Metric system and U.S. customary system

**Direct proportion:** a proportion in which the quantities being compared are directly related so that as one quantity increases (or decreases), the other quantity also increases (or decreases). This relationship is also called the direct variation, 297

Direct variation, 297

Direction, 571

Direction of vector, 571

Directional sign, 201

**Discriminant:** the radicand of the radical portion of the quadratic formula, $b^2 - 4ac$, 459

Distributive property, 17, 249, 259, 410

**Distributive property of multiplication:** the property stating that multiplying a sum or difference by a factor is equivalent to multiplying each term of the sum or difference by the factor, 17, 249
- applied to equations, 258
- common factors, 410

**Dividend:** the number being divided, 21

Divisibility tests, 61

**Divisible:** a number is said to be divisible by another number if the quotient has no remainder or if the dividend is a multiple of the divisor, 61

**Division:** the inverse operation for multiplication, 21
- check, 25
- commutativity or associativity, 21
- decimals, 24
- estimation, 25
- fractions and mixed numbers, 90, 91
- importance of first digit in quotient, 23
- inverse operation of multiplication, 21
- key words, 248
- long, 24
- multiplying by reciprocal, 90
- numerical average, 26
- polynomials, 401
- powers of 10, 221, 390
- rounding quotient, 24
- scientific notation, 228
- signed numbers, 212
- symbols, 21
- U.S. customary system, 102
- whole numbers, 22
- with zero, 212

**Divisor:** the number to divide by, 21

Domain, 323

Double root, 445

Efficiency, 119

Electromotive force (emf) of energy, 56

**Element:** another name for the member of a set of data, 628

Elimination method, 369

Endpoint, 471

Energy, 56

Engine efficiency, 119

Engineering notation, 240

English rule, 94

**English system:** a system of measurement that is currently called the U.S. Customary system of measurement, 94

Equality
- addition property, 251
- multiplication property, 252
- principle, 250
- proportions, 295
- symmetric property, 245

**Equation:** a symbolic statement that two expressions or quantities are equal in value, 244

Equations
- applying distributive property, 258
- checking solutions, 266
- clearing decimals, 271
- clearing fractions, 262
- excluded roots, 264
- horizontal line, 345
- linear, 250
- parallel lines, 347
- perpendicular lines, 349
- point-slope form, 343
- quadratic, 438
- verify, 244
- vertical line, 345

Equations in two variables, *see* Graphs and Systems of equations

Equilateral triangle, 516

**Equivalent fractions:** fractions that represent the same value, 67

Error,
- absolute, 164
- greatest possible, 163
- percent, 164
- relative, 164

Outliers, 625
Outside diameter, 17

Parabola, 450
Parallel, 38
Parallel circuits, 565
**Parallel lines:** Two or more lines in the same plane that are the same distance apart everywhere. They have no points in common, 346, 472
**Parallelogram:** a four sided polygon whose opposite sides are parallel, 38, 39, 41, 480
Partial dividend, 22
**Partial product:** the product of the multiplicand and one digit of the multiplier, 16
Partial quotient, 22
Pentagon, 532
Per: divided by, 104
**Percent:** a fractional part of 100 expressed with a percent sign (%), 107
Percent error, 164
Percent of change, 124
**Percent of increase or decrease:** the rate of change from the original amount, 124
**Percentage:** in a problem involving percent, the number (P) that represents a portion of the base, 113
**Percentage formula:** Portion = Rate times Base, 113, 121
    base, 113
    portion, 113, 121
    rate, 113, 121
    solving, 114
Percents
    complement, 123
    equivalents, 107, 109
    estimating, 116
    finding new amount directly, 122
    increases and decreases, 121
    less than 1%, 117
    more than 100%, 117
    numerical equivalents, 107, 105
    percentage formula, 113
**Perfect square:** the product of a number times itself, 31
**Perfect square trinomial:** a trinomial whose first and last terms are positive perfect squares and whose middle term has an absolute value that is twice the product of the square root of the first and last terms, 416, 420
**Perimeter:** the total length of the sides of a plane figure, 38
    circle, see circumference
    composite shapes, 493
    parallelogram, 39, 481
    polygon, 38, 39, 481
    rectangle, 18, 39, 481
    square, 39, 481
    trapezoid, 481
    triangle, 481
Period, trigonometric, 581, 582
**Periods:** groups of three place values-ones, tens, hundreds-for units, thousands, millions, and so on, 2
Perpendicular, 473
**Perpendicular lines:** two lines that intersect to form right angles (90° angles). Another term for perpendicular is normal, 349, 473
    equation of, 349
Phase angle, 566
Phase shift, 581

Pi (π), 490
Pint, 94
**Place value:** the value of a digit based on its position in a number, 2, 4
Plane, 471
Plot a point, 321
Plus or minus symbol, ±, 13
Point, 471
Point, in reading decimals, 4
Point notation, 321
Point-slope form of an equation, 343
Polar notation, 565
Polygon, 38, 480
    area, 483
    perimeter, 480
    regular, 532
Polygons, inscribed in a circle, 532
**Polynomial:** an algebraic expression in which the exponents of the variables are nonnegative integers, 395, 413
    base, 480
    basic operations 399
    degree, 397
    division, 417
    factoring special products, 419
    FOIL method of multiplying binomials, 414
    identify, 396
    multiplication, 400, 413
    order, 397
    prime, 412
    special products, 415
Positive
    key words, 207
Pound, 94
**Portion:** in a problem involving percent, the number (P) that represents a portion of the base, 113
**Power:** another term for the exponent or the result of raising a value to a power, 31
    basic operations, 221
    calculator, 32
    decimal notation, 222
    divide, 389
    fraction, 88, 393
    imaginary, 418
    imaginary numbers, 458
    multiplication, 388
    negative integer exponents, 222, 230
    power of power, 392
    product, 393
    scientific notation, 232
    signed numbers, 211
    zero, 221
    *see also* Powers of 10
**Powers of 10:** numbers whose only nonzero digit is one: 10, 100, 1,000 are examples of powers of 10, 3, 32, 221
    engineering notation, 240
    power, 225
    multiplication and division, 32, 221, 222
    scientific notation, 226
Precision, 162, 163
Prefixes, 143
Prime, 64, 428
**Prime factor:** A factors that is divisible only by one and the number itself, 64
**Prime factorization:** writing a composite number as the product of only prime numbers, 64
**Prime number:** a whole number greater than 1 that has only one factor pair, the number itself and 1, 63
Prime polynomial, 428

Principal, 271
Principle of equality, 250
**Prism:** a three-dimensional figure whose polygonal bases (ends) are parallel, congruent polygons and whose faces (sides) are parallelograms, rectangles, or squares. In a right prism the faces are perpendicular to the base or bases, 507
    oblique, 507
    right, 507
    surface area, 511
    volume, 508
**Probability:** the chance of an event occurring. It is expressed as a ratio or percent of the number of possibilities for success to the total number of possibilities, 631
Problem solving, 43, 261
    guess and check, 45
    keywords, 44
    six-step strategies, 43
**Product:** the answer or result of multiplication, 15
    cross, 295
**Proper fraction:** a common fraction whose value is less than one unit; that is, the numerator is less than the denominator, 60
Property of proportions, 295
**Proportion:** a mathematical statement that shows two fractions or ratios are equal; an equation in which each side is a fraction or ratio, 295
Proportions,
    direct variation, 297
    inverse variation, 308
    problems involving similar triangles, 301
    property of, 295
Protractor, 477
**Pure quadratic equation:** a quadratic equation and that has a quadratic term and a constant term but no linear term, 439, 440
Pyramid, 508
    right or regular, 508
**Pythagorean theorem:** the theorem that states that the square of the hypotenuse of a right triangle is equal to the sum of the squares of the two legs of the triangle, 517, 546

Quadrant, 572
    related acute angles, 576
**Quadratic equation:** an equation in which at least one letter term is raised to the second power and no letter terms have a power higher than 2 or less than 0. The standard form for a quadratic equation is $ax^2 + bx + c = 0$, where a, b, and c are real numbers and a > 0, 438
    complete quadratic equations, 439, 444
    graphing in two variables, 450
    identifying coefficients, 439
    incomplete quadratic equations, 439, 442
    nature of roots, 458
    pure quadratic equations, 439, 440
    quadratic formula, 446
    standard form, 438
    use of the discriminant in solving, 458
Quadratic formula, 446
**Quadratic polynomial:** a polynomial that has degree 2, 437
**Quadratic term:** a term that has degree 2, 397, 439
Quart, 94